Lecture Notes in Computer Science 8037

Commenced Publication in 1973
Founding and Former Series Editors:
Gerhard Goos, Juris Hartmanis, and Jan van Leeuwen

Editorial Board

Frank Dehne   Roberto Solis-Oba
Jörg-Rüdiger Sack (Eds.)

Algorithms
and Data Structures

13th International Symposium, WADS 2013
London, ON, Canada, August 12-14, 2013
Proceedings

 Springer

Volume Editors

Frank Dehne
Carleton University, Ottawa, ON, Canada
E-mail: frank@dehne.net

Roberto Solis-Oba
The University of Western Ontario, London, ON, Canada
E-mail: solis@csd.uwo.ca

Jörg-Rüdiger Sack
Carleton University, Ottawa, ON, Canada
E-mail: sack@scs.carleton.ca

ISSN 0302-9743   e-ISSN 1611-3349
ISBN 978-3-642-40103-9   e-ISBN 978-3-642-40104-6
DOI 10.1007/978-3-642-40104-6
Springer Heidelberg Dordrecht London New York

Library of Congress Control Number: 2013944325

CR Subject Classification (1998): F.2, E.1, G.2, I.3.5, G.1, C.2

LNCS Sublibrary: SL 1 – Theoretical Computer Science and General Issues

Typesetting: Camera-ready by author, data conversion by Scientific Publishing Services, Chennai, India

Printed on acid-free paper

Springer is part of Springer Science+Business Media (www.springer.com)

Preface

This volume contains the papers presented at WADS 2013: Algorithms and Data Structures Symposium (formerly Workshop on Algorithms and Data Structures) held during August 11–13, 2013, in London, Ontario. WADS alternates with the Scandinavian Workshop on Algorithms Theory (SWAT), continuing the tradition of SWAT and WADS starting with SWAT 1988 and WADS 1989.

In response to the call for papers, 139 papers were submitted. From these submissions, the Program Committee selected 44 papers for presentation at WADS 2013. In addition, invited lectures were given by the following distinguished researchers: Anjul Bhambhri (IBM Silicon Valley Lab), Timothy M. Chan (University of Waterloo), and Sergei Vassilvitskii (Google).

On behalf of the Program Committee, we would like to express our appreciation to the invited speakers, reviewers, and all the authors who submitted papers.

May 2013

Frank Dehne
Roberto Solis-Oba
Joerg-Ruediger Sack

Organization

Program Committee

David Bader
Marin Bougeret
Danny Chen
Jianer Chen
Siu-Wing Cheng
Andrea Clementi
Jose Correa
Amitava Datta
Frank Dehne
Shlomi Dolev
Faith Ellen
Thomas Erlebach
Scott Hazelhurst
Klaus Jansen
Petteri Kaski
Naoki Katoh
Rolf Klein
Darek Kowalski
Mike Langston
Monaldo Mastrolilli
Friedhelm Meyer Auf der Heide
Andrew Rau-Chaplin
Joerg-Ruediger Sack   Roberto Solis-Oba
Frits Spieksma
Takeshi Tokuyama
Gottfried Vossen

Additional Reviewers

Abu-Khzam, Faisal
Albar, Boris
Aloupis, Greg
Arya, Sunil
Bae, Sang Won
Baert, Anne-Elisabeth
Barba, Luis
Barenboim, Leonid

Becchetti, Luca
Bienkowski, Marcin
Bilò, Davide
Bohler, Cecilia
Bollig, Beate
Boudet, Vincent
Bousquet, Nicolas
Brandstadt, Andreas

Brodal, Gerth Stølting
Cardinal, Jean
Castelli Aleardi, Luca
Chalermsook, Parinya
Chambers, Erin
Chan, Timothy
Chechik, Shiri
Cheilaris, Panagiotis
Cord-Landwehr, Andreas
Crama, Yves
Crescenzi, Pierluigi
Damaschke, Peter
De Marco, Gianluca
Delling, Daniel
Di Ianni, Miriam
Dietzfelbinger, Martin
Dragan, Feodor
Duckham, Matt
Durocher, Stephane
Ediger, David
Elbassioni, Khaled
Eppstein, David
Fairbanks, James
Fischer, Johannes
Fischer, Matthias
Fox, Kyle
Fujiwara, Hiroshi
Funke, Stefan
Fusy, Eric
Gagie, Travis
Gal, Avigdor
Gao, Jie
Garcia, Alfredo
Garnero, Valentin
Gasieniec, Leszek
Gaspers, Serge
Gasten, Stefan
Gawrychowski, Pawel
Gilbers, Alexander
Green, Oded
Gualà, Luciano
Gutin, Gregory
Hagan, Ron
Har-Peled, Sariel
Hauptmann, Mathias

Haverkort, Herman
Hermelin, Danny
Higashikawa, Yuya
Hong, Seok-Hee
Ikebe, Yoshiko
Imai, Keiko
Itoh, Takehiro
Jansen, Bart
Jansen, Bart M. P.
Jansson, Jesper
Jiang, Minghui
Jung, Daniel
Kamiyama, Naoyuki
Kaothanthong, Natsuda
Kaufmann, Michael
Kawamura, Akitoshi
Kedem, Klara
Kijima, Shuji
Kobourov, Stephen
Koenemann, Jochen
Koivisto, Mikko
Korman, Matias
Kortsarz, Guy
Koutsopoulos, Andreas
Kraft, Stefan
Kratochvil, Jan
Krause, Philipp Klaus
Laekhanukit, Bundit
Lampis, Michael
Land, Felix
Langetepe, Elmar
Laughon, David
Lechtenbörger, Jens
Levcopoulos, Christos
Lingas, Andrzej
Liptak, Zsuzsanna
Liu, Chih-Hung
Lopez-Ortiz, Alejandro
Lu, Xin
Lubiw, Anna
Luxen, Dennis
Löffler, Maarten
McColl, Robert
Misra, Neeldhara
Miyazaki, Shuichi

Mnich, Matthias
Molnar, Miklos
Morgenstern, Gila
Moric, Filip
Morin, Pat
Mount, David
Mulzer, Wolfgang
Narasimhan, Giri
Navarro, Gonzalo
Nederlof, Jesper
Nekrich, Yakov
Niedermann, Benjamin
Niedermeier, Rolf
Nilsson, Bengt
Nouri Baygi, Mostafa
Nussbaum, Yahav
Nöllenburg, Martin
Okamoto, Yoshio
Otachi, Yota
Otoo, Ekow
Palop, Belen
Paluch, Katarzyna
Pasquale, Francesco
Penninger, Rainer
Phillips, Charles
Pietracaprina, Andrea
Pietrzyk, Peter
Pitre, Sylvain
Polishchuk, Valentin
Potapov, Igor
Proietti, Guido
Pérez-Lantero, Pablo
Rapaport, Ivan
Riedy, Jason
Robillard, David
Roselli, Vincenzo
Räcke, Harald
S. Anders, Peter
Saitoh, Toshiki

Sanders, Ian
Sanders, Peter
Satti, Srinivasa Rao
Saurabh, Saket
Sawada, Joe
Schlotter, Ildikó
Schoenrock, Andrew
Shioura, Akiyoshi
Silvestri, Riccardo
Sinaimeri, Blerina
Skiena, Steven
Smid, Michiel
Smorodinsky, Shakhar
Soltys, Karolina
Soto, Jose A.
Tanigawa, Shin-Ichi
Thorup, Mikkel
Tsakalidis, Konstantinos
Ukkonen, Antti
Uwe Haus, Utz
Vahrenhold, Jan
van Kreveld, Marc
Vocca, Paola
Wang, Kai
Watrigant, Rémi
Widmayer, Peter
Williams, Ryan
Wismath, Steve
Wolff, Alexander
Ximing, Li
Xue, Jason
Yamanaka, Katsuhisa
Young, Maxwell
Zaboli, Hamidreza
Zakrzewska, Anita
Zarrabi-Zadeh, Hamid
Zhu, Binhai
Ziv-Ukelson, Michal

Table of Contents

On Maximum Weight Objects Decomposable into Based Rectilinear
Convex Objects.. 1
  Mahmuda Ahmed, Iffat Chowdhury, Matt Gibson,
  Mohammad Shahedul Islam, and Jessica Sherrette

Bundling Three Convex Polygons to Minimize Area or Perimeter 13
  Hee-Kap Ahn, Helmut Alt, Sang Won Bae, and Dongwoo Park

Smart-Grid Electricity Allocation via Strip Packing with Slicing 25
  Soroush Alamdari, Therese Biedl, Timothy M. Chan, Elyot Grant,
  Krishnam Raju Jampani, Srinivasan Keshav, Anna Lubiw, and
  Vinayak Pathak

On (Dynamic) Range Minimum Queries in External Memory 37
  Lars Arge, Johannes Fischer, Peter Sanders, and Nodari Sitchinava

Distance-Sensitive Planar Point Location 49
  Boris Aronov, Mark de Berg, Marcel Roeloffzen, and
  Bettina Speckmann

Time-Space Tradeoffs for All-Nearest-Larger-Neighbors Problems 61
  Tetsuo Asano and David Kirkpatrick

Coloring Hypergraphs Induced by Dynamic Point Sets and Bottomless
Rectangles 73
  Andrei Asinowski, Jean Cardinal, Nathann Cohen,
  Sébastien Collette, Thomas Hackl, Michael Hoffmann,
  Kolja Knauer, Stefan Langerman, Michał Lasoń, Piotr Micek,
  Günter Rote, and Torsten Ueckerdt

Socially Stable Matchings in the Hospitals/Residents Problem 85
  Georgios Askalidis, Nicole Immorlica, Augustine Kwanashie,
  David F. Manlove, and Emmanouil Pountourakis

Parameterized Complexity of 1-Planarity 97
  Michael J. Bannister, Sergio Cabello, and David Eppstein

On the Stretch Factor of the Theta-4 Graph........................ 109
  Luis Barba, Prosenjit Bose, Jean-Lou De Carufel,
  André van Renssen, and Sander Verdonschot

Better Space Bounds for Parameterized Range Majority
and Minority 121
    Djamal Belazzougui, Travis Gagie, and Gonzalo Navarro

Online Control Message Aggregation in Chain Networks 133
    Marcin Bienkowski, Jaroslaw Byrka, Marek Chrobak, Łukasz Jeż,
    Jiří Sgall, and Grzegorz Stachowiak

Fingerprints in Compressed Strings................................. 146
    Philip Bille, Patrick Hagge Cording, Inge Li Gørtz, Benjamin Sach,
    Hjalte Wedel Vildhøj, and Søren Vind

Beacon-Based Algorithms for Geometric Routing 158
    Michael Biro, Justin Iwerks, Irina Kostitsyna, and
    Joseph S.B. Mitchell

Interval Selection with Machine-Dependent Intervals.................. 170
    Kateřina Böhmová, Yann Disser, Matúš Mihalák, and
    Peter Widmayer

On the Spanning Ratio of Theta-Graphs 182
    Prosenjit Bose, André van Renssen, and Sander Verdonschot

Relative Interval Analysis of Paging Algorithms on Access Graphs 195
    Joan Boyar, Sushmita Gupta, and Kim S. Larsen

On Explaining Integer Vectors by Few Homogenous Segments 207
    Robert Bredereck, Jiehua Chen, Sepp Hartung,
    Christian Komusiewicz, Rolf Niedermeier, and
    Ondřej Suchý

Trajectory Grouping Structure.................................... 219
    Kevin Buchin, Maike Buchin, Marc van Kreveld,
    Bettina Speckmann, and Frank Staals

The Art of Shaving Logs 231
    Timothy M. Chan

TREEWIDTH and PATHWIDTH Parameterized by the Vertex
Cover Number 232
    Mathieu Chapelle, Mathieu Liedloff, Ioan Todinca, and
    Yngve Villanger

Visibility and Ray Shooting Queries in Polygonal Domains 244
    Danny Z. Chen and Haitao Wang

Lift-and-Project Methods for Set Cover and Knapsack................ 256
    Eden Chlamtáč, Zachary Friggstad, and Konstantinos Georgiou

Optimal Time-Convex Hull under the L_p Metrics 268
   Bang-Sin Dai, Mong-Jen Kao, and D.T. Lee

Blame Trees 280
   *Erik D. Demaine, Pavel Panchekha, David A. Wilson, and
   Edward Z. Yang*

Plane 3-trees: Embeddability and Approximation
(Extended Abstract) 291
   Stephane Durocher and Debajyoti Mondal

A Dynamic Data Structure for Counting Subgraphs in Sparse
Graphs 304
   Zdeněk Dvořák and Vojtěch Tůma

Combinatorial Pair Testing: Distinguishing Workers from Slackers 316
   David Eppstein, Michael T. Goodrich, and Daniel S. Hirschberg

Approximation Algorithms for B_1-EPG Graphs 328
   Dror Epstein, Martin Charles Golumbic, and Gila Morgenstern

Universal Point Sets for Planar Three-Trees 341
   Radoslav Fulek and Csaba D. Tóth

Planar Packing of Binary Trees 353
   *Markus Geyer, Michael Hoffmann, Michael Kaufmann,
   Vincent Kusters, and Csaba D. Tóth*

Hierarchies of Predominantly Connected Communities 365
   Michael Hamann, Tanja Hartmann, and Dorothea Wagner

Joint Cache Partition and Job Assignment on Multi-core Processors 378
   Avinatan Hassidim, Haim Kaplan, and Omry Tuval

Finding the Minimum-Weight k-Path 390
   *Avinatan Hassidim, Orgad Keller, Moshe Lewenstein, and
   Liam Roditty*

Compressed Persistent Index for Efficient Rank/Select Queries 402
   *Wing-Kai Hon, Lap-Kei Lee, Kunihiko Sadakane, and
   Konstantinos Tsakalidis*

Tight Bounds for Low Dimensional Star Stencils in the External
Memory Model 415
   Philipp Hupp and Riko Jacob

Neighborhood-Preserving Mapping between Trees 427
   Jan Baumbach, Jiong Guo, and Rashid Ibragimov

Bounding the Running Time of Algorithms for Scheduling and Packing
Problems 439
    Klaus Jansen, Felix Land, and Kati Land

When Is Weighted Satisfiability FPT? 451
    Iyad A. Kanj and Ge Xia

Two-Sided Boundary Labeling with Adjacent Sides.................... 463
    Philipp Kindermann, Benjamin Niedermann, Ignaz Rutter,
    Marcus Schaefer, André Schulz, and Alexander Wolff

Optimal Batch Schedules for Parallel Machines 475
    Frederic Koehler and Samir Khuller

Unions of Onions: Preprocessing Imprecise Points for Fast Onion Layer
Decomposition.. 487
    Maarten Löffler and Wolfgang Mulzer

Dynamic Planar Point Location with Sub-logarithmic Local Updates ... 499
    Maarten Löffler, Joseph A. Simons, and Darren Strash

Parameterized Enumeration of (Locally-) Optimal Aggregations 512
    Naomi Nishimura and Narges Simjour

MapReduce Algorithmics 524
    Sergei Vassilvitskii

The Greedy Gray Code Algorithm 525
    Aaron Williams

Author Index 537

On Maximum Weight Objects Decomposable into Based Rectilinear Convex Objects

Mahmuda Ahmed, Iffat Chowdhury, Matt Gibson,
Mohammad Shahedul Islam, and Jessica Sherrette

Department of Computer Science
University of Texas at San Antonio
San Antonio, TX USA
{mahmed,ichowdhu,gibson,msislam,jsherett}@cs.utsa.edu

Abstract. Our main concern is the following variant of the image segmentation problem: given a weighted grid graph and a set of vertical and/or horizontal base lines crossing through the grid, compute a maximum-weight object which can be decomposed into based rectilinear convex objects with respect to the base lines. Our polynomial-time algorithm reduces the problem to solving a polynomial number of instances of the maximum flow problem.

1 Introduction

An area of work that has recently attracted extensive attention in the pattern recognition and computer vision communities is *image segmentation*. It is the process of partitioning a digital image into multiple objects for better representation and analysis of an image. From another view point, image segmentation is assigning labels to the pixels of an image such that the pixels with the same label define a particular object which may have certain visual characteristics. In practice image segmentation is used to detect objects and boundaries in the image. An example, in *medical imaging*, image segmentation is used to help locate tumors and other pathologies, measure tissue volumes, computer-guided surgery, diagnosis, treatment planning, study of anatomical structure etc. There are many other applications of image segmentation including fingerprint recognition, traffic control systems and agriculture imaging.

Image Segmentation as an Optimization Problem. Finding a "good" segmentation is often treated as an optimization problem, see for example [2,12,13,4,5,9,7,1]. Using the framework of Asano et al. [2] we are given a weighted grid graph where each grid cell corresponds to a pixel in the original image and weights on the grid cells are related to the likelihood that the particular pixel is in the object we wish to identify (positive weights are assigned to grid cells whose corresponding pixel is likely in the object and negative weights are assigned to grid cells whose corresponding pixel is likely in the background). Then we attempt to find some subset of the grid that optimizes an objective function

F. Dehne, R. Solis-Oba, and J.-R. Sack (Eds.): WADS 2013, LNCS 8037, pp. 1–12, 2013.

subject to some constraints. Let G be an $\sqrt{n} \times \sqrt{n}$ four-neighborhood grid graph. For $1 \leq i \leq \sqrt{n}$ and $1 \leq j \leq \sqrt{n}$ the grid cell p at the (i, j) position in the grid has a real value $w(p)$ called the weight of p. We call i the x-coordinate of p and j the y-coordinate of p and let p_x (resp. p_y) denote the x-coordinate (resp. y-coordinate) of p. A *region* (or *object*) R will be defined as any subset of grid cells, and we define the weight of R to be $w(R) = \sum_{p \in R} w(p)$. We are interested in computing the region R with maximum weight subject to some constraints.

Research has shown that knowledge of the geometric shape of the object that you are looking for can greatly increase an algorithm's effectiveness in practice, see for example [10,19,3,18,11]. Polynomial-time algorithms have been given which identify an optimal solution for the following classes of objects: x-monotone regions, based monotone regions, rectilinear convex regions, and star-shaped regions [4,9,8].

Objects Decomposable into Elementary Shapes. Chun et al. [7] consider the maximum-weight region problem with a twist on the constraints of some previous works. They are interested in finding a maximum-weight region that may not have simple geometric structure, but can be *decomposed* into objects with simple geometric structure. A region R can be decomposed into m objects of a particular structure if and only if there exists a coloring of the grid cells of R using m colors such that each of the objects induced by the grid cells of each of the color classes have the desired structure.

This type of problem is very interesting from both a practical perspective as well as a theoretical perspective. It is interesting in practice because an algorithm for such a problem can identify more complicated objects while still allowing control of the topology of the output object. It is interesting from a theoretical perspective because the decomposition constraints of the problem poses an interesting computational challenge to overcome. If we instead consider finding objects which are the union of m objects with simple geometric structure, the problem often becomes much harder (for example, finding the maximum-weight object that is the union of two star-shaped objects is NP-hard [7]). The decomposition variant of the problems may admit polynomial-time algorithms, but it is not trivial to design such an algorithm for many classes of objects even when $m = 2$.

Chun et al.[7] give an efficient algorithm for computing the maximum-weight region that can be decomposed into two digital star-shaped regions with respect to two given "center" grid cells. Gibson et al. [14] give a maximum-flow based algorithm for the same 2-star problem and recently Gibson et al. [15] extend the result of [14] to identify the maximum-weight object decomposable into c star-shaped objects for any constant c in polynomial time.

Chun et al. [7] consider the problem of computing the maximum-weight object decomposable into *based monotone object* with respect to a set of k given base lines. A base line of the grid graph G is a vertical ($x = i$) or horizontal ($y = j$) path of grid cells across the grid for $1 \leq i \leq \sqrt{n}$ and $1 \leq j \leq \sqrt{n}$. For a given horizontal base line $l : y = i$, a *based monotone object* is a union of segments of columns intersecting the base line. See Figure 1 (a) and (b) for an illustration. They do not require a based monotone object for a particular base line to be a connected region. This allows them to use the base lines to partition the grid into $O(k^2)$ subproblems which they solve independently using dynamic programming. Recently Chun et. al [6] gave an algorithm for finding the optimal baseline locations using quadtree decomposition.

Our Contribution. Given a weighted grid graph, we are interested in identifying the maximum-weight object decomposable into *Based Rectilinear Convex* (BRC) objects with respect to c given base lines for a constant c. For a given horizontal base line $l : y = i$, a BRC object with respect to l satisfies the properties of being a based monotone object with the additional constraint that the intersection of the object with any horizontal line is always undivided (a symmetric notion is defined for vertical base lines). See Figure 1 (c) and (d) for an illustration. In contrast to a based monotone object, a BRC object is by definition a connected object. Therefore, as opposed to the based monotone case, the base lines do not decompose the grid into subproblems which can be solved independently.

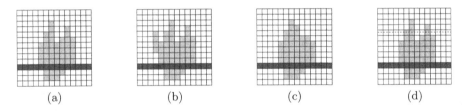

(a)   (b)   (c)   (d)

Fig. 1. Part (a) is a based monotone object with respect to the base line. Part (b) is not a based monotone object. Part (c) is a BRC object with respect to the base line. Part (d) is not a BRC object (the intersection of the object with the dotted line is not connected).

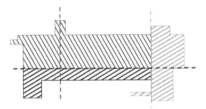

Fig. 2. A 3-BRC object with respect to 3 given base lines

We call an object which can be decomposed into c different BRC objects a *c-BRC object*. See Figure 2 for an illustration. When $c = 1$ the problem is easily solvable, but until now there has been no polynomial-time algorithm given even when $c = 2$. Our main contribution is the following theorem.

Theorem 1. *There exists a polynomial-time algorithm which computes a maximum-weight object decomposable into based rectilinear objects with respect to a set of c given base lines in a weighted grid graph for any constant $c \geq 1$.*

We prove Theorem 1 by giving a polynomial-time algorithm for computing a maximum-weight 2-BRC object for a restricted special case of the 2-BRC problem. We solve this restricted special case by observing some key geometric properties of a BRC object and show that these observations allow us to reduce the problem to computing the maximum-weight closed set in a polynomial number of appropriately defined directed graphs. It is well known that a maximum-weight closed set can be computed in polynomial time [17,16] via a reduction to the maximum flow problem. We then show how to carefully reduce the c-BRC problem to several instances of the restricted 2-BRC problem. This reduction will be done in a way so that the solution to the 2-BRC instances can be merged to obtain an optimal solution for the c-BRC instance.

To guarantee that our algorithm returns an optimal solution, our algorithm iteratively guesses the structure of an optimal solution. For each guess, we compute the maximum-weight c-BRC object which corresponds to this guess. We show that by making a polynomial number of guesses, we can guarantee that we guess the correct structure for an optimal solution; however, the polynomial is too large to be of practical interest. That being said, our result shows how the structure of a solution can be used to reduce the problem to the maximum flow problem. If this structure is given as input by a user or is found via a heuristic, then our work shows that the problem can be reduced to solving a small number of maximum flow instances which would be of practical interest. Also, our technique can easily be modified to compute the complement of a maximum weight c-BRC object (this may be more efficient for some inputs).

Organization of the Paper. In Section 2, we give an algorithm which computes a maximum-weight 2-BRC object for a restricted version of the problem. In Section 3, we extend the result to find a maximum-weight c-BRC object for any constant $c \geq 2$.

2   Algorithm for a Restricted 2-BRC Problem

In this section, we give a polynomial-time algorithm for a restricted version of 2-BRC object using an $\sqrt{n} \times \sqrt{n}$ four-neighborhood grid graph G and two base lines at the boundary of the grid. We show that this problem can be solved by computing the maximum-weight closed set for a linear number of appropriately constructed directed graphs. Given a weighted, directed graph $D = (V, E)$, a *closed set* is a subset of the vertices $C \subseteq V$ such that if $u \in C$ and $(u, v) \in E$ then $v \in C$. Intuitively, if C is a closed set then there is no edge from a vertex in C to a vertex in $V \setminus C$. The weight of a closed set C is simply the sum of the weights of the vertices in C.

Preliminaries. Initially, we assume that the base lines are parallel and without loss of generality the base lines are at $y = 1$ and $y = \sqrt{n}$. At the end of this section, we show how to handle the case where base lines are perpendicular. We view each grid cell as having a x-coordinate and a y-coordinate (the grid cell in the lower left corner has x-coordinate = y-coordinate = 1 and the grid cell in the upper right corner has x-coordinate = y-coordinate = $\sqrt{n}$). If O is a BRC object with respect to the top of the grid, we say that O is a type-N BRC object (its base line is the "northern" base line). Similarly, if O is a BRC object with respect to the bottom of the grid, we say that O is a type-S BRC object.

Let O be a BRC object and without loss of generality assume it is a type-S object. A peak of O is a grid cell $p \in O$ for which no other grid cells $p' \in O$ have y-coordinate greater than the y-coordinate of p. Similarly, for type-N object, a peak is a pixel with minimum y-coordinate over all pixels in the object. We define a peak line of O to be a vertical line through the grid which contains a peak. See Figure 3 (a) for an illustration.

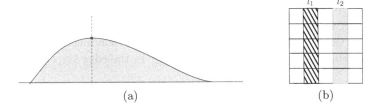

Fig. 3. Peak lines: (a) A peak line where the shaded region is a type-S BRC object. (b) The patterned and shaded portion of the grid are the peak lines l_1 and l_2 respectively.

The following observation is the key idea that allows us to reduce the restricted 2-BRC problem to a maximum-weight closed set problem. The proof has been omitted due to lack of space.

Observation 2. *Let O be a subset of grid cells in the grid, and let l be the vertical line through the grid at $x = \alpha$. Then O is a type-S (resp. type-N) BRC object with respect to peak line l if and only if the following properties hold:*

1. *for each $o \in O$ such that $o_x \leq \alpha$, each grid cell p such that $p_x = o_x$ and $p_y \leq o_y$ (resp. $p_y \geq o_y$) is in O and each grid cell q such that $q_y = o_y$ and $o_x \leq q_x \leq \alpha$ we have $q \in O$.*
2. *for each $o' \in O$ such that $o'_x > \alpha$, each grid cell p' such that $p'_x = o'_x$ and $p'_y \leq o'_y$ (resp. $p'_y \geq o'_y$) is in O and each grid cell q' such that $q'_y = o'_y$ and $o'_x \geq q'_x \geq \alpha$ we have $q' \in O$.*

The consequence of Observation 2 is that if we know a peak line for each BRC object, then we can compute them via a single maximum-weight closed set computation in an appropriately defined directed graph (we can guess all possible pairs of peak lines using $\sqrt{n} \times \sqrt{n} = n$ guesses).

Construction of the Directed Graph. We now describe the construction of the directed graph to find the 2-BRC object with respect to two peak lines l_1 and l_2. See Figure 3 (b) which shows the peak line l_1 for type-S object and l_2 for type-N object. For the remainder section, we assume when we mention a 2-BRC object, we refer a 2-BRC object with respect to l_1 and l_2.

We call our graph $D_{\{l_1,l_2\}}$. There are two "sections" of vertices in $D_{\{l_1,l_2\}}$, and each grid cell in G has exactly one vertex in each of these sections. The vertices in a closed set from the first section will determine what grid cells are in the type-S BRC object in G, and the vertices in a closed set from the second section will determine what grid cells are in the type-N BRC object in G. Let V_1 denote the vertices in the section for the type-S BRC object, and let us define V_2 similarly for type-N BRC object. For a grid cell g, let v_g^1 denote its corresponding vertex in V_1 and let v_g^2 denote its corresponding vertex in V_2. For ease of description, we view V_1 and V_2 being embedded in the same layout as the grid cells in G.

We will now define three edge sets E_1, E_2, and E_3. E_1 will consist of edges with both endpoints in V_1, E_2 will consist of edges with both endpoints in V_2, and E_3 will consist of edges with their tail in V_1 and their head in V_2. Let us now define the edge set E_1. See Figure 4(a) for an illustration. In V_1, every vertex has an edge to a vertex directly 'below' it (if it exists). And all the "horizontally adjacent" vertices have an edge between the corresponding vertices directed towards the peak line l_1. These are the all edges in the edge set E_1.

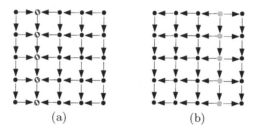

(a)   (b)

Fig. 4. (a) The arrangement of edge set of E_1 where patterned line is the peak line l_1 and this vertex set is in V_1. (b) The arrangement of edge set of E_2 where lightly shaded line is the peak line l_2 and this vertex set is in V_2.

We now describe the edge set E_2. See Figure 4(b) for an illustration. Similarly, in V_2, every vertex has an edge to a vertex directly below it (if it exists). But all the horizontally adjacent vertices have an edge between the corresponding vertices directed away from the peak line. These are all the edges in the set E_2.

The edge set E_3 consists of the directed edges (v_g^1, v_g^2) for each grid cell g. This completes the construction of the edge sets E_1, E_2, and E_3.

Our directed graph $D_{\{l_1,l_2\}}$ has vertex set $V := V_1 \cup V_2$ and edge set $E := E_1 \cup E_2 \cup E_3$. We assign weights on the vertices as follows. The weight of each vertex $v_g^1 \in V_1$ is set to be $w(g)$. The weight of each vertex $v_g^2 \in V_2$ is set to be $-w(g)$. This completes the construction of the graph.

Relationship between a Closed Set and a 2-BRC Object. We now describe a function T which will take as input a subset of vertices in $D_{\{l_1,l_2\}}$ and outputs a subset of grid cells in G. Fix any subset $V' \subseteq V$ of $D_{\{l_1,l_2\}}$. For any vertex $v_g^1 \in V' \cap V_1$, the corresponding grid cell g is in $T(V')$. For any vertex $v_g^2 \in V_2 \setminus V'$, the corresponding grid cell g is in $T(V')$. In other words, a grid cell g is in $T(V')$ if v_g^1 is in V' or if v_g^2 *is not* in V'. If v_g^1 is not in V' and v_g^2 is in V', then g is not in $T(V')$. We will prove in Lemma 1 that if V' is a closed set of $D_{\{l_1,l_2\}}$ then $T(V')$ is a 2-BRC object whose weight is the same as the weight of V' (minus a constant).

We now define another function T' which takes as input a 2-BRC object and returns a set of vertices in $D_{\{l_1,l_2\}}$. T' is the inverse of T. Fix R to be any subset of grid cells that can be decomposed into a type-S BRC object and a type-N BRC object. Fix such a decomposition, and color the grid cells in the type-S BRC object red and the cells in the type-N BRC object blue. Let us call the red grid cells R_1 and the blue grid cells R_2. For each red cell $r \in R_1$ we have that $v_r^1 \in T'(R)$ and $v_r^2 \in T'(R)$. For each blue cell $b \in R_2$ we have $v_b^1 \notin T'(R)$ and $v_b^2 \notin T'(R)$. For all uncolored cells g we have $v_g^1 \notin T'(R)$ and $v_g^2 \in T'(R)$. This concludes the definition of the function $T'(R)$, and in Lemma 2 we will prove that $T'(R)$ is a closed set in $D_{\{l_1,l_2\}}$ and has weight equal to R (minus a constant).

Note that we have $T'(T(C)) = C$ for every closed set C and $T(T'(R)) = R$ for every 2-BRC object. Thus proving Lemma 1 and Lemma 2 will complete the proof that the maximum-weight region in G that is decomposable into two BRC objects with respect to the peak lines can be computed by finding a maximum-weight closed set in $D_{\{l_1,l_2\}}$. The proof of Lemma 2 is similar to the proof of Lemma 1 and is omitted due to lack of space.

Lemma 1. *Fix any closed set C of $D_{\{l_1,l_2\}}$. Then $T(C)$ is a 2-BRC object and has weight equal to C (minus a constant).*

Proof. We first show that $T(C)$ is a 2-BRC object. Let C_1 be $C \cap V_1$, and abusing notation let $T(C_1) \subseteq T(C)$ be the grid cells g such that $v_g^1 \in C_1$. We will argue that $T(C_1)$ is a type-S BRC object by showing $T(C_1)$ satisfies properties 1 and 2 of Observation 2. We can show this is true by considering the construction of $D_{\{l_1,l_2\}}$. There is an edge in $D_{\{l_1,l_2\}}$ from v_g^1 to the vertex corresponding to the grid cell towards l_1 on the same horizontal line and the vertex directly below it. Since C is a closed set, it follows that both of these vertices must also be in the closed set. It follows from a simple inductive argument that for any $v_g' \in C_1$, all of the vertices v_c^1 which are between v_g' and the peak line l_1 on the same horizontal line and all of the vertices $v_{c'}^1$ which are below v_g' on the same vertical line will be in C. See Figure 4 (a). By the definition of T, it must be that all such grid cells c and c' are in $T(C)$. We thus have by Observation 2 that $T(C_1)$ is a type-S BRC object.

Now let C_2 be $C \cap V_2$, and abusing notation let $T(C_2) \subseteq T(C)$ be the grid cells g such that $v_g^2 \notin C_2$. We will now show that $T(C_2)$ is a type-N BRC object. Let α_2 denote the x-coordinate of the points on l_2. We remind the reader that by the definition of T, vertices in $V_2 \setminus C_2$ correspond with the grid cells that are in $T(C_2)$. Again, to show that $T(C_2)$ is type-N BRC object, we will show that properties 1 and 2 of Observation 2 hold for $T(C_2)$. Suppose for the sake of contradiction that $g \in T(C_2)$ (without loss of generality assume $g_x \leq \alpha_2$) but there is a grid cell g' such that $g_x = g'_x$ and $g_y < g'_y$ and g' is not in $T(C_2)$. Since g' is not in $T(C_2)$, we have $v^2_{g'} \in C_2$. According to the construction of $D_{\{l_1,l_2\}}$, there must be an edge from $v^2_{g'}$ to the vertex below it. Since C is a closed set, we must have that these vertices are in C_2. An inductive argument follows that all of the vertices corresponding to grid cells below g' on the same vertical line must be in C_2. This of course implies that $g \notin T(C_2)$, a contradiction. We have the similar argument for a grid cell g'' such that $g''_y = g_y$ and $g_x \leq g''_x \leq \alpha_2$. We thus prove the properties 1 and 2 of Observation 2 and hence $T(C_2)$ is a type-N BRC object.

We will now argue that $T(C)$ is a 2-BRC object. We will prove this by showing that $T(C_1)$ and $T(C_2)$ are disjoint. This is easy to see from the definition of the edge set E_3. Let g be some grid cell in $T(C_1)$. By definition, this implies that $v^1_g \in C_1$. The edge (v^1_g, v^2_g) is in E_3, and since C_1 is a closed set it must be that $v^2_g \in C_2$. This implies that for any $g \in T(C_1)$, we have $g \notin T(C_2)$. This completes the proof that $T(C_1)$ and $T(C_2)$ are disjoint and therefore $T(C)$ can be decomposed into two BRC objects.

This concludes the proof that $T(C)$ is a 2-BRC object, and we will now prove that C and $T(C)$ have the same weight (minus a constant). First let w_1 be the sum of the weights of the vertices in C_1, and let w_2 be the sum of the weights of the vertices in C_2. The weight of the closed set is exactly $w_1 + w_2$. The corresponding grid cell for each vertex in C_1 is also in $T(C)$, and moreover has the exact same weight. So the sum of the weights of the grid cells in $T(C_1)$ is w_1. Recall that the vertices in C_2 correspond to the exact set of grid cells that are not in $T(C_2)$, and thus the weight of the grid cells in $T(C_2)$ is $w(V_2) + w_2$ (we remind the reader that the weight of a vertex in C_2 is the negative of the weight of its corresponding grid cell). Therefore, the weight of the grid cells in $T(C)$ is $w_1 + w_2 + w(V_2)$. Since $w_1 + w_2$ is the weight of C, we conclude that the weight of C is equal to the weight of the grid cells in $T(C)$ minus $w(V_2)$. This concludes the proof of the lemma. □

Lemma 2. *Fix any subset R of grid cells in G that is a 2-BRC object. Then $T'(R)$ is a closed set in $D_{\{l_1,l_2\}}$ and has weight equal to R (minus a constant).*

So now we have that if C is a maximum-weight closed set of $D_{\{l_1,l_2\}}$, then $T(C)$ is a maximum-weight 2-BRC object. There are n total pairs of peak lines, so we can check all possible pairs. One of those pairs will correspond with the peak lines for the maximum-weight 2-BRC object, and therefore the maximum-weight 2-BRC object obtained for these peak lines will be the maximum-weight 2-BRC object for the entire problem.

Handling Perpendicular Base Lines. Now we will assume that the base lines can be perpendicular. Since we now have a vertical base line, the "sides" of the grid can be base lines. In this setting, we say an object O is a type-W (resp. type-E)

BRC object if O is a BRC object with respect to the "western" (resp. "eastern") base line.

Without loss of generality, assume we have the southern base line and the western base line and we wish to find the maximum-weight 2-BRC object decomposable into a type-S BRC object and a type-W BRC object. We can compute this object using a similar approach as to what we used in Section 2 by slightly changing the construction of the directed graph. Note that a peak line for a type-W object is a horizontal line (perpendicular to the base line). Suppose we are given a vertical peak line l_1 and a horizontal peak line l_2. We will construct the directed graph $D_{\{l_1, l_2\}}$ slightly differently. The vertex set will again be $V_1 \cup V_2$ where V_1 will be used to identify the type-S object and V_2 will be used to identify the type-W object. The edge sets E_1 and E_3 will be exactly as defined above, but the edge set E_2 will change. The edge set E_2 will consist of horizontal edges directed towards the base line and vertical edges directed away from the peak line. The weights are assigned the same way as before. We can argue similarly as we did in Lemma 1 and Lemma 2 that for a maximum-weight closed set C of $D_{\{l_1, l_2\}}$, we have $T(C)$ is a maximum-weight 2-BRC object with respect to l_1 and l_2, and by considering all $\sqrt{n} \times \sqrt{n}$ possible pairs of base lines we can compute in polynomial time the maximum-weight 2-BRC object with respect to the southern and western base lines. We conclude that for any two base lines, we can compute in polynomial-time the maximum-weight object for the restricted 2-BRC problem.

3   Extension to the c-BRC Problem

We now give a polynomial-time algorithm for the original problem in which we are given a weighted grid graph G and c base lines, and we wish to compute a maximum-weight c-BRC object. Our algorithm iteratively makes guesses about the structure of an optimal solution OPT. Using this structure, we reduce the problem to several instances of the restricted 2-BRC problem. The reduction is handled in two parts. First we decompose the grid into $O(c^2)$ rectangular-subgrid instances of a restricted version of the 4-BRC problem. This restricted version will be similar to the restricted 2-BRC problem considered in Section 2 (base lines are at the boundary of the grid). The key property of this instance is that for each instance I of the restricted 4-BRC problem, we have that $I \cap OPT$ can be decomposed into at most 4 BRC objects with respect to the base lines at the boundary of I. We then use a digital Voronoi diagram to break the restricted 4-BRC problem into at most 5 instances of the restricted 2-BRC problem, which we solve using the algorithm given in Section 2. The reduction is carefully done so that the merging of the solutions will be a feasible c-BRC object. When we correctly guess the structure of OPT, we show that the merged solutions will be an optimal c-BRC object. This approach is similar in flavor to the approach of Gibson et al. [15] for computing the maximum-weight object decomposable into c star-shaped objects for any fixed c. An overview of our reduction is now given. Further details and the algorithm have been omitted due to lack of space.

Reduction to the Restricted 4-BRC Problem. We will now reduce the c-BRC problem into $O(c^2)$ instances of the *restricted 4-BRC problem* in which there are at most four base lines, each of which are at the boundary of the grid. Let $B_1, B_2, \ldots, B_c$ be the c disjoint BRC objects that OPT decomposes into. Consider some B_i, and without loss of generality assume that the base line of B_i is horizontal. Let b_i denote the intersection of B_i with its base line. Let ℓ_i and r_i denote the leftmost and rightmost grid cell of b_i respectively, and consider the vertical paths through ℓ_i and r_i. Note that by the definition of a BRC object, any $g \in B_i$ cannot be "outside" of these vertical paths. Similarly a B_i with a vertical base line has a vertical b_i and must be between two horizontal paths.

Now consider any grid cell $g \in G$, and consider shooting four axis-parallel rays from g in all four directions until it hits the boundary of G or hits a b_i. We can hit at most four b_is, and g can only be in a B_i such that a ray shot from g hit b_i. To see this, first note that g can only be in a B_i such that the vertical or horizontal line through g hits b_i (otherwise g would be outside of the "vertical paths" described above). Now suppose for the sake of contradiction that an axis-parallel ray from $g \in B_i$ to b_i pierces through a $b_{i'}$. By the definition of BRC object, we have that every grid cell along this ray is in B_i including the grid cell $g' \in b_{i'}$ pierced by the ray. That implies $g' \in B_i$ and $g' \in B_{i'}$, a contradiction.

The subset of grid cells whose axis-parallel rays hit the same set of b_i (from the same "ray direction") induce the desired rectangular instances of the restricted 4-BRC problem. Let I be one such instance, and consider $I \cap OPT$. As we just argued, there will be grid cells from c' different B_i in I for some $1 \le c' \le 4$ (at most one from each "direction"). Clearly, $B_i \cap I$ for each of these B_i will be a BRC object with respect to a unique "side" of I. Thus we can view I as an instance of the restricted 4-BRC problem where we have c' base lines, each on a different "side" of I. We note that the optimal 4-BRC object for this problem may not be $OPT \cap I$. We can fix this issue by modifying the weights of the grid cells along the base line.

Reduction to the Restricted 2-BRC Problem. We now suppose we are given an instance I of the restricted 4-BRC problem and we give an overview of how to reduce it to at most 5 instances of the restricted 2-BRC problem. Due to lack of space, the details have been omitted.

First note that if there are at most 2 base lines in I, then the problem is already an instance of the 2-BRC problem. It remains to show how to handle the cases in which we have three or four base lines in a single instance. For the rest of the paper, we will assume that our instance will have all four base lines (it will be clear how to handle the case when we have three base lines).

We will now give a high level overview of the details of the decomposition of the restricted 4-BRC problem into several instances of the 2-BRC problem. Let $OPT(I)$ denote $OPT \cap I$. $OPT(I)$ can be decomposed into type-N, a type-E, a type-S, and a type-W BRC objects, so fix such a decomposition and let N, E, S, and W respectively denote each of these objects. We will consider the *digital Voronoi diagram* for these four sets. That is, we will partition the grid cells of G into four Voronoi regions $V(N), V(E), V(S)$, and $V(W)$ such that any grid cell in a Voronoi region is "closer" to that particular BRC object than it is to any of the other three. See Figure 5 for an illustration. We will show that we can use the vertices of the Voronoi diagram to help us partition the problem into instances of the 2-BRC problem. Intuitively, a vertex of the Voronoi diagram occurs where three or four different Voronoi regions "touch each other". Consider a vertex of the Voronoi diagram, and suppose this is a vertex where exactly three of the Voronoi regions "come together". For this vertex, we will find three paths in the grid. Each path will begin at this vertex and will end at one of the base lines (one path per base line). The paths will be chosen in a way such that they will consist of grid cells which all belong to the same Voronoi region (the Voronoi region associated with the base line at which the path ends). We find these paths for each of the vertices of the Voronoi diagram, and we will show that if we remove the grid cells in these paths then we are left with a constant number of connected components, each of which contains grid cells from at most 2 Voronoi regions. This allows us to use the algorithm of Section 2 to compute the maximum-weight 2-BRC object from these components and merge the solutions together to obtain $OPT(I)$.

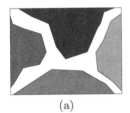   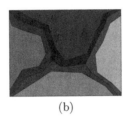   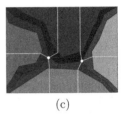

(a)   (b)   (c)

Fig. 5. Decomposing into 2-BRC instances. (a) Suppose this is $OPT(I)$. (b) The Voronoi diagram associated with $OPT(I)$. (c) The vertices and paths used to decompose into 2-BRC instances.

References

1. Anzai, S., Chun, J., Kasai, R., Korman, M., Tokuyama, T.: Effect of corner information in simultaneous placement of k rectangles and tableaux. Discrete Mathematics, Algorithms and Applications 2(4), 527–537 (2010)
2. Asano, T., Chen, D.Z., Katoh, N., Tokuyama, T.: Efficient algorithms for optimization-based image segmentation. Int. J. Comput. Geometry Appl. 11(2), 145–166 (2001)
3. Chan, T.F., Zhu, W.: Level set based shape prior segmentation. In: CVPR (2), pp. 1164–1170 (2005)
4. Chen, D.Z., Chun, J., Katoh, N., Tokuyama, T.: Efficient algorithms for approximating a multi-dimensional voxel terrain by a unimodal terrain. In: Chwa, K.-Y., Munro, J.I. (eds.) COCOON 2004. LNCS, vol. 3106, pp. 238–248. Springer, Heidelberg (2004)

5. Chen, D.Z., Hu, X.S., Luan, S., Wu, X., Yu, C.X.: Optimal terrain construction problems and applications in intensity-modulated radiation therapy. In: Möhring, R.H., Raman, R. (eds.) ESA 2002. LNCS, vol. 2461, pp. 270–283. Springer, Heidelberg (2002)

6. Chun, J., Horiyama, T., Ito, T., Kaothanthong, N., Ono, H., Otachi, Y., Tokuyama, T., Uehara, R., Uno, T.: Base location problems for base-monotone regions. In: Ghosh, S.K., Tokuyama, T. (eds.) WALCOM 2013. LNCS, vol. 7748, pp. 53–64. Springer, Heidelberg (2013)

7. Chun, J., Kasai, R., Korman, M., Tokuyama, T.: Algorithms for computing the maximum weight region decomposable into elementary shapes. In: Dong, Y., Du, D.-Z., Ibarra, O. (eds.) ISAAC 2009. LNCS, vol. 5878, pp. 1166–1174. Springer, Heidelberg (2009)

8. Chun, J., Korman, M., Nöllenburg, M., Tokuyama, T.: Consistent digital rays. Discrete & Computational Geometry 42(3), 359–378 (2009)

9. Chun, J., Sadakane, K., Tokuyama, T.: Efficient algorithms for constructing a pyramid from a terrain. IEICE Transactions 89-D(2), 783–788 (2006)

10. Das, P., Veksler, O., Zavadsky, V., Boykov, Y.: Semiautomatic segmentation with compact shape prior. Image Vision Comput. 27(1-2), 206–219 (2009)

11. Freedman, D., Zhang, T.: Interactive graph cut based segmentation with shape priors. In: CVPR (1), pp. 755–762 (2005)

12. Fukuda, T., Morimoto, Y., Morishita, S., Tokuyama, T.: Data mining using two-dimensional optimized accociation rules: Scheme, algorithms, and visualization. In: Jagadish, H.V., Mumick, I.S. (eds.) SIGMOD Conference, pp. 13–23. ACM Press (1996)

13. Fukuda, T., Morimoto, Y., Morishita, S., Tokuyama, T.: Data mining with optimized two-dimensional association rules. ACM Trans. Database Syst. 26(2), 179–213 (2001)

14. Gibson, M., Han, D., Sonka, M., Wu, X.: Maximum weight digital regions decomposable into digital star-shaped regions. In: Asano, T., Nakano, S.-I., Okamoto, Y., Watanabe, O. (eds.) ISAAC 2011. LNCS, vol. 7074, pp. 724–733. Springer, Heidelberg (2011)

15. Gibson, M., Varadarajan, K., Wu, X.: On a planar segmentation problem (2012)

16. Hochbaum, D.S.: A new - old algorithm for minimum-cut and maximum-flow in closure graphs. Networks 37(4), 171–193 (2001)

17. Picard, J.-C.: Maximal closure of a graph and applications to combinatorial problems. Management Science 22(11), 1268–1272 (1976)

18. Thiruvenkadam, S.R., Chan, T.F., Hong, B.-W.: Segmentation under occlusions using selective shape prior. SIAM J. Imaging Sciences 1(1), 115–142 (2008)

19. Veksler, O.: Star shape prior for graph-cut image segmentation. In: Forsyth, D., Torr, P., Zisserman, A. (eds.) ECCV 2008, Part III. LNCS, vol. 5304, pp. 454–467. Springer, Heidelberg (2008)

Bundling Three Convex Polygons to Minimize Area or Perimeter[*]

Hee-Kap Ahn[1], Helmut Alt[2], Sang Won Bae[3,**], and Dongwoo Park[1]

[1] POSTECH, South Korea
{heekap,dwpark}@postech.ac.kr
[2] Freie Universität Berlin, Germany
alt@mi.fu-berlin.de
[3] Kyonggi University, South Korea
swbae@kgu.ac.kr

Abstract. Given a set $\mathcal{P} = \{P_0, \ldots, P_{k-1}\}$ of k convex polygons having n vertices in total in the plane, we consider the problem of finding k translations $\tau_0, \ldots, \tau_{k-1}$ of $P_0, \ldots, P_{k-1}$ such that the translated copies $\tau_i P_i$ are pairwise disjoint and the area or the perimeter of the convex hull of $\bigcup_{i=0}^{k-1} \tau_i P_i$ is minimized. When $k = 2$, the problem can be solved in linear time but no previous work is known for larger k except a hardness result: it is NP-hard if k is part of input. We show that for $k = 3$ the translation space of P_1 and P_2 can be decomposed into $O(n^2)$ cells in each of which the combinatorial structure of the convex hull remains the same and the area or perimeter function can be fully described with $O(1)$ complexity. Based on this decomposition, we present a first $O(n^2)$-time algorithm that returns an optimal pair of translations minimizing the area or the perimeter of the corresponding convex hull.

1 Introduction

We consider the problem of finding translations of k convex polygons such that they are contained in a smallest possible convex region while their interiors are disjoint. This problem can be modelled as follows: given a set $\mathcal{P} = \{P_0, \ldots, P_{k-1}\}$ of k convex polygons in the plane with n vertices in total, find k translations $\tau_0, \ldots, \tau_{k-1}$ of $P_0, \ldots, P_{k-1}$ such that the translated copies $\tau_i P_i$'s, for $0 \leqslant i \leqslant k - 1$, do not overlap each other and the area or the perimeter of the convex hull of $\bigcup_{i=0}^{k-1} \tau_i P_i$ is minimized.

This problem can be seen as a generalization of a *packing problem* of finding a smallest region, called a *container*, of a given shape (such as a disk, a square, or a rectangle) that packs the input objects under translations. Packing problems have received significant attention in a number of disciplines. For instance, it goes back to Kepler's conjecture on sphere packing in three-dimensional Euclidean space. Sugihara et al. considered a related problem of minimizing the

---

[*] The research by H.-K. Ahn and D. Park was supported by NRF grant 2011-0030044 (SRC-GAIA) funded by the government of Korea.

[**] Corresponding author.

F. Dehne, R. Solis-Oba, and J.-R. Sack (Eds.): WADS 2013, LNCS 8037, pp. 13–24, 2013.
© Springer-Verlag Berlin Heidelberg 2013

disk bundling a set of disks [8] with applications to minimizing the sizes of holes through which sets of electric wires are to pass. They proposed a heuristic method that makes use of the Voronoi diagram of circles. Milenkovic studied the packing of a set of polygons into another polygon container with applications in the apparel industry [7]. He gave a $O(n^{k-1} \log n)$ time algorithm for packing k convex n-gons under translations into a minimum area axis-parallel rectangle container. Later, Alt and Hurtado [3] presented an algorithm for packing two convex polygons into a minimum area or perimeter rectangle whose running time is close to linear.

Much less is known about the case when the container has no restriction on its shape, for instance, when the convex hull of (the translated copies of) the input objects forms a container. For $k = 2$, Lee and Woo [6] presented a linear time algorithm for finding a translation that minimizes the area of the convex hull. The same algorithm works for minimizing the perimeter of the convex hull. Tang et al. [9] gave an $O(n^3)$ time algorithm for finding a rigid motion that minimizes the area of the convex hull. When k is part of input, the problem is known to be NP-hard even if the polygons are rectangles [4]: The proof is done by reducing the partition problem [5] into this problem. Very recently, Ahn and Cheong [1,2] presented a near-linear time approximation algorithm for finding a rigid motion that minimizes either the perimeter or the area of the convex hull.

Our Results. We consider the problem of bundling three convex polygons under translations only. We show that the translation space of P_1 and P_2 can be decomposed into $O(n^2)$ cells in each of which the combinatorial structure of the convex hull remains the same and the area or perimeter function can be fully described with $O(1)$ complexity. This is shown to be possible by a careful analysis on all event configurations at which the combinatorial structure of the convex hull changes. We then present an $O(n^2)$ time algorithm that returns an optimal pair of translations that minimizes the area or the perimeter of the convex hull of the union.

2   Preliminaries

Let $P_0, \ldots, P_{k-1}$ be k convex polygons in $\mathbb{R}^2$ with n vertices in total. For a vector $\tau \in \mathbb{R}^{2k}$, we write $\tau = (\tau_0, \ldots, \tau_{k-1})$, where $\tau_i \in \mathbb{R}^2$. The *translate* of P_i by τ_i, denoted by $\tau_i P_i$, is $\{a + \tau_i \mid a \in P_i\}$. We let $U(\tau) = \bigcup_{i=0}^{k-1} \tau_i P_i$ and let $\mathrm{conv}(\tau) := \mathrm{conv}(U(\tau))$.

Our problem can be viewed as an optimization problem of minimizing the area $\|\mathrm{conv}(\tau)\|$ or the perimeter $|\mathrm{conv}(\tau)|$ over $\tau \in \mathbb{R}^{2k}$ subject to $\tau_i P_i \cap \tau_j P_j = \emptyset$ for all $0 \leqslant i < j \leqslant k - 1$. Ahn and Cheong [2] studied the area and perimeter functions and observed the following.

Lemma 1 (Ahn and Cheong [2]). *The function $f : \mathbb{R}^{2k} \mapsto \mathbb{R}$ with $f(\tau) = |\mathrm{conv}(\tau)|$ is convex for any $k \geqslant 2$. The function $g : \mathbb{R}^{2k} \mapsto \mathbb{R}$ with $g(\tau) = \|\mathrm{conv}(\tau)\|$ is convex and piecewise linear for $k = 2$, but this is not necessarily the case for $k > 2$.*

In the bundling problem, one can reduce the search space by a simple observation.

Lemma 2. *For the bundling problem with respect to either area or perimeter, there is an optimal translation vector $\tau^* \in \mathbb{R}^{2k}$ such that the union $U(\tau^*)$ is connected, that is, every translate touches another translate under τ^*.*

We can thus concentrate only on the cases where the k polygons are connected. We shall call $\tau \in \mathbb{R}^{2k}$ a *configuration* if $U(\tau)$ is connected and each translate touches another translate under τ. A configuration τ is *feasible* if and only if the interiors of the translates are disjoint under τ. Thus, our goal is to find an optimal feasible configuration with respect to area or perimeter.

Let $\mathcal{K}$ be the set of configurations for given k polygons $P_0, \ldots, P_{k-1}$. Each configuration $\tau \in \mathcal{K}$ is associated with several properties describing the structure of the convex hull $\mathrm{conv}(\tau)$. If $\tau_i P_i$ and $\tau_j P_j$ are touching, then a vertex v of P_i lies on an edge e of P_j under τ, or vice versa. We call the pair (v, e) a *contact* induced by τ. Let $C(\tau)$ be the set of contacts induced by a configuration $\tau \in \mathcal{K}$. Note that Lemma 2 implies that $|C(\tau)| \geqslant k - 1$. The convex hull $\mathrm{conv}(\tau)$ is a closed polygonal curve consisting of the boundaries of P_i's and edges connecting the boundaries. We call such an edge a *bridge* between two polygons appearing consecutively along the boundary of the convex hull. More specifically, a vertex v_i of P_i and another v_j of P_j form a bridge (v_i, v_j) induced by τ if P_i and P_j appear consecutively along the boundary of the convex hull. Let $H(\tau)$ be the set of those pairs of vertices induced by $\tau \in \mathcal{K}$. Two configurations $\tau, \tau' \in \mathcal{K}$ are said to have the same combinatorial structure if $C(\tau) = C(\tau')$ and $H(\tau) = H(\tau')$.

In the following sections, we will show that the configuration space can be decomposed into a number of cells in which the configurations have the same combinatorial structure, so that the area or the perimeter function is described and minimized. For convenience of elaboration, we make a general position assumption on the input polygons in the sense that no two edges from the k input polygons are parallel.

3   The Configuration Space for Three Polygons

3.1   Parametrization of Configurations

As a warm-up exercise, consider the case of $k = 2$ where two convex polygons P_0 and P_1 are given. By Lemma 2, any configuration $\tau \in \mathcal{K}$ requires P_1 to touch P_0. Imagine that P_0 is stationary and P_1 translates around P_0 in the counterclockwise direction, keeping touching each other, until P_1 then reaches back to the initial position. The set $\mathcal{K}$ of configurations thus forms a space homeomorphic to a unit circle. This motion of P_1 around P_0 is piecewise affine, and the total distance that P_1 travels is exactly $|P_0| + |P_1|$. Therefore, letting $L := |P_0| + |P_1|$, the interval $[0, L)$ fully describes the configuration space $\mathcal{K}$: For any $\lambda \in [0, L)$, let $\tau(\lambda)$ be the configuration whose corresponding translated copy of P_1 is a snapshot at a moment when P_1 travels a distance of exactly λ around P_0 from its initial position.

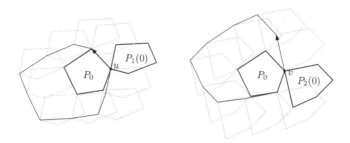

Fig. 1. Sliding P_1 and P_2 around P_0: we parameterize the configuration space $\mathcal{K}_0$ by a pair of parameters (λ_1, λ_2) for $\lambda_1 \in [0, L_1)$ and $\lambda_2 \in [0, L_2)$

We now turn to the case of $k = 3$, where three convex polygons P_0, P_1, and P_2 are given as input. Lemma 2 implies that in any configuration $\tau \in \mathcal{K}$, at least one of the three polygons touches the other two, simultaneously. Without loss of generality, we assume that both P_1 and P_2 translate around P_0 keeping touching P_0, while P_0 remains stationary. Let $\mathcal{K}_0 \subset \mathcal{K}$ be the space of configurations in which $\tau_0 = (0, 0)$ and P_0 touches both of P_1 and P_2. As discussed above for $k = 2$, the distance that each of P_1 and P_2 travels around P_0 is exactly L_1 and L_2, respectively, where $L_1 := |P_0| + |P_1|$ and $L_2 := |P_0| + |P_2|$. See Fig. 1. Then, any pair $(\lambda_1, \lambda_2) \in [0, L_1) \times [0, L_2)$ corresponds to a configuration $\tau(\lambda_1, \lambda_2)$ when P_1 and P_2 travel around P_0 by distance exactly λ_1 and λ_2, respectively, from their initial positions.

Notice that the definition of configurations do not prevent P_1 and P_2 from overlapping each other; rather, the translates of P_1 and P_2 around P_0 are independent, and are determined independently by two different parameters λ_1 and λ_2, respectively. We denote by $P_1(\lambda_1)$ and $P_2(\lambda_2)$ the translated copy of P_1 and P_2, respectively, corresponding to the parameters λ_1 and λ_2, respectively. By abuse of notation, we shall call a pair (λ_1, λ_2) a configuration in $\mathcal{K}_0$ and regard $\mathcal{K}_0$ to be $[0, L_1) \times [0, L_2)$.

3.2   Events and Event Curves

Recall that any configuration $\tau \in \mathcal{K}_0$ is associated with the set $C(\tau)$ of contacts and the set $H(\tau)$ of bridges of the corresponding convex hull. These two combinatorial associates determine the structure of the convex container and the motion of the polygons, thus being helpful in describing the objective function on the configuration space as will be shown in next sections. One natural approach would decompose the configuration space $\mathcal{K}_0$ into cells in each of which $C(\tau)$ and $H(\tau)$ remain the same for all configurations τ in the cell.

We call a configuration $\tau = (\lambda_1, \lambda_2) \in \mathcal{K}_0$ an *event* if it is one of the following cases:

C0 event. A vertex of $P_1(\lambda_1)$ or $P_2(\lambda_2)$ reaches a vertex of P_0; that is, a vertex-vertex contact occurs between P_0 and one of the others.

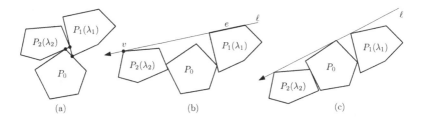

Fig. 2. Corresponding translates of the three polygons at events of different types: (a) C2 event, (b) H1 event, and (c) H2 event

C1 event. $P_1(\lambda_1)$ and $P_2(\lambda_2)$ touch each other at a vertex of $P_1(\lambda_1)$ and a vertex of $P_2(\lambda_2)$; that is, a vertex-vertex contact occurs between P_1 and P_2.

C2 event. $P_1(\lambda_1)$ and $P_2(\lambda_2)$ touch each other; that is, the three polygons are pairwise touching and it holds that $|C(\tau)| = 3$. See Fig. 2(a).

H0 event. $P_i(\lambda_i)$, for $i = 1$ or 2, is tangent to the supporting line of an edge of P_0 from the side containing P_0, or vice versa.

H1 event. $P_1(\lambda_1)$ is tangent to the supporting line of an edge of $P_2(\lambda_2)$ from the side containing $P_2(\lambda_2)$, or vice versa. See Fig. 2(b).

H2 event. The three polygons P_0, $P_1(\lambda_1)$, and $P_2(\lambda_2)$ have a common tangent line ℓ and the three lie in the same side of ℓ. See Fig. 2(c).

Remark that $\mathcal{K}_0$ includes configurations whose corresponding translates of P_1 and P_2 may overlap each other; the set of C2 events indeed form the borderline between configurations causing overlap and those not causing overlap. Note, however, that all the changes of $C(\tau)$ and $H(\tau)$ can be captured by a series of the events, when τ continuously moves inside $\mathcal{K}_0$ while it avoids overlap between P_1 and P_2. In particular, although some portions of H1 events indeed imply an overlap between P_1 and P_2, it suffices to track all the changes of $H(\tau)$ by H0, H1, and H2 events if τ continuously moves without any overlap. On the other hand, events of type C1 and C2 by definition imply no overlap between P_1 and P_2. Also, any C1 event is a C2 event by definition.

The set of all events forms a set of curves in the configuration space $\mathcal{K}_0 = [0, L_1) \times [0, L_2)$, and thus decomposes it into cells. To see this more precisely, we partition the set of events into subsets as follows:

Curves of C0 events. Any C0 event corresponds to a vertex-vertex contact, involving a pair (v, v') of vertices, exactly one of which belongs to P_0. We denote by $\gamma^{C0}_{vv'} = \gamma^{C0}_{v'v}$ the set of all C0 events with the involved pair (v, v').

Curves of C2 events. For any C2 event $\tau = (\lambda_1, \lambda_2)$, $P_1(\lambda_1)$ and $P_2(\lambda_2)$ touch each other. We have two cases: either $P_2(\lambda_2)$ is ahead of $P_1(\lambda_1)$ (in the sense that $P_1(\lambda_1 + \epsilon)$ overlaps $P_2(\lambda_2)$ for arbitrarily small $\epsilon > 0$) as depicted in Fig. 2(a), or vice versa. We denote the the set of C2 events corresponding to the former by γ^{C2}_1 and the set of C2 events corresponding to the latter one by γ^{C2}_2. Note that every C1 event coincides with a C2 event by definition, and thus all the C1 events are included in $\gamma^{C2}_1 \cup \gamma^{C2}_2$.

Curves of H0 events. Any H0 event τ corresponds to a collinearity of an edge e and a vertex v, one of which belongs to P_0. Let ℓ be the supporting line of e and assume that ℓ is directed so that the two polygons that each of v and e belongs to lie on its left side. There are two cases: the vertex v is ahead of e or behind e, along the directed line ℓ. We denote the set of all H0 events corresponding to the former by γ_{ve}^{H0} and the set of all H0 events corresponding to the latter by γ_{ev}^{H0}.

Curves of H1 events. Any H1 event $\tau = (\lambda_1, \lambda_2)$ corresponds to a collinearity of an edge e and a vertex v, each of which belongs mutually to P_1 or P_2. Let ℓ be the supporting line of e and assume that ℓ is directed so that both $P_1(\lambda_1)$ and $P_2(\lambda_2)$ lie on its left side. Note that ℓ translates as e (and the polygon containing e) translates. Observe that there are two different translations for P_1 around P_0 such that ℓ keeps being tangent to the translate of P_1; P_1 is ahead of P_0 along the directed line ℓ or vice versa. The analogue also holds for P_2. Thus, $\tau = (\lambda_1, \lambda_2)$ falls into one of the four cases. We denote by $\gamma_{ve,11}^{H1}$ the set of H1 events (λ_1, λ_2) defined by (v, e) such that $P_1(\lambda_1)$ is ahead of P_0 and $P_2(\lambda_2)$ is also ahead of P_0. Similarly, define the other three $\gamma_{ve,12}^{H1}$, $\gamma_{ve,21}^{H1}$, and $\gamma_{ve,22}^{H1}$. Fig. 2(b) shows a H1 event in the set $\gamma_{ve,21}^{H1}$.

Curves of H2 events. Any H2 event $\tau = (\lambda_1, \lambda_2)$ is associated with a line ℓ commonly tangent to the three polygons P_0, $P_1(\lambda_1)$, and $P_2(\lambda_2)$. We assume that ℓ is always directed so that P_0, together with the other two, lies on its left side. We have again four cases as we have for H1 event curves; either $P_1(\lambda_1)$ (or $P_2(\lambda_2)$) is ahead of P_0 along ℓ or is behind P_0. We divide the set of H2 events into four subsets as we did for H1 events and denote them by γ_{11}^{H2}, γ_{12}^{H2}, γ_{21}^{H2}, and γ_{22}^{H2}, respectively. Fig. 2(c) shows a H2 event in the set $\gamma_{ve,21}^{H2}$.

We let Γ be the family of those nonempty subsets of events defined above. We show in the following that every $\gamma \in \Gamma$ forms a monotone curve (or a set of monotone curves) in $\mathcal{K}_0$ with several nice behaviors.

Lemma 3. *Any set $\gamma \in \Gamma$ is monotone in both the λ_1-axis and the λ_2-axis, and consists of at most three curves on the configuration space $\mathcal{K}_0$. In addition, γ has following properties according to its type. The asterisks below mean "any."*

- $\gamma = \gamma_*^{C0}$ *or* γ_*^{H0}: γ *is a line parallel to the λ_1-axis or the λ_2-axis.*
- $\gamma = \gamma_*^{C2}$: γ *is the graph of a function from $[0, L_1)$ to $[0, L_2)$ that is increasing and piecewise linear, each of whose breakpoints coincides with a C0 or C1 event.*
- $\gamma = \gamma_*^{H1}$: γ *is the graph of a partial function on $[0, L_1)$ that is monotone and piecewise linear, each of whose breakpoints coincides with a C0 event.*
- $\gamma = \gamma_*^{H2}$: γ *is the graph of a function from $[0, L_1)$ to $[0, L_2)$ that is increasing and piecewise hyperbolic, each of whose breakpoints coincides with a C0 or H0 event.*

Each $\gamma \in \Gamma$ thus consists of one, two, or three curves unless it is axis-parallel. As shown in the proof of Lemma 3, the endpoints of γ occur when $\lambda_1 \in \{0, L_1\}$

or $\lambda_2 \in \{0, L_2\}$, except the endpoints of H1 event curves that lie on a C2 event curve. This discontinuity is because the configuration space $\mathcal{K}_0$ is indeed periodic; if we extend $P_1(\lambda_1)$ and $P_2(\lambda_2)$ for $\lambda_1 > L_1$ and $\lambda_2 > L_2$, then we have $P_1(\lambda_1 + L_1) = P_1(\lambda_1)$ and $P_2(\lambda_2 + L_2) = P_2(\lambda_2)$, and therefore γ becomes connected. We thus call each $\gamma \in \Gamma$ an *event curve* of type C0, C2, H0, H1, or H2 according to its type.

3.3   Complexity of Event Curves

We now discuss the complexity of event curves and of their arrangement $\mathcal{A}(\Gamma)$.

Lemma 4. *The family Γ consists of $O(n)$ event curves and the number of C1 events is bounded by $O(n)$. Also, each event curve in Γ consists of either $O(n)$ line segments or $O(n)$ hyperbolic segments.*

We now consider the arrangement $\mathcal{A}(\Gamma)$ of the event curves Γ.

Lemma 5. *The complexity of the arrangement $\mathcal{A}(\Gamma)$ is $O(n^3)$, and each of its edges is either a line segment or a hyperbolic arc. More specifically, the number of crossings between any two event curves in Γ is $O(n)$.*

Proof. We first show that the number of crossings between any two event curves in Γ is bounded by $O(n)$, which implies that the combinatorial complexity of the arrangement $\mathcal{A}(\Gamma)$ is bounded by $O(n^3)$ since Γ consists of $O(n)$ event curves by Lemma 4. By Lemma 3, any C0 or H0 event curve is axis-parallel and any $\gamma \in \Gamma$ is monotone in both axes. Thus, any C0 or H0 event curve intersects any other event curve at most once.

Consider two event curves $\gamma_1, \gamma_2 \in \Gamma$ of type C2, H1, or H2. Let f_1 and f_2 be the functions from $[0, L_1)$ to $[0, L_2)$ such that γ_i is the graph of f_i for $i = 1, 2$. By Lemmas 3 and 4, each f_i is monotone and has $O(n)$ breakpoints. And f_i is either linear or hyperbolic on any interval of $[0, L_1)$ between two consecutive breakpoints of f_i. This implies that there are at most two values of λ_1 in the interval such that $f_1(\lambda_1) = f_2(\lambda_1)$. Therefore, there are at most $O(n)$ crossings between γ_1 and γ_2. Since there are only $O(n)$ event curves of type C2, H1, or H2, there are at most $O(n^2)$ such combinations of (γ_1, γ_2). We thus have at most $O(n^3)$ crossings in this case. ☐

Note that the complexity of $\mathcal{A}(\Gamma)$ can be $\Omega(n^3)$ by a concrete construction of input polygons, so the bound of Lemma 5 is shown to be tight. Nonetheless, we prove a better bound if we focus on the feasible configurations, which imply no overlap between P_1 and P_2. We can easily see that the $O(n^3)$ complexity of $\mathcal{A}(\Gamma)$ is completely due to crossings among H1 event curves. By Lemma 3, any C0 or H0 curve crosses any other curve at most once. Since there are only six curves of type C2 and H2, the number of combinations (γ_1, γ_2) of any two curves of type C2, H1, or H2 but not both of H1 is $O(n)$, which implies that the total number of crossings between such combinations of curves is at most $O(n^2)$ by Lemma 5. Fortunately, the number of H1–H1 crossings that are feasible is shown to be much smaller.

Recall that the two C2 curves divide $\mathcal{K}_0$ into two regions, one consisting of all feasible configurations and the other of all infeasible configurations in $\mathcal{K}_0$. We denote by $\mathcal{F} \subset \mathcal{K}_0$ the former region. Since we want to find an optimal feasible configuration, we are mostly interested in the feasible region $\mathcal{F}$ and how it is decomposed.

Lemma 6. *For any two H1 event curves $\gamma_1, \gamma_2 \in \Gamma$, the number of crossings between γ_1 and γ_2 that lie in $\mathcal{F}$ is at most two. Therefore, the arrangement $\mathcal{A}(\Gamma)$ consists of $O(n^2)$ vertices and edges in $\mathcal{F}$.*

Proof. Consider two distinct H1 curves $\gamma_1, \gamma_2 \in \Gamma$, and suppose that $\gamma_1 \cap \gamma_2 \neq \emptyset$ and $(\lambda_1, \lambda_2) \in \gamma_1 \cap \gamma_2 \cap \mathcal{F}$. Without loss of generality, assume that γ_1 is defined by a collinearity of a vertex v of P_1 and an edge e of P_2, and γ_2 is by a vertex v' and an edge e' whichever of P_1 and P_2 they belong to. Let ℓ be the line supporting e of $P_2(\lambda_2)$ and ℓ' be the line supporting e' at this configuration. Then, we have $v \in \ell$ and $v' \in \ell'$. Also, let d be the distance between v and the closer endpoint of e along ℓ. We then observe that for any crossing in $\gamma_1 \cap \gamma_2$ the distance between v and the closer endpoint of e must be exactly d. This can be seen by simple geometry: Imagine that P_1 moves along ℓ towards e of P_2 from infinity, and see the distance between the line supporting e' and the vertex v'. There is at most one instance where e' and v' are aligned, and the distance between v and the closer endpoint of e is exactly d at the moment.

Now, consider the location of P_1 and P_2 as above. Since $(\lambda_1, \lambda_2) \in \mathcal{F}$, they do not overlap each other. We then have at most two possible position of P_0 that touches both P_1 and P_2. This means that there are at most two such coordinates (λ_1, λ_2), and thus two H1 curves can intersect at most twice in $\mathcal{F}$. Since there are $O(n)$ H1 curves in Γ, this suffices to show that the number of crossings in $\mathcal{F}$ among all H1 curves is $O(n^2)$. From Lemmas 4 and 5, we know that all the event curves consist of $O(n^2)$ line and hyperbolic segments and the number of crossings between $\gamma_1 \in \Gamma$ and $\gamma_2 \in \Gamma$ is $O(n)$ if γ_1 and γ_2 are of type C2, H1, or H2. Since the number of event curves of type C2 and H2 is only six, it suffices to show that the number of crossing among all H1 curves that lie in $\mathcal{F}$ is $O(n^2)$. $\boxdot$

Fig. 3 shows the arrangement $\mathcal{A}(\Gamma)$ of the event curves for the three input polygons depicted in Fig. 1. Although we insist to decompose $\mathcal{K}_0$ into cells in each of which the contacts $C(\tau)$ and the bridges $H(\tau)$ stay constant, remark that some cells of $\mathcal{A}(\Gamma)$ are in fact not the case. For our purpose, however, it suffices to well decompose the feasible region $\mathcal{F}$, which imply no overlap between P_1 and P_2.

Lemma 7. *The arrangement $\mathcal{A}(\Gamma)$ of the event curves decomposes the feasible region $\mathcal{F} \subset \mathcal{K}_0$ into cells σ such that both $C(\tau)$ and $H(\tau)$ remain constant over all $\tau \in \sigma$.*

Recall that all configurations in $\mathcal{K}_0$ assume P_0 to keep contact with both P_1 and P_2. Alternating the role of P_0 by P_1 or P_2, we achieve a complete description of the configuration space $\mathcal{K}$. Letting $\mathcal{K}_1$ and $\mathcal{K}_2$ be the analogous configuration space for P_1 and P_2, respectively, we have $\mathcal{K} = \mathcal{K}_0 \cup \mathcal{K}_1 \cup \mathcal{K}_2$.

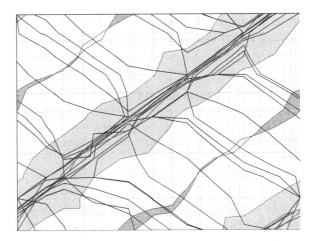

Fig. 3. The arrangement $\mathcal{A}(\Gamma)$ of the event curves in the configuration space $\mathcal{K}_0$: event curves of type C0 (light blue), H0 (light green), H1 (black), C2 (red), and H2 (purple). Any configuration in the gray region is infeasible, so the feasible region $\mathcal{F}$ is the complement of the gray region. For any configuration τ in the purple region enclosed by H2 event curves, we have $|H(\tau)| = 4$.

4   Algorithms

In this section, we present an algorithm that computes an optimal feasible configuration that minimizes the area or the perimeter of the convex hull of the three convex polygons under translation. The arrangement $\mathcal{A}(\Gamma)$ of the event curves is indeed sufficient to deal with the area or perimeter function in each feasible cell. Note that for any feasible configuration $\tau \in \mathcal{F}$, we have $2 \leqslant |H(\tau)| \leqslant 4$.

Lemma 8. *Let σ be any cell of $\mathcal{A}(\Gamma)$ with $\sigma \subset \mathcal{F}$. The area function is hyperbolic paraboloidal on σ if $|H(\tau)| = 3$ for $\tau \in \sigma$, or linear otherwise; the perimeter function is convex with $O(1)$ complexity on σ and on any edge incident to σ.*

An outline of our algorithm is as follows. We simply consider every feasible cell σ of $\mathcal{A}(\Gamma)$ in a certain order. Lemma 8 implies that the area or the perimeter function is fully described and can be minimized in constant time. For the purpose, we perform two phases: Firstly, compute the arrangement $\mathcal{A}(\Gamma)$ in the feasible region $\mathcal{F}$ only, and secondly traverse all of its cells that are feasible to minimize the area or the perimeter function restricted in each of those.

By Lemma 8, the second phase is relatively easy once the cells and the edges of $\mathcal{A}(\Gamma)$ lying in $\mathcal{F}$ are fully specified. At this phase, we visit every cell in $\mathcal{F}$ by crossing over an incident edge and thus moving to a neighboring cell. Then, by coherence, the description of the objective function restricted in the next cell can be obtained in constant time. Lemma 8 guarantees that the area or the perimeter function can be minimized in $O(1)$ time in a cell or on each of its bounding edges. Hence, the total time complexity of the second phase is bounded by $O(n^2)$ time by Lemma 6.

Computing the arrangement can be easily done in $O(n^2 \log n)$ time by a typical plane-sweep algorithm. In the following, we focus on improving the time bound to $O(n^2)$ for the task.

4.1 Computing the Arrangement $\mathcal{A}(\Gamma)$ in $\mathcal{F}$

In order to compute the arrangement $\mathcal{A}(\Gamma)$, we first compute all the event curves in Γ with full description, and then identify all the intersections among them that lie in $\mathcal{F}$.

Preprocessing. As a preprocessing, we take any two polygons P_i and P_j for $0 \leqslant i < j \leqslant 2$ and move P_j around P_i keeping a contact to P_i in the counterclockwise direction. During this motion, we gather all occurrences of vertex-vertex contact in order and store them into a sorted list C_{ij} with the corresponding pair of vertices. In addition, we maintain two external common tangents of P_i and P_j and gather all occurrences at which one of the two tangents supports an edge of P_i or P_j. We also store them into a sorted list H_{ij} with the corresponding pair of vertex and edge. Let us make C_{ij} and H_{ij} to be a circular list for later use. This preprocessing can be handled in $O(n)$ time as done in [6]. Observe that each member of C_{01} and C_{02} describes a C0 event curve in $\mathcal{K}_0$, and each of H_{01} and H_{02} describes an H0 event curve. We thus find all C0 and H0 event curves by traversing these lists.

Computing the Event Curves. Let G be the grid on $\mathcal{K}_0$ induced by all the C0 and H0 event curves. The other event curves can be obtained by tracing each across the grid cells of G. Consider the four H2 curves. By Lemma 3, each H2 event curve γ appears to be a hyperbolic segment in each grid cell σ intersected by itself, and the equation of each segment in σ can be described with help of the lists C_{ij} and H_{ij}. Its starting point at $\lambda_1 = 0$ can be found in $O(n)$ time, and then we trace γ cell by cell. As we walk along γ and move to the next grid cell σ', we immediately know the change of the contacts and the bridges so that the equation of γ in σ' can be updated in $O(1)$ time. Hence, tracing γ spends time proportional to the number of grid cells of G that are intersected by γ. Lemma 4 tells us that the number of such grid cells, and thus the cost of tracing an H2 curve is $O(n)$.

The other event curves of different types can be traced in the same fashion, taking $O(n)$ time for each. While tracing a C2 event curve, we can also specify all C1 events: this can be done by looking up the list C_{12} with a pointer that indicates the current contact between P_1 and P_2. Tracing an H1 event curve needs to look up the list H_{12}; in fact, only the members of H_{12} can determine an H1 event curve. We hence can compute all the event curves in Γ with their full description in $O(n^2)$ time.

Specifying All Necessary Crossings. We then compute the arrangement $\mathcal{A}(\Gamma)$ in $\mathcal{F}$ by specifying all necessary crossings among the event curves in Γ.

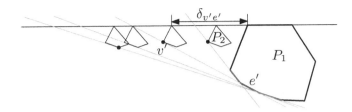

Fig. 4. Illustration to the list Δ

Note that for any two event curves $\gamma_1, \gamma_2 \in \Gamma$, all the crossings between them can be computed in $O(n)$ time by Lemmas 3 and 5. For all pairs (γ_1, γ_2) of event curves such that γ_1 is of type C2 or H2 and γ_2 is of type C2, H1, or H2, we are thus able to specify all the crossings between γ_1 and γ_2 in $O(n^2)$ time, since the number of such pairs (γ_1, γ_2) is $O(n)$. What remains is to specify the crossings among the H1 event curves.

For the last task, we take only feasible portions of every event curve into account. Let $\Gamma_{\mathcal{F}} := \{\gamma \cap \mathcal{F} \mid \gamma \in \Gamma\}$. Computing $\Gamma_{\mathcal{F}}$ can be done by cutting each $\gamma \in \Gamma$ by the C2 curves and discard its infeasible portions. The type of $\gamma' \in \Gamma_{\mathcal{F}}$ is inherited from $\gamma \in \Gamma$ such that $\gamma' = \gamma \cap \mathcal{F}$. Fortunately, this cutting does not increase the number of curves much, especially, H1 event curves.

Lemma 9. *The number of H1 event curves in $\Gamma_{\mathcal{F}}$ is $O(n)$*

Lemma 6 implies that the number of crossings in $\mathcal{F}$ between a fixed H1 curve γ and all the other H1 curves is $O(n)$. In the following, we show that all the crossings on a fixed γ with other H1 curves can be specified in $O(n)$ time. By Lemma 9, it suffices to conclude the total $O(n^2)$ time.

For the purpose, we need some more observations. Let $\gamma \in \Gamma_{\mathcal{F}}$ be an H1 event curve defined by a pair $(v, e) \in H_{12}$ of vertex v and edge e of P_1 and P_2. For any configuration $(\lambda_1, \lambda_2) \in \gamma$, $P_1(\lambda_1)$ and $P_2(\lambda_2)$ have a common external tangent that supports e. Let g be a function partially defined on $[0, L_1)$ whose graph is γ, and define $d_\gamma(\lambda_1)$ to be the distance between v and the endpoint of e that are the closer to v in the corresponding translates $P_1(\lambda_1)$ and $P_2(g(\lambda_1))$. Observe that d_γ is linear in a grid cell σ of G since P_1 and P_2 move in a linear way in σ along γ.

On the other hand, we consider the other external common tangent $\ell(\lambda_1)$ of $P_1(\lambda_1)$ and $P_2(g(\lambda_1))$. When an edge e' of P_1 or P_2 lies on $\ell(\lambda_1)$, we have a crossing between γ and another H1 event curve γ' defined by (v', e') for the vertex v' lying on $\ell(\lambda_1)$; we let $\delta_{v'e'} := d_\gamma(\lambda_1)$ at such a value of λ_1. By a geometric observation, at such a crossing, $P_1(\lambda_1)$ and $P_2(g(\lambda_1))$ have two external common tangents, one supporting e and the other supporting e'; this fixes a unique value of $d_\gamma(\lambda_1)$ to be $\delta_{v'e'}$. This implies that for any λ_1, γ' crosses γ at $(\lambda_1, g(\lambda_1))$ if and only if $d_\gamma(\lambda_1) = \delta_{v'e'}$.

We thus perform the following procedure. (See Fig. 4 for an illustration.) We compute the value $\delta_{v'e'}$ for all $(v', e') \in H_{12}$ and store them into a sorted array Δ with corresponding label (v', e'). Then, we walk along γ cell by cell to find all occurrences such that $d_\gamma(\lambda_1) = \delta$ for some $\delta \in \Delta$. This completely specifies all crossings between γ and all other H1 event curves in $\Gamma_{\mathcal{F}}$. To compute Δ, initially make P_1 and P_2 touch each other, keeping a common tangent going through v and e, and consider the other common tangent line ℓ. If we move P_2 in the direction parallel to e and away from P_1, then the tangent line ℓ will rotate monotonously in one direction. This implies that the order of $\delta_{v'e'}$ follows from the order of (v', e') in the list H_{12}. Thus, we can compute Δ in $O(n)$ time.

Let $\delta_1, \ldots, \delta_m$ be the members of Δ in the order. Once we compute Δ, we walk along γ by increasing λ_1 to find all λ_1 such that $d_\gamma(\lambda_1) = \delta$ holds for some $\delta \in \Delta$. Since d_γ is linear in each cell σ of G intersected by γ, the task is not difficult if we maintain a variable a such that $\delta_a \leqslant d_\gamma(\lambda_1) < \delta_{a+1}$ for the current value of λ_1. Hence, we can find all crossings on $\gamma \cap \sigma$ with the other H1 event curves in time $O(1 + c)$, where c is the number of the reported crossings in σ. If we sum up this over all grid cells intersected by γ, we obtain $O(n)$ time bound.

Putting it all together, we can specify all intersections among curves in $\Gamma_{\mathcal{F}}$ in $O(n^2)$ time, which are the vertices of $\mathcal{A}(\Gamma_{\mathcal{F}})$. We then cut each curve $\gamma \in \Gamma_{\mathcal{F}}$ by the crossings on γ to obtain the edges of $\mathcal{A}(\Gamma_{\mathcal{F}})$. As a result, we can build the underlying graph of the arrangement $\mathcal{A}(\Gamma_{\mathcal{F}})$, and then the arrangement $\mathcal{A}(\Gamma_{\mathcal{F}})$ can be built in the same time bound $O(n^2)$. We finally conclude our main result.

Theorem 1. *Given three convex polygons P_0, P_1, and P_2 having a total of n vertices, one can find in $O(n^2)$ time using $O(n^2)$ space an optimal pair (τ_1, τ_2) of translation vectors such that the area $\|\mathrm{conv}(P_0 \cup \tau_1 P_1 \cup \tau_2 P_2)\|$ or the perimeter $|\mathrm{conv}(P_0 \cup \tau_1 P_1 \cup \tau_2 P_2)|$ is minimized.*

References

1. Ahn, H.-K., Cheong, O.: Stacking and bundling two convex polygons. In: Deng, X., Du, D.-Z. (eds.) ISAAC 2005. LNCS, vol. 3827, pp. 882–891. Springer, Heidelberg (2005)
2. Ahn, H.K., Cheong, O.: Aligning two convex figures to minimize area or perimeter. Algorithmica 62, 464–479 (2012)
3. Alt, H., Hurtado, F.: Packing convex polygons into rectangular boxes. In: Akiyama, J., Kano, M., Urabe, M. (eds.) JCDCG 2000. LNCS, vol. 2098, pp. 67–80. Springer, Heidelberg (2001)
4. Daniels, K., Milenkovic, V.: Multiple translational containment, part i: An approximation algorithm. Algorithmica 19, 148–182 (1997)
5. Garey, M., Johnson, D.: Computers and Intractability: A Guide to the Theory of NP-Completeness. W.H. Freeman and Co., San Francisco (1979)
6. Lee, H., Woo, T.: Determining in linear time the minimum area convex hull of two polygons. IIE Trans. 20, 338–345 (1988)
7. Milenkovic, V.: Translational polygon containment and minimum enclosure using linear programming based restriction. In: Proc. 28th Annual ACM Symposium on Theory of Computation (STOC 1996), pp. 109–118 (1996)
8. Sugihara, K., Sawai, M., Sano, H., Kim, D.S., Kim, D.: Disk packing for the estimation of the size of a wire bundle. Japan J. Industrial and Applied Math. 21, 259–278 (2004)
9. Tang, K., Wang, C., Chen, D.: Minimum area convex packing of two convex polygons. Internat. J. Comput. Geom. Appl. 16, 41–74 (2006)

Smart-Grid Electricity Allocation
via Strip Packing with Slicing[*]

Soroush Alamdari[1], Therese Biedl[1], Timothy M. Chan[1], Elyot Grant[2],
Krishnam Raju Jampani[3], Srinivasan Keshav[1],
Anna Lubiw[1], and Vinayak Pathak[1]

[1] Cheriton School of Computer Science, University of Waterloo, Waterloo, Canada
{s26hosseinialamdari,biedl,tmchan,alubiw,keshav,vpathak}@uwaterloo.ca
[2] Massachusetts Institute of Technology, Cambridge, USA
elyot@mit.edu
[3] University of Guelph, Guelph, Canada
rjampani@uoguelph.ca

Abstract. One advantage of smart grids is that they can reduce the peak load by distributing electricity-demands over multiple short intervals. Finding a schedule that minimizes the peak load corresponds to a variant of a strip packing problem. Normally, for *strip packing problems*, a given set of axis-aligned rectangles must be packed into a fixed-width strip, and the goal is to minimize the height of the strip. The electricity-allocation application can be modelled as *strip packing with slicing:* each rectangle may be cut vertically into multiple slices and the slices may be packed into the strip as individual pieces. The *stacking constraint* forbids solutions in which a vertical line intersects two slices of the same rectangle.

We give a fully polynomial time approximation scheme for this problem, as well as a practical polynomial time algorithm that slices each rectangle at most once and yields a solution of height at most 5/3 times the optimal height.

1   Introduction

The conventional approach to generating and distributing electricity relies on sizing infrastructure to support the peak load, when demand for electricity is highest. However, this peak is rarely reached, so much of the expensive infrastructure is idle most of the time. For example, in 2009, 15% of the generation capacity in Massachusetts was used less than 88 hours per year [7]. Reducing the infrastructure size is not practical since unsupported demand can cause blackouts. Therefore, there is considerable benefit to reducing the peak load itself.

Peak load occurs when many consumers use power-hungry appliances simultaneously. However, there is often flexibility in scheduling the use of particular appliances. For example, a water heater requires a certain amount of electricity to heat the water, but can equally well heat the water in one continuous interval

---

[*] This work was done as part of an Algorithms Problem Session at the University of Waterloo. Research of TB, TC, SK and AL supported by NSERC.

F. Dehne, R. Solis-Oba, and J.-R. Sack (Eds.): WADS 2013, LNCS 8037, pp. 25–36, 2013.
© Springer-Verlag Berlin Heidelberg 2013

or in multiple short intervals.[1] It is anticipated that future smart grids would obtain (at each substation) daily "demand schedules" for appliance use from the consumers in its local area, and then automatically re-schedule appliance use to minimize peak load [18].

The demand schedule can be modelled as a set of rectangles, one for each appliance, with power consumption as height, and desired running time as width. The re-scheduling should cover a given length of time, which corresponds to a strip of given width. The objective is then to pack slices of the rectangles into the strip so as to minimize the maximum power consumption, i.e., the maximum height of the packing. Because appliances cannot be powered at double the usual power, we have the additional *stacking constraint* requiring that no vertical line may intersect two slices from the same rectangle. Slicing with the stacking constraint is new, but strip packing has been well-studied, as we review in the following section.

Strip Packing Problems. In the *two-dimensional strip packing problem* (abbreviated 2SP), a set of axis-aligned rectangles of specified dimensions must be packed, without rotation, into a rectangular strip of fixed width, with the goal of minimizing the height of the strip. The 2SP problem is very well-studied [14], and generalizes the *bin packing problem*, which is equivalent to the case in which all rectangles have unit height. The current best approximation algorithm for 2SP has an approximation factor of $5/3 + \varepsilon$ for any $\varepsilon > 0$ [9], and was achieved after a long sequence of successive improvements [1,16,17,19]. This is an *absolute performance bound*, i.e., the height achieved is at most $5/3 + \varepsilon$ times the optimal height. Many other authors have proposed algorithms with *asymptotic* performance guarantees [4,13,11] where an additive term is allowed.

Motivated by the electricity-allocation problem, we study a variant called *two-dimensional strip packing with slicing* (hereafter 2SP-S). In 2SP-S, we are allowed to cut each rectangle vertically into multiple slices, which may be packed into the strip as individual rectangles. Formally, the input consists of a number W and a set of rectangles $r_1, r_2, \ldots, r_n$. Here W is the width of the strip, which consists of two vertical sides at $x = 0$ and $x = W$, and the "base" at $y = 0$. Rectangle r_i has width w_i and height h_i; let $h_{\max} = \max_{i=1}^n h_i$ be the maximum height. A solution to 2SP-S consists of a partition of each rectangle r_i into vertical slices and an assignment of positions to the slices so that the interiors of the slices are pairwise disjoint. Slices must not be rotated. The *height* of a solution, denoted by H, is the minimum y-coordinate above which the strip is empty. The objective is to find a solution with minimal possible height H_{OPT}.

Related Results: Strip packing with slicing has been studied for a variant in which the width of each rectangle represents a demand for a number of concurrently running processors [2]. However, this problem differs substantially from 2SP-S because the slices must have integer widths and must be horizontally aligned due to concurrency, and results for it do not carry over.

---

[1] To simplify the modelling we presume that no extra electricity is needed for restarting the appliance.

One may observe that if each rectangle is pre-cut into slices of some small width $\delta = W/m$ (where m is a positive integer), then solving the resulting 2SP problem is precisely equivalent to solving the minimum makespan scheduling problem on m parallel machines (the $P_m||C_{max}$ problem, in three-field notation). Unfortunately, the known approximation schemes for this problem all have a running time that is either exponential in m [15] or in $\frac{1}{\varepsilon}$ [10]. In fact, when the number of machines m is an input to the problem, minimum makespan scheduling on m parallel machines admits no FPTAS unless P=NP [3], so it appears hopeless to find an FPTAS for 2SP-S using such an approach.

A second reason that existing results do not apply to electricity-allocation (at least not as far as we can prove) is that in our application we have the additional stacking constraint requiring that no vertical line may intersect two slices from the same rectangle. The version of 2SP-S with the stacking constraint is denoted by 2SP-SSC. Many of our results in this paper hold for both 2SP-S and 2SP-SSC; we shall note situations in which there are differences.

Results For 2SP-S and 2SP-SSC. The freedom to slice rectangles can be highly beneficial. It is easy to construct an example where slicing reduces the required height by a factor of $2 - \varepsilon$. Slicing also makes a difference in the complexity of the problem. Standard 2SP generalizes bin packing and is thus strongly NP-complete. Also, a simple reduction from the *Partition problem* [5] shows that 2SP admits no $(3/2 - \varepsilon)$-approximation for any $\varepsilon > 0$ unless P=NP. In contrast, 2SP-S and 2SP-SSC are easily shown to be NP-hard, but not hard to approximate: we will give a fully polynomial-time approximation scheme (FPTAS).

The FPTAS is based on solving a linear program with exponentially many constraints, and hence mostly of theoretical interest. We also develop simpler, more practical algorithms, and also limit the number of times a rectangle may be sliced (which is of interest in the electricity-allocation problem to avoid start-up costs for the appliance).[2] We give two simple 2-approximation algorithms based on the well-known First Fit and Shelf paradigms. In fact, these algorithm achieve height $H_{OPT} + h_{max}$, hence they achieve optimal height up to an additive term. Then, building on these algorithms, and splitting the problem into two halves, we give two other polynomial-time algorithms that perform no worse than First Fit and Shelf. One has absolute performance bound $3/2$. The other uses at most one cut per rectangle and has absolute performance bound $5/3$.

Our paper is organized as follows. The First Fit and Shelf algorithms are in Section 2. Section 3 contains the FPTAS, and Section 4 develops practical algorithms. We conclude in Section 5.

2   Basic Algorithms

This section describes the *First Fit* and *Shelf* heuristics for 2SP-S and 2SP-SSC. Both algorithms achieve an approximation factor of 2, which is noteworthy

---

[2] More precisely, we want to minimze the times a rectangle is *interrupted*, i.e., one slice ends and no other slice starts. The number of times a rectangle is sliced is certainly an upper bound for this.

given that, for the standard strip packing problem, 2-approximation algorithms are difficult to obtain [16,19]. Both algorithms in fact achieve a height of $H_{OPT} + h_{\max}$, and hence have asymptotically performance bound 1.

First Fit Algorithm. Given a list of rectangles $r_1, r_2, \ldots, r_n$, the *First Fit* algorithm processes them in order, repeatedly finds the lowest column in the current solution where a slice of r_i can be placed, and places the widest possible slice of r_i there, breaking ties arbitrarily. Repeat with the remainder of r_i, and continue until all rectangles have been processed. In the case of 2SP-SSC, the stacking constraint must be respected when placing slices. See Figure 1.

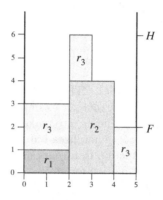

Fig. 1. An execution of the First Fit algorithm on a 2SP-SSC instance. Note that r_3 is sliced twice, and a smaller height would be achieved without the stacking constraint.

It is not hard to show that after placing each rectangle, the difference between the maximum height H and the floor F (the maximum height to which the entire strip is filled) is at most $h_{\max}$. Since by area-consideration $H_{OPT} \geq F$, First Fit achieves height at most $H_{OPT} + h_{\max}$ and is a 2-approximation since $h_{\max} \leq H_{OPT}$.

Unfortunately there are instances where First Fit needs nearly twice the optimal height. A natural improvement to First Fit is to sort the rectangles by decreasing heights first; we call this variant First Fit Decreasing. One can easily construct an instance on which even First Fit Decreasing is a factor of $\frac{4}{3}$ away from the optimum. We do not know whether this is tight, but we can show:

Lemma 1. *First Fit Decreasing is a $\frac{3}{2}$-approximation.*

Proof. (Sketch) We prove this by defining another algorithm that packs (until a certain condition is fulfilled) as much as possible into columns that contain only rectangles of height at most $H_{OPT}/2$. One can show that the "tall" (in some sense) columns of this algorithm have the same heights as the "tall" columns of First Fit Decreasing. One can also show that this other algorithm is a $\frac{3}{2}$-approximation. Putting the two together shows that First Fit Decreasing is a $\frac{3}{2}$-approximation. □

Shelf Algorithm. Given a set of rectangles, the Shelf algorithm works as follows. Sort the rectangles by decreasing height so that $h_1 \geq h_2 \geq \ldots \geq h_n$. Pack the rectangles in this order on "shelves" (also called "levels"). The first shelf is the base of the strip. Place rectangles on the current shelf from left to right. When we reach a rectangle r_i that is too wide for the remaining space, we pack the widest possible slice of r_i. The rest of r_i goes back in the list of remaining rectangles. Then we place a horizontal line across the strip to form a new shelf at the current maximum height of the packing, and continue on the new shelf with the remaining rectangles. See Figure 2. Note that the stacking constraint is automatically satisfied, and each rectangle is sliced at most once.

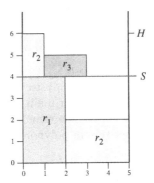

Fig. 2. An execution of the Shelf algorithm on the same instance as Figure 1 (but rectangles have been sorted by height)

The Shelf algorithm is the same as the Next-Fit-Decreasing-Height algorithm for strip packing [4], except that we fill the entire width of the shelf immediately because we can slice rectangles. It is known that Next-Fit-Decreasing-Height achieves height $2 \cdot H_{\mathrm{OPT}} + h_{\max}$ even for strip packing without slicing [4]. As we will show now, permitting to slice allows to decrease this bound to $H_{\mathrm{OPT}} + h_{\max}$.

Observe that (with $h_{n+1} := 0$) the empty space below height H has area at most $\sum_{i=1}^{n}(h_i - h_{i+1}) \cdot W$. To see this, partition the empty space into rectangles by cutting it horizontally, and assign each empty rectangle to the rectangle r_i that has a slice below it in the same shelf. Therefore, the empty space is at most $h_1 \cdot W = h_{\max} \cdot W$, which proves that the Shelf algorithm achieves height at most $H_{\mathrm{OPT}} + h_{\max}$.

3   Approximation Schemes

In this section, we sketch the FPTAS for 2SP-S and 2SP-SSC.

Theorem 1. *For any $\varepsilon > 0$, there exist $(1 + \varepsilon)$-approximation algorithms for 2SP-S and 2SP-SSC, assuming input numbers are rationals represented explicitly in binary. Their run-time is polynomial in the input size and $\frac{1}{\varepsilon}$.*

The approach uses a linear programming relaxation and is relatively standard in the literature; in particular it resembles the classic work of Karmarkar and Karp concerning the bin-packing problem [12]. The linear program we solve is similar to the one used to obtain fractional strip packings in [13], though our full algorithm requires different searching and rounding routines since the variables in our linear program must correspond to vertical configurations rather than horizontal ones.

In the remainder of this section, we prove Theorem 1 for the case of 2SP-SSC; we omit the (minor) changes that must be done for 2SP-S.

Step 1: Reducing the general problem to a decision version

Given a guess H_{GUESS} for the optimal height H_{OPT}, the main algorithm that we describe in steps 2 through 5 is capable of establishing one of the following:

(YES) There is a solution of value at most $H_{\text{GUESS}}(1 + \frac{\varepsilon}{2})$.

(NO) There is no solution of value less than or equal to H_{GUESS}.

Since the optimal height H_{OPT} is at most $\sum_{i=1}^{n} h_i$ and at least $\frac{1}{n} \sum_{i=1}^{n} h_i$, it is possible, via binary search, to establish H_{OPT} to within a multiplicative factor of $1 + \varepsilon$ using only $O(\log(\frac{n}{\varepsilon}))$ queries to our main algorithm. This then yields a $(1 + \varepsilon)$-approximation for the problem. The remaining steps describe how such queries can be answered constructively in polynomial time.

Step 2: Rounding the heights

Our linear programming method will require us to solve an instance of the knapsack problem to obtain a solution to the separation problem for the dual linear program. To render these knapsack instances tractable, we must round the heights of the rectangles in the input to multiples of an appropriate value h_0.

For 2SP-SSC, given a value of H_{GUESS}, we round all of the heights of the input rectangles down to the nearest multiple of $h_0 = \frac{\varepsilon}{2n} H_{\text{GUESS}}$. We will subsequently solve the resulting instance exactly using linear programming, obtaining a solution $\mathcal{S}$ of height H^*. It is immediate that $H^* \leq H_{\text{OPT}}$, and thus if $H^* \geq H_{\text{GUESS}}$, then there is no solution of value less than or equal to H_{GUESS}. Conversely, the stacking constraint implies that each vertical line passes through the interior of at most n rectangles in $\mathcal{S}$, so after undoing the rounding, the height of $\mathcal{S}$ increases by at most $\frac{\varepsilon}{2} H_{\text{GUESS}}$. Thus if $H^* \leq H_{\text{GUESS}}$, then there exists a solution to the original (unrounded) problem of value at most $H_{\text{GUESS}}(1 + \frac{\varepsilon}{2})$. Consequently, we can answer (YES) or (NO) depending on whether or not $H^* \leq H_{\text{GUESS}}$.

Step 3: Linear programming formulation

After rounding, each rectangle's height is a multiple of h_0, and we attempt to pack all rectangles into a strip of height at most H_{GUESS}. We define a *pattern* to be any subset of $\{r_1, \ldots, r_n\}$ whose total height is at most H_{GUESS}, and let $\mathcal{P}$ denote the set of all patterns. We observe that if arbitrary vertical slicing is permitted, then a solution to the strip packing problem can be exhibited by specifying, for each pattern $P \in \mathcal{P}$, the total width of pattern P used in the arrangement. This idea motivates our formulation.

For each pattern P, we define the variable x_P to represent the total width of pattern P used in a solution. It follows that determining the minimum strip *width* required to pack all of the rectangles into a strip of height H_{GUESS} is equivalent to solving the following linear program:

$$
\begin{aligned}
\text{minimize:} \quad & \sum_{P \in \mathcal{P}} x_P \\
\text{subject to:} \quad & \sum_{P \in \mathcal{P} | r_i \in P} x_P \geq w_i \text{ for all } 1 \leq i \leq n \qquad \text{(LP)} \\
& x_P \geq 0 \text{ for all } P \in \mathcal{P}
\end{aligned}
$$

It is immediate that upon solving this exactly, we may answer (YES) if and only if the optimal objective value W^* is at most W.

Step 4: Solving the linear program

We provide a polynomial algorithm for finding the optimal objective value W^* to our linear program. To do this, we examine the following dual of (LP):

$$
\begin{aligned}
\text{maximize:} \quad & \sum_{i=1}^{n} w_i y_i \\
\text{subject to:} \quad & \sum_{i | r_i \in P} y_i \leq 1 \text{ for all } P \in \mathcal{P} \qquad \text{(LP*)} \\
& y_i \geq 0 \text{ for all } 1 \leq i \leq n
\end{aligned}
$$

Despite this linear program having exponentially many constraints, we can tackle it using the ellipsoid algorithm. Specifically, since we assumed that the widths of the rectangles in the input are rational numbers represented explicitly in binary, we can find the *exact* optimal objective value of (LP*) in time polynomial in the input size and $\frac{1}{\varepsilon}$, provided that we can solve the corresponding *separation problem* in time polynomial in the input size and $\frac{1}{\varepsilon}$.

The separation problem for this linear program asks the following: Given values of y_i, either find a pattern P such that $\sum_{i | r_i \in P} y_i > 1$, or determine that no such pattern exists. If we regard each rectangle as having height h_i and value y_i, then this essentially asks if there is any set of rectangles of total height less than H_{GUESS} having total value greater than 1, and to return such a pattern if one exists. This can be answered by solving a knapsack instance having weight-value pairs (h_i, y_i) and maximum weight H_{GUESS}. Since each height in the rounded problem is a multiple of h_0 and $H_{\text{GUESS}} = \frac{2n}{\varepsilon} h_0$, this can be done in $O(\frac{n^2}{\varepsilon})$ time using standard dynamic programming methods.

If one wishes to achieve a more practical running time, it is feasible to replace the ellipsoid algorithm with the simplex algorithm, using the column generation technique of Gilmore and Gomory [6].

Step 5: Returning the solution

We observe that it is possible to reconstruct an optimum solution to (LP) while solving (LP*) using this technique (see [12] for details). Consequently, we can not only approximate the optimum height of a packing, but can in fact return a packing having that height. We already argued with Step 2 that this satisfies the approximation bound. Moreover, since a *basic* solution to (LP) is obtained, there are at most n patterns P for which the primal variables x_P are non-zero in the solution, implying that our algorithm returns a solution in which each rectangle is sliced at most $n - 1$ times.

We also observe that in the solution produced by the FPTAS, the number of cuts per rectangle can be further reduced to a constant that depends only on ε. More precisely, we can show (details are omitted) that any feasible solution can be modified so that each rectangle is sliced at most $(1/\varepsilon)^{O(1/\varepsilon)}$ times, without increasing the height by more than a factor of $1 + O(\varepsilon)$.

4   Algorithms with Few Slices

Although the approximation scheme from the previous section may be more practical if the simplex method is used, it is still unsuitable for electricity-allocation applications both due to its runtime and because it may result in rectangles that have been sliced numerous times. In fact, we can create instances where some rectangle must be sliced $\Omega(n/\log n)$ times in any optimal solution. For practical purposes, it would be worth sacrificing some height if in exchange we can guarantee that rectangles are not sliced too often. We develop such algorithms now.

The approach is to partition the bin vertically into two parts, slice each rectangle once, and pack the two slices in the two parts with Shelf. With a judicious choice of where to partition and slice, this results in a 5/3-approximation that slices each rectangle at most three times. (We note that this result is achieved with Shelf already if $h_{\max} \leq \frac{2}{3} H_{\mathrm{OPT}}$.) With some more work we can align rectangle slices so that each rectangle is sliced at most once.

The algorithm assumes that the value of H_{OPT} is known (we will find H_{OPT} with binary search as explained later). It depends on some parameter $t > H_{\mathrm{OPT}}/2$; using $t = 2H_{\mathrm{OPT}}/3$ gives the best approximation bound.

Step 1. Assuming the rectangles have been sorted in decreasing order of heights, find the largest k such that $w_1 + \cdots + w_k \leq W$. By $t > H_{\mathrm{OPT}}/2$ we have $h_{k+1} \leq t$. Find the largest $j \leq k$ such that $h_j \geq t$. (In case $h_{\max} < t$, we define j to be 0; the algorithm becomes identical to Shelf in this case.) Call $r_1, \ldots, r_j$ the *left floor* rectangles (of heights $\geq t$) and $r_{j+1}, \ldots, r_k$ the *right floor* rectangles (of heights $< t$). We divide the strip into two parts, where the left side has width $w_1 + \cdots + w_j$. Define α to be $(w_1 + \cdots + w_j)/W$, so the left side has width αW and the right side has width $(1 - \alpha)W$. Note that we may have $\alpha = 0$ or $\alpha = 1$, but $\alpha \leq 1$ since in any optimal packing no two floor rectangles may overlap vertically by $t > H_{\mathrm{OPT}}/2$.

Step 2. We split each rectangle into left and right pieces, subject to the constraint that the width of the left (resp. right) piece is at most αW (resp. $(1 - \alpha)W$). Either piece is allowed to be empty. The splitting procedure is described below. In the following, *shifting* a rectangle *rightward* means enlarging the width of its right piece by δ and shrinking the width of its left piece by δ for some amount $\delta > 0$. Shifting a rectangle leftward is similarly defined.

We say that a rectangle is shifted *completely* rightward if the left piece is empty or the right piece has width $(1 - \alpha)W$. We say that a rectangle is shifted *completely* leftward if the right piece is empty or the left piece has width αW.

All left floor rectangles are shifted completely leftward and all right floor rectangles are shifted completely rightward. All non-floor rectangles are initially shifted completely rightward. Let A_L^0 and A_R^0 be the total area of all left (resp. right) pieces after this initialization. We now shift rectangles so that A_R (the area of the current right pieces) equals $(1 - \alpha)H_{\text{OPT}}W$, if possible, and do this using a greedy procedure:

- If $A_R^0 \le (1 - \alpha)H_{\text{OPT}}W$, then stop.
- Otherwise, for each non-floor rectangle *from minimum to maximum height* while $A_R > (1 - \alpha)H_{\text{OPT}}W$, decrease A_R by shifting the rectangle leftward either completely or until $A_R = (1 - \alpha)H_{\text{OPT}}W$.

Observe that except for one *critical rectangle*, which we denote by r_x, all rectangles are either shifted completely leftward or completely rightward. We also claim that the above procedure ends with either $A_R = A_R^0$ or $A_R = (1 - \alpha)H_{\text{OPT}}W$, whichever is smaller. For assume that all non-floor rectangles have been shifted completely leftward. The left floor rectangles have total area at least $t\alpha W$ and the right floor rectangles have total area less than $t(1 - \alpha)W$, so among the floor rectangles, at least an α-fraction of the area has been assigned to the left. Each non-floor rectangle (shifted completely leftward) has at least an α-fraction of the area on the left. Therefore the area to the left is at least α-fraction of the total area, or $A_L \ge \alpha A$, which implies $A_R = A - A_L \le (1 - \alpha)A \le (1 - \alpha)H_{\text{OPT}}W$. So if $A_R^0 > (1 - \alpha)H_{\text{OPT}}W$, then at some point during the shifting process we reach a moment when $A_R = (1 - \alpha)H_{\text{OPT}}W$ as desired.

Step 3. Pack the left pieces into the left strip and the right pieces into the right strip, using Shelf on both sides.

Theorem 2. *There exists a 5/3-approximation for 2SP-SSC that slices every rectangle at most three times and runs in time $O(n\log(nM))$, where M is an upper bound on the integer heights of the rectangles.*

Proof. Assume for now that H_{OPT} is known and apply the above algorithm. This gives a packing with at most 3 cuts per rectangle (one to partition it into the left and right piece, and one by each application of Shelf), and the only rectangle that may have 3 cuts is r_x.

We first analyze the height of the left strip. The bottommost shelf has height $h_{\max}$, and (by definition) contains the left floor rectangles whose total area is at least $t\alpha W$. The left pieces of non-floor rectangles hence have total area at most $A_L - t\alpha W =: A_{NFL}$. Let ℓ be the tallest height of a non-floor rectangle whose left piece is non-empty ($\ell = 0$ if there is no such piece.) By the same analysis as for Shelf, the shelves for the left pieces of non-floor rectangles have empty space at most $\ell\alpha W$, hence they contribute height at most $(A_{NFL})/(\alpha W) + \ell = A_L/\alpha W - t + \ell$.

Claim: $\ell + t \le H_{\text{OPT}}$. Clearly this holds if $\ell = 0$, so assume $\ell > 0$. To prove the claim, consider an optimal packing S^* and assume that its height is less than

$\ell + t$. Then no vertical line intersects two left floor rectangles in S^*, since these all have height $\geq t$ and $\geq \ell$. Rearrange S^* so that all vertical lines containing left floor rectangles appear at the left end, and call this part the *left strip of S^**, which has width αW. Again, since the height of S^* is less than $\ell + t$, no right floor rectangles and no non-floor rectangle of height $\geq \ell$ can appear in the left strip of S^*. Thus the right floor rectangles and non-floor rectangles of height $\geq \ell$ all fit in the right strip of S^*. Also, all such non-floor rectangles have width at most $(1 - \alpha)W$. Finally, for any non-floor rectangle of height $< \ell$, at most a slice of width αW can be in the left strip, so if the rectangle has width $> \alpha W$, then its right piece, even when entirely shifted leftward, also fits within the right strip of S^*.

Recall that the greedy procedure for splitting rectangles processes rectangles in increasing height. Since we have a left piece of a non-floor rectangle of height ℓ, all non-floor rectangles of height $< \ell$ must be completely shifted leftward. So by the time the procedure reaches the rectangle of height ℓ, the right pieces consists of the right floor rectangles, the minimum possible right pieces of non-floor rectangles of height $< \ell$, and the entire non-floor rectangles of height $\geq \ell$. By the above discussion, all these pieces fit into the right strip of S^*, which has area $(1 - \alpha)W H_{\mathrm{OPT}}$. But then $A_R < (1 - \alpha)W H_{\mathrm{OPT}}$ already and the greedy procedure would have stopped and not shifted the rectangle of height ℓ leftward. This is a contradiction, so the optimal height is at least $t + \ell$, which proves the claim.

Putting all of this together and using $h_{\max} \leq H_{\mathrm{OPT}}$, the left strip has height at most

$$h_{\max} + \frac{A_L}{\alpha W} - t + \ell \leq H_{\mathrm{OPT}} + \frac{A_L}{\alpha W} - t + H_{\mathrm{OPT}} - t = 2H_{\mathrm{OPT}} - 2t + \frac{A_L}{\alpha W}.$$

In the right strip, all rectangles have height at most t. Hence the right strip has at most $t(1 - \alpha)W$ empty area, and its height is at most

$$\frac{A_R}{(1 - \alpha)W} + t \leq \frac{(1 - \alpha)H_{\mathrm{OPT}}W}{(1 - \alpha)W} + t = H_{\mathrm{OPT}} + t \leq \frac{5}{3}H_{\mathrm{OPT}}$$

by choice of t. We have chosen the partition into left and right pieces carefully to ensure that A_L is "just right". In the first case, $A_R = (1 - \alpha)H_{\mathrm{OPT}}W$, which implies that $A_L \leq \alpha H_{\mathrm{OPT}}W$. In this case the left strip has height at most $3H_{\mathrm{OPT}} - 2t \leq \frac{5}{3}H_{\mathrm{OPT}}$ by choice of t. To prove the bound for the left strip in the second case where $A_R = A_R^0 < (1 - \alpha)H_{\mathrm{OPT}}W$, we need the following result:

Claim: $\frac{A_L^0}{\alpha W} \leq H_{\mathrm{OPT}}$. To prove this claim, let R_L^0 denote the set of left pieces (including the left floor rectangles $r_1, \ldots, r_j$) when all non-floor rectangles are shifted completely rightward. Consider again an optimal solution S^* (with an unbounded number of slices) and rearrange the columns in S^* so that the left floor rectangles form a left strip of width αW. Then the left side of the strip in S^* must contain at least the pieces in R_L^0 and thus must contain a total area of at least A_L^0. It follows that $A_L^0 \leq H_{\mathrm{OPT}}\alpha W$, and hence the claim holds.

With this claim, the left strip has height $2H_{\mathrm{OPT}} - 2t + \frac{A_L^0}{\alpha W} \leq 3H_{\mathrm{OPT}} - 2t \leq \frac{5}{3}H_{\mathrm{OPT}}$ by choice of t, and we have now proved the approximation bound.

Lastly, we remove the assumption that the value of H_{OPT} is given. By replacing H_{OPT} with a user-supplied value H_{GUESS} in the algorithm, it is easy to check that the algorithm has the following behavior: if $H_{\text{GUESS}} \geq H_{\text{OPT}}$, the solution returned has height at most $(5/3)H_{\text{GUESS}}$. Thus, if the solution returned has height at most $(5/3)H_{\text{GUESS}}$, we can conclude that $H_{\text{OPT}} \leq (5/3)H_{\text{GUESS}}$, otherwise $H_{\text{OPT}} > H_{\text{GUESS}}$.

We can apply a binary search to find an approximation to H_{OPT}. Start with $X = \frac{1}{2}H_S$, the height computed by the Shelf algorithm. We know $X \leq H_{\text{OPT}} \leq (5/3)cX$ with $c = 6/5$, so this is a $(5/3)c$-approximation. Now run the above algorithm with $H_{\text{GUESS}} = \sqrt{c}X$, and conclude either that $H_{\text{OPT}} \leq (5/3)\sqrt{c}X$ or $H_{\text{OPT}} > \sqrt{c}X$. In either case, we obtain a $((5/3)\sqrt{c})$-approximation. Repeating for $O(\log(1/\varepsilon))$ iterations, we then obtain a $(5/3 + \varepsilon)$-approximation. Assuming that all rectangle heights are integers bounded by M, we can set $\varepsilon = 1/(4nM)$, for example, and a $(5/3 + \varepsilon)$-approximation becomes a $5/3$-approximation; the running time increases by an $O(\log(nM))$ factor only. $\qquad\square$

We can reduce the number of slices even further by packing them into the shelves carefully so that pieces become aligned on the shelf, and (in some cases) apply Steinberg's algorithm for 2SP [19] on one of the sides. Details are omitted.

Theorem 3. *There exists a $5/3$-approximation for 2SP-SSC that slices every rectangle at most once and runs in time $O(n \log^2 n \log(nM)/\log\log n)$.*

5   Conclusions

Motivated by an application in electricity allocation, this paper explored variants of the strip packing problem in which rectangles could be sliced vertically as long as no two slices of the same rectangle are stacked atop each other. We provided simple 2-approximation algorithms, an FPTAS of mostly theoretical interest, and practical approximation algorithms that slice rectangles only a few times.

The main remaining open problem is to find practical algorithms with better approximation factors. For example we conjecture that First Fit Decreasing is actually a $4/3$-approximation. Without the stacking constraint, this follows from Graham's $4/3$-approximation bounds for multiprocessor scheduling [8], but with the stacking constraint the best bound we can prove is $3/2$. Also, can we find a $\frac{3}{2}$-approximation (or even $\frac{4}{3}$-approximation) that slices every rectangle at most once? Finally, is there a simple PTAS for strip packing with slicing (with or without the stacking constraint)?

References

1. Baker, B.S., Coffman, E.G., Rivest, R.L.: Orthogonal packings in two dimensions. SIAM Journal on Computing 9(4), 846–855 (1980)
2. Bougeret, M., Dutot, P.F., Jansen, K., Otte, C., Trystram, D.: Approximating the non-contiguous multiple organization packing problem. In: Calude, C.S., Sassone, V. (eds.) TCS 2010. IFIP AICT, vol. 323, pp. 316–327. Springer, Heidelberg (2010)

3. Chen, B., Potts, C.N., Woeginger, G.J.: A review of machine scheduling: complexity, algorithms and approximability. In: Handbook of Combinatorial Optimization, vol. 3, pp. 21–169. Kluwer Acad. Publ., Boston (1998)

4. Coffman Jr., E.G., Garey, M.R., Johnson, D.S., Tarjan, R.E.: Performance bounds for level-oriented two-dimensional packing algorithms. SIAM J. Comput. 9(4), 808–826 (1980)

5. Garey, M.R., Johnson, D.S.: Computers and Intractability: A Guide to the Theory of NP-Completeness. Freeman & Co. Ltd. (1979)

6. Gilmore, P.C., Gomory, R.E.: A linear programming approach to the cutting-stock problem. Operations Res. 9, 849–859 (1961)

7. Giudice, P.: Our energy future and smart grid communications. Testimony before the FCC Field Hearing on Energy and Environment (2009), http://www.broadband.gov/fieldevents/fh_energy_environment/giudice.pdf

8. Graham, R.: Bounds on multiprocessing timing anomalies. SIAM J. Appl. Math. 17, 416–429 (1969)

9. Harren, R., Jansen, K., Prädel, L., van Stee, R.: A $(5/3 + \varepsilon)$-approximation for strip packing. In: Dehne, F., Iacono, J., Sack, J.-R. (eds.) WADS 2011. LNCS, vol. 6844, pp. 475–487. Springer, Heidelberg (2011)

10. Hochbaum, D.S., Shmoys, D.B.: Using dual approximation algorithms for scheduling problems: theoretical and practical results. J. Assoc. Comput. Mach. 34(1), 144–162 (1987)

11. Jansen, K., Solis-Oba, R.: New approximability results for 2-dimensional packing problems. In: Kučera, L., Kučera, A. (eds.) MFCS 2007. LNCS, vol. 4708, pp. 103–114. Springer, Heidelberg (2007)

12. Karmarkar, N., Karp, R.M.: An efficient approximation scheme for the one-dimensional bin-packing problem. In: Symposium on Foundations of Computer Science, pp. 312–320. IEEE (1982)

13. Kenyon, C., Rémila, E.: A near-optimal solution to a two-dimensional cutting stock problem. Math. Oper. Res. 25(4), 645–656 (2000)

14. Lodi, A., Martello, S., Monaci, M.: Two-dimensional packing problems: A survey. European Journal of Operational Research 141(2), 241–252 (2002)

15. Sahni, S.K.: Algorithms for scheduling independent tasks. J. Assoc. Comput. Mach. 23(1), 116–127 (1976)

16. Schiermeyer, I.: Reverse-Fit: A 2-optimal algorithm for packing rectangles. In: van Leeuwen, J. (ed.) ESA 1994. LNCS, vol. 855, pp. 290–299. Springer, Heidelberg (1994)

17. Sleator, D.D.: A 2.5 times optimal algorithm for packing in two dimensions. Information Processing Letters 10(1), 37–40 (1980)

18. Srikantha, P., Rosenberg, C., Keshav, S.: An analysis of peak demand reductions due to elasticity of omestic appliances. In: Proc. Energy-Efficient Computing and Networking (e-Energy 2012), p. 28. ACM (2012)

19. Steinberg, A.: A strip-packing algorithm with absolute performance bound 2. SIAM Journal on Computing 26(2), 401–409 (1997)

On (Dynamic) Range Minimum Queries in External Memory

Lars Arge[1], Johannes Fischer[2], Peter Sanders[2], and Nodari Sitchinava[2]

[1] MADALGO*, Aarhus University, Aarhus, Denmark
large@madalgo.au.dk
[2] Karlsruhe Institute of Technology, Karlsruhe, Germany
{johannes.fischer,sanders}@kit.edu, nodari@ira.uka.de

Abstract. We study the one-dimensional range minimum query (RMQ) problem in the external memory model. We provide the first space-optimal solution to the batched static version of the problem. On an instance with N elements and Q queries, our solution takes $\Theta(\text{sort}(N + Q)) = \Theta\left(\frac{N+Q}{B} \log_{M/B} \frac{N+Q}{B}\right)$ I/O complexity and $O(N + Q)$ space, where M is the size of the main memory and B is the block size. This is a factor of $O(\log_{M/B} N)$ improvement in space complexity over the previous solutions. We also show that an instance of the batched *dynamic* RMQ problem with N updates and Q queries can be solved in $O\left(\frac{N+Q}{B} \log^2_{M/B} \frac{N+Q}{B}\right)$ I/O complexity and $O(N + Q)$ space.

1 Introduction

Static one-dimensional range minimum queries (RMQs) are defined as follows: *Given an array $A[1, N]$ of N elements from a totally ordered universe $\mathcal{U}$, and a query in the form* RANGEMIN(i, j), *return the index of the smallest element in the subarray $A[i, j]$.*

RMQs have a wide range of applications in many areas of data structures and algorithms, for example in data compression, text indexing, and graph algorithms. For more applications, we refer the interested reader to a recent article [8], which also contains optimal solutions for one-dimensional static RMQs in the RAM model. Research on RMQs remains a hot topic, see for example a recent invited talk by Raman [12].

In this paper we consider the external memory (EM) model [1], where performance is measured by the number of block transfers (I/Os) of B words each, and where the internal memory size is limited to M words. In this setting, in case of the *online* RMQ problem (meaning that each query must be answered immediately as it arrives) all previous $O(1)$ time, linear space *static* internal memory solutions [8] are also optimal in the EM model, since each query must spend at least one I/O to report the output. However, if we consider the *offline* version of the problem (also often referred to as *batched*), where the initial array and all

---

* MADALGO is a center of the Danish National Research Foundation.

F. Dehne, R. Solis-Oba, and J.-R. Sack (Eds.): WADS 2013, LNCS 8037, pp. 37–48, 2013.
© Springer-Verlag Berlin Heidelberg 2013

queries are specified in advance and queries can be answered in arbitrary order, we can achieve better amortized bounds. In particular, Chiang et al. [7] showed that Q queries in the *static* batched range minima problem can been answered with $O((n + q) \log_m(n + q)) = O(\text{sort}(N + Q))$ I/Os. (Following the common notation in the EM literature, we define $n = N/B$, $q = Q/B$ and $m = M/B$.) Their solution takes $O(Q + N \log_m N)$ space. Batched range minima arise in suffix sorting [11] (if one also wants to find longest common prefixes) and other stringology problems, for example in string mining tasks [9].

We also consider a *dynamic* scenario, where N updates (insertions and deletion of elements) to the underlying array are arbitrarily interspersed with the Q queries. In this setting, it is impossible to achieve constant time per operation [10]: we can sort N items by first inserting them into the array, and then repeatedly querying for and removing the minimum of the entire array. This shows that online dynamic RMQs are at least as hard as priority queues.

1.1   Our Contributions

We formulate the *dynamic batched range minima problem* as follows: *Given a batch of N updates (INSERT/DELETE operations) affecting a dynamically changing "structure" over a universe $\mathcal{U}$ and Q RANGEMIN operations, report answers to all RANGEMIN queries. Answers to queries may arrive in arbitrary order, but must be correct based on the state of the "structure" at the time of the query.*

Because there is no definite concept on what the dynamization of a static array should be, in Section 2 we explore three possible versions of the underlying "structure" in the above problem definition: a linked list, a dynamic array and a point set. In section 4.2 we present a solution to the third version, which, at a first glance, seems to be the simplest to reason about in the EM model. However, in Section 5 we show that we can reduce one version to another in $O(\text{sort}(N+Q))$ I/Os. We conjecture that $\Omega(\text{sort}(N + Q))$ I/Os is the lower bound even for the static batched range minima problem, i.e., that our reductions are tight and that the three versions are equivalent.

In addition, in Section 4.1 we improve the solution of Chiang et al. [7] and present the first *linear* space solution to the *static* batched RMQ problem. Our solution also implies a linear space *parallel* I/O-efficient solution to the RMQ problem in the parallel external memory (PEM) model [3].

2   Different Scenarios for Dynamic RMQ

In general, there are different ways to think about how the INSERT/DELETE operations affect the array, and what the answers to RANGEMIN queries should be. This section is meant to explain their differences. See also Fig. 1.

Dynamic Array Version. This version most closely resembles the static version: we identify elements by their rank (number of elements, including itself, to the left) in the array at the time of the operation. Hence, although the elements are stored in a dynamic data structure, we can treat them as if they were stored in an array, and consequently denote them by $A[i]$.

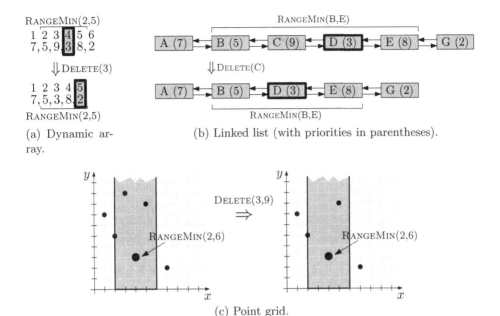

Fig. 1. Different versions of dynamic range minimum queries. Range Minima are shown in bold.

INSERT(i, x): insert a new element with value $x \in \mathcal{U}$ right after position i into A; all positions originally after i will be shifted right by one element: $A \leftarrow A[1, i] \cdot [x] \cdot A[i + 1, N]$.

DELETE(i): delete the element at position i from A; all positions originally after i will be shifted left by one element: $A \leftarrow A[1, i - 1] \cdot A[i + 1, N]$.

RANGEMIN(i, j): return (the position of) the minimum in $A[i, j]$, as in the static setting.

Though closely matching the static scenario, this version actually turns out to be most difficult to deal with: the intuitive reason for this is that the indices specify *ranks* in a dynamically changing array, and hence first need to be *unranked* (selected) before knowing which object they affect. This issue will be addressed in Sect. 5.2.

Linked List Version. Elements are part of a doubly linked list, and are identified by handles (pointers to the elements). Each list element v has an associated priority (value) $p_v \in \mathcal{U}$.

INSERT(v, u, p_v): Insert a new element v with priority $p_v \in \mathcal{U}$ right after element u: if the list A was originally $A_1 \rightsquigarrow u \rightarrow w \rightsquigarrow A_2$, it now becomes $A_1 \rightsquigarrow u \rightarrow v \rightarrow w \rightsquigarrow A_2$.

DELETE(v): Delete the element v from A: if the list A was originally $A_1 \rightsquigarrow u \rightarrow v \rightarrow w \rightsquigarrow A_2$, it now becomes $A_1 \rightsquigarrow v \rightarrow w \rightsquigarrow A_2$.

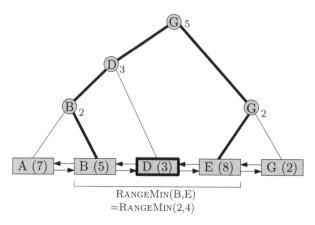

Fig. 2. Data structure to support dynamic RMQs in internal memory. Internal nodes store pointers (depicted within the nodes) to an element below them with minimum priority. The path tracked by the query RANGEMIN(B, E) is shown in bold. To the right of each internal node v is the number of leaves in the subtree rooted at v, needed for the dynamic array version (e.g., query RANGEMIN(2, 4)).

RANGEMIN(u, v): return (a pointer to the element with) the minimum priority among all elements that are currently between u and v (both inclusively).

Geometric Version. Elements are pairs of the form (x, y), and we think of them as being points on the plane. We denote the set of all points currently stored as S.

INSERT(x, y): Insert a new point (x, y) into S.
DELETE(x, y): Delete the point (x, y) from S.
RANGEMIN(x_1, x_2): return the point (x, y) such that $x_1 \leq x \leq x_2$, and y is minimum among all those points (or return only the y-value of that point).

3   Simple Dynamic Internal Memory Algorithms

As a warmup, in this section we present a fully dynamic data structure for RMQs in the internal memory (RAM/comparison) models. We show it primarily because it forms the basis of all our EM data structures. Note that distinguishing between offline and online queries is irrelevant in internal memory.

Linked List Version. We maintain a balanced binary tree over the linked list. The leaves of the tree store the elements of the linked list, and internal nodes correspond to ranges of consecutive elements of the list. Each internal node v stores a pointer to a leaf μ in the subtree rooted at v with minimum priority; we denote that pointer by μ_v. At any time, the size of the data structure is linear with the number of nodes present in the list. See also Fig. 2.

To perform INSERT or DELETE operations, we go directly to the node specified by the operation, and traverse the path up to the root, updating the values μ_x

for each node x on the path bottom up. To perform a query RANGEMIN(u, v), we traverse the two paths from u (resp. v) to the root until they meet. From there, we go back down to u (resp. v) and report the leaf μ_x with the smallest priority among those nodes x hanging off to the right (resp. left) of those paths.

The time for each operation is O($\log N$), where N is the number of elements in the linked list at the time of the operation.

Dynamic Array Version. The difference to the list version is that the queries do not specify pointers to the leaves of the tree, but rather ranks in the linked list. So the elements need to be first unranked in the list, which can be achieved by a top-down search in the tree. To support this search, we augment each node v with the size of the subtree rooted at v. Then a top-down traversal easily locates an element in O($\log N$) time given its rank. Having identified the elements, queries are processed exactly as in the linked list version. If the RANGEMIN query asks for the *position* of the minimum, we finally have to rank the minimum element by a bottom-up traversal. The subtree sizes are also easily updated during insertions/deletions and when doing rebalancing operations.

Geometric Version. We maintain a balanced search tree over the x coordinates of the elements in S. In addition to storing minimum y-values, each internal node stores the largest x-coordinate of an element in its left subtree. Then we locate the elements as in the dynamic array version (by finding predecessors of the x-coordinates), and answer queries as in the linked list version.

4 External Memory Solutions to the Geometric Version

In this section we present a solution to the geometric version of the RMQ problem. We start with a data structure for the static case, as it will form the basis of our dynamic solution.

4.1 Batched Static RMQ

In this section we present a solution to the static RMQ problem that uses only linear space but still incurs only O(sort(N)) I/O complexity.

We sort the set S of N points by the x-coordinates and build a fully balanced k-ary search tree T with $\Theta(N/M)$ leaves on top of them. Each leaf of T is associated with a contiguous x-range of $\Theta(M)$ elements. We set $k = \Theta(m)$, thus, the height of the tree is $\Theta(\log_k(N/M)) = \Theta(\log_m(N/M))$. Each internal node of the tree is associated with the range that is the union of the ranges of its children. Each node v of the tree stores $\mathcal{M}_v$ – a set of k minima of the ranges associated with the k children of v. (In contrast, the solution of Chiang et al. [7] stores prefix minima for every element of the child subtrees.) We assume that $\mathcal{M}_v$ also stores the relevant output information, e.g. the index of the corresponding element within the array. There are a total of O($\frac{N}{kM}$) internal nodes each storing $\Theta(k)$ elements. Thus, the size of the tree is O(N).

We populate the sets $\mathcal{M}_v$ bottom up. Each entry $\mathcal{M}_v[i]$ at the parent v of the leaf nodes is set to the minimum of the elements stored at each leaf w_i. The entries at each remaining internal node v are constructed by scanning the entries $\mathcal{M}_w$ of the children nodes w and computing their minima. Thus, the tree T can be built in $O(\log_m(N/M))$ rounds by scanning each level of $O(N)$ elements for a total of $O(\text{sort}(N))$ I/O complexity.

We process the queries in rounds, in each round processing all queries on a single level (starting with the root level) and propagating them down to the next level. For each query at node v, if the range of the query fully falls within the range of some child w of v, we associate the query with w for further processing and continue with the next query. If the endpoints of the query fall within ranges of two different children w_l and w_r, we can compute the value of the range minima query for the portion of the query that spans the children $w_{l+1}, \ldots, w_{r-1}$ by computing the minimum among the subset of values $\mathcal{M}_v$ and any partial answer stored with the query up to this point. We save this value with the query as a tentative answer and associate the query with the two children w_l and w_r containing the endpoints of the query for further processing. At the leaf nodes, we load the leaf of size $\Theta(M)$ into internal memory and answer the query in internal memory. Once all nodes of the tree are processed, for each query we might have up to two potential answers, each stored at the leaves containing the two endpoints of the query. The final answer is the minimum value of the two copies and can be found I/O-efficiently as follows. We sort the answers by the query identifiers (e.g., lexicographically by the left and right indices of the ranges). This process places the two copies of each query in contiguous memory and we can compute the minimum among all pairs of potential answers by scanning the sorted list.

Let Q be the total number of queries. Each query is associated with at most two nodes of the tree. And since we process all queries through each level, we use only $O(Q)$ space for the queries. Together with $O(N)$ space for the data structure, we use a total of $O(N + Q)$ space. The I/O complexity is $O((n + q)\log_m n)$: there are $O(\log_m(N/M))$ internal levels of the tree and at each level of the tree we scan the set of queries (and the internal structures of the tree, which are at most $O(N/M)$). At the leaves we load each leaf into internal memory once, thus, spending $O(n)$ I/Os and scan the set of queries. Finally, the sorting step takes $O(\text{sort}(Q))$ I/Os. Thus, the total I/O complexity is bounded by $O(\text{sort}(Q + N))$.

Parallel Extensions in the PEM Model. The algorithm of Chiang et al. [7] has been used by Arge et al. [4] to develop a parallel solution to the static batched RMQ problem in the parallel external memory (PEM) model [3]. Our new linear space EM solution immediately leads to a linear space PEM solution.

4.2   Batched Dynamic RMQ

Arge et al. [5] present a framework for solving dynamic problems in the external memory model for the so-called *external decomposable* batched problems. Let us review a few definitions from [5]. (Since the RMQ problem has constant output size per query, we omit the parts relevant for the output-sensitive problems).

Definition 1. *[5] Let $\mathcal{P}$ be a searching problem and let $\mathcal{P}(x, V)$ denote the answer to $\mathcal{P}$ with respect to a set of objects V and a query object x. $\mathcal{P}$ is called external decomposable, if for any partition $A \cup B$ of the set V and for every query x, $\mathcal{P}(x, V)$ can be computed in $\mathrm{O}(1)$ additional I/Os given $\mathcal{P}(x, A)$ and $\mathcal{P}(x, B)$ in appropriate form.*

Definition 2. *[5] Let $\mathcal{P}$ be an external decomposable batched searching problem. Consider the problem $\mathcal{P}_C$ where a color chosen from a set C is associated with each query x, and where a set of colors C_v is associated with each object $v \in V$. Only objects where $color(x) \in C_v$ are considered when answering x. Problem $\mathcal{P}$ is called $(I(N, K), S(N, K))$ $m^{1/c}$-colorable if the following two conditions hold:*

1. *For all colorings where $|C| = \Theta(\sqrt{m^{1/c}})$ and where the number of different color sets C_v is $\mathrm{O}(m^{1/c})$, for some constant $c \geq 1$, $\mathcal{P}_C$ can be solved in $\mathrm{O}(I(N, K))$ I/O operations and $\mathrm{O}(S(N, K))$ space after an initial sorting step, and*
2. *If (V_1, Q_1) and (V_2, Q_2) are two valid instances of $\mathcal{P}$ then $(V_1 \cup V_2, Q_1 \cup Q_2)$ is also a valid instance.*

Lemma 1. *The static range minima problem is $((n + q) \log_m (n + q), N + Q)$ colorable.*

Proof. Obviously, the RMQ problem is external decomposable by Definition 1. The solution to static RMQ is similar to the solution presented in Section 4.1. We build a full balanced $k = \Theta(\sqrt{m})$-ary tree T on the elements ordered by the x-coordinate and each leaf node containing $\Theta(M)$ elements. Thus, the height of the tree is still $\mathrm{O}(\log_k (N/M)) = \mathrm{O}(\log_m (N/M))$. At each internal node v, instead of storing k minima (one for each subtree rooted at the k children of v), we store $|C| = \Theta(\sqrt{m})$ sets of such k minima: one set for each color.

We populate these sets bottom up. To construct the sets of minima at the parent node v of the leaves, we load $\Theta(M)$ elements of each leaf into internal memory and construct the set of the $|C|$ minima for that leaf, one leaf at a time. We construct the sets of minima for the remaining internal nodes by computing the minima for each color from the ones stored at the children nodes bottom up.

We process the queries as before, top-down level by level, except for each query we use the minima among the ones with the same color as the query.

There are a total of $\mathrm{O}(N/M)$ leaves in the tree, and, consequently, $\mathrm{O}(\frac{N}{kM})$ internal nodes, each storing $k|C|$ minima. Thus, the total space used by the minima is $\mathrm{O}(N|C|/M)$. Each node can be at most $\mathrm{O}(m)$ in size to perform the construction of the tree and processing of the queries I/O-efficiently, i.e., $k|C| = \mathrm{O}(m)$. Thus, if we set $k = |C| = \sqrt{m}$, the tree uses linear space and each node stores $\mathrm{O}(m)$ items. $\qquad\square$

Theorem 1. *The batched dynamic range minima problem can be solved in $\mathrm{O}((n + q) \log_m^2 (n + q))$ I/Os and $\mathrm{O}(N + Q)$ space.*

Proof. Arge et al. [5] proved that the batched dynamic version of a static problem $\mathcal{P}$ that is $(I(N, Q), S(N, Q))$ colorable, can be solved in $\mathrm{O}(I(N, Q) \cdot \log_m (n + q))$ I/Os using $\mathrm{O}(S(N, Q))$ space. Combined with Lemma 1, the proof follows. $\qquad\square$

5   Reductions between Dynamic Array, Linked List, and Geometric Versions in the EM Model

In this section we reduce the linked list version of the problem to the geometric version, and the dynamic array version to the linked list version. All the reductions take sorting complexity in the EM model, which is less than the complexity of our solution to the geometric version (Section 4). This implies that Theorem 1 also applies to the linked list and dynamic array versions of the problem.

In addition, we show that the geometric version can in turn easily be reduced to the dynamic array version. This would show the equivalence of the three versions if there were a sorting lower bound to any of the versions. However, the sorting lower bound mentioned in the introduction only applies to the online dynamic case, but *not* to the batched case.

5.1   Linked List to Geometric Reduction

We show how to transform a sequence of $N+Q$ INSERT(v, u, p_v), DELETE(v), and RANGEMIN(u, v) linked list operations into a sequence of $N + Q$ INSERT(x, y), DELETE(x, y), RANGEMIN(x_1, x_2) geometric operations, by mapping each linked list node v to a point (x_v, y_v) on a plane.

For each linked list node v, we set $y_v = p_v$ by scanning the INSERT(v, u, p_v) operations. Finding the x_v's is harder, as described next.

The INSERT and DELETE operations in the linked list define the following partially persistent linked list (see Fig. 3a for an illustration). The structure is a DAG $\mathcal{G}$, initially consisting of a dummy node $\perp$, representing an artificial list head. INSERT(v, u) adds the edges $u \to v$ and $v \to w$, where w is the most recent successor of v in the structure. Likewise, DELETE(v) creates the edge $u \to w$, where u and w are the predecessor and successor of v, respectively, at the time of the operation.

To compute the values x_v, we embed $\mathcal{G}$ in one-dimensional space (a line) according to a topological ordering of $\mathcal{G}$. That is, we set x_v equal to the position of the node v in a topological ordering of $\mathcal{G}$. Note that in such an embedding any range $[x_u, x_v]$ includes the nodes of all versions of the linked lists $u \rightsquigarrow v$, and that every node of a linked list $u \rightsquigarrow v$ is contained within the range $[x_u, x_v]$. Therefore, x_v is the correct assignment of the x coordinate to node v.

As $\mathcal{G}$ is planar, we could compute its topological order in O(sort(N)) I/O complexity [6]. However, we do not know how to construct $\mathcal{G}$ I/O-efficiently. Instead, we consider the following subgraph T of $\mathcal{G}$, which will be enough for our purposes.

For each INSERT(v, u, p_v) operation we create an edge (u, v) annotated with the timestamp of the operation. Graph T is defined by the adjacency list representation with an ordered set of neighbors for each node u by sorting the generated edges (u, v) lexicographically, primarily by the first node u and secondarily by the edges' timestamps in decreasing order. The following lemma shows that we can use T to compute a topological order of $\mathcal{G}$.

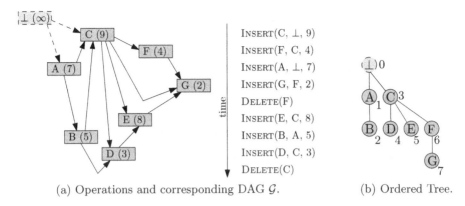

INSERT(C, ⊥, 9)
INSERT(F, C, 4)
INSERT(A, ⊥, 7)
INSERT(G, F, 2)
DELETE(F)
INSERT(E, C, 8)
INSERT(B, A, 5)
INSERT(D, C, 3)
DELETE(C)

(a) Operations and corresponding DAG $\mathcal{G}$.   (b) Ordered Tree.

Fig. 3. Reduction from linked list to geometry version

Lemma 2. *Graph T defines a tree rooted at $\perp$. Moreover, T contains all nodes of $\mathcal{G}$, and the preorder traversal of T defines a valid topological order on $\mathcal{G}$.*

Proof. Imagine the DAG $\mathcal{G}$ embedded on a plane, such that the y-coordinate of a node's embedding equals to the timestamp of the insertion operation. Since an INSERT(v, u) operation creates a node v below all other nodes, INSERT(v, u) creates a downward pointing edge (u, v) and an upward pointing edge (v, w). We draw each edge (u, w) inserted due to DELETE(v) operation as a zig-zag: first pointing downward then upward, with the kink at the y-coordinate equal to the timestamp of the DELETE operation. We draw $\mathcal{G}$ such that node v corresponding to INSERT(v, u) and the kinks in edges corresponding to DELETE(v) at time t are drawn to the left of all siblings of v up to time t. Then all edges (u, v_i) for node u (in left to right order) point to nodes v_i with decreasing timestamps. The version of the list at time t is then defined by the traversal of $\mathcal{G}$ starting at node $\perp$ and at each node following the leftmost edge that does not cross the horizontal line $y = t$.

Graph T consists of all nodes of $\mathcal{G}$ and only the downward pointing edges due to insertions, i.e., it is $\mathcal{G}$ with the zig-zag and "up-edges" removed. Thus, it is a tree. The preorder traversal of T implies a valid topological order of $\mathcal{G}$ for the following two reasons: (1) The zig-zag edges introduced by deletions do not affect the topological ordering of $\mathcal{G}$. This is simply because adding an edge $u \to w$ upon a deletion already implies the existence of the path $u \to v \to w$, and because there cannot occur a subsequent INSERT$(\cdot, v)$ operation after v was deleted. (2) The "up-edges" created by INSERT(v, u, p_v) always point to the elements inserted earlier, i.e. to the nodes in the subtrees rooted at siblings of v that appeared before v. Since v appears to the left of those siblings, the preorder traversal of the tree places v before any node it might point to.   □

The preorder traversal of T can be accomplished I/O-efficiently by building an Euler tour on T and performing list ranking with the appropriate weights in $O(\text{sort}(N))$ I/Os [7].

Thus, we successfully mapped the linked list nodes to points on the plane. In other words, we computed the mapping $v \to (x,y)$ for every node in the linked list. We use these mappings to replace INSERT(v, u, p_v), DELETE(v) and RANGEMIN(u, v) with the corresponding INSERT(x_v, y_v), DELETE(x_v, y_v), and RANGEMIN(x_1, x_2). The replacement can be performed I/O-efficiently, by sorting the operations and the mapping by the node names and simultaneously scanning the operations and the mapping, generating the geometric operations.

Given the answers to the geometric queries (namely x-coordinates of the answers), we can map them back to the linked list nodes similarly by sorting the answers and the mapping by the x-coordinates.

5.2 Dynamic Array to Linked List Reduction

In this section we show how to map each index and the corresponding timestamp to a node identifier in the linked list.

We will use the *buffer tree* technique of Arge [2] to perform this conversion. The buffer tree is an (a, b)-tree with branching parameters $a, b \in \Theta(m)$ and each leaf of size $\Theta(B)$. I.e., buffer tree is a balanced m-ary search tree of depth $O(\log_m(n+q))$. In addition, each internal node is augmented with a *buffer* of size $\Theta(M)$. All updates and queries are simply inserted into the buffer of the root, annotated with the timestamp of the operation. When a node's buffer is full, the buffer is emptied by loading its elements into the internal memory, processing the elements and pushing them to the buffers of the children nodes. If this causes some child node's buffer to become full, it is processed and emptied recursively. Rebalancing of the tree is performed in a bottom-up way, as in an ordinary (a, b)-tree.

In our case, we use the buffer tree to represent the elements of the underlying dynamic array as a linked list. Each leaf stores $\Theta(B)$ consecutive alive elements of the linked list, i.e., elements that haven't been deleted yet. Each index (in the dynamic array) is given a *name* that serves as an identifier in the linked list. Those names are stored at the leaves of the buffer tree until the element is deleted: when an INSERT operation reaches a leaf, we take the next free name from a pool of names and associate it with the element.

The goal of the conversion is that for each of the operations INSERT(i, x), DELETE(i), and RANGEMIN(i, j) we identify the name of the element i (and also j in the case of RMQs) at the current time. To this end, every internal node v with children $w_1, \ldots, w_m$ stores a list $x_1, \ldots, x_m$ of m numbers such that x_i is the number of *alive* elements in the subtree rooted at w_i, i.e., the difference between the number of INSERT and DELETE operations that passed w_i. Note that x_i is *not* the number of elements currently stored in the leaves of the subtree rooted at w_i, because there could still be some inserts and/or deletes waiting in the buffers between w_i and the leaves of the subtree rooted at w_i.[1]

Hence, during the buffer emptying process, we can route the operations to the correct children when emptying a buffer: at the root v with children $w_1, \ldots, w_m$

---

[1] However, the tree *balancing* is still performed based on the actual number of elements stored in the leaves.

and routing elements $x_1, \ldots, x_m$, we process the operations in the buffer in the order of increasing timestamps, and route an operation referring to index i to the child w_j such that $x_{j-1} < i < x_j$. Before passing the operation to the buffer of w_j, we update it to refer to the array index $i - \sum_{k<j} x_k$; by maintaining prefix sums we can keep track of these subtree ranks easily. Note that we have to update the routing elements x_i immediately when processing an update operation (i.e., increasing/decreasing x_i by one when handling an insertion/deletion, respectively), such that future operations get routed to the correct element.

This way, all INSERT and DELETE operation reach the leaf where the element they are referring to is stored. There, they are scanned by increasing timestamp, thereby identifying the name of the element they are referring to in the linked list. Corresponding operations in the linked list version are subsequently generated (together with their original timestamps), and the insertion or deletion is actually applied on the linked list stored at the leaf.

The RANGEMIN queries are handled similarly, with only one difference: for a query RANGEMIN(i, j) at time t, we create two tuples (i, t) and (j, t) and insert these tuples at the root. As before, those tuples will be routed to a leaf (using the first components of the tuples for the search), where they will be associated with a name. A final sort by the second component of the tuples (time) will bring the query endpoints back to each other; they are then translated into corresponding queries in the linked list.

A standard analysis on buffer trees concludes that in $O(\text{sort}(N))$ I/Os we create linked list operations whose answers correspond to the dynamic array queries. It remains to map the answers (elements in a linked list) back to the dynamic array version (indices in the array). We accomplish this by processing all insertions, deletions and the *answers* to the queries using a buffer tree again. However, this time we use the buffer tree in its "native" setting: from the solution to the geometric version we know the x-coordinates of all elements; therefore, we can use x-coordinates as the routing elements. As before, each node stores the number of alive elements in each of its subtrees, and these numbers are updated immediately upon handling an operation waiting in the node's buffer. If a query answer (a linked list element) is routed to child w_j of v, we need to count the sum of alive elements in v's children to the left of w_j and propagate this information down with the query. Finally, when a query answer reaches a leaf of the buffer tree, we count its relative rank within the $\Theta(B)$ elements currently stored at the leaf; this results in the final rank of the query answer.

5.3   Geometric to Dynamic Array Reduction

To convert a batch of operation from the geometric setting to the dynamic array setting, we use the ranking mechanism explained at the end of Section 5.2 (using buffer trees) to determine each point's rank. We also store the original x-coordinates (of the geometric setting) along with the new batch of operations.

6   Conclusions and Open Questions

We addressed the problem of static and dynamic batched range minimum queries in external memory and presented several algorithms for solving the problem. Since the sorting lower bound only applies to the online dynamic case, and not to batched queries, the following open questions for future research immediately come to mind: (i) what is the optimal I/O complexity of static batched range minima? We conjecture that it is the sorting bound. (ii) Likewise, what is the optimal I/O complexity of the dynamic batched version of the problem?

Other open questions concern dynamic solutions in more advanced models, like the cache oblivious model or the parallel external memory (PEM) model. The difficulty in addressing the dynamic setting in parallel lies in the seemingly sequential relationship between past updates and future queries.

References

1. Aggarwal, A., Vitter, J.S.: The input/output complexity of sorting and related problems. Communications of the ACM 31(9), 1116–1127 (1988)
2. Arge, L.: The buffer tree: A technique for designing batched external data structures. Algorithmica 37(1), 1–24 (2003)
3. Arge, L., Goodrich, M.T., Nelson, M.J., Sitchinava, N.: Fundamental parallel algorithms for private-cache chip multiprocessors. In: SPAA, pp. 197–206 (2008)
4. Arge, L., Goodrich, M.T., Sitchinava, N.: Parallel external memory graph algorithms. In: IPDPS, pp. 1–11 (2010)
5. Arge, L., Procopiuc, O., Ramaswamy, S., Suel, T., Vitter, J.S.: Theory and practice of I/O-efficient algorithms for multidimensional batched searching problems. In: SODA, pp. 685–694 (1998)
6. Arge, L., Toma, L., Zeh, N.: I/O-efficient topological sorting of planar DAGs. In: SPAA, pp. 85–93. ACM Press (2003)
7. Chiang, Y.J., Goodrich, M.T., Grove, E.F., Tamassia, R., Vengroff, D.E., Vitter, J.S.: External-memory graph algorithms. In: SODA, pp. 139–149 (1995)
8. Fischer, J., Heun, V.: Space efficient preprocessing schemes for range minimum queries on static arrays. SIAM J. Comput. 40(2), 465–492 (2011)
9. Fischer, J., Mäkinen, V., Välimäki, N.: Space efficient string mining under frequency constraints. In: Proc. ICDM, pp. 193–202. IEEE Computer Society (2008)
10. Franceschini, G., Grossi, R.: A general technique for managing strings in comparison-driven data structures. In: Díaz, J., Karhumäki, J., Lepistö, A., Sannella, D. (eds.) ICALP 2004. LNCS, vol. 3142, pp. 606–617. Springer, Heidelberg (2004)
11. Kärkkäinen, J., Sanders, P., Burkhardt, S.: Linear work suffix array construction. J. ACM 53(6), 1–19 (2006)
12. Raman, R.: Range extremum queries. In: Smyth, B. (ed.) IWOCA 2012. LNCS, vol. 7643, pp. 280–287. Springer, Heidelberg (2012)

Distance-Sensitive Planar Point Location[*]

Boris Aronov[1], Mark de Berg[2], Marcel Roeloffzen[2], and Bettina Speckmann[2]

[1] Dept. of Computer Science and Engineering, Polytechnic Institute of NYU, USA
aronov@poly.edu
[2] Dept. of Computer Science, TU Eindhoven, The Netherlands
{mdberg,mroeloff,speckman}@win.tue.nl

Abstract. Let $\mathcal{S}$ be a connected planar polygonal subdivision with n edges and of total area 1. We present a data structure for point location in $\mathcal{S}$ where queries with points far away from any region boundary are answered faster. More precisely, we show that point location queries can be answered in time $O(1 + \min(\log \frac{1}{\Delta_p}, \log n))$, where Δ_p is the distance of the query point p to the boundary of the region containing p. Our structure is based on the following result: any simple polygon P can be decomposed into a linear number of convex quadrilaterals with the following property: for any point $p \in P$, the quadrilateral containing p has area $\Omega(\Delta_p^2)$.

1 Introduction

Point location is one of the most fundamental problems in computational geometry. Given a subdivision $\mathcal{S}$ the goal is to preprocess it so that we can determine efficiently which region of $\mathcal{S}$ contains a query point p. Many different variants of the point-location problem exist; in our work we focus on planar point location in a connected polygonal subdivision $\mathcal{S}$.

There are several different solutions for planar point location that are worst-case optimal, that is, that require $O(n \log n)$ preprocessing, use $O(n)$ space, and can answer a point-location query in $O(\log n)$ time; see the survey of Snoeyink [17] for an overview. A query time of $O(\log n)$ is optimal in the worst case, but it may be possible to do better for certain types of query points. For example, if the query points are not distributed uniformly among the regions of $\mathcal{S}$, then it may be desirable to reduce the query time for points in frequently queried regions. Iacono [11] showed that this is indeed possible: given a triangulation $\mathcal{S}$ where each triangular region R_i has a probability γ_i associated with it—the probability that the query point p falls in R_i—then one can answer a point-location query in expected time $O(H(\mathcal{S}))$, where

$$H(\mathcal{S}) := \sum_{R_i \in \mathcal{S}} \gamma_i \log(1/\gamma_i),$$

[*] M. Roeloffzen and B. Speckmann were supported by the Netherlands' Organisation for Scientific Research (NWO) under project no. 600.065.120 and 639.022.707, respectively. B. Aronov has been supported by grant No. 2006/194 from the U.S.-Israel Binational Science Foundation, by NSF Grants CCF-08-30691, CCF-11-17336, and CCF-12-18791, and by NSA MSP Grant H98230-10-1-0210.

F. Dehne, R. Solis-Oba, and J.-R. Sack (Eds.): WADS 2013, LNCS 8037, pp. 49–60, 2013.
© Springer-Verlag Berlin Heidelberg 2013

is the *entropy* of $\mathcal{S}$. This result is optimal, because the entropy is a lower bound on the expected query time [14,16]. Several other point-location structures have been proposed that answer queries in $O(H(\mathcal{S}))$ (expected) time [1,2]. The structure presented by Arya, Malamatos, and Mount [1] is relatively simple and efficient in practice. It works for subdivisions with constant-complexity regions and, for any region R_i the worst-case query time for points inside R_i is $O(1 + \min(\log(1/\gamma_i), \log n))$. The results mentioned so far assume that the distribution is known in advance. Recently Iacono [12] proposed an algorithm that eventually achieves $O(H(\mathcal{S}))$ query time, but does not need any knowledge of the query distribution. Instead, the algorithm changes the structure according to the queries received. The results mentioned above require the regions of the input subdivision $\mathcal{S}$ to have constant complexity. This requirement is necessary. Indeed, if a subdivision with n edges has only two regions, each with associated probability $1/2$, then we cannot hope to achieve $O(1)$ query time. Recently, however, Collette *et al.* [8] gave an interesting extension to subdivisions with regions of non-constant complexity. They showed how to compute, for any simple polygon P and any probability distribution over P, a Steiner triangulation with near-optimal entropy, and they proved that the minimum entropy of any triangulation is a lower bound on the point-location time (in the linear decision-tree model). By applying their Steiner triangulation to every region in the given subdivision, and using the resulting triangles as input for an entropy-based point-location structure, near-optimal expected query time is achieved.

We also want to develop a point-location structure that is faster for certain query points than for others. However, instead of assuming a probability distribution over the regions, we take a more geometric approach, based on the intuition that answering a query for a point far from the boundary of its containing region should be easier than answering a query for a point near the region boundary. To make this more precise, let $\mathcal{S}$ be a planar connected polygonal subdivision with n edges in total, and assume without loss of generality that the total area of the subdivision is equal to 1. For a query point p, let Δ_p denote the Euclidean distance from p to the boundary of the region containing it. We want to create a data structure where the query time for a point p is sensitive to Δ_p. More precisely, our goal is to obtain $O(1 + \min(\log(1/\Delta_p), \log n))$ query time. When the regions in the subdivision have constant complexity, then this can be achieved using, for instance, the entropy-based point-location structure of Arya, Malamatos, and Mount [1]. To this end we simply define the probability of each region R_i to be $\gamma_i := \text{area}(R_i)$. Since for any point $p \in R_i$ the distance to the boundary of R_i is $O(\sqrt{\text{area}(R_i)})$, this gives the desired query bound.

Entropy-based point-location structures require each region to have constant complexity. So the main question that we wish to answer is this: is it possible to obtain $O(1 + \min(\log(1/\Delta_p), \log n))$ query time for subdivisions in which the regions do not have constant complexity? One may hope to use the already mentioned result of Collette *et al.* [8] to this end, perhaps by defining a probability distribution where points far away from the boundary get a higher probability. However, it is unclear how to do this, especially since their Steiner triangulation guarantees only an overall expected bound. Furthermore it may generate some small triangles in areas of high probability. Hence, the main objective of our paper is to compute a suitable decomposition directly. Our requirements are as follows: Let P be a simple polygon with n_P edges. We want to compute a decomposition of P into $O(n_P)$ subpolygons with the following properties:

- each subpolygon R is convex and has constant complexity;
- for some absolute constant α the decomposition has the α-*distance property*: for any point $p \in P$, the subpolygon R containing p has area at least $\alpha \cdot \Delta_p^2$, where Δ_p is the distance from p to the boundary of P.

We do not require the decomposition to be conforming, that is, we allow T-junctions. When we measure the complexity of a subpolygon, we disregard interior vertices on the edges (that is, T-junctions "on the opposite side" of the edge). For instance, the shaded polygon to the right is considered to have five vertices.

The problem of computing a decomposition with these properties can be considered a mesh-generation problem. Many different types of meshes exist; see the survey by Bern [4] for an overview. Several of these meshes guarantee that the number of mesh elements (subpolygons) is linear in the complexity of the polygon. For example the meshing algorithm proposed by Bern *et al.* [6] produces triangles with angles of at most 90°. However, these meshes do not guarantee any relation between the distance of a point to the boundary and the area of the containing mesh element. There are also meshes that are designed to be more detailed near the polygon boundary and coarser further away from the boundary. These meshes, however, do not guarantee a relation between the distance to the boundary and the size of mesh elements [15] or they do not have a bound on the number of mesh elements [5]. To the best of our knowledge no published mesh generation method guarantees that the mesh consists of $O(n_P)$ elements that have the required distance property.

Our Results. We start by giving a simple decomposition algorithm for convex polygons. Here we actually do not need to use non-conforming subdivisions: we show that any convex polygon can be *triangulated* in such a way that the resulting triangulation has the α-distance property for $\alpha = 1$. For possibly non-convex simple polygons we investigate several different settings that have different restrictions on the resulting subdivision. We show that it is not always possible to create a triangulation with the α-distance property without using Steiner points, and that the number of Steiner points needed in such a triangulation cannot be bounded as a function of the complexity of the polygon P. We then turn our attention to non-conforming decompositions. We prove that any simple polygon P admits a decomposition into $O(n_P)$ convex quadrilaterals and triangles that has the α-distance property for some absolute constant $\alpha > 0$. The decomposition can be computed in $O(n_P \log n_P)$ time. This result is used to obtain a linear-size data structure for point location in a planar connected polygonal subdivision $\mathcal{S}$ of area 1, such that the query time is $O(1 + \min(\log(1/\Delta_p), \log n))$, where Δ_p is the distance from the query point p to the boundary of its containing region.

2   Convex Polygons

As a warm-up exercise, we start with the problem of decomposing a convex polygon P so that the decomposition has the α-distance property for $\alpha = 1$. For this case the decomposition will actually be a triangulation.

Our algorithm is quite simple. First we split P by adding a diagonal between the vertices defining the diameter of P. We further decompose each of the two resulting subpolygons using a recursive algorithm, which we describe next. We call the edges of the input polygon P *polygon edges* and the edges created by the subdivision process *subdivision edges*. The boundary of each subpolygon we recurse on consists of one subdivision edge and a convex chain of polygon edges, where the angles between the chain and the subdivision edge are acute. Let Q be such a subpolygon and e the corresponding subdivision edge. We construct the largest triangle T contained in Q that has e as an edge by finding the vertex v on the convex chain that is farthest from e. Because the chain is convex this vertex can be found in $O(\log n_Q)$ time, where n_Q is the number of vertices of Q.

Theorem 1. *For any convex polygon P with n_P vertices we can compute in $O(n_P \log n_P)$ time a triangulation that has the 1-distance property.*

Proof. Consider the algorithm described above. To prove the resulting triangulation has the 1-distance property, consider a triangle T created by the algorithm. Suppose that T is created when a subpolygon Q is handled. Let e be

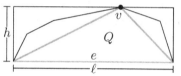

the subdivision edge of Q. Without loss of generality assume that e is horizontal and let v denote the highest point of Q. Since the angles at e's endpoints are acute, Q must be contained in an $\ell \times h$ rectangle where $\ell = |e|$ and $h = \text{dist}(v, e)$. It follows that for any point $p \in T$ we have

$$\Delta_p^2 \leq \min(h, \ell/2)^2 \leq h\ell/2 = \text{area}(T).$$

The diameter of a convex polygon can be computed in $O(n_P)$ time and the creation of each triangle takes $O(\log n_P)$ time. Since there are $n_P - 2$ triangles it follows that the algorithm takes $O(n_P \log n_P)$ time. □

3   Arbitrary Polygons

We now consider non-convex polygons. We wish to compute a decomposition of a simple polygon P into constant-complexity subpolygons that have the α-distance property. This is not always possible with a triangulation: any triangulation of the polygon in Fig. 1 must include triangle uvw or uvz, and when ε tends to zero the α-distance property is violated for points in the middle of these triangles. A Steiner triangulation with the α-distance property always exists: the quadtree-based mesh of Bern *et al.* [5] can be adapted to have the α-distance property—the (small) adaptations are required only around acute angles. However, the number of Steiner points and, hence, the number of triangles cannot by bounded as a function of the number of vertices of P. Next we show that this is necessarily so.

Theorem 2. *For any constant $\alpha > 0$ and any $m > 0$, there is a simple polygon P with eight vertices such that any Steiner triangulation of P with the α-distance property uses at least m Steiner points.*

Proof Sketch. Let P be the polygon shown in Fig. 1. Consider a Steiner triangulation $\mathcal{T}$ of P with the α-distance property. Let T_0 be the triangle in $\mathcal{T}$ that has uv as an edge.[1] If ε is very small then the other two edges of T_0 cannot be very long either, otherwise the α-distance property is violated inside T_0. This in turn implies that the neighboring triangles of T_0 cannot be very large. The idea is to repeat this argument to show that many triangles are needed to cover P.

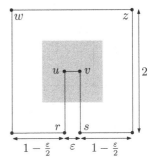

Fig. 1. Any triangulation of P with the α-distance property requires many Steiner points

Specifically, we define a sequence of triangles $T_0, T_1, T_2, \ldots$, as follows. Suppose we are given a triangle T_i and an edge e_i bounding T_i from below. (For $i = 0$, we have $e_i = uv$.) Consider the other two edges of T_i. We select one of these two edges as e_{i+1} and define T_{i+1} as the triangle directly above e_{i+1}. We select e_{i+1} as follows. If only one of the edges bounds T_i from above, then this edge is selected. If both edges bound T_i from above, then we select the edge with the smaller absolute slope. This selection guarantees that for every edge e_i at least one endpoint is above e_0.

Our goal is now to prove that the size of the triangles $T_0, T_1, \ldots$ does not increase too rapidly—more precisely, that T_{i+1} cannot be arbitrarily larger than T_i. This requires an invariant on the length of the edges e_i, but also on their absolute slope. We denote the absolute slope of e_i by σ_i. Thus $\sigma_i = |e_i|_y / |e_i|_x$, where $|e_i|_x$ and $|e_i|_y$ denote the lengths of the projection of e_i on the x- and y-axis. Let P^- denote the square with edge length 1 centered at the midpoint of uv. In Fig. 1 this square is shaded. Our argument will use the fact that for T_i inside P^- the nearest boundary point for any $p \in T_i$ lies on uv, ur, or vs. We show that both the slope and length of edge e_i are bounded as a function of i, and that e_i remains inside P^-, until $\sigma_i \cdot |e_i|$ is large enough. More precisely, we can prove that as long as $\max(4, \sigma_i^2) \cdot |e_i| < \frac{\alpha}{8\sqrt{2}}$ the following three properties hold, where (i) and (ii) are needed to prove (iii):

(i) edge e_i is contained in P^-;
(ii) the slope σ_i of e_i satisfies $\sigma_i \leq (2^{i+1} - 2)/\alpha$;
(iii) edge e_i has length at most $8\varepsilon \cdot 2^{(i+1)(i+7)}/(\alpha^{3i})$.

It follows from property (iii) that we can always choose ε small enough that we need at least m Steiner points before T_i can leave the square P^-. Property (i) is proven using an induction on i. Clearly edge e_0 is contained in P^-. For edge e_i

---

[1] The edge uv can contain Steiner vertices in its interior, as the only requirement we have for the Steiner triangulation is that any two triangles either meet in a single vertex, along a complete edge, or not at all. When uv contains Steiner vertices, we can replace uv by any subedge of uv, and the argument still holds.

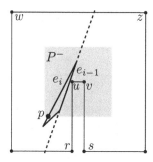

Fig. 2. Triangle T_{i-1} intersects the boundary of P^- in p

with $i > 0$ we can assume that e_{i-1} is contained in P^- and that $\max(4, \sigma_{i-1}^2) \cdot |e_{i-1}| < \frac{\alpha}{8\sqrt{2}}$. We assume for a contradiction that e_i extends outside of P^- and show that if this is the case, then T_{i-1} does not have the α-distance property for the given α. The area of T_{i-1} is upper bounded by $|e_{i-1}| \cdot |e_i| \le |e_{i-1}| \cdot 2\sqrt{2}$. Since e_i extends outside of P^- and e_{i-1} is inside it there must be a point $p \in T_{i-1}$ that is on the boundary of P^-. If p is on the left, top or right edge of P^- then $\Delta_p \ge (1 - \varepsilon)/2 \ge 1/4$. If p is on the bottom edge of P^- we can use the slope of e_{i-1} and the fact that its top endpoint is above e_0 to bound the distance from p to the boundary of P. Without loss of generality assume that p is to the left of ur. Since one endpoint of e_{i-1} is above e_0 and e_{i-1} cannot intersect e_0 the distance from p to ur (the nearest boundary edge) is at least $1/(2\sigma_{i-1})$, see also Fig. 2. We get that $\Delta_p \ge 1/(2\max(2, \sigma_{i-1}))$. This would imply that

$$\text{area}(T_i) \le |e_{i-1}| \cdot 2\sqrt{2} < \frac{\alpha}{8\sqrt{2}\max(4, \sigma_{i-1}^2)} \cdot 2\sqrt{2} \le \alpha \cdot \Delta_p^2,$$

contradicting that T_i has the α-distance property. Hence, we can conclude that e_i must be contained in P^-. Using (i) we can prove (ii), which can be used to prove (iii) (proof omitted here). The lemma follows then from (iii).   □

Theorem 2 implies that we cannot restrict ourselves to triangulations if we want a linear-size decomposition with the α-distance property. We hence consider possibly non-conforming decompositions using convex k-gons (that is, we allow T-junctions). We first show how to compute a linear-size decomposition with the α-distance property that uses convex k-gons for $k \le 7$, and then we argue that each k-gon can be further subdivided into convex quadrilaterals and triangles.

A Decomposition with 7-Gons. We assume without loss of generality that no two vertices of the input polygon P have the same x- or y-coordinates. We describe a recursive algorithm that computes in each step a single 7-gon[2] of the subdivision and then recurses on up to four smaller polygons. In a generic step of the recursive procedure, we are given a polygon bounded by a chain of edges from the original polygon and by two subdivision edges, one vertical and one horizontal; see Fig. 3a. (In our figures we use gray lines for subdivision edges,

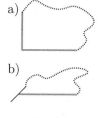

Fig. 3.

solid black lines for polygon edges, and dotted black lines to indicate an unspecified continuation of the boundary of the input polygon. Black circles mark vertices of the input polygon.) The subdivision edges meet in a vertex, the *corner* of the polygon. One of the subdivision edges can have zero length (see Fig. 3b).

---

[2] From now on, when we use the term 7-gon, we mean a convex k-gon for $k \le 7$.

Without loss of generality we assume that the horizontal subdivision edge, e_h, is the longer of the two subdivision edges, and that the vertical subdivision edge e_v extends upward from the left endpoint of e_h. Initially, P does not have the right form as there are no subdivision edges. Hence we first pick an arbitrary point in the interior of P and shoot axis-aligned rays in all four directions. This subdivides P into four polygons that each have exactly two subdivision edges that meet in a vertex.

We now describe how we generate a 7-gon of the decomposition in a recursive step on input polygon $Q \subset P$ with two subdivision edges, e_h and e_v, meeting in corner v. We first grow a square with v as lower-left corner, until the square hits the boundary of Q. (This could be immediately, if the vertical subdivision edge has zero length.) If one of the edges of the square hits a vertex of the original polygon P, we stop. Otherwise a vertex of the square hits an edge of P. We then start pushing the square along the edge, meanwhile growing it so that it remains in contact with the subdivision edge. This again continues until the boundary of P is hit, which may either terminate the process (when a vertex of P is hit) or not (when an edge is hit), and so on. The 7-gon will be the union (swept volume) of all squares generated during the entire process. Fig. 4 gives an overview of the cases that can arise, with A being the start configuration. Thick arrows indicate a transition from one case to another. As mentioned, we stop pushing a square when a new vertex of P occurs on the boundary. Cases where this happens are given a number (A1, B1, B2, ...). Next we provide more details on how to push the squares in each of the cases and when one case transitions to another. The top left, top right, bottom left, and bottom right vertex of a square will be denoted by $p_\mathrm{nw}, p_\mathrm{ne}, p_\mathrm{sw}, p_\mathrm{se}$, respectively, and the top, right, bottom, and left edge by $e_\mathrm{n}, e_\mathrm{e}, e_\mathrm{s}, e_\mathrm{w}$. In each case the process ends when a vertex of P is hit.

A We grow a square from the corner while keeping e_s on e_h and e_w on e_v until it hits an edge or vertex of P. We go into case B if p_nw hits an edge e_nw of P or into case E and F if p_ne hits an edge e_ne. Note that p_se cannot hit an edge of the polygon before p_nw, since e_h is at least as long as e_v.

B The vertex p_nw is on an edge e_nw of P and e_s is on e_h. We push the square to the right while maintaining these contacts. We go into case C if p_se hits an edge e_se of P or into case D and F if p_ne hits an edge e_ne of P.

C The vertex p_nw is on an edge e_nw of P and p_se is on an edge e_se of P. We push the square up and to the right maintaining these contacts. We go into case D and G if p_ne hits an edge e_ne of P.

D The vertex p_nw is on an edge e_nw and p_ne is on an edge e_ne of P. We push the square upward while maintaining these contacts.

E The vertex p_ne is on an edge e_ne of P and e_w is on e_v. We push the square upward while maintaining these contacts. We go into case D if p_nw hits an edge of P.

F The vertex p_ne is on an edge e_ne of P and e_s is on e_h and we push the square to the right while maintaining these contacts. We go into case G if p_se hits an edge of P.

G The vertex p_ne is on an edge e_ne and p_se is on an edge e_se of P and we push the square to the right while maintaining these contacts.

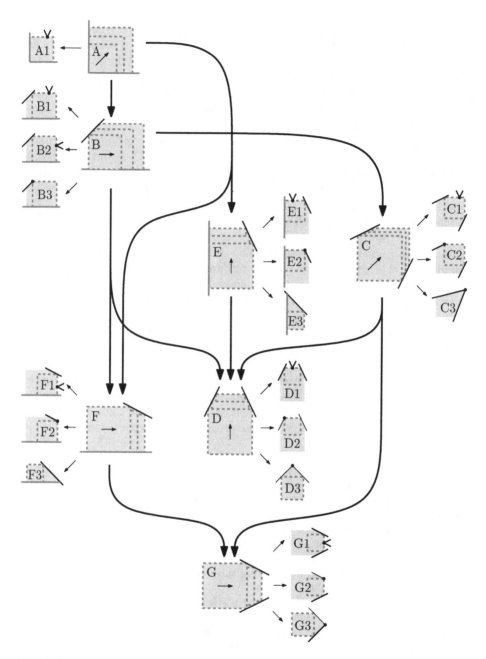

Fig. 4. We construct 7-gons by pushing squares through the polygon according to cases A to G. Fat arrows indicate a transition from one case to another and a split means that we continue in two separate directions. Note that cases E and F, and D and G are symmetric.

Lemma 1. *The process above generates a convex k-gon C with $k \leq 7$. Moreover, for any $p \in C$ we have $\mathrm{area}(C) \geq \frac{1}{2} \cdot \Delta_p^2$, where Δ_p denotes the distance from p to the boundary of the original polygon P.*

Proof. A straightforward case analysis of the different paths that the process may follow in Fig. 4—note that we can actually follow several paths, since sometimes we continue pushing in two separate directions—shows that C is a convex 7-gon. The construction guarantees that C is the union of a (possibly infinite) set of squares that each touch the boundary of P. Let σ be such a square containing p. Then $\Delta_p \leq \sqrt{2} \cdot \mathrm{length}(\sigma)$, where $\mathrm{length}(\sigma)$ denotes the edge length of σ. It follows that $\mathrm{area}(C) \geq \mathrm{area}(\sigma) = \mathrm{length}(\sigma)^2 \geq \frac{1}{2} \cdot \Delta_p^2$. □

After constructing the 7-gon C, we should recurse on the remaining parts of the polygon. The parts we can recurse on must have at most two orthogonal subdivision edges that meet in a point, as in Fig. 3. Parts for which this is not yet the case are first subdivided further by shooting horizontal and/or vertical rays from certain vertices of C 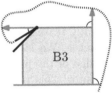 so that the required property holds for the resulting subparts. Which rays to shoot depends on the final case in the construction of C. The figure to the right shows case B3; the corners of the parts on which we recurse are also indicated. In total, we may get up to four parts in which we recursively construct new 7-gons.

Lemma 2. *The algorithm described above creates $O(n_P)$ 7-gons in total, when applied to a polygon P with n_P vertices.*

Proof Sketch. Let V_Q denote the subset of vertices of P that are on the boundary of a polygon $Q \subset P$ on which we recurse, excluding the possible vertices of P that are the endpoints of the subdivision edges of Q. Recall that after we construct a 7-gon C inside Q, the remainder of Q is subdivided into at most four parts on which we recurse again. At least one vertex of V_Q is on the boundary of C, so each part has strictly fewer vertices of P on its boundary. We also know that each vertex of V_Q can be on the boundary of at most one part (recall that vertices on endpoints of subdivision edges are not considered). It follows that only $O(n_P)$ 7-gons are constructed. □

Next we describe how to implement the algorithm in $O(n_P \log n_P)$ time. Each of the cases A to G can be viewed as moving a square from a start location to an end location such that all intermediate squares have specific contacts to the polygon Q as detailed in each case description. To find the swept volume of this sequence of squares it suffices to know at which squares we start and end. To find these start and end squares we need some supporting data structures.

We use the medial axis $\mathcal{M}$ of P, with the following asymmetric distance measure. Let p and q be two points in the plane. The distance from p to q is the scaling factor of a unit square with its lower left corner on p, that has q on its boundary. This distance is defined only if q lies to the north-east of p. However, for any point inside P the distance to the boundary of P and the nearest point on the boundary are well defined, which is sufficient for our purpose. Conceptually, one can also set the undefined distances to infinity. We compute four such medial axes, one for each direction in which we can grow squares, in $O(n \log n)$ time [10]. We then construct the following data structures:

- We preprocess each medial axis so that we can do point location in $O(\log n)$ time. Since the medial axis is a connected subdivision this can be done in $O(n)$ time [13].
- We also preprocess each medial axis so that we can answer horizontal and vertical ray shooting queries in $O(\log n)$ time. This can again be done in $O(n)$ time by first computing the horizontal and vertical decomposition of P [7], and then preprocessing these trapezoidal maps for point location.
- Finally, we preprocess P itself in $O(n)$ time such that we can do horizontal and vertical ray shooting in $O(\log n)$ time.

Initially (case A) we want to find the largest square that we can grow from the corner v. We locate the cell of $\mathcal{M}$ that contains v, which gives us the vertex or edge of the polygon, say edge e, that is closest to v in the specified distance measure. This implies that e is the first edge hit by the boundary of a square grown from v. In this way we determine in $O(\log n)$ time if we are in case A, B, or E. Next we push the square upward or to the right. We then have to determine the final square for that movement and in which case we should continue. We distinguish two different types of movement for the square. Either the square has one edge on one of the vertical or horizontal subdivision edges (case B, E, and F), or it has two corners on polygon edges of P (case C, D, and G).

If one edge of the square stays on a subdivision edge then specifically the lower left vertex stays on the subdivision edge and the series of squares that we create are exactly the largest squares with their lower left corners on the subdivision edge. Recall that we stop moving the square when another edge or vertex of P hits the boundary of the square. Let q denote the lower left corner of this square. By definition of $\mathcal{M}$ the point q has to be on a bisector of $\mathcal{M}$ as there are two different features (edges or vertices) of P that are at equal distance. Hence, the process of moving a square along a subdivision edge is essentially the same as moving its lower left corner point until it hits an edge of the medial axis (or P). We can use horizontal or vertical ray shooting to find in $O(\log n)$ time the point q where we end the movement along the subdivision edge.

When we move a square while keeping two vertices on edges of P it follows from the definition of the medial axis and our distance measure that the lower left vertex of the square remains on the bisector of the two edges of P. The movement ends when a third edge or vertex of P is on the boundary of the square, so at a vertex of the medial axis. Specifically the vertex where the bisector along which the lower left vertex was moving, ends. To find the final square of the movement we have to find the bisector, determine which endpoint of the bisector we need and find the three edges or vertices of P that determine that vertex. Since we already found the right bisector in the previous case, each of these steps can be done in $O(1)$ time after which we can determine in $O(1)$ time how to continue. To summarize, we obtain the following lemma.

Lemma 3. *Computing the 7-gon in a recursive step of the algorithm takes $O(\log n)$ time, after $O(n)$ preprocessing.*

From 7-gons to quadrilaterals and triangles. As a last step we can convert the 7-gons from our decomposition into convex quadrilaterals and triangles. The resulting decomposition still has the α-distance property, although the value for α will decrease from $1/2$ to $1/8$. Let Q denote a convex polygon with n_Q vertices. By the ham-sandwich theorem [9], there exists a line cutting Q into two portions of equal area with at most $\lfloor n_Q/2 \rfloor$ vertices of Q strictly on each side of the line. Cutting along this line, we obtain two polygons with half the area and at most $\lfloor n_Q/2 \rfloor + 2$ vertices each. By repeating this process, if necessary, we obtain either triangles or quadrilaterals. We hence arrive at the following theorem.

Theorem 3. *Given a simple polygon P we can compute in $O(n \log n)$ time a subdivision of P consisting of $O(n)$ triangles and convex quadrilaterals with the $(1/8)$-distance property.*

Proof. By Lemmas 2 and 3 we can compute in $O(n \log n)$ time a decomposition of P into $O(n)$ convex k-gons, for $k \leq 7$, that has the $(1/2)$-distance property. We further subdivide each k-gon using ham-sandwich cuts, as explained above. In the worst case we start with a 7-gon that is split into two 5-gons by the first ham-sandwich cut, after which each 5-gon is split into two quadrilaterals. We then get four quadrilaterals each having $1/4$ of the area of the 7-gon. Since the decomposition into 7-gons had the $(1/2)$-distance property, the new decomposition has the $(1/8)$-distance property. $\square$

Distance-sensitive point location. Let us return to the problem of point-location in a subdivision $\mathcal{S}$, where the goal is to obtain faster query time for query points far away from the boundary of the region they are contained in. We decompose each region of $\mathcal{S}$ using Theorem 3. Next we assign each quadrilateral or triangle Q in the new decomposition a probability of area(Q), and we apply the entropy-based point-location method of Arya *et al.* [2]. (Recall that we assume that area$(\mathcal{S}) = 1$.) This method requires each face of the decomposition to have constant complexity. More precisely, since they use a randomized incremental method that adds the maximal segments defining the subdivision one by one, it is sufficient if each face is bounded by a constant number of such maximal segments. Thus it is not a problem that our subdivision is non-conforming. We can therefore conclude with the following theorem.

Theorem 4. *Given a subdivision $\mathcal{S}$ with area$(\mathcal{S}) = 1$ we can compute in $O(n \log n)$ time a data structure for point location, such that for any query point p the query time is $O(1 + \min(\log(1/\Delta_p), \log n))$, where Δ_p is the distance from p to the boundary of the region containing p.*

4   Conclusions

We presented a data structure for distance-sensitive point location in a connected planar polygonal subdivision $\mathcal{S}$. The key ingredient is a non-conforming decomposition of each region $P \in S$ into a linear number of quadrilaterals and triangles with the (1/8)-distance property. We also proved that if we insist on decomposing each region using a conforming Steiner triangulation, then the number of Steiner vertices cannot be bounded as a function of the region complexity. This poses obvious open problems: can we obtain a linear-size decomposition with the α-distance property (for some constant α) that uses only triangles, or one that uses both triangles and quadrilaterals but is conforming? Another interesting question is whether the decomposition can be generalized to $\mathbb{R}^3$. Note, however, that this does not directly lead to distance-sensitive point location since, to our knowledge, no 3D entropy-based point location structures are known.

References

1. Arya, S., Malamatos, T., Mount, D.M.: A simple entropy-based algorithm for planar point location. ACM Trans. Algorithms 3, article 17 (2007)
2. Arya, S., Malamatos, T., Mount, D.M., Wong, K.C.: Optimal expected-case planar point location. SIAM J. Comput. 37, 584–610 (2007)
3. de Berg, M., Cheong, O., van Kreveld, M., Overmars, M.: Computational Geometry: Algorithms and Applications, 3rd edn. Springer (2008)
4. Bern, M.: Triangulations and mesh generation. In: Goodman, J.E., O'Rourke, J. (eds.) Handbook of Discrete and Computational Geometry, 2nd edn., ch. 25. Chapman & Hall/CRC (2004)
5. Bern, M., Eppstein, D., Gilbert, J.: Provably good mesh generation. J. of Computer and System Sciences 48(3), 384–409 (1994)
6. Bern, M., Mitchell, S., Ruppert, J.: Linear-size nonobtuse triangulation of polygons. Discrete & Computational Geometry 14(1), 411–428 (1995)
7. Chazelle, B.: Triangulating a simple polygon in linear time. Discrete & Computational Geometry 6, 485–524 (1991)
8. Collette, S., Dujmović, V., Iacono, J., Langerman, S., Morin, P.: Entropy, triangulation, and point location in planar subdivisions. ACM Trans. Algorithms 8(3), 1–18 (2012)
9. Edelsbrunner, H.: Algorithms in Combinatorial Geometry. Springer (1987)
10. Fortune, S.: A fast algorithm for polygon containment by translation. In: Brauer, W. (ed.) ICALP 1985. LNCS, vol. 194, pp. 189–198. Springer, Heidelberg (1985)
11. Iacono, J.: Expected asymptotically optimal planar point location. Computational Geometry 29(1), 19–22 (2004)
12. Iacono, J., Mulzer, W.: A static optimality transformation with applications to planar point location. Int. J. of Comput. Geom. and Appl. 22(4), 327–340 (2012)
13. Kirkpatrick, D.: Optimal search in planar subdivisions. SIAM J. Comput. 12(1), 28–35 (1983)
14. Knuth, D.E.: Sorting and Searching, 2nd edn. The Art of Computer Programming, vol. 3. Addison-Wesley (1998)
15. Ruppert, J.: A Delaunay refinement algorithm for quality 2-dimensional mesh generation. J. Algorithms 18(3), 548–585 (1995)
16. Shannon, C.E.: A mathematical theory of communication. Bell Sys. Tech. Journal 27, 379–423, 623–656 (1948)
17. Snoeyink, J.: Point location. In: Goodman, J.E., O'Rourke, J. (eds.) Handbook of Discrete and Computational Geometry, 2nd edn., ch. 34. Chapman & Hall/CRC (2004)

Time-Space Tradeoffs
for All-Nearest-Larger-Neighbors Problems

Tetsuo Asano[1] and David Kirkpatrick[2]

[1] School of Information Science, JAIST, Japan
[2] Department of Computer Science, University of British Columbia, Canada

Abstract. This paper addresses two versions of a fundamental problem, referred to as the All-Nearest-Larger-Neighbors (ANLN) problem, defined as follows: given a one-dimensional array A of n real-valued keys, find, for each array element $A[i]$, the index of a nearest array element, if one exists, whose key is strictly larger than $A[i]$.

We develop algorithms for one- and two-sided versions of the ANLN problem that run in $O(n \log_b n)$ time, using $\Theta(b)$ work-space, for all $b = O(n)$, exhibiting a full time-space tradeoff that subsumes all known (memory-restricted) special cases. In addition, a non-trivial lower bound is developed for the time complexity of solving both versions on a pointer machine with limited work-space. This lower bound matches the time complexity of our algorithms, when restricted to constant space.

The fundamental nature of ANLN problems make them intrinsically interesting to study. They also capture the essence of a variety of other familiar problems, such as determining the forest structure associated with a given string of nested parentheses, and triangulating monotone polygons. For both of these, we describe reductions to versions of the ANLN problem, achieving the same time-space tradeoffs.

1 Introduction

We consider the following *All-Nearest-Larger-Neighbor (ANLN) problem*: Given an array $A[0 : n-1]$ of n real values, determine an array $NLN[0 : n-1]$, where $NLN[i]$ is the index of a *nearest larger neighbor* of $A[i]$ in $A[0 : n-1]$, i.e. an index j such that $A[j] > A[i]$ and $|j - i|$ is minimized. (If $A[i]$ is a maximum element in $A[0 : n-1]$, i.e. it has no larger neighbor, this is signified by assigning $NLN[i] \notin [0 : n-1]$.)

Of course, the nearest larger neighbor of $A[i]$ is the closer of its *nearest larger left-neighbor* and its *nearest larger right-neighbor* (with the obvious definitions). We refer to the variant of the ANLN problem in which neighbors are restricted to right-neighbors as the *All-Nearest-Larger-Right-Neighbor (ANLRN) problem*. It is straightforward to design a linear-time algorithm for solving the ANLRN problem, if we are free to use linear work-space: simply scan the input array A from right to left while keeping a monotonically decreasing subsequence in a stack to find, for each successive element of A, the nearest larger neighbor to its right. Combining this with the symmetric left to right scan, constructing all

F. Dehne, R. Solis-Oba, and J.-R. Sack (Eds.): WADS 2013, LNCS 8037, pp. 61–72, 2013.
© Springer-Verlag Berlin Heidelberg 2013

nearest larger left-neighbors, gives an equally straightforward solution to the full ANLN problem; we will refer to this as the *double-scan algorithm*.

We are interested in understanding the complexity of solving the ANLN problem with sub-linear work space. For this reason, we assume that the input array A is read-only, and the output array NLN is write-once. In this setting, the complexity of the ANLN problem turns out to depend on (1) whether or not we allow duplicate keys, and (2) whether we require the outputs to be constructed in some specified order (think of this as an *on-line* version); in the easiest case keys are distinct and arbitrary output order is permitted.

Throughout this paper, we measure work-space by the number of words of $O(\log n)$ bits used in an algorithm in addition to input data stored in a read-only array. Thus, by a $\Theta(b)$-work-space algorithm we mean an algorithm that uses work-space consisting of $\Theta(b)$ words of $O(\log n)$ bits, which can be used as counters, indices and pointers, with a range of $\Theta(n)$ values. This generalizes the constant work-space model introduced by Asano et al [1,3,2] for the study of a variety of geometric problems. It is also consistent with the framework of *memory-adjustable* algorithms, and the time-space tradeoffs, described in [2] and [4] respectively. Of course, memory-constrained computational models and time-space tradeoffs have been the subject of study for a long time (see, for example, [7,8]); we refer the reader to [2] for a succinct overview of this background work.

In an earlier paper [1], we showed that the ANLN problem can be solved in $O(n \log n)$ time using only constant work-space. In section 2, we give a full space-time tradeoff: an algorithm that runs in $O(n \log_b n)$ time with $\Theta(b)$ work-space, in particular, $O(n/\varepsilon)$ time, if $b = \Theta(n^\varepsilon)$. In section 3, we present an algorithm for the ANLRN problem with a similar tradeoff. (It should be noted that the close relationship between these problems, in the absence of space constraints, cannot be exploited in general; athough they have some common features, the algorithms that we develop for the two problems are fundamentally different.) Of course, the same results follow if we restrict attention to left-neighbors, or if we want to determine smaller, rather than larger neighbors; we assume that the reader will interpret the acronyms ANSN, ANSLN, etc. appropriately.

Section 4 describes a lower bound on the product of time and work-space that applies to both the ANLN and ANLRN problems. This bound is based on a pointer machine model that is sufficiently powerful to realize all existing algorithms for these problems. This lower bound matches the time complexity of our algorithms, when restricted to constant space.

Our algorithms for ANLN problems provide a foundation for the solution of other problems such as extracting the forest structure implicit in a given string of nested parentheses, and constructing a triangulation of a monotone polygon. In sections 5 and 6, we present algorithms for these problems that run in $O(n \log_b n)$ time using $\Theta(b)$ work-space, using direct reductions to variants of the ANLN problem. This improves the earlier $O(n^2)$-time constant work space algorithm of [2] for triangulating mountain polygons (a special sub-class of monotone polygons), and the algorithm of [4], for b in the range $[1, \log n]$ (including their $O(n^2)$-time constant work-space and $O(n \log n)$-time $\Theta(\log n)$ work-space algorithms).

2   The All-Nearest-Larger-Neighbor Problem

2.1   Background

Let $A[0 : n-1]$ be an array of (not necessarily distinct) real-valued keys. As previously noted, a straightforward double-scan algorithm provides a linear-time solution for the ANLN problem on A, if we are free to use linear work-space. It was shown in [1] that, *assuming distinctness of inputs*, the problem can be solved in $O(n \log n)$ time, using only constant work-space, by using a simple *bidirectional search* from each element of $A[0 : n-1]$. In fact, Bose and Moran [5] showed even earlier that a problem, equivalent to the ANLN Problem for circular lists of length n containing *distinct numbers*, can be solved using bidirectional search, using $\Theta(1)$ pointers, in $O(n \log n)$ time.

Theorem 1. *[5,1] There is an algorithm using bidirectional search that solves the All-Nearest-Larger-Neighbor problem for n distinct numbers, in $O(n \log n)$ time in the worst case, using only $\Theta(1)$ work-space.*

2.2   A New Time-Space Tradeoff Using a Hierarchical Decomposition

We have noted the existence of two different algorithms for the general ANLN problem (with duplicate inputs permitted). The first one runs in $O(n)$ time using $O(n)$ work space, and the other takes $O(n \log n)$ time but uses only $O(1)$ work-space. In this section, we describe a unified algorithm with space-time tradeoffs matching the results above. More precisely, the algorithm is characterized by the amount b of work-space, and runs in $O(n \log_b n)$ time. In particular, it runs in $O(n/\varepsilon)$ time, if $b = \Theta(n^\varepsilon)$ for any $\varepsilon \in (0, 1]$, and $O(n \log n)$ time if $b = \Theta(1)$.

Our algorithm is based on an implicit hierarchical decomposition of the input array. For ease of description, we let $k = \lceil \log_b n \rceil$ and we assume that the input

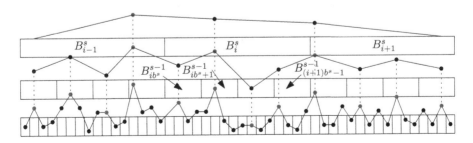

Fig. 1. Computing the approximate NLN values for the maximum elements of the $3b$ sub-blocks of B_{i-1}^s, B_i^s and B_{i+1}^s

array $A[0 : n-1]$ has been augmented appropriately so that $A[i] = -\infty$, for indices i outside the range $[0 : n-1]$, except for $A[b^{k+1}-1]$ and $A[-b^k]$, which take the value ∞. Of course, this requires no additional space, since accesses outside of $A[0 : n-1]$ can be handled by index checking. By this convention, we ensure that all elements in $A[0 : b^k-1]$ have a well-defined nearest larger left- and right-neighbors in the augmented array, and reduce our problem to the case where $n = b^k$, for some integer k.

For each s, $0 \leq s \leq k$, and i, $-1 \leq i < b^{k-s}$, we denote by B_i^s the sub-array $A[ib^s : (i+1)b^s-1]$, which we refer to as the i-th b^s-block of A. For $s > 0$, the block B_i^s is the concatenation of b^{s-1}-blocks $B_{ib}^{s-1}, B_{ib+1}^{s-1}, \ldots, B_{(i+1)b-1}^{s-1}$, which we refer to as the sub-blocks of B_i^s. (See Figure 1.)

For each block B_i^s, with $0 \leq s \leq k$, and $0 \leq i \leq b^{k-s}$, we determine the NLN-indices for the maximum elements in each of the sub-blocks of B_i^s, with the exception of those one or more sub-blocks that contain the maximum value in the entire block B_i^s. We exploit the (easily confirmed) fact that the NLN-indices of each of these sub-block maxima must lie outside of its own sub-block, but within one of the blocks B_{i-1}^s, B_i^s or B_{i+1}^s. For this purpose, we first construct an array $M[-b : 2b-1]$, where $M[t]$ contains the maximum element in the sub-block B_{ib+t}^{s-1}, that is $M[-b : 2b-1]$ records the sub-block maxima for all $3b$ sub-blocks of B_{i-1}^s, B_i^s and B_{i+1}^s. Using the double-scan algorithm, we determine the NLN-indices for all elements in $M[0 : b-1]$ within $M[-b : 2b-1]$.

The NLN-indices for the elements in $M[0 : b]$ provide approximate NLN-indices for the corresponding elements of A in the sense that if t' is the NLN-index of $M[t]$ in $M[-b : 2b-1]$, then the exact NLN-index of the corresponding elements of A (i.e. the maximum elements in B_{ib+t}^{s-1}), must lie either in $B_{ib+t-\Delta}^{s-1}$ or $B_{ib+t+\Delta}^{s-1}$, where $\Delta = |t' - t|$.

It remains to determine p_l, the index of the rightmost element of $B_{ib+t-\Delta}^{s-1}$ that is larger than $M[t]$ (choosing $p_l = -b^k$, if such an element does not exist), and the index of the leftmost element of $B_{ib+t+\Delta}^{s-1}$ that is larger than $M[t]$ (choosing $p_r = 2b^k - 1$, if such an element does not exist). We then assign the exact NLN-index to all maxima in B_{ib+t}^{s-1} by assigning the closer of $A[p_l]$ and $A[p_r]$. See Algorithm 1 below.

The following captures the essence of our algorithm:

Lemma 1. *The loop at line 2 of Algorithm 1 constructs the NLN-values for the maximum elements of all b principal sub-blocks of B_i^s that are not maximum elements of B_i^s, in $O(b^s)$ time, using $\Theta(b)$ work-space.*

Proof. In the first step (lines 3-4), we scan $3 \cdot b^s$ elements, and determine (and record) $3b$ sub-block maxima. Lines 5-9 determine the NLN-values restricted to these sub-block maxima (using the double-scan algorithm). Finally, lines 10-18 convert each of these approximate NLN-values to actual NLN-values, by searching exhaustively in the appropriate sub-block of size b^{s-1}. □

Since the loop at line 2 is repeated for all b^{k-s} b^s-blocks, for all s, $1 \leq s \leq k$, the following is an immediate consequence of Lemma 1:

Algorithm 1. Two-sided All-Nearest-Larger-Neighbors

Input: Read-only array $A[0 : b^k - 1]$ of keys.
Output: Write-once array $NLN[0 : b^k - 1]$, where $A[NLN[p]]$ is the
    nearest larger neighbor of $A[p]$.

1 **for** $s \leftarrow 1$ **to** k **do**
2   **for** $i \leftarrow 0$ **to** $b^{k-s}-1$ **do**   //determine NLN-values for all
      //sub-block maxima that are not block maxima of B_i^s;
3   **for** $t \leftarrow -b$ **to** $2b - 1$ **do**
4       $M[t] \leftarrow \max\{B_{ib+t}^{s-1}\}$ //find maximum elements in each of;
      //the sub-blocks of B_{i-1}^s, B_i^s, and B_{i+1}^s;
5   **for** $t \leftarrow 0$ **to** $b - 1$ **do**
6       $\Delta \leftarrow \infty$;
7   **if** $M[t] \neq \max\{M[0 : b - 1]\}$ **then**
8        $\Delta \leftarrow$ distance to NLN of $M[t]$ within $M[-b : 2b - 1]$
9   (using the double-scan algorithm)
10   ;
11   **for** $t \leftarrow 0$ **to** $b-1$ **do**
12   **if** $\Delta < \infty$ **then**
13        $p_l \leftarrow$ index of rightmost element of $B_{ib+t-\Delta}^{s-1}$ with
14   value greater than $M[t]$ ($p_l \leftarrow -b^k$, otherwise) ;
15        $p_r \leftarrow$ index of leftmost element of $B_{ib+t+\Delta}^{s-1}$ with
16   value greater than $M[t]$ ($p_r \leftarrow 2b^k - 1$, otherwise);
17        $p_m \leftarrow \lfloor (p_l + p_r)/2 \rfloor$;
18   **for** each element $A[u]$ of B_{ib+t}^{s-1} with $A[u] = M[t]$ **do**
19   **if** $u \leq p_m$ **then** $NLN[u] \leftarrow p_l$ **else** $NLN[u] \leftarrow p_r$;

Theorem 2. *We can find the NLN-values of all elements in a given array of n (not-necessarily-distinct) elements using $\Theta(b)$ work-space and $O(n \log_b n)$ time.*

Remark. We note that our algorithm has a straightforward implementation, with the same asymptotic cost, in which the only index operations used are (i) increment, (ii) decrement, and (iii) copy. Thus, it can be applied directly to a variant of the ANLN problem in which the input is given as an immutable list, and the output is provided as a sequence of pointer pairs.

3   The All-Nearest-Larger-Right-Neighbor Problem

Suppose that we are given n real numbers in an array $A[0 : n-1]$. For each $i \in [0 : n-1]$, we wish to determine the nearest larger right-neighbor of $A[i]$, $NLRN[i]$, that is the smallest index $j \in [i+1 : n-1]$ such that $A[j] > A[i]$. Note that there may be many elements $A[i]$ that do not have such a nearest larger right-neighbor; this is signified by assigning $NLRN[i] \notin [0 : n-1]$.

In this section, we describe an algorithm that solves this ANLRN problem, within the same $O(n \log_b n)$ time and $\Theta(b)$ work-space constraints as the two-sided variant. As in the previous section, we find it convenient to assume that $n = b^k$ and that the array A has been augmented appropriately so that indices outside the range $[0 : n-1]$ return the value $-\infty$, except for $A[n]$, which takes the value ∞. By this convention, we ensure that all elements in $A[0 : n-1]$ have a well-defined nearest larger right-neighbor in the augmented array.

We exploit the same hierarchical decomposition of A as before. Specifically, for each s, $0 \le s \le k$, and i, $-1 \le i \le b^{k-s}$, B_i^s denotes the sub-array $A[ib^s : (i+1)b^s-1]$ (the i-th b^s-block of A) and for $s > 0$, the block B_i^s is the concatenation of the b^{s-1}-blocks $B_{ib}^{s-1}, B_{ib+1}^{s-1}, \dots, B_{(i+1)b-1}^{s-1}$, (the b sub-blocks of B_i^s).

We say that a pair (p, q) is a NLRN-pair if $q = NLRN[p]$. Such a pair is said to have span s if p and q belong to the same b^s-block, but different b^{s-1}-blocks. The s-profile of block B_{ib+r}^{s-1}, the r-th sub-block of block B_i^s, is defined to be the smallest interval in B_{ib+r}^{s-1}, possibly empty, that contains all of the left endpoints of NLRN-pairs whose right endpoints lie outside of B_i^s. (More generally, for $r \le t$, the t-intermediate s-profile of B_{ib+r}^{s-1} is defined to be the smallest interval in B_{ib+r}^{s-1}, possibly empty, that contains all of the left endpoints of NLRN-pairs whose right endpoints lie to the right of B_{ib+t}^{s-1}; the $b-1$-intermediate s-profile of B_{ib+r}^{s-1} is exactly the same as the s-profile of B_{ib+r}^{s-1}.) The s-profile of B_i^s is just the sequence of the at most b non-empty s-profiles associated with its sub-blocks.

We construct the s-profile of B_i^s iteratively, introducing sub-blocks from left to right. As a by-product we identify and output all NLRN-pairs with span s whose left endpoints lie in B_i^s. After introducing sub-block B_{ib+t}^{s-1} we have the invariant that (i) the non-empty t-intermediate s-profiles of $B_{ib}^{s-1}, \dots, B_{ib+t}^{s-1}$ appear, from bottom to top, on a stack S (initialized to contain the interval $[-1, -1]$), as interval endpoint pairs, and (ii) all NLRN-pairs with span s with both endpoints in $B_{ib}^{s-1}, \dots, B_{ib+t}^{s-1}$ have been detected and output.

Phase t, for $0 \le t \le b - 1$, introduces the sub-block B_{ib+t}^{s-1}; this involves a simultaneous left-to-right scan of B_{ib+t}^{s-1} with a right-to-left scan of the $(t-1)$-intermediate s-profiles on S. We advance the scan on B_{ib+t}^{s-1} until the maximum element discovered so far in that block, $A[j_r]$, dominates the current maximum element discovered so far, $A[j_l]$, in the profile scan (at which point we set $NLRN[j_l] = j_r$). While this latter condition holds we advance the profile scan, updating j_l when a newly scanned element exceeds the current $A[j_l]$ and either setting $NLRN[j_l] = j_r$, if $A[j_l]$ continues to be dominated by $A[j_r]$, or switching back to the scan on B_{ib+t}^{s-1}, otherwise. If the profile scan exhausts the current $(t-1)$-intermediate s-profile at the top of S, it pops S and continues with the next $(t-1)$-intermediate s-profile. The process continues until the scan of B_{ib+t}^{s-1} is completed, since the stack contains a reference to an element with value ∞ at it bottom, in which case the interval $[j_r, (ib+t+1)b^{s-1}-1]$, the elements of B_{ib+t}^{s-1} to the right of its leftmost maximum element, is pushed on to S, thereby completing the phase.

Once again, our algorithm has a straightforward implementation, with the same asymptotic cost, in which the only index operations used are (i) increment, (ii) decrement, and (iii) copy. Thus, it is directly applicable to a variant of the ANLRN problem in which the input is given as an immutable list, and the output is provided as a sequence of pointer pairs.

4   A Lower Bound via Interval-Acknowledgment

In the preceding two sections we have seen that the ANLN and ANLRN problems for arrays of size n both have has an $O(n \log n)$-time solution with $O(1)$ work space, even if duplicate elements exist. It is natural to ask if it is possible to reduce the time complexity without increasing the work-space.

Suppose we are given an array $A[0 : n-1]$ of n keys drawn from some key space, and a collection of k registers (pointers) whose values range over the set $\{0, 1, \ldots, n-1\}$. We consider a computation model that permits registers to be incremented or decremented, compared to one another or to any integer constant in the range $[0, n-1]$, or copied. (As we have noted, our algorithms, in fact all algorithms that we can imagine for addressing ANLN and ANLRN problems, can be realized in this model.)

We present a nontrivial lower bound for the time complexity of algorithms in this model that solve instances of the following fundamental array processing problem.

Interval-Acknowledgement Problem: Let $\mathcal{I}$ be a set of intervals $[a_i, b_i]$, $i = 0, 1, \ldots, n-1$ such that $a_i, b_i \in \{0, 1, \ldots, n-1\}$. Initially all registers have the value 0. An interval $[a_i, b_i]$ is said to be *acknowledged* at time t if at that time some register has value a_i and some other register has value b_i. The problem is to acknowledge all of the intervals in $\mathcal{I}$.

It should be clear that for any set $\mathcal{I}$ of *nested* intervals (where no pair of intervals $[a_i, b_i]$ and $[a_j, b_j]$, satisfies $a_i < a_j < b_i < b_j$), there is an array $A[0 : n-1]$ such that $A[b_i]$ is the nearest larger neighbor of $A[a_i]$ or $A[a_i]$ is the nearest larger neighbor of $A[b_i]$, for $i = 0, 1, \ldots, n-1$. For this reason, lower bounds on the (space-constrained) time-complexity of Interval-Acknowledgement for nested intervals carry over to the ANLN (and ANLRN) problems.

A set of nested intervals $\mathcal{I}$ is said to be *well-nested* if for every non-disjoint pair $I_1, I_2 \in \mathcal{I}$ either (i) I_1 is a subset of either the first or second half of I_2, or (ii) I_2 is a subset of either the first or second half of I_1.

An interval $[a, b]$ has two *endpoint neighborhoods* both of size $(b-a)/2$: the left (resp., right) endpoint neighborhood is the interval $[(5a-b)/4, (3a+b)/4]$ (resp., $[(3b+a)/4, (5b-a)/4]$). It is straightforward to confirm that if $\mathcal{I}$ is well-nested then the following *endpoint neighborhood property* must hold: for any pair of intervals $I_1, I_2 \in \mathcal{I}$, if $|I_1| \geq |I_2|$ then one of the endpoint neighborhoods of I_1 must be disjoint from both of the endpoint intervals of I_2.

Theorem 3. *Any k-register algorithm $\mathcal{A}$ that solves the Interval-Acknowledgment problem for a well-nested set $\mathcal{I}$ of intervals must make a total of $\Omega((1/k) \sum_{I \in \mathcal{I}} |I|)$ register modifications.*

Proof. We say that two registers p and q *approximate* the interval I if p has a value in I's left endpoint interval and q has a value in I's right endpoint interval. It follows directly from the endpoint neighborhood property that at any fixed time any pair of registers approximates at most one interval in $\mathcal{I}$.

Let $[a, b]$ be any interval in $\mathcal{I}$. For $0 \leq y \leq (b - a)/4$, we define $t_y^{[a,b]}$ to be the *first* time step when, for some pair of registers p, q, $|a - p| \leq y$ and $|b - q| \leq y$. Note that, since $\mathcal{A}$ acknowledges all intervals in $\mathcal{I}$, $t_0^{[a,b]}$ (and hence $t_y^{[a,b]}$, for all $0 \leq y \leq (b-a)/4$) is well defined. Furthermore, since register values can only be incremented, decremented or copied, it is clear that at time $t_y^{[a,b]}$ we must have either $|a - p| = y$ or $|b - q| = y$.

So, we *charge* interval $[a, b]$ for each time step $t_y^{[a,b]}$, where $0 \leq y \leq (b - a)/4$. There are clearly $\frac{1}{4} \sum_{I \in \mathcal{I}} |I|$ charges in total. But no time step is charged to more than $k - 1$ intervals, since only one register is modified at time $t_y^{[a,b]}$ and hence there are only $k - 1$ register pairs that can generate charges (and, as previously observed, each pair can generate at most one charge). Thus, there must be at least $\frac{1}{4k} \sum_{I \in \mathcal{I}} |I|$ distinct steps of $\mathcal{A}$ at which charges are made.   □

Consider an array $A[0 : n{-}1]$ whose elements form a recursively defined sequence $\sigma_{\log n}$. The sequence σ_0 is defined as $< 0 >$; to form the sequence σ_{i+1} we concatenate the sequence σ_i^+, formed from σ_i by incrementing the first element, and σ_i^r, formed by reversing the sequence σ_i. Thus, we have
$\sigma_1 = < 1, 0 >$, $\sigma_2 = < 2, 0, 0, 1 >$, $\sigma_3 = < 3, 0, 0, 1, 1, 0, 0, 2 >$,
$\sigma_4 = < 4, 0, 0, 1, 1, 0, 0, 2, 2, 0, 0, 1, 1, 0, 0, 3 >$, ...

Let $\mathcal{I}$ be the set of intervals $[a, b]$ corresponding to a pair of indices a, b in the array $A[0 : n{-}1]$ for which either (i) b is the index of the NLN of $A[a]$ or (ii) a is the index of the NLN of $A[b]$. It is easy to confirm that $\sum_{I \in \mathcal{I}} |I| = \Omega(n \log n)$.

In summary, we have:

Corollary 1. *Any k-register algorithm that solves either the ANLN problem requires $\Omega((n/k) \log n)$ time in the worst case.*

A similar construction gives the same worst case lower bound for the ANLRN problem. Note that the lower bound for Interval-Acknowledgement applies even to non-deterministic algorithms, and counts only increment and decrement operations on registers. Nevertheless, we conjecture that the deterministic complexity of our ANLN problems is fully captured by the non-deterministic complexity of Interval-Acknowledgment, and that both will be shown to coincide with our upper bounds, in the worst case.

5   Extracting Forest Structure from a Parenthesis String

A *well-formed parenthesis string* has the familiar recursive definition: (i) the string () is well-formed, and (ii) if X and Y are well-formed strings, then so are (X) and XY. Let Z be any well-formed parenthesis string of length $2n$. We associate an n-node forest with Z as follows: (i) if Z has the form () then its corresponding forest is a single node, (ii) if Z has the form (X), then its forest is the tree whose root has, as children, the root nodes of the forest associated with X, and (iii) if Z has the form XY, then its forest is the union of the forests associated with X and Y. There is an obvious association between the matching pairs of parentheses in Z and the nodes of its associated forest. (See Figure 2 for an example.)

It is straightforward to identify the matching pairs of parentheses in a given well-formed parenthesis string in linear time, using a stack that may consist of linearly many indices. In fact, if we are only interested in verifying the well-formed property, then by simply maintaining a depth-of-nesting count, initially zero, of the excess of left parentheses over right parentheses, a single counter suffices.

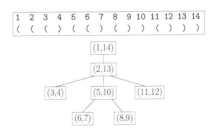

Fig. 2. An example of a well-formed parenthesis string and its associated forest expression

While some underlying structure, for example left-child and next-sibling associations, is readily available from the parenthesis string, it is not so straightforward to identify matching parenthesis pairs, and the parent structure in the associated forest, if work-space is limited. We note, however, that the single counter algorithm associates with the input string a depth-of-nesting sequence (the counter values) that embodies information that can be exploited in this task. Fortunately, this information does not need to be recorded; the nesting depth associated with every index used is easily maintained by any algorithm whose indexing operations are restricted to increment, decrement and copy. Thus we can view this depth-of-nesting sequence as a read-only input sequence D suitable for input to our ANLN algorithms.

By the definition of the depth-of-nesting sequence, if the i-th parenthesis is a left parenthesis and $D[i]$ has $D[j]$ as its nearest smaller right neighbor, then the parentheses in positions i and j form a matching pair. Moreover, if the i-th parenthesis is a right parenthesis and $D[i]$ has $D[j]$ as its nearest smaller right neighbor, then the pair of parentheses ending at the i-th parenthesis has, as its parent, the pair ending at the j-th parenthesis. It follows that, using $O(b)$ work-space, we can compute left-child, next-sibling, partner and parent relations for all parentheses in an array (or list) representation of a well-formed string of parentheses, in $O(n \log_b n)$ time.

6   Triangulation of Monotone Polygons

A simple polygonal chain is *monotone* with respect to a line ℓ if the orthogonal projection of its vertices onto ℓ, in order, form a monotonically increasing (or decreasing) sequence on ℓ. A simple polygon is *monotone* with respect to a line ℓ if it can be split into two chains both of which are monotone with respect to ℓ. Monotone polygons arise frequently in the decomposition of more general simple polygons; in particular, a simple approach to the triangulation of simple polygons exploits the fact that monotone polygons can be triangulated in linear time [6].

This section presents an algorithm for triangulating an arbitrary x-monotone polygon P with n vertices, using $\Theta(s)$ work-space. We assume that P is given by two non-intersecting x-monotone chains, the *upper chain* $U = (u_1, u_2, \ldots, u_{n_1})$ and the *lower chain* $V = (v_1, v_2, \ldots, v_{n_2})$, whose first and last vertices coincide.

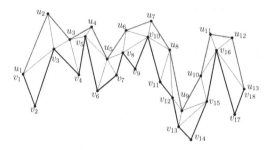

Fig. 3. Triangulation of an x-monotone polygon

A special class of x-monotone polygons is called a *mountain* when the lower chain consists of exactly one edge. Let P be a mountain polygon with upper chain $U = (u_1, u_2, \ldots, u_n)$. We associate with each vertex u_i, $1 \le i \le n$, a value h_i corresponding to the distance from u_i to the line through u_1 and u_n. We treat the sequence $h_1, h_2, \ldots, h_n$ as an all nearest smaller neighbor input sequence. It is straightforward to confirm (using the nesting properties of NSN-pairs) that: (i) if (h_i, h_j) is a nearest smaller right-neighbor pair, with $j > i+1$, then (u_i, u_j) is a chord of P; and (ii) if (h_i, h_j) is a nearest smaller left-neighbor pair, with $j < i-1$, then (u_i, u_j) is a chord of P.

In fact, the NSN-chords defined in (i) and (ii) serve to fully triangulate P:

Lemma 2. *A mountain $P = (u_1, u_2, \ldots, u_n)$ with a single edge (u_1, u_n) on the lower boundary can be triangulated by repeatedly adding NSN-chords according to rule (i) or (ii) above, whenever they are applicable.*

Proof. For simplicity we assume all vertices have different h-values. It suffices to observe that (a) no two chords selected by rules (i) and (ii) cross, and (b) the faces that remain after all applications of rules (i) and (ii) are all triangles.

The first follows immediately from the nesting properties of NSN-pairs. We prove the second by contradiction. Suppose that there is a face bounded by four or more edges of P and NSN-chords that contains no internal NSN-chord. Let u_i be the vertex on this face with the maximum h-value, and let u_j and u_k be its adjacent vertices (on the face). If $h_j > h_k$ then u_j must have an NSN-chord to u_k, or to some other vertex on P between u_i and u_k. But the latter would intersect the chord (or polygon edge) (u_i, u_k), contradicting (a).   □

Since the h_i-values need not be explicitly stored (they can be computed, when needed, from u_i, u_1 and u_n), it follows from the results of section 3 that:

Lemma 3. *A mountain polygon consisting of n vertices can be triangulated in $O(n \log_s n)$ time using $O(s)$ work-space.*

We can extend the above result to arbitrary monotone polygons as follows. Given an x-monotone polygon P, we choose the s-th vertex in the increasing order of their x-coordinates. Assume that a vertex u_i on the upper chain is the s-th vertex. Let v_{j-1} and v_j be the vertices on the lower chain such that $v_{j-1}.x \leq u_i.x \leq v_j.x$, where $v.x$ denotes the x-coordinate of vertex v (see Figure 4).

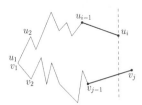

Fig. 4. The s-th vertex u_i from the left and its associated vertices u_{i-1}, v_j and v_{j-1}

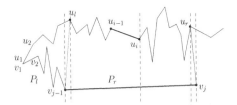

Fig. 5. Two vertices u_l and u_r associated with v_{j-1} and v_j

Let u_l be the first vertex on the upper boundary with $u_l.x > v_{j-1}.x$. and let u_r be the last vertex on the upper boundary with $u_r.x < v_j.x$. These vertices can be found by traversing the upper boundary from u_i; the cost of this traversal can be charged to the following triangulation.

It follows that (u_l, v_{j-1}) is a chord since there is no vertex with x-coordinate between $u_l.x$ and $v_{j-1}.x$. For the same reason, (u_r, v_j) is also a chord. The monotone sub-polygon $P_l = (u_1, \ldots, u_l; v_1, \ldots, v_{j-1})$ to the left of the chord (u_l, v_{j-1}) contains $O(s)$ vertices, and hence it can be triangulated in $O(s)$ time using $O(s)$ space. The right sub-polygon $P_r = (v_{j-1}, u_1, \ldots, u_r; v_{j-1}, v_j)$ is a mountain, and hence it can be triangulated in $O(m \log_s m)$ time in $O(s)$ space if it has m vertices. By the definition of P_l and P_r, the number of vertices in P_l and P_r is $\Omega(s)$. The remaining monotone polygon $P' = (u_r, u_{r+1}, \ldots, u_{n_1}; v_j, v_{j+1}, \ldots, v_{n_2})$ is triangulated in the same manner.

Thus, we have the following theorem:

Theorem 4. *Any n-vertex x-monotone polygon can be triangulated in $O(n \log_s n)$ time using $\Theta(s)$ work-space.*

7 Concluding Remarks and Future Work

We have seen in this paper that variants of the All-Nearest-Larger-Neighbors problem admit a rich variety of memory-constrained algorithms. Our central result is a full time-space tradeoff for the solution of two these variants; a lower bound demonstrates that these are asymptotically optimal when the work-space is restricted to a constant number of registers, on a computational model that permits such registers to be used as pointer/indices into the input array. It would be very interesting to strengthen our lower bound results, when the number of pointers grows as a function of the input size.

It would be desirable to characterize, in the spirit of time-space tradeoff results for stack-based algorithms in [4], those applications for which time-space tradeoffs can be achieved by reduction to variants of the ANLN problem.

Acknowledgment. The work of T.A. was supported in part by KAKENHI No. 24106004 and No. 23300001, and that of D.K. was partially by the Natural Sciences and Engineering Research Council of Canada. The authors would like to thank Jun Tarui, for suggesting the one-sided ANLN problem and for preliminary results associated with it, and Sergey Bereg and Faith Ellen, for helpful discussions, particularly concerning lower bounds.

References

1. Asano, T., Bereg, S., Kirkpatrick, D.: Finding Nearest Larger Neighbors: A Case Study in Algorithm Design and Analysis. In: Albers, S., Alt, H., Näher, S. (eds.) Efficient Algorithms. LNCS, vol. 5760, pp. 249–260. Springer, Heidelberg (2009)
2. Asano, T., Buchin, K., Buchin, M., Korman, M., Mulzer, W., Rota, G., Schultz, A.: Memory-constrained algorithms for simple polygons. In: 28th European Workshop on Computational Geometry (EuroCG), Booklet of Abstracts, pp. 49–52 (2012)
3. Asano, T., Mulzer, W., Rote, G., Wang, Y.: Constant-work-space algorithms for geometric problems. Journal of Computational Geometry 2(1), 46–68 (2011)
4. Barba, L., Korman, M., Langerman, S., Sadakane, K., Silveira, R.I.: Space-time trade-offs for stack-based algorithms. Proc. STACS, pp. 281–292 (2013)
5. Bose, P., Morin, P.: An improved algorithm for subdivision traversal without extra storage. International Journal of Computational Geometry & Applications 12(4), 297–308 (2002)
6. Garey, M.R., Johnson, D.S., Preparata, F.P., Tarjan, R.E.: Triangulating a simple polygon. Information Processing Letters 7(4), 175–179 (1978)
7. Munro, J.I., Raman, V.: Selection from read-only memory and sorting with minimum data movement. Theoretical Computer Science 165, 311–323 (1996)
8. Munro, J.I., Paterson, M.S.: Selection and sorting with limited storage. Theoretical Computer Science 12, 315–323 (1980)

Coloring Hypergraphs Induced by Dynamic Point Sets and Bottomless Rectangles

Andrei Asinowski[1], Jean Cardinal[2], Nathann Cohen[3], Sébastien Collette[4], Thomas Hackl[5], Michael Hoffmann[6], Kolja Knauer[7], Stefan Langerman[2], Michał Lasoń[8], Piotr Micek[8], Günter Rote[9], and Torsten Ueckerdt[10]

[1] Freie Universität Berlin
asinowski@mi.fu-berlin.de
[2] Université Libre de Bruxelles
{jcardin,slanger}@ulb.ac.be
[3] Université Paris-Sud 11
nathann.cohen@gmail.com
[4] Université Libre de Bruxelles
me@scollette.com
[5] TU Graz
thackl@ist.tugraz.at
[6] ETH Zürich
hoffmann@inf.ethz.ch
[7] Université Montpellier 2
kolja.knauer@gmail.com
[8] Jagiellonian University in Krakow
{mlason,piotr.micek}@tcs.uj.edu.pl
[9] Freie Universität Berlin
rote@inf.fu-berlin.de
[10] Karlsruhe Institute of Technology
torsten.ueckerdt@kit.edu

Abstract. We consider a coloring problem on dynamic, one-dimensional point sets: points appearing and disappearing on a line at given times. We wish to color them with k colors so that at any time, any sequence of $p(k)$ consecutive points, for some function p, contains at least one point of each color.

We prove that no such function $p(k)$ exists in general. However, in the restricted case in which points appear gradually, but never disappear, we give a coloring algorithm guaranteeing the property at any time with $p(k) = 3k - 2$. This can be interpreted as coloring point sets in $\mathbb{R}^2$ with k colors such that any bottomless rectangle containing at least $3k-2$ points contains at least one point of each color. Here a bottomless rectangle is an axis-aligned rectangle whose bottom edge is below the lowest point of the set. For this problem, we also prove a lower bound $p(k) > ck$, where $c > 1.67$. Hence, for every k there exists a point set, every k-coloring of which is such that there exists a bottomless rectangle containing ck points and missing at least one of the k colors.

Chen *et al.* (2009) proved that no such function $p(k)$ exists in the case of general axis-aligned rectangles. Our result also complements recent results from Keszegh and Pálvölgyi on cover-decomposability of octants (2011, 2012).

F. Dehne, R. Solis-Oba, and J.-R. Sack (Eds.): WADS 2013, LNCS 8037, pp. 73–84, 2013.
© Springer-Verlag Berlin Heidelberg 2013

1   Introduction

It is straightforward to color n points lying on a line with k colors in such a way that any set of k consecutive points receive different colors; just color them cyclically with the colors $1, 2, \ldots, k, 1, \ldots$. What can we do if points can appear and disappear on the line, and we wish a similar property to hold at any time? More precisely, we fix the number k of colors, and wish to maintain the property that at any given time, any sequence of $p(k)$ consecutive points, for some function p, contains at least one point of each color.

We show that in general, such a function does not exist: there are dynamic point sets on a line that are impossible to color with two colors so that monochromatic subsequences have bounded length. This holds even if the whole schedule of appearances and disappearances is known in advance. This family of point sets is described in Section 2.

We prove, however, that there exists a linear function p in the case where points can appear on the line at any time, but *never disappear*. Furthermore, this is achieved in a constructive, *semi-online* fashion: the coloring decision for a point can be delayed, but at any time the currently colored points yield a suitable coloring of the set. The algorithm is described in Section 3.

In Section 4, we restate the result in terms of a coloring problem in $\mathbb{R}^2$: for any integer $k \geq 1$, every point set in $\mathbb{R}^2$ can be colored with k colors so that any *bottomless* rectangle containing at least $3k - 2$ points contains one point of each color. Here, an axis-aligned rectangle is said to be bottomless whenever the y-coordinate of its bottom edge is $-\infty$.

In Section 5, we give lower bounds on the problem of coloring points with respect to bottomless rectangles. We show that the number of points $p(k)$ contained in a bottomless rectangle must be at least $1.67k$ in order to guarantee the presence of at least one point of each color.

Finally, in Section 6, we consider an alternative problem in which we fix the size of the sequence to k, but we are allowed to increase the number of colors.

Motivation and previous works. The problem is motivated by previous intriguing results in the field of geometric hypergraph coloring. Here, a geometric hypergraph is a set system defined by a set of points and a set of geometric ranges, typically polygons, disks, or pseudodisks. Every hyperedge of the hypergraph is the intersection of the point set with a range.

It was shown recently [7] that for every convex polygon P, there exists a constant c, such that any point set in $\mathbb{R}^2$ can be colored with k colors in such a way that any translation of P containing at least $p(k) = ck$ points contains at least one point of each color. This improves on several previous intermediate results [15,17,2]. Similar positive results for other families of geometric hypergraphs are given by Aloupis et al. [3,1], and Smorodinsky and Yuditsky [18]. Discussions on the relation between this coloring problem and ε-nets can be found in Pach and Tardos [13].

The problem for translates of polygons can be cast in its dual form as a covering decomposition problem: given a set of translates of a polygon P, we wish to color them with k colors so that any point covered by at least $p(k)$ of them is covered by at least one of each color. The two problems can be seen to be equivalent by replacing the points by translates of a symmetric image of P centered on these points. The covering decomposition problem has a long history that dates back to conjectures by János Pach in the early 80s (see for instance [11,4], and references therein). The decomposability of coverings by unit disks was considered in a seemingly lost unpublished manuscript by Mani and Pach in 1986. Up to recently, however, surprisingly little was known about this problem.

For other classes of ranges, such as axis-aligned rectangles, disks, translates of some concave polygons, or arbitrarily oriented strips [5,12,14,16], such a coloring does not always exists, even when we restrict ourselves to two colors.

Keszegh [8] showed in 2007 that every point set could be 2-colored so that any bottomless rectangle containing at least 4 points contains both colors. Our positive result on bottomless rectangles (Corollary 2) is a generalization of Keszegh's results to k-colorings. Later, Keszegh and Pálvölgyi [9] proved the following cover-decomposability property of octants in $\mathbb{R}^3$: every collection of translates of the positive octant can be 2-colored so that any point of $\mathbb{R}^3$ that is covered by at least 12 octants is covered by at least one of each color. This result generalizes the previous one (with a looser constant), as incidence systems of bottomless rectangles in the plane can be produced by restricted systems of octants in $\mathbb{R}^3$. It also implies similar covering decomposition results for homothetic copies of a triangle. More recently, they generalized their result to k-colorings, and proved an upper bound of $p(k) < 12^{2^k}$ on the corresponding function $p(k)$ [10].

2   Coloring Dynamic Point Sets

A *dynamic point set* S in $\mathbb{R}$ is a collection of triples $(v_i, a_i, d_i) \in \mathbb{R}^3$, with $d_i \geq a_i$, that is interpreted as follows: the point $v_i \in \mathbb{R}$ appears on the real line at time a_i and disappears at time d_i. Hence, the set $S(t)$ of points that are present at time t are the points v_i with $t \in [a_i, d_i)$. A k-coloring of a dynamic point set assigns one of k colors to each such triple.

We now show that it is not possible to find a 2-coloring of such a point set while avoiding long monochromatic subsequences at any time.

Theorem 1. *For every $p \in \mathbb{N}$, there exists a dynamic point set S with the following property: for every 2-coloring of S, there exists a time t such that $S(t)$ contains p consecutive points of the same color.*

Proof. In order to prove this result, we work on an equivalent two-dimensional version of the problem. From a dynamic point set, we can build n horizontal segments in the plane, where the ith segment goes from (a_i, v_i) to (d_i, v_i). At any time t the visible points $S(t)$ correspond to the intervals that intersect the line $x = t$. It is therefore equivalent, in order to obtain our result, to build a collection of horizontal segments in the plane that cannot be 2-colored in such a way that any set of p segments intersecting some vertical segment contains one element of each color.

Our construction borrows a technique from Pach, Tardos, and Tóth [14]. In this paper, the authors provide an example of a set system whose base set cannot be 2-colored without leaving some set monochromatic. This set system $\mathcal{S}$ is built on top of the $1 + p + \cdots + p^{p-1} = \frac{1-p^p}{1-p}$ vertices of a p-regular tree $\mathcal{T}^p$ of depth p, and contains two kinds of sets :

- the $1 + p + \cdots + p^{p-2}$ sets of *siblings*: the sets of p vertices having the same father,
- the p^{p-1} sets of p vertices corresponding to a path from the root vertex to one of the leaves in $\mathcal{T}^p$.

It is not difficult to realize that this set system is not 2-colorable: by contradiction, if every set of siblings is non-monochromatic, we can greedily construct a monochromatic path from the root to a leaf.

We now build a collection $\mathcal{S}$ of horizontal segments corresponding to the vertices of $\mathcal{T}^p$, in such a way that for any set $E \in \mathcal{S}$ there exists a time t at which the elements of E are consecutive among those that intersect the line $x = t$. For any p (see Fig. 1), the construction starts with a building block B_p^1 of p horizontal segments, the ith segment going from $(-\frac{i}{p}, i)$ to $(0, i)$. Because these p segments represent *siblings* in $\mathcal{T}^p$, they are consecutive on the vertical line that goes through their rightmost endpoint, and hence cannot all receive the same color.

Block B_p^{j+1} is built from a copy of B_p^1 to which are added p resized and translated copies of B_p^j : the ith copy lies in the rectangle with top-right corner $(-\frac{i-1}{p}, i+1)$ and bottom-left corner $(-\frac{i}{p}, i)$. By adding to B_p^{p-1} a last horizontal segment below all others, corresponding to the root of $\mathcal{T}^p$, the ancestors of a segment are precisely those that are below it on the vertical line that goes through its leftmost point. When such sets of ancestors are of cardinality $p - 1$, which only happens when one considers the set of ancestors of a leaf, then the set formed by the leaf and its ancestors is required to be non-monochromatic.

With this construction we ensure that a feasible 2-coloring of the segments would yield a proper 2-coloring of $\mathcal{S}$, which we know does not exist.   □

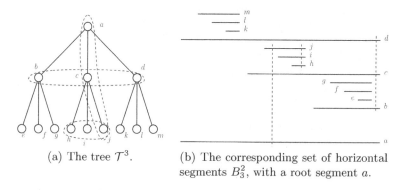

(a) The tree $\mathcal{T}^3$.   (b) The corresponding set of horizontal segments B_3^2, with a root segment a.

Fig. 1. The recursive construction of Theorem 1, for $p = 3$

The above result implies that no function $p(k)$ exists for any k that answers the original question. If it were the case, then we could simply merge color classes of a k-coloring into two groups and contradict the above statement.

Theorem 1 can also be interpreted as the indecomposability of coverings by a specific class of unbounded polytopes in $\mathbb{R}^3$. We define a *corner* with coordinates (a, b, c) as the following subset of $\mathbb{R}^3$: $\{(x, y, z) \in \mathbb{R}^3 : a \le x \le b, y \le c \le z\}$. An example is given in Fig. 2. One can verify that a point (x, y, z) is contained in a corner a, b, c if and only if the vertical line segment with endpoints (x, y) and (x, z) intersects the horizontal line segment with endpoints (a, c) and (b, c). The corollary follows.

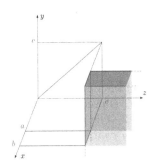

Fig. 2. A corner with coordinates (a, b, c)

Corollary 1. *For every $p \in \mathbb{N}$, there exists a collection S of corners with the following property: for every 2-coloring of S, there exists a point $x \in \mathbb{R}^3$ contained in exactly p corners of S, all of the same color. In other words, corners are not cover-decomposable.*

3   Coloring Point Sets under Insertion

Since we cannot bound the function $p(k)$ in the general case, we now consider a simple restriction on our dynamic point sets: we let the deletion times d_i be infinite for every i. Hence, points appear on the line, but never disappear.

A natural idea to tackle this problem is to consider an online coloring strategy, that would assign a color to each point in order of their arrival times a_i, without any knowledge of the points appearing later. However, we cannot guarantee any bound on $p(k)$ unless we delay some of the coloring decisions. To see this, consider the case $k = 2$, and call the two colors red and blue. An online algorithm must color each new point in red or blue as soon as it is presented. We can design an adversary such that the following invariant holds: at any time, the set of points is composed of a sequence of consecutive red points, followed by a sequence of consecutive blue points. The adversary simply chooses the new point to lie exactly between the two sequences at each step.

Our computation model will be *semi-online*: The algorithm considers the points in their order of the arrival time a_i. At any time, a point in the sequence either has one of the k colors, or is uncolored. Uncolored points can be colored later, but once a point is colored, it keeps its color for the rest of the procedure. At any time, the colors that are already assigned suffice to satisfy the property that any subsequence of $3k - 2$ points has one point of each color, i.e., $p(k) \le 3k - 2$.

Theorem 2. *Every dynamic point set without disappearing points can be k-colored in the semi-online model such that at any time, every subsequence of at least $3k - 2$ consecutive points contains at least one point of each color.*

Proof. We define a *gap for color i* as a maximal interval (set of consecutive points) containing no point of color i, that is, either between two successive occurrences of color i, or before the first occurrence (first gap), or after the last occurrence (last gap), or the whole line if no point has color i. A *gap* is simply a gap for color i, for some $1 \leq i \leq k$. We propose an algorithm for the semi-online model keeping the sizes of all gaps to be at most $3k - 3$. This means every set of $3k - 2$ consecutive points contains each color at least once and implies $p(k) \leq 3k - 2$. The algorithm maintains two invariants:
(a) every gap contains at most $3k - 3$ points; (b) if there is some point colored with i then every gap for color i, except the first and the last gap, contains at least $k - 1$ points.

The two invariants are vacuous when the set of points is empty. Now, suppose that the invariants hold for an intermediate set of points and consider a new point on the line presented by an adversary. Clearly, invariant *(b)* cannot be violated in the extended set as no gaps decrease in size. However, there may arise some gaps of size $3k - 2$ violating *(a)*. If not then the invariants hold for the extended set and the algorithm does not color any point in this step. Suppose there are some gaps of size $3k - 2$. Consider one of them, say a gap of color i, and denote the points in the gap in their natural ordering on the line from left to right as $(\ell_1, \ldots, \ell_{k-1}, m_1, \ldots, m_k, r_1, \ldots r_{k-1})$. Now, color i does not appear among these points. Invariant *(b)* yields that none of the $k - 1$ remaining colors appears twice among $m_1, \ldots, m_k$. Thus, there is some m_j, which is uncolored and the algorithm colors it with i. This splits the large gap into two smaller gaps. Moreover, since there are $k - 1$ ℓ-points and $k - 1$ r-points invariant *(b)* is maintained for both new i-gaps. The algorithm repeats that process until all gaps are of size at most $3k - 3$.

This concludes the proof, as after the algorithm ends all remaining uncolored points can be arbitrary colored. □

4 Coloring Points with Respect to Bottomless Rectangles

A *bottomless rectangle* is a set of the form $\{(x, y) \in \mathbb{R}^2 : a \leq x \leq b, y \leq c\}$, for a triple of real numbers (a, b, c) with $a \leq b$. We consider the following geometric coloring problem: given a set of points in the plane, we wish to color them with k colors so that any bottomless rectangle containing at least $p(k)$ points contains at least one point of each color. It is not difficult to realize that the problem is equivalent to that of the previous section.

Corollary 2. *Every point set $S \subset \mathbb{R}^2$ can be colored with k colors so that any bottomless rectangle containing at least $3k - 2$ points of S contains at least one point of each color.*

Proof. The algorithm proceeds by sweeping S vertically in increasing y-coordinate order. This defines a dynamic point set S' that contains at time t the x-coordinates of the points below the horizontal line of equation $y = t$. The set of points of S that are contained in a bottomless rectangle $\{(x,y) \in \mathbb{R}^2 : a \leq x \leq b, y \leq t\}$ correspond to the points in the interval $[a, b]$ in $S'(t)$. Hence, the two coloring problems are equivalent, and Theorem 2 applies. □

5 Lower Bound

We now give a lower bound on the smallest possible value of $p(k)$.

Theorem 3. *For any k sufficiently large, there exists a point set P such that for any k-coloring of P, there exists a color $i \in [k]$ and a bottomless rectangle containing at least $1.677k - 2.5$ points, none of which are colored with color i.*

Proof. Fix $k \geq 100$. For $n \in \mathbb{N}$ and $0 \leq a < k$ we define the point set $P = P(n, a)$ to be the union of point sets L, R and B (standing for left, right and bottom, respectively) as follows:

$$L := \{(i - n, 2i - 1) \in \mathbb{R}^2 \mid i \in [n]\}$$
$$B := \{(i, 0) \in \mathbb{R}^2 \mid i \in [a]\}$$
$$R := \{(a + i, 2n + 2 - 2i) \in \mathbb{R}^2 \mid i \in [n]\}$$

See Figure 3(a) for an illustration. Note that $|L| = |R| = n$ and $|B| = a$. Consider

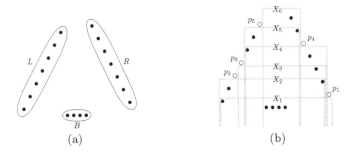

(a)   (b)

Fig. 3. (a) The point set $P = P(n, a)$ with $n = 7$ and $a = 4$, and (b) the bottomless rectangles $X_1, \ldots, X_6$ corresponding to the color class $P(c^*) = \{p_1, \ldots, p_5\}$

any coloring of the points in P with colors from $[k]$. For a color $i \in [k]$ we define $P(i)$ to be the subset of points of P colored with i. We assume for the sake of contradiction that every bottomless rectangle that contains $b := \lfloor 1.677k - 2.5 \rfloor$ points, contains one point of each color. In the remainder of the proof we will identify a bottomless rectangle containing b' points but no point of one particular color. We give a lower bound for b' depending on n and a, but independent of the fixed coloring under consideration. Taking sufficiently large n and choosing $a = \lfloor 0.655k \rfloor$ we will prove $b' > b$, which contradicts our assumption and hence concludes the proof.

A color used at least once for the points in B is called a *low color* and a point colored with a low color is a *low point*. Note that there are low points outside of the set B. Let ℓ be the number of low colors. Clearly, $\ell \le |B| = a$.

Claim 1.

(i) *For every non-low color c there are at least $\left\lfloor \frac{n}{b-a} \right\rfloor$ points of color c in L.*

(ii) *There are at least $\sum_{i=0}^{\ell-1} \left\lfloor \frac{n}{b-i} \right\rfloor$ low points in L.*

Proof. Fix a color $c \in [k]$ and assume that the j leftmost points in B are not colored with c. Order the points in L colored with c according to their x-coordinate: p_1, $p_2, \ldots, p_m$. Now for each $1 < i \le m$ there is a bottomless rectangle containing all points in L between p_{i-1} and p_i, and the leftmost j points in B, and nothing else. Additionally, there is a bottomless rectangle containing all points in L to the left of p_1 together with j leftmost points in B, and a bottomless rectangle containing all points in L to the right of p_m together with j leftmost points in B. Note that all these rectangles are disjoint within L and each point from L not colored with c lies in exactly one such rectangle. Since each such rectangle X avoids the color c we get that $|X \cap P| \le b - 1$ and $|X \cap L| \le b - 1 - j$ and therefore

$$m + (m+1)(b-1-j) = m(b-j) + b - j - 1 \ge |L| = n,$$

$$m \ge \left\lfloor \frac{n}{b-j} \right\rfloor. \tag{1}$$

In order to prove (i) consider a non-low color c. As c is not used on points in B at all we can put $j = a$ in (1) and the statement of (i) follows. Now, if c is a low color, then j defined as the maximum number of leftmost points in B avoiding c is always less than a. However, for each low color c we obtain a different j. Thus the sum of inequality (1) over all low colors is minimized by $\sum_{i=0}^{\ell-1} \lfloor \frac{n}{b-i} \rfloor$, which gives (ii). $\qquad\square$

By Claim 1 (i) and (ii) combined we get that there is a set S of $k - a$ non-low colors such that at most $n - \sum_{i=0}^{a-1} \lfloor \frac{n}{b-i} \rfloor$ points in L have a color from S. Analogously, at most $n - \sum_{i=0}^{a-1} \lfloor \frac{n}{b-i} \rfloor$ points in R have a color from S. Summing up we get:

$$\sum_{c \in S} |P(c)| = \sum_{c \in S} (|P(c) \cap L| + |P(c) \cap R|)$$

$$\leq 2n - 2 \sum_{i=0}^{a-1} \left\lfloor \frac{n}{b-i} \right\rfloor \leq 2n - 2 \sum_{i=0}^{a-1} \left(\frac{n}{b-i} - 1 \right)$$

$$= 2n \left(1 - \sum_{i=b-a+1}^{b} \frac{1}{i} \right) + 2a$$

$$= 2n \left(1 - \sum_{i=1}^{b} \frac{1}{i} + \sum_{i=1}^{b-a} \frac{1}{i} \right) + 2a.$$

Using that $\sum_{i=1}^{x} \frac{1}{i} = \ln(x+1) - \sum_{j=1}^{\infty} \frac{B_j}{j(x+1)^j} + \gamma$ for every $x \geq 1$, where B_j are the second Bernoulli numbers and γ is the Euler-Mascheroni constant, we obtain

$$\sum_{c \in S} |P(c)| < 2n \left(1 - \ln(b+1) + \ln(b-a+1) \right) + 2a$$

$$= 2n \left(1 - \ln \left(\frac{b+1}{b-a+1} \right) \right) + 2a.$$

From the pigeonhole principle we know that there has to exist a color $c^* \in S$, such that

$$q := |P(c^*)| \leq \left\lfloor \frac{2n(1 - \ln(\frac{b+1}{b-a+1})) + 2a}{k-a} \right\rfloor. \tag{2}$$

Enumerate the points in $P(c^*)$ by $p_1, p_2, \ldots, p_q$ according to their increasing y-coordinates, i.e., we have $i < j$ iff p_i has smaller y-coordinate than p_j. Now we consider all maximal bottomless rectangles that completely contain B and contain no point of color c^*. There are exactly $q+1$ such rectangles: For every point $p_i \in P(c^*)$ there is a bottomless rectangle X_i whose top side lies immediately below p_i. And one further bottomless rectangle X_{q+1} containing the entire strip between L and R, and with sides bounded by the point in $P(c^*) \cap L$ and the point in $P(c^*) \cap R$ with the highest index. See Figure 3(b) for an illustration.

Claim 2. $\sum_{i=1}^{q} |X_i \cap (L \cup R)| \geq \frac{3}{2}(2n - q - b + a).$

Proof. Let Y_1 and Y_{q+1} be the sets of points in $L \cup R$ with y-coordinate smaller than p_1 and larger than p_q, respectively. Let Y_i, $2 \leq i \leq q$, be the set of points with y-coordinate between p_{i-1} and p_i. Note that $Y_i \subset X_i \cap (L \cup R)$ for all $1 \leq i \leq q+1$, and that the $q+1$ sets $Y_1, \ldots, Y_{q+1}$ partition the points of $L \cup R$ that are not colored with c^*. Clearly, $|X_i \cap Y_i| = |Y_i|$. We claim that $|X_{i+1} \cap Y_i| \geq \frac{1}{2}|Y_i|$, for $i = 1, \ldots, q$.

Without loss of generality, let us assume that $p_i \in L$. Then either $Y_i = \emptyset$ or the point in Y_i with largest y-coordinate lies in R. Since points from L and R alternate in the ordering of $L \cup R$ with respect to increasing y-coordinate it follows that Y_i is almost equally partitioned into its left part $Y_i \cap L$ and its right part $Y_i \cap R$. Since the topmost point in Y_i lies in R we have $|Y_i \cap R| \geq \frac{1}{2}|Y_i|$. Now since $p_i \in L$ we have $X_{i+1} \supset Y_i \cap R$, and thus

$$|X_{i+1} \cap Y_i| \geq |Y_i \cap R| \geq \frac{1}{2}|Y_i|. \tag{3}$$

Note also that $|X_{q+1} \cap Y_q| + |Y_{q+1}| \leq |X_{q+1} \cap (L \cup R)| < b - a$ as X_{q+1} avoids color c^*, so $|X_{q+1}| < b$, and contains all a points in B.

Now we calculate

$$\sum_{i=1}^{q} |X_i \cap (L \cup R)| \geq \left(\sum_{i=1}^{q} |X_i \cap Y_i| + |X_{i+1} \cap Y_i|\right) - |X_{q+1} \cap Y_q|$$

$$\overset{(3)}{\geq} \sum_{i=1}^{q} \frac{3}{2}|Y_i| - |X_{q+1} \cap Y_q|$$

$$= \frac{3}{2}\left(2n - |P(c^*)| - |Y_{q+1}|\right) - |X_{q+1} \cap Y_q|$$

$$\geq \frac{3}{2}\left(2n - q - (|Y_{q+1}| + |X_{q+1} \cap Y_q|)\right) \geq \frac{3}{2}\left(2n - q - (b - a)\right).$$

$\square$

From Claim 2 we get from the pigeonhole principle that there is a bottomless rectangle $X^* \in \{X_1, \ldots, X_q\}$ with

$$|X^*| \geq \frac{\frac{3}{2}(2n - q - b + a)}{q} + a = \frac{3n}{q} - \frac{3}{2} - \frac{3(b-a)}{2q} + a$$

$$\overset{(2)}{\geq} \frac{3(k-a)}{2\left(1 - \ln\left(\frac{b+1}{b-a+1}\right) + \frac{2a}{n}\right)} + a - \frac{3}{2} - \frac{3(b-a)}{2q}$$

Now, if we increase n, then $q = |P(c^*)|$ increases as well, and for sufficiently large n the terms $\frac{2a}{n}$ in the denominator and the additive term $\frac{3(b-a)}{2q}$ become negligible. In particular, with $a := \lfloor 0.655k \rfloor$ and $b = \lfloor 1.677k - 2.5 \rfloor$ and sufficiently large n we have

$$|X^*| \geq \frac{3(k-a)}{2\left(1 - \ln\left(\frac{b+1}{b-a+1}\right)\right)} + a - \frac{3}{2}$$

$$= \frac{3(k - \lfloor 0.655k \rfloor)}{2\left(1 - \ln\left(\frac{\lfloor 1.677k - 2.5 \rfloor + 1}{\lfloor 1.677k - 2.5 \rfloor - \lfloor 0.655k \rfloor + 1}\right)\right)} + \lfloor 0.655k \rfloor - \frac{3}{2}$$

$$\sim \left(\frac{1.035}{2\left(1 - \ln\left(\frac{1.677}{1.022}\right)\right)} + 0.655\right)k > 1.68k.$$

Hence if k is big enough ($k \geq 100$ is actually enough) the bottomless rectangle X^* contains strictly more than $1.677k - 2.5$ points but no point of color c^*, which is a contradiction and concludes the proof. $\square$

6   Increasing the Number of Colors

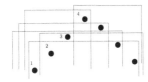

Fig. 4. A point set witnessing $c(k) \geq 2k-1$ for $k = 4$

There is another problem which can be tackled this time in an *online* model. The number $c(k)$ is the minimum number of colors needed to color the points on a line such that any set of at most k consecutive points is completely colored by distinct colors. The same problem has been considered for other types of geometric hypergraphs by Aloupis et al. [3]. Again, the algorithm considers the points in their order of the arrival time a_i but now colors them immediately.

Proposition 1. *Every dynamic point set without disappearing points can be $(2k - 1)$-colored in the online model such that at any time, every subsequence of at least k consecutive points contains no color twice.*

Proof. At the arrival of a new point p denote by $(\ell_1, \ldots, \ell_{k-1})$ and $(r_1, \ldots, r_{k-1})$ the $k - 1$ points to its left and to its right, respectively. Together they have at most $2k - 2$ colors, Thus, there is at least one of the $2k - 1$ colors unused among these points. The algorithm colors p with this color.   □

Corollary 3. *Every point set $S \subset \mathbb{R}^2$ can be colored with $2k - 1$ colors so that any bottomless rectangle containing at least k points of S contains no color twice.*

The number of colors used in Corollary 3 is smallest possible. This is witnessed by a point set S consisting of k points of the form $\{(i, 2i) \mid 0 \leq i \leq k - 1\}$ and $k - 1$ points of the form $\{(2k - i, 2i - 1) \mid 1 \leq i \leq k - 1\}$, see Fig. 4 for an example. It is easy to see that every pair of points in such a point set is in a common bottomless rectangle of size at most k. Finally, let us remark that an upper bound on $c(k)$ for dynamic point sets in which points can both appear and disappear, as in Section 2, can be obtained by bounding the chromatic number of the corresponding so-called *bar k-visibility graph*, as defined by Dean et al. [6]. In particular, they show that those graphs have $O(kn)$ edges, yielding $c(k) = O(k)$ for that case.

Acknowledgments. This research is supported by the ESF EUROCORES programme EuroGIGA, CRP ComPoSe, the Austrian Science Fund (FWF): P23629-N18 "Combinatorial Problems on Geometric Graphs" (Thomas Hackl), CRP GraDR and the Swiss National Science Foundation, SNF Project 20GG21-134306 (Michael Hoffmann). It was initiated at the ComPoSe kickoff meeting held at CIEM in Castro de Urdiales (Spain) on May 23–27, 2011, and pursued at the 2nd ComPoSe Workshop held at TU Graz (Austria) on April 16–20, 2012. The authors warmly thank the organizers of these two meetings as well as all the other participants. Part of this work was done during a stay of Kolja Knauer,

Piotr Micek, and Torsten Ueckerdt at ULB (Brussels) and supported as EURO-CORES short-term visit. A preliminary version was presented by a subset of the authors at EuroCG'12 in Assisi (Italy).

References

1. Aloupis, G., Cardinal, J., Collette, S., Imahori, S., Korman, M., Langerman, S., Schwartz, O., Smorodinsky, S., Taslakian, P.: Colorful strips. Graphs and Combinatorics 27(3), 327–339 (2011)
2. Aloupis, G., Cardinal, J., Collette, S., Langerman, S., Orden, D., Ramos, P.: Decomposition of multiple coverings into more parts. Discrete & Computational Geometry 44(3), 706–723 (2010)
3. Aloupis, G., Cardinal, J., Collette, S., Langerman, S., Smorodinsky, S.: Coloring geometric range spaces. Discrete & Computational Geometry 41(2), 348–362 (2009)
4. Brass, P., Moser, W.O.J., Pach, J.: Research Problems in Discrete Geometry. Springer (2005)
5. Chen, X., Pach, J., Szegedy, M., Tardos, G.: Delaunay graphs of point sets in the plane with respect to axis-parallel rectangles. Random Struct. Algorithms 34(1), 11–23 (2009)
6. Dean, A.M., Evans, W., Gethner, E., Laison, J.D., Safari, M.A., Trotter, W.T.: Bar k-visibility graphs. J. Graph Algorithms Appl. 11(1), 45–59 (2007)
7. Gibson, M., Varadarajan, K.R.: Optimally decomposing coverings with translates of a convex polygon. Discrete & Computational Geometry 46(2), 313–333 (2011)
8. Keszegh, B.: Weak conflict-free colorings of point sets and simple regions. In: CCCG, pp. 97–100 (2007)
9. Keszegh, B., Pálvölgyi, D.: Octants are cover-decomposable. Discrete & Computational Geometry 47(3), 598–609 (2012)
10. Keszegh, B., Pálvölgyi, D.: Octants are cover-decomposable into many coverings. CoRR, abs/1207.0672 (2012)
11. Pach, J.: Covering the plane with convex polygons. Discrete & Computational Geometry 1, 73–81 (1986)
12. Pach, J., Tardos, G.: Coloring axis-parallel rectangles. J. Comb. Theory, Ser. A 117(6), 776–782 (2010)
13. Pach, J., Tardos, G.: Tight lower bounds for the size of epsilon-nets. In: Proceedings of the 27th Annual ACM Symposium on Computational Geometry, SoCG 2011, pp. 458–463 (2011)
14. Pach, J., Tardos, G., Tóth, G.: Indecomposable coverings. In: CJCDGCGT, pp. 135–148 (2005)
15. Pach, J., Tóth, G.: Decomposition of multiple coverings into many parts. Comput. Geom. 42(2), 127–133 (2009)
16. Pálvölgyi, D.: Indecomposable coverings with concave polygons. Discrete & Computational Geometry 44(3), 577–588 (2010)
17. Pálvölgyi, D., Tóth, G.: Convex polygons are cover-decomposable. Discrete & Computational Geometry 43(3), 483–496 (2010)
18. Smorodinsky, S., Yuditsky, Y.: Polychromatic coloring for half-planes. J. Comb. Theory, Ser. A 119(1), 146–154 (2012)

Socially Stable Matchings
in the Hospitals/Residents Problem

Georgios Askalidis[1], Nicole Immorlica[1,2], Augustine Kwanashie[3],
David F. Manlove[3], and Emmanouil Pountourakis[1]

[1] Dept. of Electrical Engineering and Computer Science, Northwestern University
[2] Microsoft Research New England
[3] School of Computing Science, University of Glasgow

Abstract. In the Hospitals/Residents (HR) problem, agents are partitioned into hospitals and residents. Each agent wishes to be matched to
an agent (or agents) in the other set and has a strict preference over these
potential matches. A matching is stable if there are no blocking pairs,
i.e., no pair of agents that prefer each other to their assigned matches.
Such a situation is undesirable as it could lead to a deviation in which
the blocking pair form a private arrangement outside the matching. This
however assumes that the blocking pair have social ties or communication channels to facilitate the deviation. Relaxing the stability definition
to take account of the potential lack of social ties between agents can
yield larger stable matchings.

In this paper, we define the Hospitals/Residents problem under Social
Stability (HRSS) which takes into account social ties between agents by
introducing a *social network graph* to the HR problem. Edges in the social
network graph correspond to resident-hospital pairs in the HR instance
that know one another. Pairs that do not have corresponding edges in
the social network graph can belong to a matching M but they can never
block M. Relative to a relaxed stability definition for HRSS, called *social
stability*, we show that socially stable matchings can have different sizes
and the problem of finding a maximum socially stable matching is NP-
hard, though approximable within $3/2$. Furthermore we give polynomial
time algorithms for special cases of the problem.

1   Introduction

Matching problems generally involve the assignment of a set (or sets) of agents
to one another. Agents may be required to list other agents they find acceptable
in order of preference, either explicitly or implicitly through a list of desirable
characteristics. Agents may also be subject to capacity constraints, indicating
the maximum number of assignments they are allowed to be involved in.

An example of such a matching problem that has received much attention in
literature is the *Hospitals/Residents problem (HR)* [9,10,18,15]. An HR instance
consists of a set of *residents* seeking to be matched to a set of *hospitals*. Each
resident ranks a subset of the hospitals in strict order of preference, and vice
versa. Further, each resident forms an *acceptable pair* with every hospital on his

F. Dehne, R. Solis-Oba, and J.-R. Sack (Eds.): WADS 2013, LNCS 8037, pp. 85–96, 2013.
© Springer-Verlag Berlin Heidelberg 2013

preference list. Finally, each hospital has a *capacity*, indicating the maximum number of residents that it can be assigned. A *matching* is a set assignments among acceptable pairs such that no resident is assigned to more than one hospital, and no hospital exceeds its capacity. An acceptable pair forms a *blocking pair* with respect to a matching, or *blocks* a matching, if both agents would rather be assigned to each other than remain with their assignees (if any) in the matching. A matching is *stable* if it admits no blocking pair. HR has a wide range of applications including traditional markets like the assignment of graduating medical students (residents) to hospitals [13,17] and students to high schools [1,2], and online markets like oDesk (an online labour market), AirBnB (an online short-term housing rental market), and Match.com/OkCupid/etc. (online dating markets). In applications such as these, it has been convincingly argued that stability is a desirable property of a matching [17].

Although the concept of stability is important in many applications of matching problems, there are classes of matching problems (such as the Stable Roommates problem) for which an instance is not guaranteed to admit a stable matching [9]. Moreover, enforcing the stability requirement tends to reduce the size of matchings discovered [6]. This is an issue particularly in the case of applications where it is desirable to find the largest possible matching. Also, it is generally assumed that a resident-hospital pair that blocks a matching in theory will also block the matching in practice. However this assumption is not always true in some real-life applications, as resident-hospital pairs are more likely to form blocking pairs in practice if social ties exist between them. These factors have motivated studies into alternative, weaker stability definitions that still aim to prevent a given matching from being subverted in practice while increasing the number of agents involved in the matchings.

Arcaute and Vassilvitskii [3] described the Hospitals/Residents problem in the context of assigning job applicants to company positions. They observed that applicants are more likely to be employed by a company if they are recommended by their friends who are already employees of that company. In their model, an applicant-company pair (a, c) may block a matching M if (a, c) blocks it in the traditional sense (as described in the analogous HR context) and *a is friends with* another applicant a' assigned to c in M. Thus their problem incorporates both the traditional HR problem and additionally an underlying social network, represented as an undirected graph consisting of applicants as nodes and edges between nodes where the corresponding applicants have some social ties (e.g., are friends). Matchings that admit no blocking pair in this context are called *locally stable* due to the addition of the informational constraint on blocking pairs. Cheng and McDermid [8] investigated the problem (which they called HR+SN) further and established various algorithmic properties and complexity results. They showed that locally stable matchings can be of different sizes and the problem of finding a maximum locally stable matching is NP-hard. They identified special cases where the problem is polynomially solvable and gave upper and lower bounds on the approximability of the problem.

While the HR+SN model is quite natural in the job market, it makes an assumption that the employed applicant a' will always be willing to make a recommendation. This however may not be the case as a recommendation may in practice lead to a' being rejected by his assigned company. Ultimately this may lead to a reassignment for a' to a worse company or indeed a' may end up unmatched. While it is true that a scenario may arise where these social ties between applicants may lead to a blocking pair of a matching, it is arguably equally likely that social ties between an applicant and the company itself will exist. That is, an applicant need not know another applicant who was employed by the company in order to block a matching; it is enough for him to know *any* employee in the company (for example the Head of Human Resources). Such a model could also be natural in many applications both within and beyond the job market context.

Additionally, many matching markets are cleared by a centralised clearinghouse. While more traditional markets require agents to explicitly list potential matches, many online markets ask agents to list desirable characteristics and then use software to infer the preference lists of the agents. In these markets, communication between agents is facilitated by the centralised clearinghouse. Some agent pairs in the market may have social ties outside the clearinghouse. Often these social ties are due to past interactions within the marketplace and so the clearinghouse is aware of them. These pairs can communicate outside the clearinghouse and might block proposed matchings. Most pairs, however, only become aware of each other when the clearinghouse proposes them as a match. Thus even if they prefer each other to their assigned matches, they will not be able to discover each other and deviate from the matching.

Based on these ideas, we present a variant of HR called the *Hospitals / Residents problem under Social Stability (HRSS)*. In this model, which we describe in the context of assigning graduating medical residents to hospital positions, we assume that a resident-hospital pair will only form a blocking pair in practice if there exists some social relationship between them. Two agents that have such a social relationship are called an *acquainted pair*, and this is represented by an edge in a *social network graph*. We call a pair of agents that do not have such a social relationship an *unacquainted pair*. Such a pair may be part of a matching M (given that M is typically constructed by a trusted third party, i.e., a centralised clearinghouse) but cannot form a blocking pair with respect to M. As a consequence, although a resident-hospital pair may form a blocking pair in the classical sense, if they are an unacquainted pair, they will not form a blocking pair in the HRSS context. A matching that admits no blocking pair in this new context is said to be *socially stable*. We denote the one-to-one restriction of HRSS as the *Stable Marriage problem with Incomplete lists under Social Stability (SMISS)*.

Hoefer and Wagner [11,12] studied a problem that generalises both HR+SN and HRSS. In their model, the social network graph involves all agents and need not be bipartite. A pair *locally blocks* a given matching M if (i) it blocks in the classical sense, and (ii) the agents involved are at most l edges apart in the social network graph augmented by M. This scenario can be viewed as a generalisation of the HR+SN ($l = 2$) and HRSS ($l = 1$) models. They studied the convergence time for better-response dynamics that converge to locally stable matchings, and also established a lower bound for the approximabiliy of the problem of finding a maximum locally stable matching (for the case that $l \leq 2$).

Locally stable matchings have also been investigated in the context of the Stable Roommates problem (a non-bipartite generalisation of the Stable Marriage problem) in [7]. Here, the *Stable Roommates problem with Free edges (SRF)* as introduced was motivated by the observation that, in kidney exchange matching schemes, donors and recipients do not always have full information about others and are more likely to have information only on others in the same transplant centre as them. The problem is defined by the traditional Stable Roommates problem together with a set of *free edges*. These correspond to pairs of agents in different transplant centres that do not share preference information; such pairs may be involved in stable matchings, but cannot block any matching. It is shown in [7] that the problem of determining whether a stable matching exists, given an SRF instance, is NP-complete.

In this paper, we present some algorithmic results for the HRSS model described above. In Section 2, we present some preliminary definitions and observations. In Section 3, we consider the approximability of MAX HRSS, the problem of finding a maximum socially stable matching in an HRSS instance. We give a 3/2-approximation algorithm for the problem, and also show that it is not approximable within $21/19 - \varepsilon$, for any $\varepsilon > 0$, unless P=NP, and not approximable within $3/2 - \varepsilon$, for any $\varepsilon > 0$, assuming the Unique Games Conjecture. In Section 4 we present polynomial-time algorithms for two special cases of MAX HRSS where (i) the number of unacquainted pairs is constant, and (ii) the number of acquainted pairs is constant. Finally some open problems are given in Section 5. All proofs for this paper are omitted for space reasons but can be found in [5].

2   Preliminary Definitions and Results

An instance I of the Hospitals/Residents problem (HR), as defined in [9], contains a set $R = \{r_1, r_2, ..., r_{n_1}\}$ of residents, a set $H = \{h_1, h_2, ..., h_{n_2}\}$ of hospitals. Each resident $r_i \in R$ ranks a subset of H in strict order of preference; each hospital $h_j \in H$ ranks a subset of R, consisting of those residents who ranked h_j, in strict order of preference. Each hospital h_j also has a capacity $c_j \in \mathbb{Z}^+$ indicating the maximum number of residents that can be assigned to it. A pair (r_i, h_j) is called an *acceptable pair* if h_j appears in r_i's preference list. We denote by $\mathcal{A}$ the set of all acceptable pairs. A *matching* M is a set of acceptable pairs such that each resident is assigned to at most one hospital and the number of residents assigned to each hospital does not exceed its capacity. If r_i is matched in M, we denote the hospital assigned to resident r_i in M by $M(r_i)$. We denote the set of residents assigned to hospital h_j in M as $M(h_j)$. A resident r_i is *unmatched* in M if no pair in M contains r_i. A hospital h_j is *undersubscribed* in M if $|M(h_j)| < c_j$. A pair (r_i, h_j) is said to *block* a matching M, or form a

men's preferences   women's preferences
m_1: $\underline{w_1}$ w_1: m_2 $\underline{m_1}$
m_2: $\underline{w_1}$ $\underline{w_2}$ w_2: $\underline{m_2}$

Fig. 1. SMISS instance (I, G)

blocking pair with respect to M, in the classical sense, if (i) r_i is unmatched in M or prefers h_j to $M(r_i)$ and (ii) h_j is undersubscribed in M or prefers r_i to some resident in $M(h_j)$. A matching that admits no blocking pair is *stable*.

We define an instance (I, G) of the *Hospitals/Residents Problem under Social Stability (HRSS)* as consisting of an HR instance I (as defined above) and a bipartite graph $G = (R \cup H, A)$, where $A \subseteq \mathcal{A}$. A pair (r_i, h_j) belongs to A if and only if r_i has social ties with h_j. We call (r_i, h_j) an *acquainted pair*. We also define the set of *unacquainted* pairs (which cannot block any matching) to be $U = \mathcal{A} \backslash A$. A pair (r_i, h_j) *socially blocks* a matching M, or forms a *social blocking pair* with respect to M, if (r_i, h_j) blocks M in the classical sense in the underlying HR instance I and $(r_i, h_j) \in A$. A matching M is said to be *socially stable* if there exists no social blocking pair with respect to M. If we restrict the hospitals' capacities to 1, we obtain the *Stable Marriage problem with Incomplete lists under Social Stability (SMISS)*.

Clearly every instance of HRSS admits a socially stable matching. This is because the underlying HR instance is bound to admit a stable matching [9] which is also socially stable. However socially stable matchings could be larger than stable matchings. Consider the SMISS instance (I, G) shown in Figure 1, where the acquainted pairs in the social network graph are underlined in the preference lists. Matchings $M_1 = \{(m_1, w_1), (m_2, w_2)\}$ and $M_2 = \{(m_2, w_1)\}$ are both socially stable in (I, G) and M_2 is the unique stable matching. Thus an instance of SMISS (and hence HRSS) can admit a socially stable matching that is twice the size of a stable matching. Clearly the instance shown in Figure 1 can be replicated to give an arbitrarily large SMISS instance with a socially stable matching that is twice the size of a stable matching. This, and applications where we seek to match as many agents as possible, motivates MAX HRSS.

There is also a strong relationship between HRSS and the HR+SN problem described in [8]. We have shown (following an idea of Király) in [5] that an instance (I, G) of HRSS can be transformed in polynomial time to an instance (I', G') of HR+SN such that a socially stable matching M in (I, G) is locally stable in (I', G') and a complete locally stable matching (one in which all the residents are matched) in (I', G') is a complete socially stable matching in (I, G).

3   Approximating MAX HRSS

We begin this section by noting that MAX HRSS is NP-hard even in a very restricted setting. Let MAX SMISS denote the restriction of MAX HRSS in which all hospitals have capacity 1.

Theorem 1. *MAX SMISS is NP-hard even if each list is of length at most 3.*

In order to deal with this hardness, polynomial-time approximation algorithms can be developed for MAX HRSS. In this section we present a 3/2-approximation algorithm for MAX HRSS. We show this is tight assuming the Unique Games Conjecture (UGC), and also show a $21/19 - \varepsilon$ lower bound assuming $P \neq NP$. The lower bounds hold even for MAX SMISS. We start by giving the inapproximability result assuming P$\neq$NP.

Theorem 2. *MAX SMISS is not approximable within $21/19 - \varepsilon$, for any $\varepsilon > 0$, unless P=NP.*

We can obtain a better lower bound of $3/2 - \varepsilon$, for any $\varepsilon > 0$, if we strengthen our assumption from $P \neq NP$ to the truth of the UGC.

Theorem 3. *Assuming the UGC, MAX SMISS cannot be approximated within $3/2 - \varepsilon$, for any $\varepsilon > 0$.*

For the upper bound for MAX HRSS, we observe that a technique known as *cloning* has been described in literature [10,18], which may be used to convert an HR instance I into an instance I' of the Stable Marriage problem with Incomplete lists in polynomial time, such that there is a one-to-one correspondence between the set of stable matchings in I and I'. A similar technique can be used to convert an HRSS instance to an SMISS instance in polynomial time.

Theorem 4. *Given an instance (I, G) of HRSS, we may construct in $O(n_1 + c_{max}m)$ time an instance (I', G') of SMISS such that a socially stable matching M in (I, G) can be transformed in $O(c_{max}m)$ time to a socially stable matching M' in (I', G') with $|M'| = |M|$ and conversely, where n_1 is the number of residents, c_{max} is the maximum hospital capacity and m is the number of acceptable resident-hospital pairs in I.*

Due to Theorem 4, an approximation algorithm α for MAX SMISS with performance guarantee c (for some constant $c > 0$) can be used to obtain an approximation for MAX HRSS with the same performance guarantee. This can be done by cloning the HRSS instance (I, G) to form an SMISS instance (I', G'), and applying α to (I', G') to obtain a matching M'. This matching can then be transformed to a matching M in (I, G) such that $|M| = |M'|$. Our first upper bound for MAX HRSS is an immediate consequence of the fact that any stable matching is at least half the size of a maximum socially stable matching.

Proposition 5. *MAX HRSS is approximable within a factor of 2.*

We now present a 3/2-approximation algorithm for MAX SMISS. The algorithm relies on the principles outlined in the 3/2-approximation algorithms for the general case of MAX HRT, the problem of finding a maximum cardinality stable matching given an instance of the Hospitals / Residents problem with Ties, as presented by Király [14] and McDermid [16]. Given an instance (I, G) of SMISS, the algorithm works by running a modified version of the extended Gale-Shapley algorithm [9] where unmatched men are given a chance to propose again by promoting them on all the preference lists on which they appear. Let A and U denote the sets of acquainted and unacquainted pairs in (I, G) respectively.

Consider a woman w_j in (I, G). We denote an *unacquainted man* m_i *on* w_j's *preference list* as one where $(m_i, w_j) \in U$. Similarly we denote an *acquainted man* m_i *on* w_j's *preference list* as one where $(m_i, w_j) \in A$. For a man m_i, we denote $next(m_i)$ as the next woman on m_i's list succeeding the last woman to whom he proposed to or the first woman on m_i's list if he has been newly promoted or is proposing for the first time. During the execution of the algorithm if a man runs out of women to propose to on his list for the first time, he is *promoted*, thus allowing him to propose to the remaining women on his list beginning from the first. A man can only be promoted once during the execution of the algorithm. If a promoted man still remains unmatched after proposing to all the women on his preference list, he is removed from the instance and will not be part of the final matching.

In the classical Gale-Shapley algorithm [9], a woman w_j prefers a man m_i to another m_k if $rank(w_j, m_i) < rank(w_j, m_k)$. We define a modified version of the extended Gale-Shapley algorithm [10], *mod-EXGS*, where a woman does not accept or reject proposals from men solely on the basis of their positions on her preference list, but also on the basis of their status as to whether they are acquainted or unacquainted men on her list and whether they have been promoted. Given two men m_i and m_k on a woman w_j's preference list, we define the relations $\lhd_{w_j}$, $\lhd'_{w_j}$ and $\prec_{w_j}$ as follows:

Definition 6. *Let m_i and m_k be any two men on a woman w_j's list. Then*
1. *$m_i \lhd_{w_j} m_k$ if either*
   (i) $(m_i, w_j) \in U$, $(m_k, w_j) \in U$, m_i is promoted and m_k is unpromoted or
   (ii) $(m_i, w_j) \in A$, $(m_k, w_j) \in U$ and m_k is unpromoted.
2. *$m_i \lhd'_{w_j} m_k$ if $m_i \not\lhd_{w_j} m_k$, $m_k \not\lhd_{w_j} m_i$ and w_j prefers m_i to m_k in the classical sense.*
We define $\prec_{w_j} = \lhd_{w_j} \cup \lhd'_{w_j}$.

The relation $\prec_{w_j}$ will be used to determine whether a proposal from a man is accepted or rejected by w_j.

The main algorithm *approx-SMISS* (as shown in Algorithm 1) starts by calling *mod-EXGS* (as shown in Algorithm 2) where a *proposal sequence* is started by allowing each man to propose to women beginning from the first woman on his preference list. If a man m_i proposes to a woman w_j on his list and w_j is matched and $m_i \prec_{w_j} M(w_j)$, then w_j is unmatched from her partner m_k, and m_k will be allowed to continue proposing to other women on his list. w_j is then assigned to m_i. On the other hand, if $M(w_j) \prec_{w_j} m_i$ then w_j rejects m_i's proposal. Also if w_j is unmatched when m_i proposes, she is assigned to m_i. Irrespective of whether the proposal from m_i is accepted or rejected, if $(m_i, w_j) \in A$ then all pairs (m_k, w_j) such that $rank(w_j, m_k) > rank(w_j, m_i)$ are deleted from the instance. However if $(m_i, w_j) \in U$ no such deletions take place. This proposal sequence continues until every man is either matched or has exhausted his preference list.

After each proposal sequence (where control is returned to the *approx-SMISS* algorithm), if a promoted man still remains unmatched after proposing to all the women on his preference list, he is removed from the instance. Also if a previously unpromoted man exhausts his preference lists and is still unmatched,

---
Algorithm 1. approx-SMISS

1: initial matching $M = \emptyset$;
2: **while** some unmatched man with a non-empty preference list exists **do**
3:     call mod-EXGS;
4:     **for** all m_i such that m_i is unmatched and promoted **do**
5:         remove m_i from instance;
6:     **end for**
7:     **for** all m_i such that m_i is unmatched, unpromoted and has a non-empty preference list **do**
8:         promote m_i;
9:     **end for**
10: **end while**
11: return the resulting matching M;

Algorithm 2. mod-EXGS

1: **while** some man m_i is unmatched and still has a woman left on his list **do**
2:     $w_j = next(m_i)$;
3:     **if** w_j is matched in M and $m_i \prec_{w_j} M(w_j)$ **then**
4:         $M = M \setminus \{(M(w_j), w_j)\}$;
5:     **end if**
6:     **if** w_j is unmatched in M **then**
7:         $M = M \cup \{(m_i, w_j)\}$;
8:     **end if**
9:     **if** $(m_i, w_j) \in A$ **then**
10:         **for** each m_k such that $(m_k, w_j) \in \mathcal{A}$ and $rank(w_j, m_k) > rank(w_j, m_i)$ **do**
11:             delete (m_k, w_j) from instance;
12:         **end for**
13:     **end if**
14: **end while**

---

he is promoted and a new proposal sequence initiated (by calling *mod-EXGS*). The algorithm terminates when each man either (i) is assigned a partner, (ii) has no woman on his preference list or (iii) has been promoted and has proposed to all the women on his preference list for a second time.

Lemma 7. *If algorithm* approx-SMISS *is executed on an SMISS instance* (I, G), *it terminates with a socially stable matching* M *in* (I, G).

The execution of the *mod-EXGS* algorithm takes $O(m)$ time where $m = |\mathcal{A}|$ is the number of acceptable pairs. These executions can be performed at most $2n_1$ times, where n_1 is the number of men, as a man is given at most two chances to propose to the women on his list. Thus the overall time complexity of the algorithm is $O(n_1 m)$. The above results, together with Theorem 4, lead us to state the following theorem concerning the performance guarantee of the approximation algorithm for MAX HRSS.

Theorem 8. *MAX HRSS is approximable within a factor of 3/2.*

men's preferences	women's preferences
m_1: $\underline{w_1}$ $\underline{w_3}$	w_1: m_2 $\underline{m_1}$
m_2: $\underline{w_1}$ $\underline{w_2}$	w_2: $\underline{m_2}$ m_3
m_3: w_2	w_3: $\underline{m_1}$

Fig. 2. $|M_{opt}| = (3/2).|M|$

The SMISS instance shown in Figure 2 (where the acquainted pairs in the social network graph are underlined in the preference lists) shows that the 3/2 bound for the algorithm is tight. Here $M_{opt} = \{(m_1, w_3), (m_2, w_1), (m_3, w_2)\}$ is the unique maximum socially stable matching. Also the approximation algorithm outputs $M = \{(m_1, w_1), (m_2, w_2)\}$ irrespective of the order in which proposals are made. Clearly this instance can be replicated to obtain an arbitrarily large SMISS instance for which the performance guarantee is tight.

We remark that a similar 3/2-approximation algorithm for MAX HRSS was presented independently by Askalidis et al. in [4].

4   Some Special Cases of HRSS

Given the hardness results obtained for the problem of finding a maximum socially stable matching in a general HRSS instance, the need arises to investigate special cases of the problem that are tractable. This section describes some polynomial-time solvability results for two special cases of HRSS.

Before presenting the two main results of this section, we first note that we have given an $O(n^{3/2} \log n)$ algorithm for finding a maximum socially stable matching, given an instance of MAX SMISS where each man is allowed to have at most two women on his preference list and n is the total number of men and women involved. This algorithm, presented in [5], is omitted for space reasons.

4.1   HRSS with a Constant Number of Unacquainted Pairs

It is easy to see that in the special case where the set U of unacquainted pairs is exactly the set $\mathcal{A}$ of acceptable pairs in the underlying HR instance, then the set A of acquainted pairs satisfies $A = \emptyset$ and every matching found is a socially stable matching. Also if the instance contains no unacquainted pairs (i.e., $A = \mathcal{A}$ and $U = \emptyset$), then only stable matchings in the classical sense are socially stable. In both these cases, a maximum socially stable matching can be generated in polynomial time. The case may however arise where the number of unacquainted pairs is constant. In this case, we show that it is also possible to generate a maximum socially stable matching in polynomial time.

Let (I, G) be an instance of HRSS and let $S \subseteq \mathcal{A}$ be a subset of the acceptable pairs in I. We denote $I \backslash S$ as the instance of HR obtained from I by deleting the pairs in S from the preference lists in I. The following proposition plays a key role in establishing the correctness of the algorithm.

Proposition 9. *Let (I, G) be an instance of HRSS. Let M be a socially stable matching in (I, G). Then there exists a set of unacquainted pairs $U' \subseteq U$ such that M is stable in $I' = I \backslash U'$. Conversely suppose that M is a stable matching in $I' = I \backslash U'$ for some $U' \subseteq U$. Then M is socially stable in (I, G).*

By considering all subsets $U' \subseteq U$, forming I', finding a stable matching in each such I' and keeping a record of the maximum stable matching found, we obtain a maximum socially stable matching in (I, G). This discussion leads to the following theorem.

Theorem 10. *Given an instance (I, G) of HRSS where the set U of unacquainted pairs is of constant size, a maximum socially stable matching can be generated in $O(m)$ time, where $m = |\mathcal{A}|$ is the number of acceptable pairs.*

4.2 HRSS with a Constant Number of Acquainted Pairs

We now consider the restriction of HRSS in which the set A of acquainted pairs is of constant size k. Given an instance (I, G) of this problem we show that a maximum socially stable matching can be found in polynomial time. Let $A = \{e_1, e_2, ..., e_k\}$ where e_i represents an acquainted pair (r_{s_i}, h_{t_i}) $(1 \leq i \leq k)$. A tree T of depth k is constructed with all nodes at depth i labelled e_{i+1} $(i \geq 0)$. There are left and right branches below e_i. Each branch corresponds to a condition placed on r_{s_i} or h_{t_i} with respect to a matching M. The left branch below e_i (i.e., a resident condition branch) corresponds to the condition that r_{s_i} is matched in M and prefers his partner to h_{t_i}. The right branch below e_i (i.e., a hospital condition branch) corresponds to the condition that h_{t_i} is fully subscribed in M and has a partner no worse than r_{s_i}. Satisfying at least one of these conditions ensures that M admits no blocking pair involving (r_{s_i}, h_{t_i}). The tree is constructed in this manner with the nodes at depth $k - 1$, labelled e_k, branching in the same way to dummy leaf nodes e_{k+1} (not representing acquainted pairs).

A path P from the root node e_1 to a leaf node e_{k+1} will visit all pairs in A exactly once. Every left branch in P gives a resident condition and every right branch gives a hospital condition. Let R' and H' be the set of residents and hospitals involved in resident and hospital conditions in P respectively. Given a matching M, enforcing all the conditions along P can be achieved by first deleting all pairs from the instance I that could potentially violate these conditions. So if some resident condition along P states that a resident r_{s_i} must be matched in M to a hospital he prefers to h_{t_i} then r_{s_i}'s preference list is truncated starting with h_{t_i}. If some hospital condition states that a hospital h_{t_i} must be fully subscribed in M and must not be matched to a resident worse than r_{s_i} then h_{t_i}'s preference list is truncated starting from the resident immediately following r_{s_i}. After performing these truncations based on the conditions along P, a new HR instance I' is obtained.

Proposition 11. *If M is a matching in I' that is computed at the leaf node of a path P and all residents in R' are matched in M and all hospitals in H' are fully subscribed in M then M is a socially stable matching in (I, G).*

By finding a maximum weight matching in a suitable weighted graph (see [5] for further details), we may in polynomial time find a largest matching M satisfying the constraints of Proposition 11 or report that no such matching exists. In the latter case P is ruled as infeasible and another path is considered, otherwise P is called feasible.

There are 2^k paths from the root node to leaf nodes in the tree T. The following proposition is important to our result.

Proposition 12. *There must exist at least one feasible path in T.*

To generate a maximum socially stable matching M in an instance (I, G) of HRSS, all 2^k paths through T from the root node to leaf nodes are considered with a record kept of the largest matching M (satisfying the constraints of Proposition 11) computed at the leaf node of each feasible path. M is then the desired matching as the following proposition shows

Proposition 13. *If M is a matching obtained from the process described above, M is a maximum socially stable matching in (I, G).*

The above proposition leads to the following main result of this subsection.

Theorem 14. *Given an instance (I, G) of HRSS where the set A of acquainted pairs satisfies $|A| = k$ for some constant k, a maximum socially stable matching can be generated in $O(c_{max} m \sqrt{n_1 + C})$ time where n_1 is the number of residents, m is the number of acceptable pairs, c_{max} is the largest capacity of any hospital and C is the total capacity of all the hospitals in the problem instance.*

Following the results in Theorems 10 and 14, we conclude this section with the theorem below showing the existence of FPT algorithms for MAX HRSS under two different parameterisations.

Theorem 15. *MAX HRSS is in FPT with parameter k, where either $k = |A|$ or $k = |U|$, and A and U are the sets of acquainted and unacquainted pairs respectively.*

5   Open Problems

The study of the Hospitals / Residents problem under Social Stability is still at an early stage, and some interesting open problems remain. Firstly it is worth considering the scenario where ties exist in the preference lists of agents. Also it could be argued that information about undersubscribed hospitals would be in the public domain, and hence an undersubscribed hospital may form an acquainted pair with all the residents on its preference list. It would be interesting to investigate algorithmic aspects of this variant of HRSS.

Acknowledgments. The fourth author is supported by grant EP/K010042/1 from the Engineering and Physical Sciences Research Council. We would like to thank Martin Hoefer for making the third and fourth authors aware of [4]; Zoltán Király for observing Theorem 15 and for other valuable comments; and Rob Irving and an anonymous referee for further valuable suggestions concerning this paper.

References

1. Abdulkadiroğlu, A., Pathak, P.A., Roth, A.E.: The Boston public school match. American Economic Review 95(2), 368–371 (2005)
2. Abdulkadiroğlu, A., Pathak, P.A., Roth, A.E.: The New York City high school match. American Economic Review 95(2), 364–367 (2005)
3. Arcaute, E., Vassilvitskii, S.: Social networks and stable matchings in the job market. In: Leonardi, S. (ed.) WINE 2009. LNCS, vol. 5929, pp. 220–231. Springer, Heidelberg (2009)
4. Askalidis, G., Immorlica, N., Pountourakis, E.: Socially stable matchings. CoRR Technical Report 1302.3309, http://arxiv.org/abs/1302.3309
5. Askalidis, G., Immorlica, N., Kwanashie, A., Manlove, D.F., Pountourakis, E.: Socially stable matchings in the Hospitals / Residents problem. CoRR Technical Report 1303.2041, http://arxiv.org/abs/1303.2041
6. Biró, P., Manlove, D.F., Mittal, S.: Size versus stability in the marriage problem. Theoretical Computer Science 411, 1828–1841 (2010)
7. Cechlárová, K., Fleiner, T.: Stable roommates with free edges. Technical Report 2009-01, Egerváry Research Group on Combinatorial Optimization, Budapest (2009)
8. Cheng, C., McDermid, E.: Maximum locally stable matchings. In: Proc. MATCH-UP 2012, pp. 51–62 (2012)
9. Gale, D., Shapley, L.S.: College admissions and the stability of marriage. American Mathematical Monthly 69, 9–15 (1962)
10. Gusfield, D., Irving, R.W.: The Stable Marriage Problem: Structure and Algorithms. MIT Press (1989)
11. Hoefer, M.: Local matching dynamics in social networks. Information and Computation 222, 20–35 (2013)
12. Hoefer, M., Wagner, L.: Locally stable marriage with strict preferences. In: Smotrovs, J., Yakaryilmaz, A. (eds.) ICALP 2013, Part II. LNCS, vol. 7966, pp. 620–631. Springer, Heidelberg (2013)
13. Irving, R.W.: Matching medical students to pairs of hospitals: A new variation on a well-known theme. In: Bilardi, G., Pietracaprina, A., Italiano, G.F., Pucci, G. (eds.) ESA 1998. LNCS, vol. 1461, pp. 381–392. Springer, Heidelberg (1998)
14. Király, Z.: Linear time local approximation algorithm for maximum stable marriage. In: Proc. MATCH-UP 2012, pp. 99–110 (2012)
15. Manlove, D.F.: Algorithmics of Matching Under Preferences. World Scientific (2013)
16. McDermid, E.: A 3/2-approximation algorithm for general stable marriage. In: Albers, S., Marchetti-Spaccamela, A., Matias, Y., Nikoletseas, S., Thomas, W. (eds.) ICALP 2009, Part I. LNCS, vol. 5555, pp. 689–700. Springer, Heidelberg (2009)
17. Roth, A.E.: The evolution of the labor market for medical interns and residents: a case study in game theory. Journal of Political Economy 92(6), 991–1016 (1984)
18. Roth, A.E., Sotomayor, M.A.O.: Two-sided matching: a study in game-theoretic modeling and analysis. Cambridge University Press (1990)

Parameterized Complexity of 1-Planarity

Michael J. Bannister[1], Sergio Cabello[2], and David Eppstein[1]

[1] Department of Computer Science, University of California, Irvine, USA
[2] Faculty of Mathematics and Physics, University of Ljubljana, Slovenia

Abstract. We consider the problem of finding a 1-planar drawing for a general graph, where a 1-planar drawing is a drawing in which each edge participates in at most one crossing. Since this problem is known to be NP-hard we investigate the parameterized complexity of the problem with respect to the vertex cover number, tree-depth, and cyclomatic number. For these parameters we construct fixed-parameter tractable algorithms. However, the problem remains NP-complete for graphs of bounded bandwidth, pathwidth, or treewidth.

1 Introduction

1-planar graphs (the graphs that can be drawn in the plane with at most one crossing per edge) were introduced by Ringel in 1965 [23] and have since been extensively studied from the point of view of basic properties such as their colorings [2,6], edge density [3,22,24], characterization by forbidden subgraphs [18,19], and embeddings on nonplanar surfaces [26]. In graph drawing, 1-planarity has more recently become of interest, as a way of generalizing planar drawings in a controlled way that does not lead to too much visual complexity. Works in this area have compared 1-planarity to other forms of controlled crossings such as RAC (right-angle-crossing) graphs [11], found an algorithmic characterization of the 1-planar drawings that can be straightened to have all edges represented by straight line segments [17], and studied the transformation of rotation systems into 1-planar drawings [10]. However, until now there have been no published algorithms for finding 1-planar drawings of arbitrary graphs. Unfortunately, testing 1-planarity is NP-hard in general [14,19], even for graphs obtained from planar graphs by adding a single edge [4], so we cannot expect it to be solved by an algorithm whose running time is a polynomial of the input size.

Because of the difficulty of recognizing 1-planar drawings, and their usefulness in graph drawing, it becomes of interest to study the complexity of algorithms for testing 1-planarity that are not fully polynomial. An important tool for this sort of study is *parameterized complexity* [9,13], according to which we seek additional numeric parameters (other than the numbers of edges and vertices) that measure the complexity of an input graph, and seek algorithms whose running time is the product of a polynomial in the input size and a non-polynomial function of the other parameter or parameters. If this can be accomplished, the result will in general be an algorithm that solves the problem correctly on all graphs, that can

F. Dehne, R. Solis-Oba, and J.-R. Sack (Eds.): WADS 2013, LNCS 8037, pp. 97–108, 2013.
© Springer-Verlag Berlin Heidelberg 2013

be relied on to be efficient for graphs that have small values of the parameter, and that has a performance that degrades gracefully as the parameter increases.

In this paper we study for the first time the parameterized complexity of 1-planarity. We provide a fixed-parameter tractable algorithm for the problem when it is parameterized by the cyclomatic number (the minimum number of edges that must be removed from the graph to make a forest) or the tree-depth. For a third parameter, the vertex cover number, we show that even more efficient FPT algorithms are possible, based on a polynomial kernel (a transformation of any instance to an equivalent instance with size polynomial in the parameter). However, as we show in the full version of this paper, the problem remains NP-complete for graphs of bounded bandwidth; therefore, it is unlikely that there exists a fixed-parameter tractable algorithm for 1-planarity when parameterized by bandwidth, pathwidth, treewidth, or clique-width.

Although our primary motivation is in understanding the complexity of 1-planarity, our research on the vertex cover and tree-depth parameters has a secondary purpose as well, in exploring the circumstances in which general theorems that guarantee the existence of an inexplicit FPT algorithm (with unknown dependence on the parameter) can be made explicit. It is known that the graphs of bounded vertex cover number, and the graphs of bounded tree-depth, are well-quasi-ordered under induced subgraphs [21]. This means that for any graph recognition problem closed under induced subgraphs (as 1-planarity is), and for any fixed bound on vertex cover or tree-depth, there is a finite set of forbidden induced subgraphs that can be used to characterize the problem, and a linear time recognition algorithm. However, the theorems that prove these results do not imply any computable bound on the size of these forbidden subgraphs or on the dependence on the parameter of these linear time algorithms. In contrast, for 1-planarity with these parameters we provide algorithms whose dependence on the parameter is known, explicit, and computable (albeit impractically large).

2   Vertex Cover Number

The *vertex cover number* k of an undirected graph G is the minimum number of vertices needed to touch all of the edges of G. This number is central to the theory of parameterized complexity, to the point where Guo et al. call it "the *Drosophila* of fixed-parameter algorithmics" [15]. After much earlier work on the problem, the best fixed-parameter tractable algorithms for computing the vertex cover number, parameterized by this number, take time $O(1.2738^k + kn)$ [5]. We will show that, when parameterized by vertex cover number, 1-planarity is also fixed-parameter tractable, using a standard technique, kernelization, whereby we replace an instance graph by an equivalent instance of size bounded by a function of the kernel. Although the vertex cover number is a weaker parameter than the tree-depth that we consider later (a graph of vertex cover number k has tree-depth at most $k + 1$), we begin with this parameter for two reasons: (1) for this parameter we achieve stronger results, namely a polynomial kernel, than we do for the other parameters that we consider, and (2) the simplicity of this case makes it an appropriate warm-up for the other parameters.

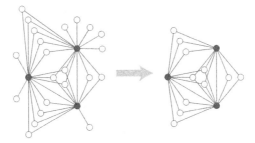

Fig. 1. Kernelization for vertex cover number k: remove degree-one vertices, and reduce each $K_{2,i}$ subgraph (with two cover vertices on one side of the bipartition) to $K_{2,\min\{i,2k-3\}}$. Here $k = 3$, so the $K_{2,i}$ subgraphs are reduced to $K_{2,3}$.

Lemma 1 (Czap and Hudák [7]). *A complete bipartite graph is 1-planar if and only if it is of the form* $K_{1,n}$, $K_{2,n}$, $K_{3,i}$ *for* $i \in \{3,4,5,6\}$, *or* $K_{4,4}$.

Lemma 2. *Testing 1-planarity of an n-vertex graph G takes time $2^{O(n)}$.*

Proof. If G has more than $4n$ edges, we return that it is not 1-planar [22]. Otherwise, we proceed with a divide and conquer algorithm: the existence of cycle separators [20] implies that in any 1-planar drawing there is a curve passing through $O(\sqrt{n})$ vertices that separates the drawing into two balanced parts. We can then solve the problem by solving $2^{O(n)}$ subproblems, each of them smaller by a constant fraction. $\qquad\square$

Lemma 3. *Let G be a 1-planar graph, with a subgraph H of the form $K_{2,i}$ formed by i vertices of degree two, all with the same two neighbors. Then G has a 1-planar drawing in which the induced drawing of H is planar.*

Proof. If two edges of H that share an endpoint cross then we can uncross them, resulting in a drawing with fewer crossings, and if two non-incident edges of H cross each other then we can redraw all of H without crossings near the previous position of these two crossed edges, again reducing the total number of crossings. Therefore, a 1-planar drawing of G that minimizes the total number of crossings has the desired property. $\qquad\square$

Lemma 4. *Let a graph G have a known vertex cover C of size $|C| = k$. Then in time $O(n)$ we can transform G into an kernel G_C of size $O(k^2)$ such that G is 1-planar if and only if G_C is 1-planar. A 1-planar drawing of G_C may be transformed into a 1-planar drawing of G in linear time.*

Proof. Delete any vertices of degree one in $G \setminus C$; this cannot change 1-planarity. If G is to be 1-planar, there are at most $5k$ two-edge paths connecting distinct pairs of vertices of C through a different vertex of $G \setminus C$; otherwise smoothing out the internal vertices of those $5k$ paths we would contradict the bound on [22] for drawings with at most two crossings per edge. Moreover, in $G \setminus C$ there are at most $6k$ vertices of degree three or more sharing any two fixed neighbors in C, as otherwise G contains a $K_{3,7}$. It follows that $G \setminus C$ has $O(k^2)$ vertices of degree three or more, if G is 1-planar.

The vertices of degree two in $G \setminus C$ can be grouped by radix sort according to the identities of their two neighbors in C, forming a collection of $K_{2,i}$ subgraphs. If G is 1-planar, there are $O(k)$ such subgraphs $K_{2,i}$ because each of them gives a two-edge path connecting two distinct vertices of C. If one of these $K_{2,i}$ subgraphs has $i > 2k-3$ then we claim that G is 1-planar if and only if the subgraph G' formed by deleting $i - (2k-3)$ vertices within this subgraph to form a smaller $K_{2,2k-3}$ subgraph is also 1-planar. In one direction, if G is 1-planar, then clearly so is G'. In the other direction, suppose G' is 1-planar; then by Lemma 3 it has a 1-planar drawing in which the given $K_{2,2k-3}$ subgraph is drawn planarly, with $2k-3$ quadrilateral faces. Two adjacent faces among this set of $2k-3$ must be empty of the $k-2$ vertices of C that are not part of the $K_{2,2k-3}$ subgraph. Therefore, the two edges e and f separating these two faces cannot be crossed by any edge of the 1-planar drawing, for any crossing edge would either have to cross entirely across one of these two faces (violating 1-planarity) or have an endpoint in each of the two faces (violating the assumption that neither of these faces contains a vertex of C). The remaining vertices and edges of G that were deleted to form the $K_{2,2k-3}$ subgraph may be added to the drawing, near path ef, without violating 1-planarity, showing as desired that G is 1-planar.

Performing this replacement of $K_{2,i}$ by $K_{2,\min(i,2k-3)}$ separately for each of the groups of vertices in $G \setminus C$ results in the desired kernel G_C. G_C has $O(k^2)$ vertices of high degree and $O(k)$ groups of $O(k)$ vertices in $K_{2,i}$ subgraphs, for a total of $O(k^2)$ vertices.

If a drawing of G_C is found, a corresponding drawing of G may be found by eliminating crossings between pairs of edges belonging to the same $K_{2,i}$ subgraphs in G_C, finding an uncrossed length-two path with two vertices in C as path endpoints within each $K_{2,2k-3}$ subgraph, expanding each of these $K_{2,2k-3}$ subgraphs to $K_{2,i}$ for the correct value of i from the original graph G (placing the restored vertices near the uncrossed path), and finally adding back any deleted degree-one vertices of G. ☐

An example of this kernelization is depicted in Figure 1, for a graph with vertex cover number three.

Theorem 1. *We can test the 1-planarity of a given n-vertex graph, parameterized by its vertex cover number k, in time $O(n + 2^{O(k^2)})$.*

Proof. Apply an FPT algorithm to find an optimal vertex cover C, apply Lemma 4 to replace G with a kernel G_C of size $O(k^2)$, in linear time, and then apply Lemma 2 to this kernel.

To reduce the dependence on n in the time for the initial vertex cover step from $O(kn)$ to $O(n)$, we abort the algorithm if the input has more than $4n-8$ edges, and otherwise apply a standard kernelization for vertex cover: find a maximal matching M in G, and find all vertices of degree greater than $2|M|$; these must all belong to the optimal vertex cover, and can be removed from G, leaving a smaller graph G' that has $O(k^2)$ edges (otherwise it could not be covered by the remaining low-degree vertices). Apply the vertex cover algorithm to G' instead of to G.   □

Corollary 1. *We can test 1-planarity for split graphs in time $O(n)$.*

Proof. If a given split graph has a clique of size seven, it is not 1-planar, and otherwise, it has a vertex cover of size six and we use the above algorithm.   □

3   Tree-Depth

As we now show, 1-planarity parameterized by tree-depth may be tested by an FPT algorithm. The *tree-depth* of a graph G is the smallest depth of a forest F on the same vertex set as G such that every edge of G connects an ancestor-descendant pair in F, where we measure the depth of a tree as the maximum number of vertices on a root-leaf path [21]. Equivalently, it is the size of a maximum clique in a trivially perfect supergraph of G chosen to minimize this clique size; here, a trivially perfect graph is the graph of ancestor-descendant pairs in a forest. A graph G with vertex cover number k has tree-depth at most $k + 1$, for we may find a tree T of depth $k + 1$ that has the k vertices of the cover on a path, from which all other vertices descend as leaves; all edges of G connect ancestor-descendant pairs in G. For this reason, in some sense the result of this section is stronger than that of Theorem 1, although the dependence on the parameter is worse.

An n-vertex path has tree-depth $\lceil \log_2(n + 1) \rceil$. It follows that an arbitrary depth-first search tree for a given graph G has a depth that is at most $2^d - 1$ (because otherwise it would contain a path that is too long for the given depth) and at least the tree-depth d (because the DFS tree has the ancestor-descendant property from which tree-depth is defined). Based on this observation, one can derive an FPT algorithm for computing the tree-depth, by finding a DFS tree, using it to construct a tree decomposition, and applying standard dynamic programming techniques to this decomposition [21].

Lemma 5. *Let G be a graph with tree-depth at most d, as witnessed by a forest F of depth d for which all edges of G connect ancestor-descendant pairs. Then in linear time it is possible to replace G by an equivalent kernel for 1-planarity consisting of a collection of disconnected subgraphs with $O(2^{2d^2+O(d)})$ vertices each.*

Proof. If G is not biconnected we may test 1-planarity on each biconnected component of G separately; therefore, we can assume without loss of generality that the given graph G is 2-connected, and that we have a tree T of depth d such that every edge of G connects an ancestor-descendant pair in T. We can also assume without loss of generality that each node of T is adjacent to at least one node in each of its child subtrees (because otherwise we could move those children up to be siblings of the node, which does not increase the depth) and that each child subtree induces a connected subgraph (because otherwise we could split it into two separate children). Because the tree-depth is d, the longest path in G has length less than 2^d.

Now consider how many children a node v in T can have. For each child subtree T_i, consider the set S_i of v and ancestors of v that are connected to nodes in T_i. And for each subset S of v and its ancestors, let $C(S)$ be the set of child subtrees T_i of v for which $S_i = S$. There are at most 2^d different sets S, and we want to show that for each of them, $C(S)$ has bounded size. If $|S| = 1$, this is easy: then $S = \{v\}$ (because otherwise there is no node in T_i adjacent to v) and v is an articulation point, violating the assumption of 2-connectivity.

Next, consider the case that $|S| \geq 3$. That is, we have a set S consisting of v and two or more of its ancestors, and a set $C(S)$ of child subtrees of v that are each connected to all of the nodes in S. Choose exactly three nodes of S and, for each child subtree T_i in $C(S)$, let X_i be a smallest subgraph connecting the three chosen nodes in the subgraph of G induced by $T_i \cup S$. By the bound on the length of paths in G, $|X_i| = O(2^d)$. Note that, among any three of these trees X_i, X_j, and X_k (all for members of $C(S)$) there must be at least one crossing, because contracting each tree to a single node produces a $K_{3,3}$ subgraph. There are $\Omega(|C(S)|^3)$ triples of trees, at least one crossing per triple, and at most $|C(S)|$ triples that involve any single crossing, so there are $\Omega(|C(S)|^2)$ crossings altogether, among a set of only $O(|C(S)|2^d)$ edges. In order to prevent the pigeonhole principle from forcing some edge to be crossed twice, we must have $|C(S)| = O(2^d)$.

Finally, consider the case that $|S| = 2$. In this case, $|C(S)|$ can be unbounded (e.g. consider the graph $K_{2,n}$, which has tree-depth three). But, if it is greater than 2^d, then it does not matter how much greater it is: no cycle in the drawing can separate the two vertices in S, because the minimal such cycle would have to have length at most 2^d but would have to cross each of the subgraphs T_i, a contradiction. So in this case we can split the graph into subgraphs formed from each child T_i together with an uncrossable edge between the two nodes in S, and test 1-planarity separately for each of these subgraphs. When $C(S)$ is small enough that no such split is possible, $|C(S)| = O(2^d)$.

After performing any splits from the $|S| = 2$ case, the remaining graph has its nodes arranged into a tree of height d in which each node has $O(2^{2d})$ children. Therefore, the total number of nodes in the tree is $O(2^{2d^2 + O(d)})$.   □

By combining this kernelization with the known FPT algorithm for computing tree-depth and with Lemma 2 for testing the 1-planarity of the kernel, we obtain

Theorem 2. *The 1-planarity of a given graph, with tree-depth d, may be computed in time $O(n2^{2^{2d^2 + O(d)}})$.*

As an example of the power of this approach, we show how to use it to recognize 1-planar cographs. Cographs are well-quasi-ordered by induced subgraphs [8], from which it follows that there is an algorithm for testing 1-planarity by checking for the existence of a finite set of forbidden induced subgraphs; however, we

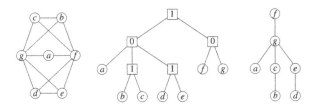

Fig. 2. Finding a low-tree-depth representation of a cograph by forming a path for each 1-labeled cotree node, consisting of the cotree leaves that descend from it but are not in its heaviest child. Left: a cograph. Center: its cotree. Right: the tree formed by connecting together the paths L_x. Each cotree node has the same color as its corresponding path.

do not know how to explicitly list these forbidden subgraphs nor do we know how to turn a recognition algorithm along such lines into an algorithm for finding a 1-planar drawing. In contrast, the algorithm outlined below for recognizing 1-planar cographs is explicit (albeit with impractically large constants) and constructs a drawing of the graph.

Lemma 6. *Let $C_{a,b}$ denote the class of cographs that do not contain K_a nor $K_{b,b}$ as subgraphs. Then the graphs in $C_{a,b}$ have tree-depth at most $1 + (a-1)(b-1)$.*

Proof. For any graph G in this class, we use a cotree representing G, and use it to guide the construction of a forest F on the nodes of G. A cotree has the vertices of G as its leaves; every internal node is labeled either 0 or 1, and two vertices of G are adjacent if and only if their lowest common ancestor is labeled 1. We assume that this tree is in canonical form meaning that no two adjacent internal nodes have the same label as each other and that each internal node has at least two children.

For each node x labeled 1 in the cotree, let H_x denote the subtree descending from a child of x that contains the largest number of leaves (breaking ties arbitrarily) and let L_x denote the set of leaf descendants of x that are not in H_x. For each maximal set L_x (not contained in L_y for some 1-labeled node y), we form a path, which will form a subgraph of F. If the closest 1-labeled ancestor of cotree node x is node y, we set the parent of the top node of path L_x to be the bottom node of path L_y. In addition, if any vertex v of G does not belong to a set L_x, we make it a leaf of the forest F, and we set the parent of v to be the bottom node of the path for the lowest 1-labeled ancestor of v in the cotree. The forest constructed in this way (shown in Figure 2) will necessarily have the defining property of tree-depth that every edge in G connects an ancestor-descendant pair in F.

If L_x has at least $2b - 1$ leaves, then the leaf descendants of x contain a $K_{b,b}$ subgraph. For this reason, every path L_x has at most $2(b - 1)$ vertices of G in it. Additionally, on any path from the root to a leaf in the cotree, at most one of the 1-labeled cotree nodes can have more than $b - 1$ nodes in L_x, for if one such node does, then each of its ancestors must have at most $b - 1$ nodes in L_x, or else we would again have a $K_{b,b}$ subtree. Finally, observe that a path from the root to a leaf in the cotree that has $a - 1$ 1-labeled nodes would give rise to a K_a subgraph; therefore, every such path has at most $a - 2$ 1-labeled nodes. By this analysis, the longest path from leaf to root that could exist in the forest F consists of one vertex of G that does not belong to a set L_x, one set L_x of size $2(b - 1)$, and $a - 3$ sets L_x of size $b - 1$, matching the depth given in the statement of the lemma.   □

Corollary 2. *We can recognize 1-planar cographs, and find 1-planar drawings of them, in $O(n)$ time.*

Proof. We first test whether the given cograph contains K_7 or $K_{5,5}$ as a subgraph. If it does, it is not 1-planar. If it does not, we may apply Lemma 6 and Theorem 2.   □

4   Cyclomatic Number

We say that a graph G has *cyclomatic number* k if k is the smallest number of edges that must be removed from G to yield a forest; equivalently $k = m - n + c$, where c is the number of connected components in G. By a *maximal degree two path* we shall mean a path between two vertices each of degree greater than two such that all vertices in the interior of the path have degree two. For technical reasons, an edge between vertices each having degree greater than two will also be considered a maximal degree two path. Gurevich et al. define a k-*almost-tree* to be a graph G such that given a spanning tree T of G every biconnected component of G has at most k edges not in T [16]. This is equivalent to each biconnected component having cyclomatic number k.

The cyclomatic number and k-almost-tree parameter have previously been used as parameters in fixed parameter algorithms. For example, in biology, gene expression can be represented as a Boolean network in which individual genes are represented as vertices and edges represent correlations between pairs of genes. Fixed parameter tractable algorithms have been designed for the control problem, which involves finding sequences of valid labelings of genes as being active or inactive [1]. In operations research, fixed-parameter algorithms for the continuous facility location problem have been constructed, where weighted edges represent a road network on which to efficiently place facilities serving clients in the network [16]. Intraprogram communication networks in distributed systems use vertices to represent modules of a program to be computed in parallel and edges to represent communicating pairs of modules; they also have structure yielding fixed-parameter algorithms [12] with respect to this parameter.

Lemma 7. *If G is a graph with cyclomatic number k and no degree one vertices, then G has at most $2k - 2$ vertices of degree greater than two. Furthermore, this bound is tight. Also, the number of maximal degree two paths is at most $3k - 3$.*

Proof. Double counting edges yields $2(n - c + k) \geq 2a + 3b$, where a is number of degree two vertices and b is the number of vertices of degree greater than two.

Fig. 3. Removing a crossing in a degree two path

Using $n = a + b$ and $c \geq 1$ we obtain $b \leq 2k - 2$, establishing the upper bound. For the upper bound consider any biconnencted cubic graph with $2k - 2$ vertices, e.g., a cubic Halin graph whose characteristic tree has k leaves.

For the bound on the maximal degree two paths consider the graph G' where each maximal degree two path is reduced to a single edge. The graph G' has cyclomatic number k and at most $2k - 2$ vertices. This implies that G' has at most $3k - 3$ edges, establishing the bound. □

Lemma 8. *If G is a 1-planar, then there is a 1-planar drawing of G such that maximal degree two paths do not self intersect.*

Proof. It suffices to show that a self crossing in a maximal degree two path can be removed without increasing the number of crossings on any edges. We can locally uncross a self intersection changing the drawing within a circular region $\mathcal{R}$ around the intersection that is not crossed by other edges. See Figure 3 for an example of this operation. □

Lemma 9. *Every word on $n > 1$ symbols, without consecutive equal symbols, of length greater than $2n! - 1$ has a subword on $k > 1$ symbols, for some $k \leq n$, such that each symbol appears at least k times in the subword. Furthermore, this bound is tight, i.e., there exists a word w of length $2n! - 1$ on n symbols such that for every $1 < k \leq n$, w has no subword on k symbols in which each symbol appears at least k times.*

Proof. Let w be a word on n symbols of length at least $2(n!) - 1$, and let σ be the symbol appearing least often in w. If σ occurs more than n times in w, then we are done. So assume that σ occurs at most $n - 1$ times. Removing σ from w leaves us with at least $2(n!) - n$ symbols split into at most n subwords. Thus, the longest of these subwords, call it u, has length at least $(2(n!) - n)/n = 2(n - 1)! - 1$. Since u contains at most $n - 1$ unique symbols we are done by induction on n.

To construct a word on n symbols of length $2(n!) - 1$ with no reducible subword, let $\sigma_0, \sigma_1, \ldots, \sigma_n$ be our n symbols. Now recursively define the words by $w_k = (w_{k-1}\sigma_k)^{k-1}w_{k-1}$ and $w_2 = \sigma_0\sigma_1\sigma_0$. A simple induction argument shows that the length of w_k is $2(k!) - 1$. □

Lemma 10. *If G is a 1-planar graph with p maximal degree two paths, then G has a 1-planar drawing such that every maximal degree two path is crossed at most $2p! - 1$ times.*

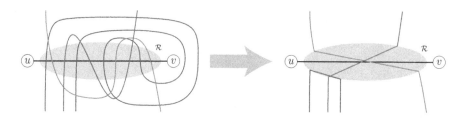

Fig. 4. Left crossing sequence `rgbrbrgbrg`; Right crossing sequence `bg`

Proof. We need only show that given a maximal degree two path from u to v with more than $2p! - 1$ crossings, we can reduce the number of times that it is crossed without increasing the crossing count on other degree two paths.

First, we continuously deform the plane such that the path from u to v is a straight line. This is possible since we may assume that maximal degree two paths do not self intersect by Lemma 8. Now we consider the sequence of crossings through the path from u to v my other maximal degree two paths. In this sequence there are at most p symbols. So if the number of crossings on the path from u to v is greater than $2p! - 1$, Lemma 9 implies that there is a subword on p' symbols such that every symbol appears at least p' times.

Now, we construct a strictly convex region $\mathcal{R}$ around the crossings represented by this word such that only paths represented in the word intersect the region, and such that a path does not reintersect the path from u to v without first leaving $\mathcal{R}$. For every path we shortcut it from the first time it intersects $\mathcal{R}$ to the last time it intersects $\mathcal{R}$, in path order, with a straight line. So now each path in $\mathcal{R}$ is a straight line, and therefore they can only intersect each other at most once. So, we have reduced the number of crossings on the path from u to v, without increasing the crossings on the other paths. □

Lemma 11. *Let G be a graph with cyclomatic number k. Then in linear time we can transform G into a kernel G_C of size $O((3k - 3)(3k - 3)!)$ such that G is 1-planar if and only if G_C is 1-planar. In addition, a 1-planar drawing of G_C may be transformed into a 1-planar drawing of G in linear time.*

Proof. We remove degree one vertices from G until no more are left, producing the 2-*core* of G [25]. This process can be done in linear time by maintaining a queue of degree one vertices. A degree one vertex may be added to any drawing without introducing crossings, so a graph has a 1-planar drawing if and only if its 2-core has a 1-planar drawing.

Lemma 7 implies that we have at most $p = 3k - 3$ maximal degree two paths. For each of these maximal degree two paths we reduce the number of degree two vertices to $2p! + 1$ if they exceed this amount. Since Lemma 10 guarantees that, if G is 1-planar, then it has a drawing such that no maximal degree two path is crossed more than $2p! - 1$ times, this reduction does not change the 1-planarity of the graph. Thus, we have a kernel G_C of size $O((3k - 3)(3k - 3)!)$ such that G is 1-planar if and only if G_C is 1-planar. □

Theorem 3. *We can test the 1-planarity of a graph with cyclomatic number k in time $O\left(n + 2^{O((3k)!)}\right)$.*

Since a graph can be decomposed into its biconnected components in linear time and edges in separate biconnected components need not cross we have the following corollary to Theorem 3.

Corollary 3. *We can test the 1-planarity of a k-almost tree in time $O(n2^{O((3k)!)})$.*

Acknowledgements. The research of Bannister and Eppstein was supported in part by the National Science Foundation under grants 0830403 and 1217322, and by the Office of Naval Research under MURI grant N00014-08-1-1015. The research of Cabello was supported in part by the Slovenian Research Agency, program P1-0297, project J1-4106, and within the EUROCORES Programme EUROGIGA (project GReGAS) of the European Science Foundation. We also gratefully acknowledge the Slovenian Research Agency for travel funds allowing the authors to meet and perform this research.

References

[1] Akutsu, T., Hayashida, M., Ching, W.-K., Ng, M.K.: Control of Boolean networks: Hardness results and algorithms for tree structured networks. J. Theor. Biol. 244(4), 670–679 (2007), doi:10.1016/j.jtbi.2006.09.023

[2] Borodin, O.V.: Solution of the Ringel problem on vertex-face coloring of planar graphs and coloring of 1-planar graphs. Metody Diskret. Analiz. (41), 12–26, 108 (1984)

[3] Brandenburg, F.J., Eppstein, D., Gleißner, A., Goodrich, M.T., Hanauer, K., Reislhuber, J.: On the density of maximal 1-planar graphs. In: Didimo, W., Patrignani, M. (eds.) GD 2012. LNCS, vol. 7704, pp. 327–338. Springer, Heidelberg (2013)

[4] Cabello, S., Mohar, B.: Adding one edge to planar graphs makes crossing number and 1-planarity hard. CoRR abs/1203.5944 (2012)

[5] Chen, J., Kanj, I.A., Xia, G.: Improved upper bounds for vertex cover. Theoretical Computer Science 411(40-42), 3736–3756 (2010), doi:10.1016/j.tcs.2010.06.026

[6] Chen, Z.-Z., Kouno, M.: A linear-time algorithm for 7-coloring 1-plane graphs. Algorithmica 43(3), 147–177 (2005), doi:10.1007/s00453-004-1134-x

[7] Czap, J., Hudák, D.: 1-planarity of complete multipartite graphs. Discrete Applied Mathematics 160(4-5), 505–512 (2012), doi:10.1016/j.dam.2011.11.014

[8] Damaschke, P.: Induced subgraphs and well-quasi-ordering. J. Graph Th. 14(4), 427–435 (1990), doi:10.1002/jgt.3190140406

[9] Downey, R.G., Fellows, M.R.: Parameterized Complexity. Monographs in Computer Science. Springer (1999), doi:10.1007/978-1-4612-0515-9

[10] Eades, P., Hong, S.-H., Katoh, N., Liotta, G., Schweitzer, P., Suzuki, Y.: Testing maximal 1-planarity of graphs with a rotation system in linear time. In: Didimo, W., Patrignani, M. (eds.) GD 2012. LNCS, vol. 7704, pp. 339–345. Springer, Heidelberg (2013)

[11] Eades, P., Liotta, G.: Right angle crossing graphs and 1-planarity. In: Speckmann, B. (ed.) GD 2011. LNCS, vol. 7034, pp. 148–153. Springer, Heidelberg (2011)

[12] Fernandez-Baca, D.: Allocating modules to processors in a distributed system. IEEE Transactions on Software Engineering 15(11), 1427–1436 (1989), doi:10.1109/32.41334

[13] Flum, J., Grohe, M.: Parameterized Complexity Theory. Texts in Theoretical Computer Science. Springer (2006)

[14] Grigoriev, A., Bodlaender, H.L.: Algorithms for graphs embeddable with few crossings per edge. Algorithmica 49(1), 1–11 (2007), doi:10.1007/s00453-007-0010-x

[15] Guo, J., Niedermeier, R., Wernicke, S.: Parameterized complexity of generalized vertex cover problems. In: Dehne, F., López-Ortiz, A., Sack, J.-R. (eds.) WADS 2005. LNCS, vol. 3608, pp. 36–48. Springer, Heidelberg (2005)

[16] Gurevich, Y., Stockmeyer, L., Vishkin, U.: Solving NP-Hard Problems on Graphs That Are Almost Trees and an Application to Facility Location Problems. J. ACM 31(3), 459–473 (1984), doi:10.1145/828.322439

[17] Hong, S.-H., Eades, P., Liotta, G., Poon, S.-H.: Fáry's theorem for 1-planar graphs. In: Gudmundsson, J., Mestre, J., Viglas, T. (eds.) COCOON 2012. LNCS, vol. 7434, pp. 335–346. Springer, Heidelberg (2012)

[18] Korzhik, V.P.: Minimal non-1-planar graphs. Discrete Mathematics 308(7), 1319–1327 (2008), doi:10.1016/j.disc.2007.04.009

[19] Korzhik, V.P., Mohar, B.: Minimal Obstructions for 1-Immersions and Hardness of 1-Planarity Testing. J. Graph Th. 72(1), 30–71 (2013), doi:10.1002/jgt.21630

[20] Miller, G.L.: Finding Small Simple Cycle Separators for 2-Connected Planar Graphs. J. Comput. Syst. Sci. 32(3), 265–279 (1986), doi:10.1016/0022-0000(86)90030-9

[21] Nešetřil, J., Ossona de Mendez, P.: Sparsity: Graphs, Structures, and Algorithms. Algorithms and Combinatorics 28, 115–144 (2012), doi:10.1007/978-3-642-27875-4

[22] Pach, J., Tóth, G.: Graphs drawn with few crossings per edge. Combinatorica 17(3), 427–439 (1997), doi:10.1007/BF01215922

[23] Ringel, G.: Ein Sechsfarbenproblem auf der Kugel. Abhandlungen aus dem Mathematischen Seminar der Universität Hamburg 29, 107–117 (1965), doi:10.1007/BF02996313

[24] Schumacher, H.: Zur Struktur 1-planarer Graphen. Mathematische Nachrichten 125, 291–300 (1986)

[25] Seidman, S.B.: Network structure and minimum degree. Social Networks 5(3), 269–287 (1983), doi:10.1016/0378-8733(83)90028-X

[26] Suzuki, Y.: Optimal 1-planar graphs which triangulate other surfaces. Discrete Mathematics 310(1), 6–11 (2010), doi:10.1016/j.disc.2009.07.016

On the Stretch Factor of the Theta-4 Graph[*]

Luis Barba[1,2,**], Prosenjit Bose[1], Jean-Lou De Carufel[1], André van Renssen[1], and Sander Verdonschot[1]

[1] School of Computer Science, Carleton University, Ottawa, Canada
jit@scs.carleton.ca,
{andre,jdecaruf,sander}@cg.scs.carleton.ca
[2] Département d'Informatique, Université Libre de Bruxelles, Brussels, Belgium
lbarbafl@ulb.ac.be

Abstract. In this paper we show that the θ-graph with 4 cones has constant stretch factor, i.e., there is a path between any pair of vertices in this graph whose length is at most a constant times the Euclidean distance between that pair of vertices. This is the last θ-graph for which it was not known whether its stretch factor was bounded.

Keywords: computational geometry, geometric spanners, θ-graphs.

1 Introduction

A c-spanner of a weighted graph G is a connected sub-graph H with the property that for all pairs of vertices u and v, the weight of the shortest path between u and v in H is at most c times the weight of the shortest path between u and v in G, for some fixed constant $c \geq 1$. The smallest constant c for which H is a c-spanner of G is referred to as the *stretch factor* or *spanning ratio* of the graph.

The graph G is referred to as the *underlying graph*. In our setting, the underlying graph is the complete graph on a set of n points in the plane and the weight of an edge is the Euclidean distance between its endpoints. A c-spanner of such a graph is called a *geometric c-spanner*. For a comprehensive overview of geometric c-spanners, see the book by Narasimhan and Smid [10].

In this paper, we focus on θ-graphs. Introduced independently by Clarkson [7] and Keil [9], the θ_m-graph is constructed as follows. Given a set P of points in the plane, we consider each point $p \in P$ and partition the plane into m cones (regions in the plane between two rays originating from the same point) with apex p, each defined by two rays at consecutive multiples of $\theta = 2\pi/m$ radians from the negative y-axis. We label the cones $C_0(p)$ through $C_{m-1}(p)$, in counterclockwise order around p, starting from the negative y-axis; see Fig. 1. In each cone $C_i(p)$, we add an edge between p and p_i, the point in $C_i(p)$ nearest to p. However, instead of using the Euclidean distance, we measure distance in $C_i(p)$ by projecting each point onto the angle bisector of this cone. Formally, p_i is the point in $C_i(p)$ such that for every other point $w \in C_i(p)$, the projection of p_i

---

[*] Research supported in part by NSERC and FQRNT.
[**] Boursier FRIA du FNRS.

F. Dehne, R. Solis-Oba, and J.-R. Sack (Eds.): WADS 2013, LNCS 8037, pp. 109–120, 2013.
© Springer-Verlag Berlin Heidelberg 2013

onto the angle bisector of $C_i(p)$ lies closer to p than the projection of w. For simplicity, we assume that no two points of P lie on a line parallel to either the boundary or the angle bisector of a cone.

It has been shown that θ_m-graphs are geometric spanners for $m \geq 7$, and their stretch factor approaches 1 as m goes to infinity [4,6,11]. The proofs crucially rely on the fact that, given two points p and q such that $q \in C_i(p)$, the distance between p_i and q is always less than the distance between p and q. This property does not hold for $m \leq 6$ and indeed, the path obtained by starting at p and repeatedly following the edge in the cone that contains q, is not necessarily a spanning path. The main motivation for using spanners is usually to reduce the number of edges in the graph without increasing the length of shortest paths

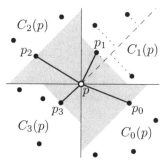

Fig. 1. The neighbors of p in the θ_4-graph of P. Each edge supports an empty isosceles triangle.

too much. Thus, θ-graphs with fewer cones are more interesting in practice, as they have fewer edges. This raises the following question: "What is the smallest m for which the θ_m-graph is a geometric spanner?" Bonichon et al. [1] showed that the θ_6-graph is a geometric 2-spanner. Recently, Bose et al. [5] proved that the θ_5-graph is a geometric 9.96-spanner. Coming from the other side, El Molla [8] showed that there is no constant c for which the θ_2- and θ_3-graphs are geometric c-spanners. This leaves the θ_4-graph as the only open question. Moreover, its resemblance to graphs like the Yao$_4$-graph [3] and the L_∞-Delaunay triangulation [2], both of which are geometric c-spanners, makes this question more tantalizing. In this paper we establish an upper bound of approximately 237 on the stretch factor of the θ_4-graph. In Section 5, we present a lower bound of 7 that we believe is closer to the true stretch factor of the θ_4-graph.

2   Existence of a Spanning Path

Let P be a set of points in the plane. In this section, we prove that the θ_4-graph of P is a spanner. We do this by showing that the θ_4-graph approximates the L_∞-Delaunay triangulation. The L_∞-Delaunay triangulation of P is a geometric graph with vertex set P, and an edge between two points of P whenever there exists an empty axis-aligned square having these two points on its boundary.

Bonichon et al. [2] showed that the L_∞-Delaunay triangulation has a stretch factor of $c^* = \sqrt{4 + 2\sqrt{2}}$, i.e., there is a path between any two vertices whose length is at most c^* times their Euclidean distance. We approximate this path in the L_∞-Delaunay triangulation by showing the existence of a spanning path in the θ_4-graph of P joining the endpoints of every edge in the L_∞-Delaunay triangulation. The main ingredient to obtain this approximation is Lemma 1 whose proof is presented in Section 4. Before stating this result, we need a few more definitions. Given two points s and t, their L_1 distance $d_{L_1}(s,t)$ is the sum of the absolute differences of their x- and y-coordinates.

Let $S_t(s)$ be the smallest axis-aligned square centered on t that contains s. Let ℓ_t^- and ℓ_t^+ be the lines with slope -1 and $+1$ passing through t, respectively.

Throughout this paper, we repeatedly use t to denote a *target* point of P that we want to reach via a path in the θ_4-graph. Therefore, we typically omit the reference to t and write ℓ^-, ℓ^+ and $S(s)$ when referring to ℓ_t^-, ℓ_t^+ and $S_t(s)$, respectively.

We say that an object is *empty* if its interior contains no point of P. An *s-t-path* is a path with endpoints s and t.

Lemma 1. *Let s and t be two points of P such that t lies in $C_0(s)$. If the top-right quadrant of $S(s)$ is empty and $C_1(s)$ contains no point of P below ℓ^-, then there is an s-t-path in the θ_4-graph of P of length at most $18 \cdot d_{L_1}(s,t)$.*

For ease of readability, the proof of Lemma 1 is deferred to Section 4.

Given a path φ, let $|\varphi|$ denote the sum of the lengths of the edges in φ. Using Lemma 1, we obtain the following.

Lemma 2. *Let s and t be two points of P. If the smallest axis-aligned square enclosing s and t, that has t as a corner, is empty, then there is an s-t-path in the θ_4-graph of P of length at most $(\sqrt{2}+36) \cdot |st|$.*

Proof. Assume without loss of generality that s lies in $C_1(t)$. Then, the top-right quadrant of $S(s)$ is empty as it coincides with the smallest axis-aligned square enclosing s and t that has t as a corner; see Fig. 2(a). Recall that s_3 is the neighbor of s in the θ_4-graph inside the cone $C_3(s)$. Assume that $s_3 \neq t$ as otherwise the result follows trivially. Consequently, s_3 must lie either in $C_0(t)$ or in $C_2(t)$. Assume without loss of generality that s_3 lies in the top-left quadrant of $S(s)$. As s_3 lies in the interior of $S(s)$, $S(s_3) \subset S(s)$ and hence, the top-right quadrant of $S(s_3)$ is empty. Moreover, s_3 lies above ℓ^- and hence $C_1(s_3)$ contains no point of P below ℓ^-. Therefore, by Lemma 1 there is an s_3-t-path φ of length at most $18 \cdot d_{L_1}(s_3,t)$. Since s_3 lies inside $S(s)$, $|s_3t| \leq \sqrt{2} \cdot |st|$ and hence $|\varphi| \leq 18 \cdot d_{L_1}(s_3,t) \leq 18\sqrt{2} \cdot |s_3t| \leq 18\sqrt{2}\sqrt{2} \cdot |st| = 36 \cdot |st|$. Moreover, the length of edge ss_3 is at most $d_{L_1}(s,t) \leq \sqrt{2} \cdot |st|$ since s_3 must lie above ℓ^-. Thus, $ss_3 \cup \varphi$ is an s-t-path of length $|ss_3| + |\varphi| \leq (\sqrt{2}+36) \cdot |st|$. □

The following observation is depicted in Fig. 2(b).

Observation 1 *Let S be an axis-aligned square. If two points a and b lie on consecutive sides along the boundary of S, then there is a square S_{ab} containing the segment ab such that $S_{ab} \subseteq S$ and either a or b lies on a corner of S_{ab}.*

Lemma 3. *Let ab be an edge of the L_∞-Delaunay triangulation of P. There is an a-b-path φ_{ab} in the θ_4-graph of P such that $|\varphi_{ab}| \leq (1+\sqrt{2}) \cdot (\sqrt{2}+36) \cdot |ab|$.*

Proof. Let $T = (a,b,c)$ be a triangle in the L_∞-Delaunay triangulation of P. By definition of this triangulation, there is an empty square S such that every vertex of T lies on the boundary of S. By the general position assumption, a, b and c must lie on different sides of S. If a and b lie on consecutive sides of the boundary

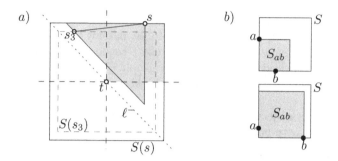

Fig. 2. *a)* Configuration used in the proof of Lemma 2, grey areas represent empty regions. *b)* If a and b lie on consecutive sides of a square S, there is a square S_{ab} such that $ab \subset S_{ab} \subseteq S$ and either a or b lies on a corner of S_{ab}

of S, then by Observation 1 and Lemma 2 there is a path φ_{ab} contained in the θ_4-graph of P such that $|\varphi_{ab}| \leq (\sqrt{2} + 36) \cdot |ab|$.

If a and b lie on opposite sides of S, then both ac and cb have their endpoints on consecutive sides along the boundary of S. Let S_{ac} be the square contained in S existing as a consequence of Observation 1 when applied on the edge ac. Thus, either a or c lies on a corner of S_{ac}. Furthermore, as S_{ac} is contained in S, it is also empty. Consequently, by Lemma 2, there is a a-c-path φ_{ac} such that $|\varphi_{ac}| \leq (\sqrt{2}+36) \cdot |ac|$. Analogously, there is a path φ_{cb} such that $|\varphi_{cb}| \leq (\sqrt{2}+36) \cdot |cb|$. Using elementary geometry, it can be shown that since a and b lie on opposite sides of S, $|ac| + |cb| \leq (1 + \sqrt{2}) \cdot |ab|$. Therefore, the path $\varphi_{ab} = \varphi_{ac} \cup \varphi_{cb}$ is an a-b-path such that $|\varphi_{ab}| \leq (1 + \sqrt{2}) \cdot (\sqrt{2} + 36) \cdot |ab|$. □

Theorem 1. *The θ_4-graph of P is a spanner whose stretch factor is at most*

$$(1 + \sqrt{2}) \cdot (\sqrt{2} + 36) \cdot \sqrt{4 + 2\sqrt{2}} \approx 237$$

Proof. Let ν be the shortest path joining s with t in the L_∞-Delaunay triangulation of P. Bonichon *et al.* [2] proved that the length of ν is at most $\sqrt{4 + 2\sqrt{2}} \cdot |st|$. By replacing every edge in ν with the path in the θ_4-graph of P that exists by Lemma 3, we obtain an s-t-path of length at most

$$(1 + \sqrt{2}) \cdot (\sqrt{2} + 36) \cdot |\nu| \leq (1 + \sqrt{2}) \cdot (\sqrt{2} + 36) \cdot \sqrt{4 + 2\sqrt{2}} \cdot |st| \qquad □$$

3   Light Paths

We introduce some tools that will help us prove Lemma 1 in Section 4.

For a point p in the θ_4-graph, recall that there is an edge between p and p_i, the point in $C_i(p)$ nearest to p. We call the edge pp_i an *i-edge*. Notice that every i-edge is associated with an empty isosceles right triangle. For a point p, the empty triangle generated by its i-edge is denoted by $\Delta_i(p)$.

Let φ be a path that follows only 0- and 1-edges. A 0-edge pp_0 of φ is *light* if no edge of φ crosses the horizontal ray shooting to the right from p. We say that φ is a *light* path if all its 0-edges are light. In this section we show how to bound the length of a light path with respect to the Euclidean distance between its endpoints.

Lemma 4. *Given a light path φ, every pair of 0-edges of φ has disjoint orthogonal projection on the line defined by the equation $y = -x$.*

Proof. Let s and t be the endpoints of φ. Let pp_0 be any 0-edge of φ and let ν_{p_0} be the *diagonal line* extending the hypotenuse of $\Delta_0(p)$, i.e., ν_{p_0} is a line with slope $+1$ passing through p_0. Let γ be the path contained in φ that joins p_0 with t. We claim that every point in γ lies below ν_{p_0}. If this claim is true, the diagonal lines constructed from the empty triangles of every 0-edge in φ split the plane into disjoint slabs, each containing a different 0-edge of φ. Thus, their projection on the line defined by the equation $y = -x$ must be disjoint.

   To prove that every point in γ lies below ν_{p_0}, notice that every point in γ must lie to the right of p since φ is x-monotone, and below p since pp_0 is light, i.e., γ is contained in $C_0(p)$. Since $\Delta_0(p)$ is empty, no point of γ lies above ν_{p_0} and inside $C_0(p)$ yielding our claim.   □

Given a point w of P, we say that a point p of P is *w-protected* if $C_1(p)$ contains no point of P below or on ℓ_w^-, recall that ℓ_w^- is the line with slope -1 passing through w. In other words, a point p is w-protected if either $C_1(p)$ is empty or p_1 lies above ℓ_w^-. Moreover, every point lying above ℓ_w^- is w-protected and no point in $C_3(w)$ is w-protected.

   Given two points s and t such that s lies to the left of t, we aim to construct a path joining s with t in the θ_4-graph of P. The role of t-protected points will be central in this construction. However, as a first step, we relax our goal and prove instead the existence of a light path $\sigma_{s \to t}$ going from s towards t that does not necessarily end at t.

   To construct $\sigma_{s \to t}$, start at a point $z = s$ and repeat the following steps until reaching either t or a t-protected point w lying to the right of t.

   – If z is not t-protected, then follow its 1-edge, i.e., let $z = z_1$.
   – If z is t-protected, then follow its 0-edge, i.e., let $z = z_0$.

The pseudocode of this algorithm can be found in Algorithm 1.

Algorithm 1. Given two points s and t of P such that s lies to the left of t, algorithm to compute path $\sigma_{s \to t}$

1: Let $z = s$.
2: Append s to $\sigma_{s \to t}$.
3: **while** $z \neq t$ and z is not a t-protected point lying to the right of t **do**
4:     **if** z is t-protected **then** $z = z_0$ **else** $z = z_1$
5:     Append z to $\sigma_{s \to t}$.
6: **end while**
7: **return** $\sigma_{s \to t}$

Lemma 5. *Let s and t be two points of P such that s lies to the left of t. Algorithm 1 produces a light x-monotone path $\sigma_{s \to t}$ joining s with a t-protected point w such that either $w = t$ or w lies to the right of t. Moreover, every edge on $\sigma_{s \to t}$ is contained in $S(s)$.*

Proof. By construction, Algorithm 1 finishes only when reaching either t or a t-protected point lying to the right of t. Since every edge of $\sigma_{s \to t}$ is either a 0-edge or a 1-edge traversed from left to right, the path $\sigma_{s \to t}$ is x-monotone. The left endpoint of every 0-edge in $\sigma_{s \to t}$ lies in $C_2(t)$ as it most be t-protected and no t-protected point lies in $C_3(t)$. Thus, if vv_0 is a 0-edge, then v lies in $C_2(t)$ and hence, v_0 lies inside $S(s)$ and above ℓ^+. Otherwise t would lie inside $\Delta_0(v)$. Therefore, every 0-edge in $\sigma_{s \to t}$ is contained in $S(s)$.

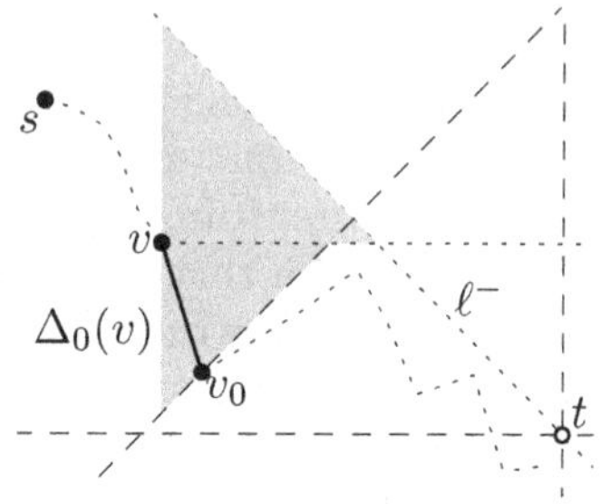

Fig. 3. If v is a t-protected point, then edge vv_0 is light in any path $\sigma_{s \to t}$ that contains it

Every 1-edge in $\sigma_{s \to t}$ has its two endpoints lying below ℓ^-; otherwise, we followed the 1-edge of a t-protected point which is not allowed by Step 4 of Algorithm 1. Thus, every 1-edge in $\sigma_{s \to t}$ lies below ℓ^- and to the right of s. As 1-edges are traversed from bottom to top and the 0-edges of $\sigma_{s \to t}$ are enclosed by $S(s)$, every 1-edge in $\sigma_{s \to t}$ is contained in $S(s)$.

Let vv_0 be any 0-edge of $\sigma_{s \to t}$. Since we followed the 0-edge of v, we know that v is t-protected and hence no point of P lies in $C_1(v)$ and below ℓ^-. As every 1-edge has its two endpoints lying below ℓ^- and $\sigma_{s \to t}$ is x-monotone, no 1-edge in $\sigma_{s \to t}$ can have an endpoint in $C_1(v)$. In addition, every 0-edge of $\sigma_{s \to t}$ joins its left endpoint with a point below it. Thus, no 0-edge of $\sigma_{s \to t}$ can cross the ray shooting to the right from v. Consequently, vv_0 is light and hence $\sigma_{s \to t}$ is a light path; see Fig 3. $\square$

Given two points p and q, let $|pq|_x$ and $|pq|_y$ be the absolute differences between their x- and y-coordinates, respectively, i.e., $d_{L_1}(p, q) = |pq|_x + |pq|_y$.

Lemma 6. *Let s and t be two points of P such that s lies to the left of t. If s is t-protected, then $|\sigma_{s \to t}| \le 3 \cdot d_{L_1}(s, t)$.*

Proof. To bound the length of $\sigma_{s \to t}$, we bound the length of its 0-edges and the length of its 1-edges separately. Let Z be the set of all 0-edges in $\sigma_{s \to t}$ and consider their orthogonal projection on ℓ^-. By Lemma 4, all these projections are disjoint. Moreover, the length of every 0-edge in Z is at most $\sqrt{2}$ times the length of its projection. Let $s_\perp$ be the orthogonal projection of s on ℓ^- and let δ be the segment joining $s_\perp$ with t. Since s is t-protected and $\sigma_{s \to t}$ is x-monotone, the orthogonal projection of every 0-edge of Z on ℓ^- is contained in δ and hence $\sum_{e \in Z} |e| \le \sqrt{2} \cdot |\delta|$. Since $|\delta| = d_{L_1}(s, t)/\sqrt{2}$ as depicted in Fig. 4(a), we conclude that $\sum_{e \in Z} |e| \le d_{L_1}(s, t)$.

Let O be the set of all 1-edges in $\sigma_{s \to t}$ and let η be the horizontal line passing through t. Since $\sigma_{s \to t}$ is x-monotone, the orthogonal projections of all edges in

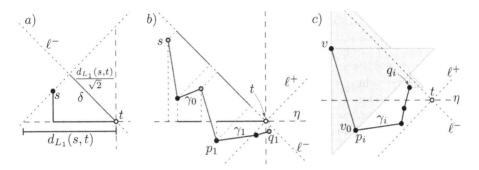

Fig. 4. *a)* The segment δ having length $d_{L_1}(s,t)/\sqrt{2}$. *b)* The 0-edges of $\sigma_{s\to t}$ have disjoint projections on ℓ^- and the 1-edges have disjoint projections on the horizontal line passing through t. The slope between the endpoints of the maximal paths γ_0 and γ_1 is less than 1. *c)* The slope between p^i with q^i is smaller than 1.

O on η are disjoint. Let $\gamma_0, \ldots, \gamma_k$ be the connected components induced by O, i.e., the set of maximal connected paths that can be formed by the 1-edges in O; see Fig. 4(b). We claim that the slope of the line joining the two endpoints p^i, q^i of every γ_i is smaller than 1. If this claim is true, the length of every γ_i is bounded by $|p^i q^i|_x + |p^i q^i|_y \leq 2 \cdot |p^i q^i|_x$ as each γ_i is x- and y-monotone.

To prove that the slope between p^i and q^i is smaller than 1, let $v v_0$ be the 0-edge of $\sigma_{s\to t}$ such that $v_0 = p^i$. Since $v v_0$ is in $\sigma_{s\to t}$, v is t-protected by Step 4 of Algorithm 1 and hence, as $\Delta_0(v)$ is empty, q^i must lie below the line with slope $+1$ passing through p^i yielding our claim; see Fig. 4(c) for an illustration.

Let ω be the segment obtained by shooting a ray from t to the left until hitting the boundary of $S(s)$. We bound the length of all edges in O using the length of ω. Notice that the orthogonal projection of every γ_i on η is contained in ω, except maybe for γ_k whose right endpoint q^k could lie below and to the right of t. Two cases arise: If the projection of γ_k on η is contained in ω, then $\sum_{i=0}^{k} |\gamma_i| \leq \sum_{i=0}^{k} 2 \cdot |p^i q^i|_x \leq 2 \cdot |\omega|$. Otherwise, since q_k is t-protected, q_k lies below ℓ^- and hence $d_{L_1}(p^k, q^k) \leq d_{L_1}(p^k, t)$. Moreover, p^k must lie above ℓ^+ as p^k is reached by a 0-edge coming from above η, i.e., $|p^k t|_y < |p^k t|_x$. Therefore,

$$|\gamma_k| \leq d_{L_1}(p^k, q^k) \leq d_{L_1}(p^k, t) = |p^k t|_x + |p^k t|_y \leq 2 \cdot |p^k t|_x$$

Consequently, $\sum_{i=0}^{k} |\gamma_i| \leq 2 \cdot |p^k t|_x + \sum_{i=0}^{k-1} 2 \cdot |p^i q^i|_x \leq 2 \cdot |\omega|$. Since $|\omega| \leq d_{L_1}(s,t)$, we get that $\sum_{e \in O} |e| = \sum_{i=0}^{k} |\gamma_i| \leq 2 \cdot d_{L_1}(s,t)$. Thus, $\sigma_{s\to t}$ is a light path of length at most $\sum_{e \in O} |e| + \sum_{e \in Z} |e| \leq 3 \cdot d_{L_1}(s,t)$. □

By the construction of the light path in Algorithm 1, we observe the following.

Lemma 7. *Let s and t be two points of P such that s lies to the left of t. If the right endpoint w of $\sigma_{s\to t}$ is not equal to t, then w lies either above ℓ^+ if $w \in C_1(t)$, or below ℓ^- if $w \in C_0(t)$.*

Proof. If w lies in $C_1(t)$, then by Step 4 of Algorithm 1, w was reached by a 0-edge pw such that p is a t-protected point lying above and to the left of t. As $\Delta_0(p)$ is empty, t lies below the hypotenuse of $\Delta_0(p)$ and hence w lies above ℓ^+.

Assume that w lies in $C_0(t)$. Notice that w is the only t-protected point of $\sigma_{s \to t}$ that lies to the right of t; otherwise, Algorithm 1 finishes before reaching w. By Step 4 of Algorithm 1, every 0-edge of $\sigma_{s \to t}$ needs to have a t-protected left endpoint. Moreover, every t-protected point of $\sigma_{s \to t}$, other that w, lies above and to the left of t. Therefore, w is not reached by a 0-edge of $\sigma_{s \to t}$, i.e., w must be the right endpoint of a 1-edge pw of $\sigma_{s \to t}$. Notice that w cannot lie above ℓ^- since otherwise p is t-protected and hence Algorithm 1 finishes before reaching w yielding a contradiction. Thus, w lies below ℓ^-. $\qquad\square$

4   One Empty Quadrant

Before stepping into the proof of Lemma 1, we need one last definition. Given a point p of P, the $\max_1$-*path* of p is the longest x-monotone path having p as an endpoint that consists only of 1-edges and contains edge pp_1. We restate Lemma 1 using the notions of t-protected and s-t-path.

Lemma 1. *Let s and t be two points of P such that t lies in $C_0(s)$. If the top-right quadrant of $S(s)$ is empty and s is t-protected, then there is an s-t-path in the θ_4-graph of P of length at most $18 \cdot d_{L_1}(s,t)$.*

Proof. Since s is t-protected, no point of P lies above s, to the right of s and below ℓ^-; see the dark-shaded region in Fig. 5. Let R be the smallest axis-aligned rectangle enclosing s and t and let k be the number of t-protected points inside R, by the general position assumption, these points are strictly contained in R. We prove the lemma by induction on k.

Base Case: Assume that R contains no t-protected point, i.e., $k = 0$. We claim that R must be empty and we prove it by contradiction. Let q be a point in R and note that q cannot lie above ℓ^- as it would be t-protected yielding a contradiction. If q lies below ℓ^-, we can follow the $\max_1$-path from q until reaching a t-protected point p lying below ℓ^-. Since s is t-protected, p must lie inside R which is also a contradiction. Thus, R must be empty.

Assume that $s_0 \neq t$ since otherwise the result is trivial. As R is empty and $s_0 \neq t$, s_0 lies below t and above ℓ^+. Moreover, no point of P lies

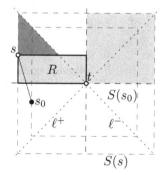

Fig. 5. Base case

above t, below ℓ^- and inside $S(s_0)$ since s is t-protected. Thus, if we think of the set of points P rotated 90 degrees clockwise around t, Lemma 6 and Lemma 7 guarantee the existence of an s_0-t-path γ of length at most $3 \cdot d_{L_1}(s_0,t)$. Since s_0 lies above ℓ^+, $d_{L_1}(s,s_0) \leq d_{L_1}(s,t)$. Furthermore, $d_{L_1}(s_0,t) \leq 2 \cdot d_{L_1}(s,t)$ as s_0

lies inside $S(s)$. Thus, by joining ss_0 with γ, we obtain an s-t-path of length at most $7 \cdot d_{L_1}(s,t)$.

Inductive Step: We aim to show the existence of a path γ joining s with a t-protected point $w \in R$ such that the length of γ is at most $18 \cdot d_{L_1}(s,w)$. If this is true, we can merge γ with the w-t-path φ existing by the induction hypothesis to obtain the desired s-t-path with length at most $18 \cdot d_{L_1}(s,t)$. We analyze two cases depending on the position of s_0 with respect to R.

Case 1. Assume that s_0 lies inside R. If s_0 lies above ℓ^-, then s_0 is t-protected and hence we are done after applying our induction hypothesis on s_0. If s_0 lies below ℓ^-, then we can follow its $\max_1$-path to reach a t-protected point w that must lie inside R as s is t-protected. By running Algorithm 1 on s and w, we obtain a path $\sigma_{s \to w}$ that goes through the edge ss_0 and then follows the $\max_1$-path of s_0 until reaching w; see Fig. 6.

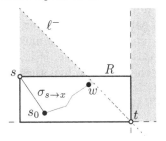

Fig. 6. Case 1

Since s is t-protected and w lies below ℓ^-, s is also w-protected. Therefore, Lemma 6 guarantees that $|\sigma_{s \to w}| \leq 3 \cdot d_{L_1}(s,w)$. By induction hypothesis on w, there is a w-t-path φ such that $|\varphi| \leq 18 \cdot d_{L_1}(w,t)$. As w lies in R, by joining $\sigma_{s \to w}$ with φ we obtain the desired s-t-path of length at most $18 \cdot d_{L_1}(s,t)$.

Case 2. Assume that s_0 does not lie in R. This implies that s_0 lies below t. Assume also that $\sigma_{s \to t}$ does not reach t; otherwise we are done since $|\sigma_{s \to t}| \leq 3 \cdot d_{L_1}(s,t)$. Thus, as the top-right quadrant of $S(s)$ is empty, $\sigma_{s \to t}$ ends at a t-protected point z lying in the bottom-right quadrant of $S(s)$. We consider two sub-cases depending on whether $\sigma_{s \to t}$ contains a point inside R or not.

Case 2.1. If $\sigma_{s \to t}$ contains a point inside R, let w be the first t-protected point of $\sigma_{s \to t}$ after s and note that w also lies inside R since s is t-protected. Notice that the part of $\sigma_{s \to t}$ going from s to w is in fact equal to $\sigma_{s \to w}$ since w lies above t and only 1-edges were followed after s_0 by Step 4 of Algorithm 1; see Fig. 7. Thus, as s is also w-protected, the length of $\sigma_{s \to w}$ is bounded by $3 \cdot d_{L_1}(s,w)$ by Lemma 6. Hence, we can apply the induction hypothesis on w as before and obtain the desired s-t-path.

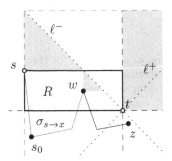

Fig. 7. Case 2.1

Case 2.2. If $\sigma_{s \to t}$ does not contain a point inside R, then $\sigma_{s \to t}$ follows only 1-edges from s_0 until reaching z in the bottom-right quadrant of $S(s)$; see Fig. 8(a) for an illustration of this case.

Let P^* be the set of points obtained by reflecting P on line ℓ^+. Since this reflection preserves 1-edges, it preserves t-protected points. Therefore, if z^* is the reflection of z, then z^* lies in $C_2(t)$ and is also t-protected. Hence, we can use Algorithm 1 to produce a path $\sigma_{z^* \to t}$ in the θ_4-graph of P^*. Let $\gamma_{z \to t}$ be the path in the θ_4-graph of P obtained by reflecting $\sigma_{z^* \to t}$ on ℓ^+. Note that $\gamma_{z \to t}$ is a path that uses only 1- and 2-edges. Because the top-right quadrant of $S(s)$ is empty, $\gamma_{z \to t}$ ends at a point w such that w is either equal to t or w lies in the top-left quadrant of $S(s)$; see Fig. 8(a). Since z lies inside $S(s)$, $d_{L_1}(z,t) \le 2 \cdot d_{L_1}(s,t)$. Hence, by Lemma 6, the length of $\sigma_{s \to t} \cup \gamma_{z \to t}$ is at most

$$|\sigma_{s \to t}| + |\gamma_{z \to t}| \le 3 \cdot d_{L_1}(s,t) + 3 \cdot d_{L_1}(z,t) \le 3 \cdot d_{L_1}(s,t) + 6 \cdot d_{L_1}(s,t) = 9 \cdot d_{L_1}(s,t).$$

Two cases arise: If $\gamma_{z \to t}$ reaches t ($w = t$), then we are done since $\sigma_{s \to t} \cup \gamma_{z \to t}$ joins s with t through z and its length is at most $18 \cdot d_{L_1}(s,t)$.

If $\gamma_{z \to t}$ does not reach t ($w \ne t$), then w lies below ℓ^- by Lemma 7 applied on path $\sigma_{z^* \to t}$. Moreover, as s is t-protected, no point in $C_1(s)$ can be reached by $\gamma_{z \to t}$ and hence w must lie inside R. We claim that $d_{L_1}(s,t) \le 2 \cdot d_{L_1}(s,w)$. If this claim is true, $|\sigma_{s \to t} \cup \gamma_{z \to t}| \le 9 \cdot d_{L_1}(s,t) \le 18 \cdot d_{L_1}(s,w)$. Furthermore, by the induction hypothesis, there is a path φ joining w with t of length at most $18 \cdot d_{L_1}(w,t)$. Consequently, by joining $\sigma_{s \to t}, \gamma_{z \to t}$ and φ, we obtain an s-t-path of length at most $18 \cdot d_{L_1}(s,w) + 18 \cdot d_{L_1}(w,t) = 18 \cdot d_{L_1}(s,t)$.

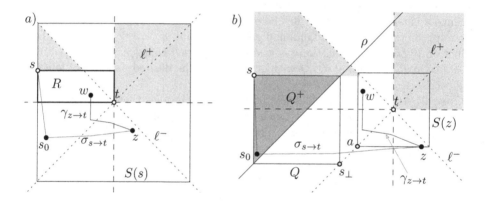

Fig. 8. *a)* Case 2.2 in the proof of Lemma 1, path $\sigma_{s \to t}$ has no point inside R and reaches a point z lying in the bottom-right quadrant of $S(s)$. *b)* The inductive argument proving that the point w, reached after taking the path $\gamma_{z \to t}$, lies outside of the triangle Q^+ containing all the points above ρ and below s. As s is t-protected, the region above s and below ρ is empty.

To prove that $d_{L_1}(s,t) \le 2 \cdot d_{L_1}(s,w)$, let $s_\perp$ be the orthogonal projection of s on ℓ^+. Let ρ be the perpendicular bisector of the segment $s s_\perp$ and notice that for every point y in $C_0(s)$, $d_{L_1}(s,t) \le 2 \cdot d_{L_1}(s,y)$ if and only if y lies below ρ.

Let Q be the minimum axis-aligned square containing s and $s_\perp$. Note that ρ splits Q into two equal triangles Q^+ and Q^- as one diagonal of Q is contained in ρ. Assume that Q^+ is the triangle that lies above ρ. Notice that all points lying in $C_0(s)$ and above ρ are contained in Q^+; see Fig. 8(b). We prove that w lies outside of Q^+ and hence, that w must lie below ρ.

If s_0 lies below ρ, then the empty triangle $\Delta_0(s)$ contains Q^+ forcing w to lie below ρ. Assume that s_0 lies above ρ. In this case, z lies above s_0 as we only followed 1-edges to reach z in the construction of $\sigma_{s \to t}$ by Step 4 of Algorithm 1. Let a be the intersection of ℓ^+ and the ray shooting to the left from z. Notice that w must lie to the right of a as the path $\gamma_{z \to t}$ is contained in the square $S(z)$ and a is one of its corners. As z lies above s_0 and s_0 lies above $s_\perp$, we conclude that a is above $s_\perp$ and both lie on ℓ^+. Therefore, a lies to the right of $s_\perp$, implying that w lies to the right of $s_\perp$ and hence outside of Q^+. As we proved that w lies below ρ, we conclude that $d_{L_1}(s,t) \leq 2 \cdot d_{L_1}(s,w)$.   □

5   Lower Bound

We show how to construct a lower bound of 7 for the θ_4-graph. We start with two points u and w such that w lies in $C_2(u)$ and the difference of their x-coordinates is arbitrarily small. To construct the lower bound, we repeatedly replace a single edge of the shortest u-w-path by placing points in the corners of the empty triangle(s) associated with that edge. The final graph is shown in Fig. 9.

We start out by removing the edge between u and w by placing two points, one inside $\Delta_2(u)$ and one inside $\Delta_0(w)$, both arbitrarily close to the corner that does not contain u nor w. Let v_1 be the point placed in $\Delta_2(u)$. Placing v_1 and the other point in $\Delta_0(w)$ removed edge uw, but created two new shortest paths, uv_1w being one of them. Hence, our next step is to extend this path.

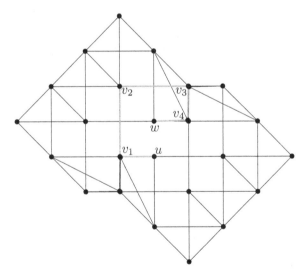

Fig. 9. A lower bound for the θ_4-graph. One of the shortest paths from u to w goes via v_1, v_2, v_3, and v_4.

We remove edge v_1w (and its equivalent in the other path) by placing a point arbitrarily close to the corner of $\Delta_1(v_1)$ and $\Delta_3(w)$ that is farthest from u. Let v_2 be the point placed inside $\Delta_1(v_1)$. Hence, edge v_1w is replaced by the path v_1v_2w. Next, we extend the path again by removing edge v_2w (and its equivalent edge in the other paths). Like before, we place a point arbitrarily close to the corner of $\Delta_0(v_2)$ and $\Delta_2(w)$ that is farthest from u. Let v_3 be the point placed in $\Delta_0(v_2)$. Hence, edge v_2w is replaced by v_2v_3w.

Finally, we replace edge v_3w (and its equivalent edge in the other paths). For all paths for which this edge lies on the outer face, we place a point in the corner of the two empty triangles defining that edge. However, for edge v_3w which does not lie on the outer face, we place a single point v_4 in the intersection of $\Delta_3(v_3)$ and $\Delta_1(w)$. In this way, edge v_3w is replaced by v_3v_4w. When placing v_4, we need to ensure that no edge uv_4 is added as this would created a shortcut. This is easily achieved by placing v_4 such that it is closer to v_3 than to w. The resulting graph is shown in Fig. 9.

Lemma 8. *The stretch factor of the θ_4-graph is at least 7.*

Proof. We look at path $uv_1v_2v_3v_4w$ from Fig. 9. Edges uv_1, v_3v_4, and v_4w have length $|uw| - \varepsilon$ and edges v_1v_2 and v_2v_3 have length $2 \cdot |uw| - \varepsilon$, where ε is positive and arbitrarily close to 0. Hence, the stretch factor of this path is arbitrarily close to 7. □

References

1. Bonichon, N., Gavoille, C., Hanusse, N., Ilcinkas, D.: Connections between Theta-Graphs, Delaunay Triangulations, and Orthogonal Surfaces. In: Thilikos, D.M. (ed.) WG 2010. LNCS, vol. 6410, pp. 266–278. Springer, Heidelberg (2010)
2. Bonichon, N., Gavoille, C., Hanusse, N., Perković, L.: The stretch factor of L_1- and L_∞-Delaunay triangulations. In: Epstein, L., Ferragina, P. (eds.) ESA 2012. LNCS, vol. 7501, pp. 205–216. Springer, Heidelberg (2012)
3. Bose, P., Damian, M., Douïeb, K., O'Rourke, J., Seamone, B., Smid, M., Wuhrer, S.: $\pi/2$-angle Yao graphs are spanners. International Journal of Computational Geometry & Applications 22(1), 61–82 (2012)
4. Bose, P., De Carufel, J.-L., Morin, P., van Renssen, A., Verdonschot, S.: Optimal bounds on theta-graphs: More is not always better. In: Proceedings of CCCG, pp. 305–310 (2012)
5. Bose, P., Morin, P., van Renssen, A., Verdonschot, S.: The θ_5-graph is a spanner. To appear in the proceedings of WG 2013 (2013)
6. Bose, P., van Renssen, A., Verdonschot, S.: On the spanning ratio of theta-graphs. In: Dehne, F., Solis-Oba, R., Sack, J.-R. (eds.) WADS 2013. LNCS, vol. 8037, pp. 182–194. Springer, Heidelberg (2013)
7. Clarkson, K.: Approximation algorithms for shortest path motion planning. In: Proceedings of STOC, pp. 56–65 (1987)
8. El Molla, N.M.: Yao spanners for wireless ad hoc networks. Master's thesis, Villanova University (2009)
9. Keil, J.: Approximating the complete Euclidean graph. In: Karlsson, R., Lingas, A. (eds.) SWAT 1988. LNCS, vol. 318, pp. 208–213. Springer, Heidelberg (1988)
10. Narasimhan, G., Smid, M.: Geometric Spanner Networks. Cambridge University Press (2007)
11. Ruppert, J., Seidel, R.: Approximating the d-dimensional complete Euclidean graph. In: Proceedings of CCCG, pp. 207–210 (1991)

Better Space Bounds
for Parameterized Range Majority and Minority

Djamal Belazzougui[1], Travis Gagie[1,2], and Gonzalo Navarro[3]

[1] Department of Computer Science, University of Helsinki
[2] Helsinki Institute for Information Technology
[3] Department of Computer Science, University of Chile

Abstract. Karpinski and Nekrich (2008) introduced the problem of parameterized range majority, which asks to preprocess a string of length n such that, given the endpoints of a range, one can quickly find all the distinct elements whose relative frequencies in that range are more than a threshold τ. Subsequent authors have reduced their time and space bounds such that, when τ is given at preprocessing time, we need either $\mathcal{O}(n \lg(1/\tau))$ space and optimal $\mathcal{O}(1/\tau)$ query time or linear space and $\mathcal{O}((1/\tau) \lg \lg \sigma)$ query time, where σ is the alphabet size. In this paper we give the first linear-space solution with optimal $\mathcal{O}(1/\tau)$ query time. For the case when τ is given at query time, we significantly improve previous bounds, achieving either $\mathcal{O}(n \lg \lg \sigma)$ space and optimal $\mathcal{O}(1/\tau)$ query time or compressed space and $\mathcal{O}\left((1/\tau) \lg \frac{\lg(1/\tau)}{\lg \lg n}\right)$ query time. Along the way, we consider the complementary problem of parameterized range minority that was recently introduced by Chan et al. (2012), who achieved linear space and $\mathcal{O}(1/\tau)$ query time even for variable τ. We improve their solution to use either nearly optimally compressed space with no slowdown, or optimally compressed space with nearly no slowdown. Some of our intermediate results, such as density-sensitive query time for one-dimensional range counting, may be of independent interest.

1 Introduction

Finding frequent elements in a dataset is a fundamental operation in data mining. Finding the most frequent elements can be challenging when all the distinct elements have nearly equal frequencies and we do not have the resources to compute all their frequencies exactly. In some cases, however, we are interested in the most frequent elements only if they really are frequent. For example, Misra and Gries [20] showed how, given a string and a threshold τ with $0 < \tau \leq 1$, with two passes and $\mathcal{O}(1/\tau)$ words of space we can find all the distinct elements in a string whose relative frequencies are at least τ. These elements are called the τ-majorities of the string. Misra and Gries' algorithm was rediscovered by Demaine, López-Ortiz and Munro [9], who noted it can be made to run in $\mathcal{O}(1)$ time per element on a word RAM with $\Omega(\lg n)$-bit words, where n is the length of the string, which is the model we use; it was then rediscovered again by Karp,

F. Dehne, R. Solis-Oba, and J.-R. Sack (Eds.): WADS 2013, LNCS 8037, pp. 121–132, 2013.
© Springer-Verlag Berlin Heidelberg 2013

Shenker and Papadimitriou [16]. As Cormode and Muthukrishnan [8] put it, "papers on frequent items are a frequent item!"

Krizanc, Morin and Smid [18] introduced the problem of preprocessing the string such that later, given the endpoints of a range, we can quickly return the mode of that range (i.e., the most frequent element). They gave two solutions, one of which takes $\mathcal{O}(n^{2-2\epsilon})$ space for any fixed positive $\epsilon \leq 1/2$, and answers queries in $\mathcal{O}(n^\epsilon \lg \lg n)$ time; the other takes $\mathcal{O}(n^2 \lg \lg n / \lg n)$ space and answers queries in $\mathcal{O}(1)$ time. Petersen [22] reduced Krizanc et al.'s first time bound to $\mathcal{O}(n^\epsilon)$ for any fixed non-negative $\epsilon < 1/2$, and Petersen and Grabowski [23] reduced the second space bound to $\mathcal{O}(n^2 \lg \lg n / \lg^2 n)$. Chan et al. [6] recently gave a linear-space solution that answers queries in $\mathcal{O}\left(\sqrt{n/\lg n}\right)$ time. They also gave evidence suggesting we cannot easily achieve query time substantially smaller than $\sqrt{n}$ using linear space; however, the best known lower bound, by Greve et al. [15], says only that we cannot achieve query time $o\left(\lg(n)/\lg(sw/n)\right)$ using s words of w bits each. Because of the difficulty of supporting range mode queries, Bose et al. [5] and Greve et al. [15] considered the problem of approximate range mode, for which we are asked to return an element whose frequency is at least a constant fraction of the mode's frequency.

Karpinski and Nekrich [17] took a different direction, analogous to Misra and Gries' approach, when they introduced the problem of preprocessing the string such that later, given the endpoints of a range, we can quickly return the τ-majorities of that range. We refer to this problem as parameterized range majority. Assuming τ is given when we are preprocessing the string, they showed how we can store the string in $\mathcal{O}(n(1/\tau))$ space and answer queries in $\mathcal{O}\left((1/\tau)(\lg \lg n)^2\right)$ time. They also gave bounds for dynamic and higher-dimensional versions. Durocher et al. [10] independently posed the same problem and showed how we can store the string in $\mathcal{O}(n \lg(1/\tau + 1))$ space and answer queries in $\mathcal{O}(1/\tau)$ time. Notice that, because there can be up to $1/\tau$ distinct elements to return, this time bound is worst-case optimal. Gagie et al. [14] showed how to store the string in compressed space — i.e., $\mathcal{O}(n(H + 1))$ bits, where H is the entropy of the distribution of elements in the string — such that we can answer queries in $\mathcal{O}((1/\tau) \lg \lg n)$ time. They also showed how to drop the assumption that τ is fixed and simultaneously achieve optimal query time, at the cost of increasing the space bound by a $(\lg n)$-factor. That is, they gave a data structure that stores the string in $\mathcal{O}(n(H + 1))$ space such that later, given the endpoints of a range and τ, we can return the τ-majorities of that range in $\mathcal{O}(1/\tau)$ time. Chan et al. [7] recently gave another solution for variable τ, which also has $\mathcal{O}(1/\tau)$ query time but uses $\mathcal{O}(n \lg n)$ space. As far as we know, these are all the relevant bounds for Karpinski and Nekrich's original exact, static, one-dimensional problem, both for fixed and variable τ; they are summarized in Table 1 together with our own results. Related work includes Elmasry et al.'s [11] solution for the dynamic version and Lai, Poon and Shi's [19] and Wei and Yi's [26] approximate solutions for the dynamic version.

In this paper we first consider the complementary problem of parameterized range minority, which was recently introduced by Chan et al. [7]. For this problem we are asked to preprocess the string such that later, given the endpoints of a

Table 1. Results for the problem of parameterized range majority on a string of length n over an alphabet of size σ in which the distribution of the elements has entropy H

source	space	time	variable τ
Karpinski and Nekrich [17]	$\mathcal{O}(n(1/\tau))$ words	$\mathcal{O}((1/\tau)(\lg\lg n)^2)$	no
Durocher et al. [10]	$\mathcal{O}(n\lg(1/\tau))$ words	$\mathcal{O}(1/\tau)$	no
Gagie et al. [14]	$\mathcal{O}(n(H+1))$ bits	$\mathcal{O}((1/\tau)\lg\lg\sigma)$	no
Theorem 3	$\mathcal{O}(n)$ words	$\mathcal{O}(1/\tau)$	no
Gagie et al. [14]	$\mathcal{O}(n(H+1))$ words	$\mathcal{O}(1/\tau)$	yes
Chan et al. [7]	$\mathcal{O}(n\lg n)$ words	$\mathcal{O}(1/\tau)$	yes
Theorem 4	$\mathcal{O}(n\lg\lg\sigma)$ words	$\mathcal{O}(1/\tau)$	yes
Theorem 5	$nH + o(n\lg\sigma)$ bits	$\mathcal{O}((1/\tau)\lg\lg\sigma)$	yes
Theorem 7	$(1+\epsilon)nH + o(n\lg\sigma)$ bits	$\mathcal{O}\!\left((1/\tau)\lg\frac{\lg(1/\tau)}{\lg\lg n}\right)$	yes

range, we can return (if one exists) a distinct element that occurs in that range but is not one of its τ-majorities. Such an element is called a τ-minority for the range. At first, finding a τ-minority might seem harder than finding a τ-majority because, e.g., we are less likely to find a τ-minority by sampling. Nevertheless, Chan et al. gave a linear-space solution with $\mathcal{O}(1/\tau)$ query time even when τ is given at query time. In Section 3 we give two results, also for the case of variable τ:

1. for any positive constant ϵ, a solution with $\mathcal{O}(1/\tau)$ query time that takes $(1+\epsilon)nH + \mathcal{O}(n)$ bits;
2. for any function $f(n) = \omega(1)$, a solution with $\mathcal{O}((1/\tau)f(n))$ query time that takes $nH + \mathcal{O}(n) + o(nH)$ bits.

In the full version of this paper we will reduce the space bound in point 2 above to $nH + o(n(H+1))$ bits. That is, we improve Chan et al.'s solution to use either nearly optimally compressed space with no slowdown, or optimally compressed space with nearly no slowdown. We reuse ideas from this section in our solutions for parameterized range majority.

In Section 4 we return to Karpinski and Nekrich's original problem of parameterized range majority with fixed τ and give the first linear-space solution with worst-case optimal $\mathcal{O}(1/\tau)$ query time. In Section 5 we adapt this solution to the more challenging case of variable τ and give three results:

1. a solution with $\mathcal{O}(1/\tau)$ query time that takes $\mathcal{O}(n\lg\lg\sigma)$ space, where σ is the size of the alphabet;
2. a solution with $\mathcal{O}((1/\tau)\lg\lg\sigma)$ query time that takes $nH + o(n\lg\sigma)$ bits;
3. for any positive constant ϵ, a solution with $\mathcal{O}\!\left((1/\tau)\lg\frac{\lg(1/\tau)}{\lg\lg n}\right)$ query time that takes $(1+\epsilon)nH + o(n\lg\sigma)$ bits.

With (2), we can support $\mathcal{O}(1)$-time access to the string and $\mathcal{O}(\lg\lg\sigma)$-time rank and select (see definitions in Section 2.1); with (3), select also takes $\mathcal{O}(1)$ time.

In the full version of this paper we will reduce the space bounds in (2) and (3) to $nH + o(n(H+1))$ and $(1 + \epsilon)nH + \mathcal{O}(n)$ bits, respectively. While proving (3) we introduce a compressed data structure with density-sensitive query time for one-dimensional range counting, which may be of independent interest; due to space constraints, however, we leave the description of this data structure to the full version of this paper. We will also show in the full version how to use our data structures for (2) or (3) to find a range mode quickly when it is actually reasonably frequent. We leave as an open problem reducing the space bound in (1) or the time bound in (2) or (3), to obtain linear or compressed space with optimal query time.

2   Preliminaries

2.1   Access, Select and (Partial) Rank

Let $S[1..n]$ be a string over an alphabet of size σ and let H be the entropy of the distribution of elements in S. An access query on S takes a position k and returns $S[k]$; a rank query takes a distinct element a and a position k and returns the number of occurrences of a in $S[1..k]$; a select query takes a distinct element a and a rank r and returns the position of the rth occurrence of a in S. A partial rank query is a rank query with the restriction that the given distinct element must occur in the given position; i.e., $S[k] = a$. These are among the most well-studied operations on strings, so we state here only the results most relevant to this paper.

For $\sigma = 2$ and any constant c, Pătraşcu [24] showed how we can store S in $nH + \mathcal{O}(n/\lg^c n)$ bits. For $\sigma = \lg^{\mathcal{O}(1)} n$, Ferragina et al. [12] showed how we can store S in $nH + o(n)$ bits and support access, rank and select in $\mathcal{O}(1)$ time. For $\sigma < n$, Barbay et al. [1] showed how, for any positive constant ϵ, we can store S in $(1 + \epsilon)nH + o(n)$ bits and support access and select in $\mathcal{O}(1)$ time and rank in $\mathcal{O}(\lg \lg \sigma)$ time. Belazzougui and Navarro [3] showed how to support $\mathcal{O}(1)$-time partial rank using $\mathcal{O}(n(\lg H + 1))$ bits; in the full version of their paper [2] they reduced that space bound to $o(n)(H + 1)$ bits. In another paper, Belazzougui and Navarro [4] showed how, for any function $f(n) = \omega(1)$, we can store S in $nH + o(n(H+1)$ bits and support access in $\mathcal{O}(1)$ time, select in $\mathcal{O}(f(n))$ time and rank in $\mathcal{O}(\lg \lg \sigma)$ time. They also proved, via a reduction from the predecessor problem, that we cannot support general rank queries in $o(\lg(\lg \sigma / \lg \lg n))$ time while using $n \lg^{\mathcal{O}(1)} n$ space.

2.2   Coloured Range Listing

Motivated by the problem of document listing, Muthukrishnan [21] showed how we can store $S[1..n]$ such that, given the endpoints of a range, we can quickly list the distinct elements in that range and the positions of their leftmost occurrences therein. This is the special case of one-dimensional coloured range listing in which the points' coordinates are the integers from 1 to n. Let $C[1..n]$ be the array in which $C[k]$ is the position of the last occurrence of the distinct element $S[k]$ in $S[1..k-1]$ — i.e., the last occurrence before $S[k]$ itself — or 0 if there is no such occurrence. Notice $S[k]$ is the first occurrence of that distinct element in a range $S[i..j]$ if and only if $i \le k \le j$ and $C[k] < i$. We store C, implicitly or explicitly, and a data structure supporting $\mathcal{O}(1)$-time range-minimum queries on C that return the position of the leftmost occurrence of the minimum in the range.

To list the distinct elements in a range $S[i..j]$ given i and j, we find the position m of the leftmost occurrence of the minimum in the range $C[i..j]$; check whether $C[m] < i$; and, if so, output $S[m]$ and m and recurse on $C[i..m-1]$ and $C[m+1..j]$. This procedure is online — i.e., we can stop it early if we want only a certain number of distinct elements — and the time it takes per distinct element is $\mathcal{O}(1)$ plus the time to access C.

Suppose we already have data structures supporting access, select and partial rank queries on S, all in $\mathcal{O}(t)$ time. Notice $C[k] = S.\text{select}_{S[k]}\big(S.\text{rank}_{S[k]}(k) - 1\big)$, so we can also support access to C in $\mathcal{O}(t)$ time. Sadakane [25] and Fischer [13] gave $\mathcal{O}(n)$-bit data structures supporting $\mathcal{O}(1)$-time range-minimum queries. Therefore, we can implement Muthukrishnan's solution using $\mathcal{O}(n)$ extra bits such that it takes $\mathcal{O}(t)$ time per distinct element listed.

3   Parameterized Range Minority

Recall from Section 1 that a τ-minority for a range is a distinct element that occurs in that range but is not one of its τ-majorities. The problem of parameterized range minority is to preprocess a string such that later, given the endpoints of a range and τ, we can quickly return a τ-minority for that range if one exists. Chan et al. gave a linear-space solution with $\mathcal{O}(1/\tau)$ query time even for the case of variable τ. They first build a list of $\lfloor 1/\tau \rfloor + 1$ distinct elements that occur in the given range (or as many as there are, if fewer) and then check those elements' frequencies to see which are τ-minorities. There cannot be more than $\lfloor 1/\tau \rfloor$ τ-majorities so, if there exists a τ-minority for that range, then at least one must be in the list. In this section we show how to implement this idea using compressed space.

To support parameterized range minority on $S[1..n]$ in $\mathcal{O}(1/\tau)$ time, we store data structures supporting $\mathcal{O}(1)$-time access, select and partial rank queries on S and a data structure supporting $\mathcal{O}(1)$-time range-minimum queries on C. For any positive constant ϵ, we can store these data structures in a total of $(1 + \epsilon)nH + \mathcal{O}(n)$ bits. Given τ and endpoints i and j, in $\mathcal{O}(1/\tau)$ time we use Muthukrishnan's algorithm to build a list of $\lfloor 1/\tau \rfloor + 1$ distinct elements that occur in $S[i..j]$ (or as many as there are, if fewer) and the positions of their leftmost occurrences therein. We check whether these distinct elements are τ-minorities using the following lemma:

Lemma 1. *Suppose we know the position of the leftmost occurrence of a distinct element in a range. We can check whether that distinct element is a τ-minority or a τ-majority using a partial rank query and a select query on S.*

Proof. Let k be the position of the first occurrence of a in $S[i..j]$. If $S[k]$ is the rth occurrence of a in S, then a is a τ-minority for $S[i..j]$ if and only if the $(r + \lceil \tau(j - i + 1) \rceil - 1)$th occurrence of a in S is strictly after $S[j]$; otherwise a is a τ-majority. That is, we can check whether a is a τ-minority for $S[i..j]$ by checking whether

$$S.\text{select}_a \left(S.\text{rank}_a(k) + \lceil \tau(j - i + 1) \rceil - 1 \right) > j ;$$

since $S[k] = a$, computing $S.\text{rank}_a(k)$ is only a partial rank query.   □

This gives us the following theorem, which improves Chan et al.'s solution to use nearly optimally compressed space with no slowdown.

Theorem 1. *For any positive constant ϵ, we can store S in $(1+\epsilon)nH+\mathcal{O}(n)$ bits such that later, given the endpoints of a range and τ, we can return a τ-minority for that range (if one exists) in $\mathcal{O}(1/\tau)$ time.*

Alternatively, for any function $f(n) = \omega(1)$, we can store our data structures for access, select and partial rank on S and range-minimum queries on C in a total of $nH + \mathcal{O}(n) + o(nH)$ at the cost of select queries taking $\mathcal{O}(f(n))$ time.

Theorem 2. *For any function $f(n) = \omega(1)$, we can store S in $nH + \mathcal{O}(n) + o(nH)$ bits such that later, given the endpoints of a range and τ, we can return a τ-minority for that range (if one exists) in $\mathcal{O}((1/\tau) f(n))$ time.*

In the full version of this paper we will reduce the space bound of Theorem 2 to $nH + o(n(H + 1))$ bits. That is, we improve Chan et al.'s solution to use optimally compressed space with nearly no slowdown.

4   Parameterized Range Majority with Fixed τ

The standard approach to finding τ-majorities, going back to Misra and Gries' work, is to build a list of $\mathcal{O}(1/\tau)$ candidate elements and then verify them. For parameterized range majority, an obvious way to verify candidates is to use rank queries. The problem with this approach is that, as noted in Subsection 2.1, we cannot support general rank queries in $o(\lg(\lg \sigma / \lg \lg n))$ time while using $n \lg^{\mathcal{O}(1)} n$ space; e.g., with only linear space, we cannot support general rank queries in $\mathcal{O}(1)$ time when the alphabet is super-polylogarithmic. If we can find the position of candidates' first occurrences in the range, however, then by Lemma 1 we can check them using only partial rank and select queries.

Suppose we want to support parameterized range majority on $S[1..n]$ for a fixed threshold τ. We first store data structures that support access, select and partial rank on S in $\mathcal{O}(1)$ time, which takes $\mathcal{O}(n)$ space. For $0 \leq b \leq \lfloor \lg n \rfloor$, let $F_b[1..n]$ be the binary string in which $F_b[k] = 1$ if the distinct element $S[k]$ occurs at least $\tau 2^b$ times in $S[k..k + 2^{b+1} - 1]$; and let S_b and C_b be the subsequences of S and C, respectively, consisting of those elements flagged by 1s in F_b. We store

F_b in $\mathcal{O}(n)$ bits such that we can support access, rank and select queries on F_b in $\mathcal{O}(1)$ time. Notice we can implement an access query on S_b or C_b as a select query on F_b and access queries on S or C, respectively. As described in Subsection 2.2, we can implement an access query to C as access, select and partial rank queries on S. We also store an $\mathcal{O}(1)$-time range-minimum data structure for C_b, which takes $\mathcal{O}(|S_b|)$ bits.

With these data structures, given endpoints i and j with $\lfloor \lg(j-i+1) \rfloor = b$, we use Muthukrishnan's algorithm to list the distinct elements in $S_b[F_b.\mathrm{rank}_1(i)..$ $F_b.\mathrm{rank}_1(j)]$ and the positions of their leftmost occurrences therein; we then use select queries on F_b to find the positions of those elements in S. That is, we list the distinct elements in $S[i..j]$ that are flagged by 1s in F_b and the positions of their leftmost flagged occurrences therein. We then apply Lemma 1 to each of these elements, treating the positions of their leftmost flagged occurrences as the positions of their leftmost occurrences. Since each distinct element in $S[i..j]$ that is flagged in F_b occurs at least $\tau 2^b$ times in $S[i..j + 2^{b+1} - 1] \subset S[i..i + 2^{b+2}]$, there are $\mathcal{O}(1/\tau)$ of them and we use a total of $\mathcal{O}(1/\tau)$ time.

Notice that the leftmost flagged occurrences of a distinct element a in $S[i..j]$ may not necessarily be the leftmost occurrence therein. However, if a is a τ-majority in $S[i..j]$ then, by definition, a occurs at least $\tau(j - i + 1) \geq \tau 2^b$ times in $S[i..j] \subset S[i..i + 2^{b+1} - 1]$, so a's leftmost occurrence in $S[i..j]$ is flagged by a 1 in F_b and, therefore, we apply Lemma 1 to it. It follows that we return each τ-majority in $S[i..j]$.

We store only one set of data structures supporting access, select and partial rank on S. Summing over b from 0 to $\lfloor \lg n \rfloor$, the data structures for access, select, partial rank and range-minimum queries take a total of $\mathcal{O}(n \lg n)$ bits, which is $\mathcal{O}(n)$ words. Therefore, we have the first linear-space data structure with worst-case optimal $\mathcal{O}(1/\tau)$ query time for Karpinski and Nekrich's original problem of parameterized range majority with fixed τ.

Theorem 3. *Given a threshold τ, we can store a string in linear space and support parameterized range majority in $\mathcal{O}(1/\tau)$ time.*

5   Parameterized Range Majority with Variable τ

5.1   Nearly Linear Space with Optimal Query Time

Suppose we have an instance of the data structure from Theorem 3 for each threshold $1, 1/2, 1/4, \ldots, 1/2^{\lceil \lg n \rceil}$, which takes a total of $\mathcal{O}(n \lg n)$ space. Given endpoints i and j and a threshold τ, we can use the instance for threshold $1/2^{\lceil \lg(1/\tau) \rceil}$ to build a list of $\mathcal{O}(1/\tau)$ candidate elements and then check them with Lemma 1; this takes a total of $\mathcal{O}(1/\tau)$ time and returns all the τ-majorities in $S[i..j]$. Gagie et al. used a variant of this idea to obtain the first data structure for variable τ. We can easily reduce our space bound to $\mathcal{O}(n \lg \sigma)$ because, if $1/\tau \geq \sigma$, then we can simply use Muthukrishnan's algorithm with S and C to list in $\mathcal{O}(\sigma) = \mathcal{O}(1/\tau)$ time all the distinct elements in $S[i..j]$ and the positions of their leftmost occurrences therein, then check them with Lemma 1.

Notice that we need store only one set of data structures supporting access, select and partial rank on S. Also, if $S[k]$ is a $(1/2^t)$-majority in a range, then it is also a $(1/2^{t'})$-majority for all $t' \geq t$. It follows that if, instead of querying only the instance for the threshold $1/2^{\lceil \lg(1/2) \rceil}$, we query the instances for all the thresholds $1, 1/2, 1/4, \ldots, 1/2^{\lceil \lg(1/\tau) \rceil}$ — which still takes $\mathcal{O}\left(\sum_{t=0}^{2^{\lceil \lg(1/\tau) \rceil}} 2^t\right) = \mathcal{O}(1/\tau)$ time — then we can modify the instances to reduce the total number of 1s in their binary strings. Specifically, for $0 \leq t \leq \lceil \lg \sigma \rceil$, let F_b^t be the binary string F_b in the instance for threshold $1/2^t$; we modify F_b^t such that $F_b^t[k] = 1$ if and only if the number of occurrences of the distinct element $S[k]$ in $S[k..k + 2^{b+1} - 1]$ is at least 2^{b-t} times but less than 2^{b-t+1}.

For $0 \leq b \leq \lfloor \lg n \rfloor$ and $1 \leq k \leq n$, we have $F_b^t[k] = 1$ for at most one value of t. Therefore, all the binary strings contain a total of at most $n(\lfloor \lg n \rfloor + 1)$ copies of 1, so all the range-minimum data structures take a total of $\mathcal{O}(n \lg n)$ bits. Since the binary strings have total length $n \lceil \lg n \rceil \lceil \lg \sigma \rceil$, we can use Pătraşcu's data structure to store them in a total of $\mathcal{O}(n \lg(n) \lg \lg \sigma)$ bits. A slightly neater approach is to represent all the binary strings $F_b^0, \ldots, F_b^{\lceil \lg \sigma \rceil}$ as a single string $T_b[1..n]$ in which $T_b[k] = t$ if $F_b^t[k] = 1$, and ∞ if there is no such value t. We can implement access, rank and select queries on $F_b^0, \ldots, F_b^{\lceil \lg \sigma \rceil}$ by access, rank and select queries on T_b. Since T_b is an alphabet of size $\mathcal{O}(\lg \sigma)$, we can use Ferragina et al.'s data structure to store it in $\mathcal{O}(n \lg \lg \sigma)$ bits and support access, rank and select queries in $\mathcal{O}(1)$ time. Either way, in total we use $\mathcal{O}(n \lg \lg \sigma)$ space.

Theorem 4. *We can store S in $\mathcal{O}(n \lg \lg \sigma)$ space such that later, given the endpoints of a range and τ, we can return the τ-majorities for that range in $\mathcal{O}(1/\tau)$ time.*

5.2 Optimally Compressed Space with Nearly Optimal Query Time

To be able to apply Lemma 1, we must be able to find the leftmost occurrence of each τ-majority in a range. For this reason, we may flag many occurrences of the same distinct element even when they appear in close succession, because we cannot know in advance where the query range will start. As discussed in Section 4, however, if we have a data structure that supports rank queries on S, then it is sufficient for us to build a list of $\mathcal{O}(1/\tau)$ candidate elements that includes all the τ-majorities — without any information about positions — and then verify them using rank queries. This lets us flag fewer elements and so reduce our space bound, at the cost of using slightly suboptimal query time.

We store an instance of Belazzougui and Navarro's data structure supporting access on S in $\mathcal{O}(1)$ and rank and select on S in $\mathcal{O}(\lg \lg \sigma)$ time, which takes $nH + o(n(H + 1))$ bits. For $0 \leq t \leq \lceil \lg \sigma \rceil$ and $\lfloor \lg(2^t \lg \lg \sigma) \rfloor \leq b \leq \lfloor \lg n \rfloor$, we divide S into blocks of length 2^{b-1} and store data structures supporting access, rank and select on the binary string $G_b^t[1..n]$ in which $G_b^t[k] = 1$ if, first, the distinct element $S[k]$ occurs at least 2^{b-t} times in $S[k - 2^{b+1}..k + 2^{b+1}]$ and, second, $S[k]$ is the leftmost or rightmost occurrence of that distinct element in its block. We also store an $\mathcal{O}(1)$-time range-minimum data structure for the subsequence of C consisting of elements flagged by 1s in G_b^t.

The number of distinct elements that occur at least 2^{b-t} times in a range of size $\mathcal{O}(2^b)$ is $\mathcal{O}(2^t)$, so there are $\mathcal{O}(2^t)$ elements in each block flagged by 1s in G_b^t. It follows that we can store an instance of Pătraşcu's data structure supporting $\mathcal{O}(1)$-time access, rank and select on G_b^t in $\mathcal{O}(n2^{t-b}(b-t) + n/\lg^3 n)$ bits; we need $\mathcal{O}(2^t)$ bits for the corresponding range-minimum data structure. Summing over t from 0 to $\lceil \lg \sigma \rceil$ and over b from $\lfloor \lg(2^t \lg \lg \sigma) \rfloor$ to $\lfloor \lg n \rfloor$, calculation shows we use a total of $\mathcal{O}\left(\frac{n \lg \sigma \lg \lg \lg \sigma}{\lg \lg \sigma} + \frac{n}{\lg n}\right) = o(n \lg \sigma)$ bits for the binary strings and range-minimum data structures. Therefore, including the instance of Belazzougui and Navarro's data structure for S, we use $nH + o(n \lg \sigma)$ bits altogether.

Given endpoints i and j and a threshold τ, if

$$\lfloor \lg(j - i + 1) \rfloor < \left\lfloor \lg\left(2^{\lceil \lg(1/\tau) \rceil} \lg \lg \sigma\right) \right\rfloor,$$

then we simply run Misra and Gries' algorithm on $S[i..j]$ in $\mathcal{O}(j - i) = \mathcal{O}((1/\tau) \lg \lg \sigma)$ time. Otherwise, we use Muthukrishnan's algorithm to list the distinct elements flagged by 1s in G_b^t, where $t = \lceil \lg(1/\tau) \rceil$ and $b = \lfloor \lg(j-i+1) \rfloor \geq \lfloor \lg(2^t \lg \lg \sigma) \rfloor$, and use rank queries on S to check whether each of them is a τ-majority in $S[i..j]$. Since $S[i..j]$ overlaps at most 5 blocks of length 2^{b-1}, it contains $\mathcal{O}(1/\tau)$ distinct elements flagged by 1s in G_b^t; therefore, Muthukrishnan's algorithm takes $\mathcal{O}(1/\tau)$ time and we use a total of $\mathcal{O}((1/\tau) \lg \lg \sigma)$ time for all the rank queries on S.

Since $S[i..j]$ cannot be completely contained in a block of length 2^{b-1}, if $S[i..j]$ overlaps a block then it includes one of that block's endpoints. Therefore, if $S[i..j]$ contains an occurrence of a distinct element a, then it includes the leftmost or rightmost occurrence of a in some block. Suppose a is a τ-majority in $S[i..j]$. For $i \leq k \leq j$, a occurs at least 2^{b-t} times in $S[k - 2^{b+1}..k + 2^{b+1}]$, so some occurrence of a in $S[i..j]$ is flagged by a 1 in G_b^t. Therefore, we return a.

Theorem 5. *We can store S in $nH + o(n \lg \sigma)$ bits such that later, given the endpoints of a range and τ, we can return the τ-majorities for that range in $\mathcal{O}((1/\tau) \lg \lg \sigma)$ time.*

Since our solution includes an instance of Belazzougui and Navarro's data structure, we can also support $\mathcal{O}(1)$-time access to S and $\mathcal{O}(\lg \lg \sigma)$-time rank and select. In the full version of this paper we will reduce the space bound of Theorem 5 to $nH + o(n(H + 1))$ bits.

5.3 Nearly Optimally Compressed Space with Very Nearly Optimal Query Time

Recall from Subsection 5.1 that, if $1/\tau \geq \sigma$, then we can simply use Muthukrishnan's algorithm to list all the distinct elements in a range and then check them with Lemma 1; therefore, we can assume $1/\tau < \sigma$. In this subsection we use a new data structure with density-sensitive query time for one-dimensional range counting, which may be of independent interest, to obtain a nearly optimally compressed data structure for parameterized range majority with $\mathcal{O}\left((1/\tau) \lg \frac{\lg(1/\tau)}{\lg \lg n}\right)$

query time. Due to space constraints, however, we leave the description of our range-counting data structure to the full version of this paper and merely state our result here:

Theorem 6. *For any positive constant ϵ, we can store S in $(1 + \epsilon)nH + \mathcal{O}(n)$ bits such that later, given endpoints i and j and a distinct element a, we can return $\mathrm{occ}(a, S[i..j])$ in $\mathcal{O}\left(\lg \frac{\lg \frac{j-i+1}{\mathrm{occ}(a, S[i..j])}}{\lg \lg n}\right)$ time. We can also support access and select in $\mathcal{O}(1)$ time and rank in $\mathcal{O}(\lg \lg \sigma)$ time.*

To obtain a compressed data structure for parameterized range majority with $\mathcal{O}\left((1/\tau) \lg \frac{\lg(1/\tau)}{\lg \lg n}\right)$ query time, we combine our solution from Theorem 5 with Theorem 6. Instead of using $\mathcal{O}(\lg \lg \sigma)$-time rank queries to check each of the $\mathcal{O}(1/\tau)$ candidate elements returned by Muthukrishnan's algorithm, we use range-counting queries. We can make all $\mathcal{O}(1/\tau)$ range-counting queries each take $\mathcal{O}\left(\lg \frac{\lg(1/\tau)}{\lg \lg n}\right)$ time because, if one starts taking too much time, then the distinct element we are checking cannot be a τ-majority and we can stop the query early. (In fact, as we will show in the full version of this paper, our data structure from Theorem 6 does not need such intervention.) This gives us our final result:

Theorem 7. *We can store S in $(1 + \epsilon)nH + o(n \lg \sigma)$ bits such that later, given the endpoints of a range and τ, we can return the τ-majorities for that range in $\mathcal{O}\left((1/\tau) \lg \frac{\lg(1/\tau)}{\lg \lg n}\right)$ time.*

Notice our solution in Theorem 7 takes optimal $\mathcal{O}(1/\tau)$ time when $1/\tau = \lg^{\mathcal{O}(1)} n$. Again, we can also support access and select in $\mathcal{O}(1)$ time and rank in $\mathcal{O}(\lg \lg \sigma)$ time. In the full version of this paper we will reduce the space bound in Theorem 7 to $(1 + \epsilon)nH + \mathcal{O}(n)$ bits, and show how to use our data structures from Theorems 5 and 7 to find a range mode quickly when it is actually reasonably frequent.

6   Conclusions

We have given the first linear-space data structure for parameterized range majority with query time $\mathcal{O}(1/\tau)$, which is worst-case optimal in terms of n and τ. Moreover, we have improved the space bounds for parameterized range majority and minority in the important case of variable τ. For parameterized range majority with variable τ, we have achieved nearly linear space and worst-case optimal query time, or compressed space with a slight slowdown. For parameterized range minority, we have improved Chan et al.'s solution to use nearly compressed space with no slowdown or compressed space with nearly no slowdown. In the full version of this paper we will also reduce the lower-order terms in our compressed space bounds to $o(n(H + 1))$ with the same slowdowns. We leave as an open problem achieving linear or compressed space with $\mathcal{O}(1/\tau)$ query time for variable τ, or showing that this is impossible.

Acknowledgments. Many thanks to Patrick Nicholson for helpful comments.

References

1. Barbay, J., Claude, F., Gagie, T., Navarro, G., Nekrich, Y.: Efficient fully-compressed sequence representations. Algorithmica (to appear)
2. Belazzougui, D., Navarro, G.: Alphabet-independent compressed text indexing. ACM Transactions on Algorithms (to appear)
3. Belazzougui, D., Navarro, G.: Alphabet-independent compressed text indexing. In: Demetrescu, C., Halldórsson, M.M. (eds.) ESA 2011. LNCS, vol. 6942, pp. 748–759. Springer, Heidelberg (2011)
4. Belazzougui, D., Navarro, G.: New lower and upper bounds for representing sequences. In: Epstein, L., Ferragina, P. (eds.) ESA 2012. LNCS, vol. 7501, pp. 181–192. Springer, Heidelberg (2012)
5. Bose, P., Kranakis, E., Morin, P., Tang, Y.: Approximate range mode and range median queries. In: Diekert, V., Durand, B. (eds.) STACS 2005. LNCS, vol. 3404, pp. 377–388. Springer, Heidelberg (2005)
6. Chan, T.M., Durocher, S., Larsen, K.G., Morrison, J., Wilkinson, B.T.: Linear-space data structures for range mode query in arrays. In: Proceedings of the 29th Symposium on Theoretical Aspects of Computer Science (STACS), pp. 290–301 (2012)
7. Chan, T.M., Durocher, S., Skala, M., Wilkinson, B.T.: Linear-space data structures for range minority query in arrays. In: Fomin, F.V., Kaski, P. (eds.) SWAT 2012. LNCS, vol. 7357, pp. 295–306. Springer, Heidelberg (2012)
8. Cormode, G., Muthukrishnan, S.: Data stream methods. Lecture 3 of Rutger's 198:671 Seminar on Processing Massive Data Sets (2003), http://www.cs.rutgers.edu/~muthu/198-3.pdf
9. Demaine, E.D., López-Ortiz, A., Munro, J.I.: Frequency estimation of internet packet streams with limited space. In: Möhring, R.H., Raman, R. (eds.) ESA 2002. LNCS, vol. 2461, pp. 348–360. Springer, Heidelberg (2002)
10. Durocher, S., He, M., Munro, J.I., Nicholson, P.K., Skala, M.: Range majority in constant time and linear space. Information and Computation 222, 169–179 (2013)
11. Elmasry, A., He, M., Munro, J.I., Nicholson, P.K.: Dynamic range majority data structures. In: Asano, T., Nakano, S.-I., Okamoto, Y., Watanabe, O. (eds.) ISAAC 2011. LNCS, vol. 7074, pp. 150–159. Springer, Heidelberg (2011)
12. Ferragina, P., Manzini, G., Mäkinen, V., Navarro, G.: Compressed representations of sequences and full-text indexes. ACM Transactions on Algorithms 3(2) (2007)
13. Fischer, J.: Optimal succinctness for range minimum queries. In: López-Ortiz, A. (ed.) LATIN 2010. LNCS, vol. 6034, pp. 158–169. Springer, Heidelberg (2010)
14. Gagie, T., He, M., Munro, J.I., Nicholson, P.K.: Finding frequent elements in compressed 2D arrays and strings. In: Grossi, R., Sebastiani, F., Silvestri, F. (eds.) SPIRE 2011. LNCS, vol. 7024, pp. 295–300. Springer, Heidelberg (2011)
15. Greve, M., Jørgensen, A.G., Larsen, K.D., Truelsen, J.: Cell probe lower bounds and approximations for range mode. In: Abramsky, S., Gavoille, C., Kirchner, C., Meyer auf der Heide, F., Spirakis, P.G. (eds.) ICALP 2010. LNCS, vol. 6198, pp. 605–616. Springer, Heidelberg (2010)
16. Karp, R.M., Shenker, S., Papadimitriou, C.H.: A simple algorithm for finding frequent elements in streams and bags. ACM Transactions on Database Systems 28(1), 51–55 (2003)

17. Karpinski, M., Nekrich, Y.: Searching for frequent colors in rectangles. In: Proceedings of the 20th Canadian Conference on Computational Geometry (CCCG), pp. 11–14 (2008)
18. Krizanc, D., Morin, P., Smid, M.H.M.: Range mode and range median queries on lists and trees. Nordic Journal of Computing 12(1), 1–17 (2005)
19. Lai, Y.K., Poon, C.K., Shi, B.: Approximate colored range and point enclosure queries. Journal of Discrete Algorithms 6(3), 420–432 (2008)
20. Misra, J., Gries, D.: Finding repeated elements. Science of Computer Programming 2(2), 143–152 (1982)
21. Muthukrishnan, S.: Efficient algorithms for document retrieval problems. In: Proceedings of the 13th Symposium on Discrete Algorithms (SODA), pp. 657–666 (2002)
22. Petersen, H.: Improved bounds for range mode and range median queries. In: Geffert, V., Karhumäki, J., Bertoni, A., Preneel, B., Návrat, P., Bieliková, M. (eds.) SOFSEM 2008. LNCS, vol. 4910, pp. 418–423. Springer, Heidelberg (2008)
23. Petersen, H., Grabowski, S.: Range mode and range median queries in constant time and sub-quadratic space. Information Processing Letter 109(4), 225–228 (2009)
24. Pătraşcu, M.: Succincter. In: Proceedings of the 49th Symposium on Foundations of Computer Science (FOCS), pp. 305–313 (2008)
25. Sadakane, K.: Succinct data structures for flexible text retrieval systems. Journal of Discrete Algorithms 5(1), 12–22 (2007)
26. Wei, Z., Yi, K.: Beyond simple aggregates: indexing for summary queries. In: Proceedings of the 30th Symposium on Principles of Database Systems (PODS), pp. 117–128 (2011)

Online Control Message Aggregation
in Chain Networks[*]

Marcin Bienkowski[1], Jaroslaw Byrka[1], Marek Chrobak[2], Łukasz Jeż[1,3],
Jiří Sgall[4], and Grzegorz Stachowiak[1]

[1] Institute of Computer Science, University of Wroclaw, Poland
[2] Department of Computer Science, University of California at Riverside, USA
[3] Dept. of Computer, Control, and Management Engineering,
Sapienza University of Rome, Italy
[4] Computer Science Institute, Faculty of Mathematics and Physics, Charles
University, Czech Republic

Abstract. In the Control Message Aggregation (CMA) problem, control
packets are generated over time at the nodes of a tree T and need to
be transmitted to the root of T. To optimize the overall cost, these
transmissions can be delayed and different packets can be aggregated,
that is a single transmission can include all packets from a subtree rooted
at the root of T. The cost of this transmission is then equal to the total
edge length of this subtree, independently of the number of packets that
are sent. A sequence of transmissions that transmits all packets is called
a *schedule*. The objective is to compute a schedule with minimum cost,
where the cost of a schedule is the sum of all the transmission costs and
delay costs of all packets. The problem is known to be $\mathbb{NP}$-hard, even for
trees of depth 2. In the online scenario, it is an open problem whether
a constant-competitive algorithm exists.

We address the special case of the problem when T is a chain network.
For the online case, we present a 5-competitive algorithm and give a
lower bound of $2 + \phi \approx 3.618$, improving the previous bounds of 8 and 2,
respectively. Furthermore, for the offline version, we give a polynomial-
time algorithm that computes the optimum solution.

1 Introduction

In the Control Message Aggregation (CMA) problem, introduced in [6], we are
given a tree T whose edges have positive *lengths*. Over time, packets are generated at the nodes of T. Each packet is specified by a pair (t, v), where t is the
injection time of this packet and $v \in T$ the vertex where the packet is injected.
All packets must be transmitted to the root of T, although not necessarily immediately; to reduce cost, packets can be delayed and different packets can be
aggregated into a single transmission. A transmission is defined as a subtree

---

[*] Research partially supported by NSF grants CCF-1217314 and OISE-1157129,
MNiSW grant no. N N206 368839, 2010-2013, EU ERC project 259515 PAAl, CE-ITI
(project P202/12/G061 of GA ČR), and grant IAA100190902 of GA AV ČR.

F. Dehne, R. Solis-Oba, and J.-R. Sack (Eds.): WADS 2013, LNCS 8037, pp. 133–145, 2013.

rooted at the root of T that transports *all* the packets currently contained in its nodes. The cost of this transmission is then equal to the total edge length of this subtree, independently of the number of packets that are sent. Delaying a transmission of a packet incurs cost equal to the time it waits for the transmission. A *schedule* is specified by a sequence of transmissions that transmit all packets to the root. The cost of a schedule is the sum of its transmission costs and the waiting costs of all packets. The objective is to find a schedule of minimum cost.

In reality, what we refer to as *packets* in the paper are abstractions of control messages acknowledging the receipt of some network packets from a communication stream. This motivates the assumptions of our cost model, in particular the fact that the transmission cost does not depend on the number of packets involved. The reason is that in practice acknowledgement messages are very small, so the cost of including them in the transmission is negligibly small in comparison to the overhead of sending it. More generally, the CMA problem can model transportation problems where large quantities of small items need to be shipped to a common destination, incurring both the transportation and delay costs.

While it is possible to consider the problem in the *offline* scenario, where the algorithm knows the whole input sequence in advance, the *online* model more accurately reflects the constraints that arise in practice. In the online model, packets arrive as time passes, with each packet (t, v) arriving at time t, and at each time step the algorithm needs to decide whether and what to transmit. In the online model, one can further distinguish two versions: distributed algorithms, where the nodes make local decisions independently, and centralized, *full information* algorithms [2], that have complete knowledge about the input sequence so far and the current state.

Previous Work. Khanna et al. [6] defined the CMA problem and gave an $O(\log \alpha)$-competitive online algorithm, where α is the sum of all edge lengths. (Their algorithm works under an additional technical assumption that each packet has to wait at least one time unit.) The CMA problem on a single-edge tree is equivalent to the prominent TCP acknowledgement problem. The online version of TCP acknowledgement has been essentially solved: the optimal competitive ratios for deterministic and randomized algorithms are, respectively, 2 and $e/(e - 1) \approx 1.582$ [4,5].

While finding better bounds on the competitive ratio remains an open problem, $O(1)$-competitive online algorithms were presented for some special classes of trees. For *flat trees*, where the tree has depth two and the root has only one child, the problem becomes equivalent to the Joint Replenishment Problem (JRP) with linear penalties, well studied in Operations Research. For this version, Brito et al. [2] gave a 5-competitive algorithm. The ratio was later improved to 3 by Buchbinder et al. [3], who also presented a lower bound of 2.64. These results apply in fact to arbitrary trees of depth two, as such trees can be decomposed into flat trees that are processed independently. In the offline case, Arkin et al. [1] proved that JRP is $\mathbb{NP}$-hard and a 1.8-approximation algorithm was given by Levi et al. [7,8,9].

Another previously studied case is that of chain networks, where each non-leaf node has exactly one child. This version was introduced by Brito et al. [2], who presented an 8-competitive online algorithm. No online algorithm can be better than 2-competitive in this case, due to the lower bound for the TCP-acknowledgement problem [4].

Our Contributions. Following the work in [2], we study online and offline algorithms for Control Message Aggregation on chain networks. Three results are provided. In the online case, we focus on centralized algorithms. For this model we give a 5-competitive algorithm and we prove a lower bound of $2 + \phi \approx 3.618$ (where $\phi = (1 + \sqrt{5})/2$ is the golden ratio). Both results improve the previously known bounds, described above. As we show, the analysis of our online algorithm is tight. For the offline case, we provide an algorithm that computes an optimal schedule in polynomial time.

2   Preliminaries

Throughout the paper, we think of a chain network as a half-line $\mathbb{R}^+$, consisting of all non-negative real numbers. Slightly abusing the terminology, we will refer to it as *the line*. Any $x \in \mathbb{R}^+$ represents a node of the network at distance x from the origin. Packets arrive over time and need to be transmitted to the destination at point 0. A transmission from x at time t sends all packets from the interval $(0, x]$, and the cost of this transmission is x. (Without loss of generality, we may assume that no packet ever arrives at point 0 as they might be transmitted immediately at zero cost.) We consider continuous time, that is packets can be injected and transmitted at arbitrary real-valued times. We also allow packets to have non-negative weights. The penalty function for waiting is assumed to be linear, that is a packet of weight w injected at time a and transmitted at time t pays the cost $w(t - a)$ for waiting. The weights do not affect the cost of transmissions. Thus, for the purpose of computing cost, w unit-weight packets arriving at the same time are equivalent to one packet of weight w. Note that our lower bounds use only integer weights, and thus can be trivially simulated using non-weighted packets.

3   An Online 5-Competitive Algorithm

In this section, we present our 5-competitive algorithm for Control Message Aggregation on chain networks.

For any $x \in \mathbb{R}^+$ and time t, denote by $\mathsf{twc}_t[x]$ the total waiting cost of the packets currently at x, starting at their arrival until time t. We generalize this notation to intervals: for example $\mathsf{twc}_t(x, y]$ is the total waiting cost in the interval $(x, y]$, that includes y but not x, etc. We will drop subscript t if its value is clear from the context or not relevant. The *length* of a transmission from point x is equal to x. For any time t, $\mathsf{trl}(t)$ denotes the length of a transmission at time t; we let $\mathsf{trl}(t) = 0$ if there is no transmission at t. For the analysis, we define an analogous notion of $\mathsf{trl}_{\mathrm{ADV}}(t)$ for adversarial transmissions.

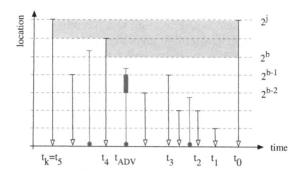

Fig. 1. An example for the analysis of BALANCE's cost at time t_0. The line (the vertical axis) is shown in the logarithmic scale. The algorithm's transmissions end with empty triangles. The adversary's transmissions end with circles. The algorithm's cost at t_0 is charged to the waiting cost of packets within the shaded area and the 2^{b-1}-to-2^{b-2} segment of the adversarial transmission (marked with thick line) at t_{ADV}.

Algorithm BALANCE. At every time t, transmit from the maximum point 2^j such that $\mathsf{twc}_t(0, 2^j] = 2^{j-2}$, if such j exists; otherwise stay idle. (Note that every packet will eventually be transmitted.)

Theorem 1. BALANCE *is 5-competitive.*

Proof. Assume that BALANCE transmits from 2^j at time t_0. Then, it pays 2^j for the transmission and it has paid $\mathsf{twc}(0, 2^j] = 2^{j-2}$ for waiting of the packets it has just transmitted. Altogether, these terms contribute $5 \cdot 2^{j-2}$ to the total cost of BALANCE. We show that it is possible to charge this cost to adversary's actions (waiting or transmitting), whose cost is at least 2^{j-2}, assuring that no action will be charged more than once. This implies that BALANCE is 5-competitive.

To this end, we choose the sequence of BALANCE's transmissions at times $t_1, t_2, \ldots, t_k$, going back in time, with k largest possible, where t_1 is the last transmission time before t_0, each other t_i has a transmission longer than the one at t_{i-1}, and the transmission at t_k has length at least 2^j. (To avoid boundary cases, we assume that there is an artificial transmission at time -1 of infinite length and zero cost, so t_k is always well defined.) Formally, we find a sequence of transmission times $t_k < t_{k-1} < \ldots < t_2 < t_1 < t_0$, such that

1. $\mathsf{trl}(t_k) \geq 2^j$,
2. $\mathsf{trl}(t_{i+1}) > \mathsf{trl}(t_i)$ for $1 \leq i \leq k-1$,
3. $\mathsf{trl}(t) \leq \mathsf{trl}(t_i)$ for $0 \leq i \leq k-1$ and $t \in (t_{i+1}, t_i)$, and $\mathsf{trl}(t) = 0$ for $t \in (t_1, t_0)$.

We call these transmissions the *cover sequence* for t_0, cf. Fig. 1.

Then, we consider adversarial transmissions occurring within the time interval $(t_k, t_0]$. An adversarial transmission occurring at time $t \in (t_{\ell+1}, t_\ell]$ is called *unobstructed* if its length is at least $\mathsf{trl}(t_\ell)$. Let $t_{ADV} \in (t_k, t_0]$ be the time of the longest unobstructed adversary's transmission (with ties broken in favor of later transmissions), and let $A = \mathsf{trl}_{\mathrm{ADV}}(t_{\mathrm{ADV}})$; if there was no unobstructed adversary's transmission in $(t_k, t_0]$, then let $t_{\mathrm{ADV}} = t_0$ and $A = 0$. Finally, let $b = \min\{j, \lfloor \log_2 A \rfloor + 1\}$, where $b = -\infty$ if $A = 0$. Thus, for $A \in (0, 2^j)$, it holds that $2^{b-1} \leq A < 2^b$.

Now we focus on packets at points from $(2^b, 2^j]$ that are transmitted at t_0; their waiting periods are contained in the shaded area in Fig. 1. Let us denote these packets by $W(t_0)$. Note that the waiting cost of $W(t_0)$ is $\mathsf{twc}_{t_0}(2^b, 2^j]$, and that, by the definition of t_{ADV}, the adversary pays for the waiting of these packets at least as much as BALANCE does.

We will charge BALANCE's cost of $5 \cdot 2^{j-2}$ to two actions of the adversary: its waiting of the packets in $W(t_0)$ and the movement of packets from 2^{b-1} to 2^{b-2} during the transmission at t_{ADV} (if there was any). In other words, we charge to the segment $[2^{b-2}, 2^{b-1}]$ of the adversary's transmission at t_{ADV}. As mentioned at the beginning of the proof, it is sufficient to show two properties: (i) the total cost of these two actions is at least 2^{j-2} and (ii) none of these actions is charged again when we analyze another transmission of algorithm BALANCE.

For property (i), we claim that $\mathsf{twc}_{t_0}(2^b, 2^j] \geq 2^{j-2} - 2^{b-2}$. Clearly, this is the case for $b = j$ or $b = -\infty$. For $b < j$, recall that $\mathsf{twc}_{t_0}(0, 2^j] = 2^{j-2}$ and furthermore, $\mathsf{twc}_{t_0}(0, 2^b] \leq 2^{b-2}$ as otherwise the algorithm would have transmitted from 2^b earlier. The adversary cost of transmitting across the segment $[2^{b-1}, 2^{b-2}]$ at time t_{ADV} is 2^{b-2}. Together, these costs add up to at least $2^{j-2} - 2^{b-2} + 2^{b-2} = 2^{j-2}$, as claimed in (i).

To show property (ii), consider a transmission of BALANCE at some time $t'_0 > t_0$ with $\mathsf{trl}(t'_0) = 2^{j'}$. We also consider the corresponding cover sequence for t'_0, the time t'_{ADV} of the longest unobstructed adversarial transmission within that cover sequence, and the corresponding values of $A' = \mathsf{trl}_{\mathrm{ADV}}(t'_{\mathrm{ADV}})$ and b'. $W(t_0)$ and $W(t'_0)$ are disjoint sets of packets since BALANCE transmits them at distinct times t_0 and t'_0. Thus, it suffices to prove that transmissions at t_0 and t'_0 charge their costs to different parts of the adversarial transmissions. Clearly, it is the case when they are charged to different transmissions, so in the following we assume that $t'_{\mathrm{ADV}} = t_{\mathrm{ADV}}$ (and hence $A = A'$). Note that $t_{\mathrm{ADV}} \leq t_0 < t'_0$. Then $j < j'$, as otherwise the cover sequence for t'_0 would end at some time point after t_0. Furthermore, the adversarial transmission at t_{ADV} is unobstructed in the cover sequence for t'_0, which means that $A \geq 2^j$, and hence $b = j$. But then, as $j' > j$, by the definition of b', we have $b' > j$. This means that $b' \neq b$, i.e., the transmissions at t_0 and t'_0 are charged to different parts of the adversarial transmission at $t'_{\mathrm{ADV}} = t_{\mathrm{ADV}}$. □

The analysis of our algorithm can be shown to be tight. In fact, we can prove (the proof will be given in the full version) an even stronger tightness result, namely the following: *Every deterministic algorithm that transmits only from integer powers of 2 has competitive ratio at least 5.*

4   A Lower Bound of $2 + \phi \approx 3.618$

To prove the lower bound of $2+\phi$, it is sufficient to show that for any $R < 2+\phi$ there is a strategy for the adversary that forces any deterministic algorithm ALG to pay at least R times the cost of an optimal solution OPT.

In the next section we show how to construct an adversarial strategy for a slightly modified version of the problem, called the *single-phase game*. In such a game, the adversary injects (weighted) packets only at time 0, at points $b_1 < b_2 < \ldots < b_m$ that will be specified later. Additionally, the adversary has the capability to end the game at an arbitrary time τ. As an algorithm may finish with non-transmitted packets, the definition of cost associated with such a packet q has to be adapted: it is simply the waiting time of q from time 0 till time τ. We define *phase ratio* as the ALG-to-OPT cost ratio with the waiting costs modified as described above. Our single-phase construction has two additional properties, namely that there exist absolute lower and upper bounds on the duration of a game, and there exists an absolute lower bound on the cost of ALG in a single phase.

While the actual adversarial construction of a single-phase game and its analysis are given in the subsequent subsections, here we argue that if the adversary can force the phase ratio to be at least R, then R is a lower bound on the competitive ratio for CMA on chain networks. To this end, the adversary chooses a large integer ℓ, and the actual input sequence consist of $\ell + 1$ phases, numbered $0, 1, \ldots, \ell$. In a phase p, the adversary plays the single-phase game, but with the weights of the packets multiplied by $\ell^{p \cdot \ell}$ and all the time values used in his strategy divided by $\ell^{p \cdot \ell}$. Intuitively, increased weights cause waiting costs to accumulate faster, but we compensate for it by "shrinking" the time. With this rescaling, the adversarial single-phase strategy will also force ratio R in the single-phase game played in phase p. Clearly, the cost of ALG in the whole CMA instance is at least the sum of its costs in the individual single-phase games. (It could be larger if some packets are not transmitted in the phase when they are issued.) On the other hand, since the time intervals of consecutive phases decrease so fast, if a packet is injected at the beginning of phase p, and is not transmitted by the adversary within phase p, then its remaining waiting cost, in phases $p+1, p+2, \ldots, \ell$, is negligible. Finally, all the packets not sent by the adversary by the end of phase ℓ can be sent at that time at the cost of at most b_m, and this cost's contribution is also negligible in comparison to the total cost. Thus, except for a negligible low-order term, the adversary's cost is also the sum of his costs in the individual single-phase games. Therefore the overall cost of ALG is at least R times the adversary's cost, minus a low-order term. (A similar reduction to a single-phase game was used in [3].)

The rest of this section is organized in the following way. In Section 4.1, we present the strategy of the adversary for a single phase. For the construction, the adversary has to carefully choose the number of injected packets m, their injection points $b_1 < b_2 < \ldots < b_m$, and some waiting thresholds $w_1, \ldots, w_m$. We list the desired properties of these sequences and show that if these properties are satisfied, then the phase ratio is at least R. Finally, in Section 4.2, we prove the existence of such sequences.

4.1   Construction of a Single Phase

For the construction, we define the infinite sequences of reals $\{b_i\}$ and $\{w_i\}$, where $b_0 = w_0 = 0$ and $b_1 = 1$. Using notation $B_j = \sum_{i=1}^{j} b_i$ and $W_j = \sum_{i=1}^{j} w_i$ (thus, $B_0 = W_0 = 0$), the two sequences are defined by:

$$w_j = \frac{1}{R-1} \cdot (W_{j-1} + B_j - Rb_{j-1}) \text{ for } j \geq 1 , \qquad (1)$$

$$b_j = R \cdot b_{j-1} + b_{j-2} - B_{j-1} - W_{j-2} \text{ for } j \geq 2 . \qquad (2)$$

From (2),

$$W_{j-1} = Rb_j + b_{j-1} - B_{j+1} \text{ for } j \geq 1 . \qquad (3)$$

Plugging this into (1) and rearranging, we obtain the following useful identity:

$$(R-1)w_j = -b_{j+1} + Rb_j - (R-1)b_{j-1} \quad \text{for } j \geq 1 \qquad (4)$$

The crux of our construction lies in the algebraic property that, for $R \in (2, 2+\phi)$, the sequence $\{b_i\}$ stops increasing at some point. In particular, in Section 4.2, we show the following crucial lemma.

Lemma 1. *For $R \in (2, 2+\phi)$ and the sequences $\{b_j\}, \{B_j\}, \{w_j\}, \{W_j\}$ defined by equations (1) and (2), there exists an integer $m \geq 1$, such that the following properties hold.*

(i) $1 = b_1 < b_2 < \ldots < b_m$ and $b_{m+1} \leq b_m$.
(ii) $w_j \geq 0$ for all $0 \leq j \leq m$.

Note that it may happen that $b_{m+1} < 0$, but this does not affect the validity of our proof, since we do not use b_{m+1} as a packet injection point in the lower bound strategy; in the argument below we only need that $b_{m+1} \leq b_m$.

The adversary chooses m, whose existence is guaranteed by Lemma 1 and a very large integer K. At time 0 the adversary injects a packet of weight K^{m-j} at point b_j, for $j = 1, 2, \ldots, m$. K is chosen to be at least $\max_j(w_{j+1}/w_j)$, which guarantees that the time when the waiting cost for the packet in b_j reaches w_j is an increasing function of j. Given that no further packets are injected within the phase, within this phase ALG will execute a sequence of transmissions from increasing points. Suppose that k is the largest integer such that ALG transmitted from b_k and $k < m$. Then ALG pays for the waiting cost of the packets at $b_{k+1}, ..., b_m$. We will not charge ALG for the waiting cost of packets at $b_{k+2}, ..., b_m$. In fact, in the calculations we will also not charge the adversary for the waiting cost of these packets. This decreases the adversary cost, but since the waiting cost of packets at $b_{k+2}, ..., b_m$ is negligible, this decrease is negligible as well. (A rigorous limit argument will appear in the full version of the paper.)

The Adversarial Strategy. We can assume that ALG transmits each packet if the phase is long enough, for otherwise the cost of ALG would be unbounded. With this assumption, the adversarial strategy is this: Suppose that the last transmission from ALG was from b_{j-1} and let ω be the waiting cost of the packets in b_j when ALG makes the next transmission, from some $b_{j'}$ (where $j' \geq j$). Then, if $j' = j < m$ and $\omega \geq w_j$, the adversary continues the phase. Otherwise, the phase ends.

The intuition is that the adversary tries to force ALG to transmit from points $b_1, b_2, ...$, one by one. Suppose that the adversary ends the game at step j. The algorithm paid both the waiting cost and the transmission cost of the packets at $b_1, ..., b_{j-1}$, while the adversary can perform significantly better by transmitting from b_{j-1} at the beginning of the phase and not paying their waiting cost at all. If ALG skips b_j and transmits from $b_{j'}$, for $j' > j$, then it pays extra transmission cost. On the other hand, if it transmits from b_j before the threshold w_j, the phase will be short, so the adversary can save cost by not transmitting from b_j. The formal argument follows the theorem below.

Theorem 2. *Every deterministic algorithm has competitive ratio at least* $2 + \phi$.

Proof. As explained at the beginning of Section 4, it is sufficient to show that for any $R \in (2, 2 + \phi)$, the phase ratio is at least R. The adversary uses the strategy described above.

Suppose that the phase ends at step j. Then, up to this step, ALG made separate transmissions from $b_1, b_2, ..., b_{j-1}$, and for each of these b_i the waiting cost of the packet at b_i was at least w_i. So the cost of ALG so far is at least $W_{j-1} + B_{j-1}$, plus the cost associated with the packet at b_j. For determining the phase ratio, we consider three possibilities of finishing the phase. Notice that, by Lemma 1, in each case both the enumerator and the denominator are positive. Recall also that, as explained earlier, we can ignore the waiting costs for packets at $b_{j+1}, ..., b_m$.

<u>Case 1</u>: $j' > j$. Note that this implies $j < m$. ALG's cost is $W_{j-1} + B_{j-1} + \omega + b_{j'}$. We consider two options for the adversary: he can transmit from b_{j-1} at time 0 and pay ω for waiting, or he can transmit from b_j. So the ratio is

$$\frac{W_{j-1} + B_{j-1} + \omega + b_{j'}}{\min(b_{j-1} + \omega, b_j)} \geq \frac{W_{j-1} + B_{j-1} + \omega + b_{j+1}}{\min(b_{j-1} + \omega, b_j)}$$

$$\geq \frac{1}{b_j}\left(W_{j-1} + B_{j-1} + b_j - b_{j-1} + b_{j+1}\right)$$

$$= \frac{1}{b_j}\left(W_{j-1} + B_{j+1} - b_{j-1}\right)$$

$$= \frac{1}{b_j}\left(b_{j+1} - b_{j-1} + B_j + W_{j-1}\right) = R,$$

where the last equality follows from (2).

<u>Case 2</u>: $j' = j$ and $\omega < w_j$. ALG's cost is at least $W_{j-1} + B_j + \omega$. The adversary will transmit from b_{j-1} at time 0 and pay the cost of waiting at b_j, paying $b_{j-1} + \omega$. So the ratio is

$$\frac{W_{j-1} + B_j + \omega}{b_{j-1} + \omega} \geq \frac{W_{j-1} + B_j + w_j}{b_{j-1} + w_j} = R,$$

where the last equality follows from (1).

<u>Case 3:</u> $j' = j = m$ and $\omega \geq w_j$. The cost of ALG is $W_{m-1} + B_m + \omega$. The adversary will transmit from b_m at time 0, paying b_m. Note that, by the choice of m we have $b_{m+1} \leq b_m$. By (4), $(R-1)w_m = -b_{m+1} + Rb_m - (R-1)b_{m-1} \geq (R-1)(b_m - b_{m-1})$, that is $w_m \geq b_m - b_{m-1}$. So the ratio is

$$\frac{W_{m-1} + B_m + \omega}{b_m} \geq \frac{W_{m-1} + B_m + w_m}{b_{m-1} + w_m} = R,$$

where the last equality again follows from (1).   □

4.2   Proof of Lemma 1

Proof of Part (i). We first derive a single recurrence for the sequence $\{b_j\}$. Assuming $j \geq 1$, plugging (3) into (2) and simplifying, we obtain the recurrence

$$(R-1)b_{j+2} - (R^2 - R + 1)b_{j+1} + (R^2 - R + 1)b_j = 0 \quad \text{for } j \geq 1. \quad (5)$$

with $b_0 = 0$, $b_1 = 1$ and $b_2 = R - 1$. (Note that an initial condition for $j = 2$ is not covered by (5).) The characteristic equation of (5) is $(R-1)x^2 - (R^2 - R + 1)x + (R^2 - R + 1) = 0$ with discriminant $\Delta = (R^2 - R + 1)(R^2 - 5R + 5)$. In the interval $(2, 2 + \phi)$ the value of Δ is negative, so the characteristic equation has two imaginary conjugate roots

$$\beta_{1,2} = \frac{R^2 - R + 1 \pm \sqrt{\Delta}}{2(R-1)},$$

and consequently

$$b_j = \alpha_1 \cdot \beta_1^j + \alpha_2 \cdot \beta_2^j, \quad j \geq 1, \quad (6)$$

for some complex numbers $\alpha_1, \alpha_2 \neq 0$. From the theory of recurrence equations of order 2 (the case of conjugate imaginary roots), the sequence of $\{b_j\}$ cannot increase infinitely (cf. Appendix A), which proves Part (i) of Lemma 1.

Proof of Part (ii). It is sufficient to show that the right hand side of (4) is non-negative, i.e., that

$$-b_{j+1} + Rb_j - (R-1)b_{j-1} \geq 0 \quad \text{for } j \geq 1. \quad (7)$$

For $j = 1$, the left-hand-side of (7) is $-b_2 + Rb_1 - (R-1)b_0 = -(R-1) + R \cdot 1 - (R-1) \cdot 0 = 1 \geq 0$. Suppose $j \geq 2$ and that the claim holds for $j - 1$. After multiplying the left-hand side of (7) by $R - 1$ and rearranging it, we get

$$(R-1)[-b_{j+1} + Rb_j - (R-1)b_{j-1}]$$
$$= [-(R-1)b_{j+1} + (R^2 - R + 1)b_j - (R^2 - R + 1)b_{j-1}]$$
$$+ [-b_j + Rb_{j-1} - (R-1)b_{j-2}] + (R-1)b_{j-2} \geq 0,$$

The last inequality follows because the first term is 0 by (5), the second one is non-negative by the inductive assumption, and the third one is positive by the choice of m.

5   Polynomial-Time Offline Solution

The input is a sequence of packets numbered $1, 2, ..., n$, where packet j is specified as a triple (t_j, x_j, w_j). In this triple, t_j is the injection time, x_j is the point of injection, and w_j is the weight of packet j. For simplicity, we assume that $t_1 < t_2 < ... < t_n$ and that all x_j are different. Any instance can be modified to have this form by infinitesimal perturbations on the time and space axes.

Without loss of generality, we can assume that in an optimal solution each transmission occurs at some time t_k and it includes the packet injected at this time, that is, it transmits from some point x_j where $x_j \geq x_k$. We can further assume that $j \leq k$, since otherwise x_j itself does not yet have a packet. We call a transmission from x_j at time t_k satisfying these conditions a (j, k)-*transmission*.

For $i < k$ and any j we define a sub-instance $\mathcal{I}_{i,j,k}$ that consists of the triples (t_a, x_a, w_a) such that $i < a < k$ and $x_a < x_j$. We consider the quantity $F_{i,j,k}$ that represents the minimum cost of sub-instance $\mathcal{I}_{i,j,k}$ accrued in the time window (t_i, t_k). To define it formally, we relax the rules to allow some packets in $\mathcal{I}_{i,j,k}$ not to be transmitted. The cost of transmissions is defined as before. The waiting cost of each packet is either the cost of waiting until its transmission, if it gets transmitted, or until time t_k, if it's not. Then $F_{i,j,k}$ is the minimum cost of $\mathcal{I}_{i,j,k}$ under these rules.

We now derive the recurrence equation for the $F_{i,j,k}$'s. For $k = i+1$, $\mathcal{I}_{i,j,k} = \emptyset$, so we have $F_{i,j,k} = 0$. Let $k > i+1$. For any h, let $G^h_{i,j,k}$ be the waiting time for packets from $\mathcal{I}_{i,j,k}$ that are beyond x_h, assuming that they are not transmitted before t_k, cf. Figure 2. Thus

$$G^h_{i,j,k} = \sum_{\substack{i < \ell < k \\ x_h < x_\ell < x_j}} w_\ell(t_k - t_\ell).$$

Slightly abusing notation, we will allow $h = 0$ in the above formula, with x_0 understood to be 0. Thus $G^0_{i,j,k}$ is the total waiting cost of packets in $\mathcal{I}_{i,j,k}$ if there are no transmissions. We then claim that

$$F_{i,j,k} = \min \begin{cases} G^0_{i,j,k} \\ \min_{\substack{i < h \leq g < k \\ x_g \leq x_h < x_j}} \left\{ F_{i,h,g} + x_h + w_h(t_g - t_h) + F_{g,h,k} + G^h_{i,j,k} \right\} \end{cases}$$

The recurrence is illustrated in Fig. 2. To show correctness, we argue as follows. Consider the optimum schedule for $\mathcal{I}_{i,j,k}$. If no transmissions occur in this schedule, then $F_{i,j,k} = G^0_{i,j,k}$. If there is at least one transmission, choose the (h, g)-transmission with maximum x_h. Then all packets ℓ in $\mathcal{I}_{i,j,k}$ above x_h pay the waiting cost for the interval $[t_\ell, t_k]$. The total of these waiting costs is $G^h_{i,j,k}$, the last term in the formula. The cost of the (h, g)-transmission is x_h and we need to pay for the waiting cost of x_h, which is equal $w_h(t_g - t_h)$. These are the second and third terms in the formula. We then need to add the cost of serving the packets of $\mathcal{I}_{i,j,k}$ that are below x_h. These packets constitute two

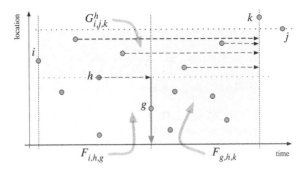

Fig. 2. The idea behind the recurrence for $F_{i,j,k}$

sub-instances, $\mathcal{I}_{i,h,g}$ and $\mathcal{I}_{g,h,k}$, with respective costs $F_{i,h,g}$ and $F_{g,h,k}$, which are the first and the forth terms in the formula.

It remains to show how to use the recurrence to compute the solution of the whole instance. To this end, we modify the instance by adding two new packets of weight 0, injected first and last. Denoting by n the number of packets in this new instance, the original instance will consist of packets injected into $x_2, ..., x_{n-1}$ at times $t_2, ..., t_{n-1}$. We set $t_1 < t_2$, $t_n > t_{n-1}$ with the difference $t_n - t_{n-1}$ large enough, so that in the optimum solution no packets from the original instance will wait until time t_n. Also, $x_1 < x_n$ are both larger than all $x_2, ..., x_{n-1}$. Then, $F_{1,n,n}$ is the same as the cost of the optimum solution of the original instance.

Summarizing, we proved the following result:

Theorem 3. *The optimum solution for message aggregation on chain networks can be computed in time $O(n^5)$.*

6   Final Comments

For the online Control Message Aggregation problem on chain networks we proved that the optimal competitive ratio is between 3.681 and 5. Closing or tightening this gap remains an open problem, although we showed that to improve the upper bound a new approach is needed. We have not addressed the case of randomized algorithms and we leave it for future work. It is intuitively clear that randomization should help to reduce the upper bound. In fact, there are at least two ways to take advantage of randomization: one, by choosing the cutoff points (a sequence other than the powers of 2) randomly, and two, by choosing the transmissions at random.

References

1. Arkin, E., Joneja, D., Roundy, R.: Computational complexity of uncapacitated multi-echelon production planning problems. Operations Research Letters 8(2), 61–66 (1989)

2. Brito, C., Koutsoupias, E., Vaya, S.: Competitive analysis of organization networks or multicast acknowledgement: How much to wait? Algorithmica 64(4), 584–605 (2012)
3. Buchbinder, N., Kimbrel, T., Levi, R., Makarychev, K., Sviridenko, M.: Online make-to-order joint replenishment model: primal dual competitive algorithms. In: Proc. of the 19th ACM-SIAM Symp. on Discrete Algorithms (SODA), pp. 952–961 (2008)
4. Dooly, D.R., Goldman, S.A., Scott, S.D.: On-line analysis of the TCP acknowledgment delay problem. Journal of the ACM 48(2), 243–273 (2001)
5. Karlin, A.R., Kenyon, C., Randall, D.: Dynamic TCP acknowledgement and other stories about e/(e - 1). Algorithmica 36(3), 209–224 (2003)
6. Khanna, S., Naor, J(S.), Raz, D.: Control message aggregation in group communication protocols. In: Widmayer, P., Triguero, F., Morales, R., Hennessy, M., Eidenbenz, S., Conejo, R. (eds.) ICALP 2002. LNCS, vol. 2380, pp. 135–146. Springer, Heidelberg (2002)
7. Levi, R., Roundy, R., Shmoys, D.B.: A constant approximation algorithm for the one-warehouse multi-retailer problem. In: Proc. of the 16th ACM-SIAM Symp. on Discrete Algorithms (SODA), pp. 365–374 (2005)
8. Levi, R., Roundy, R., Shmoys, D.B., Sviridenko, M.: A constant approximation algorithm for the one-warehouse multiretailer problem. Management Science 54(4), 763–776 (2008)
9. Levi, R., Sviridenko, M.: Improved approximation algorithm for the one-warehouse multi-retailer problem. In: Díaz, J., Jansen, K., Rolim, J.D.P., Zwick, U. (eds.) APPROX and RANDOM 2006. LNCS, vol. 4110, pp. 188–199. Springer, Heidelberg (2006)

A   Order 2 Recurrence Equations with Non-Real Base Solutions

For completeness, we prove here that for every real-valued sequence $\{b_i\}$ defined by a recurrence equation of order 2 with non-real base solutions, there exists $n > 1$ such that $b_n \leq 0$. Note that for the sequence from our lower bound construction, this trivially implies the existence of m.

Let $\alpha_1, \alpha_2, \beta_1, \beta_2$ be as defined in (6), and let us assume for convenience that (6) holds for $j = 0$ as well. Since β_1 and β_2 are non-real roots of a quadratic equation, it follows that $\beta_2 = \overline{\beta_1}$, i.e., they are complex conjugates.

In the following, we reason about the imaginary part of (6), which is 0 for all $j \in \mathbb{N}$ by our assumption. Laying $j = 0$, we have

$$\Im(\alpha_1) + \Im(\alpha_2) = 0 \ , \tag{8}$$

and laying $j = 1$, we have

$$\Im(\alpha_1) \cdot \Re(\beta_1) + \Re(\alpha_1) \cdot \Im(\beta_1) + \Im(\alpha_2) \cdot \Re(\beta_2) + \Re(\alpha_2) \cdot \Im(\beta_2) = 0 \ . \tag{9}$$

As $\beta_2 = \overline{\beta_1}$, we have $\Re(\beta_1) = \Re(\beta_2)$. Together with (8) this implies $\Im(\alpha_1) \cdot \Re(\beta_1) + \Im(\alpha_2) \cdot \Re(\beta_2) = 0$, which subtracted from (9) yields

$$\Re(\alpha_1) \cdot \Im(\beta_1) + \Re(\alpha_2) \cdot \Im(\beta_2) = 0 \ .$$

As $\Im(\beta_1) + \Im(\beta_2) = 0$, this implies $\Re(\alpha_1) = \Re(\alpha_2)$, which together with (8) means that $\alpha_2 = \overline{\alpha_1}$.

Thus to get $b_n \leq 0$, it suffices to pick n such that $\arg(\alpha_1 \cdot \beta_1^n) \in [\pi/2, 3\pi/2]$, which is equivalent to $\arg(\alpha_1) + n \cdot \arg(\beta_1) \in [\pi/2 + 2k\pi, 3\pi/2 + 2k\pi]$ for some $k \in \mathbb{N}$. This is possible, since $\beta_1 \notin \mathbb{R}$ means that $\arg(\beta_1)$ is not an integer multiple of π.

Fingerprints in Compressed Strings

Philip Bille[1], Patrick Hagge Cording[1], Inge Li Gørtz[1,*], Benjamin Sach[2],
Hjalte Wedel Vildhøj[1], and Søren Vind[1,**]

[1] Technical University of Denmark, DTU Compute
{phbi,phaco,inge,hwvi,sovi}@dtu.dk
[2] University of Warwick, Department of Computer Science
sach@dcs.warwick.ac.uk

Abstract. The Karp-Rabin fingerprint of a string is a type of hash
value that due to its strong properties has been used in many string
algorithms. In this paper we show how to construct a data structure
for a string S of size N compressed by a context-free grammar of size
n that answers fingerprint queries. That is, given indices i and j, the
answer to a query is the fingerprint of the substring $S[i, j]$. We present the
first $O(n)$ space data structures that answer fingerprint queries without
decompressing any characters. For Straight Line Programs (SLP) we
get $O(\log N)$ query time, and for Linear SLPs (an SLP derivative that
captures LZ78 compression and its variations) we get $O(\log \log N)$ query
time. Hence, our data structures has the same time and space complexity
as for random access in SLPs. We utilize the fingerprint data structures to
solve the longest common extension problem in query time $O(\log N \log \ell)$
and $O(\log \ell \log \log \ell + \log \log N)$ for SLPs and Linear SLPs, respectively.
Here, ℓ denotes the length of the LCE.

1 Introduction

Given a string S of size N and a Karp-Rabin fingerprint function ϕ, the answer
to a FINGERPRINT(i, j) query is the fingerprint $\phi(S[i, j])$ of the substring $S[i, j]$.
We consider the problem of constructing a data structure that efficiently answers
fingerprint queries when the string is compressed by a context-free grammar of
size n.

The fingerprint of a string is an alternative representation that is much shorter
than the string itself. By choosing the fingerprint function randomly at runtime
it exhibits strong guarantees for the probability of two different strings having
different fingerprints. Fingerprints were introduced by Karp and Rabin [20] and
used to design a randomized string matching algorithm. Since then, they have
been used as a central tool to design algorithms for a wide range of problems
(see e.g., [2, 3, 11–13, 15, 16, 19, 22]).

* Supported by a grant from the Danish Council for Independent Research | Natural
Sciences.
** Supported by a grant from the Danish National Advanced Technology Foundation.

F. Dehne, R. Solis-Oba, and J.-R. Sack (Eds.): WADS 2013, LNCS 8037, pp. 146–157, 2013.
© Springer-Verlag Berlin Heidelberg 2013

A fingerprint requires constant space and it has the useful property that given the fingerprints $\phi(S[1, i-1])$ and $\phi(S[1, j])$, the fingerprint $\phi(S[i, j])$ can be computed in constant time. By storing the fingerprints $\phi(S[1, i])$ for $i = 1 \ldots N$ a query can be answered in $O(1)$ time. However, this data structure uses $O(N)$ space which can be exponential in n. Another approach is to use the data structure of Gąsieniec et al. [17] which supports linear time decompression of a prefix or suffix of the string generated by a node. To answer a query we find the deepest node that generates a string containing $S[i]$ and $S[j]$ and decompress the appropriate suffix of its left child and prefix of its right child. Consequently, the space usage is $O(n)$ and the query time is $O(h + j - i)$, where h is the height of the grammar. The $O(h)$ time to find the correct node can be improved to $O(\log N)$ using the data structure by Bille et al. [8] giving $O(\log N + j - i)$ time for a FINGERPRINT(i, j) query. Note that the query time depends on the length of the decompressed string which can be large.

We present the first data structures that answers fingerprint queries on grammar compressed strings without decompressing any characters, and improve all of the above time-space trade-offs. Assume without loss of generality that the compressed string is given as a Straight Line Program (SLP). An SLP is a grammar in Chomsky normal form, i.e., each nonterminal has exactly two children. A Linear SLP is an SLP where the root is allowed to have more than two children, and for all other internal nodes, the right child must be a leaf. Linear SLPs capture the LZ78 compression scheme [27] and its variations. Our data structures give the following theorem.

Theorem 1. *Let S be a string of length N compressed into an SLP G of size n. We can construct data structures that support FINGERPRINT queries in:*

(i) $O(n)$ space and query time $O(\log N)$
(ii) $O(n)$ space and query time $O(\log \log N)$ if G is a Linear SLP

Hence, we show a data structure for fingerprint queries that has the same time and space complexity as for random access in SLPs.

Our fingerprint data structures are based on the idea that a random access query for i produces a path from the root to a leaf labelled $S[i]$. The concatenation of the substrings produced by the left children of the nodes on this path produce the prefix $S[1, i]$. We store the fingerprints of the strings produced by each node and concatenate these to get the fingerprint of the prefix instead. For Theorem 1(i), we combine this with the fast random access data structure by Bille et al. [8]. For Linear SLPs we use the fact that the production rules form a tree to do large jumps in the SLP in constant time using a level ancestor data structure. Then a random access query is dominated by finding the node that produces $S[i]$ among the children of the root, which can be modelled as the predecessor problem.

Furthermore, we show how to obtain faster query time in Linear SLPs using finger searching techniques. Specifically, a finger for position i in a Linear SLP is a pointer to the child of the root that produces $S[i]$.

Theorem 2. *Let S be a string of length N compressed into an SLP G of size n. We can construct an $O(n)$ space data structure such that given a finger f for position i or j, we can answer a* FINGERPRINT(i, j) *query in time $O(\log \log D)$ where $D = |i - j|$.*

Along the way we give a new and simple reduction for solving the finger predecessor problem on integers using any predecessor data structure as a black box.

In compliance with all related work on grammar compressed strings, we assume that the model of computation is the RAM model with a word size of $\log N$ bits.

Longest Common Extension in Compressed Strings. As an application we show how to efficiently solve the longest common extension problem (LCE). Given two indices i, j in a string S, the answer to the LCE(i, j) query is the length ℓ of the maximum substring such that $S[i, i + \ell] = S[j, j + \ell]$. The compressed LCE problem is to preprocess a compressed string to support LCE queries. On uncompressed strings this is solvable in $O(N)$ preprocessing time, $O(N)$ space, and $O(1)$ query time with a nearest common ancestor data structure on the suffix tree for S [18]. Other trade-offs are obtained by doing an exponential search over the fingerprints of strings starting in i and j [7]. Using the exponential search in combination with the previously mentioned methods for obtaining fingerprints without decompressing the entire string we get $O((h + \ell) \log \ell)$ or $O((\log N + \ell) \log \ell)$ time using $O(n)$ space for an LCE query. Using our new (finger) fingerprint data structures and the exponential search we obtain Theorem 3.

Theorem 3. *Let G be an SLP of size n that produces a string S of length N. The SLP G can be preprocessed in $O(N)$ time into a Monte Carlo data structure of size $O(n)$ that supports LCE queries on S in*

(i) $O(\log \ell \log N)$ time
(ii) $O(\log \ell \log \log \ell + \log \log N)$ time if G is a Linear SLP.

Here ℓ denotes the LCE value and queries are answered correctly with high probability. Moreover, a Las Vegas version of both data structures that always answers queries correctly can be obtained with $O(N^2/n \log N)$ preprocessing time with high probability.

We furthermore show how to reduce the Las Vegas preprocessing time to $O(N \log N \log \log N)$ when all the internal nodes in the Linear SLP are children of the root (which is the case in LZ78).

The following corollary follows immediately because an LZ77 compression [26] consisting of n phrases can be transformed to an SLP with $O(n \log \frac{N}{n})$ production rules [9, 23].

Corollary 1. *We can solve the LCE problem in $O(n \log \frac{N}{n})$ space and query time $O(\log \ell \log N)$ for LZ77 compression.*

Finally, the LZ78 compression can be modelled by a Linear SLP G_L with constant overhead. Consider an LZ78 compression with n phrases, denoted $r_1, \ldots, r_n$. A terminal phrase corresponds to a leaf in G_L, and each phrase $r_j = (r_i, a)$, $i < j$, corresponds to a node $v \in G_L$ with r_i corresponding to the left child of v and the right child of v being the leaf corresponding to a. Therefore, we get the following corollary.

Corollary 2. *We can solve the* LCE *problem in* $O(n)$ *space and query time* $O(\log \ell \log \log \ell + \log \log N)$ *for LZ78 compression.*

2   Preliminaries

Let $S = S[1, |S|]$ be a string of length $|S|$. Denote by $S[i]$ the character in S at index i and let $S[i, j]$ be the substring of S of length $j - i + 1$ from index $i \geq 1$ to $|S| \geq j \geq i$, both indices included.

A Straight Line Program (SLP) G is a context-free grammar in Chomsky normal form that we represent as a node-labeled and ordered directed acyclic graph. Each leaf in G is labelled with a character, and corresponds to a terminal grammar production rule. Each internal node in G is labeled with a nonterminal rule from the grammar. The unique string $S(v)$ of length $size(v) = |S(v)|$ is *produced* by a depth-first left-to-right traversal of $v \in G$ and consist of the characters on the leafs in the order they are visited. We let $root(G)$ denote the root of G, and $left(v)$ and $right(v)$ denote the left and right child of an internal node $v \in G$, respectively.

A Linear SLP G_L is an SLP where we allow $root(G_L)$ to have more than two children. All other internal nodes $v \in G_L$ have a leaf as $right(v)$. Although similar, this is not the same definition as given for the Relaxed SLP by Claude and Navarro [10]. The Linear SLP is more restricted since the right child of any node (except the root) must be a leaf. Any Linear SLP can be transformed into an SLP of at most double size by adding a new rule for each child of the root.

We extend the classic *heavy path decomposition* of Harel and Tarjan [18] to SLPs as in [8]. For each node $v \in G$, we select one edge from v to a child with maximum size and call it the *heavy edge*. The remaining edges are *light edges*. Observe that $size(u) \leq size(v)/2$ if v is a parent of u and the edge connecting them is light. Thus, the number of light edges on any path from the root to a leaf is at most $O(\log N)$. A *heavy path* is a path where all edges are heavy. The heavy path of a node v, denoted $H(v)$, is the unique path of heavy edges starting at v. Since all nodes only have a single outgoing heavy edge, the heavy path $H(v)$ and its leaf $leaf(H(v))$, is well-defined for each node $v \in G$.

A *predecessor data structure* supports predecessor and successor queries on a set $R \subseteq U = \{0, \ldots, N - 1\}$ of n integers from a universe U of size N. The answer to a *predecessor query* PRED(q) is the largest integer $r^- \in R$ such that $r^- \leq q$, while the answer to a *successor query* SUCC(q) is the smallest integer $r^+ \in R$ such that $r^+ \geq q$. There exist predecessor data structures achieving a query time of $O(\log \log N)$ using space $O(n)$ [21, 24, 25].

Given a rooted tree T with n vertices, we let $depth(v)$ denote the length of the path from the root of T to a node $v \in T$. A *level ancestor data structure* on T supports *level ancestor queries* $\text{LA}(v, i)$, asking for the ancestor u of $v \in T$ such that $depth(u) = depth(v) - i$. There is a level ancestor data structure answering queries in $O(1)$ time using $O(n)$ space [14] (see also [1,5,6]).

Fingerprinting. The Karp-Rabin fingerprint [20] of a string x is defined as $\phi(x) = \sum_{i=1}^{|x|} x[i] \cdot c^i \mod p$, where c is a randomly chosen positive integer, and $2N^{c+4} \le p \le 4N^{c+4}$ is a prime. Karp-Rabin fingerprints guarantee that given two strings x and y, if $x = y$ then $\phi(x) = \phi(y)$. Furthermore, if $x \ne y$, then with high probability $\phi(x) \ne \phi(y)$. Fingerprints can be composed and subtracted as follows.

Lemma 1. *Let $x = yz$ be a string decomposable into a prefix y and suffix z. Let N be the maximum length of x, c be a random integer and $2N^{c+4} \le p \le 4N^{c+4}$ be a prime. Given any two of the Karp-Rabin fingerprints $\phi(x)$, $\phi(y)$ and $\phi(z)$, it is possible to calculate the remaining fingerprint in constant time as follows:*

$$\phi(x) = \phi(y) \oplus \phi(z) = \phi(y) + c^{|y|} \cdot \phi(z) \mod p$$

$$\phi(y) = \phi(x) \ominus_s \phi(z) = \phi(x) - \frac{c^{|x|}}{c^{|z|}} \cdot \phi(z) \mod p$$

$$\phi(z) = \phi(x) \ominus_p \phi(y) = \frac{\phi(x) - \phi(y)}{c^{|y|}} \mod p$$

In order to calculate the fingerprints of Lemma 1 in constant time, each fingerprint for a string x must also store the associated exponent $c^{|x|} \mod p$, and we will assume this is always the case. Observe that a fingerprint for any substring $\phi(S[i, j])$ of a string can be calculated by subtracting the two fingerprints for the prefixes $\phi(S[1, i - 1])$ and $\phi(S[1, j])$. Hence, we will only show how to find fingerprints for prefixes in this paper.

3   Basic Fingerprint Queries in SLPs

We now describe a simple data structure for answering $\text{FINGERPRINT}(1, i)$ queries for a string S compressed into a SLP G in time $O(h)$, where h is the height of the parse tree for S. This method does not unpack the string to obtain the fingerprint, instead the fingerprint is generated by traversing G.

The data structure stores $size(v)$ and the fingerprint $\phi(S(v))$ of the string produced by each node $v \in G$. To compose the fingerprint $f = \phi(S[1, i])$ we start from the root of G and do the following. Let v' denote the currently visited node, and let $p = 0$ be a variable denoting the size the concatenation of strings produced by left children of visited nodes. We follow an edge to the right child of v' if $p + size(left(v')) < i$, and follow a left edge otherwise. If following a right edge, update $f = f \oplus \phi(S(left(v')))$ such that the fingerprint of the full string generated by the left child of v' is added to f, and set $p = p + size(left(v'))$.

When following a left edge, f and p remains unchanged. When a leaf is reached, let $f = f \oplus \phi(S(v'))$ to include the fingerprint of the terminal character. Aside from the concatenation of fingerprints for substrings, this procedure resembles a random access query for the character in position i of S.

The procedure correctly composes $f = \phi(S[1, i])$ because the order in which the fingerprints for the substrings are added to f is identical to the order in which the substrings are decompressed when decompressing $S[1, i]$.

Since the fingerprint composition takes constant time per addition, the time spent generating a fingerprint using this method is bounded by the height of the parse tree for $S[i]$, denoted $O(h)$. Only constant additional space is spent for each node in G, so the space usage is $O(n)$.

4   Faster Fingerprints in SLPs

Using the data structure of Bille et al. [8] to perform random access queries allows for a faster way to answer FINGERPRINT$(1, i)$ queries.

Lemma 2 ([8]). *Let S be a string of length N compressed into a SLP G of size n. Given a node $v \in G$, we can support random access in $S(v)$ in $O(\log(size(v)))$ time, at the same time reporting the sequence of heavy paths and their entry- and exit points in the corresponding depth-first traversal of $G(v)$.*

The main idea is to compose the final fingerprint from substring fingerprints by performing a constant number of fingerprint additions per heavy path visited.

In order to describe the data structure, we will use the following notation. Let $V(v)$ be the left children of the nodes in $H(v)$ where the heavy path was extended to the right child, ordered by increasing depth. The order of nodes in $V(v)$ is equal to the sequence in which they occur when decompressing $S(v)$, so the concatenation of the strings produced by nodes in $V(v)$ yields the prefix $P(v) = S(v)[1, L(v)]$, where $L(v) = \sum_{u \in V(v)} size(u)$. Observe that $P(u)$ is a suffix of $P(v)$ if $u \in H(v)$. See Figure 1 for the relationship between u, v and the defined strings.

Let each node $v \in G$ store its unique outgoing heavy path $H(v)$, the length $L(v)$, $size(v)$, and the fingerprints $\phi(P(v))$ and $\phi(S(v))$. By forming heavy path trees of total size $O(n)$ as in [8], we can store $H(v)$ as a pointer to a node in a heavy path tree (instead of each node storing the full sequence).

The fingerprint $f = \phi(S[1, i])$ is composed from the sequence of heavy paths visited when performing a single random access query for $S[i]$ using Lemma 2. Instead of adding all left-children of the path towards $S[i]$ to f individually, we show how to add all left-children hanging from each visited heavy path in constant time per heavy path. Thus, the time taken to compose f is $O(\log N)$.

More precisely, for the pair of entry- and exit-nodes v, u on each heavy path H traversed from the root to $S[i]$, we set $f = f \oplus (\phi(P(v)) \ominus_s \phi(P(u)))$ (which is allowed because $P(u)$ is a suffix of $P(v)$). If we leave u by following a right-pointer, we additionally set $f = f \oplus \phi(S(left(u)))$. If u is a leaf, set $f = f \oplus \phi(S(u))$ to include the fingerprint of the terminal character.

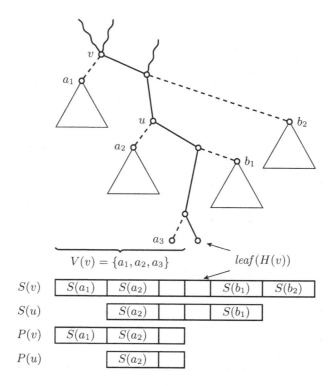

Fig. 1. Figure showing how $S(v)$ and its prefix $P(v)$ is composed of substrings generated by the left children a_1, a_2, a_3 and right children b_1, b_2 of the heavy path $H(v)$. Also illustrates how this relates to $S(u)$ and $P(u)$ for a node $u \in H(v)$.

Remember that $P(v)$ is exactly the string generated from v along H, produced by the left children of nodes on H where the heavy path was extended to the right child. Thus, this method corresponds exactly to adding the fingerprint for the substrings generated by all left children of nodes on H between the entry- and exit-nodes in depth-first order, and the argument for correctness from the slower fingerprint generation also applies here.

Since the fingerprint composition takes constant time per addition, the time spent generating a fingerprint using this method is bounded by the number of heavy paths traversed, which is $O(\log N)$. Only constant additional space is spent for each node in G, so the space usage is $O(n)$. This concludes the proof of Theorem 1(i).

5   Faster Fingerprints in Linear SLPs

In this section we show how to quickly answer FINGERPRINT$(1, i)$ queries on a Linear SLP G_L. In the following we denote the sequence of k children of $root(G_L)$

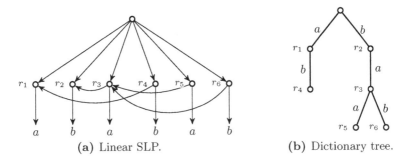

(a) Linear SLP.   (b) Dictionary tree.

Fig. 2. A Linear SLP compressing the string abbaabbaabab and the dictionary tree obtained from the Linear SLP

from left to right by $r_1, \ldots, r_k$. Also, let $R(j) = \sum_{m=1}^{j} size(r_m)$ for $j = 0, \ldots, k$. That is, $R(j)$ is the length of the prefix of S produced by G_L including r_j (and $R(0)$ is the empty prefix).

We also define the dictionary tree F over G_L as follows. Each node $v \in G_L$ corresponds to a single vertex $v^F \in F$. There is an edge (u^F, v^F) labeled c if $u = left(v)$ and $c = S(right(v))$. If v is a leaf, there is an edge $(root(F), v^F)$ labeled $S(v)$. That is, a left child edge of $v \in G_L$ is converted to a parent edge of $v^F \in F$ labeled like the right child leaf of v. Note that for any node $v \in G_L$ except the root, producing $S(v)$ is equivalent to following edges and reporting edge labels on the path from $root(F)$ to v^F. Thus, the prefix of length a of $S(v)$ may be produced by reporting the edge labels on the path from $root(F)$ until reaching the ancestor of v^F at depth a.

The data structure stores a predecessor data structure over the prefix lengths $R(j)$ and the associated node r_j and fingerprint $\phi(S[1, R(j)])$ for $j = 0, \ldots, k$. We also have a doubly linked list of all r_j's with bidirectional pointers to the predecessor data structure and G_L. We store the dictionary tree F over G_L, augment it with a level ancestor data structure, and add bidirectional pointers between $v \in G_L$ and $v^F \in F$. Finally, for each node $v \in G_L$, we store the fingerprint of the string it produces, $\phi(S(v))$.

A query FINGERPRINT$(1, i)$ is answered as follows. Let $R(m)$ be the predecessor of i among $R(0), R(1), \ldots, R(k)$. Compose the answer to FINGERPRINT$(1, i)$ from the two fingerprints $\phi(S[1, R(m)]) \oplus \phi(S[R(m) + 1, i])$. The first fingerprint $\phi(S[1, R(m)])$ is stored in the data structure and the second fingerprint $\phi(S[R(m) + 1, i])$ can be found as follows. Observe that $S[R(m) + 1, i]$ is fully generated by r_{m+1} and hence a prefix of $S(r_{m+1})$ of length $i - R(m)$. We can get r_{m+1} in constant time from r_m using the doubly linked list. We use a level ancestor query $u^F = LA(r_{m+1}^F, i - R(m))$ to determine the ancestor of r_{m+1}^F at depth $i - R(m)$, corresponding to a prefix of r_{m+1} of the correct length. From u_F we can find $\phi(S(u)) = \phi(S[R(m) + 1, i])$.

It takes constant time to find $\phi(S[R(m) + 1, i])$ using a single level ancestor query and following pointers. Thus, the time to answer a query is bounded by the time spent determining $\phi(S[1, R(m)])$, which requires a predecessor query among k elements (i.e. the number of children of $root(G_L)$) from a universe of size N. The data structure uses $O(n)$ space, as there is a bijection between nodes in G_L and vertices in F, and we only spend constant additional space per node in G_L and vertex in F. This concludes the proof of Theorem 1(ii).

6   Finger Fingerprints in Linear SLPs

The $O(\log \log N)$ running time of a FINGERPRINT$(1, i)$ query is dominated by having to find the predecessor $R(m)$ of i among $R(0), R(1), \ldots, R(k)$. Given $R(m)$ the rest of the query takes constant time. In the following, we show how to improve the running time of a FINGERPRINT$(1, i)$ query to $O(\log \log |j - i|)$ given a finger for position j. Recall that a finger f for a position j is a pointer to the node r_m producing $S[j]$. To achieve this, we present a simple linear space finger predecessor data structure that is interchangeable with any other predecessor data structure.

Finger Predecessor. Let $R \subseteq U = \{0, \ldots, N - 1\}$ be a set of n integers from a universe U of size N. Given a finger $f \in R$ and a query point $q \in U$, the *finger predecessor problem* is to answer finger predecessor or successor queries in time depending on the universe distance $D = |f - q|$ from the finger to the query point. Belazzougui et al. [4] present a succinct solution for solving the finger predecessor problem relying on a modification of z-fast tries. Here, we use a simple reduction for solving the finger predecessor problem using any predecessor data structure as a black box. The proof is omitted due to lack of space.

Lemma 3. *Let $R \subseteq U = \{0, \ldots, N - 1\}$ be a set of n integers from a universe U of size N. Given a predecessor data structure with query time $t(N, n)$ using $s(N, n)$ space, we can solve the finger predecessor problem in time $O(t(D, n))$ using space $O(s(N, \frac{n}{\log N}) \log N)$.*

Using the van Emde Boas predecessor data structure [21, 24, 25] with $t(N, n) = O(\log \log N)$ query time using $s(N, n) = O(n)$ space, we obtain the following corollary.

Corollary 3. *Let $R \subseteq U = \{0, \ldots, N - 1\}$ be a set of n integers from a universe U of size N. Given a finger $f \in R$ and a query point $q \in U$, we can solve the finger predecessor problem in time $O(\log \log |f - q|)$ and space $O(n)$.*

Finger Fingerprints. We can now prove Theorem 2. Assume wlog that we have a finger for i, i.e., we are given a finger f to the node r_m generating $S[i]$. From this we can in constant time get a pointer to r_{m+1} in the doubly linked list and from this a pointer to $R(m + 1)$ in the predecessor data structure. If $R(m+1) > j$ then $R(m)$ is the predecessor of j. Otherwise, using Corollary 3 we can in time $O(\log \log |R(m+1) - j|)$ find the predecessor of j. Since $R(m+1) \geq i$ and the rest of the query takes constant time, the total time for the query is $O(\log \log |i - j|)$.

7   Longest Common Extensions in Compressed Strings

Given an SLP G, the longest common extension (LCE) problem is to build a data structure for G that supports longest common extension queries $LCE(i, j)$. In this section we show how to use our fingerprint data structures as a tool for doing LCE queries,and hereby obtain Theorem 3.

7.1   Computing Longest Common Extensions with Fingerprints

We start by showing the following general lemma that establishes the connection between LCE and fingerprint queries.

Lemma 4. *For any string S and any partition $S = s_1 s_2 \cdots s_t$ of S into k non-empty substrings called phrases, $\ell = \mathrm{LCE}(i, j)$ can be found by comparing $O(\log \ell)$ pairs of substrings of S for equality. Furthermore, all substring comparisons $x = y$ are of one of the following two types:*

Type 1 *Both x and y are fully contained in (possibly different) phrase substrings.*

Type 2 *$|x| = |y| = 2^p$ for some $p = 0, \ldots, \log(\ell) + 1$ and for x or y it holds that*
  (a) The start position is also the start position of a phrase substring, or
  (b) The end position is also the end position of a phrase substring.

Proof. Let a position of S be a *start* (*end*) position if a phrase starts (ends) at that position. Moreover, let a comparison of two substrings be of *type 1* (*type 2*) if it satisfies the first (second) property in the lemma. We now describe how to find $\ell = \mathrm{LCE}(i, j)$ by using $O(\log \ell)$ type 1 or 2 comparisons.

If i or j is not a start position, we first check if $S[i, i+k] = S[j, j+k]$ (type 1), where $k \geq 0$ is the minimum integer such that $i + k$ or $j + k$ is an end position. If the comparison fails, we have restricted the search for ℓ to two phrase substrings, and we can find the exact value using $O(\log \ell)$ type 1 comparisons.

Otherwise, $\mathrm{LCE}(i, j) = k + \mathrm{LCE}(i + k + 1, j + k + 1)$ and either $i + k + 1$ or $j + k + 1$ is a start position. This leaves us with the task of describing how to answer $\mathrm{LCE}(i, j)$, assuming that either i or j is a start position.

We first use $p = O(\log \ell)$ type 2 comparisons to determine the biggest integer p such that $S[i, i + 2^p] = S[j, j + 2^p]$. It follows that $\ell \in [2^p, 2^{p+1}]$. Now let $q < 2^p$ denote the length of the longest common prefix of the substrings $x = S[i + 2^p + 1, i + 2^{p+1}]$ and $y = S[j + 2^p + 1, j + 2^{p+1}]$, both of length 2^p. Clearly, $\ell = 2^p + q$. By comparing the first half x' of x to the first half y' of y, we can determine if $q \in [0, 2^{p-1}]$ or $q \in [2^{p-1} + 1, 2^p - 1]$. By recursing we obtain the exact value of q after $\log 2^p = O(\log \ell)$ comparisons.

However, comparing $x' = S[a_1, b_1]$ and $y' = S[a_2, b_2]$ directly is not guaranteed to be of type 1 or 2. To fix this, we compare them indirectly using a type 1 and type 2 comparison as follows. Let $k < 2^p$ be the minimum integer such that $b_1 - k$ or $b_2 - k$ is a start position. If there is no such k then we can compare x' and y' directly as a type 1 comparison. Otherwise, it holds that $x' = y'$ if and only if $S[b_1 - k, b_1] = S[b_2 - k, b_2]$ (type 1) and $S[a_1 - k - 1, b_1 - k - 1] = S[a_2 - k - 1, b_2 - k - 1]$ (type 2). $\qquad\square$

Theorem 3 follows by using fingerprints to perform the substring comparisons. In particular, we obtain a Monte Carlo data structure that can answer a LCE query in $O(\log \ell \log N)$ time for SLPs and in $O(\log \ell \log \log N)$ time for Linear SLPs. In the latter case, we can use Theorem 2 to reduce the query time to $O(\log \ell \log \log \ell + \log \log N)$ by observing that for all but the first fingerprint query, we have a finger into the data structure.

7.2   Verifying the Fingerprint Function

Since the data structure is Monte Carlo, there may be collisions among the fingerprints used to determine the LCE, and consequently the answer to a query may be incorrect. We describe how to obtain a Las Vegas data structure that always answers LCE queries correctly by efficiently verifying that the fingerprint function ϕ is collision-free on all substrings compared in the computation of $LCE(i, j)$. We give two verification algorithms. One that works for LCE queries in SLPs, and a faster one that works for Linear SLPs where all internal nodes are children of the root (e.g. LZ78). Due to lack of space, the details of the algorithms are omitted. They will appear in a full version of this paper.

The verification algorithm for SLPs has running time $O(N^2/n \log N)$ and uses $O(n)$ space. It uses the fingerprints of substrings of size 2^{p-1} to verify fingerprints of substrings of size 2^p similarly to the verification algorithm in [7]. For Linear SLPs where all internal nodes are children of the root, the running time is reduced to $O(N \log N \log \log N)$ while using $O(n)$ space.

References

1. Alstrup, S., Holm, J.: Improved algorithms for finding level ancestors in dynamic trees. In: Welzl, E., Montanari, U., Rolim, J.D.P. (eds.) ICALP 2000. LNCS, vol. 1853, pp. 73–84. Springer, Heidelberg (2000)
2. Amir, A., Farach, M., Matias, Y.: Efficient randomized dictionary matching algorithms. In: Apostolico, A., Galil, Z., Manber, U., Crochemore, M. (eds.) CPM 1992. LNCS, vol. 644, pp. 262–275. Springer, Heidelberg (1992)
3. Andoni, A., Indyk, P.: Efficient algorithms for substring near neighbor problem. In: Proc. 17th SODA, pp. 1203–1212 (2006)
4. Belazzougui, D., Boldi, P., Vigna, S.: Predecessor search with distance-sensitive query time. arXiv:1209.5441 (2012)
5. Bender, M., Farach-Colton, M.: The level ancestor problem simplified. Theoret. Comput. Sci. 321, 5–12 (2004)
6. Berkman, O., Vishkin, U.: Finding level-ancestors in trees. J. Comput. System Sci. 48(2), 214–230 (1994)
7. Bille, P., Gørtz, I.L., Sach, B., Vildhøj, H.W.: Time-space trade-offs for longest common extensions. In: Kärkkäinen, J., Stoye, J. (eds.) CPM 2012. LNCS, vol. 7354, pp. 293–305. Springer, Heidelberg (2012)
8. Bille, P., Landau, G., Raman, R., Sadakane, K., Satti, S., Weimann, O.: Random access to grammar-compressed strings. In: Proc. 22nd SODA, pp. 373–389 (2011)
9. Charikar, M., Lehman, E., Liu, D., Panigrahy, R., Prabhakaran, M., Sahai, A., Shelat, A.: The smallest grammar problem. IEEE Trans. Inf. Theory 51(7), 2554–2576 (2005)

10. Claude, F., Navarro, G.: Self-indexed grammar-based compression. Fundamenta Informaticae 111(3), 313–337 (2011)
11. Cole, R., Hariharan, R.: Faster suffix tree construction with missing suffix links. SIAM J. Comput. 33(1), 26–42 (2003)
12. Cormode, G., Muthukrishnan, S.: Substring compression problems. In: Proc. 16th SODA, pp. 321–330 (2005)
13. Cormode, G., Muthukrishnan, S.: The string edit distance matching problem with moves. ACM Trans. Algorithms 3(1), 2 (2007)
14. Dietz, P.F.: Finding level-ancestors in dynamic trees. In: Dehne, F., Sack, J.-R., Santoro, N. (eds.) WADS 1991. LNCS, vol. 519, pp. 32–40. Springer, Heidelberg (1991)
15. Farach, M., Thorup, M.: String matching in Lempel–Ziv compressed strings. Algorithmica 20(4), 388–404 (1998)
16. Gąsieniec, L., Karpinski, M., Plandowski, W., Rytter, W.: Randomized efficient algorithms for compressed strings: The finger-print approach. In: Hirschberg, D.S., Meyers, G. (eds.) CPM 1996. LNCS, vol. 1075, pp. 39–49. Springer, Heidelberg (1996)
17. Gąsieniec, L., Kolpakov, R., Potapov, I., Sant, P.: Real-time traversal in grammar-based compressed files. In: Proc. 15th DCC, p. 458 (2005)
18. Harel, D., Tarjan, R.E.: Fast algorithms for finding nearest common ancestors. SIAM J. Comput. 13(2), 338–355 (1984)
19. Kalai, A.: Efficient pattern-matching with don't cares. In: Proc. 13th SODA, pp. 655–656 (2002)
20. Karp, R.M., Rabin, M.O.: Efficient randomized pattern-matching algorithms. IBM J. Res. Dev. 31(2), 249–260 (1987)
21. Mehlhorn, K., Näher, S.: Bounded ordered dictionaries in $O(\log \log N)$ time and $O(n)$ space. Inform. Process. Lett. 35(4), 183–189 (1990)
22. Porat, B., Porat, E.: Exact and approximate pattern matching in the streaming model. In: Proc. 50th FOCS, pp. 315–323 (2009)
23. Rytter, W.: Application of Lempel–Ziv factorization to the approximation of grammar-based compression. Theoret. Comput. Sci. 302(1), 211–222 (2003)
24. van Emde Boas, P., Kaas, R., Zijlstra, E.: Design and implementation of an efficient priority queue. Theory Comput. Syst. 10(1), 99–127 (1976)
25. Willard, D.: Log-logarithmic worst-case range queries are possible in space $\Theta(N)$. Inform. Process. Lett. 17(2), 81–84 (1983)
26. Ziv, J., Lempel, A.: A universal algorithm for sequential data compression. IEEE Trans. Inf. Theory 23(3), 337–343 (1977)
27. Ziv, J., Lempel, A.: Compression of individual sequences via variable-rate coding. IEEE Trans. Inf. Theory 24(5), 530–536 (1978)

Beacon-Based Algorithms
for Geometric Routing*

Michael Biro**, Justin Iwerks**,
Irina Kostitsyna***, and Joseph S.B. Mitchell**

Stony Brook University

Abstract. We consider beacons, an analog of geographical greedy routing, motivated by sensor network applications. A beacon b is a point object that can be activated to create a 'magnetic pull' towards itself everywhere in a polygonal domain P. We explore the properties of beacons and their effect on points in polygons, as well as demonstrate polynomial-time algorithms to compute a variety of structures defined by the action of beacons on P. We establish a polynomial-time algorithm for routing from a point s to a point t using a discrete set of candidate beacons, as well as a 2-approximation and a PTAS for routing between beacons placed without restriction in P.

1 Introduction

We consider a model of *beacon-based routing* that generalizes geographical greedy routing in sensor networks. In geographical routing [1, 2], each node is given a Euclidean coordinate and a message is transmitted to the neighbor whose Euclidean distance to the destination is a minimum. When the distribution of sensors is very dense, i.e. close to infinity, the route a message takes under geographical routing will follow a straight line towards the destination, or, when the message hits the network boundary, may follow a boundary edge to greedily minimize the distance to the destination. This is precisely the model of beacon-based routing in this paper (see also [3–6]), where the destination is a beacon. In this context, we demonstrate algorithms to compute all nodes that can transmit a message to a given beacon, to compute all nodes that a given node can transmit to, as well as algorithms to compute an optimal, or nearly-optimal, sequence of beacon locations to transmit messages between two given nodes in the network.

Other routing schemes in sensor networks that are related to beacons include a family of routing methods that use *landmarks*, as in Fang et al., Fonseca et al., and Nguyen et al., [7–9]. In this type of routing, a collection of nodes, called landmarks, first transmit throughout the entire network so that each node

---

* This research was partially supported by the National Science Foundation (CCF-1018388) and the US-Israel Binational Science Foundation (project 2010074).
** Department of Applied Mathematics and Statistics, Stony Brook University, {mbiro,jiwerks,jsbm}@ams.stonybrook.edu
*** Department of Computer Science, Stony Brook University, ikost@cs.stonybrook.edu

F. Dehne, R. Solis-Oba, and J.-R. Sack (Eds.): WADS 2013, LNCS 8037, pp. 158–169, 2013.
© Springer-Verlag Berlin Heidelberg 2013

can record its distance to each landmark. Then, in order to route towards a destination, a function based on the distance vector to the landmarks is used for selecting the neighbor to which to transmit the message next. The paper containing a model most similar to ours is the one adopted in Nguyen et al., [9]. In their paper, the message is routed directly towards a single landmark until the current node is at an equal distance away from the landmark as the destination. At that point another landmark is selected. The paper shows that by choosing the landmarks carefully, the message path's length is within a constant factor of the shortest path.

In our model, a beacon can occupy a point location on the interior or the boundary of P, ∂P. When a beacon is *activated*, we imagine that an object starting at a point $p \in P$ moves along a straight line toward b until it either reaches b or makes contact with ∂P. If contact is made with ∂P, the object will follow along ∂P as long as its straight-line distance to b decreases monotonically. Following the path determined by the beacon, the object may alternate between moving in a straight-line path toward b on the interior of P and following along ∂P. If there is no infinitesimal movement that an object at p can make so that its distance to b (strictly) decreases, we say that the object is 'stuck' and has reached a local minimum or *dead point* on ∂P (see Figure 3). If an object starting at p eventually reaches b we say that b *attracts* p. Two points are *routed* if there is a sequence of beacons that can be activated and then deactivated, one at a time and in order, so that an object beginning at a starting point s would visit each beacon in the sequence after it is activated and terminate at a destination point t, which we will always require to be a beacon itself.

2   Properties of Beacons

We examine the effect of beacons among obstacles in the plane. Our terminology describes the attracted components to be moving objects, however, autonomous robots, message delivery, or geographic routing interpretations are equally valid. We begin with some definitions describing the structures and behavioral properties of beacons in polygons.

We define a *beacon* b as an transmitter-like object that is placed at a point in a polygon P and can be *activated* to effect a pull on objects in P. When b is activated, objects in P move to greedily minimize their Euclidean distance to b, while being constrained to remain interior to P. A beacon b *attracts* a point p if, under the action of b, an object starting at p moves so that its Euclidean distance to b eventually decreases to 0. In this case, we also say that p is attracted to b.

Using these definitions, we may want to determine the set of points that are attracted to a beacon b, called the *attraction region of b*, $A(b)$ or determine the set of beacons that attract a point p, called the *inverse attraction region of p*, $IA(p)$ (See Figure 1). If a beacon b is activated, objects at the points that are attracted to b will reach b, but objects at the points not attracted to b will reach a local minimum with respect to distance to b in P and remain there under the influence of b. We are interested in determining the classification of the points of

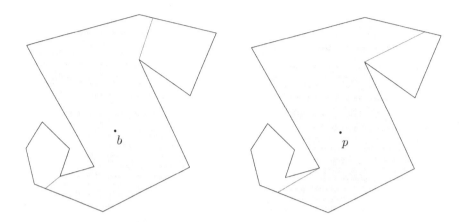

Fig. 1. (Left) The attraction region of a beacon b. (Right) The inverse attraction region of a point p.

P based on the final position of the points under the action of b. These questions motivate the following definitions.

We say a point d in P is a *dead point* with respect to a beacon b if d is not b, and an object at d remains stationary under the influence of the beacon b. That is, d is a point such that the Euclidean distance from b to d is a non-zero local minimum inside of P. For a given beacon b in a polygon P, let $D(b)$ be the set of dead points with respect to b in P.

Then, for each dead point of a beacon b in a polygon P, $d \in D(b)$, define the *dead region* of d with respect to b, $DR_b(d)$, to be the set of points of P that reach d if the beacon at b is activated. Since d is at a local minimum with respect to Euclidean distance to b, if a point reaches d it can never leave d under the action of b.

We can bound the number of dead points a given beacon may have in a polygon P.

Theorem 1. *Let P be a simple polygon on n vertices. If $D(b)$ is the set of dead points with respect to b, then $0 \leq |D(b)| \leq n - 3$. Similarly, let P be a polygon with n vertices and h holes. If $D(b)$ is the set of dead points with respect to b, then $0 \leq |D(b)| \leq n - h - 3$. Furthermore, these bounds are tight.*

Proof. For simple P: If P is convex, then there are no dead points, satisfying the lower bound. Since $b \in P$, at least 3 edges of P must have a point visible to b, which implies that these three edges cannot have any dead points. Since no edge can have more than 1 dead point, this implies that the upper bound of $n - 3$ cannot be exceeded. An example achieving the upper bound is shown in Figure 2.

For arbitrary P: If P is convex and the holes are triangles oriented so they lack dead points, then there are no dead points, satisfying the lower bound. Let H be the set of holes, and let n_i be the number of vertices of the ith hole. First, examine the polygon, $P - H$, with all the holes removed. Polygon $P - H$ has at most $n - \sum_i n_i$ vertices, and since $b \in P$, $P - H$ can have at most $n - \sum_i n_i - 3$ dead points. Now examine a hole $i \in H$, with n_i vertices. Take a planar arrangement consisting of only b and i and note that at least one edge of i must be visible to b, so it does not contribute a dead point. Therefore, i can contribute at most $n_i - 1$ dead points, and all holes together contribute at most $\sum_i (n_i - 1) = (\sum_i n_i) - h$. Adding the two contributions yields $n - \sum_i n_i - 3 + \sum_i n_i - h = n - h - 3$ dead points. An example achieving the upper bound is shown in Figure 2. □

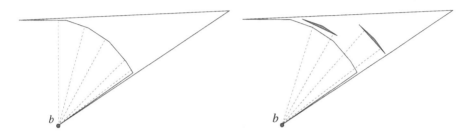

Fig. 2. (Left) A simple polygon with $n = 8$ vertices and $n - 3 = 5$ dead points. (Right) A polygon with $n = 14$ vertices, $h = 2$ holes and $n - h - 3 = 14 - 2 - 3 = 9$ dead points. The four additional dead points are shown.

Lemma 1. *The set of dead regions, $D(b)$, along with the attraction region of b, $A(b)$, forms a partition of the polygon P.*

Proof. We see that every point must eventually either reach b or be forced to stop at a dead point, so these sets cover P. We remove any ambiguity about the movement of a point on a *reflex* vertex (having internal angle greater than π) by assuming it always falls to the left of $\overrightarrow{bp}$. Then, every point follows a unique path induced by the beacon b, as the rules for all possible positions are fixed. Therefore a point cannot end up at two different dead points d_1 and d_2, and so the dead regions and attraction region subdivide the polygon into disjoint regions. Since each point is in a region, this is a partition of the polygon P. □

Using local criteria (see [6] for details), we can determine special *cut vertices*, *split vertices*, and *split edges* that are vital in determining the boundary edges of the partition of a polygon under the action of a beacon. Cut vertices are reflex vertices of the polygon such that the ray emanating from the vertex oriented away from b lies interior to the polygon. There are three classes of cut vertices, depicted in Figure 4, corresponding to the different ways the ray may lie in the polygon. Furthermore, we examine the different cut vertices for situations where the vertex acts as a separator, i.e., where the edges are angled so that points on the left of the ray from b slide away from points on the right of the ray and vice

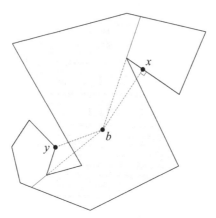

Fig. 3. The partition of P with respect to b. Highlighted is the attraction region $A(b)$ of beacon b; x and y are dead points with respect to b.

versa. These special vertices are the split vertices, and the line segment from them to the first intersection of the ray with the polygon is called the split edge of that split vertex. The far vertex of a split edge is called the *ray vertex* of that split edge (split vertex, respectively). Using these special classes of vertices, we may classify the boundary of the partition of a simple polygon P into an attraction region and dead regions.

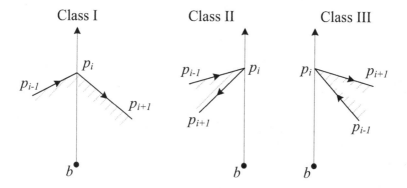

Fig. 4. Three classes of cut vertices

Theorem 2. *If P is a simple polygon, and $e = \overline{p_i q_i}$ is a split edge of b, then e is the boundary between two regions of the partition of P with respect to b. If P has holes, then e may lie entirely interior to a region of the partition, but cannot intersect more than one region of the partition due to b.*

Proof. In the case where P is simple, e is a diagonal of the polygon $P + q_i$ and therefore splits it into two pieces, P_L and P_R. The convention mentioned in Lemma 1 means that the points on e move to the left and so are part of P_L. Since e is parallel to $\overline{bp_i}$, the direct unconstrained action of b can never pull a point from P_L to P_R or vice versa. Therefore, the only possible way for a point to move from one side of e to the other is to move unconstrained until reaching ∂P and then slide along an edge. In order to slide along an edge across e, it must pass p_i, and therefore must slide along $p_i p_{i+1}$, or $p_{i-1} p_i$, depending on whether it started in P_L or P_R. Since p_i is a split vertex, then regardless of which class cut vertex it is, due to the angles defined for a split vertex, a point on edge $p_i p_{i+1}$ or $p_{i-1} p_i$ is pulled away from p_i, and can never reach it. Therefore, since the polygon is simple, the points on one side of e cannot end in the same location as the points on the right side and so are in different regions. Furthermore, e is the boundary of at most two regions as the unconstrained points all travel parallel to their ray from b and end at the same point, meaning that a given ray lies in the same region. Therefore, they are in different regions and e is their boundary.

If P has holes, then the same argument holds, except for the fact that some split edges may have points arbitrarily close on either side that end up in the same location due to points sliding around holes. This corresponds to the split edge lying entirely inside a given region.   □

Conversely, we can classify the boundary edges of the partition of P with respect to b.

Theorem 3. *If a curve is a boundary edge of a region in the partition of a polygon P defined by a beacon b then it is either a part of the boundary of P or a split edge of b.*

Proof. Take a boundary component c of a region that is not part of an edge of P. If some length of c is not parallel to the ray from b, then the unconstrained attraction from b will pull points across c, implying that the two sides share a dead region. Therefore, c must be a straight segment parallel to the ray from b. Now, c must intersect the polygon at two locations, say s_1 and s_2, with s_1 closer to b. All points on c slide down c to s_1 under the influence of b. If s_1 is on the interior of an edge, there are two cases. If s_1 is a dead point, then points on both sides of the edge of s_1 slide to s_1, implying that c is in the interior of the dead region of s_1, contradiction. If s_1 is not a dead point, then it will slide along the edge, either left or right. In both cases, points from both sides of c end at the same dead point, so c is not on a boundary. Therefore, s_1 is a vertex. We see that the conditions that force all points from one side of c to a different region than all points on the other side of c are exactly the conditions that make c a split vertex.

Therefore the boundary edges of regions in the attraction arrangement are exactly the edges of P, dead edges, or edges of the form (p_k, q_i) or (q_i, q_j) for some pair of adjacent ray vertices q_i, q_j.   □

These theorems form the idea for the attraction-region algorithms for points given in the next section. We first find the split vertices of the polygon with respect to the beacon b, then propagate the split edges to find the ray vertices. In simple polygons, this immediately gives the attraction partition of the polygon P with respect to a point beacon b. For polygons with holes, some of the split edges may not be relevant, so we then walk along the boundary of each region, deleting edges seen twice, as they are interior edges. Then, the attraction arrangement is exactly a decomposition of P into polygons each of which either contains a single dead point or, in the case of the attraction region, b. The arrangement therefore consists of the dead regions and the attraction region.

In the following, we give some additional global properties of the attraction region of a point in a polygon P. Specifically, properties of connectedness, convexity, simplicity, and complexity are given for both simple polygons and polygons with holes.

Proposition 1. *Given a beacon b, $b \in A(b)$. Furthermore, the visibility polygon of b, $V(b)$, is a subset of $A(b)$.*

Proof. Each point in the visibility polygon moves to greedily minimize its Euclidean distance to b. Since the straight line path connecting them to b lies in the polygon, that is the path they take, and they are attracted to b. There are also examples where equality is achieved. □

Proposition 2. *The attraction region of a beacon b in a polygon P is connected.*

Proof. Take two arbitrary points in $A(b)$, say p_1 and p_2. Then, b attracts both p_1 and p_2, as well as the entirety of the paths each take under the action of b. The concatenation of these two paths gives a path from p_1 to b to p_2, showing that $A(b)$ is connected. □

Theorem 4. *The attraction region of a beacon b in a simple polygon P is convex with respect to P. This is not necessarily the case if P has holes.*

Proof. A subset R of a polygon P is convex with respect to P if, for any two points p_1 and p_2 in R, either p_1 is not visible to p_2, or the line segment $\overline{p_1 p_2}$ lies entirely in R. See Figure 5. Take a line segment that does not intersect ∂P with endpoints p_1 and p_2 that are both attracted to b. Suppose there exists a point on the segment, p_3, that is not attracted to b. Since p_3 is not attracted to b, p_3 lies outside $A(b)$ and so there exists a split edge that separates p_3 from b. Since the line segment is convex, that split edge must also separate one of p_1 or p_2 from b, and that is a contradiction, as both p_1 and p_2 are attracted to b. In polygons with holes, the line segment may be intersected by many split edges, instead of at most 2 as in simple polygons. Therefore, there can be points on the line segment that are not attracted to b. □

Corollary 1. *The attraction region of a beacon b in a simple polygon P is simple, i.e. it has no holes.*

Theorem 5. *The partition of a polygon P with respect to a beacon b has boundary complexity $O(n)$. Specifically, $A(b)$ has complexity $O(n)$. Furthermore, there are cases with $\Omega(n)$ complexity. See Figure 2.*

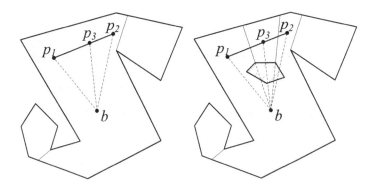

Fig. 5. A beacon's attraction region is convex with respect to a simple polygon and is not necessarily convex with respect to a polygon with holes

Proof. P has n vertices and there are at most $n-3$ additional ray vertices added to make P'. Split edges cannot cross, and so there are at most $2n - 3$ vertices in the partition of P. Therefore, the partition has complexity $O(n)$. Since $A(b)$ is a subset of the partition, it must also have at most $O(n)$ complexity.   □

3   Algorithms for Computing Attraction Regions

Recall the definition of the attraction region of a point: The *attraction region*, $A(b)$, of a beacon b in a polygon P is defined as the set of points p in P that are attracted by b.

We can compute the attraction region, $A(b)$, by various methods; details can be found in [6]. One method is based on a simple rotational sweep, taking time $O(n \log n)$, and space $O(n)$, in simple polygons and polygons with holes. Another method is based on preprocessing P for ray-shooting queries, resulting in time $O(n \log n)$ in a simple polygon and $O(\sqrt{hn} \log n)$ in a polygon with h holes. Alternatively, the ray shooting approach can be done in time $O(n)$ in a simple polygon and $O(nh)$ in a polygon with h holes by using a triangulation to propagate only the split edges relevant to the attraction region, exploiting the connectedness and convexity properties of the attraction region.

Our overall most efficient method of computing $A(b)$ yields a running-time bound of $O(T(n) + n)$, where $T(n)$ is the time to triangulate the given polygon ($O(n)$ in simple polygons, and $O(n + h \log^{1+\epsilon} h)$, respectively for simple polygons and polygons with holes [10,11]). This is (nearly) optimal time, yet the difficulty of implementation may make the prior algorithms more suitable for practical use.

This algorithm computes the necessary split edges by finding a *radial trapezoidization* of the polygon P emanating from the beacon b. This trapezoidization may be computed from a triangulation of P in linear time, by a modification of the result on parallel trapezoidization by Fournier and Montuna [12]. See also the thesis of Mouawad [13] for a description of the modification from parallel to radial trapezoidization. Once the trapezoidization is found, the split edges with respect to b may be found quickly, and after a linear amount of additional work as described above, we compute the full partition of P with respect to b in $O(T(n) + n)$ time and $O(n)$ space.

Theorem 6. *The partition of a polygon P with respect to a beacon b can be computed in $O(n)$ time and space if P is simple, and $O(n + h \log^{1+\epsilon} h)$ time, $O(n)$ space if P has holes.*

4 Algorithms for Computing Inverse Attraction Regions

Recall the definition of the inverse attraction region of a point or region: The *inverse attraction region*, $IA(p)$, of a point p in a polygon P is defined as the set of beacon locations b in P that attract p. Similarly, the *inverse attraction region* of a subset R, $IA(R)$, in a polygon P is defined as the set of points p in P such that a beacon b at p attracts at least one point of R.

In this section we discuss the computation of the inverse attraction region of a point p in a polygon P, as well as the inverse attraction region of a subset of P. Note that inverse attraction regions, unlike attraction regions, may have $\Omega(n)$ connected components ($\Omega(n^2)$ components in polygons with holes) and the components may be free-floating in the interior of the polygon, with boundary edges defined by non-local conditions. This makes their computation more difficult, and we resort to a decomposition approach that determines an arrangement that contains the inverse attraction, then test each face of the arrangement for attraction, using the algorithms in the preceding section. (see [6] for details).

4.1 Algorithm for the Inverse Attraction Region of a Point

The algorithm begins by constructing an arrangement $\mathcal{A}_p$ made from taking the arrangement of lines defined by each edge of the polygon and the lines through each reflex vertex that are perpendicular to the edges incident on the reflex vertex. Then, using the properties of split vertices, we can prove the following results.

Lemma 2. *If b_1 and b_2 are two points in a face F of the arrangement $\mathcal{A}_p$ and p_i is a split vertex relative to b_1, then p_i is a split vertex relative to b_2.*

This allows us to show that the faces of the arrangement are constant with respect to attracting p.

Theorem 7. *If b_1 and b_2 are two points in a face F of the arrangement $\mathcal{A}_p$ and $p \in A(b_1)$, then $p \in A(b_2)$.*

Therefore, we can test a candidate point from each face of the constructed arrangement, using the algorithms discussed in the previous section, and determine which faces make up the inverse attraction region. Walking through the arrangement allows us to update the attraction regions quickly, so this algorithm runs in $O(n^2)$ time.

Theorem 8. *The inverse attraction region of a point p in a polygon P can be computed in $O(n^2)$ time.*

4.2   Algorithm for the Inverse Attraction Region of a Region R

The following algorithm is a modification of the preceding algorithm, and works to compute the inverse attraction region of a polygonal region R with $|R| = m$. It uses the same decomposition idea, but with a slightly more refined arrangement, $\mathcal{A}_R$, made of the lines defined by each edge of the polygon, the lines through each reflex vertex perpendicular to the edges incident to the reflex vertex, and the lines from each vertex of R through each reflex vertex of P.

This allows us to show that the faces of the arrangement $\mathcal{A}_R$ are constant with respect to attracting a point from R.

Theorem 9. *If b_1 and b_2 are two points in a face F of the arrangement $\mathcal{A}_R$ and $R \cap A(b_1) \neq \emptyset$, then $R \cap A(b_2) \neq \emptyset$.*

Therefore, we can test a candidate point from each face of the constructed arrangement, using the attraction region algorithms from the previous section, and determine which faces make up the attraction region of R. Walking through the arrangement allows us to update the attraction regions quickly, so this algorithm runs in $O(m^2 n^2)$ time.

Theorem 10. *The inverse attraction region of a region R with $|R| = m$, in a polygon P, can be computed in $O(m^2 n^2)$ time.*

Later, we will use this algorithm for computing the inverse attraction region of a triangle, which takes $O(n^2)$ time.

Corollary 2. *The inverse attraction region of a triangle in a polygon P can be computed in $O(n^2)$ time.*

5   Beacon Routing

We are interested in finding a *minimum beacon path* between two points s, t, in a polygon P. A minimum beacon path from s to t is the smallest possible collection of points $b_1, b_2, \ldots, b_k$ in P with the property that b_1 attracts s, b_{i+1} attracts b_i for $i = 1, \ldots, k-1$, and t attracts b_k. Specifically, in this section, we solve the minimum beacon path problem for a special case where we are given a set of m candidate beacon locations, and we also approximate the solution in the general case.

5.1   Algorithm for Minimum Beacon Routing with Candidate Beacons

If we are given a collection of m candidate beacon locations, the minimum beacon path algorithm constructs a digraph G whose vertices are the candidate locations and which has the edge $\overrightarrow{(u,v)}$ if $u \in A(v)$. The minimum beacon path is then given by the shortest $s - t$ path in G.

Theorem 11. *A minimum beacon path from s to t, chosen from a set of m candidate locations in a polygon P, can be found in time $O(mn + m^2)$ for simple polygons, and $O(m(n + h \log^{1+\epsilon} h + m \log h))$ for polygons with holes.*

Proof. Correctness follows from the one-to-one correspondence between $s - t$ beacon paths and $s - t$ paths in the graph G.

For simple polygons, the main contributor to the running-time is constructing G, where for each of the $m + 2$ point in C, we spend $O(n)$ computing the attraction region, and spend $O(n + m)$ determining which edges to include in G. This yields a total running-time of $O(m(n+m)) = O(nm + m^2)$. Computing the triangulation, the point location, and the shortest-path algorithm are all dominated by this running-time, so the total running-time is $O(nm + m^2)$.

For polygons with holes, the running time is increased, as for each of the $m+2$ points in C, we spend $O(n + h \log^{1+\epsilon} h)$ computing the attraction regions, and then spend $O(n + m \log h)$ to locate the candidates in the triangles. Again, the triangulation, point-location, and shortest-path algorithms are dominated by this running-time, so the total is $O(m(n + h \log^{1+\epsilon} h + m \log h))$ □

5.2   Approximation Algorithm for Minimum Beacon Routing

We also approximate the minimum beacon path by finding the inverse attraction regions of the set of triangles in a triangulation, then building a digraph G whose vertices correspond to triangles (along with s and t) and which has edge $\overrightarrow{(u,v)}$ if $u \cap IA(v) \neq \emptyset$. Then, a minimum s, t path in G corresponds to a sequence of triangles where a point in each triangle is attracted by a point in its successor triangle in the path. These points may not be the same, but a single additional beacon per triangle allows us to link the sequence of points together to yield a 2-approximation to the minimum beacon path.

Theorem 12. *A 2-approximation for the minimum beacons path from s to t can be found in time $O(n^3)$. This procedure can be iterated to achieve a polynomial-time approximation scheme for minimum beacon paths.*

Proof. Since the minimum beacon $s - t$ path is at least as long as the minimum path in G, and we use at most two beacons in our beacon path for each beacon in the minimum path in G, we have at most twice as many beacons as necessary. The running time is dominated by the computing of the inverse attraction region of triangles. By Corollary 2, this takes $O(n^2)$ per triangle, so $O(n^3)$ total. We can then find the attracted/attracting points by walking through the path starting from t, computing attraction regions for the points already determined. The minimum beacon path has length at most $O(n)$, [3], and so we spend at most $O(n^2)$, or $O(n^2 + nh \log^{1+\epsilon} h)$, to do so.

By increasing the number of iterations, (i.e $IA(IA(\dots(IA(a)))) = IA^k(a)$) we modify the above algorithm to find a beacon path that has at most $k + 1$ beacons for every k beacons in the minimum beacon path, yielding a PTAS.   □

References

1. Bose, P., Morin, P., Stojmenović, I., Urrutia, J.: Routing with guaranteed delivery in ad hoc wireless networks. Wireless Networks 7(6), 609–616 (2001)
2. Karp, B., Kung, H.: GPSR: Greedy perimeter stateless routing for wireless networks. In: Proceedings of the 6th Annual International Conference on Mobile Computing and Networking, pp. 243–254. ACM (2000)
3. Biro, M., Gao, J., Iwerks, J., Kostitsyna, I., Mitchell, J.: Beacon-based routing and coverage. In: 21st Fall Workshop on Computational Geometry (2011)
4. Biro, M., Gao, J., Iwerks, J., Kostitsyna, I., Mitchell, J.: Beacon-based structures in polygonal domains. In: CG:YRF 2012, Abstracts of the 1st Computational Geometry: Young Researchers Forum (2012)
5. Iwerks, J.: Combinatorics and complexity in geometric visibility problems. Dissertation, Stony Brook University (2012)
6. Biro, M.: Beacon-based routing and guarding. Dissertation, Stony Brook University (2013)
7. Fang, Q., Gao, J., Guibas, L., de Silva, V., Zhang, L.: GLIDER: Gradient landmark-based distributed routing for sensor networks. In: Proceedings of the IEEE 24th Annual Joint Conference of the IEEE Computer and Communications Societies, INFOCOM 2005, vol. 1, pp. 339–350. IEEE (2005)
8. Fonseca, R., Ratnasamy, S., Zhao, J., Ee, C., Culler, D., Shenker, S., Stoica, I.: Beacon vector routing: Scalable point-to-point routing in wireless sensornets. In: Proceedings of the 2nd Conference on Symposium on Networked Systems Design & Implementation, vol. 2, pp. 329–342. USENIX Association (2005)
9. Nguyen, A., Milosavljevic, N., Fang, Q., Gao, J., Guibas, L.: Landmark selection and greedy landmark-descent routing for sensor networks. In: 26th IEEE International Conference on Computer Communications, INFOCOM 2007, pp. 661–669. IEEE (2007)
10. Chazelle, B.: Triangulating a simple polygon in linear time. Discrete & Computational Geometry 6(1), 485–524 (1991)
11. Bar-Yehuda, R., Chazelle, B.: Triangulating disjoint jordan chains. International Journal of Computational Geometry and Applications 4(4), 475–481 (1994)
12. Fournier, A., Montuno, D.Y.: Triangulating simple polygons and equivalent problems. ACM Transactions on Graphics (TOG) 3(2), 153–174 (1984)
13. Mouawad, N.: Minimal obscuring sets. Master's thesis, McGill University (1990)

Interval Selection
with Machine-Dependent Intervals

Kateřina Böhmová[1], Yann Disser[2], Matúš Mihalák[1], and Peter Widmayer[1]

[1] Institute of Theoretical Computer Science, ETH Zürich, Zürich, Switzerland
{katerina.boehmova,matus.mihalak,widmayer}@inf.ethz.ch
[2] Department of Mathematics, TU Berlin, Berlin, Germany
disser@math.tu-berlin.de

Abstract. We study an offline interval scheduling problem where every job has exactly one associated interval on every machine. To schedule a set of jobs, exactly one of the intervals associated with each job must be selected, and the intervals selected on the same machine must not intersect. We show that deciding whether all jobs can be scheduled is NP-complete already in various simple cases. In particular, by showing the NP-completeness for the case when all the intervals associated with the same job end at the same point in time (also known as just-in-time jobs), we solve an open problem posed by Sung and Vlach (J. Sched., 2005). We also study the related problem of maximizing the number of scheduled jobs. We prove that the problem is NP-hard even for two machines and unit-length intervals. We present a 2/3-approximation algorithm for two machines (and intervals of arbitrary lengths).

Keywords: Intervals, Scheduling, Complexity, Approximation.

1 Introduction

We consider an interval scheduling problem with m machines and n jobs. A job consists of m open intervals—each associated with exactly one machine. In other words, each job has exactly one interval on each machine. To *schedule* a job, exactly one of its intervals must be selected. To schedule several jobs, no two selected intervals on the same machine may intersect. The goal is to schedule the maximum number of jobs. We will refer to this problem as INTERVALSELECTION.

The presented problem (much like general interval scheduling problems) is motivated by several applications, see, e.g., [2,5,6]. Our motivation comes from the area of car-sharing where a set of users (jobs) wish to reserve a car (machine) for a certain amount of time (interval), sufficiently large to drive to an appointment location (specific to each user) and back. The distance of the parking place of each car to the destination may vary, and this results, for each user, in various time intervals for the cars.

In the special case of a single machine, our problem becomes the classical interval scheduling problem which is solvable in polynomial time by a simple greedy algorithm that considers the intervals in increasing order of their right

F. Dehne, R. Solis-Oba, and J.-R. Sack (Eds.): WADS 2013, LNCS 8037, pp. 170–181, 2013.
© Springer-Verlag Berlin Heidelberg 2013

end-points. For the case of two machines, it can be decided in polynomial time whether all jobs can be scheduled (by a reduction to 2-SAT). In contrast to this, in the present paper we show that the same question is NP-complete for the case of three machines. Moreover, we show that the problem of maximizing the number of scheduled jobs is NP-hard already for two machines. Both results hold even if all the intervals have unit length.

We also consider variants of INTERVALSELECTION where all intervals of the same job, when seen on the real line, have a non-empty intersection (e.g., this would be the time around the user's appointment in the mentioned car-sharing application). We call such a non-empty intersection a *core* of a job. We refer to INTERVALSELECTION where each job has a core as INTERVALSELECTION *with cores*. A special case of such a variant is when all intervals of a job have the same end-point (so called just-in-time jobs [16]). We show that, in this setting, the problem of deciding whether all jobs can be scheduled is NP-complete. This solves an open problem posed by Sung and Vlach [14,16]. If the cores do not have to be at the right-end of the intervals, we show that deciding whether all jobs can be scheduled is NP-complete already when all intervals have unit length.

Our problem can be seen as a special case of the *job interval selection problem*, denoted as JISP_k, where each job has k associated intervals on the real line. To see the relation, consider the machines of an instance of INTERVALSELECTION in any order, and just concatenate the intervals for the machines along the real line, thus creating an instance of JISP_m. JISP_k is APX-hard for any $k \geq 2$, and only a deterministic $1/2$-approximation algorithm is known (in fact, a simple greedy algorithm) [15], and a randomized $\approx \frac{e-1}{e}$-approximation algorithm [5] that gives a $3/4$-approximation for JISP_2. We present a simple deterministic $2/3$-approximation algorithm for INTERVALSELECTION with two machines. Thus, our algorithm is the first deterministic algorithm for a non-trivial special case of JISP_2 that beats the barrier of 2.

Table 1 provides an overview of the known (white background) and new (grey background) complexity results for INTERVALSELECTION and related problems. The columns distinguish three basic computational goals: scheduling *all* jobs, the *maximum number* of jobs, or jobs of *maximum weight*. Each row, from top to bottom, is a generalization of the problem in the previous row, starting with INTERVALSELECTION on a single machine, and ending with JISP_k. As can be seen from the table, the (general) INTERVALSELECTION, denoted as "no core required" in the table, is closely related to well-known and studied problems: it offers a natural generalization of the setting "with cores" [14,16], and it is an interesting special case of JISP_k [5,6,15]. Previous work left a gap in the understanding of the complexity of the problems (the grey areas in the table), which we address and completely close in this paper. To achieve tight hardness results for the boundary cases of 2 and 3 machines (for the decision variant), or 1 and 2 machines (for the maximization objective), we devise gadgets that we plug together using known results on a specific graph coloring problem (solvable in polynomial time), which might be of independent interest. Notably, where meaningful, the hardness results hold even if all intervals are of unit length.

Table 1. Summary of the complexity of INTERVALSELECTION problems with n jobs, and m machines. The cells in gray indicate our contribution.

	Schedule all jobs	Max # jobs	Max $\sum$ weights
single machine	$O(n \log n)$	$O(n \log n)$	$O(n \log n)$
identical intervals per job	$O(n \log n)$	$O(n \log n)$	$O(n^2 \log n)$
with cores, any m	**NP-complete** † §	NP-hard † §	NP-hard † §
	$O(mn^{m+1})$	$O(mn^{m+1})$	$O(mn^{m+1})$
no core required	**NP-complete** †	NP-hard †	NP-hard †
2 machines	$O(n^2)$	**NP-hard** †	NP-hard †
≥ 3 machines	**NP-complete** †	NP-hard †	NP-hard †
JISP$_k$ (single machine)			
2 intervals per job	$O(n^2)$	NP-hard†	NP-hard†
≥ 3 intervals per job	NP-complete†	NP-hard†	NP-hard†

§ even if   † even if all intervals have unit length
 – all cores at the end, or
 – all cores in the middle

Related Work. The general interest in interval scheduling problems dates back to the 1950s. The classical variant, in which each job has associated an interval and can be scheduled on any of the machines (i.e., in our setting, each job has exactly the same interval on every machine) and the goal is to decide whether all the jobs can be scheduled, is polynomially solvable [1]. The maximization version is polynomially solvable as well, even if the jobs are weighted [4]. However, Arkin and Silverberg [1] showed that if each job can only be scheduled on a subset of the m machines, the problem becomes NP-hard (even in the unweighted case). They also gave a $O(n^{m+1})$-time algorithm (i.e., polynomial for a constant m).

The special case of our problem with just-in-time jobs (i.e., where all intervals of a job have the same right end point) has been studied by Sung and Vlach [16]. They showed that the weighted version is NP-hard and presented a dynamic programming algorithm that solves the problem in time $O(m \cdot n^{m+1})$. Settling the complexity of the problem with unit-weight jobs was posed as an open problem [16]; this open problem has also been stated in a recent survey on just-in-time job scheduling [14].

As outlined beforehand, our problem is a special class of JISP$_k$ (job interval scheduling problem on a single machine with k intervals per job). Nakajima and Hakimi [11] showed that the decision version of JISP$_3$ is NP-complete. Keil [8] showed that this is the case even if the intervals have the same length, while the general decision version of JISP$_2$ can be solved in polynomial time. The maximization version has been studied as outlined earlier by Spieksma [15] and Chuzhoy [5]. Erlebach and Spieksma [6] consider the weighted JISP$_k$ with more than one machine (every job has the same set of k intervals on every machine) and they study myopic (single-pass) greedy algorithms.

JISP$_k$ is, in some sense, a discrete variant of the throughput-maximization problem (also known as the time-constrained scheduling problem, or the real-time scheduling problem), in which each job has a length, a release time, and a deadline, and a job is associated with the (infinite) set of intervals of given length lying between the job's release time and the deadline. Bar-Noy et al. [2] study this problem and give the currently best approximation algorithms for most of the existing variants of the problem.

There are many other, for the scope of the paper less relevant variants of scheduling where intervals "come into play". We refer to the survey by Kolen et al. [9] for more information on the topic. We also stress that online variants of the presented problems have been studied as well, see e.g., the recent paper of Sgall [13] on online throughput maximization.

2 Approximation of Interval Selection on Two Machines

In this section we present a 2/3-approximation algorithm for INTERVALSELEC-
TION with two machines. We stress that by *interval* we understand a time interval associated with both a job, and a machine. Recall that INTERVALSELECTION on one machine is solvable by a simple greedy algorithm that considers all intervals on the machine sorted by the right end-points in the ascending order and selects each considered interval if it does not intersect any of the previously selected intervals. We denote this algorithm by A^1. We can also apply the greedy algorithm in the setting with two machines M_A and M_B. More formally, let $A^2(M_A, M_B)$ be the algorithm that first runs A^1 on machine M_A, removes from M_B the intervals for jobs whose intervals were selected on machine M_A, and runs A^1 on M_B. This algorithm gives a 1/2-approximation [15], which is tight for the algorithm.

Obviously, we can run the greedy algorithm in the other direction, i.e., first on M_B and then on M_A (denoted by $A^2(M_B, M_A)$), which again gives a 1/2-approximation. Perhaps surprisingly, the algorithm that chooses the better solution of the two provided by $A^2(M_A, M_B)$ and $A^2(M_B, M_A)$ is a 2/3-approximation. Even though the algorithm, let us call it A^3, is extremely simple, the analysis thereof is more interesting.

Consider an optimum solution O where O_A denotes the intervals selected on M_A and O_B the intervals selected on M_B. Consider $A^2(M_A, M_B)$ and let S_A be the intervals selected by $A^2(M_A, M_B)$ on M_A. Obviously, $A^2(M_A, M_B)$ selects on M_A at least $|O_A|$ intervals (which follows from the fact that A^1 finds an optimum on a single machine). The only reason that A^2 selects less than $|O_B|$ intervals on M_B is that it cannot select intervals that correspond to jobs already scheduled on M_A (see Figure 1 for illustration). In fact, every job scheduled on machine M_A prevents selecting one interval on M_B (the one that corresponds to the same job) and each such selected interval on M_A can cause that we can select one interval less on M_B. We introduce the following definition to measure how a selection S_A on M_A reduces the size of the solution on M_B with respect to O. We say that a set I of intervals *reduces the selection on M_B by k* if after selecting the intervals I on M_A the algorithm A^1 selects $|O_B| - k$ intervals on M_B. Note that a set I can never reduce the selection by more than $|I|$ intervals; in particular, a single interval can reduce the selection by at most one.

In Figure 1, the interval for job β_1 on M_A reduces the selection on M_B by one, but the interval for job α_1 on M_A reduces the selection on M_B by one

Fig. 1. Instance where A^3 returns exactly $2/3 \cdot |O|$ jobs: O contains all jobs α_i and β_i for $i = 1, 2, 3$ (in grey), but both $A^2(M_A, M_B)$ and $A^2(M_B, M_A)$ schedule only the jobs $\alpha_1, \alpha_2, \beta_1, \beta_2$.

only with the help of β_2 on M_A. That is, sometimes we need more than one interval to reduce the selection by one. Accordingly, we will further distinguish the intervals in S_A as follows. S_A^O are the intervals that are both in S_A and in O_A. Observe that every interval $i_O \in O_A \setminus S_A$ has an interval $i_A \in S_A$ such that its right end-point intersects i_O. For each such i_O we place the leftmost such interval i_A in the set $S_A^\cap$. We define $S_A^\emptyset$ to be the remaining intervals of S_A. Note that, by definition, $|S_A^O \cup S_A^\cap| = |O_A|$. Similarly, we define S_B to be the intervals scheduled by the "reverse" algorithm $A^2(M_B, M_A)$ on M_B, and we analogically define the sets S_B^O, $S_B^\cap$, $S_B^\emptyset$.

Intuitively, if $S_A^\cap$ or $S_B^\cap$ is small, then the choice of $A^2(M_A, M_B)$ or $A^2(M_B, M_A)$ on the first machine reduces the selection on the second machine only a little (and thus it schedules many jobs). On the other hand, if both $S_A^\cap$ and $S_B^\cap$ are large, we need to select twice as many jobs to reduce the selection. We will show that the trade-off between these constraints lies at $|S_A^\cap| = 1/3 \cdot |O|$. To make this formal, we analyze how much the selection S_A reduces the selection on M_B.

Lemma 1. *Assume that $A^2(M_A, M_B)$ selects r intervals on M_A corresponding to jobs from S_B^O, s intervals corresponding to jobs from $S_B^\cap$, and t intervals corresponding to jobs from $O_B \setminus S_B^O$. Then the selection on M_B is reduced by at most $r + \min\{s, t\}$.*

Proof. Observe that O_B and $S_B^O \cup S_B^\cap$ are two selections of size $|O_B|$ having exactly S_B^O in common. Now, it is enough to realize that, after removing the intervals corresponding to jobs in S_A, we can select $|O_B| - r - t$ intervals from O_B, and we can select $|O_B| - r - s$ intervals from $S_B^O \cup S_B^\cap$.   □

Theorem 1. *A^3 is a 2/3-approximation algorithm. This bound is tight for the algorithm.*

Proof. Without loss of generality, we assume that $|S_A^\cap| \leq |S_B^\cap|$. We distinguish two cases. First, assume that $|S_A^\cap| \leq 1/3 \cdot |O|$. Since S_A^O are the intervals from O, they correspond to different jobs than the jobs to which the intervals in O_B correspond. Thus, on M_A, at most $|S_A^\cap| + |S_A^\emptyset|$ intervals corresponding to jobs in O_B are selected, and the selection on M_B is reduced by at most this amount. Therefore, among the intervals in O_B, algorithm $A^2(M_A, M_B)$ selects at least $|O_B| - |S_A^\cap| - |S_A^\emptyset|$ intervals. In total, algorithm $A^2(M_A, M_B)$ selects at least $|S_A^O| + |S_A^\cap| + |S_A^\emptyset| + |O_B| - |S_A^\cap| - 1/3 \cdot |O| = 2/3 \cdot |O|$ jobs.

Now, assume that $|S_B^\cap| \geq |S_A^\cap| > 1/3 \cdot |O|$. We analyze how much the intervals S_A can reduce the selection on M_B. At most $|S_B^O|$ intervals corresponding to jobs in S_B^O can be selected on M_A. By Lemma 1, the selection on M_B is reduced at the maximum possible way if in S_A there is the same number of intervals corresponding to jobs in $S_B^\cap$ as the number of intervals corresponding to jobs in $O_B \setminus S_B^O$. Thus, the selection on M_B will be reduced the most, if $|S_B^O|$ intervals in S_A correspond to jobs in S_B^O, and the rest of S_A is split evenly between $S_B^\cap$ and $O_B \setminus S_B^O$. The selection on M_B can thus be reduced by at most

$$|S_B^O| + \frac{|S_A| - |S_B^O|}{2} = \frac{|S_A^\emptyset| + |O_A| + |S_B^O|}{2} = \frac{|S_A^\emptyset| + |O| - |S_B^\cap|}{2} \le \frac{|S_A^\emptyset| + 2/3 \cdot |O|}{2}.$$

Thus, also in this case, algorithm $A^2(M_A, M_B)$ schedules at least $|S_A^O| + |S_A^\cap| + |S_A^\emptyset| + |O_B| - \frac{|O|}{3} - \frac{|S_A^\emptyset|}{2} \ge |O_A| + |O_B| - \frac{|O|}{3} = \frac{2}{3}|O|$ jobs.

Therefore, in every case, the algorithm A^3 schedules at least $2/3 \cdot |O|$ jobs. The analysis is tight, as the example from Figure 1 shows.   □

As an obvious future work, we want to analyze the natural generalization of the algorithm to $m \ge 3$ machines.

3   Hardness Results

In this section we study the complexity of INTERVALSELECTION and show that most of the natural variants are NP-complete or NP-hard. We first describe generic gadgets that we will use as building blocks in our hardness proofs. In the subsequent sections we give the actual hardness proofs. Some of the proofs, as well as the parts discussing the correctness of the reductions, are omitted due to space constraints. They can be found in full version in technical report [3].

Recall that by an *interval* we understand a time interval associated with both a job and a machine. In the following, we will also use time intervals not associated with a job or a machine. To avoid confusion, we use the following terminology. When we consider a time interval with respect to a single machine, but independently of the jobs, we call it a *slot*. And when considering a time interval independently of machines and jobs, we call it a *window*.

We will also use the notion of *blocking*. We say that an interval i *blocks a slot* s if i intersects s and both are associated with the same machine. We say that a set of intervals I *blocks window w on a set of machines* $\mathcal{M}$ if for each machine M in $\mathcal{M}$ there is an interval in I that blocks the slot corresponding to w on M. We say that a set of intervals I *completely blocks a window* w if each slot that intersects the window w is blocked by some interval in I.

We call a schedule in which all jobs are scheduled a *complete schedule*.

Our hardness results are shown by a reduction from variants of the NP-complete *satisfiability problem* (SAT). SAT is the problem of finding, for a given a set of r clauses $C = \{c_1, c_2, \ldots, c_r\}$ over a set of Boolean variables $X = \{x_1, x_2, \ldots, x_s\}$, a truth assignment such that every clause is satisfied, i.e., at least one literal in every clause evaluates to TRUE (see, e.g., [7] for an exact definition of the problem). SAT is NP-complete, even if every clause is restricted to have at most three literals (denoted as 3-SAT) [7], and even, if each clause

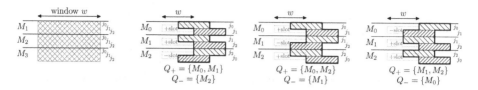

Fig. 2. The first drawing illustrates how we depict a blocking gadget for a window w. The last three drawings illustrate decision gadgets on three machines M_0, M_1, and M_2. Each of the decision gadgets has two positive slots on machines in Q_+ and one negative slot on the machine in Q_-. The crucial intervals constituting the gadget are depicted by the shaded boxes (always one interval spans the respective box). The associated jobs of the intervals are indicated on the sides. The remaining intervals of the jobs are blocked by a blocking gadget, and thus never selected. These intervals and the blocking gadget are for simplicity not displayed. The two different shades in the boxes depict the only two possibilities how to select the intervals in the decision gadget.

contains at most three literals and each variable appears in the formula at most three times, once as a negative literal and at most twice as a positive literal (denoted as ($\leq$3,3)-Sat) [7]. The problem of finding a truth assignment that maximizes the number of satisfied clauses is NP-hard, even if each clause contains two literals and each variable appears at most three times in the formula (denoted as (2,3)-MaxSat) [12].

Building Blocks for Hardness Proofs. In order to simplify the explanation of the hardness proofs, we define the following two gadgets (specific sets of jobs) and use them as building blocks in our reductions.

The purpose of the *blocking gadget* is to completely block a certain window w, i.e., to make sure that in any complete schedule no interval that intersects w is ever scheduled, with the exception of the intervals of the jobs that constitute the gadget itself. Let w be a window (that we want to completely block). The gadget consists of m jobs, each having w as their interval on every machine. We visually depict a blocking gadget as in Figure 2.

Lemma 2. *In any complete schedule for* INTERVALSELECTION *that contains the blocking gadget B for window w, no selected interval outside B intersects w.*

The purpose of the *decision gadget* is to mimic a truth assignment to a variable in a boolean formula of 3-SAT. This is done by blocking a certain window either on one set of machines or on another disjoint set. Given a window w and two disjoint subsets Q_-, Q_+ of machines, we will call the window w on the machines in Q_+ the *positive slots* and w on Q_- the *negative slots* of the gadget (cf. Figure 2). With our gadget we want to achieve that in any complete schedule either all the positive slots of the gadget are free and all the negative slots are blocked by the schedule, or vice versa. Let us refer to the former situation as the *positive decision* of the gadget and to the latter as the *negative decision*. Intuitively, we achieve this effect by using jobs with intervals placed so that we have exactly two ways how to schedule all jobs. To ensure that there is no other way to schedule the jobs of the gadget, we may need to block some intervals of these jobs. For this purpose we use a blocking gadget.

Formally, we construct the decision gadget as follows. We denote by Q the union of Q_-, Q_+, by k the size of Q, and by $M_0, M_1, \ldots, M_{k-1}$ the machines in Q. Without loss of generality, we assume that w has unit length. We use k jobs $j_0, j_1, \ldots, j_{k-1}$, one job per machine in Q. The intervals for all these jobs have unit length $|w|$. There is a blocking gadget B such that all intervals of the decision gadget except for intervals of j_i on M_i, M_{i-1} intersect B (we write M_{i-1} instead of $M_{i-1 \bmod k}$ for simplicity). The exact placement of j_i and j_{i+1} on M_i depends on whether the window w is supposed to be a positive or a negative slot on M_i. In particular, if M_i is in Q_- (w is a negative slot on M_i), the interval for j_i is placed directly to the right of w and the interval for j_{i+1} is placed so that its left end is at the center of w. Otherwise, if M_i is in Q_+, the left end of the interval for j_i is at the center of w and the interval for j_{i+1} is directly to the right of w. Note that the intervals constituting the gadget occupy a window of length 2 (excluding the intervals that are blocked by the blocking gadget).

Lemma 3. *In any complete schedule for an instance of* INTERVALSELECTION *that contains the decision gadget D for window w and subsets Q_-, Q_+ of machines, either D blocks w on all machines in Q_- and leaves it free on all machines in Q_+, or vice versa.*

Corollary 1. *Given a window w and subsets Q_-, Q_+ of machines, in any complete schedule, the intervals of the decision gadget as constructed above enforce the following. Either on all the positive slots of the gadget intervals can be scheduled and all the negative slots are blocked, or vice versa.*

3.1 Interval Selection with Shared Cores

In this section we analyze the complexity of INTERVALSELECTION with cores. We study two variants. First, we consider the case when every job has a core at the end, i.e., all intervals of a job end at the same point in time. We show that deciding whether there is a complete schedule for this variant is NP-complete. By this we resolve an open problem posed by Sung and Vlach [14,16]. Afterwards, we consider the case where every job has a core at an arbitrary position and show that this variant is NP-complete even if all intervals have unit length. We note that both variants are solvable in time $O(m \cdot n^{m+1})$, and thus in polynomial time if m is constant [16].

Theorem 2. *The problem of deciding whether there exists a complete schedule in* INTERVALSELECTION *with cores at the end is NP-complete.*

Proof. The problem is in NP, since the completeness of a given schedule can be checked in linear time. To show the hardness, we present a reduction from 3-SAT.

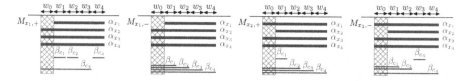

Fig. 3. Example of the construction of INTERVALSELECTION with cores at the end for an instance Φ of the 3-SAT problem (each figure shows the intervals on a single machine, the figures of the machines $M_{x_3,+}$, $M_{x_3,-}$, $M_{x_4,+}$, $M_{x_4,-}$ are not displayed), where $\Phi = (x_1 \vee x_2 \vee \overline{x_3}) \wedge (x_1 \vee x_3 \vee \overline{x_4}) \wedge (\overline{x_2} \vee \overline{x_3} \vee x_4) \wedge (x_1 \vee \overline{x_3} \vee x_4)$

Let us consider an arbitrary instance Φ of 3-SAT given by a set of clauses $C = \{c_1, c_2, \ldots, c_r\}$ over a set of Boolean variables $X = \{x_1, x_2, \ldots, x_s\}$. We construct the following instance $\mathcal{S}$ of the INTERVALSELECTION problem (cf. Figure 3 along with the construction). We use two machines for each variable x_i, denoted by $M_{x_i,+}$ and $M_{x_i,-}$. The machine $M_{x_i,+}$ corresponds to the positive literal of x_i, whereas $M_{x_i,-}$ corresponds to the negative literal of x_i. On the machines we consider a window of $r + 1$ units and we denote the unit windows constituting it by $w_0, w_1, w_2, \ldots, w_r$. We place a blocking gadget over all machines on the window w_0. Next, for each variable x_i we add a job α_{x_i} with two possible ways of scheduling it (in any complete schedule). This mimics a truth assignment to the variable x_i. We call these jobs the *variable jobs*. We place the intervals of a variable job α_{x_i} as follows. On $M_{x_i,+}$ and $M_{x_i,-}$ we place an interval such that it covers $w_1, w_2, \ldots, w_r$, and on every other machine we place an interval such that it covers $w_0, w_1, w_2, \ldots, w_r$. Note that the blocking gadget ensures that in any complete schedule each job α_{x_i} is scheduled on one of the machines $M_{x_i,+}$, $M_{x_i,-}$, and no other job is scheduled on that machine on any window $w_1, w_2, \ldots, w_r$. Intuitively, by scheduling α_{x_i}, one of the two literals of x_i is selected and thus set to FALSE, implicitly setting a truth assignment for variable x_i. Lastly, we add r jobs linked to the clauses so that the actual scheduling of these jobs is related to the way how the clauses of Φ are satisfied. For each clause c_j we have one *clause job* denoted by β_{c_j}. We place the intervals for the job β_{c_j} on window w_j on those machines that correspond to literals that appear in the clause c_j, and on the windows $w_0, w_1, \ldots, w_j$ on the other machines. In other words, in any complete schedule, a job β_{c_j} can only be scheduled on a machine that corresponds to a literal that appears in clause c_j, since on all other machines the intervals for β_{c_j} intersect the blocking gadget. Moreover, if the same literal appears in clauses c_j and $c_{j'}$, $j \neq j'$, then the intervals for jobs β_{c_j} and $\beta_{c_{j'}}$ do not intersect on the machine that corresponds to this literal.

Note that the constructed instance of INTERVALSELECTION has the property that all the intervals corresponding to one job have at their end a unit window in common. Obviously, the above construction can be done in polynomial time. $\square$

The presented hardness implies the hardness of other variants of INTERVALSE-LECTION, such as that of cores at arbitrary positions, or with no required core at all. Similarly, the presented hardness implies the hardness of the maximization versions of these variants. Moreover, using the gadgets described before, we can construct a reduction from ($\leq$3,3)-SAT to prove the following theorem concerning INTERVALSELECTION with arbitrary cores and unit length intervals.

Theorem 3. *The problem of deciding whether there exists a complete schedule in* INTERVALSELECTION *with cores is NP-complete even if all intervals have unit length.*

3.2   Interval Selection with Restricted Number of Machines

In this section we consider the complexity of non-restricted INTERVALSELECTION. We show that, in contrast to INTERVALSELECTION with cores, the problem is NP-hard even if the number of machines is constant. In particular, we prove that deciding whether there is a complete schedule is NP-complete already for three machines. In contrast, the problem is polynomially solvable for two machines [8]. We show that the problem of maximizing the number of scheduled intervals, on the other hand, is NP-hard already for two machines (while polynomially solvable for one machine). Moreover, all these hardness results hold even when all intervals have the same length.

We believe that the techniques used in the proofs may be of independent interest. The decision gadgets capture the relation between a schedule and an assignment. However, we also use properties of edge coloring that provide us with a mapping that lets us put the pieces together and finalize the construction of a scheduling problem under the required, rather restrictive conditions.

Unit Interval Selection with Three Machines. We consider INTERVALSELECTION with three machines and unit length intervals, with the objective of deciding whether there is a complete schedule. We will present a reduction from ($\leq$3,3)-SAT. We will use the following lemma in the subsequent hardness result.

Lemma 4. *Let Φ be an instance of* ($\leq$3,3)-SAT, *given by a set of clauses C over a set of Boolean variables X. Then, there exists a mapping p from $E = \{(x,c) \in X \times C \mid x \in c\}$ to the set $\{M_1, M_2, M_3\}$, such that $p(x,c) \neq p(x,c')$ for $c \neq c'$ and $p(x,c) \neq p(x',c)$ for $x \neq x'$. Moreover, such a mapping p can be found in polynomial time.*

Proof. We prove the statement by edge-coloring the bipartite graph $G = (X \cup C, E)$. The structure of ($\leq$3,3)-SAT implies that all vertices of the constructed graph G have a degree at most 3. A bipartite graph is Δ-edge-colorable in polynomial time, where Δ is the maximum degree [10]. Therefore, the graph G is 3-edge-colorable, with colors from $\{M_1, M_2, M_3\}$. This coloring gives us the desired mapping from E to $\{M_1, M_2, M_3\}$. $\square$

Theorem 4. *The problem of deciding whether there exists a complete schedule in* INTERVALSELECTION *is NP-complete even for three machines and unit length intervals.*

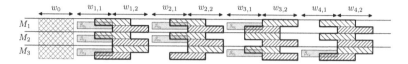

Fig. 4. Instance $\Phi = (x_1 \vee x_2 \vee x_3) \wedge (x_1 \vee x_2 \vee \overline{x_4}) \wedge (\overline{x_1} \vee \overline{x_3})$ of ($\leq$3,3)-SAT and the corresponding instance of INTERVALSELECTION with three machines. Intervals intersecting the blocking gadget are not shown in the figure.

Proof. The problem is obviously in NP. To show the hardness, we reduce ($\leq$3,3)-SAT to it. Let Φ be an instance of ($\leq$3,3)-SAT, given by a set of clauses $C = \{c_1, c_2, \ldots, c_r\}$ over a set of Boolean variables $X = \{x_1, x_2, \ldots, x_s\}$. We construct from Φ the following instance $\mathcal{S}$ of the INTERVALSELECTION problem (cf. Figure 4 along with the construction), using three machines M_1, M_2, M_3. We use a window of $2s + 1$ units, and denote the unit windows constituting it by $w_0, w_{1,1}, w_{1,2}, w_{2,1}, w_{2,2}, \ldots, w_{s,1}, w_{s,2}$. We introduce jobs with unit length intervals as follows. We place a blocking gadget on window w_0 over all machines. For each variable x_i we place a decision gadget D_{x_i} on the machines such that it has two positive and one negative slot on window $w_{i,1}$, in an arrangement that we will specify later. The gadget D_{x_i} occupies windows $w_{i,1}$ and $w_{i,2}$ and uses internally the blocking gadget on w_0. The positive/negative decision of gadget D_{x_i} corresponds to the truth assignment of the variable x_i and the decision of D_{x_i} is independent of the other decision gadgets. We introduce a *clause job* β_{c_j} for each clause c_j. To place the intervals for β_{c_j}, we look at the literals that appear in c_j. For each appearance of a positive literal of some variable x_i in c_j we place an interval for β_{c_j} on a positive slot of D_{x_i}, and for each appearance of a negative literal of $x_{i'}$ in c_j we place an interval for β_{c_j} on the negative slot of $D_{x_{i'}}$. If c_j contains only two literals, we place one interval for β_{c_j} on the window w_0 so that it intersects the blocking gadget and cannot be selected in any complete schedule.

To obtain a valid construction, we need to ensure that all the intervals for each clause job β_{c_j} are placed on different machines, and at the same time, we require that each positive/negative slot of the decision gadgets is occupied by at most one interval. We now explain the exact placement of the positive/negative slots, as well as the distribution of the clause jobs over the slots that achieve this. We have three machines and we need to place each decision gadget so that it has its negative slot on some machine and its positive slots on the other two machines. Finding a way to arrange the decision gadgets and distribute their slots is equivalent to finding a mapping from a set of pairs (variable x, clause c containing x) to the set $\{M_1, M_2, M_3\}$ that assigns different machines to the variables in each clause and different machines to the clauses containing a fixed variable. Such a mapping can be efficiently constructed due to Lemma 4.   □

Unit Interval Selection with Two Machines. The maximization variant of INTERVALSELECTION turns to be NP-hard already for two machines. The proof is similar to that of Theorem 4, but uses a reduction from (2,3)-MAXSAT.

Theorem 5. *Maximizing the number of scheduled intervals in* INTERVALSELEC-TION *is NP-hard, even for two machines and unit length intervals.*   .

Acknowledgements. This work was partially supported by the EU FP7/2007-2013 (DG CONNECT.H5-Smart Cities and Sustainability), under grant agreement no. 288094 (project eCOMPASS). Kateřina Böhmová is a recipient of the Google Europe Fellowship in Optimization Algorithms, and this research is supported in part by this Google Fellowship. We would like to thank Thomas Graffagnino from Swiss Federal Railways (SBB) and Rastislav Šrámek for pointing out optimization problems with applications in car sharing. The first author would like to thank Petr Škovroň for feedback on preliminary versions of the paper.

References

1. Arkin, E.M., Silverberg, E.B.: Scheduling jobs with fixed start and end times. Discrete Applied Mathematics 18(1), 1–8 (1987)
2. Bar-Noy, A., Guha, S., Naor, J., Schieber, B.: Approximating the throughput of multiple machines in real-time scheduling. SIAM J. Comput. 31(2), 331–352 (2001)
3. Böhmová, K., Disser, Y., Mihalák, M., Widmayer, P.: Interval selection with machine-dependent intervals. Tech. Rep. 786, Institute of Theoretical Computer Science, ETH Zurich (2013)
4. Bouzina, K.I., Emmons, H.: Interval scheduling on identical machines. Journal of Global Optimization 9, 379–393 (1996)
5. Chuzhoy, J., Ostrovsky, R., Rabani, Y.: Approximation algorithms for the job interval selection problem and related scheduling problems. In: Proc. of the 42nd IEEE Symp. on Foundations of Computer Science (FOCS), pp. 348–356 (2001)
6. Erlebach, T., Spieksma, F.C.R.: Interval selection: applications, algorithms, and lower bounds. Journal of Algorithms 46(1), 27–53 (2003)
7. Garey, M.R., Johnson, D.S.: Computers and intractability: a guide to the theory of NP-completeness. W. H. Freeman & Co., New York (1979)
8. Keil, J.M.: On the complexity of scheduling tasks with discrete starting times. Operations Research Letters 12(5), 293–295 (1992)
9. Kolen, A.W.J., Lenstra, J.K., Papadimitriou, C.H., Spieksma, F.C.R.: Interval scheduling: a survey. Naval Research Logistics (NRL) 54(5), 530–543 (2007)
10. König, D.: Über Graphen und ihre Anwendung auf Determinantentheorie und Mengenlehre. Mathematische Annalen 77, 453–465 (1916)
11. Nakajima, K., Hakimi, S.L.: Complexity results for scheduling tasks with discrete starting times. Journal of Algorithms 3(4), 344–361 (1982)
12. Raman, V., Ravikumar, B., Rao, S.S.: A simplified NP-complete MAXSAT problem. Information Processing Letters 65(1), 1–6 (1998)
13. Sgall, J.: Open problems in throughput scheduling. In: Epstein, L., Ferragina, P. (eds.) ESA 2012. LNCS, vol. 7501, pp. 2–11. Springer, Heidelberg (2012)
14. Shabtay, D., Steiner, G.: Scheduling to maximize the number of just-in-time jobs: a survey. In: Just-in-Time Systems. Springer Optimization and Its Applications, vol. 60, pp. 3–20. Springer, New York (2012)
15. Spieksma, F.C.R.: On the approximability of an interval scheduling problem. Journal of Scheduling 2(5), 215–227 (1999)
16. Sung, S.C., Vlach, M.: Maximizing weighted number of just-in-time jobs on unrelated parallel machines. Journal of Scheduling 8, 453–460 (2005)

On the Spanning Ratio of Theta-Graphs*

Prosenjit Bose, André van Renssen, and Sander Verdonschot

School of Computer Science, Carleton University, Ottawa, Canada
jit@scs.carleton.ca, {andre,sander}@cg.scs.carleton.ca

Abstract. We present improved upper bounds on the spanning ratio of a large family of θ-graphs. A θ-graph partitions the plane around each vertex into m disjoint cones, each having aperture $\theta = 2\pi/m$. We show that for any integer $k \geq 1$, θ-graphs with $4k + 4$ cones have spanning ratio at most $1 + 2\sin(\theta/2)/(\cos(\theta/2) - \sin(\theta/2))$. We also show that θ-graphs with $4k + 3$ and $4k + 5$ cones have spanning ratio at most $\cos(\theta/4)/(\cos(\theta/2) - \sin(3\theta/4))$. This is a significant improvement on all families of θ-graphs for which exact bounds are not known. For example, the spanning ratio of the θ-graph with 7 cones is decreased from at most 7.5625 to at most 3.5132. We also improve the upper bounds on the competitiveness of the θ-routing algorithm for these graphs to $1 + 2\sin(\theta/2)/(\cos(\theta/2) - \sin(\theta/2))$ on θ-graphs with $4k + 4$ cones and to $1 + 2\sin(\theta/2) \cdot \cos(\theta/4)/(\cos(\theta/2) - \sin(3\theta/4))$ on θ-graphs with $4k + 3$ and $4k + 5$ cones. For example, the routing ratio of the θ-graph with 7 cones is decreased from at most 7.5625 to at most 4.0490.

Keywords: computational geometry, spanners, θ-graphs, spanning ratio.

1   Introduction

In a weighted graph G, let the distance $\delta_G(u, v)$ between two vertices u and v be the length of the shortest path between u and v in G. A subgraph H of G is a *t-spanner* of G if for all pairs of vertices u and v, $\delta_H(u, v) \leq t \cdot \delta_G(u, v), t \geq 1$. The *spanning ratio* of H is the smallest t for which H is a t-spanner. The graph G is referred to as the *underlying graph* [7]. A routing strategy is said to be *c-competitive* with respect to G if the length of the path returned by the routing strategy is not more than c times the length of the shortest path in G [3].

We consider the situation where the underlying graph G is a straightline embedding of K_n, the complete graph on a set of n points in the plane. The weight of each edge (u, v) is the Euclidean distance $|uv|$ between u and v. A spanner of such a graph is called a *geometric spanner*. We look at a specific type of geometric spanner: θ-graphs.

Introduced independently by Clarkson [5] and Keil [6], θ-graphs are constructed as follows (a more precise definition follows in the next section): for each vertex u, we partition the plane into m disjoint cones with apex u, each having aperture $\theta = 2\pi/m$. When m cones are used, we denote the resulting

---

* Research supported in part by NSERC.

F. Dehne, R. Solis-Oba, and J.-R. Sack (Eds.): WADS 2013, LNCS 8037, pp. 182–194, 2013.
© Springer-Verlag Berlin Heidelberg 2013

θ-graph as θ_m. The θ-graph is constructed by, for each cone with apex u, connecting u to the vertex v whose projection along the bisector of the cone is closest. Ruppert and Seidel [8] showed that the spanning ratio of these graphs is at most $1/(1-2\sin(\theta/2))$, when $\theta < \pi/3$, i.e. there are at least seven cones. This proof also showed that the θ-*routing* algorithm (defined in the next section) is $1/(1-2\sin(\theta/2))$-competitive on these graphs.

Bonichon *et al.* [1] showed that the θ_6-graph has spanning ratio 2. This was done by dividing the cones into two sets, positive and negative cones, such that each positive cone is adjacent to two negative cones and vice versa. It was shown that when edges are added only in the positive cones, in which case the graph is called the half-θ_6-graph, the resulting graph is equivalent to the TD-Delaunay triangulation (the Delaunay triangulation where the empty region is an equilateral triangle) whose spanning ratio is 2, as shown by Chew [4]. An alternative, inductive proof of the spanning ratio of the half-θ_6-graph was presented by Bose *et al.* [3] along with an optimal local competitive routing algorithm on the half-θ_6-graph. Recently, Bose *et al.* [2] generalized this inductive proof to show that the $\theta_{(4k+2)}$-graph has spanning ratio $1 + 2\sin(\theta/2)$, where k is an integer and at least 1. This spanning ratio is exact, i.e. there is a matching lower bound.

In this paper, we generalize the results from Bose *et al.* [2]. We look at the three remaining families of θ-graphs: the $\theta_{(4k+3)}$-graph, the $\theta_{(4k+4)}$-graph, and the $\theta_{(4k+5)}$-graph, where k is an integer and at least 1. We show that the $\theta_{(4k+4)}$-graph has a spanning ratio of at most $1 + 2\sin(\theta/2)/(\cos(\theta/2) - \sin(\theta/2))$. We also show that the $\theta_{(4k+3)}$-graph and the $\theta_{(4k+5)}$-graph have spanning ratio at most $\cos(\theta/4)/(\cos(\theta/2) - \sin(3\theta/4))$. We also improve the competitiveness of θ-routing on these graphs. The θ-routing algorithm is the standard routing algorithm on all θ-graphs having at least seven cones. For both the spanning ratio and the routing ratio, the best known bound was $1/(1-2\sin(\theta/2))$ by Ruppert and Seidel [8].

Table 1. An overview of current and previous spanning and routing ratios

	Current Spanning	Current Routing	Previous Spanning/Routing
$\theta_{(4k+2)}$-graph	$1 + 2\sin(\theta/2)$ [2]	$\frac{1}{1-2\sin(\theta/2)}$ [8]	$\frac{1}{1-2\sin(\theta/2)}$ [8]
$\theta_{(4k+3)}$-graph	$\frac{\cos(\theta/4)}{\cos(\theta/2)-\sin(3\theta/4)}$	$1 + \frac{2\sin(\theta/2)\cos(\theta/4)}{\cos(\theta/2)-\sin(\theta/2)}$	$\frac{1}{1-2\sin(\theta/2)}$ [8]
$\theta_{(4k+4)}$-graph	$1 + \frac{2\sin(\theta/2)}{\cos(\theta/2)-\sin(\theta/2)}$	$1 + \frac{2\sin(\theta/2)}{\cos(\theta/2)-\sin(\theta/2)}$	$\frac{1}{1-2\sin(\theta/2)}$ [8]
$\theta_{(4k+5)}$-graph	$\frac{\cos(\theta/4)}{\cos(\theta/2)-\sin(3\theta/4)}$	$1 + \frac{2\sin(\theta/2)\cos(\theta/4)}{\cos(\theta/2)-\sin(\theta/2)}$	$\frac{1}{1-2\sin(\theta/2)}$ [8]

2 Preliminaries

Let a *cone* C be the region in the plane between two rays originating from the same point (referred to as the apex of the cone). For ease of exposition, we only consider point sets in general position: no two vertices lie on a line parallel to one of the rays that define the cones and no two vertices lie on a line perpendicular to the bisector of one of the cones.

When constructing a θ_m-graph, for each vertex u of K_n consider the rays originating from u with the angle between consecutive rays being $\theta = 2\pi/m$. Each pair of consecutive rays defines a cone. The cones are oriented such that the bisector of some cone coincides with the vertical halfline through u that lies above u. Let this cone be C_0 of u and number the cones in clockwise order around u. The cones around the other vertices have the same orientation as the ones around u. We write C_i^u to indicate the i-th cone of a vertex u.

The θ_m-graph is constructed as follows: for each cone C of each vertex u, add an edge from u to the closest vertex in that cone, where distance is measured along the bisector of the cone. More formally, we add an edge between two vertices u and v if $v \in C$ and for all vertices $w \in C$ $(v \neq w)$, $|uv'| \leq |uw'|$, where v' and w' denote the orthogonal projection of v and w on the bisector of C. Note that our general position assumption implies that each vertex adds at most one edge per cone to the graph.

Given a vertex w in cone C of vertex u, we define the *canonical triangle* T_{uw} as the triangle defined by the borders of C and the line through w perpendicular to the bisector of C. We use m to denote the midpoint of the side of T_{uw} opposite u and α to denote the smaller unsigned angle between uw and um (see Figure 1). Note that for any pair of vertices u and w, there exist two canonical triangles: T_{uw} and T_{wu}.

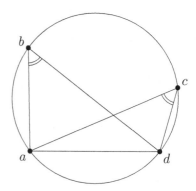

Fig. 1. The canonical triangle T_{uw} **Fig. 2.** Four points a, b, c, d on a circle

Using the structure of the θ_m-graph, *θ-routing* is defined as follows. From the current vertex u, follow the edge to the closest vertex in T_{ut}, where t is the destination. This step is repeated until the destination is reached.

Next, we prove a few geometric lemmas that will be useful when bounding the spanning ratios of the graphs. We use $\angle xyz$ to denote the smaller angle between line segments xy and yz.

Lemma 1. *Let a, b, c, and d be four points on a circle such that $\angle cad \le \angle bad \le \angle adc$. It holds that $|ac| + |cd| \le |ab| + |bd|$ and $|cd| \le |bd|$.*

Proof. Since b and c lie on the same circle and $\angle abd$ and $\angle acd$ are the angle opposite to the same chord ad, the inscribed angle theorem implies that $\angle abd = \angle acd$ (see Figure 2). First, we show that $|ac| + |cd| \le |ab| + |bd|$.

We look at the function $\sin \alpha + \sin(\pi - \gamma - \alpha)$, where γ is a fixed constant and $\gamma + \alpha \le \pi$. Using elementary calculus, it can be shown that this function has a maximum at $\alpha = (\pi - \gamma)/2$ and is strictly unimodal for $\alpha \in (0, \pi - \gamma)$. Next, we note that $|ac| + |cd| \le |ab| + |bd|$ can be rewritten as $2 \cdot r \cdot (\sin \angle adc + \sin \angle cad) \le 2 \cdot r \cdot (\sin \angle adb + \sin \angle bad)$, where r is the radius of the circle. Since we can express $\angle adc$ and $\angle adb$ as $\pi - \angle acd - \angle cad$ and $\pi - \angle abd - \angle bad$, both sides of the inequality have the form $\sin \alpha + \sin(\pi - \gamma - \alpha)$, with $\gamma = \angle abd = \angle acd$. Hence, since $\angle cad \le \angle bad \le \pi - \angle acd - \angle cad = \angle adc$, we have that $|ac| + |cd| \le |ab| + |bd|$ and $|cd| \le |bd|$. $\qquad\square$

Lemma 2. *Let u, v and w be three vertices in the $\theta_{(4k+x)}$-graph, $x \in \{3, 4, 5\}$, such that $w \in C_0^u$ and $v \in T_{uw}$, to the left of uw. Let a be the intersection of the side of T_{uw} opposite u and the left boundary of C_0^u. Let C_i^v denote the cone of v that contains w and let c and d be the upper and lower corner of T_{vw}. If $1 \le i \le k - 1$, or $i = k$ and $|cw| \le |dw|$, then $\max\{|vc| + |cw|, |vd| + |dw|\} \le |va| + |aw|$ and $\max\{|cw|, |dw|\} \le |aw|$.*

Proof. This situation is illustrated in Figure 3. We perform case distinction on $\max\{|cw|, |dw|\}$.

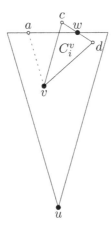

Fig. 3. The situation where we apply Lemma 1

Case 1: If $|cw| > |dw|$, we need to show that when $1 \le i \le k - 1$, $|vc| + |cw| \le |va| + |aw|$ and $|cw| \le |aw|$. Since angles $\angle vaw$ and $\angle vcw$ are both angles between the boundary of a cone and the line perpendicular to its bisector, $\angle vaw = \angle vcw$. Thus, c lies on the circle through a, v, and w. Therefore, if we can show that $\angle cvw \leq \angle avw \leq \angle vwc$, Lemma 1 proves this case.

We show $\angle cvw \leq \angle avw \leq \angle vwc$ in two steps. Since $w \in C_i^v$ and $i \geq 1$, we have that $\angle avc = i \cdot \theta \geq \theta$. Hence, since $\angle avw = \angle avc + \angle cvw$, $\angle cvw \leq \angle avw$. It remains to show that $\angle avw \leq \angle vwc$. We note that $\angle avw \leq (i+1) \cdot \theta$ and $(\pi - \theta)/2 \leq \angle vwc$, since $|cw| > |dw|$. Using that $\theta = 2\pi/(4k+x)$ and $x \in \{3, 4, 5\}$, we compute the maximum value of i for which $\angle avw \leq \angle vwc$:

$$\angle avw \leq \angle vwc$$
$$(i+1) \cdot \theta \leq \frac{\pi - \theta}{2}$$
$$i \leq \frac{\pi}{2\theta} - \frac{3}{2}$$
$$i \leq \frac{\pi \cdot (4k+x)}{4\pi} - \frac{3}{2}$$
$$i \leq k + \frac{x}{4} - \frac{3}{2}$$
$$i \leq k - 1$$

Hence, $\angle avw \leq \angle vwc$ when $i \leq k - 1$.

Case 2: If $|cw| \leq |dw|$, we need to show that when $1 \leq i \leq k$, $|vd| + |dw| \leq |va| + |aw|$ and $|dw| \leq |aw|$. Since angles $\angle vaw$ and $\angle vdw$ are both angles between the boundary of a cone and the line perpendicular to its bisector, $\angle vaw = \angle vcw$. Thus, when we reflect d around vw, the resulting point d' lies on the circle through a, v, and w. Therefore, if we can show that $\angle d'vw \leq \angle avw \leq \angle vwd'$, Lemma 1 proves this case.

We show $\angle d'vw \leq \angle avw \leq \angle vwd'$ in two steps. Since $w \in C_i^v$ and $i \geq 1$, we have that $\angle avw \geq \angle avc = i \cdot \theta \geq \theta$. Hence, since $\angle d'vw \leq \theta$, $\angle d'vw \leq \angle avw$. It remains to show that $\angle avw \leq \angle vwd'$. We note that $\angle vwd' = \angle dwv = \pi - (\pi - \theta)/2 - \angle dvw$ and $\angle avw = \angle avd - \angle dvw = (i+1) \cdot \theta - \angle dvw$. Using that $\theta = 2\pi/(4k+x)$ and $x \in \{3, 4, 5\}$, we compute the maximum value of i for which $\angle avw \leq \angle vwd'$:

$$\angle avw \leq \angle vwd'$$
$$(i+1) \cdot \theta - \angle dvw \leq \frac{\pi + \theta}{2} - \angle dvw$$
$$i \leq \frac{\pi}{2\theta} - \frac{1}{2}$$
$$i \leq \frac{\pi \cdot (4k+x)}{4\pi} - \frac{1}{2}$$
$$i \leq k + \frac{x}{4} - \frac{1}{2}$$
$$i \leq k$$

Hence, $\angle avw \leq \angle vwd'$ when $i \leq k$.   □

Lemma 3. *Let u, v and w be three vertices in the $\theta_{(4k+x)}$-graph, such that $w \in C_0^u$, $v \in T_{uw}$ to the left of uw, and $w \notin C_0^v$. Let a be the intersection of the side of T_{uw} opposite u and the line through v parallel to the left boundary of T_{uw}. Let y and z be the corners of T_{vw} opposite to v. Let $\beta = \angle awv$ and let γ be the unsigned angle between vw and the bisector of T_{vw}. Let c be a positive constant. If*

$$c \geq \frac{\cos \gamma - \sin \beta}{\cos \left(\frac{\theta}{2} - \beta \right) - \sin \left(\frac{\theta}{2} + \gamma \right)},$$

then

$$|vp| + c \cdot |pw| \leq |va| + c \cdot |aw|,$$

where p is y if $|yw| \geq |zw|$ and z if $|yw| < |zw|$.

Proof. Using that the angle between the bisector of a cone and its boundary is $\theta/2$, we first express the four line segments in terms of β and γ (see Figure 4):

$$|vp| = |vw| \cdot \cos \gamma / \cos(\theta/2)$$
$$|pw| = |vw| \cdot (\sin \gamma + \cos \gamma \cdot \tan(\theta/2))$$
$$|va| = |vw| \cdot \sin \beta / \cos(\theta/2)$$
$$|aw| = |vw| \cdot (\cos \beta + \sin \beta \cdot \tan(\theta/2))$$

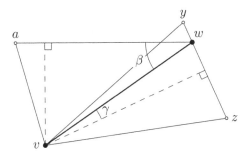

Fig. 4. Finding a constant c such that $|vz| + c \cdot |zw| \leq |va| + c \cdot |aw|$

To compute for which values of c the inequality $|vp| + c \cdot |pw| \leq |va| + c \cdot |aw|$ holds, we first multiply both sides by $\cos(\theta/2)/|vw|$ and rewrite as follows:

$$\frac{\cos(\theta/2)}{|vw|} \cdot (|vp| + |pw| \cdot c) = \cos \gamma + c \cdot (\sin \gamma \cdot \cos(\theta/2) + \cos \gamma \cdot \sin(\theta/2))$$
$$= \cos \gamma + c \cdot \sin(\theta/2 + \gamma)$$

$$\frac{\cos(\theta/2)}{|vw|} \cdot (|va| + |aw| \cdot c) = \sin \beta + c \cdot (\cos \beta \cdot \cos(\theta/2) + \sin \beta \cdot \sin(\theta/2))$$
$$= \sin \beta + c \cdot \cos(\theta/2 - \beta)$$

We can now calculate for which values of c the inequality holds:

$$\cos \gamma + c \cdot \sin(\theta/2 + \gamma) \leq \sin \beta + c \cdot \cos(\theta/2 - \beta)$$
$$\cos \gamma - \sin \beta \leq c \cdot (\cos(\theta/2 - \beta) - \sin(\theta/2 + \gamma))$$
$$c \geq \frac{\cos \gamma - \sin \beta}{\cos(\theta/2 - \beta) - \sin(\theta/2 + \gamma)}$$

It remains to show that $c > 0$. Since $w \notin C_0^v$, we have that $\beta \in (0, (\pi - \theta)/2)$, and by definition $\gamma \in [0, \theta/2)$. This implies that $\sin(\pi/2 + \gamma) > \sin \beta$ or equivalently $\cos \gamma - \sin \beta > 0$. Thus, we need to show that $\cos(\theta/2 - \beta) - \sin(\theta/2 + \gamma) > 0$ or equivalently $\sin(\pi/2 + \theta/2 - \beta) > \sin(\theta/2 + \gamma)$. It suffices to show that $\theta/2 + \gamma < \pi/2 + \theta/2 - \beta < \pi - \theta/2 - \gamma$. This follows from $\beta \in (0, (\pi - \theta)/2)$, $\gamma \in [0, \theta/2)$, and the fact that $\theta \leq 2\pi/7$.   □

3   Generic Framework for the Spanning Proof

Using the lemmas from the previous section, we provide a generic framework for the spanning proof for the three families of θ-graphs. After providing this framework, we fill in the blanks for the individual families.

Theorem 1. *Let u and w be two vertices in the plane. Let m be the midpoint of the side of T_{uw} opposite u and let α be the unsigned angle between uw and um. There exists a path connecting u and w in the $\theta_{(4k+x)}$-graph of length at most*

$$\left(\frac{\cos \alpha}{\cos \left(\frac{\theta}{2}\right)} + \left(\cos \alpha \cdot \tan \left(\frac{\theta}{2}\right) + \sin \alpha \right) \cdot c \right) \cdot |uw|,$$

where $c \geq 1$ is a constant that depends on $x \in \{3, 4, 5\}$. For the $\theta_{(4k+4)}$-graph, c equals $1/(\cos(\theta/2) - \sin(\theta/2))$ and for the $\theta_{(4k+3)}$-graph and $\theta_{(4k+5)}$-graph, c equals $\cos(\theta/4)/(\cos(\theta/2) - \sin(3\theta/4))$.

Proof. We assume without loss of generality that $w \in C_0^u$. We prove the theorem by induction on the area of T_{uw} (formally, induction on the rank, when ordered by area, of the canonical triangles for all pairs of vertices). Let a and b be the upper left and right corners of T_{uw}. Our inductive hypothesis is the following, where $\delta(u, w)$ denotes the length of the shortest path from u to w in the $\theta_{(4k+x)}$-graph: $\delta(u, w) \leq \max\{|ua| + |aw| \cdot c, |ub| + |bw| \cdot c\}$.

We first show that this induction hypothesis implies the theorem. Basic trigonometry gives us the following equalities: $|um| = |uw| \cdot \cos \alpha$, $|mw| = |uw| \cdot \sin \alpha$, $|am| = |bm| = |uw| \cdot \cos \alpha \cdot \tan(\theta/2)$, and $|ua| = |ub| = |uw| \cdot \cos \alpha / \cos(\theta/2)$. Thus the induction hypothesis gives that $\delta(u, w)$ is at most $|ua| + (|am| + |mw|) \cdot c = |uw| \cdot (\cos \alpha / \cos(\theta/2) + (\cos \alpha \cdot \tan(\theta/2) + \sin \alpha) \cdot c)$.

Base Case: T_{uw} has rank 1. Since the triangle is a smallest triangle, w is the closest vertex to u in that cone. Hence the edge (u, w) is part of the $\theta_{(4k+x)}$-graph, and $\delta(u, w) = |uw|$. From the triangle inequality and the fact that $c \geq 1$, we have $|uw| \leq \max\{|ua| + |aw| \cdot \boldsymbol{c}, |ub| + |bw| \cdot \boldsymbol{c}\}$, so the induction hypothesis holds.

Induction Step: We assume that the induction hypothesis holds for all pairs of vertices with canonical triangles of rank up to j. Let T_{uw} be a canonical triangle of rank $j + 1$.

If (u, w) is an edge in the $\theta_{(4k+x)}$-graph, the induction hypothesis follows by the same argument as in the base case. If there is no edge between u and w, let v be the vertex closest to u in T_{uw}, and let a' and b' be the upper left and right corners of T_{uv} (see Figure 5). By definition, $\delta(u, w) \leq |uv| + \delta(v, w)$, and by the triangle inequality, $|uv| \leq \min\{|ua'| + |a'v|, |ub'| + |b'v|\}$.

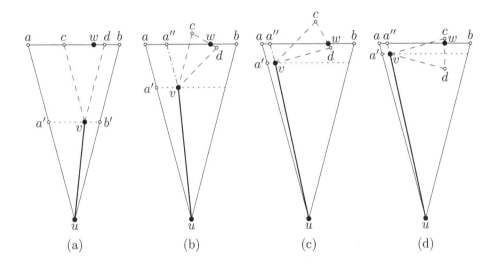

Fig. 5. The four cases based on the cone of v that contains w, in this case for the θ_{12}-graph

Without loss of generality, we assume that v lies to the left of w. We perform a case analysis based on the cone of v that contains w, where c and d are the left and right corners of T_{vw}, opposite to v: (a) $w \in C_0^v$, (b) $w \in C_i^v$ where $1 \leq i \leq k - 1$, or $i = k$ and $|cw| \leq |dw|$, (c) $w \in C_k^v$ and $|cw| > |dw|$, (d) $w \in C_{k+1}^v$.

Case (a): Vertex w lies in C_0^v (see Figure 5a). Since T_{vw} has smaller area than T_{uw}, we apply the inductive hypothesis to T_{vw}. Hence we have $\delta(v, w) \leq \max\{|vc| + |cw| \cdot \boldsymbol{c}, |vd| + |dw| \cdot \boldsymbol{c}\}$. Since v lies to the left of w, the maximum of the left hand side is attained by its second argument $|vc| + |cw| \cdot \boldsymbol{c}$. Since vertices $v, c, a,$ and a' form a parallelogram and $c \geq 1$, we have that:

$$\delta(u, w) \leq |uv| + \delta(v, w)$$
$$\leq |ua'| + |a'v| + |vc| + |cw| \cdot \boldsymbol{c}$$
$$\leq |ua| + |aw| \cdot \boldsymbol{c}$$
$$\leq \max\{|ua| + |aw| \cdot \boldsymbol{c}, |ub| + |bw| \cdot \boldsymbol{c}\},$$

which proves the induction hypothesis.

Case (b): Vertex w lies in C_i^v, where $1 \leq i \leq k-1$, or $i = k$ and $|cw| \leq |dw|$. Let a'' be the intersection of the side of T_{uw} opposite u and the left boundary of C_0^v. Since T_{vw} is smaller than T_{uw}, by induction we have $\delta(v, w) \leq \max\{|vc| + |cw| \cdot \boldsymbol{c}, |vd| + |dw| \cdot \boldsymbol{c}\}$ (see Figure 5b). Since $w \in C_i^v$ where $1 \leq i \leq k-1$, or $i = k$ and $|cw| \leq |dw|$, we can apply Lemma 2. Note that point a in Lemma 2 corresponds to point a'' in this proof. Hence, we get that $\max\{|vc| + |cw|, |vd| + |dw|\} \leq |va''| + |a''w|$ and $\max\{|cw|, |dw|\} \leq |a''w|$. Since $\boldsymbol{c} \geq 1$, this implies that $\max\{|vc| + |cw| \cdot \boldsymbol{c}, |vd| + |dw| \cdot \boldsymbol{c}\} \leq |va''| + |a''w| \cdot \boldsymbol{c}$. Since $|uv| \leq |ua'| + |a'v|$ and v, a'', a, and a' form a parallelogram, we have that $\delta(u, w) \leq |ua| + |aw| \cdot \boldsymbol{c}$, proving the induction hypothesis for T_{uw}.

Case (c) and (d) Vertex w lies in C_k^v and $|cw| > |dw|$, or w lies in C_{k+1}^v. Let a'' be the intersection of the side of T_{uw} opposite u and the left boundary of C_0^v (see Figures 5c and d). Since T_{vw} is smaller than T_{uw}, we can apply induction on it. The precise application of the induction hypothesis varies for the three families of θ-graphs and, using Lemma 3, determines the value of $\boldsymbol{c}$. Hence, these cases are discussed in the spanning proofs of the three families.   □

4   The $\theta_{(4k+4)}$-Graph

In this section, we give improved upper bounds on the spanning ratio of the $\theta_{(4k+4)}$-graph, for any integer $k \geq 1$.

Theorem 2. *Let u and w be two vertices in the plane. Let m be the midpoint of the side of T_{uw} opposite u and let α be the unsigned angle between uw and um. There exists a path connecting u and w in the $\theta_{(4k+4)}$-graph of length at most*

$$\left(\frac{\cos \alpha}{\cos \left(\frac{\theta}{2}\right)} + \frac{\cos \alpha \cdot \tan \left(\frac{\theta}{2}\right) + \sin \alpha}{\cos \left(\frac{\theta}{2}\right) - \sin \left(\frac{\theta}{2}\right)} \right) \cdot |uw|.$$

Proof. We apply Theorem 1 using $\boldsymbol{c} = 1/(\cos(\theta/2) - \sin(\theta/2))$. It remains to handle Case (c), where $w \in C_k^v$ and $|cw| > |dw|$, and Case (d), where $w \in C_{k+1}^v$.

Recall that c and d are the left and right corners of T_{vw}, opposite to v, and a'' is the intersection of aw and the line through v, parallel to ua. Let β be $\angle a''wv$ and let γ be the angle between vw and the bisector of T_{vw}. Since T_{vw} is smaller than T_{uw}, the induction hypothesis gives a bound on $\delta(v, w)$. Since $|uv| \leq |ua'| + |a'v|$ and v, a'', a, and a' form a parallelogram, we need to show that $\delta(v, w) \leq |va''| + |a''w| \cdot \boldsymbol{c}$ for both cases in order to complete the proof.

Case (c): When w lies in C_k^v and $|cw| > |dw|$, the induction hypothesis for T_{vw} gives $\delta(v, w) \leq |vc| + |cw| \cdot \boldsymbol{c}$. We note that $\gamma = \theta - \beta$. Hence Lemma 3 gives that the inequality holds when $c \geq (\cos(\theta - \beta) - \sin\beta)/(\cos(\theta/2 - \beta) - \sin(3\theta/2 - \beta))$. As this function is decreasing in β for $\theta/2 \leq \beta \leq \theta$, it is maximized when β equals $\theta/2$. Hence c needs to be at least $(\cos(\theta/2) - \sin(\theta/2))/(1 - \sin\theta)$, which can be rewritten to $1/(\cos(\theta/2) - \sin(\theta/2))$.

Case (d): When w lies in C_{k+1}^v, w lies above the bisector of T_{vw} and the induction hypothesis for T_{vw} gives $\delta(v, w) \leq |wd| + |dv| \cdot c$. We note that $\gamma = \beta$. Hence Lemma 3 gives that the inequality holds when $c \geq (\cos\beta - \sin\beta)/(\cos(\theta/2 - \beta) - \sin(\theta/2 + \beta))$. As this function is decreasing in β for $0 \leq \beta \leq \theta/2$, it is maximized when β equals 0. Hence c needs to be at least $1/(\cos(\theta/2) - \sin(\theta/2))$. □

Since $\cos\alpha/\cos(\theta/2) + (\cos\alpha \cdot \tan(\theta/2) + \sin\alpha)/(\cos(\theta/2) - \sin(\theta/2))$ is increasing for $\alpha \in [0, \theta/2]$, for $\theta \leq \pi/4$, it is maximized when $\alpha = \theta/2$, and we obtain the following corollary:

Corollary 1. *The $\theta_{(4k+4)}$-graph is a* $\left(1 + \dfrac{2 \cdot \sin\left(\frac{\theta}{2}\right)}{\cos\left(\frac{\theta}{2}\right) - \sin\left(\frac{\theta}{2}\right)}\right)$ *-spanner of K_n.*

Furthermore, we observe that the proof of Theorem 2 follows the same path as the θ-routing algorithm follows: if the direct edge to the destination is part of the graph, it follows this edge, and if it is not, it follows the edge to the closest vertex in the cone that contains the destination.

Corollary 2. *The θ-routing algorithm is* $\left(1 + \dfrac{2 \cdot \sin\left(\frac{\theta}{2}\right)}{\cos\left(\frac{\theta}{2}\right) - \sin\left(\frac{\theta}{2}\right)}\right)$ *-competitive on the $\theta_{(4k+4)}$-graph.*

5   The $\theta_{(4k+3)}$-Graph and the $\theta_{(4k+5)}$-Graph

In this section, we give improved upper bounds on the spanning ratio of the $\theta_{(4k+3)}$-graph and the $\theta_{(4k+5)}$-graph, for any integer $k \geq 1$.

Theorem 3. *Let u and w be two vertices in the plane. Let m be the midpoint of the side of T_{uw} opposite u and let α be the unsigned angle between uw and um. There exists a path connecting u and w in the $\theta_{(4k+3)}$-graph of length at most*

$$\left(\frac{\cos\alpha}{\cos\left(\frac{\theta}{2}\right)} + \frac{(\cos\alpha \cdot \tan\left(\frac{\theta}{2}\right) + \sin\alpha) \cdot \cos\left(\frac{\theta}{4}\right)}{\cos\left(\frac{\theta}{2}\right) - \sin\left(\frac{3\theta}{4}\right)}\right) \cdot |uw|.$$

Proof. We apply Theorem 1 using $c = \cos(\theta/4)/(\cos(\theta/2) - \sin(3\theta/4))$. It remains to handle Case (c), where $w \in C_k^v$ and $|cw| > |dw|$, and Case (d), where $w \in C_{k+1}^v$.

Recall that c and d are the left and right corners of T_{vw}, opposite to v, and a'' is the intersection of aw and the line through v, parallel to ua. Let β be $\angle a''wv$ and let γ be the angle between vw and the bisector of T_{vw}. Since T_{vw} is smaller than T_{uw}, the induction hypothesis gives a bound on $\delta(v, w)$. Since $|uv| \leq |ua'| + |a'v|$ and v, a'', a, and a' form a parallelogram, we need to show that $\delta(v, w) \leq |va''| + |a''w| \cdot c$ for both cases in order to complete the proof.

Case (c): When w lies in C_k^v and $|cw| > |dw|$, the induction hypothesis for T_{vw} gives $\delta(v, w) \leq |vc| + |cw| \cdot c$. We note that $\gamma = 3\theta/4 - \beta$. Hence Lemma 3 gives that the inequality holds when $c \geq (\cos(3\theta/4 - \beta) - \sin\beta)/(\cos(\theta/2 - \beta) - \sin(5\theta/4 - \beta))$. As this function is decreasing in β for $\theta/4 \leq \beta \leq 3\theta/4$, it is maximized when β equals $\theta/4$. Hence c needs to be at least $c \geq (\cos(\theta/2) - \sin(\theta/4))/(\cos(\theta/4) - \sin\theta)$, which is equal to $\cos(\theta/4)/(\cos(\theta/2) - \sin(3\theta/4))$.

Case (d): When w lies in C_{k+1}^v, w lies above the bisector of T_{vw} and the induction hypothesis for T_{vw} gives $\delta(v, w) \leq |wd| + |dv| \cdot c$. We note that $\gamma = \theta/4 + \beta$. Hence Lemma 3 gives that the inequality holds when $c \geq (\cos(\theta/4 + \beta) - \sin\beta)/(\cos(\theta/2 - \beta) - \sin(3\theta/4 + \beta))$, which is equal to $\cos(\theta/4)/(\cos(\theta/2) - \sin(3\theta/4))$. □

Theorem 4. *Let u and w be two vertices in the plane. Let m be the midpoint of the side of T_{uw} opposite u and let α be the unsigned angle between uw and um. There exists a path connecting u and w in the $\theta_{(4k+5)}$-graph of length at most*

$$\left(\frac{\cos\alpha}{\cos\left(\frac{\theta}{2}\right)} + \frac{(\cos\alpha \cdot \tan\left(\frac{\theta}{2}\right) + \sin\alpha) \cdot \cos\left(\frac{\theta}{4}\right)}{\cos\left(\frac{\theta}{2}\right) - \sin\left(\frac{3\theta}{4}\right)} \right) \cdot |uw|.$$

Proof. We apply Theorem 1 using $c = \cos(\theta/4)/(\cos(\theta/2) - \sin(3\theta/4))$. It remains to handle Case (c), where $w \in C_k^v$ and $|cw| > |dw|$, and Case (d), where $w \in C_{k+1}^v$.

Recall that c and d are the left and right corners of T_{vw}, opposite to v, and a'' is the intersection of aw and the line through v, parallel to ua. Let β be $\angle a''wv$ and let γ be the angle between vw and the bisector of T_{vw}. Since T_{vw} is smaller than T_{uw}, the induction hypothesis gives a bound on $\delta(v, w)$. Since $|uv| \leq |ua'| + |a'v|$ and v, a'', a, and a' form a parallelogram, we need to show that $\delta(v, w) \leq |va''| + |a''w| \cdot c$ for both cases in order to complete the proof.

Case (c): When w lies in C_k^v and $|cw| > |dw|$, the induction hypothesis for T_{vw} gives $\delta(v, w) \leq |vc| + |cw| \cdot c$. We note that $\gamma = 5\theta/4 - \beta$. Hence Lemma 3 gives that the inequality holds when $c \geq (\cos(5\theta/4 - \beta) - \sin\beta)/(\cos(\theta/2 - \beta) - \sin(5\theta/4 - \beta))$. As this function is decreasing in β for $3\theta/4 \leq \beta \leq 5\theta/4$, it is maximized when β equals $3\theta/4$. Hence c needs to be at least $c \geq (\cos(\theta/2) - \sin(3\theta/4))/(\cos(\theta/4) - \sin\theta)$, which is less than $\cos(\theta/4)/(\cos(\theta/2) - \sin(3\theta/4))$.

Case (d): When w lies in C_{k+1}^v, the induction hypothesis for T_{vw} gives $\delta(v, w) \leq \max\{|vc| + |cw| \cdot c, |vd| + |dw| \cdot c\}$. If $\delta(v, w) \leq |vc| + |cw| \cdot c$, we note that $\gamma = \theta/4 - \beta$. Hence Lemma 3 gives that the inequality holds when $c \geq (\cos(\theta/4 - \beta) - \sin\beta)/(\cos(\theta/2 - \beta) - \sin(3\theta/4 - \beta))$. As this function is decreasing in β for $0 \leq \beta \leq \theta/4$, it is maximized when β equals 0. Hence c needs to be at least $c \geq \cos(\theta/4)/(\cos(\theta/2) - \sin(3\theta/4))$.

If $\delta(v, w) \leq |vd| + |dw| \cdot c$, we note that $\gamma = \theta/4 + \beta$. Hence Lemma 3 gives that the inequality holds when $c \geq (\cos(\beta - \theta/4) - \sin\beta)/(\cos(\theta/2 - \beta) - \sin(\theta/4 + \beta))$, which is equal to $\cos(\theta/4)/(\cos(\theta/2) - \sin(3\theta/4))$. □

When looking at two vertices u and w in the $\theta_{(4k+3)}$-graph and the $\theta_{(4k+5)}$-graph, we notice that when the angle between uw and the bisector of T_{uw} is α, the angle between wu and the bisector of T_{wu} is $\theta/2 - \alpha$. Hence the worst case spanning ratio becomes the minimum of the spanning ratio when looking at T_{uw} and the spanning ratio when looking at T_{wu}.

Theorem 5. *The $\theta_{(4k+3)}$-graph and $\theta_{(4k+5)}$-graph are $\dfrac{\cos\left(\frac{\theta}{4}\right)}{\cos\left(\frac{\theta}{2}\right)-\sin\left(\frac{3\theta}{4}\right)}$-spanners of K_n.*

Proof. The spanning ratio of the $\theta_{(4k+3)}$-graph and the $\theta_{(4k+5)}$-graph is at most:

$$\min \left\{ \begin{array}{l} \dfrac{\cos\alpha}{\cos\left(\frac{\theta}{2}\right)} + \dfrac{\left(\cos\alpha\cdot\tan\left(\frac{\theta}{2}\right)+\sin\alpha\right)\cdot\cos\left(\frac{\theta}{4}\right)}{\cos\left(\frac{\theta}{2}\right)-\sin\left(\frac{3\theta}{4}\right)}, \\[3mm] \dfrac{\cos\left(\frac{\theta}{2}-\alpha\right)}{\cos\left(\frac{\theta}{2}\right)} + \dfrac{\left(\cos\left(\frac{\theta}{2}-\alpha\right)\cdot\tan\left(\frac{\theta}{2}\right)+\sin\left(\frac{\theta}{2}-\alpha\right)\right)\cdot\cos\left(\frac{\theta}{4}\right)}{\cos\left(\frac{\theta}{2}\right)-\sin\left(\frac{3\theta}{4}\right)} \end{array} \right\}$$

Since $\cos\alpha/\cos(\theta/2)+(\cos\alpha\cdot\tan(\theta/2)+\sin\alpha)\cdot c$ is increasing for $\alpha \in [0, \theta/2]$, for $\theta \leq 2\pi/7$, the minimum of these two functions is maximized when the two functions are equal, i.e. when $\alpha = \theta/4$. Thus the $\theta_{(4k+3)}$-graph and the $\theta_{(4k+5)}$-graph has spanning ratio at most:

$$\frac{\cos\left(\frac{\theta}{4}\right)}{\cos\left(\frac{\theta}{2}\right)} + \frac{\left(\cos\left(\frac{\theta}{4}\right)\cdot\tan\left(\frac{\theta}{2}\right)+\sin\left(\frac{\theta}{4}\right)\right)\cdot\cos\left(\frac{\theta}{4}\right)}{\cos\left(\frac{\theta}{2}\right)-\sin\left(\frac{3\theta}{4}\right)} = \frac{\cos\left(\frac{\theta}{4}\right)\cdot\cos\left(\frac{\theta}{2}\right)}{\cos\left(\frac{\theta}{2}\right)\cdot\left(\cos\left(\frac{\theta}{2}\right)-\sin\left(\frac{3\theta}{4}\right)\right)}$$

$\square$

Furthermore, we observe that the proofs of Theorem 3 and Theorem 4 follow the same path as the θ-routing algorithm follows.

Theorem 6. *The θ-routing algorithm is $1 + \dfrac{2\cdot\sin\left(\frac{\theta}{2}\right)\cdot\cos\left(\frac{\theta}{4}\right)}{\cos\left(\frac{\theta}{2}\right)-\sin\left(\frac{3\theta}{4}\right)}$-competitive on the $\theta_{(4k+3)}$-graph and the $\theta_{(4k+5)}$-graph.*

References

1. Bonichon, N., Gavoille, C., Hanusse, N., Ilcinkas, D.: Connections between theta-graphs, Delaunay triangulations, and orthogonal surfaces. In: Thilikos, D.M. (ed.) WG 2010. LNCS, vol. 6410, pp. 266–278. Springer, Heidelberg (2010)
2. Bose, P., De Carufel, J.L., Morin, P., van Renssen, A., Verdonschot, S.: Optimal bounds on theta-graphs: More is not always better. In: CCCG, pp. 305–310 (2012)
3. Bose, P., Fagerberg, R., van Renssen, A., Verdonschot, S.: Competitive routing in the half-θ_6-graph. In: SODA 2012, pp. 1319–1328 (2012)
4. Chew, P.: There are planar graphs almost as good as the complete graph. Journal of Computer and System Sciences 39(2), 205–219 (1989)
5. Clarkson, K.: Approximation algorithms for shortest path motion planning. In: STOC, pp. 56–65 (1987)
6. Keil, J.: Approximating the complete Euclidean graph. In: Karlsson, R., Lingas, A. (eds.) SWAT 1988. LNCS, vol. 318, pp. 208–213. Springer, Heidelberg (1988)
7. Narasimhan, G., Smid, M.: Geometric Spanner Networks. Cambridge University Press (2007)
8. Ruppert, J., Seidel, R.: Approximating the d-dimensional complete Euclidean graph. In: CCCG, pp. 207–210 (1991)

A   Approximate Values

The following table shows approximate values of our improved spanning and routing ratios, compared to the existing spanning ratios by Ruppert and Seidel [8] (for the $\theta_{(4k+3)}$-graph, the $\theta_{(4k+4)}$-graph, and the $\theta_{(4k+5)}$-graph) and Bose *et al.* [2] (for the $\theta_{(4k+2)}$-graph) and the existing routing ratios by Ruppert and Seidel [8].

Table 2. Approximate spanning and routing ratios

m	New Spanning	Previous Spanning	New Routing	Previous Routing
7	3.5136	7.5625	4.0490	7.5625
8	2.4143	4.2620	2.4143	4.2620
9	2.2398	3.1650	2.5321	3.1650
10	-	1.6181	-	2.6181
11	1.8193	2.2908	2.0251	2.2908
12	1.7321	2.0732	1.7321	2.0732
13	1.6107	1.9181	1.7710	1.9181
14	-	1.4451	-	1.8020
15	1.4863	1.7119	1.6181	1.7119
16	1.4967	1.6399	1.4967	1.6399
17	1.4039	1.5811	1.5159	1.5811
18	-	1.3473	-	1.5321
19	1.3452	1.4908	1.4429	1.4908
20	1.3764	1.4554	1.3764	1.4554
21	1.3014	1.4247	1.3879	1.4247
22	-	1.2847	-	1.3979
23	1.2674	1.3743	1.3452	1.3743
24	1.3033	1.3533	1.3033	1.3533
25	1.2402	1.3346	1.3109	1.3346

Relative Interval Analysis of Paging Algorithms on Access Graphs[*]

Joan Boyar, Sushmita Gupta, and Kim S. Larsen

University of Southern Denmark, Odense, Denmark
{joan,sgupta,kslarsen}@imada.sdu.dk

Abstract. Access graphs, which have been used previously in connection with competitive analysis and relative worst order analysis to model locality of reference in paging, are considered in connection with relative interval analysis. The algorithms LRU, FIFO, FWF, and FAR are compared using the path, star, and cycle access graphs. In this model, some of the expected results are obtained. However, although LRU is found to be strictly better than FIFO on paths, it has worse performance on stars, cycles, and complete graphs, in this model. We solve an open question from [Dorrigiv, López-Ortiz, Munro, 2009], obtaining tight bounds on the relationship between LRU and FIFO with relative interval analysis.

1 Introduction

The paging problem is the problem of maintaining a subset of a potentially very large set of pages from memory in a significantly smaller cache. When a page is requested, it may already be in cache (called a "hit"), or it must be brought into cache (called a "fault"). The algorithmic problem is the one of choosing an eviction strategy, i.e., which page to evict from cache in the case of a fault, with the objective of minimizing the total number of faults.

Many different paging algorithms have been considered in the literature, many of which can be found in [3,13]. Among the best known are LRU (least-recently-used), which always evicts the least recently used page, and FIFO (first-in-first-out), which evicts pages in the order they entered the cache. We also consider a known bad algorithm, FWF (flush-when-full), which is often used for reference, since quality measures ought to be able to determine at the very least that it is worse than the other algorithms. If FWF encounters a fault with a full cache, it empties its cache, and brings the new page in. Finally, we consider a more involved algorithm, FAR, which works with respect to a known access graph. Whenever a page is requested, it is marked. When it is necessary to evict a page, it always evicts an unmarked page. If all pages are marked in such a situation, FAR first unmarks all pages. The unmarked page it chooses to evict is the one

---

[*] Supported in part by the Danish Council for Independent Research. Part of this work was carried out while the first and third authors were visiting the University of Waterloo, Ontario, Canada.

F. Dehne, R. Solis-Oba, and J.-R. Sack (Eds.): WADS 2013, LNCS 8037, pp. 195–206, 2013.
© Springer-Verlag Berlin Heidelberg 2013

farthest from any marked page in the access graph. For breaking possible ties, we assume the LRU strategy in this paper.

Understanding differences in paging algorithms' behavior under various circumstances has been a topic for much research. The most standard measure of quality of an online algorithm, competitive analysis [18,15], cannot directly distinguish between most of them. It deems LRU, FIFO, and FWF equivalent, with a competitive ratio of k, where k denotes the size of the cache. Other measures, such as relative worst order analysis [5,6], can be used to obtain more separations, including that LRU and FIFO are better than FWF and that look-ahead helps. No techniques have been able to separate LRU and FIFO, without adding some modelling of locality of reference.

Although LRU performs better than FIFO in some practical situations [19], if one considers all sequences of length n for any n, bijective/average analysis shows that their average number of faults on these sequences is identical [2], which basically follows from LRU and FIFO being demand paging algorithms. Thus, it is not surprising that some assumptions involving locality of reference are necessary to separate them.

A separation between FIFO and LRU was established quite early using access graphs for modelling locality of reference [10], showing that under competitive analysis, no matter which access graph one restricts to, LRU always does at least as well as FIFO. This proved a conjecture in [4], where the access graph model was introduced. Another way to restrict the input sequences was investigated in [1]. Using Denning's working set model [11,12] as an inspiration, sequences were limited with regards to the number of distinct pages in a sliding window of size k. This also favors LRU, as does bijective analysis [2], using the same locality of reference definition as [1]. There has also been work in the direction of probabilistic models, including the diffuse adversary model [17] and Markov chain based models [16].

The earlier successes and the generality of access graphs, together with the possibilities the model offers with regards to investigating specific access patterns, makes it an interesting object for further studies. In the light of the recent focus on development of new performance measures, together with the comparative studies initiated in [9], exploring access graphs results in the context of new performance measures seems like a promising direction for expanding our understanding of performance measures as well as concrete algorithms.

One step in that direction was carried out in [7], where more nuanced results were demonstrated, showing that restricting input sequences using the access graph model, while applying relative worst order analysis, LRU is strictly better than FIFO on paths and cycles. The question as to whether or not LRU is at least as good as FIFO on all finite graphs was left as an open problem, but it was shown that there exists a family of graphs which grows with the length of the corresponding request sequence, where LRU and FIFO are incomparable. Since LRU is optimal on paths, it is not surprising that both competitive analysis and relative worst order analysis find that LRU is better than FIFO on paths. Any "reasonable" analysis technique should give this result. Under competitive analysis, LRU and FIFO are equivalent on cycles. The separation by relative worst order analysis occurs because cycles contain paths, LRU is better on paths, and relative worst order analysis can reflect this. The fact that there exists an infinite family of graphs which grows with the length of the sequence where LRU and FIFO are incomparable may or may not be interesting. There are many sequences were FIFO is better than LRU; they just seem to occur less often in real applications.

Comparing two algorithms under almost any analysis technique is generally equivalent to considering them with the complete graph as an access graph, since the complete graph does not restrict the request sequence in any way. Thus, LRU and FIFO are equivalent on complete graphs under both competitive analysis and relative worst order analysis, since they are equivalent without considering access graphs.

In this paper, we consider relative interval analysis [14]. In some ways relative interval analysis is between competitive analysis and relative worst order analysis. As with relative worst order analysis, two algorithms are compared directly to each other, rather than compared to OPT. This gives the advantage that, when one algorithm dominates another in the sense that it is at least as good as the other on every request sequence and better on some, the analysis will reflect this. However, it is similar to competitive analysis in that the two algorithms are always compared on exactly the same sequence. To compare two algorithms, LRU and FIFO for example, one considers the difference between LRU's and FIFO's performance on any sequence, divided by the length of that sequence. The range that these ratios can take is the "interval" for that pair of algorithms. For FIFO and LRU, [14] found two families of sequences I_n and J_n such that $\lim_{n\to\infty} \frac{\text{FIFO}(I_n) - \text{LRU}(I_n)}{n} = -1 + \frac{1}{k}$ and $\lim_{n\to\infty} \frac{\text{FIFO}(J_n) - \text{LRU}(J_n)}{n} = \frac{1}{2} - \frac{1}{4k-2}$. They left it as an open problem to determine if worse sequences exist, making the interval even larger. In their notation, they proved: $[-1 + \frac{1}{k}, \frac{1}{2} - \frac{1}{4k-2}] \subseteq \mathcal{I}(\text{FIFO}, \text{LRU})$. We start by proving that this is tight: $\mathcal{I}(\text{FIFO}, \text{LRU}) = [-1 + \frac{1}{k}, \frac{1}{2} - \frac{1}{4k-2}]$. These results would be interpreted as saying that FIFO has *better performance* than LRU, since the absolute value of the minimum value in the interval is larger than the maximum, but also that they have different strengths, since zero is contained in the interior of the interval. We obtain more nuanced results by considering various types of access graphs: complete graphs (K_N), paths (P_N), stars (S_N), and cycles (C_N). This splits the interval of $[-1 + \frac{1}{k}, \frac{1}{2} - \frac{1}{4k-2}]$ into subintervals for the respective graph classes. Table 1 shows our results.

Comparing these results with the results from competitive analysis and relative worst order analysis, both with respect to access graphs, it becomes clear that different measures highlight different aspects of the algorithms. Both measures show that LRU is strictly better than FIFO on paths, which is not surprising since it is in fact optimal on paths and FIFO is not. On the other access graphs considered here, relative interval analysis gives results which can be interpreted as incomparability, but leaning towards deeming FIFO the better algorithm. Relative worst order analysis, on the other hand, shows that on cycles,

Table 1. Summary of Results: $\mathcal{A} \in \{\text{FAR}, \text{LRU}\}$, $\mathcal{B} \in \{\text{FAR}, \text{FIFO}, \text{LRU}\}$, $N = k + r$, with $1 \le r \le k - 1$, $X_r = r(x-1) + \lceil \frac{N}{2^x} \rceil$ with $x = \lfloor \log \frac{N}{r} \rfloor$, and $\hat{N}$ denotes N if N is even, and $N - 1$ otherwise.

Lower Bound	Relative Interval	Upper Bound	Theorem
	$\mathcal{I}^{K_N}[\text{FIFO}, \text{LRU}] =$	$\left[-1 + \frac{1}{k}, \; \frac{1}{2} - \frac{1}{4k-2}\right]$	1
	$\mathcal{I}^{K_N}[\text{FWF}, \mathcal{A}] \quad =$	$\left[0, \; 1 - \frac{1}{k}\right]$	2
$\left[0, \; 1 - \frac{k+1}{k^2}\right]$	$\subseteq \mathcal{I}^{K_N}[\text{FWF}, \text{FIFO}] \subseteq$	$\left[0, \; 1 - \frac{1}{k}\right]$	3
	$\mathcal{I}^{P_N}[\text{FIFO}, \mathcal{A}] \quad =$	$\left[0, \; \frac{1}{2} - \frac{1}{2k}\right]$	4
	$\mathcal{I}^{P_N}[\text{FWF}, \mathcal{A}] \quad =$	$\left[0, \; 1 - \frac{1}{k}\right]$	2
$\left[0, \; 1 - \frac{k+1}{k^2}\right]$	$\subseteq \mathcal{I}^{P_N}[\text{FWF}, \text{FIFO}] \subseteq$	$\left[0, \; 1 - \frac{1}{k}\right]$	3
$\left[-\frac{1}{2} + \Theta(\frac{1}{k}), \; \frac{1}{4} + \Theta(\frac{1}{k})\right]$	$\subseteq \mathcal{I}^{S_N}[\text{FIFO}, \mathcal{A}] \subseteq$	$\left[-\frac{1}{2} + \Theta(\frac{1}{k}), \; \frac{1}{4} + \Theta(\frac{1}{k})\right]$	5
	$\mathcal{I}^{S_N}[\text{FWF}, \mathcal{B}] \quad =$	$\left[0, \; \frac{1}{2}\right]$	6
$\left[-1 + \frac{r}{k}, \; \frac{1}{2} - \frac{1}{4k-2}\right]$	$\subseteq \mathcal{I}^{C_N}[\text{FIFO}, \text{LRU}] \subseteq$	$\left[-1 + \frac{1}{k}, \; \frac{1}{2} - \frac{1}{4k-2}\right]$	7
	$\mathcal{I}^{C_N}[\text{FWF}, \text{LRU}] \quad =$	$\left[0, \; 1 - \frac{1}{k}\right]$	8
$\left[-\frac{r\left(\lfloor \log \frac{\hat{N}}{r} \rfloor - 1\right)}{N-1}, \; 1 - \frac{X_r}{k}\right]$	$\subseteq \mathcal{I}^{C_N}[\text{LRU}, \text{FAR}] \subseteq$	$\left[-\frac{X_r-1}{k}, \; 1 - \frac{1}{k}\right]$	9
$\left[-\frac{X_r-r}{k}, \; 1 - \frac{X_r}{k}\right]$	$\subseteq \mathcal{I}^{C_N}[\text{FIFO}, \text{FAR}] \subseteq$	$\left[-\frac{X_r-1}{k}, \; 1 - \frac{1}{k}\right]$	9
$\left[0, \; 1 - \frac{X_r}{k}\right]$	$\subseteq \mathcal{I}^{C_N}[\text{FWF}, \text{FAR}] \subseteq$	$\left[0, \; 1 - \frac{1}{k}\right]$	9
$\left[0, \; 1 - \frac{k+1}{k^2}\right]$	$\subseteq \mathcal{I}^{C_N}[\text{FWF}, \text{FIFO}] \subseteq$	$\left[0, \; 1 - \frac{1}{k}\right]$	3

LRU is strictly better than FIFO, and on complete graphs, they are equivalent. It has not yet been studied on stars, but an incomparability result for LRU and FIFO has been found for a family of graphs growing with the length of the input.

2   Preliminaries

We have defined the paging algorithms in the introduction. If more detail is desired, the algorithms are described in [3].

An *access graph* for paging models the access patterns, i.e., which pages can be requested after a given page. Thus, the vertices are pages, and after a page p has been requested, the next request is to p or one of its neighbors in the access graph. We let N denote the number of vertices of the access graph under consideration at a given time. This is the same as the number of different pages we consider. We will always assume that $N > k$, since otherwise the problem is trivial, and let $r = N - k$. A requests sequence is a sequence of pages and the sequence *respects* a given access graph if any two consecutive requests are either identical or neighbors in the access graph. We let $\mathcal{L}(G)$ denote the set of all request sequences respecting G.

We use the definition of k-phases from [3]:

Definition 1. *A request sequence can be divided recursively into a number of k-phases as follows: Phase 0 is the empty sequence. For every $i \ge 1$, Phase i is a maximal sequence following Phase $i - 1$ containing at most k distinct requests.*

Thus, Phase i begins on the $(k + 1)$st distinct page requested since the start of Phase $i - 1$, and the last phase may contain fewer than k different pages. We generally want to ignore Phase 0, and refer to Phase 1 as the first phase.

Similarly, we can define x-blocks, for some integer x, focusing on when a given algorithm $\mathcal{A}$ has faulted x times.

Definition 2. *A request sequence can be divided recursively into a number of x-blocks with respect to an algorithm $\mathcal{A}$ as follows: The 0th x-block is the empty sequence. For every $i \geq 1$, the ith x-block is a maximal sequence following the $(i - 1)$st x-block for which $\mathcal{A}$ faults at most x times.*

The complete *blocks are defined to be the ones with x faults, i.e., excluding the 0th block and possibly the last.*

There are some well-known and important classifications of paging algorithms, which are used here and in most other papers on paging [3]: A paging algorithm is called *conservative* if it incurs at most k page faults on any consecutive subsequence of the input containing k or fewer distinct page references. LRU and FIFO belong to this class. Similarly, a paging algorithm is called a *marking* algorithm if for any k-phase, once a page has been requested in that phase, it is not evicted for the duration of that phase. LRU, FAR, and FWF are marking algorithms.

If $\mathcal{A}$ is a paging algorithm, we let $\mathcal{A}(I)$ denote $\mathcal{A}$'s cost (number of faults) on the input (request) sequence I. We adapt relative interval analysis from [14] to access graphs. Let $\mathcal{A}$ and $\mathcal{B}$ be two algorithms. We define the following notation:

$$\text{Min}^G(\mathcal{A}, \mathcal{B}) = \lim_{n \to \infty} \inf \frac{\min_{|I|=n, I \in \mathcal{L}(G)} \{\mathcal{A}(I) - \mathcal{B}(I)\}}{n} \text{ and}$$

$$\text{Max}^G(\mathcal{A}, \mathcal{B}) = \lim_{n \to \infty} \sup \frac{\max_{|I|=n, I \in \mathcal{L}(G)} \{\mathcal{A}(I) - \mathcal{B}(I)\}}{n}$$

Definition 3. *The relative interval of two algorithms $\mathcal{A}$ and $\mathcal{B}$ with respect to the access graph, G, is*

$$\mathcal{I}^G(\mathcal{A}, \mathcal{B}) = [\text{Min}^G(\mathcal{A}, \mathcal{B}), \text{Max}^G(\mathcal{A}, \mathcal{B})]$$

$\mathcal{B}$ has better performance than $\mathcal{A}$ *if* $\text{Max}^G(\mathcal{A}, \mathcal{B}) > |\text{Min}^G(\mathcal{A}, \mathcal{B})|$. $\mathcal{B}$ dominates $\mathcal{A}$ *if* $\mathcal{I}^G(\mathcal{A}, \mathcal{B}) = [0, \beta]$ *for some* $\beta > 0$. *Note that* $\text{Max}^G(\mathcal{A}, \mathcal{B}) = -\text{Min}^G(\mathcal{B}, \mathcal{A})$.

This definition generalizes the one from [14] in that the original definition is the special case where G is the complete graph, which is the same as saying that there are no restrictions on the sequences.

Note that if $\mathcal{B}$ *dominates* $\mathcal{A}$, this means that $\mathcal{A}$ does not outperform $\mathcal{B}$ on any sequence (asymptotically), while there are sequences on which $\mathcal{B}$ outperforms $\mathcal{A}$. Also, when $\text{Max}^G(\mathcal{A}, \mathcal{B})$ is close to 0, this indicates that $\mathcal{A}$'s performance is not much worse than that of $\mathcal{B}$'s.

Due to space limitations, most proofs and the statements of most lemmas have been omitted. Refer to [8] for all the details.

3   Complete Graphs

As remarked earlier, if the access graph is complete, it incurs no restrictions, so the result of this section is in the same model as [14]. In [14], it is shown that $[-\frac{k-1}{k}, \frac{k-1}{2k-1}] \subseteq \mathcal{I}(\text{FIFO}, \text{LRU})$. Below, we answer an open question from [14], proving that this is tight. The full version of the paper [8] contains a more detailed proof.

Lemma 1. *For any access graph* G,

$$-1 + \frac{1}{k} \leq \text{Min}^G(\text{FIFO}, \text{LRU}) \ \text{and} \ \text{Max}^G(\text{FIFO}, \text{LRU}) \leq \frac{1}{2} - \frac{1}{4k - 2}.$$

Proof. We first consider the Min value. Suppose that a sequence I has b complete k-phases. Since LRU is conservative and a complete k-phase contains k distinct pages, it cannot fault more than $bk + k - 1$ times [3]. With b complete k-phases, $\text{FIFO}(I) \geq k + b - 1$, so $\text{FIFO}(I) - \text{LRU}(I) \geq k + b - 1 - (bk + k - 1) = -b(k - 1)$. Each k-phase must have length at least k, so $|I| \geq bk$. Thus, $\text{Min}^G(\text{FIFO}, \text{LRU}) \geq -\frac{b(k-1)}{bk} = -1 + \frac{1}{k}$.

We now consider the Max value. Given a request sequence I, we let B_i denote the ith k-block for FIFO. Assume that there are b complete k-blocks. FIFO faults k times per complete k-block and up to $k - 1$ times for the possible final k-block. Thus, $\text{FIFO}(I) \leq bk + (k - 1)$. Assume that LRU faults α_i times in B_i. With b complete k-blocks, which are at least as long as k-phases, LRU faults at least $b + k - 1$ times. Thus, $\Sigma_{i=1}^{b} \alpha_i \geq b + k - 1$.

We now compute a lower bound on the length of the request sequence I based on the number of complete k-blocks in it and the algorithms' behavior on it.

As a first step, with every request on which FIFO faults and LRU has a hit, we associate a distinct request where FIFO has a hit. Let r be such a request to a page p in B_i. Since it is a hit for LRU, p must have been requested in the maximal subsequence of requests I' consisting of k distinct pages and ending just before r. Consider the first such request, r', in I'. If it were a fault for FIFO, FIFO could not have faulted again on r. Thus, r' was a hit for FIFO and we associate r' with r.

To establish that the association is distinct, assume that r' also gets associated with a request r''. Without loss of generality, assume that r'' is later than r. For FIFO to fault on both r and r'', there must be at least k distinct pages different from p in between r and r''. However, since we are assuming that LRU has a hit on r'', by the property of LRU, the page requested by r'' must have been requested during the same k distinct pages. Thus, by the construction above, the page that gets associated with r'' (and r) will be later than r, which is a contradiction.

Thus, if LRU faults α_i times in B_i, by the procedure above, we identify at least $k - \alpha_i$ distinct requests. In total, there are at least $\Sigma_{i=1}^{b}(k - \alpha_i) = bk - \Sigma_{i=1}^{b}\alpha_i$

distinct hits for FIFO in I and, since there are b complete k-blocks, at least bk faults. Thus, the length of I is at least $2bk - \Sigma_{i=1}^{b}\alpha_i$, and

$$\frac{\text{FIFO}(I) - \text{LRU}(I)}{|I|} \leq \frac{bk + k - 1 - \Sigma_{i=1}^{b}\alpha_i}{2bk - \Sigma_{i=1}^{b}\alpha_i}.$$

By the lower bound on $\Sigma_{i=1}^{b}\alpha_i$ above, and the arithmetic observation that $\frac{u-y}{v-y} < \frac{u-x}{v-x}$, if $u < v$ and $x < y < v$, we have that

$$\frac{bk + k - 1 - \Sigma_{i=1}^{b}\alpha_i}{2bk - \Sigma_{i=1}^{b}\alpha_i} \leq \frac{bk + k - 1 - (b + k - 1)}{2bk - (b + k - 1)} = \frac{b(k-1)}{b(2k-1) - k + 1}.$$

Clearly, $\max_{|I|=n, I \in \mathcal{L}(G)}\{\text{FIFO}(I) - \text{LRU}(I))\}$ is unbounded as a function of n. Since $\lim_{b \to \infty} \frac{b(k-1)}{b(2k-1)-k+1} = \frac{k-1}{2k-1}$, we have $\text{Max}^G(\text{FIFO}, \text{LRU}) \leq \frac{k-1}{2k-1} = \frac{1}{2} - \frac{1}{4k-2}$. □

From [14] and Lemma 1, we have the following:

Theorem 1. $\mathcal{I}(\text{FIFO}, \text{LRU}) = [-1 + \frac{1}{k}, \frac{1}{2} - \frac{1}{4k-2}]$.

3.1   FWF

FWF performs very badly compared to the other algorithms considered here, LRU, FAR, and FIFO. The following is folklore:

Lemma 2. *For any sequence I and any conservative or marking algorithm $\mathcal{A}$, we have $\mathcal{A}(I) \leq \text{FWF}(I)$.*

Thus, for any graph G, $\text{Min}^G[\text{FWF}, \mathcal{A}] = 0$, where $\mathcal{A}$ is either FAR, LRU, or FIFO. Hence, LRU, FIFO, and FAR all dominate FWF.

Theorem 2. *For the path access graph P_N, where $N \geq k + 1$ (for LRU, for any graph containing P_{k+1}), $\mathcal{I}^{P_N}[\text{FWF}, \mathcal{A}] = [0, 1 - \frac{1}{k}]$, where $\mathcal{A} \in \{\text{LRU}, \text{FAR}\}$.*

For FWF versus FIFO, a result almost as tight holds:

Theorem 3. *For any graph G containing a path with $k + 1$ vertices, if k is odd, then $\mathcal{I}^G[\text{FWF}, \text{FIFO}] = [0, 1 - \frac{1}{k}]$, and if k is even, then $[0, \frac{k^2-k-1}{k^2}] \subseteq \mathcal{I}^G[\text{FWF}, \text{FIFO}] \subseteq [0, \frac{k-1}{k}]$.*

4   Path Graphs

Lemma 3. *For the path access graph P_N, we have $\text{Max}^{P_N}(\text{FIFO}, \text{LRU}) \leq \frac{1}{2} - \frac{1}{2k}$.*

Proof. Consider any request sequence I. We divide the sequence up into phases as described now (these are *not* k-phases). Initially, define a direction by where LRU makes its kth fault compared with its cache content. Without loss of generality, we assume this happens going to the right on the path.

We start the first phase with the first request and later explain how subsequent phases are started. In all the phases, we start to the left (relatively). In all phases, except the first, LRU has the first $k - 1$ distinct pages that will be requested during that phase in cache. In all phases, the first fault by LRU in the phase, after having processed the first $k - 1$ distinct pages, is to the right. We maintain this as an invariant that holds at the start of any phase, though the direction can change, as we will get back to at the end of the proof. The exception in the first phase, adding an extra $k - 1$ faults to the cost of LRU as compared with the analysis below, will not influence the result in the the limit for the length of the request sequence going towards infinity.

We want to analyze a phase where LRU faults to the right before it faults to the left again. These faults to the right may not appear consecutively. There may be some faults in a row, but then there may be hits and then faults again, etc. Thus, assume that there are m maximal subsequences of requests to the right where LRU faults—all of this before LRU faults going to the left again. Assume further that these maximal subsequences of requests give rise to $s_1, s_2, \ldots, s_m$ faults, respectively, where, by definition, $m \geq 1$, and let $s = \Sigma_{i=1}^m s_i$.

For now, we assume that for all i, $s_i < k$. Thus, LRU moves left and right at least m times; maybe more times where it does not give rise to faults. Since it does not fault going to the left during these turns, the faults are to pages further and further to the right. Let E_{right} denote the extreme rightmost position it reaches during these faults to the right.

When LRU faults again to the left after having processed E_{right}, we consider the leftmost node E_{left}, where LRU faults after the s faults described above, but before it faults to the right again. We end the phase with the first request to E_{left} after the s faults. We define subsequent phases inductively in the same way, starting with the first request not included in the previous phase, possibly leaving an incomplete phase at the end.

We now consider the costs of the algorithms and the length of the sequence per phase. LRU faults s times going to the right during the m turns in the phase. Additionally, LRU must fault at least t times going from E_{right} to E_{left}, where t is defined by there being $k + t$ nodes between E_{left} and E_{right}, including both endpoints. This sums up to $s + t$ faults.

For FIFO, we postpone the discussion of the first s_1 distinct pages seen in a phase. Just to avoid any confusion, note that these pages are immediately to the right of E_{left} (the endpoint of the previous phase) and thus not the pages that LRU faults on. After that, consider the maximal subsequence of at most k distinct pages. This subsequence starts with the $(s_1 + 1)$st distinct request (the last request to it before the s_2 faults) and continues up to, but not including the first request that LRU has one of its s_2 faults on. We know that there are at most k pages there, because LRU only faults s_1 times there. Assume that FIFO faults f_1 times on this subsequence. Since FIFO is conservative, $f_1 \leq k$.

We define more such subsequences repeatedly, the $(m-1)$st of these ending just before LRU's first fault of the s_m faults, and the mth including the s_m faults and k of the $k+t$ nodes before we reach E_{left}. Finally, we return to the question of the first s_1 distinct pages seen in the phase. These overlap with the "t pages" from the previous phase; otherwise we would not have started the phase where we did. If FIFO faults on one of these pages when going through the t pages in the previous phase, it will not fault on them again in this phase. Thus, we only have to count them in one phase, and choose to do this in the previous phase. In total, FIFO faults at most $(\Sigma_{i=1}^{m} f_i) + t$ times, and for all i, $f_i \leq k$.

The difference between the cost of FIFO and LRU is then at most $(\Sigma_{i=1}^{m} f_i) + t - (s+t) = (\Sigma_{i=1}^{m} f_i) - s = (\Sigma_{i=1}^{m} (f_i - 1)) - (s - m)$.

From the analysis of FIFO above, knowing that on a subsequence of length at most k, FIFO can fault at most once on any given page, if it faults f_i times, the subsequence has at least f_i distinct pages. Given that the subsequence starts at the left end of the "s_i pages" and ends at the right end of the "s_i pages", all pages that FIFO faults on, except possibly the leftmost, must be requested at least twice, giving at least $2f_i - 1$ requests. So, the length of the sequence is at least $(\Sigma_{i=1}^{m} (2f_i - 1)) + t$. We now sum up over all phases, equipping each variable with a superscript denoting the phase number.

First, the total length, L, is at least

$$L \geq \Sigma_j(\Sigma_{i=1}^{m^j}(2f_i^j - 1)) + t^j = \Sigma_j(\Sigma_{i=1}^{m^j} 2f_i^j) - m^j + t^j.$$

Since s expresses how far we move to the right and t how far we move to the left, and the whole path has a bounded number of nodes N, we have that $\Sigma_j t^j \geq \Sigma_j s^j - N$. Thus, $L \geq (\Sigma_j(\Sigma_{i=1}^{m^j} 2f_i^j) - m^j + s^j) - N$.

I has a number of complete phases and then some extra requests in addition to that. There must exist a fixed constant c independent of I such that the cost of FIFO on the extra part of any sequence is bounded by c. This follows since there is a limit of N on how far requests can move to the right. So if requests never again come so far to the left that LRU faults, all requests thereafter are to only k pages. This added constant can also take care of the initial extra cost of $k-1$. Since we are just using a lower bound on the sequence length, we can ignore the length of a possibly incomplete phase at the end. Thus,

$$\frac{\text{FIFO}(I) - \text{LRU}(I)}{|I|} \leq \frac{c + \Sigma_j \Sigma_{i=1}^{m^j}(f_i^j - 1) - (s^j - m^j)}{-N + \Sigma_j(\Sigma_{i=1}^{m^j} 2f_i^j) - m^j + s^j} \leq \frac{c + \Sigma_j \Sigma_{i=1}^{m^j}(f_i^j - 1)}{-N + \Sigma_j \Sigma_{i=1}^{m^j} 2f_i^j}$$

$$\leq \frac{c + \Sigma_j m^j(k-1)}{-N + \Sigma_j m^j 2k} = \frac{c + (k-1)\Sigma_j m^j}{-N + 2k \Sigma_j m^j}$$

The second inequality follows since $s^j \geq m^j$, and the third inequality follows because $\frac{f_i^j - 1}{2f_i^j} \leq \frac{1}{2}$ and $k \geq f_i$ implies that $\frac{f_i^j - 1}{2f_i^j} \leq \frac{k-1}{2k}$.

For sequences where the number of phases does not approach infinity, as argued above, FIFO's cost will be bounded. For the number of phases approaching infinity, $\lim_{j \to \infty} \frac{c+(k-1)\Sigma_j m^j}{-N+2k\Sigma_j m^j} = \frac{k-1}{2k} = \frac{1}{2} - \frac{1}{2k}$, which implies the result.

Now, for this proof, we assumed that $s_i < k$. If $s_i \geq k$, we simply terminate the phase after the processing of the s_i requests that LRU faults on, and continue to define phases inductively from there. All the bounds from above hold with $t = 0$ and the observation that FIFO will not fault on the first s_1 requests in the next phase. The direction of the construction is now reversed. In this process, whenever we reverse the direction as above, we also rename the variable s to t and t to s, such that s continues to keep track of movement to the right and t of movement to the left, and the inequality $\Sigma_j t^j \geq \Sigma_j s^j - N$ still holds.   □

Theorem 4. $\mathcal{I}^{P_N}[\text{FIFO}, \text{LRU}] = \left[0, \frac{1}{2} - \frac{1}{2k}\right]$, and LRU dominates FIFO on paths.

Note that FAR and LRU perform identically on paths, so FAR also dominates FIFO with the same interval.

5   Star Graphs

We let S_N denote a star graph with N vertices. A star graph has a central vertex, s, which is directly connected to $N-1$ other vertices, none of which are directly connected. Thus, we could also see a star graph as a tree with root s and $N-1$ leaves, all located at a distance one from the root. The algorithms FAR and LRU behave identically on star graphs. Neither of them ever evicts the central vertex.

Theorem 5. For $\mathcal{A} \in \{\text{LRU}, \text{FAR}\}$, we have

$$\left[-\frac{1}{2} + \frac{1}{2(k-1)}, \frac{1}{4} + \frac{1}{8k-12}\right] \subseteq \mathcal{I}^{S_N}[\text{FIFO}, \mathcal{A}]$$
$$\subseteq \left[-\frac{1}{2} + \frac{1}{2(k-1)} + \frac{1}{2k(k-1)}, \frac{1}{4} + \frac{1}{8k-12}\right]$$

In [14], it was shown that $\text{Max}^G(\text{FIFO}, \text{LRU}) \geq \frac{k-1}{2k-1} = \frac{1}{2} - \frac{1}{4k-2}$. The above result shows that for star access graphs, that bound can be decreased by a factor of approximately two.

Since LRU and FAR perform identically on stars, $\text{Min}^{S_N}(\text{FAR}, \text{LRU}) = \text{Max}^{S_N}(\text{FAR}, \text{LRU}) = 0$.

Theorem 6. For $\mathcal{A} \in \{\text{LRU}, \text{FAR}, \text{FIFO}\}$, we have $\mathcal{I}^{S_N}[\text{FWF}, \mathcal{A}] = \left[0, \frac{1}{2}\right]$.

6   Cycle Graphs

We consider graphs consisting of exactly one cycle, containing $N \geq k+1$ vertices, and define $r = N - k$. We will concentrate on the case where $r < k$, since otherwise the cycle is so large that for the algorithms considered here, it works as if it were an infinite path. Thus, for example, there are sequences where FIFO performs worse than LRU, but on worst case sequences, simply going around the cycle, the algorithms perform identically.

In the following statements of theorems, we use $N = k + r$ with $1 \leq r \leq k - 1$, and $X_r = r(x - 1) + \lceil \frac{N}{2^x} \rceil$, where $x = \lfloor \log \frac{N}{r} \rfloor$, and $\hat{N}$ to denote N if N is even and $N - 1$ otherwise.

Theorem 7. *For the cycle access graph C_N,*

$$\left[-1 + \frac{r}{k}, \frac{1}{2} - \frac{1}{4k - 2} \right] \subseteq \mathcal{I}^{C_N} [\text{FIFO}, \text{LRU}] \subseteq \left[-1 + \frac{1}{k}, \frac{1}{2} - \frac{1}{4k - 2} \right]$$

Theorem 8. *For the cycle access graph C_N,*

$$\mathcal{I}^{C_N} [\text{FWF}, \text{LRU}] = \left[0, 1 - \frac{1}{k} \right]$$

Theorem 9. *For the cycle access graph C_N,*

$$\left[-\frac{X_r - r}{k}, 1 - \frac{X_r}{k} \right] \subseteq \mathcal{I}^{C_N} [\text{FIFO}, \text{FAR}] \subseteq \left[-\frac{X_r - 1}{k}, 1 - \frac{1}{k} \right],$$

$$\left[-\frac{r \left(\lfloor \log \frac{\hat{N}}{r} \rfloor - 1 \right)}{N - 1}, 1 - \frac{X_r}{k} \right] \subseteq \mathcal{I}^{C_N} [\text{LRU}, \text{FAR}] \subseteq \left[-\frac{X_r - 1}{k}, 1 - \frac{1}{k} \right], \ and$$

$$\left[0, 1 - \frac{X_r}{k} \right] \subseteq \mathcal{I}^{C_N} [\text{FWF}, \text{FAR}] \subseteq \left[0, 1 - \frac{1}{k} \right].$$

7   Concluding Remarks

Relative interval analysis has the advantage that it can separate algorithms properly when one algorithm is at least as good as another on every sequence and is better on some. This was reflected in the results concerning FWF which is dominated by the other algorithms considered for all access graphs. It was also reflected by the result showing that LRU and FAR have better performance than FIFO on paths. The analysis also found the expected result that FAR, which is designed to perform well on access graphs, performs better than both LRU and FIFO on cycles.

However, it is disappointing that the relative interval analysis of LRU and FIFO on stars and cycles found that FIFO had the better performance, confirming the original results by [14] on complete graphs. Clearly, access graphs cannot automatically be used with arbitrary quality measures for online algorithms to show that LRU is better than FIFO. To try to understand other quality measures for online algorithms better, it would be interesting to determine on which such measures access graphs are useful for separating LRU and FIFO, and on which they are not.

References

1. Albers, S., Favrholdt, L.M., Giel, O.: On paging with locality of reference. Journal of Computer and System Sciences 70(2), 145–175 (2005)
2. Angelopoulos, S., Dorrigiv, R., López-Ortiz, A.: On the separation and equivalence of paging strategies. In: SODA 2007, pp. 229–237 (2007)
3. Borodin, A., El-Yaniv, R.: Online Computation and Competitive Analysis. Cambridge University Press (1998)
4. Borodin, A., Irani, S., Raghavan, P., Schieber, B.: Competitive paging with locality of reference. Journal of Computer and System Sciences 50(2), 244–258 (1995)
5. Boyar, J., Favrholdt, L.M.: The relative worst order ratio for on-line algorithms. ACM Transactions on Algorithms 3(2), article No. 22 (2007)
6. Boyar, J., Favrholdt, L.M., Larsen, K.S.: The relative worst order ratio applied to paging. Journal of Computer and System Sciences 73(5), 818–843 (2007)
7. Boyar, J., Gupta, S., Larsen, K.S.: Access graphs results for LRU versus FIFO under relative worst order analysis. In: Fomin, F.V., Kaski, P. (eds.) SWAT 2012. LNCS, vol. 7357, pp. 328–339. Springer, Heidelberg (2012)
8. Boyar, J., Gupta, S., Larsen, K.S.: Relative interval analysis of paging algorithms on access graphs, arXiv:1305.0669 (cs.DS) (2013)
9. Boyar, J., Irani, S., Larsen, K.S.: A comparison of performance measures for online algorithms. In: Dehne, F., Gavrilova, M., Sack, J.-R., Tóth, C.D. (eds.) WADS 2009. LNCS, vol. 5664, pp. 119–130. Springer, Heidelberg (2009)
10. Chrobak, M., Noga, J.: LRU is better than FIFO. Algorithmica 23(2), 180–185 (1999)
11. Denning, P.J.: The working set model for program behaviour. Communications of the ACM 11(5), 323–333 (1968)
12. Denning, P.J.: Working sets past and present. IEEE Transactions on Software Engineering 6(1), 64–84 (1980)
13. Dorrigiv, R., López-Ortiz, A.: A survey of performance measures for on-line algorithms. SIGACT News 36(3), 67–81 (2005)
14. Dorrigiv, R., López-Ortiz, A., Munro, J.I.: On the relative dominance of paging algorithms. Theoretical Computer Science 410, 3694–3701 (2009)
15. Karlin, A.R., Manasse, M.S., Rudolph, L., Sleator, D.D.: Competitive snoopy caching. Algorithmica 3, 79–119 (1988)
16. Karlin, A.R., Phillips, S.J., Raghavan, P.: Markov paging. SIAM Journal on Computing 30(3), 906–922 (2000)
17. Koutsoupias, E., Papadimitriou, C.H.: Beyond competitive analysis. SIAM Journal on Computing 30(1), 300–317 (2000)
18. Sleator, D.D., Tarjan, R.E.: Amortized efficiency of list update and paging rules. Communications of the ACM 28(2), 202–208 (1985)
19. Young, N.: The k-server dual and loose competitiveness for paging. Algorithmica 11, 525–541 (1994)

On Explaining Integer Vectors
by Few Homogenous Segments

Robert Bredereck[1,*], Jiehua Chen[1,**], Sepp Hartung[1], Christian Komusiewicz[1],
Rolf Niedermeier[1], and Ondřej Suchý[2,***]

[1] Institut für Softwaretechnik und Theoretische Informatik, TU Berlin
{robert.bredereck,jiehua.chen,sepp.hartung,christian.komusiewicz,
rolf.niedermeier}@tu-berlin.de
[2] Faculty of Information Technology, Czech Technical University in Prague
ondrej.suchy@fit.cvut.cz

Abstract. We extend previous studies on NP-hard problems dealing with the decomposition of nonnegative integer vectors into sums of few homogeneous segments. These problems are motivated by radiation therapy and database applications. If the segments may have only positive integer entries, then the problem is called VECTOR EXPLANATION$^+$. If arbitrary integer entries are allowed in the decomposition, then the problem is called VECTOR EXPLANATION. Considering several natural parameterizations (including maximum vector entry, maximum difference between consecutive vector entries, maximum segment length), we obtain a refined picture of the computational (in-)tractability of these problems. In particular, we show that in relevant cases VECTOR EXPLANATION$^+$ is algorithmically harder than VECTOR EXPLANATION.

1 Introduction

We study two variants of a "mathematically fundamental" [4], NP-hard combinatorial problem occurring in cancer radiation therapy planning [10] and database and data warehousing applications [1, 18]:

VECTOR EXPLANATION (VECTOR EXPLANATION$^+$)
Input: A vector $A \in \mathbb{N}^n$ with $A[1] > 0$ and $A[n] > 0$ and an integer k.
Question: Can A be explained by at most k (positive) segments?

Herein, a *segment* is a $0/a$-vector, $a \in \mathbb{Z} \setminus \{0\}$, with n entries where all a-entries occur consecutively, and it is positive if a is positive. An *explanation* is a set of segments that sum up to the input vector. For instance, in case of VECTOR EXPLANATION (VE for short) the vector $(4, 3, 3, 4)$ can be explained by the segments $(4, 4, 4, 4)$ and $(0, -1, -1, 0)$, and in case of VECTOR EXPLANATION$^+$ (VE$^+$ for short) it can be explained by $(3, 3, 3, 3)$, $(1, 0, 0, 0)$, and $(0, 0, 0, 1)$.

---

* Supported by the DFG, research project PAWS, NI 369/10.
** Supported by der Studienstiftung des Deutschen Volkes.
*** The main work was done while O. Suchý was at TU Berlin, supported by the DFG, research project AREG, NI 369/9.

F. Dehne, R. Solis-Oba, and J.-R. Sack (Eds.): WADS 2013, LNCS 8037, pp. 207–218, 2013.
© Springer-Verlag Berlin Heidelberg 2013

Table 1. An overview of previous and new results

Parameters	VECTOR EXPLANATION	VECTOR EXPLANATION$^+$
max. value γ		$2^{O(\sqrt{\gamma})} \cdot \gamma n$ [6]
		no poly. kernel (Thm. 3)
max. difference δ of two consecutive entries	fpt (Thm. 2(2))	$O(n^{\delta+1} \cdot e^{\pi\sqrt{2\delta/3}})$ (Thm. 2(3))
(# of peaks p, δ)		fpt (Thm. 2(1))
number k of segments	$k^{O(k)} + n^{O(1)}$ (Thm. 4)	
	$(2k-1)$-entry kernel (Thm. 4)	
$k' = 2k - n$	$k'^{O(k')} + n^{O(1)}$ (Thm. 5(3))	$k^{O(k')} + n^{O(1)}$ (Thm. 5(1))
	$3k'$-entry kernel (Thm. 5(3))	W[1]-hard (Thm. 5(2))
$n - k$	NP-hard for $(n-k) = 1$ (Thm. 6(2))	
max. segment length l	$l \geq 3$: NP-hard (Thm. 6(1))	
	$l \leq 2$: $O(n^2)$ (Thm. 6(2))	
max. number o of overlapping segments	$o = 1$: trivial	
	$o = 2$ (and $l = 3$ and $n - k = 1$): NP-hard (Thm. 6(1))	

VE occurs in the database context and VE$^+$ occurs in the radiation therapy context. Motivated by previous work providing polynomial-time solvable special cases [1, 4], polynomial-time approximation [5, 19] and fixed-parameter tractability results [6, 8] (approximation and fixed-parameter algorithms both exploit problem-specific structural parameters), we head on a systematic parameterized and multivariate complexity analysis [13, 21] of both problems; see Table 1 for a survey of parameterized complexity results.

Previous Work. Agarwal et al. [1] studied a polynomial-time solvable variant ("tree-ordered") of VE relevant in data warehousing. Karloff et al. [18] initiated a study of (special cases of) the two-dimensional ("matrix") case of VE and provided NP-hardness results as well as polynomial-time constant-factor approximations. Parameterized complexity aspects of VE and its two-dimensional variant seem unstudied so far.

The literature on VE$^+$ is richer. For a general account on the motivation from radiation therapy refer to the survey by Ehrgott et al. [10]. Concerning computational complexity, VE$^+$ is known to be strongly NP-hard [3] and APX-hard [4]. A significant amount of work has been done to achieve approximation algorithms for minimizing the number of segments which improve on the straightforward factor of two [4] (also see Biedl et al. [5]). Improving a previous fixed-parameter tractability result for the parameter "maximum value γ of a vector entry" by Cambazard et al. [8], Biedl et al. [6] developed a fixed-parameter algorithm solving VE$^+$ in polynomial time when $\gamma = O((\log n)^2)$ with n being the number of entries in the input vector. Moreover, the parameter "maximum difference between two consecutive vector entries" has been exploited for developing polynomial-time approximation algorithms [5, 19]. Finally, we remark that most of the previous studies also looked at the two-dimensional ("matrix") case, whereas we focus on the one-dimensional ("vector") case.

Our Contributions. We observe that the combinatorial structure of the considered problems is extremely rich, opening the way to a more thorough study of the computational complexity landscape under the perspective of problem parameterization. We take a closer look at these parameterization aspects that help in better understanding and exploiting problem-specific properties. To start with, note that previous work [6, 8], motivated by the application in radiation therapy, studied the parameterization by the maximum vector entry γ. They showed fixed-parameter tractability for VE$^+$ parameterized by γ, which we complement by showing the non-existence (assuming a standard complexity-theoretic assumption) of a corresponding polynomial-size problem kernel. Using an integer linear program (ILP) formulation, we also show fixed-parameter tractability for VE parameterized by γ. Moreover, for the perhaps most obvious parameter, the number k of explaining segments, we show fixed-parameter tractability for both problems. In addition, we study the following parameters:

Definition 1. *For an input vector $A \in \mathbb{N}^n$ where, for notational convenience, additionally $A[0] = A[n+1] = 0$ define:*

- *the maximum difference δ between two consecutive vector entries ($\delta = \max_{1 \leq i \leq n+1} |A[i] - A[i-1]|$);*
- *the number p of peaks (a position $1 \leq i \leq n$ is a peak if $A[i-1] < A[i] > A[i+1]$);*
- *the maximum segment length l (number of a-entries);*
- *the maximum number o of segments having a non-zero entry at a particular vector entry;*
- *"distance from triviality"-parameter $n - k$;*
- *"distance from triviality"-parameter $k' := 2k - n$.*

Concerning the parameters $n - k$ and k', a brief discussion is appropriate. As to $n - k$, note that the problems have trivial solutions if $k = n$: just take n segments with one non-zero entry each. In this sense, $n - k$ is a parameterization by "distance from triviality" [16, 21]. We show that, somewhat surprisingly, both problems are already NP-hard for $k = n - 1$. As to k', note that by a simple preprocessing which will be explained later on, we can achieve that for every resulting instance which is not already classified as no-instance, we have that $n \leq 2k - 1$.[1] Moreover, if $k = \lfloor n/2 \rfloor + 1$, then the instances are polynomial-time solvable, motivating the "distance from triviality-parameter" k'. Interestingly, while we show that VE$^+$ is W[1]-hard for parameter k', we show that VE is fixed-parameter tractable for k'. Finally, we show NP-hardness for $l = 3$ and $o = 2$.

Table 1 summarizes our and previous results. Our work is organized as follows. In Section 2, we present a number of useful combinatorial properties of vector

---

[1] The definition of k' refers to the number n of entries in the instance after the preprocessing.

explanation problems which may be of independent interest and which are used throughout our work. In Section 3, we study the "smoothness of input vector" parameters γ, δ, and p. In Section 4, we present results for further parameters as discussed above, and we conclude with some challenges for future research. Due to the lack of space most proofs and details are deferred to a full version.

Parameterized Complexity Preliminaries. A parameterized problem is *fixed-parameter tractable* (fpt) if all instances (I, k) consisting of the "classical" problem instance I and the parameter k can be solved in $f(k) \cdot |I|^{O(1)}$ time for any function f solely depending on k. A *kernelization* algorithm is a polynomial-time algorithm that transforms each instance (I, k) for a problem L into an instance (I', k') for L such that $(I, k) \in L \Leftrightarrow (I', k') \in L$ (equivalence) and $k', |I'| \leq g(k)$ for some function g. The instance (I', k') is called a (problem) *kernel* of size $g(k)$. A kernelization algorithm is often described by a set of *data reduction rules* whose exhaustive application leads to a problem kernel. An instance is called *reduced* with respect to a data reduction rule if a further application would have no effect on the instance.

A problem that is shown to be *W[1]-hard* by means of a *parameterized reduction* from a W[1]-hard problem is not fpt, unless FPT = W[1]. A parameterized reduction maps an instance (I, k) in $f(k) \cdot |I|^{O(1)}$ time to an equivalent instance (I', k') with $k' \leq g(k)$ for some functions f and g. See [20] for a more detailed introduction to parameterized algorithmics. We assume the unit-cost RAM model where arithmetic operations on numbers count as a single computation step.

2   Combinatorial Properties

Formally, for an input vector $A \in \mathbb{N}^n$ a *segment* I is a pair written as $[l, r]$ with $l, r \in \{1, 2, \ldots, n + 1\}$ and $l < r$. We say I *starts* at position l and *ends* at positions r. A segment $[l, r]$ *covers* position i whenever $l \leq i < r$. A set of segments $\mathcal{I}$, together with a weight function $w : \mathcal{I} \to \mathbb{Z} \setminus \{0\}$, forms an *explanation* for $A \in \mathbb{N}^n$ if

$$\forall 1 \leq i \leq n : A[i] = \sum_{I \in \mathcal{I} \text{ covers } i} w(I),$$

where $A[i]$ denotes the ith entry in A. We also say $(\mathcal{I}, w)$ *explains* A. In case of VE$^+$, we only allow positive weights, that is, $w : \mathcal{I} \to \mathbb{N} \setminus \{0\}$. We refer to $|\mathcal{I}|$ as *solution size*. Segments with positive weight are called *positive segments*, those with negative weight are called *negative segments*.

Definition 2. *A position $1 \leq i \leq n + 1$ with respect to a vector $A \in \mathbb{N}^n$, is called an* uptick *if $A[i - 1] < A[i]$ and called* downtick *if $A[i - 1] > A[i]$. The size of an uptick (resp. downtick) i equals $|A[i] - A[i - 1]|$.*

By the following known data reduction rule [4], we may assume that each position is either an uptick or a downtick.

Rule 1. *If vector A has two consecutive equal entries, then remove one of them.*

Rule 1 can be applied exhaustively in $O(n)$ time. Afterwards, each position in A is either an uptick or a downtick. By the following lemma, we can assume that in solutions for VE^+ the segments start in upticks and end in downticks.

Lemma 1 ([4, Lemma 1]). *Let (A, k) be an instance of VE^+. There is a minimum-size explanation for vector A in which each segment starts at an uptick and ends at a downtick.*

For VE, we can generalize Lemma 1 to hold for negative and positive segments. Actually, one can even "reorder" all consecutive up- and downticks. This implies that for VE actually only the sizes of the upticks and downticks matter, not their order. To formalize this, we introduce the notion of *single-peakedness*.

Definition 3. *A vector is* single-peaked *if it contains only one peak. A single-peaked instance is an instance with a single-peaked vector.*

The following theorem summarizes combinatorial properties of VE and VE^+ which are used throughout the paper and which may be of independent interest.

Theorem 1. *Let (A, k) be an instance of VE. Then, the following holds.*

1. *There is a minimum-size explanation for (A, k) in which each positive segment starts at an uptick and ends at a downtick, and each negative segment starts at a downtick and ends at an uptick.*
2. *For any position $1 \leq i \leq n$ setting $A[i] \leftarrow A[i-1] + A[i+1] - A[i]$ results in an equivalent instance.*
3. *If (A, k) is single-peaked, then (A, k) is an equivalent VE^+ instance.*
4. *The instance (A, k) can be reduced in $O(n + k^2)$ time to an equivalent single-peaked instance (A', k) such that the maximum difference between consecutive entries is the same in A and A'.*
5. *There is an equivalent instance (A', k) with $A' \in \{0, \dots, 2\delta - 1\}^n$ where δ is the maximum difference between consecutive entries of A.*

Further, the following holds for VE^+ and for single-peaked VE instances.

6. *There is a minimum-size explanation such that*
   (a) *there is only one segment, starting at an uptick, that covers the last entry and*
   (b) *there is only one segment, ending at a downtick, that covers the first entry.*
7. *If an instance (A, k) is a yes-instance, then A contains at most k upticks and at most k downticks.*

3   Parameterization by Input Smoothness

In this section, we examine how the computational complexity of VE and VE^+ is influenced by parameters that measure how "smooth" the input vector $A \in \mathbb{N}^n$ is. We assume that A is reduced with respect to Rule 1 and thus all consecutive positions in A have different values. We consider the following three measurements: the maximum difference δ between two consecutive values in A, the number p of peaks, that is, the number of positions $1 \le i \le n$ in A such that $A[i-1] < A[i] > A[i+1]$, and the maximum value γ occurring in A. Our main results are fixed-parameter algorithms for the combined parameter (p, δ) in the case of VE$^+$ and for the parameter δ in the case of VE. For the parameter maximum value γ, we show that VE$^+$ does not admit a polynomial-size problem kernel unless NP $\subseteq$ coNP/poly.

Theorem 2. *1. VE$^+$ parameterized by the combined parameter number p of peaks and maximum difference δ is fixed-parameter tractable.*

   2. VE parameterized by the maximum difference δ or the maximum value γ is fixed-parameter tractable.

   3. VE$^+$ is solvable in $O(n^{\delta+1} \cdot e^{\pi\sqrt{2\delta/3}})$ time.

Proof. We only prove the correctness of Theorem 2(1). This also implies Theorem 2(2): By Theorem 1(3) and Theorem 1(4), we can transform input instances of VE into single-peaked ones of VE$^+$ without increasing the maximum difference δ. Furthermore, $\delta \le \gamma$. Together with the above transformation this implies fixed-parameter tractability for δ and for γ. The proof of Theorem 2(3) is based on a dynamic programming algorithm, omitted from this extended abstract.

To show Theorem 2(1), we provide an integer linear program (ILP) formulation for VE$^+$ where the number of variables is a function of p and δ. This ILP determines whether there is a size-k solution with the properties given by Lemma 1, that is, a solution in which each segment starts at an uptick and ends at a downtick. In such a solution, the multiset of weights of segments that start at an uptick sum up to the uptick size. This analogously holds for segments ending at a downtick. Motivated by this fact, we introduce the following notion: For a positive integer x, we say that a multiset $X = \{x_1, x_2, \ldots, x_r\}$ of positive integers *partitions* x if $x = \sum_{i=1}^{r} x_i$. Similarly, we say that X *partitions* an uptick (downtick) i of size x if X partitions x. Let $\mathcal{P}(x)$ denote the set of all multisets that partition x.

In the ILP formulation, we describe a solution by "fixing" for each $i \in \{1, \ldots, n\}$ a multiset X_i of positive integers which partitions the uptick (downtick) at i. The crucial observation for our ILP is that if a set of consecutive upticks contains more than one uptick of size x, it is sufficient to fix how many of these upticks were partitioned in which way. In other words, one does not need to know the partition for each position; instead one can distribute freely the partitions of x onto the upticks of size x. This also holds for consecutive downticks. Since each peak is preceded by consecutive upticks and succeeded by consecutive downticks, and since we introduce variables in the ILP formulation to "model" how many upticks (downticks) exist between two consecutive peaks, the number of variables in the formulation is bounded by a function of p and δ. We now give the details of the formulation. Herein, we assume that the peaks are ordered from left to right; we refer to the i-th peak in this order as peak i.

For an integer $x \in \{1, \ldots, \delta\}$, let $\mathrm{occ}(x, i)$ denote the number of upticks of size x that directly precede peak i, that is, the number of upticks succeeding peak $i - 1$ and preceding peak i. Similarly, let occ$(-x, i)$ denote the number of downticks of size x that directly succeed i. For two positive integers y and x with $y \leq x$ and a multiset $P \in \mathcal{P}(x)$ let mult(y, P) denote how often y appears in P. We use mult(y, P) to "model" how many segments of weight y start (end) at some uptick (downtick) that is partitioned by P.

To formulate the ILP, we introduce for each peak i, each $x \in \{1, \ldots, \delta\}$, and each $P \in \mathcal{P}(x)$ two nonnegative variables var$_{x,P,i}$ and var$_{-x,P,i}$. The variables correspond to the number of upticks directly preceding peak i and downticks directly succeeding peak i of size x that are partitioned by P in a possible explanation of A. To enforce that a particular assignment to these variables corresponds to a valid explanation, we introduce the following constraints.

First, for each peak i and each $1 \leq x \leq \delta$ we ensure that the number of directly proceeding size-x upticks (succeeding size-x downticks) that are partitioned by some $P \in \mathcal{P}(x)$ is equal to the number of directly proceeding size-x upticks (succeeding size-x downticks):

$$\forall i \in \{1, \ldots, p\}, \ \forall x \in \{-\delta, \ldots, \delta\} \setminus \{0\} : \sum_{P \in \mathcal{P}(x)} \text{var}_{x,P,i} = \text{occ}(x, i). \quad (1)$$

Second, we ensure that for each peak i and each value $y \in \{1, \ldots, \delta\}$ the number of segments of weight y that end directly after peak i is at most the number of segments of weight y that start at positions (not necessarily directly) preceding peak i minus the number of segments of weight y that end at positions succeeding some peak $j < i$. Informally, this means that we only "use" the available number of segments of weight y. To enforce this property, for each peak $1 \leq i \leq p$ and each possible segment weight $1 \leq y \leq \delta$ we add:

$$\sum_{j=1}^{i} \sum_{x=y}^{\delta} \sum_{P \in \mathcal{P}(x)} (\ \underbrace{\text{mult}(y, P) \cdot \text{var}_{x,P,j}}_{\text{\# of opened weight-}y\text{ segments}} \ - \ \underbrace{\text{mult}(y, P) \cdot \text{var}_{-x,P,j}}_{\text{\# of closed weight-}y\text{ segments}} \ ) \geq 0 \quad (2)$$

Finally, we ensure that the total number of segments is at most k:

$$\sum_{i=1}^{p} \sum_{x=1}^{\delta} \sum_{P \in \mathcal{P}(x)} \sum_{y=1}^{x} \text{mult}(y, P) \cdot \text{var}_{x,P,i} \leq k. \quad (3)$$

Correctness: The equivalence of the ILP instance and (A, k) can be seen as follows. Assume that there is a size-at-most-k explanation S for (A, k). Recall that by definition of $\mathcal{P}(x)$, for any uptick i of size x there is a partition in $\mathcal{P}(x)$ that corresponds to the weights of the segments starting in i. For each peak i, for any value $1 \leq x \leq \delta$ and each $P \in \mathcal{P}(x)$, count how many upticks of size x that directly precede peak i are explained by segments in S (segments that start in this uptick) whose weights correspond to $\mathcal{P}(x)$ and set var$_{x,P,i}$ to this value. Symmetrically, do the same for the downticks succeeding the peak i and set var$_{-x,P,i}$ accordingly. It is straightforward to verify that eqs. (1) to (3) hold.

Reversely, assume that there is an assignment to the variables such that eqs. (1) to (3) are fulfilled. We form a set of segments S as follows: For any peak i and any value $1 \leq x \leq \delta$ with $\mathrm{occ}(x, i) > 0$ let $\mathcal{P}_{i,x}$ be the multiset of partitions of $\mathcal{P}(x)$ that contains each $P \in \mathcal{P}(x)$ exactly $\mathrm{var}_{x,P,i}$ times. By eq. (1), $|\mathcal{P}_{i,x}| = \mathrm{occ}(x, i)$. For an arbitrary ordering of $\mathcal{P}_{i,x}$ and the upticks of size x directly preceding peak i, add to S for the jth element $\mathcal{P}_j$ of $\mathcal{P}_{i,x}$ exactly $|\mathcal{P}_j|$ segments with weight corresponding to $\mathcal{P}_j$ and let them start at the jth uptick with size x that directly precedes peak i. By eq. (3) we added at most k segments. It remains to specify the end of the segments. Symmetrically to the upticks, for each downtick directly succeeding peak i of size x let $\mathcal{P}_{i,x}$ be the multiset of elements from $\mathcal{P}(x)$ containing each $P \in \mathcal{P}(x)$ exactly $\mathrm{var}_{-x,P,i}$ times. For the jth element $\mathcal{P}_j$ of $\mathcal{P}_{i,x}$ and the jth downtick directly succeeding peak i (again both with respect to any ordering) and for each $\alpha \in \mathcal{P}_j$ pick any weight-α segment from S (so far without end) and let it end directly one position behind the jth downtick. Observe that the existence of this segment is ensured by eq. (2). Finally, it remains to argue that the of end of each segment in S is determined. This follows from the fact that eqs. (1) and (2) together imply for each $1 \leq y \leq \delta$ that

$$\sum_{i=1}^{p} \sum_{x=y}^{\delta} \sum_{P \in \mathcal{P}(x)} (\mathrm{mult}(y, P) \cdot \mathrm{var}_{x,P,i} - \mathrm{mult}(y, P) \cdot \mathrm{var}_{-x,P,i}) = 0,$$

and thus the total number of opened weight-y segments is equal to the number of closed weight-y segments.

Running time: The ILP can be solved within the following time bound. The number of variables in the constructed ILP instance is

$$p \cdot \sum_{x \in \{-\delta,\ldots,\delta\} \setminus \{0\}} |\mathcal{P}(|x|)| = 2p \sum_{x=1}^{\delta} |\mathcal{P}(x)| \leq 2\delta p \cdot |\mathcal{P}(\delta)| \leq 2\delta p \cdot e^{\pi \sqrt{\frac{2}{3}\delta}} =: f(\delta, p),$$

where the last inequality is due to de Azevedo Pribitkin [2]. Thus, due to a deep result in combinatorial optimization the feasibility of the ILP can decided in $O(f(\delta, p)^{2.5f(\delta,p)+o(f(\delta,p))} \cdot |L|)$ time, where $|L|$ is the size of the instance [14, 17]. Moreover, as we have $O(\delta p)$ inequalities, we also have $|L| = O(\delta^2 p^2 \cdot e^{\pi \sqrt{\frac{2}{3}\delta}})$.   □

For the parameter maximum value γ, VE$^+$ is known to be fixed-parameter tractable [6]. We complement this result by showing a lower bound on the problem kernel size, and thus demonstrate limitations on the power of preprocessing.

Theorem 3. *Unless $NP \subseteq coNP/poly$, there is no polynomial-size problem kernel for VE$^+$ parameterized by the maximum value γ.*

Proof. We provide a so-called AND-cross-composition [7, 9] from the 3-PARTITION problem. Given a multiset $S = \{a_1, \ldots, a_{3m}\}$ of positive integers and an integer bound B with $m \cdot B = \sum_{i=1}^{3m} a_i$ and $B/4 < a_i < B/2$ for every $i \in \{1, \ldots, 3m\}$, 3-PARTITION asks whether the set S can be partitioned into m subsets $P_1, \ldots, P_m$ with $|P_j| = 3$ and $\sum_{a_i \in P_j} a_i = B$ for every $j \in \{1, \ldots, m\}$. 3-PARTITION is NP-hard even if B (and thus all a_i's) is bounded by a polynomial in m [15]. We show that this variant of 3-PARTITION AND-cross-composes to VE$^+$

parameterized by the maximum value γ. Then, results of Bodlaender et al. [7] and Drucker [9] imply that VE^+ does not have a polynomial-size problem kernel with respect to parameter γ, unless $NP \subseteq coNP/poly$.

First, let (S, B) be a single instance of 3-PARTITION. We show that it reduces to an instance $(A', 3m)$ of VE^+. This reduction is very similar to a previous NP-hardness reduction for VE^+ [4]. We define A' as length-$(4m - 1)$ vector:

$$\left(a_1, a_1 + a_2, \ldots, \overset{j}{\underset{i=1}{\sum}} a_i, \ldots, \overset{3m}{\underset{i=1}{\sum}} a_i = mB, (m - 1)B, (m - 2)B, \ldots, B \right).$$

On the one hand, if a partition $P_1, \ldots, P_m$ of S forms a solution, then the set of segments $\{[i, 3m + j] \mid a_i \in P_j\}$ each with weight $w([i, 3m + j]) = a_i$ is an explanation for the vector A'. On the other hand, let $(\mathcal{I}, w)$ be an explanation for $(A', 3m)$. By Lemma 1 we may assume that every segment starts at an uptick and ends at a downtick. Therefore, $\mathcal{I}$ contains $3m$ segments and the segment starting at position i has weight a_i. Since $B/4 < a_i < B/2$ for each integer $a_i \in S$, exactly three segments end at a downtick whose size is exactly B. Thus, grouping the segments according to the position they end at, we get the desired partition of S, solving the instance of 3-PARTITION.

Now let $(S_1, B_1), \ldots, (S_t, B_t)$ be instances of 3-PARTITION such that $S_r = \{a_1^r, \ldots, a_{3m_r}^r\}$ and $B_r \leq m_r^c$ for every $r \in \{1, \ldots, t\}$ and some constant c. We build an instance (A, k) of VE^+ by first using the above reduction for each (S_r, B_r) separately to produce a vector A'_r, and then concatenating the vectors A'_r one after another, leaving a single position of value 0 in between. The total length of the vector A is $4(\sum_{r=1}^{t} m_r) - 1$ and we set $k := 3\sum_{r=1}^{t} m_r$.

Due to the argumentation for the single instance case, on the one hand, if each of the instances is a yes-instance, then there is an explanation using $3m_r$ segments per instance (S_r, B_r), that is $3\sum_{r=1}^{t} m_r$ segments in total. On the other hand, we need at least $3m_r$ segments to explain A'_r and there is an explanation with $3m_r$ segments if and only if (S_r, B_r) is a yes-instance. Since all segments are positive and the subvectors A'_r's are separated by a position with value zero, no segment can span over two subvectors. In other words, no segment can be used to explain more than one of the A'_r's. Therefore, an explanation for A with $3\sum_{r=1}^{t} m_r$ segments implies that (S_r, B_r) is a yes-instance for every $r \in \{1, \ldots, t\}$.

Finally, observe that the maximum value γ in the vector A is equal to $\max_{r=1}^{t} m_r B_r \leq \max_{r=1}^{t} m_r^{c+1}$. Since in each input 3-PARTITION instance the maximum value $m_r B_r$ is polynomially bounded in the instance size $|S_r|$, this value is thus polynomially bounded in $\max_{r=1}^{t} |S_r|$. Hence, 3-PARTITION AND-cross-composes to VE^+ parameterized by the maximum value γ, and there is no polynomial-size problem kernel for this problem unless $NP \subseteq coNP/poly$.   □

4   Further Parameterizations

We now provide fixed-parameter tractability and (parameterized) hardness results for further natural parameters. Specifically, we consider the number k of segments in the solution, so-called "above-guarantee" and "below-guarantee" parameterizations (which are smaller than k), the maximum segment length l, and the maximum number of segments covering a position.

For the parameter k we obtain fixed-parameter tractability by using Rule 1, Theorem 1(6), and Theorem 1(7) to develop search tree algorithms for VE$^+$ and VE. The depth and the branching degree of the search tree are bounded by the solution size k. The second part of Theorem 4 follows directly from exhaustively applying Rule 1.

Theorem 4. VE$^+$ *and* VE *can be solved in* $O(k! \cdot k + n)$ *time. Any instance of* VE$^+$ *or* VE *can be reduced in* $O(n)$ *time to an equivalent one with at most* $(2k - 1)$ *entries.*

The second part of Theorem 4 implies that for a reduced instance every explanation needs at least $\lfloor n/2 \rfloor + 1$ segments. Furthermore, instances with $k = \lfloor n/2 \rfloor + 1$ are solvable in polynomial time (below, we will state a generalization of this fact). Hence, it is interesting to study parameters that measure how far we have to exceed this lower bound for the solution size; notably, such above-guarantee parameters can be significantly smaller than k. For this reason, we study a parameter that measures $k - \lfloor n/2 \rfloor - 1$. For ease of presentation, we define this parameter as $k' := 2k - n$. The concepts of "clean" and "messy" positions, which are defined as follows, are crucial for the design of our algorithms.

Definition 4. *Let* (A, k) *be an instance of* VE *or* VE$^+$ *and let* $\mathcal{I}$ *be an explaining segment set for vector* A. *A segment* $I = [i, j] \in \mathcal{I}$ *is* clean *if all other segments start and end at positions different from* i *and* j. *A position* i *is* clean *with respect to* $\mathcal{I}$ *if it is the start or endpoint of a clean segment in* $\mathcal{I}$. *A position or segment that is not clean is called* messy.

We show that clean positions can always be covered by clean segments of "minimum length": Iterate from left to right over all clean positions and for each position i (still clean) find the first clean position $j > i$ with $-(A[i] - A[i - 1]) = A[j] - A[j + 1]$ and add a segment of weight $A[i] - A[i - 1]$ from i to $j + 1$.

For every yes-instance of VE, the number of messy positions is at most $3k'$ and the number of messy segments used by an explanation is at most $2k'$ with $k' = 2k - n$. Furthermore, if there are an uptick and a downtick of same size in a single-peaked instance, then we may assume that the corresponding segment is contained in the solution.

As the following theorem shows, using the properties concerning clean and messy positions, we can replace the exponent k in the running time of Theorem 4 by the smaller k'. This also implies that VE$^+$ is polynomial-time solvable for constant k'. Unless W[1]=FPT, this result cannot be improved to fixed-parameter tractability since we can give a parameterized reduction from the W[1]-hard SUB-SET SUM problem [11] to VE$^+$. In contrast, VE$^+$ for single-peaked instances as well as VE in general are fixed-parameter tractable with respect to k' and can be efficiently reduced to equivalent instances with at most $3k'$ positions.

Theorem 5. *1.* VE$^+$ *can be solved in* $O((2k)^{3k'} \cdot (k^2 + (2k')! \cdot k') + n)$ *time.*

2. VE^+ *is W[1]-hard with respect to* k'.
3. *Any single-peaked instance of* VE^+ *and any instance of* VE *can be reduced in* $O(k^2 + n)$ *time to an equivalent one with most* $3k'$ *entries. Moreover,* VE^+ *and* VE *are solvable in* $O((2k')! \cdot k' + k^2 + n)$ *time.*

The previous parameter k' measures how far the solution exceeds the lower bound $\lfloor n/2 \rfloor + 1$. Another bound on the solution size is n: If $k = n$, then any instance of VE^+ or VE is a trivial yes-instance. Hence, it is interesting to consider the parameter $n - k$. Furthermore, it is natural to consider explanations with restricted segment length l or the maximum number o of segments overlapping at some position. The following theorem shows that VE^+ and VE are already NP-hard even if $k = n - 1$, $l \geq 3$, and $o = 2$. To this end, we reduce from the NP-hard PARTITION problem [15]. In terms of parameterized complexity this implies that, unless P=NP, VE^+ is not fixed-parameter tractable with respect to the "maximum segment length l", the "maximum number o of segments overlapping at some position", and the "below guarantee parameter" $n - k$.

We also show that, in contrast to the NP-hardness for $l \geq 3$, VE^+ and VE are polynomial-time solvable for $l \leq 2$.

Theorem 6. *1. VE^+ and VE are NP-hard even if $k = n - 1$ and every yes-instance has an explanation of at most k segments where each position is covered by at most two segments and each segment has length at most three.*
  2. Both VE^+ and VE can be solved in $O(n^2)$ time for maximum segment length $l = 2$.

5   Conclusion

It would be interesting to significantly improve on several of the running time upper bounds of our (theoretical) tractability results (cf. Table 1 for an overview). Moreover, we also left open a number of concrete problems. We conclude with three of them:

- Is VE^+ fixed-parameter tractable with respect to the maximum difference δ?
- Does VE parameterized by δ or parameterized by γ admit a polynomial-size problem kernel?
- Is VE or VE^+ fixed-parameter tractable with respect to the parameter "number of different values in the input vector A"? This parameter would be a natural version of "parameterization by the number of numbers" [12].

Acknowledgement. We are very grateful for the very detailed and constructive feedback provided by the *WADS* reviewers.

References

[1] Agarwal, D., Barman, D., Gunopulos, D., Young, N., Korn, F., Srivastava, D.: Efficient and effective explanation of change in hierarchical summaries. In: Proc. 13th KDD, pp. 6–15. ACM (2007)

[2] de Azevedo Pribitkin, W.: Simple upper bounds for partition functions. The Ramanujan Journal 18, 113–119 (2009)

[3] Baatar, D., Hamacher, H.W., Ehrgott, M., Woeginger, G.J.: Decomposition of integer matrices and multileaf collimator sequencing. Discrete Appl. Math. 152(1-3), 6–34 (2005)

[4] Bansal, N., Chen, D.Z., Coppersmith, D., Hu, X.S., Luan, S., Misiolek, E., Schieber, B., Wang, C.: Shape rectangularization problems in intensity-modulated radiation therapy. Algorithmica 60(2), 421–450 (2011)

[5] Biedl, T.C., Durocher, S., Hoos, H.H., Luan, S., Saia, J., Young, M.: A note on improving the performance of approximation algorithms for radiation therapy. Inf. Process. Lett. 111(7), 326–333 (2011)

[6] Biedl, T.C., Durocher, S., Engelbeen, C., Fiorini, S., Young, M.: Faster optimal algorithms for segment minimization with small maximal value. Discrete Appl. Math. 161(3), 317–329 (2013)

[7] Bodlaender, H.L., Jansen, B.M.P., Kratsch, S.: Cross-composition: A new technique for kernelization lower bounds. In: Proc. 28th STACS. LIPIcs, vol. 9, pp. 165–176. Schloss Dagstuhl–Leibniz-Zentrum fuer Informatik (2011)

[8] Cambazard, H., O'Mahony, E., O'Sullivan, B.: A shortest path-based approach to the multileaf collimator sequencing problem. Discrete Appl. Math. 160(1-2), 81–99 (2012)

[9] Drucker, A.: New limits to classical and quantum instance compression. In: Proc. 53rd IEEE FOCS, pp. 609–618. IEEE Computer Society (2012)

[10] Ehrgott, M., Güler, C., Hamacher, H., Shao, L.: Mathematical optimization in intensity modulated radiation therapy. Ann. Oper. Res. 175, 309–365 (2010)

[11] Fellows, M.R., Koblitz, N.: Fixed-parameter complexity and cryptography. In: Moreno, O., Cohen, G., Mora, T. (eds.) AAECC 1993. LNCS, vol. 673, pp. 121–131. Springer, Heidelberg (1993)

[12] Fellows, M.R., Gaspers, S., Rosamond, F.A.: Parameterizing by the number of numbers. Theory Comput. Syst. 50(4), 675–693 (2012)

[13] Fellows, M.R., Jansen, B.M.P., Rosamond, F.A.: Towards fully multivariate algorithmics: Parameter ecology and the deconstruction of computational complexity. European J. Combin. 34(3), 541–566 (2013)

[14] Frank, A., Tardos, É.: An application of simultaneous diophantine approximation in combinatorial optimization. Combinatorica 7(1), 49–65 (1987)

[15] Garey, M.R., Johnson, D.S.: Computers and Intractability: A Guide to the Theory of NP-Completeness. Freeman (1979)

[16] Guo, J., Hüffner, F., Niedermeier, R.: A structural view on parameterizing problems: Distance from triviality. In: Downey, R.G., Fellows, M.R., Dehne, F. (eds.) IWPEC 2004. LNCS, vol. 3162, pp. 162–173. Springer, Heidelberg (2004)

[17] Kannan, R.: Minkowski's convex body theorem and integer programming. Math. Oper. Res. 12, 415–440 (1987)

[18] Karloff, H., Korn, F., Makarychev, K., Rabani, Y.: On parsimonious explanations for 2-d tree- and linearly-ordered data. In: Proc. 28th STACS. LIPIcs, vol. 9, pp. 332–343. Schloss Dagstuhl–Leibniz-Zentrum fuer Informatik (2011)

[19] Luan, S., Saia, J., Young, M.: Approximation algorithms for minimizing segments in radiation therapy. Inf. Process. Lett. 101(6), 239–244 (2007)

[20] Niedermeier, R.: Invitation to Fixed-Parameter Algorithms. Oxford University Press (2006)

[21] Niedermeier, R.: Reflections on multivariate algorithmics and problem parameterization. In: Proc. 27th STACS. LIPIcs, vol. 5, pp. 17–32. Schloss Dagstuhl–Leibniz-Zentrum fuer Informatik (2010)

Trajectory Grouping Structure*

Kevin Buchin[1], Maike Buchin[1], Marc van Kreveld[2],
Bettina Speckmann[1], and Frank Staals[2]

[1] Dep. of Mathematics and Computer Science, TU Eindhoven
[2] Dep. of Information and Computing Sciences, Utrecht University

Abstract. The collective motion of a set of moving entities like people, birds, or other animals, is characterized by groups arising, merging, splitting, and ending. Given the trajectories of these entities, we define and model a structure that captures all of such changes using the Reeb graph, a concept from topology. The *trajectory grouping structure* has three natural parameters, namely group size, group duration, and entity inter-distance. These parameters allow us to obtain detailed or global views of the data. We prove complexity bounds on the maximum number of maximal groups that can be present, and give algorithms to compute the grouping structure efficiently. Furthermore, we showcase the results of experiments using data generated by the NetLogo flocking model and from the Starkey project. Although there is no ground truth for the groups in this data, the experiments show that the trajectory grouping structure is plausible and has the desired effects when changing the essential parameters. Our research provides the first complete study of trajectory group evolvement, including combinatorial, algorithmic, and experimental results.

1 Introduction

In recent years there has been an increase in location-aware devices and wireless communication networks. This has led to a large amount of trajectory data capturing the movement of animals, vehicles, and people. The increase in trajectory data goes hand in hand with an increasing demand for techniques and tools to analyze them, for example, in sports, ecology, transport, and social services.

An important task is the analysis of movement patterns. In particular, given a set of moving entities we wish to determine when and which subsets of entities travel together. When a sufficiently large set of entities travels together for a sufficiently long time, we call such a set a *group* (we give a more formal definition later). Groups may start, end, split and merge with other groups. Apart from the question what the current groups are, we also want to know which splits and merges led to the current groups, when they happened, and which groups they involved. We wish to capture this group change information in a model that we call the *trajectory grouping structure*.

* MB, BS & FS are supported by the Netherlands Organisation for Scientific Research (NWO) under project no. 612.001.106, 639.022.707 & 612.001.022, respectively.

F. Dehne, R. Solis-Oba, and J.-R. Sack (Eds.): WADS 2013, LNCS 8037, pp. 219–230, 2013.

The informal definition above suggests that three parameters are needed to define groups: (i) a spatial parameter for the distance between entities; (ii) a temporal parameter for the duration of a group; (iii) a count for the number of entities in a group. We will design our grouping structure definition to incorporate these parameters so that we can study grouping at different scales. We use the three parameters as follows: a small spatial parameter implies we are interested only in spatially close groups, a large temporal parameter implies we are interested only in long-lasting groups, and a large count implies we are interested only in large groups. By adjusting the parameters suitably, we can obtain more detailed or more generalized views of the trajectory grouping structure.

The use of scale parameters and the fact that the grouping structure changes at discrete events suggest the use of computational topology [6]. In particular, we use Reeb graphs to capture the grouping structure. Reeb graphs have been used extensively in shape analysis and the visualization of scientific data (see e.g. [2,5,8]). A Reeb graph captures the structure of a two- or higher-dimensional scalar function, by considering the evolution of the connected components of the level sets. The computation of Reeb graphs has received considerable attention in computational geometry and topology; an overview is given in [4]. Recently, a deterministic $O(n \log n)$ time algorithm was presented for constructing the Reeb graph of a 2-skeleton of size n [16]. Edelsbrunner et al. [5] discuss time-varying Reeb graphs for continuous space-time data. Although we also analyze continuous space-time data (2D-space in our case), our Reeb graphs are not time-varying, but time is the parameter that defines the Reeb graph.

Our research is motivated by and related to previous research on flocks [1,9,10,19], herds [11], convoys [12], moving clusters [13], mobile groups [20] and swarms [14]. These concepts differ from each other in the way in which space and time are used to test if entities form a group: do the entities stay in a single disc or are they density-connected [7], should they stay together during consecutive time steps or not, can the group members change over time, etc. Only the herds concept [11] includes the splitting and merging of groups.

Contributions. We present the first complete study of trajectory group evolvement, including combinatorial, algorithmic, and experimental results. Our research differs from and improves on previous research in the following ways. Firstly, our model is simpler than herds and thus more intuitive. Secondly, we consider the grouping structure at continuous times instead of at discrete steps (which was done only for flocks). Thirdly, we analyze the algorithmic and combinatorial aspects of groups and their changes. Fourthly, we implemented our algorithms and provide evidence that our model captures the grouping structure well and can be computed efficiently. We created videos based on our implementation showing the maximal groups we found in simulated NetLogo flocking data [21] and in real-world data from the Starkey project [15], see www.staff.science.uu.nl/~staal006/grouping.

A Definition for a Group. Let $\mathcal{X}$ be a set of entities of which we have locations over time. The ε-*disc* of an entity x (at time t) is a disc of radius ε centered at x at time t. Two entities are *directly connected* at time t if their ε-discs overlap. Two entities x and y are ε-*connected* at time t if there is a sequence $x = x_0, .., x_k = y$ of entities such that for all i, x_i and x_{i+1} are directly connected.

A subset $S \subseteq \mathcal{X}$ of entities is ε-connected at time t if all entities in S are pairwise ε-connected at time t. This means that the union of the ε-discs of entities in S forms a single connected region. The set S forms a *component* at time t if and only if S is ε-connected, and S is maximal with respect to this property. The set of components $\mathcal{C}(t)$ at time t forms a partition of the entities in $\mathcal{X}$ at time t.

Let the spatial parameter of a group be ε, the temporal parameter δ, and the size parameter m. A set G of k entities forms a *group* during time interval I if and only if the following three conditions hold: (i) G contains at least m entities, so $k \geq m$, (ii) the interval I has length at least δ, and (iii) at all times $t \in I$, there is a component $C \in \mathcal{C}(t)$ such that $G \subseteq C$.

We denote the interval $I = [t_s, t_e]$ of group G with I_G. Group H *covers* group G if $G \subseteq H$ and $I_G \subseteq I_H$. If there are no groups that cover G, we say G is *maximal* (on I_G). In Fig. 1, groups $\{x_1, x_2\}$, $\tilde{G} = \{x_3, x_4\}$, $\hat{G} = \{x_5, x_6\}$, and $G = \{x_1, .., x_4\}$ are maximal: $\tilde{G}$ and $\hat{G}$ on $[t_0, t_5]$, G on $[t_1, t_2]$. Group $\{x_1, x_3\}$ is covered by G and hence not maximal.

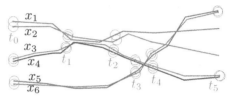

Fig. 1. For $m = 2$ and $\delta > t_4 - t_3$ there are four maximal groups: $\{x_1, x_2\}$, $\{x_3, x_4\}$, $\{x_5, x_6\}$, and $\{x_1, .., x_4\}$

Note that entities can be in multiple maximal groups at the same time. For example, entities $\{y_1, y_2, y_3\}$ can travel together for a while, then y_4, y_5 may become ε-connected, and shortly thereafter y_1, y_4, y_5 separate and travel together for a while. Then y_1 may be in two otherwise disjoint maximal groups for a short time. An entity can also be in two maximal groups where one is a subset of the other. In that case the group with fewer entities must last longer. That an entity is in more groups simultaneously may seem counterintuitive at first, but it is necessary to capture all grouping information. We will show that the total number of maximal groups is $O(\tau n^3)$, where n is the number of entities in $\mathcal{X}$ and τ is the number of edges of each input trajectory. This bound is tight in the worst case.

Our maximal group definition uses three parameters, which all allow a more global view of the grouping structure. In particular, we observe that there is *monotonicity* in the group size and the duration: If G is a group during interval I, and we decrease the minimum required group size m or decrease the minimum required duration δ, then G is still a group on time interval I. Also, if G is a maximal group on I, then it is also a maximal group for a smaller m or smaller δ. For the spatial parameter ε we observe monotonicity in a slightly different manner: if G is a group for a given ε, then for a larger value of ε there exists a group $G' \supseteq G$. The monotonicity property is important when we want to have a more detailed view of the data: we do not lose maximal groups in a more detailed view. The group may however be extended in size and/or duration.

We capture the grouping structure using a Reeb graph of the ε-connected components together with the set of all maximal groups. Parts of the Reeb graph that do not support a maximal group can be omitted. The grouping structure can help us in answering various questions. For example:

- What is the largest/longest maximal group at time t?
- How many entities are currently (not) in any maximal group?
- What is the first maximal group that starts/ends after time t?
- What is the total time that an entity was part of any maximal group?
- Which entity has shared maximal groups with the most other entities?

Furthermore, the grouping structure can be used to partition the trajectories in independent data sets, to visualize grouping aspects of the trajectories, and to compare grouping across different data sets.

Results and Organization. We discuss how to represent the grouping structure in Section 2, and prove that there are always $O(\tau n^3)$ maximal groups, which is tight in the worst case. Here n is the number of trajectories (entities) and τ the number of edges in each trajectory. We present an algorithm to compute the trajectory grouping structure and all maximal groups in Section 3. This algorithm runs in $O(\tau n^3 + N)$ time, where N is the total output size. In Section 4 we discuss robustness briefly; all details can be found in the full version of the paper [3]. In Section 5 we evaluate our methods on synthetic and real-world data.

2   Representing the Grouping Structure

Let $\mathcal{X}$ be a set of n entities, where each entity travels along a path of τ edges. To compute the grouping structure we consider a manifold $\mathcal{M}$ in $\mathbb{R}^3$, where the z-axis corresponds to time. The manifold $\mathcal{M}$ is the union of n "tubes". Each tube consists of τ skewed cylinders with horizontal radius ε that we obtain by tracing the ε-disc of an entity x over its trajectory.

Let H_t denote the horizontal plane at height t, then the set $\mathcal{M} \cap H_t$ is the *level set* of t. The connected components in the level set of t correspond to the components (maximal sets of ε-connected entities) at time t. We will assume that all trajectories have their known positions at the same times $t_0, .., t_\tau$ and that no three entities become ε-(dis)connected at the same time. Our theory does not depend on these assumptions and we could remove them, but they make the descriptions considerably more clear.

2.1   The Reeb Graph

We start out with a possibly disconnected solid that is the union of a collection of tube-like regions: a 3-manifold with boundary. Note that this manifold is not explicitly defined. We are interested in horizontal cross-sections, and the evolution of the connected components of these cross-sections defines the Reeb graph. Note that this is different from the usual Reeb graph that is obtained from the 2-manifold that is the boundary of our 3-manifold, using the level sets of the height function (the function whose level sets we follow is the height function above a horizontal plane below the manifold), see [6] for more on this topic.

To describe how the components change over time, we consider the Reeb graph $\mathcal{R}$ of $\mathcal{M}$. The Reeb graph has a vertex v at every time t_v where the components change. The vertex times are usually not at any of the given times $t_0, .., t_\tau$, but in between two consecutive time steps. The vertices of the Reeb graph can be classified in four groups. There is a *start vertex* for every component at t_0 and an *end vertex* at t_τ. A start vertex has in-degree zero and out-degree one, and an end vertex has in-degree one and out-degree zero. The remaining vertices are either *merge vertices* or *split vertices*. Since we assume that no three entities become ε-(dis)connected at exactly the same time there are no simultaneous splits and merges. This means merge vertices have in-degree two and out-degree one, and split vertices have in-degree one and out-degree two. A directed edge $e = (u, v)$ connecting vertices u and v, with $t_u < t_v$, corresponds to a set C_e of entities that form a component at any time $t \in I_e = [t_u, t_v]$. The Reeb graph is this directed graph. Note that the Reeb graph depends on the spatial parameter ε, but not on the other two parameters of maximal groups.

Theorem 1. *Given a set $\mathcal{X}$ of n entities, in which each entity travels along a trajectory of τ edges, the Reeb graph $\mathcal{R} = (V, E)$ has $O(\tau n^2)$ vertices and edges. These bounds are tight in the worst case.*

Proof. Lemma 1 in the full paper [3] gives a simple construction that shows that the Reeb graph may have $\Omega(\tau n^2)$ vertices and edges in the worst case. For the upper bound, consider a trajectory edge (v_i, v_{i+1}) of entity $x \in \mathcal{X}$. An other entity $y \in \mathcal{X}$ is directly connected to x during at most one interval $I \subseteq [t_i, t_{i+1}]$. This interval yields at most two vertices in $\mathcal{R}$. The trajectory of x consists of τ edges, hence a pair x, y produces $O(\tau)$ vertices in $\mathcal{R}$. This gives a total of $O(\tau n^2)$ vertices, each with constant degree, so there are $O(\tau n^2)$ edges. $\qquad\square$

The trajectories of entities are associated with the edges of the Reeb graph in a natural way. Each entity follows a directed path in the Reeb graph from a start vertex to an end vertex. Similarly, (maximal) groups follow a directed path from a start or merge vertex to a split or end vertex. If $m > 0$ or $\delta > 0$, there may be edges in the Reeb graph with which no group is associated. These edges do not contribute to the grouping structure, so we can discard them. The remainder of the Reeb graph we call the *reduced Reeb graph*, which, together with all maximal groups associated with its edges, forms the *trajectory grouping structure*.

2.2 Bounding the Number of Maximal Groups

To bound the total number of maximal groups, we study the case where $m = 1$ and $\delta = 0$, because larger values can only reduce the number of maximal groups. It may seem as if each vertex in the Reeb graph simply creates as many maximal groups as it has outgoing edges. However, consider for example Fig. 2. Split vertex v creates not only the maximal groups $\{1, 3, 5, 7\}$ and $\{2, 4, 6, 8\}$, but also $\{1, 3\}$, $\{5, 7\}$, $\{2, 4\}$, and $\{6, 8\}$. These last four groups are all maximal on $[t_2, t]$, for $t > t_4$.

Notice that all six newly discovered groups start strictly before t_v, but only at t_v do we realize that these groups are maximal, which is the meaning that should be understood with "creating maximal groups". This example can be extended to arbitrary size. Hence a vertex v may create many new maximal groups, some of which start before t_v. We can show that each vertex creates at most n new maximal groups, which leads to a total of $O(\tau n^3)$ maximal groups. The proof of the following theorem is given in the full paper [3].

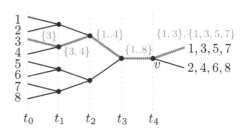

Fig. 2. The maximal groups containing entity 3 (green). Vertex v creates six new groups, including $\{1, 3\}$ and $\{1, 3, 5, 7\}$.

Theorem 2. *Let $\mathcal{X}$ be a set of n entities, in which each entity travels along a trajectory of τ edges. There are at most $O(\tau n^3)$ maximal groups, and this is tight in the worst case.*

3   Computing the Grouping Structure

To compute the grouping structure we need to compute the reduced Reeb graph and the maximal groups. We now show how to do this efficiently. Removing the edges of the Reeb graph that are not used is an easy post-processing step which we do not discuss further.

3.1   Computing the Reeb Graph

We can compute the Reeb graph $\mathcal{R} = (V, E)$ as follows. We first compute all times where two entities x and y are at distance 2ε from each other. We distinguish two types of events, *connect events* at which x and y become directly connected, and *disconnect events* at which x and y stop being directly connected.

We now process the events on increasing time while maintaining the current components. We do this by maintaining a graph $G = (\mathcal{X}, Z)$ representing the directly-connected relation, and the connected components in this graph. The set of vertices in G is the set of entities. The graph G changes over time: at connect events we insert new edges into G, and at disconnect events we remove edges. At any given time t, G contains an edge (x, y) if and only if x and y are directly connected at time t. Hence the components at t (the maximal sets of ε-connected entities) correspond to the connected components in G at time t. Since we know all times at which G changes in advance, we can use the same approach as in [16] to maintain the connected components: we assign a weight to each edge in G and we represent the connected components using a maximum weight spanning forest. The weight of edge (x, y) is equal to the time at which we remove it from G, that is, the time at which x and y become directly disconnected. We store the maximum weight spanning forest F as an ST-tree [17], which allows connectivity queries, inserts, and deletes, in $O(\log n)$ time.

We spend $O(n^2)$ time to initialize the graph G at t_0 in a brute-force manner. For each component we create a start vertex in $\mathcal{R}$. We also initialize a one-to-one mapping M from the current components in G to the corresponding vertices in $\mathcal{R}$. When we handle a connect event of entities x and y at time t, we query F to get the components C_x and C_y containing x and y, respectively. Using M we locate the corresponding vertices v_x and v_y in $\mathcal{R}$. If $C_x \neq C_y$ we create a new merge vertex v in $\mathcal{R}$ with time $t_v = t$, add edges (v_x, v) and (v_y, v) to $\mathcal{R}$ labeled C_x and C_y, respectively. If $C_x = C_y$ we do not change $\mathcal{R}$. Finally, we add the edge (x, y) to G (which may cause an update to F), and update M.

At a disconnect event we first query F to find the component C currently containing x and y. Using M we locate the vertex u corresponding to C. Next, we delete the edge (x, y) from G, and again query F. Let C_x and C_y denote the components containing x and y, respectively. If $C_x = C_y$ we are done, meaning x and y are still ε-connected. Otherwise we add a new split vertex v to $\mathcal{R}$ with time $t_v = t$, and an edge $e = (u, v)$ with $C_e = C$ as its component. We update M accordingly.

Finally, we add an end vertex v for each component C in F with $t_v = t_\tau$. We connect the vertex $u = M(C)$ to v by an edge $e = (u, v)$ and let $C_e = C$ be its component.

Analysis. We need $O(\tau n^2 \log n)$ time to compute all $O(\tau n^2)$ events and sort them according to increasing time. To handle an event we query F a constant number of times, and we insert or delete an edge in F. These operations all take $O(\log n)$ time. So the total time required for building $\mathcal{R}$ is $O(\tau n^2 \log n)$.

Theorem 3. *Given a set $\mathcal{X}$ of n entities, in which each entity travels along a trajectory of τ edges, the Reeb graph $\mathcal{R} = (V, E)$ has $O(\tau n^2)$ vertices and edges, and can be computed in $O(\tau n^2 \log n)$ time.*

3.2 Computing the Maximal Groups

We now show how to compute all maximal groups using the Reeb graph $\mathcal{R} = (V, E)$. We will ignore the requirements that each maximal group should contain at least m entities and have a minimal duration of δ. That is, we assume $m = 1$ and $\delta = 0$. It is easy to adapt the algorithm for larger values.

Labeling the Edges. Our algorithm labels each edge $e = (u, v)$ in the Reeb graph with a set of maximal groups $\mathcal{G}_e$. The groups $G \in \mathcal{G}_e$ are those groups for which we have discovered that G is a maximal group at a time $t \leq t_u$. Each maximal group G becomes maximal at a vertex, either because a merge vertex created G as a new group that is maximal, or because G is now a maximal set of entities that is still together after a split vertex. This means we can compute all maximal groups as follows.

We traverse the set of vertices of $\mathcal{R}$ in topological order. For every vertex v we compute the maximal groups on its outgoing edge(s) using the information on its incoming edge(s).

If v is a start vertex it has one outgoing edge $e = (v, u)$. We set $\mathcal{G}_e$ to $\{(C_e, t_v)\}$ where $t_v = t_0$. If v is a merge vertex it has two incoming edges, e_1 and e_2. We propagate the maximal groups from e_1 and e_2 on to the outgoing edge e, and we discover (C_e, t_v) as a new maximal group. Hence $\mathcal{G}_e = \mathcal{G}_{e_1} \cup \mathcal{G}_{e_2} \cup \{(C_e, t_v)\}$.

If v is a split vertex it has one incoming edge e, and two outgoing edges e_1 and e_2. A maximal group G on e may end at v, continue on e_1 or e_2, or spawn a new maximal group $G' \subset G$ on either e_1 or e_2. In particular, for any group G' in $\mathcal{G}_{e_i}$, there is a group G in $\mathcal{G}_e$ such that $G' = G \cap C_i \neq \emptyset$. The starting time of G' is $t' = \min\{t \mid (G, t) \in \mathcal{G}_e \wedge G' \subseteq G\}$. Thus, t' is the first time G' was part of a maximal group on e. Stated differently, t' is the first time

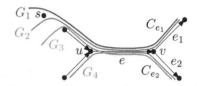

Fig. 3. After split vertex v, $\mathcal{G}_{e_1}$ contains the groups $C_{e_1} = G_1 \cup G_2$ (with starting time t_s), G_1, and G_2. Maximal groups $C_{e_2} = G_3 \cup G_4$ (with starting time t_u), G_3, and G_4 go to e_2. The maximal groups C_e and $G_1 \cup G_2 \cup G_3$ end at v.

G' was in a component on a path to v. Fig. 3 illustrates this case. If v is an end vertex it has no outgoing edges. So there is nothing to be done.

Storing the Maximal Groups. We need a way to store the maximal groups $\mathcal{G}_e$ on an edge $e = (u, v)$ in such a way that we can efficiently compute the set(s) of maximal groups on the outgoing edge(s) of a vertex v. We now show that we can use a tree $\mathcal{T}_e$ to represent $\mathcal{G}_e$, with which we can handle a merge vertex in $O(1)$ time, and a split vertex in $O(k)$ time, where k is the number of entities involved. The tree uses $O(k)$ storage.

We say a group G is a *subgroup* of a group H if and only if $G \subseteq H$ and $I_H \subseteq I_G$. For example, in Fig. 1 $\{x_1, x_2\}$ is a subgroup of $\{x_1, .., x_4\}$. Note that both G and H could be maximal. The proof of the following lemma is given in the full paper [3].

Lemma 1. *Let e be an edge of $\mathcal{R}$, and let S and T be maximal groups in $\mathcal{G}_e$ with starting times t_S and t_T, respectively. There is also a maximal group $G \supseteq S \cup T$ on e with starting time $t_G \geq \max(t_S, t_T)$, and if $S \cap T \neq \emptyset$ then S is a subgroup of T or vice versa.*

We represent the groups $\mathcal{G}_e$ on an edge $e \in E$ by a tree $\mathcal{T}_e$. We call this the *grouping tree*. Each node $v \in \mathcal{T}_e$ represents a group $G_v \in \mathcal{G}_e$. The children of a node v are the largest subgroups of G_v. From Lemma 1 it follows that any two children of v are disjoint. Hence an entity $x \in G_v$ occurs in only one child of v. Furthermore, note that the starting times are monotonically decreasing on the path from the root to a leaf: smaller groups started earlier. A leaf corresponds to a smallest maximal group on e: a singleton set with an entity $x \in C_e$. It follows that $\mathcal{T}_e$ has $O(n)$ leaves, and therefore has size $O(n)$. Note, however, that the summed sizes of all maximal groups can be quadratic.

Analysis. We analyze the time required to label each edge e with a tree $\mathcal{T}_e$ for a given Reeb graph $\mathcal{R} = (V, E)$. Topologically sorting the vertices takes linear time. So the running time is determined by the processing time in each vertex, that is, computing the tree(s) $\mathcal{T}_e$ on the outgoing edge(s) e of each vertex. Start, end, and merge vertices can be handled in $O(1)$ time: start and end vertices are trivial, and at a merge vertex v the tree $\mathcal{T}_e$ is simply a new root node with time t_v and as children the (roots of the) trees of the incoming edges. At a split vertex we have to split the tree $\mathcal{T} = \mathcal{T}_{(u,v)}$ of the incoming edge (u, v) into two trees for the outgoing edges of v. For this, we traverse $\mathcal{T}$ in a bottom-up fashion, and for each node, check whether it induces a vertex in one or both of the trees after splitting. This algorithm runs in $O(|\mathcal{T}|)$ time. Since $|\mathcal{T}| = O(n)$ the total running time of our algorithm is $O(n|V|) = O(\tau n^3)$.

Reporting the Groups. We can augment our algorithm to report all maximal groups at split and end vertices. The main observation is that a maximal group ending at a split vertex v, corresponds exactly to a node in the tree $\mathcal{T}_{(u,v)}$ (before the split) that has entities in leaves below it that separate at v. The procedures for handling split and end vertices can easily be extended to report the maximal groups of size at least m and duration at least δ by simply checking this for each maximal group. Although the number of maximal groups is $O(\tau n^3)$ (Theorem 2), the summed size of all maximal groups can be $\Theta(\tau n^4)$. The running time of our algorithm is $O(\tau n^3 + N)$, where N is the total output size.

Theorem 4. *Given a set $\mathcal{X}$ of n entities, in which each entity travels along a trajectory of τ edges, we can compute all maximal groups in $O(\tau n^3 + N)$ time, where N is the output size.*

4   Robustness

The grouping structure definition we have given and analyzed has a number of good properties. It fulfills monotonicity, and in the previous sections we showed that there are only polynomially many maximal groups, which can be computed in polynomial time as well. In this section we study the property of robustness, which our definition of grouping structure does not have yet. Intuitively, a robust grouping structure ignores short interruptions of groups, as these interruptions may be insignificant at the temporal scale at which we are studying the data. For example, if we are interested in groups that have a duration of one hour or more, we may want to consider interruptions of a minute or less insignificant.

We introduce a new temporal parameter α. Interruptions of duration at most α may be ignored, and the precise moment of events is not relevant beyond a value of α. We can incorporate α in our definition of the grouping structure and obtain (details and proofs are in the full paper [3]):

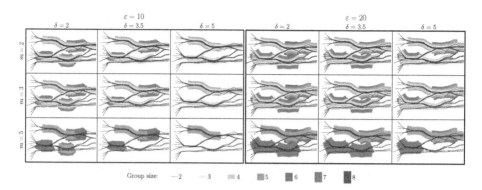

Fig. 4. The maximal groups for varying parameter values. The time associated with each trajectory vertex is proportional to its x-coordinate.

Theorem 5. *Given a set $\mathcal{X}$ of n entities, in which each entity travels along a trajectory of τ edges, we can compute all robust maximal groups in $O(\tau n^3 \log n + N)$ time, where N is the output size.*

5   Evaluation

To see if our model of the grouping structure is practical and indeed captures the grouping behavior of entities we implemented and evaluated our algorithms. We would like to visually inspect the maximal groups identified by our algorithm, and compare this to our intuition of groups. In restricted cases we can show this in a figure, see for example Fig. 4, but for a larger number of trajectories the resulting figures become too cluttered to analyze. So instead we generated short videos.[1]

We use two types of data sets to evaluate our method: a synthetic data set generated using a slightly modified version of the NetLogo Flocking model [21], and a real-world data set consisting of deer, elk, and cattle [15].

NetLogo. We generated several data sets using an adapted version of the Net-Logo Flocking model [21]. In our adapted model the entities no longer wrap around the world border, but instead start to turn when they approach the border. Furthermore, we allow small random direction changes for the entities. The data set that we consider here contains 400 trajectories, each with 818 edges. Our videos show all maximal groups for varying parameter values.

The videos show that our model indeed captures the crucial properties of grouping behavior well. We observe that the choice of parameter values is important. In particular, if we make ε too large we see that the entities are loosely coupled, and too many groups are found. Similarly, for large values of m virtually no groups are found. However, for reasonable parameter settings, for example $\varepsilon = 5.25$, $m = 4$, and $\delta = 100$, we can clearly see that our algorithm identifies

---

[1] See **www.staff.science.uu.nl/~staal006/grouping**

virtually all sets of entities that travel together. Furthermore, if we see a set of entities traveling together that is not identified as group, we indeed see that they disperse quickly after they have come together. The coloring of the line-segments also nicely shows how smaller groups merge into larger ones, and how the larger groups break up into smaller subgroups. This is further evidence that our model captures the grouping behavior well.

Starkey. We also ran our algorithms on a real-world data set, namely on tracking data obtained in the Starkey project [15]. We chose a period of 30 days for which we have the locations of most of the animals. This yields a data set containing 126 trajectories with 1264 vertices each. In the Starkey video we can see that a large group of entities quickly forms in the center, and then slowly splits into multiple smaller groups. We notice that some entities (groups) move closely together, whereas others often stay stationary, or travel separately.

Running Times. Since we are mainly interested in how well our model captures the grouping behavior, we do not extensively evaluate the running times of our algorithms. On our desktop system with a AMD Phenom II X2 CPU running at 3.2Ghz our algorithm, implemented in Haskell, computes the grouping structure for our data sets in a few seconds. Even for 160 trajectories with roughly 20 thousand vertices each we can compute and report all maximal groups in three minutes. Most of the time is spent on computing the Reeb graph, in particular on computing the connect/disconnect events.

6   Concluding Remarks

We introduced a trajectory grouping structure which uses Reeb graphs and a notion of persistence for robustness. We showed how to characterize and efficiently compute the maximal groups and group changes in a set of trajectories, and bounded their maximal number. Our paper demonstrates that computational topology provides a mathematically sound way to define grouping of moving entities. The complexity bounds, algorithms and implementation together form the first comprehensive study of grouping. Our videos show that our methods produce results that correspond to human intuition.

Further work includes more extensive experiments together with domain specialists, such as behavioral biologists, to ensure further that the grouping structure captures groups and events in a natural way, and changes in the parameters have the desired effect. Further, our research may be linked to behavioral models of collective motion [18] and provide a (quantifiable) comparison of these.

We expect that for realistic inputs the size of the grouping structure is much smaller than the worst-case bound that we proved. In almost all our initial experiments the number of maximal groups was less than τ. We plan to do further experiments to get a better estimate of this number, and to provide faster algorithms under realistic input models. We will also work on improving the visualization of the maximal groups and the grouping structure, based on the reduced Reeb graph.

References

1. Benkert, M., Gudmundsson, J., Hübner, F., Wolle, T.: Reporting flock patterns. Computational Geometry 41(3), 111–125 (2008)
2. Biasotti, S., Giorgi, D., Spagnuolo, M., Falcidieno, B.: Reeb graphs for shape analysis and applications. Theor. Comput. Sci. 392(1-3), 5–22 (2008)
3. Buchin, K., Buchin, M., van Kreveld, M.J., Speckmann, B., Staals, F.: Trajectory grouping structures. CoRR, abs/1303.6127 (2013)
4. Dey, T.K., Wang, Y.: Reeb graphs: approximation and persistence. In: Proc. 27th ACM Symp. on Computational Geometry, pp. 226–235 (2011)
5. Edelsbrunner, H., Harer, J., Mascarenhas, A., Pascucci, V., Snoeyink, J.: Time-varying Reeb graphs for continuous space-time data. Computational Geometry 41(3), 149–166 (2008)
6. Edelsbrunner, H., Harer, J.L.: Computational Topology – an introduction. American Mathematical Society (2010)
7. Ester, M., Kriegel, H., Sander, J., Xu, X.: A density-based algorithm for discovering clusters in large spatial databases with noise. In: Proc. 2nd International Conference Knowledge Discovery and Data Mining, vol. 1996, pp. 226–231. AAAI Press (1996)
8. Fomenko, A., Kunii, T. (eds.): Topological Methods for Visualization. Springer, Tokyo (1997)
9. Gudmundsson, J., van Kreveld, M.: Computing longest duration flocks in trajectory data. In: Proc. 14th ACM International Symposium on Advances in Geographic Information Systems, GIS 2006, pp. 35–42. ACM (2006)
10. Gudmundsson, J., van Kreveld, M., Speckmann, B.: Efficient detection of patterns in 2D trajectories of moving points. GeoInformatica 11, 195–215 (2007)
11. Huang, Y., Chen, C., Dong, P.: Modeling herds and their evolvements from trajectory data. In: Cova, T.J., Miller, H.J., Beard, K., Frank, A.U., Goodchild, M.F. (eds.) GIScience 2008. LNCS, vol. 5266, pp. 90–105. Springer, Heidelberg (2008)
12. Jeung, H., Yiu, M.L., Zhou, X., Jensen, C.S., Shen, H.T.: Discovery of convoys in trajectory databases. PVLDB 1, 1068–1080 (2008)
13. Kalnis, P., Mamoulis, N., Bakiras, S.: On discovering moving clusters in spatio-temporal data. In: Medeiros, C.B., Egenhofer, M., Bertino, E. (eds.) SSTD 2005. LNCS, vol. 3633, pp. 364–381. Springer, Heidelberg (2005)
14. Li, Z., Ding, B., Han, J., Kays, R.: Swarm: Mining relaxed temporal moving object clusters. PVLDB 3(1), 723–734 (2010)
15. Oregon Department of Fish and Wildlife and the USDA Forest Service. The Starkey project (2004)
16. Parsa, S.: A deterministic $O(m \log m)$ time algorithm for the Reeb graph. In: Proc. 28th ACM Symp. on Computational Geometry, pp. 269–276 (2012)
17. Sleator, D.D., Tarjan, R.E.: A data structure for dynamic trees. Journal of Computer and System Sciences 26(3), 362–391 (1983)
18. Sumpter, D.: Collective Animal Behavior. Princeton University Press (2010)
19. Vieira, M.R., Bakalov, P., Tsotras, V.J.: On-line discovery of flock patterns in spatio-temporal data. In: Proc. 17th ACM International Conference on Advances in Geographic Information Systems, GIS 2009, pp. 286–295. ACM (2009)
20. Wang, Y., Lim, E.-P., Hwang, S.-Y.: Efficient algorithms for mining maximal valid groups. The VLDB Journal 17(3), 515–535 (2008)
21. Wilensky, U.: NetLogo flocking model. Center for Connected Learning and Computer-Based Modeling. Northwestern University, Evanston, IL (1998)

The Art of Shaving Logs

Timothy M. Chan

Cheriton School of Computer Science, University of Waterloo,
Waterloo, Ontario N2L 3G1, Canada
tmchan@uwaterloo.ca

There have been many instances in the literature where an algorithm with running time of the form $O(n^a)$ is improved to an algorithm with running time $O(n^a/\log^b n)$ for some constants a and b. The "four Russians" algorithm for Boolean matrix multiplication is perhaps one of the most well known examples.

In this talk, we will look at a few selected recent examples of this phenomenon, including the *3-SUM* problem, the problem of detecting *affine degeneracy* in a point set, *Klee's measure* problem, and the *all-pairs shortest paths* problem. (The selection is not comprehensive but is biased towards the speaker's own expertise.) Bit tricks, table lookups, and constructions of small decision trees are used to achieve many of these polylogarithmic-factor speedups, but often the applications of these techniques require interesting ideas.

Some of these results hold for integer input, and some hold for arbitrary real-valued input. Most of these results work in a standard word RAM model (and some work even in a pointer machine). None of the results is obtained through "cheating", as the speaker will try to argue. Many open problems will be mentioned.

F. Dehne, R. Solis-Oba, and J.-R. Sack (Eds.): WADS 2013, LNCS 8037, p. 231, 2013.
© Springer-Verlag Berlin Heidelberg 2013

TREEWIDTH and PATHWIDTH Parameterized by the Vertex Cover Number*

Mathieu Chapelle[1], Mathieu Liedloff[2], Ioan Todinca[2], and Yngve Villanger[3]

[1] IGM-LabInfo, Universit Paris-Est Marne-la-Vallée, 5 Bd Descartes - Champs sur Marne 77454 Marne la Vallée cedex 2, France
mathieu.chapelle@univ-mlv.fr
[2] LIFO, Université d'Orléans, BP 6759, F-45067 Orléans Cedex 2, France
(mathieu.liedloff,ioan.todinca)@univ-orleans.fr
[3] Department of Informatics, University of Bergen, N-5020 Bergen, Norway
yngve.villanger@uib.no

Abstract. After the number of vertices, *Vertex Cover Number* is the largest of the classical graph parameters and has more and more frequently been used as a separate parameter in parameterized problems, including problems that are not directly related to the *Vertex Cover Number*. Here we consider the TREEWIDTH and PATHWIDTH problems parameterized by k, the size of a minimum vertex cover of the input graph. We show that the PATHWIDTH and TREEWIDTH can be computed in $O^*(3^k)$ time. This complements recent polynomial kernel results for TREEWIDTH and PATHWIDTH parameterized by the *Vertex Cover Number*.

1 Introduction

Parameterized algorithms are typically used in the setting where the provided problem is NP-hard and we want to bound the exponential part of the running time to a function of some specific parameter. This parameter can be any property related to the input, the output, or the problem itself. A classical parameter is n, the size of the input or the number of vertices in the input graph. Algorithms of this type are usually refered to as *moderately exponential time* algorithms [13], and in many cases it is non trivial to improve the exponential dependence on n to something better than the naive brute force bound.

The number of vertices is not the only natural graph parameter; there are also parameters like *treewidth, feedback vertex set*, and *vertex cover number*. For every graph, there is an increasing order on these parameters: *treewidth* is the smallest, and then *feedback vertex set, vertex cover number* and eventually n. We refer to Bodlaender et al. [5] for more parameters and the relation between them. Many moderately exponential time algorithms have an exponential dependence on n that is of the form c^n for some constant $c < 2$. When the exponential part of the running time is bounded by one of the other graph parameters, we typically see a much faster growing function than we do for parameter n. Thus, we have reached a situation where tradeoffs can be made between the size the parameter we choose and the exponential dependence on this parameter.

We use a modified big-Oh notation that suppresses all other (polynomially bounded) terms. Thus for functions f and g we write $f(n,k) = O^*(g(n,k))$ if $f(n,k) =$

---

* Partially supported by the ANR project AGAPE.

F. Dehne, R. Solis-Oba, and J.-R. Sack (Eds.): WADS 2013, LNCS 8037, pp. 232–243, 2013.
© Springer-Verlag Berlin Heidelberg 2013

$O(g(n, k) \cdot n^{\mathcal{O}(1)})$. Consider the problems of computing the TREEWIDTH or the PATH-WIDTH of a given graph G. For parameter n both these values can be computed in $O^*(2^n)$ time by a dynamic programming approach proposed by Held and Karp [16]. Currently the best moderately exponential time algorithms for these problems have running times $O(1.735^n)$ [12] and $O(1.89^n)$ [17] respectively. On the other hand if we go to the smaller parameters *treewidth* and *pathwidth* the best known running times are of the from $O^*(2^{O(k^3)})$ [2]. Thus, it is preferable to use the $O^*(2^{O(k^3)})$ algorithm parameterized by *treewidth* if the *treewidth* is $O(n^{1/3})$, and the algorithms parameterized by n otherwise. In this paper we are considering *vertex cover number* as a parameter for the TREEWIDTH or the PATHWIDTH problems. Our goal is to find the most efficient algorithm for these problems where the exponential part of the running time only depends on the *vertex cover number*, i.e. the minimum size of a vertex cover of the graph.

Using the *vertex cover number* as a parameter when analyzing algorithms and solving problems is not a new idea. Some examples from the literature are an $O^*(2^k)$ algorithm for CUTWIDTH parameterized by the *vertex cover number* [9], and different variants of graph layout problems [11] with the same parameter. Let us also mention an $O^*(2^k)$ algorithm for CHORDAL GRAPH sandwich parameterized by the *vertex cover number* of an edge set [15].

Another direction in the area of parameterized complexity is kernelization or instance compression. Recently it was shown [10] that we can not expect that the TREEWIDTH and PATHWIDTH problems have a polynomial kernel unless $NP \subseteq coNP/poly$ when parameterized by *treewidth* or *pathwidth*, but on the other hand they do have a kernel of size $O(k^3)$ when parameterized by the *vertex cover number* [5,6]. Existence of a polynomial size kernel does not necessarily imply the existence of an algorithm that has a slow growing exponential function in the size of the parameter. Indeed if we first kernelize and then use the best moderately exponential time algorithm of [12] on the kernel, we still obtain an $O^*(2^{O(k^3)})$ algorithm for TREEWIDTH parameterized by the *vertex cover number*. Hence dependence in the parameter is still similar to the algorithm parameterized by *treewidth* [2].

Our Results. We provide an $O^*(3^k)$ time algorithm for PATHWIDTH and TREEWIDTH when parameterized by vc, the *vertex cover number*. It means that this algorithm will be preferable for graphs where the *treewidth* is $\Omega(vc^{1/3})$ and the vertex cover number is at most $0.5n$ and $0.58n$ for the TREEWIDTH and PATHWIDTH problems respectively. Another consequence is that the TREEWIDTH and PATHWIDTH of a bipartite graph can be computed in $O^*(3^{n/2})$ or $O^*(1.733^n)$ time, which is better than the running time provided by the corresponding moderate exponential time algorithms ($O^*(1.735^n)$ [12] and $O^*(1.89^n)$ [17] respectively). Due to space restrictions, we only present here an $O^*(4^k)$ algorithm for TREEWIDTH, based on dynamic programming. The algorithm is then modified to obtain a running time of $O^*(3^k)$, and for this purpose we use the subset convolution technique introduced in [1]. This result is detailed in the full version of the paper [7].

In addition to this we also show in the full version that the PATHWIDTH can be computed in $O^*(2^{k'})$ time where k' is the *vertex cover number* of the complement of the graph. This matches the result of [4] for TREEWIDTH parameterized by the *vertex cover number* of the complement of the graph.

2 Preliminaries

All graphs considered in this article are simple and undirected. For a graph $G = (V, E)$ we denote by $n = |V|$ the number of vertices and by $m = |E|$ the number of edges. The neighborhood of a vertex v is defined as $N(v) = \{u \in V : \{u, v\} \in E\}$, and the closed neighborhood is defined as $N[v] = N(v) \cup \{v\}$. For a vertex set W, we define its neighborhood as $N(W) = \bigcup_{v \in W} N(v) \setminus W$, and its closed neighborhood as $N[W] = N(W) \cup W$. A vertex set $C \subseteq V$ in a graph $G = (V, E)$ is called a *vertex cover* if for every edge $uv \in E(G)$ we have that vertex u or v is in C. The minimum size of a vertex cover of G is called the *vertex cover number* of G. Vertex set X is called a clique of G if for each pair $u, v \in X$ we have that $uv \in E$.

Proposition 1 ([8]). *The minimum vertex cover problem can be solved in $O^*(1.28^k)$ time, where k is the vertex cover number of the input graph.*

We now define *tree* and *path decompositions*. A *tree decomposition* of a graph $G = (V, E)$ is a pair $(T, \mathcal{X})$ where $T = (I, F)$ is a tree and $\mathcal{X} = \{X_i \mid i \in I\}$ is a family of subsets of V, called *bags*, where

- $V = \bigcup_{i \in I} X_i$,
- for each edge $uv \in E$ there exists an $i \in I$ such that $u, v \in X_i$, and
- for each vertex $v \in V$ the nodes $\{i \in I \mid v \in X_i\}$ form a connected subtree of T.

The *width* of the tree decomposition $(T, \mathcal{X})$ is $\max_{i \in I} |X_i| - 1$ (the maximum size of a bag, minus one) and the *treewidth* of G is the minimum width over all tree decompositions of G.

A *path decomposition* of G is a tree decomposition $(T, \mathcal{X})$ such that the tree T is actually a path. The *pathwidth* of G is the minimum width over all path decomposition of G.

The following result is a straightforward consequence of Helly's property for a family of subtrees of a tree.

Proposition 2. *Let $(T, \mathcal{X})$ be a tree decomposition of graph $G = (V, E)$. Let $H = (V, F)$ be the graph such that $xy \in F$ if and only if there exists a bag of the decomposition containing both x and y. A set $W \subseteq V$ of vertices induces a clique in H if and only if there is a bag $X_i \in \mathcal{X}$ such that $W \subseteq X_i$.*

It is well-known that the graph H constructed above is a chordal graph (or an interval graph if we replace tree decomposition by path decomposition), but we will not use this here. See e.g. [14] for more details on these graphs and a proof of the previous proposition.

Let i be a node of an arbitrarily rooted tree decomposition $(T, \mathcal{X})$. Let T_i be the subtree of T rooted in i. We denote by V_i the union of bags of the subtree T_i. We let $L_i = V_i \setminus X_i$ (L like "lower") and $R_i = V \setminus V_i$ (R like "rest"). Clearly, (L_i, X_i, R_i) is a partition of V.

Proposition 3 ([3]). *Let $(T, \mathcal{X})$ be a tree decomposition of graph $G = (V, E)$. The bag X_i separates, in graph G, any two vertices $a \in L_i$ and $b \in R_i$, i.e. a and b are in different components of $G[V \setminus X_i]$.*

For our purpose, it is very convenient to use *nice* tree and path decompositions (see e.g. [3]). In a nice tree decomposition $(T, \mathcal{X})$, the tree is rooted, and has only four types of nodes :

1. Leaf nodes i, in which case $|X_i| = 1$.
2. Introduce nodes i, having a unique child j s.t. $X_i = X_j \cup \{u\}$ for some $u \in V \setminus X_j$.
3. Forget nodes i, having a unique child j s.t. $X_i = X_j \setminus \{u\}$ for some $u \in X_j$.
4. Join nodes i, having exactly two children j and k, s.t. $X_i = X_j = X_k$.

Moreover, we can assume that the root node corresponds to a bag of size 1.

Let us associate an *operation* τ_i to each node of a nice tree decomposition. If we are in the second case of the definition (introduce node i), we associate operation $\tau_i = introduce(u)$, where u is the vertex introduced in bag X_i. If we are in the third case (forget node i), we associate operation $\tau_i = forget(u)$, where u is the forgotten vertex. In the fourth case (join node), we associate operation $\tau_i = join(X_i; L_j, L_k)$. For a leaf node i with $X_i = \{u\}$, we also associate operation $\tau_i = introduce(u)$. Nice path decompositions are defined in a similar way, but of course they do not have join nodes.

It is well known [3] that any tree or path decomposition can be refined into a nice one in linear time, without increasing the width.

Proposition 4 ([3]). *Let $(T, \mathcal{X})$ be a tree decomposition of G. There exists a nice tree decomposition $(T', \mathcal{X}')$, such that*

- *each bag of $\mathcal{X}'$ is a subset of a bag in $\mathcal{X}$*
- *for each node i of T, there is a node i' of T' such that the corresponding partitions (L_i, X_i, R_i) (induced by i in $(T, \mathcal{X})$) and $(L'_{i'}, X'_{i'}, R'_{i'})$ (induced by i' in $(T', \mathcal{X}')$) are equal.*

Traces and Valid Partitions. Let C be a vertex cover of minimum size of our input graph G, and let $S = V \setminus C$ be the remaining independent set. We denote $k = |C|$. Our objective is to describe, in a first step, an $O^*(4^k)$ algorithm for treewidth and an $O^*(3^k)$ algorithm for pathwidth. Very informally, if we fix a nice tree or path decomposition of $G[C]$, then there is an optimal way of adding the vertices of S to this tree or path decomposition. Trying all nice decompositions of $G[C]$ by brute force would be too costly. Therefore we introduce the notion of *traces* and *valid partitions* of C.

Definition 1. *Consider a node i of a tree decomposition $(T, \mathcal{X})$ of G. The* trace *of node i on C is the three-partition (L_i^C, X_i^C, R_i^C) of C such that $L_i^C = L_i \cap C$, $X_i^C = X_i \cap C$ and $R_i^C = R_i \cap C$.*

A partition (L^C, X^C, R^C) of C is called a valid triple *or* valid partition *if it is the trace of some node of a tree decomposition. We say that a tree decomposition* respects *the valid partition (L^C, X^C, R^C) if some node of the tree decomposition produces this trace on C.*

The following lemma gives an easy characterization of valid partitions of C. It also proves that a partition is the trace of a node of some tree decomposition, this also holds for some path decomposition. Therefore we do not need to distinguish between partitions that would be valid for tree decompositions or valid for path decompositions.

Lemma 1. *A three-partition* (L^C, X^C, R^C) *is the trace of some tree decomposition (or path decomposition) if and only if* X^C *separates* L^C *from* R^C *in the graph* $G[C]$.

Proof. "$\Rightarrow$:" Consider a node i of a tree decomposition $(T, \mathcal{X})$ of G such that the trace of node i on C is (L^C, X^C, R^C). By Proposition 3, bag X_i separates L_i from R_i in G. Therefore $X_i \cap C = X^C$ separates $L_i \cap C = L^C$ from $R_i \cap C = R^C$ in $G[C]$.

"$\Leftarrow$:" Conversely, since X^C separates L^C from R^C in $G[C]$ and $S = V \setminus C$ is an independent set of G, note that the three bags $X^C \cup L^C \cup S$, $X^C \cup S$ and $X^C \cup R^C \cup S$ form a path decomposition of G. The trace of the middle bag is (L^C, X^C, R^C). □

By Proposition 4, for any valid partition (L^C, X^C, R^C), there exists a nice tree or path decomposition respecting it. Our algorithms will proceed by dynamic programming over valid three-partitions (L^C, X^C, R^C) of this type, for a given vertex cover C. There is a natural partial ordering on such three-partitions.

Definition 2. *We say that a valid three-partition* (L_j^C, X_j^C, R_j^C) *precedes the three-partition* (L_i^C, X_i^C, R_i^C) *if they are different and they are the respective traces of two nodes j and i of a same nice tree decomposition* $(T, \mathcal{X})$, *where i is the father of j in T.*

Observe that if (L_j^C, X_j^C, R_j^C) precedes (L_i^C, X_i^C, R_i^C) we have that $L_j^C \subsetneq L_i^C$ (if i is a *join* or *forget* node) or $L_j^C = L_i^C$ and $X_j^C \subsetneq X_i^C$ (if i is an *introduce* node). In particular, we can order the three-partitions according to a linear extension of the precedence relation. Our algorithms will proceed by dynamic programming over three-partitions of C, according to this order.

It is convenient for us to have a unique maximal three-partition w.r.t. the precedence order. Therefore, starting from graph G, we create a new graph G' by adding a universal vertex $univ$ (i.e. adjacent to all other vertices of G). Clearly, $C \cup \{univ\}$ is a vertex cover of G', of size $k + 1$. Note that the treewidth (resp. pathwidth) of G' equals the treewidth (resp. pathwidth) of G, plus one. Moreover, G has an optimal nice tree (resp. path) decomposition whose root bag only contains vertex $univ$. Therefore, it is sufficient to compute the treewidth (pathwidth) for graph G'. From now on we assume that the input graph is G', i.e. it contains a special universal vertex $univ$, and we only use nice tree (path) decompositions whose root bag is $\{univ\}$. If C denotes the vertex cover of the input graph, then the trace of the root is always $(C \setminus \{univ\}, \{univ\}, \emptyset)$.

3   TREEWIDTH Parameterized by the Vertex Cover Number

Recall that the nice tree decompositions are rooted, thus we can speak of *lower* and *upper* nodes of the decomposition tree.

Lemma 2. *Let* (L^C, X^C, R^C) *be a three-partition of C. Let* $(T, \mathcal{X})$ *be a nice tree-decomposition and consider the set of nodes of T whose trace on C is* (L^C, X^C, R^C). *If* $L^C \neq \emptyset$, *then these nodes of T induce a directed subpath in T, from a lower node* $i_{\min}$ *to an upper node* $i_{\max}$.

Proof. Consider two nodes i and j leaving this same trace (L^C, X^C, R^C) on C. We claim that one of them is ancestor of the other in the tree. By contradiction, assume there is a lowest common ancestor k of i and j, different from i, j. Let $x \in L^C$ (note that here we use the condition $L^C \neq \emptyset$). Observe that x appears in bags of both subtrees T_i and T_j of T, hence by definition of a tree decomposition it must belong to bag X_k. Since x is in X_k and in the subtree T_i, we must also have $x \in X_i$. But $X_i \cap C = X^C$, implying that x is in both X^C and L^C — contradicting the fact that the latter sets do not intersect. It follows that one of i, j must be ancestor of the other.

Let $i_{\min}$ (resp. $i_{\max}$) be the lowest (resp. highest) node whose trace on C is the three-partition (L^C, X^C, R^C). It remains to prove that any node on the path from $i_{\min}$ to $i_{\max}$ in T leaves the same trace. Let i be a node on this path. Recall that L_i denotes the set of vertices of G that appear only in bags strictly below i, and R_i denotes the vertices that do not appear in bags below i. Since i is between $i_{\min}$ and $i_{\max}$, cleary $X^C \subseteq X_i$. If X_i contains some vertex $x \in C \setminus X^C$, then either $x \in L^C$ and thus x must also appear in bag $X_{i_{\min}}$, or $x \in R^C$ and it must appear in bag $X_{i_{\max}}$. In both cases, this contradicts the trace of $i_{\min}$ and $i_{\max}$ on C. We thus have $L^C \subseteq L_i \cap C$. If $L_i \cap C$ also contains some node $x \in R^C$, as before we have that x must be in bag $X_{i_{\max}}$ — a contradiction. Eventually, observe that $R^C \subseteq R_i \cap C$, and that if $R_i \cap C$ contained some vertex $x \in L^C$, this vertex must appear in bag $i_{\min}$ — a contradiction. $\square$

In order to "glue" a valid three-partition (L^C, X^C, R^C) with the previous and next ones, into a nice tree decomposition of $G[C]$, we need to control the operation right below and right above the subpath of nodes leaving this trace. Therefore we introduce the following notion of *valid quintuples*.

Definition 3. *Let (L^C, X^C, R^C) be a valid partition of C, with $L^C \neq \emptyset$. Let τ_+ and τ_- be operations of type introduce, forget or join. We say that $(\tau_-, L^C, X^C, R^C, \tau_+)$ is a* valid quintuple *if there is a nice tree decomposition $(T, \mathcal{X})$ of G respecting the three-partition (L^C, X^C, R^C), with $i_{\min}$ and $i_{\max}$ being the lower and upper node corresponding to this trace, such that $\tau_- = \tau_{i_{\min}}$ and $\tau_+ = \tau_{i_{\max}+1}$, where $i_{\max} + 1$ is the father of $i_{\max}$. In the particular case when $i_{\max}$ is the root we assume for convenience that $\tau_{i_{\max}+1}$ is the forget operation on the unique vertex of the root bag.*

We also say that this nice tree decomposition $(T, \mathcal{X})$ respects the valid quintuple $(\tau_-, L^C, X^C, R^C, \tau_+)$.

The following result is needed for the enumeration of valid quintuples. Its (rather straightforward) proof is given in the full version of the article [7].

Lemma 3. *Given a quintuple $Q = (\tau_-, L^C, X^C, R^C, \tau_+)$, there is a linear time algorithm checking if Q is valid.*

To be able to start our dynamic programming, we introduce a new category of valid quintuples, called *degenerate*, corresponding to valid partitions of type $(\emptyset, X^C, R^C)$. Roughly, they will correspond to the leaves of our optimal tree decomposition. We point out that for degenerate quintuples, parameter τ_- is irrelevant.

Definition 4. *Let $(\emptyset, X^C, R^C)$ be a valid partition of C. Let τ_+ be an operation of type $forget(u)$, with $u \in X^C$ such that $N_G(u) \subseteq X^C$. We say that $(\tau_-, \emptyset, X^C, R^C, \tau_+)$ is a* degenerate valid quintuple *and a tree decomposition respects this quintuple if it has a node $i_{\max}$ whose trace on C is $(\emptyset, X^C, R^C)$, and whose father corresponds to operation $forget(u)$.*

Let us fix a valid quintuple $Q = (\tau_-, L^C, X^C, R^C, \tau_+)$. We want to construct a tree decomposition $(T, \mathcal{X})$ respecting Q, of minimum width. Recall that $S = V \setminus C$ denotes the independent set of the graph obtained by removing the vertices of the vertex cover C. We must understand how to place the vertices of S in the bags of $(T, \mathcal{X})$. For this purpose we define some special subsets of S w.r.t. Q, and the next lemmata describe how these subsets are forced to be in some bags on the subpath of T from $i_{\min}$ to $i_{\max}$ (cf. Lemma 2).

Notation 1. *Let $Q = (\tau_-, L^C, X^C, R^C, \tau_+)$ be a valid quintuple.*

- *We denote $XTR^S(Q) = \{x \in S \mid N(x) \cap L^C \neq \emptyset \text{ and } N(x) \cap R^C \neq \emptyset\}$.*
- *• If τ_- is of type $introduce(u)$, then we denote $XL^S(Q) = \{x \in S \mid N(x) \subseteq L^C \cup X^C \text{ and } u \in N(x) \text{ and } N(x) \cap L^C \neq \emptyset\}$.*
    - *• If τ_- is of type $join(X^C; L1^C, L2^C)$, then $XL^S(Q) = \{x \in S \mid N(x) \cap L1^C \neq \emptyset \text{ and } N(x) \cap L2^C \neq \emptyset \text{ and } N(x) \cap R^C = \emptyset\}$. In particular, the last condition ensures that $XL^S(Q)$ does not intersect $XTR^S(Q)$.*
    - *• If τ_- is a forget operation or if the quintuple is degenerate, then we let $XL^S(Q) = \emptyset$.*
- *Suppose that τ_+ is of type $forget(v)$. Then we let $XR^S(Q) = \{x \in S \mid N(x) \subseteq R^C \cup X^C \text{ and } v \in N(x) \text{ and } N(x) \cap R^C \neq \emptyset\}$. If τ_+ is a introduce or join operation, then $XR^S = \emptyset$.*

Lemma 4. *Let $Q = (\tau_-, L^C, X^C, R^C, \tau_+)$ be a valid quintuple and let $(T, \mathcal{X})$ be a nice tree decomposition respecting Q. Denote by $[i_{\min}, i_{\max}]$ the directed subpath of nodes whose trace on C is (L^C, X^C, R^C) (in the case where $L^C = \emptyset$, we take $i_{\min} = i_{\max}$). Then*

- *For any i in the subpath $[i_{\min}, i_{\max}]$, X_i contains $X^C \cup XTR^S(Q)$.*
- *$X_{i_{\min}}$ contains $X^C \cup XTR^S(Q) \cup XL^S(Q)$.*
- *$X_{i_{\max}}$ contains $X^C \cup XTR^S(Q) \cup XR^S(Q)$.*

Proof. Let $x \in XTR^S(Q)$. By definition of XTR^S, vertex x has a neighbor $a \in L^C$ and a neighbor $b \in R^C$. Since $a \in L^C$, it only appears in the bags of T strictly below $i_{\min}$. Since x is adjacent to a, it must also appear on one of these bags. Since $b \in R^C$, vertex b appears in no bag below $i_{\max}$ (included). Therefore, x must appear in some bag which is not below $i_{\max}$. Consequently, x appears in every bag of the $[i_{\min}, i_{\max}]$ subpath.

Assume that $XL^S(Q)$ is not empty. If $\tau_- = \tau_{i_{\min}} = introduce(u)$, then every vertex $x \in XL^S(Q)$ must appear in some bag strictly below $i_{\min}$ (because it has a neighbor in L^C) and in some bag containing u (because it sees u). This latter bag cannot be strictly below $i_{\min}$. Thus $x \in X_{i_{\min}}$ and $XL^S(Q)$ is contained in $X_{i_{\min}}$. When $i_{\min}$ is a *join* node, $L^C \neq \emptyset$ and we must show that $X_{i_{\min}}$ contains $XL^S(Q)$. But then each vertex $x \in XL^S(Q)$ has a neighbor which only appears in the left subtree of $i_{\min}$, strictly below $i_{\min}$, and one in the right subtree of $i_{\min}$, strictly below $i_{\min}$. Thus x must appear in the bag of $i_{\min}$.

If $XR^S(Q)$ is not empty, then $\tau_+ = forget(v)$ and v is a neighbor of each $x \in XR^S(Q)$. Hence x must appear in a bag below $i_{\max}$ (included). But x also has neighbors in R^C, thus it must appear in some bag which is not below $i_{\max}$. Consequently, $XR^S(Q)$ is contained in $X_{i_{\max}}$. $\square$

We consider now vertices of S whose neighborhood is a subset of X^C.

Notation 2. *Let $Q = (\tau_-, L^C, X^C, R^C, \tau_+)$ be a valid quintuple. Denote by $XF^S(Q)$ the set of vertices $x \in S$ such that $N(x) \subseteq X^C$.*

Let $\epsilon(Q)$ be set to 1 if there is some $x \in XF^S(Q)$ such that $N(x) = X^C$, set to 0 otherwise.

Lemma 5. *Let $Q = (\tau_-, L^C, X^C, R^C, \tau_+)$ be a valid quintuple and let $(T, \mathcal{X})$ be a tree decomposition respecting it. Then $(T, \mathcal{X})$ has a bag of size at least $|X^C| + \epsilon(Q)$.*

Proof. If $\epsilon(Q) = 0$ the claim is trivial. If $\epsilon(Q) = 1$, let $x \in S$ such that $N(x) = X^C$. By Helly's property (see Proposition 2), there must be a bag of $(T, \mathcal{X})$ containing x and X^C. □

Notation 3. *Let $Q = (\tau_-, L^C, X^C, R^C, \tau_+)$ be a valid quintuple. We define the* local treewidth *of Q as*

$$loctw(Q) = |X^C| + \max\{|XTR^S| + |XL^S|, |XTR^S| + |XR^S|, \epsilon(Q)\} - 1.$$

The -1 used above plays the same role as in the definition of treewidth. By Lemmata 4 and 5 we deduce.

Corollary 1. *Any nice tree decomposition of G respecting a valid quintuple Q is of width at least $loctw(Q)$.*

We now define the partial treewidth of a valid quintuple. Intuitively, the partial treewidth of a quintuple Q is the minimum value t such that there is a nice tree decomposition of $G[C]$, respecting Q, with all valid quintuples below Q having local treewidth at most t. We shall prove in Lemma 6 and Theorem 5 that actually the partial treewidth of Q is at most t if and only if there exists a nice tree decomposition of the whole graph G, respecting Q, such that all bags below $i_{\max}$ have size at most $t + 1$.

Notation 4. *Given a valid quintuple $Q = (\tau_-, L^C, X^C, R^C, \tau_+)$, we define the* partial treewidth *of Q, denoted $ptw(Q)$, as follows.*

- *If $Q = (\tau_-, \emptyset, X^C, R^C, forget(u))$ is a degenerate valid quintuple then*

$$ptw(Q) = loctw(Q).$$

- *If $\tau_- = introduce(u)$,*

$$ptw(Q) = \max\left\{ loctw(\tau_-, L^C, X^C, R^C, \tau_+), \min_{valid\ quintuple\ Q_-} ptw(Q_-)\right\}$$

where the minimum is taken over all valid quintuples Q_- of type $(\eta, L^C, X^C \setminus \{u\}, R^C \cup \{u\}), introduce(u))$.

- *If $\tau_- = forget(u)$,*

$$ptw(Q) = \max\left\{ loctw(\tau_-, L^C, X^C, R^C, \tau_+), \min_{valid\ quintuple\ Q_-} ptw(Q_-)\right\}$$

where the minimum is taken over all valid quintuples Q_- of type $(\eta, L^C \setminus \{u\}, X^C \cup \{u\}, R^C), forget(u))$.

– *If* $\tau_- = join(X^C; L1^C, L2^C)$,

$$ptw(Q) = \max \; (loctw(\tau_-, L^C, X^C, R^C, \tau_+),$$

$$\min_{\text{valid quintuple } Q1_-} ptw(Q1_-),$$

$$\min_{\text{valid quintuple } Q2_-} ptw(Q2_-))$$

where the minima are taken over all valid quintuples $Q1_-$ of type $(\eta 1, L1^C, X^C, R^C \cup L2^C), join(X^C; L1^C, L2^C))$ and all quintuples $Q2_-$ of type $(\eta 2, L2^C, X^C, R^C \cup L1^C), join(X^C; L1^C, L2^C))$.

Lemma 6. *Any nice tree decomposition of G respecting a valid quintuple Q is of width at least $ptw(Q)$.*

Proof. We order the three-partitions (L^C, X^C, R^C) of C according to the precedence relation (Definition 2). We prove the lemma for each valid quintuple $Q = (\tau_-, L^C, X^C, R^C, \tau_+)$, by induction (according to this order) on (L^C, X^C, R^C).

For quintuples such that $L^C = \emptyset$, the property follows directly from Corollary 1 and the base case of Notation 4.

Now take $Q = (\tau_-, L^C, X^C, R^C, \tau_+)$ with $L^C \neq \emptyset$. Let $i_{\min}$ the lowest node of the tree decomposition respecting Q, whose trace is (L^C, X^C, R^C). If $i_{\min}$ is a join node, it has two sons with traces $(L1^C, X^C, R^C \cup L2^C)$ and $(L2^C, X^C, R^C \cup L1^C)$ and the proof follows from the $join$ case of Notation 4 and the induction hypothesis on the valid quintuples preceding Q. Note that both $L1^C$ and $L2^C$ are non empty, otherwise $i_{\min}$ would not be the lowest node with trace (L^C, X^C, R^C). Similarly, if $i_{\min}$ is an $introduce(u)$ node, then we apply Corollary 1 and the $introduce$ case of Notation 4 to the quintuple preceding Q in the tree decomposition. The same holds if $i_{\min}$ is of type $forget(u)$ (using the $forget$ case of Notation 4). We point out that, if $\tau_- = forget(u)$ and $L^C = \{u\}$, the quintuple Q_- of Notation 4 corresponds to the base case of our induction. □

Theorem 5. *The treewidth of G is*

$$tw(G) = \min_{Q_{last}} ptw(Q_{last})$$

over all valid quintuples Q_{last} of the form $(\tau_-, C \setminus \{univ\}, \{univ\}, \emptyset, forget(univ))$.

Proof. First note that $tw(G) \geq \min_{Q_{last}} ptw(Q_{last})$. Indeed, an optimal tree decomposition will contain a root whose bag corresponds to a single vertex $univ$, and this root will leave a trace on C of type $(C \setminus \{univ\}, \{univ\}, \emptyset)$. The inequality follows from Lemma 6.

Conversely, let $Q_{last} = (\tau_-, C \setminus \{univ\}, \{univ\}, \emptyset, forget(univ))$ be the valid quintuple of minimum ptw, among all quintuples of this type; denote by t this minimum value. The computation of $ptw(Q_{last})$ naturally provides a tree T^C of quintuples, the root being Q_{last}, and such that for the node corresponding to quintuple Q its sons are the preceding quintuples realizing the minimum value for $ptw(Q)$ in Notation 4. The leaves of this tree correspond to the base case of Notation 4, hence to degenerate valid quintuples. By definition of ptw, all these selected quintuples have $loctw$ at most t. We construct a tree decomposition of G with bags of size at most $t + 1$.

Let $Q_i = (\tau_{-i}, L_i^C, X_i^C, R_i^C, \tau_{+i})$ be the quintuple associated to node i in T^C. Let $(T^C, \mathcal{X}^C)$ be the tree-decomposition of $G[C]$ obtained by associating to each node i of T^C the bag X^C. Each node i, except for the leaves, corresponds to an *introduce*, *forget* or *join* operation τ_{-i}.

Let T be the tree obtained from T^C by replacing each node i with a path of three nodes, denoted $i_{\min}, imid$ and $i_{\max}$ (from the bottom towards the top). Initially, we associate to the three nodes $i_{\min}, imid, i_{\max}$ the same bag X_i^C. Now, for each i,

1. add $XTR^S(Q_i)$ to all bags in the subpath $[i_{\min}, i_{\max}]$ of T;
2. add $XL^S(Q_i)$ to bag number $i_{\min}$;
3. add $XR^S(Q_i)$ to bag number $i_{\max}$;
4. For each vertex $x \in XF^S(Q_i)$, which has not yet been added to some bag of T, create a new node of T adjacent only to $imid$ and associate to this node the bag $N[x]$. These nodes are called *pending nodes*.

We claim that in this way we have obtained a tree decomposition $(T, \mathcal{X})$ of G. Clearly all bags created at step i are of size at most $loctw(Q_i) + 1$, hence at most $t + 1$. It remains to prove that these bags satisfy the conditions of a tree decomposition.

Recall that $(T^C, \mathcal{X}^C)$ is a tree decomposition of $G[C]$. By construction of $(T, \mathcal{X})$, for each vertex $y \in C$, the bags of $(T, \mathcal{X})$ containing it will form a subtree of T. Also, for each edge yz of $G[C]$, some bag of $(T, \mathcal{X})$ shall contain both y and z. It remains to verify the same type of conditions for vertices of S and edges incident to them. This part of the proof is detailed in the full version of the paper [7]. □

Theorem 6. *The* TREEWIDTH *problem can be solved in* $O^*(4^k)$ *time, where k is the size of the minimum vertex cover of the input graph.*

Proof. Given an arbitrary graph G, we compute a minimum vertex cover in $O^*(1.28^k)$ (Proposition 1). Then G is transformed into a graph G' by adding a universal vertex $univ$. Let C be the vertex cover of G' obtained by adding $univ$ to the minimum vertex cover of G (hence $|C| = k + 1$). The treewidth of G' is computed as follows.

Step 1. Compute all valid partitions (L^C, X^C, R^C), by enumerating all three-partitions of C and keeping only the valid ones (Lemma 1). This can be done in time $O^*(3^k)$. The number of valid partitions is at most 3^{k+1}.

Step 2. Compute all valid quintuples Q using Lemma 3. For $Q = (\tau_-, L^C, X^C, R^C, \tau_+)$ where τ_+ is a *join* node, the parameters of this *join* are not relevant for $loctw(Q)$ and $ptw(Q)$. Therefore, we do not need to memorize the parameters of the *join*. With this simplification, we only need to store $O^*(4^k)$ valid (simplified) quintuples, and their computation can be performed in time $O^*(4^k)$. The 4^k comes from quintuples of the type $(join(X^C; L1^C, L2^C), L^C, X^C, R^C, \tau_+)$, since $L1^C, L2^C, X^C$ and R^C form a partition of C into four parts. The quintuples are then sorted by the precedence relation on the corresponding valid three-partitions. This can be done within the same running time, the triples (L^C, X^C, R^C) being sorted by increasing size of L^C, and in case of tie-breaks by increasing size of X^C (see Definition 2 and following remarks).

Step 3. For each valid quintuple $Q = (\tau_-, L^C, X^C, R^C, \tau_+)$, according to the ordering above, compute by dynamic programming $loctw(Q)$ (Notation 3) and then $ptw(Q)$

(Notation 4). In order to process efficiently the quintuples Q_- of Notation 4, let us observe the value $\min_{Q_-} ptw(Q_-)$ over all Q_- of a given type can be updated online, as soon as we compute $ptw(Q_-)$. Indeed, for all the Q_- of a same type, the first parameter η will differ, but the four others are equal. So it is actually a minimum over all η. The same holds for the minimum over all $Q1_-$ and over all $Q2_-$. Hence, when we process quintuple Q, we have these minima at hand and the value $ptw(Q)$ is computable in polynomial time. This step can be performed in polynomial for each Q, so the overall running time is still $O^*(4^k)$.

Step 4. Compute the treewidth of G' using Theorem 5, and return $tw(G) = tw(G') - 1$. This step takes polynomial running time.

Altogether, the algorithm takes $O^*(4^k)$ running time and space. This achieves the proof of the theorem. The algorithm can also be adapted to return, within the same time bounds, an optimal tree decomposition of the input graph.

With a careful analysis, the most costly part of the algorithm is Step 2, running in $O(k4^k m)$ time, where the $O(k4^k)$ part represents the number of quintuples that are enumerated, and $O(m)$ is the time required to check whether a quintuple is valid.   □

The algorithm for pathwidth, described in the full version of the paper [7], is quite similar, with a slight difference in the definition of local pathwidth. Due to the fact that it only uses *introduce* and *forget* operations, the number of valid quintuples is $O^*(3^k)$, and so is the running time of the pathwidth algorithm. More efforts are required to transform the $O^*(4^k)$ algorithm for treewidth into an algorithm with $O^*(3^k)$ running time. For valid quintuples $Q = (\tau_-, L^C, X^C, R^C, \tau_+)$ where τ_- is a *join* node, we do not explicitly store all possible parameters of the join. Instead, we use the powerful subset convolution technique [1] to compute the best parameters of the join, optimizing $ptw(Q)$. This can be done in $O^*(3^k)$ time [7].

Theorem 7 (see [7]). *The* TREEWIDTH *and* PATHWIDTH *problems can be solved in* $O^*(3^k)$ *time, where k is the vertex cover number of the input graph.*

4   Concluding Remarks

We have shown that it is possible to obtain $O^*(3^k)$ time algorithms for computing TREEWIDTH and PATHWIDTH where parameter k is the *vertex cover number* of the graph. This puts the *vertex cover number* in the same class as parameter n as both allows an $O^*(c^k)$ time algorithm for the considered problems. It is an interesting question whether an $O^*(c^k)$ time algorithm exists when using the *feedback vertex set* of the graph as the parameter k.

References

1. Björklund, A., Husfeldt, T., Kaski, P., Koivisto, M.: Fourier meets Möbius: fast subset convolution. In: Johnson, D.S., Feige, U. (eds.) STOC, pp. 67–74. ACM (2007)
2. Bodlaender, H.L.: A linear-time algorithm for finding tree-decompositions of small treewidth. SIAM J. Comput. 25(6), 1305–1317 (1996)

3. Bodlaender, H.L.: Treewidth: Algorithmic techniques and results. In: Privara, I., Ružička, P. (eds.) MFCS 1997. LNCS, vol. 1295, pp. 19–36. Springer, Heidelberg (1997)

4. Bodlaender, H.L., Fomin, F.V., Koster, A.M.C.A., Kratsch, D., Thilikos, D.M.: On exact algorithms for treewidth. In: Azar, Y., Erlebach, T. (eds.) ESA 2006. LNCS, vol. 4168, pp. 672–683. Springer, Heidelberg (2006)

5. Bodlaender, H.L., Jansen, B.M.P., Kratsch, S.: Preprocessing for treewidth: A combinatorial analysis through kernelization. In: Aceto, L., Henzinger, M., Sgall, J. (eds.) ICALP 2011, Part I. LNCS, vol. 6755, pp. 437–448. Springer, Heidelberg (2011)

6. Bodlaender, H.L., Jansen, B.M.P., Kratsch, S.: Kernel Bounds for Structural Parameterizations of Pathwidth. In: Fomin, F.V., Kaski, P. (eds.) SWAT 2012. LNCS, vol. 7357, pp. 352–363. Springer, Heidelberg (2012)

7. Chapelle, M., Liedloff, M., Todinca, I., Villanger, Y.: Treewidth and pathwidth parameterized by vertex cover. arXiv:1305.0433 (2013)

8. Chen, J., Kanj, I.A., Xia, G.: Improved upper bounds for vertex cover. Theor. Comput. Sci. 411(40-42), 3736–3756 (2010)

9. Cygan, M., Lokshtanov, D., Pilipczuk, M., Pilipczuk, M., Saurabh, S.: On cutwidth parameterized by vertex cover. In: Marx, D., Rossmanith, P. (eds.) IPEC 2011. LNCS, vol. 7112, pp. 246–258. Springer, Heidelberg (2012)

10. Drucker, A.: New limits to classical and quantum instance compression. In: FOCS, pp. 609–618. IEEE Computer Society (2012)

11. Fellows, M.R., Lokshtanov, D., Misra, N., Rosamond, F.A., Saurabh, S.: Graph layout problems parameterized by vertex cover. In: Hong, S.-H., Nagamochi, H., Fukunaga, T. (eds.) ISAAC 2008. LNCS, vol. 5369, pp. 294–305. Springer, Heidelberg (2008)

12. Fomin, F.V., Villanger, Y.: Finding induced subgraphs via minimal triangulations. In: Marion, J.-Y., Schwentick, T. (eds.) STACS. LIPIcs, vol. 5, pp. 383–394. Schloss Dagstuhl - Leibniz-Zentrum fuer Informatik (2010)

13. Fomin, F.V., Kratsch, D.: Exact Exponential Algorithms. Texts in Theoretical Computer Science. An EATCS Series. Springer (2010)

14. Golumbic, M.C.: Algorithmic Graph Theory and Perfect Graphs. Academic Press, New York (1980)

15. Heggernes, P., Mancini, F., Nederlof, J., Villanger, Y.: A parameterized algorithm for CHORDAL SANDWICH. In: Calamoneri, T., Diaz, J. (eds.) CIAC 2010. LNCS, vol. 6078, pp. 120–130. Springer, Heidelberg (2010)

16. Held, M., Karp, R.M.: A dynamic programming approach to sequencing problems. Journal of the Society for Industrial and Applied Mathematics 10(1), 196–210 (1962)

17. Kitsunai, K., Kobayashi, Y., Komuro, K., Tamaki, H., Tano, T.: Computing directed pathwidth in $O(1.89^n)$ time. In: Thilikos, D.M., Woeginger, G.J. (eds.) IPEC 2012. LNCS, vol. 7535, pp. 182–193. Springer, Heidelberg (2012)

Visibility and Ray Shooting Queries in Polygonal Domains

Danny Z. Chen[1,*] and Haitao Wang[2,**]

[1] Department of Computer Science and Engineering
University of Notre Dame, Notre Dame, IN 46556, USA
dchen@cse.nd.edu
[2] Department of Computer Science
Utah State University, Logan, UT 84322, USA
haitao.wang@usu.edu

Abstract. Given a polygonal domain (or polygon with holes) in the plane, we study the problem of computing the visibility polygon of any query point. As a special case of visibility problems, we also study the ray-shooting problem of finding the first point on the polygon boundaries that is hit by any query ray. These are fundamental problems in computational geometry and have been studied extensively. We present new algorithms and data structures that improve the previous results.

1 Introduction

Given a set $\mathcal{P} = \{P_1, P_2, \ldots, P_h\}$ of h pairwise-disjoint polygonal obstacles of totally n vertices in the plane, the space minus the interior of all obstacles is called the *free space*. Two points are *visible* to each other if the open line segment connecting them lies entirely in the free space. For any point q in the free space, the *visibility polygon* of q, denoted by $Vis(q)$, is the set of points in the plane visible to q. The *visibility query* problem seeks an efficient data structure that allows fast computation of $Vis(q)$ for any query point q. Let $|Vis(q)|$ denote the number of vertices of $Vis(q)$. We present two new visibility query data structures. The first one uses $O(n^2)$ space and is constructed in $O(n^2 \log n)$ time; for any query point q, $Vis(q)$ can be computed in $O(\log^2 n + \min\{h, |Vis(q)|\} \log n + |Vis(q)|)$ time. Our second data structure is of size $O(n + h^2)$, and its preprocessing time and query time are $O(n + h^2 \log h)$ and $O(|Vis(q)| \log n)$, respectively. Note that in some cases the value h can be substantially smaller than n.

We also study the *ray-shooting query*, a special case of visibility problems: Given any query ray $\sigma(q)$ with its origin point q in the free space, find the first point on the obstacle boundaries or in infinity that is hit by $\sigma(q)$. We construct a data structure of size $O(n + h^2)$ in $O(n + h^2 \cdot \text{poly}(\log h))$ time that answers any query in $O(\log n)$ time, where $\text{poly}(\log h)$ is a polynomial function of $\log h$.

---

* Chen's research was supported in part by NSF under Grants CCF-0916606 and CCF-1217906.
** Corresponding author.

F. Dehne, R. Solis-Oba, and J.-R. Sack (Eds.): WADS 2013, LNCS 8037, pp. 244–255, 2013.
© Springer-Verlag Berlin Heidelberg 2013

Table 1. Summary of ray-shooting data structures in polygonal domains

Data Structure	Preprocessing Time	Size	Query Time
[5,10]	$O(n\sqrt{h} + n\log n + h^{3/2}\log h)$	$O(n)$	$O(\sqrt{h}\log n)$
[17]	$O(n^2)$	$O(n^2)$	$O(\log n)$
[1]	$O((n\log n + h^2)\log h)$	$O((n + h^2)\log h)$	$O(\log^2 n \log^2 h)$
Our Result	$O(n + h^2 \cdot \text{poly}(\log h))$	$O(n + h^2)$	$O(\log n)$

Throughout this paper, we always let k denote $|Vis(q)|$ for any query point q. We say the *complexity* of a data structure is $O(f_1(\cdot), f_2(\cdot), f_3(\cdot))$ if its preprocessing time, size, and query time are $O(f_1(\cdot)), O(f_2(\cdot))$, and $O(f_3(\cdot))$, respectively.

Previous Work. For the ray-shooting query problem, Table 1 gives a summary. Our new data structure improves the previous work for small h. For simple polygons, ray-shooting data structures of $O(n, n, \log n)$ complexity have been proposed [5,6,9,10].

For the visibility query problem, previous work has been done on both the single simple polygon case and the polygonal domain case. For a single simple polygon, Bose *et al.* [4] proposed a data structure of complexity $O(n^3 \log n, n^3, k + \log n)$. Aronov *et al.* [2] gave a smaller-size data structure with a little larger query time, with complexity $O(n^2 \log n, n^2, k + \log^2 n)$. As indicated in [2], by using a ray-shooting data structure [5,10], a visibility query data structure of complexity $O(n, n, k \log n)$ is possible. For the polygonal domain case, Zarei and Ghodsi [21] gave a data structure of complexity $O(n^3 \log n, n^3, k + \min\{h, k\} \log n)$, and Inkulu and Kapoor [12] obtained a data structure of complexity $O(n^2 \log n, n^2, k + h + \min\{h, k\} \log^2 n)$. Another data structure in [12] has complexity related to the size of the visibility graph of the polygonal domain, which is $O(n^2)$; in the worst case, its complexity is $O(n^2 h^3, n^2 h^2, k \log n)$. Nouri and Ghodsi [16] gave a data structure of complexity $O(n^4 \log n, n^4, k + \log n)$, and Lu *et al.* [15] presented a data structure of complexity $O(n^2 \log n, n^2, k + \log^2 n + h \log(n/h))$. Table 2 summarizes these results for the polygonal domain case.

Comparing with the result in [21], our first data structure is $O(n)$ smaller in space and preprocessing time, but with an additive $O(\log^2 n)$ query time, which seems difficult to improve unless the query time of the data structure for the simple polygon case [2] can be reduced (because it has the same preprocessing time and space as our data structure). Comparing with the results in [12,15], our first data structure has the same processing time and space but with smaller query time. Our second data structure, comparing with the second one in [12], has the same query time but uses much less preprocessing time and space.

In addition, our results for visibility queries can be extended to *cone visibility queries* where, in addition to a query point q, a query also includes a cone with q as the apex that delimits the visibility of q. Our first visibility query data structure can be extended to this case with the same performances; for our

Table 2. Summary of visibility query data structures in polygonal domains. The value k is the output size of the visibility polygon of the query point.

Data Structure	Preprocessing Time	Size	Query Time
[21]	$O(n^3 \log n)$	$O(n^3)$	$O(k + \min\{h, k\} \log n)$
[12]	$O(n^2 \log n)$	$O(n^2)$	$O(k + h + \min\{h, k\} \log^2 n)$
[12]	$O(n^2 h^3)$	$O(n^2 h^2)$	$O(k \log n)$
[15]	$O(n^2 \log n)$	$O(n^2)$	$O(k + \log^2 n + h \log(n/h))$
Our Result 1	$O(n^2 \log n)$	$O(n^2)$	$O(k + \log^2 n + \min\{h, k\} \log n)$
Our Result 2	$O(n + h^2 \log h)$	$O(n + h^2)$	$O(k \log n)$

second one, the extended version has the same performances as before except that the preprocessing time becomes $O(n + h^2 \text{poly}(\log h))$.

Our Approaches. A corridor structure of polygonal domains has been used for solving shortest path problems [7,11,13], and later some new concepts like "bays", "canals", and "ocean" were introduced [8], which we refer to as the "extended corridor structure". In this paper, we also use the extended corridor structure [8], which partitions the free space into an ocean $\mathcal{M}$, bays, and canals. Each bay or canal is a simple polygon. The ocean $\mathcal{M}$ is multiply connected and its boundary consists of $O(h)$ convex chains. The extended data structure was used in [8] for computing the visibility polygon from a single line segment in polygonal domains. Unfortunately, the algorithm in [8] does not work for visibility queries. The techniques given in this paper focus on visibility queries. We process each bay/canal using data structures for simple polygons, and process $\mathcal{M}$ using data structures for convex obstacles. For example, for the visibility query problem, we build the data structure [2] for each bay/canal; for $\mathcal{M}$, we utilize the visibility complex [19,20]. For any query point q, $Vis(q)$ is obtained by consulting the data structures for $\mathcal{M}$ and for bays/canals.

Note that the corridor structure [13] was also used by the visibility query data structures in [12,15]; but, their approaches are quite different from ours. For example, they do not use the extended corridor structure (i.e., they do not use the ocean, bays, and canals). As shown later, our techniques not only yields better results but also makes the solutions quite simple.

In Section 2, we review the geometric structures of $\mathcal{P}$. We present our ray-shooting data structure in Section 3. In Section 4, we give our data structures for the visibility query problems. Due to the space limit, some proofs are omitted and can be found in the full version of this paper. For ease of exposition, we assume that no three obstacle vertices of $\mathcal{P}$ are collinear.

2 Preliminaries

For completeness of this paper, we briefly review the extended corridor structure [8]. Further, the rest of this paper relies heavily on the notation related to the

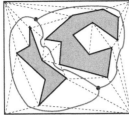
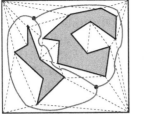
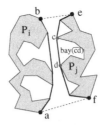
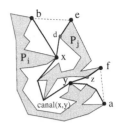

Fig. 1. Illustrating a triangulation of the free space among two obstacles and the corridors (with red solid curves). There are two junction triangles indicated by the large dots inside them, connected by three solid (red) curves. Removing the two junction triangles results in three corridors.

Fig. 2. Illustrating an open hourglass (left) and a closed hourglass (right) with a corridor path connecting the apices x and y of the two funnels. The dashed segments are diagonals. The paths $\pi(a, b)$ and $\pi(e, f)$ are marked by thick solid curves. A bay $bay(\overline{cd})$ with gate $\overline{cd}$ (left) and a canal $canal(x, y)$ with gates $\overline{xd}$ and $\overline{yz}$ (right) are also shown.

structure. For simplicity, we assume all obstacles in $\mathcal{P}$ are contained in a rectangle $\mathcal{R}$ (see Fig. 1), and we also view $\mathcal{R}$ as an obstacle in $\mathcal{P}$.

Let $\mathcal{F}$ denote the free space in $\mathcal{R}$, and $Tri(\mathcal{F})$ denote a triangulation of $\mathcal{F}$. Let $G(\mathcal{F})$ be the (planar) dual graph of $Tri(\mathcal{F})$. The degree of each node in $G(\mathcal{F})$ is at most three. Using $G(\mathcal{F})$, we compute a planar 3-regular graph, denoted by G^3 (the degree of each node in G^3 is three), possibly with loops and multi-edges, as follows. First, we remove every degree-one node from $G(\mathcal{F})$ together with its incident edge; repeat this process until no degree-one node remains in the graph. Second, remove every degree-two node from $G(\mathcal{F})$ and replace its two incident edges by a single edge; repeat this process until no degree-two node remains. The resulting graph is G^3 (see Fig. 1), which has $O(h)$ faces, nodes, and edges [13]. Each node of G^3 corresponds to a triangle in $Tri(\mathcal{F})$, which is called a *junction triangle* (see Fig. 1). The removal of all junction triangles results in $O(h)$ *corridors* (defined below), each of which corresponds to one edge of G^3.

The boundary of a corridor C consists of four parts (see Fig. 2): (1) A boundary portion of an obstacle $P_i \in \mathcal{P}$, from a point a to a point b; (2) a diagonal of a junction triangle from b to a boundary point e on an obstacle $P_j \in \mathcal{P}$ ($P_i = P_j$ is possible); (3) a boundary portion of the obstacle P_j from e to a point f; (4) a diagonal of a junction triangle from f to a. The corridor C is a simple polygon. Let $\pi(a, b)$ (resp., $\pi(e, f)$) be the shortest path from a to b (resp., e to f) inside C. The region H_C bounded by $\pi(a, b), \pi(e, f)$, and the two diagonals $\overline{be}$ and $\overline{fa}$ is called an *hourglass*, which is *open* if $\pi(a, b) \cap \pi(e, f) = \emptyset$ and *closed* otherwise (see Fig. 2). If H_C is open, then both $\pi(a, b)$ and $\pi(e, f)$ are convex chains and are called the *sides* of H_C; otherwise, H_C consists of two "funnels" [14] and a path $\pi_C = \pi(a, b) \cap \pi(e, f)$ joining the two apices of the two funnels, called the *corridor path* of C. Each funnel side is also a convex chain. We compute the hourglass of each corridor. The triangulation $Tri(\mathcal{F})$ can be computed in $O(n \log n)$ time or $O(n + h \log^{1+\epsilon} h)$ time for any constant $\epsilon > 0$ [3]. After $Tri(\mathcal{F})$ is produced, computing all hourglasses takes $O(n)$ time.

Let $\mathcal{M}$ be the union of all $O(h)$ junction triangles, open hourglasses, and funnels. We call the space $\mathcal{M}$ the *ocean*. Note that $\mathcal{M} \subseteq \mathcal{F}$. Since the sides of open hourglasses and funnels are all convex, the boundary $\partial \mathcal{M}$ of $\mathcal{M}$ consists of $O(h)$ convex chains with totally $O(n)$ vertices; further, $\partial \mathcal{M}$ has $O(h)$ reflex vertices (with respect to $\mathcal{R} \setminus \mathcal{M}$). Thus, $\mathcal{R} \setminus \mathcal{M}$ can be partitioned into a set $\mathcal{P}'$ of $O(h)$ pairwise interior-disjoint convex polygons of totally $O(n)$ vertices [13] (e.g., by extending an angle-bisecting segment inward from each reflex vertex). If we view the convex polygons in $\mathcal{P}'$ as obstacles, then the ocean $\mathcal{M}$ is the free space with respect to $\mathcal{P}'$. The set $\mathcal{P}'$ can be obtained easily in $O(n + h \log h)$ time. It should be pointed out that our algorithms given later can be applied to $\mathcal{M}$ directly without explicitly computing the convex polygons in $\mathcal{P}'$. But for ease of exposition, we always discuss our algorithms on $\mathcal{P}'$ instead of on $\mathcal{M}$.

2.1   Bays and Canals

Recall that $\mathcal{M} \subseteq \mathcal{F}$. We examine the free space of $\mathcal{F}$ not in $\mathcal{M}$, i.e., $\mathcal{F} \setminus \mathcal{M}$, which consists of two types of regions: *bays* and *canals*, as defined below.

Consider the hourglass H_C of a corridor C. We first discuss the case when H_C is open (see Fig. 2). H_C has two sides. Let $S_1(H_C)$ be an arbitrary side of H_C. The obstacle vertices on $S_1(H_C)$ all lie on the same obstacle, say $P \in \mathcal{P}$. Let c and d be any two consecutive vertices on $S_1(H_C)$ such that the line segment $\overline{cd}$ is not an edge of P (see the left figure in Fig. 2, with $P = P_j$). The free region enclosed by $\overline{cd}$ and a boundary portion of P between c and d is called the *bay* of $\overline{cd}$ and P, denoted by $bay(\overline{cd})$, which is a simple polygon. We call $\overline{cd}$ the *bay gate* of $bay(\overline{cd})$, which is a common edge of $bay(\overline{cd})$ and $\mathcal{M}$.

If the hourglass H_C is closed, then let x and y be the two apices of its two funnels. Consider two consecutive vertices c and d on a side of a funnel such that $\overline{cd}$ is not an obstacle edge. If neither c nor d is a funnel apex, then c and d must lie on the same obstacle and the segment $\overline{cd}$ also defines a bay with that obstacle. However, if c or d is a funnel apex, say, $c = x$, then c and d may lie on different obstacles. If they lie on the same obstacle, then they also define a bay; otherwise, we call $\overline{xd}$ the *canal gate* at $x = c$ (see Fig. 2). Similarly, there is also a canal gate at the other funnel apex y, say $\overline{yz}$. Let P_i and P_j be the two obstacles bounding the hourglass H_C. The free region enclosed by P_i, P_j, and the two canal gates $\overline{xd}$ and $\overline{yz}$ that contains the corridor path of H_C is the *canal* of H_C, denoted by $canal(x, y)$, which is also a simple polygon.

Clearly, all bays and canals together constitute the space $\mathcal{F} \setminus \mathcal{M}$.

The fact that each bay has only one gate allows us to process a bay easily. Intuitively, an observer outside a bay cannot see any point outside the bay "through" its gate. But, each canal has two gates, which could cause trouble. The next lemma, proved in [8], gives an important property that an observer outside a canal cannot see any point outside the canal through the canal (and its two gates); we call it *the opaque property* of canals.

Lemma 1. [8] (The Opaque Property) *For any canal, suppose a line segment $\overline{pq}$ is in $\mathcal{F}$ (i.e., p is visible to q) such that neither p nor q is in the canal. Then $\overline{pq}$ cannot contain any point of the canal that is not on its two gates.*

3   The Ray-shooting Queries

We present our ray-shooting data structure in this section. We assume that we have already computed the ocean $\mathcal{M}$, and all bays and canals. We also assume the convex obstacle set $\mathcal{P}'$ is given. Recall that $\mathcal{M}$ is the free space among $\mathcal{P}'$. The preprocessing for these takes $O(n + h \log^{1+\epsilon} h)$ time.

Consider a ray $\sigma(q)$ with its origin $q \in \mathcal{F}$. Let q^* be the outcome of the ray-shooting query of $\sigma(q)$, i.e., q^* is the point on the input obstacles of $\mathcal{P}$ or on the boundary of $\mathcal{R}$ (denoted by $\partial \mathcal{R}$) that is hit first by $\sigma(q)$. We first show how to find q^*, and then discuss the preprocessing of our data structure. For simplicity of discussion, we assume the line containing the ray $\sigma(q)$ does not contain any obstacle vertex. Note that the origin q can be in $\mathcal{M}$, a bay, or a canal.

We first consider the case of $q \in \mathcal{M}$. If $\sigma(q)$ does not hit any obstacle of $\mathcal{P}'$ before it hits $\partial \mathcal{R}$, then the portion of $\sigma(q)$ inside $\mathcal{R}$ lies entirely in $\mathcal{M}$ and thus q^* is on $\partial \mathcal{R}$. Below, we assume $\sigma(q)$ hits an obstacle of $\mathcal{P}'$. Let p be the first point on the obstacles of $\mathcal{P}'$ hit by $\sigma(q)$. Based on our discussion in Section 2, each edge of any obstacle of $\mathcal{P}'$ is either an edge of an input obstacle of $\mathcal{P}$ or a bay/canal gate. If p is not on a gate of any bay/canal, then p is on an input obstacle of $\mathcal{P}$, and hence $q^* = p$. Otherwise, p is on a gate of a bay or a canal. If p is on the gate of a bay B, then since B has only one gate, q^* must be on the boundary of B (and thus on the boundary of an input obstacle of $\mathcal{P}$). If p is on a gate of a canal C, then although C has two gates, due to the opaque property of Lemma 1, q^* must be on the boundary of C that lies on an input obstacle.

If the origin q is in a bay/canal, then we find the first point p on the boundary of the bay/canal hit by $\sigma(q)$. If p is not on a gate, then $q^* = p$; otherwise, the ray $\sigma(q)$ goes out of the bay/canal and enters $\mathcal{M}$ through that gate, and we use a procedure as for the case of $q \in \mathcal{M}$ to compute q^*.

The discussion above shows that to compute q^*, we only need to conduct at most three ray-shooting queries each of which is either on a bay/canal or on the convex obstacle set $\mathcal{P}'$. We perform the preprocessing accordingly. For a bay/canal, because it is a simple polygon, we build a data structure for simple polygons [5,10] for it. Since the total number of vertices of all bays and canals is $O(n)$, preprocessing all bays and canals takes $O(n)$ time and space, and each query inside a bay/canal takes $O(\log n)$ time.

For the convex obstacle set $\mathcal{P}'$, Pocchiola and Vegter [18] showed that by using the visibility complex, a data structure of $O(n + k')$ size can be built in $O(n + k' \cdot \text{poly}(\log h))$ time that allows to answer each ray-shooting query in $O(\log n)$ time, where $k' = O(h^2)$ is the number of common tangents of the convex obstacles in $\mathcal{P}'$ that lie in the free space of $\mathcal{P}'$ (i.e., $\mathcal{M}$).

In summary, we have the following result.

Theorem 1. *For an input polygonal domain $\mathcal{P}$, we can build a data structure of size $O(n + h^2)$ in $O(n + h^2 \cdot \text{poly}(\log h))$ preprocessing time that allows to answer each ray-shooting query in $O(\log n)$ time.*

4   The Visibility Queries

In this section, we present our two visibility query data structures. We assume that the ocean $\mathcal{M}$, and all bays and canals have been computed, and the convex obstacle set $\mathcal{P}'$ is given. The needed preprocessing takes $O(n + h \log^{1+\epsilon} h)$ time. We begin with the first data structure, described in Sections 4.1, 4.2, and 4.3. The second data structure is shown in Section 4.4, which uses some ingredients of the first data structure.

For a query point q, we seek to compute the visibility polygon $Vis(q)$. For simplicity of discussion, assume q is not collinear with any two obstacle vertices.

To provide some intuition, in Section 4.1, we sketch an algorithmic procedure for computing $Vis(q)$ without any preprocessing, and argue its correctness. Our query algorithm (with preprocessing) given later will follow this procedure. In Section 4.2, we present the preprocessing of our first data structure. Its query algorithm and time analysis are shown in Section 4.3.

4.1   The Algorithm for Computing $Vis(q)$

The query point q may be in $\mathcal{M}$, a bay, or a canal. We start with the case of $q \in \mathcal{M}$. For any subset S of the free space $\mathcal{F}$, let $Vis(q, S)$ denote the intersection of $Vis(q)$ and S. For example, $Vis(q, \mathcal{M})$ is the subpolygon of $Vis(q)$ in the ocean $\mathcal{M}$, and $Vis(q, \mathcal{F})$ is $Vis(q)$.

We first compute $Vis(q, \mathcal{M})$. Because the space $\mathcal{F} \setminus \mathcal{M}$ consists of all bays and canals, the region $Vis(q) \setminus Vis(q, \mathcal{M})$ is the union of the visibility subpolygons of $Vis(q)$ in all bays and canals. Next, we show how to compute $Vis(q) \setminus Vis(q, \mathcal{M})$.

Observation 1. *For $q \in \mathcal{M}$, if a bay/canal does not have any gate that intersects with the boundary of $Vis(q, \mathcal{M})$, then no point in that bay/canal is visible to q.*

Proof. Consider any point p in a bay $bay(\overline{cd})$. Suppose p is visible to q. Since $q \in \mathcal{M}$, $\overline{pq}$ must intersect the gate $\overline{cd}$, say at a point p'. Hence, p' is visible to q. Because $\overline{cd}$ is on $\partial\mathcal{M}$, p' is on the boundary of $Vis(q, \mathcal{M})$. Thus $\overline{cd}$ intersects the boundary of $Vis(q, \mathcal{M})$. The case for canals can be proved similarly.

Suppose for a bay $bay(\overline{cd})$ with gate $\overline{cd}$, we want to compute $Vis(q, bay(\overline{cd}))$. If its gate $\overline{cd}$ does not intersect the boundary $\partial Vis(q, \mathcal{M})$ of $Vis(q, \mathcal{M})$, then by Observation 1, $Vis(q, bay(\overline{cd})) = \emptyset$. If $\overline{cd}$ has a single sub-segment on $\partial Vis(q, \mathcal{M})$, then q can see part of $bay(\overline{cd})$ through the cone delimited by this sub-segment and with q as the apex, and we compute $Vis(q, bay(\overline{cd}))$ "seeing through" this cone. The general case is when multiple disjoint sub-segments of $\overline{cd}$ are on $\partial Vis(q, \mathcal{M})$ (e.g., see Fig. 3). In this case, some interior points of $bay(\overline{cd})$ are visible to q

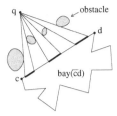

Fig. 3. Three sub-segments (the thick ones) of $\overline{cd}$ are visible to the point q

through multiple cones. We compute the visible region of q in $bay(\overline{cd})$ for each such cone. It is easy to see that the visible regions for these cones are mutually disjoint. Therefore, $Vis(q, bay(\overline{cd}))$ is the union of them.

Next, suppose for a canal $canal(x,y)$ with two gates $\overline{xd}$ and $\overline{yz}$ (as in Fig. 2), we want to compute $Vis(q, canal(x,y))$. Similarly, if its two gates do not intersect $\partial Vis(q, \mathcal{M})$, then $Vis(q, canal(x,y)) = \emptyset$. Otherwise, let $Vis(\overline{xd})$ denote the region of $canal(x,y)$ visible to q through the gate $\overline{xd}$ and $Vis(\overline{yz})$ denote the region of $canal(x,y)$ visible to q through the gate $\overline{yz}$. Clearly, $Vis(q, canal(x,y)) = Vis(\overline{xd}) \cup Vis(\overline{yz})$. We compute $Vis(\overline{xd})$ and $Vis(\overline{yz})$ separately using our above approach for the bay case. Note that $x = y$ is possible, in which case the two gates share a common vertex x but we view x as belonging only to $\overline{xd}$ (i.e., $\overline{yz}$ is viewed as a half-open segment). In this way, the two gates never intersect. Lemma 2 below shows that $Vis(\overline{xd})$ and $Vis(\overline{yz})$ are mutually disjoint. Thus, once $Vis(\overline{xd})$ and $Vis(\overline{yz})$ are available, computing $Vis(\overline{xd}) \cup Vis(\overline{yz})$ is trivial. The proof of Lemma 2 is omitted.

Lemma 2. *For $q \in \mathcal{M}$, the visibility polygons $Vis(\overline{xd})$ and $Vis(\overline{yz})$ in $canal(x,y)$ do not intersect with each other.*

Based on the above, after we obtain $Vis(q, \mathcal{M})$, to compute $Vis(q) \setminus Vis(q, \mathcal{M})$, we can simply check the boundary $\partial Vis(q, \mathcal{M})$. For each sub-segment of a bay/canal gate on $\partial Vis(q, \mathcal{M})$, we compute the region in the bay/canal visible to q through that sub-segment. All these regions are pairwise disjoint and $Vis(q) \setminus Vis(q, \mathcal{M})$ is a trivial union of them. We hence finish the discussion of our procedure for computing $Vis(q)$ in the case of $q \in \mathcal{M}$.

Next, we consider the case when the query point q is in a bay, say $bay(\overline{cd})$. In this case, we first compute the visibility polygon of q in $bay(\overline{cd})$, i.e., $Vis(q, bay(\overline{cd}))$. If the gate $\overline{cd}$ does not intersect the boundary of $Vis(q, bay(\overline{cd}))$, then $Vis(q) = Vis(q, bay(\overline{cd}))$ because q is not visible to any point outside $bay(\overline{cd})$. Otherwise, there must be a single (maximal) sub-segment of $\overline{cd}$ on the boundary of $Vis(q, bay(\overline{cd}))$ through which q can see the outside of $bay(\overline{cd})$ (in the cone delimited by that sub-segment). In other words, $Vis(q) \setminus Vis(q, bay(\overline{cd}))$ is the visible region in the space $\mathcal{F} \setminus bay(\overline{cd})$ visible to q through the cone. To compute $Vis(q) \setminus Vis(q, bay(\overline{cd}))$, we use a procedure similar to that for the case of $q \in \mathcal{M}$. The difference is that here the visibility is through a cone.

The remaining case is when the query point q is in a canal. This case is very similar to the bay case above. The difference is that we consider the two canal gates separately, using the procedure for the bay case. We omit the details.

4.2   The Preprocessing

We discuss the preprocessing for our algorithm in Section 4.1, in which we need to compute the visibility polygons of q in $\mathcal{M}$ or in a bay/canal.

We first discuss the preprocessing for computing $Vis(q, \mathcal{M})$ when $q \in \mathcal{M}$. Recall that we have a set $\mathcal{P}'$ of $O(h)$ convex obstacles of totally $O(n)$ vertices and its free space is the ocean $\mathcal{M}$. By using the visibility complex [19,20], we have the following lemma with proof omitted.

Lemma 3. *We can build a data structure of size $O(n + h^2)$ in $O(n + h^2 \log h)$ time that allows to compute $Vis(q, \mathcal{M})$ in $O(|Vis(q, \mathcal{M})| + h' \log n)$ time for any query point $q \in \mathcal{M}$, where h' is the number of obstacles in $\mathcal{P}'$ visible to q.*

Further, recall that in our algorithm discussed in Section 4.1, when the query point q is in a bay (or canal), $Vis(q, \mathcal{M})$ is the visible region of q in $\mathcal{M}$ through a cone (or a sub-segment of the bay gate). Therefore, we need to deal with the cone visibility in $\mathcal{M}$. For this, we extend the result in Lemma 3.

Corollary 1. *We can build a data structure of size $O(n+h^2)$ in $O(n+h^2 \log h)$ time that allows to compute $Vis(q, \mathcal{M})$ in $O(|Vis(q, \mathcal{M})| + h' \log n)$ time for any query point q in a bay or canal within its visibility cone, where h' is the number of obstacles in $\mathcal{P}'$ visible to q.*

Next, we discuss the preprocessing for bays and canals. Recall that there are two types of query situations on a bay/canal. The first type is that the query point q is inside a bay/canal and we need to compute the visibility polygon of q in that bay/canal. The second type is that q is outside a bay/canal along with a sub-segment of a gate of that bay/canal and we need to compute the visibility polygon of q in the bay/canal through that sub-segment.

For the first type, we simply use the data structure by Aronov *et al.* [2] for simple polygons. Since all bays and canals have totally at most n vertices, the preprocessing time and space for all bays and canals are $O(n^2 \log n)$ and $O(n^2)$, respectively. After that, for any query point q in a bay/canal, the visibility polygon P of q in the bay/canal can be computed in $O(\log^2 n + |P|)$ time.

For the second type, we do the following preprocessing. Consider a convex obstacle $P \in \mathcal{P}'$. Let $BayCanal(P)$ (or $BC(P)$) denote the set of bays and canals each of which has a gate lying on the boundary of P. For any query point $q \notin P$, let $\mathcal{C}_q$ be a cone with apex q. Denote by $Vis(q, BC(P))$ the union of the visibility polygons of q in all bays and canals of $BC(P)$, and here all other obstacles in $\mathcal{P}'$ are ignored (i.e., we assume they are transparent and do not block the view of q). Let $Vis(\mathcal{C}_q, BC(P)) = \mathcal{C}_q \cap Vis(q, BC(P))$, i.e., $Vis(\mathcal{C}_q, BC(P))$ is the union of the visibility polygons of q in all bays and canals of $BC(P)$ through the cone $\mathcal{C}_q$. Using the techniques in [2], we have Lemma 4, with proof omitted.

Lemma 4. *For a convex obstacle P, suppose the total number of vertices in all bays and canals of $BC(P)$ is m. We can build a data structure of size $O(m^2)$ in $O(m^2 \log m)$ time such that for any query point $q \notin P$, in $O(\log m)$ time, we can obtain (a pointer to) a data structure storing $Vis(q, BC(P))$, and if needed, report $Vis(q, BC(P))$ explicitly in additional $O(|Vis(q, BC(P))|)$ time. Further, given any cone C_q with apex q, from the above data structure, we can obtain $Vis(C_q, BC(P))$ in additional $O(\log m + |Vis(C_q, BC(P))|)$ time.*

We compute the data structure for Lemma 4 for each convex obstacle in $\mathcal{P}'$. Since all bays and canals have $O(n)$ vertices, the total preprocessing time is $O(n^2 \log n)$ and the space is $O(n^2)$.

In summary, our preprocessing includes: (1) preprocessing $\mathcal{M}$ (or $\mathcal{P}'$) using Lemma 3 and Corollary 1, (2) preprocessing all bays and canals for the first type query situation using the data structure in [2], and (3) preprocessing all bays and canals for the second type query situation using Lemma 4. The overall preprocessing time is $O(n^2 \log n)$ and the space is $O(n^2)$.

4.3   The Query Algorithm

Consider a query point q. Our query algorithm for computing $Vis(q)$ follows the same procedure as given in Section 4.1. We first discuss the case of $q \in \mathcal{M}$.

In Step (1), we compute $Vis(q, \mathcal{M})$ using the data structure for Lemma 3, which takes $O(|Vis(q, \mathcal{M})| + h' \log n)$ time. In Step (2), for each obstacle $P \in \mathcal{P}'$ visible to q, by Lemma 4, we obtain the data structure for storing $Vis(q, BC(P))$, in $O(\log n)$ time. In Step (3), we check the boundary of $Vis(q, \mathcal{M})$; for every obstacle P visible to q, if q's view of P is blocked partially by some other obstacles of $\mathcal{P}'$, i.e., there are some cones through which q is visible to one or more portions of P, then for each such cone C_q, by Lemma 4, we compute $Vis(C_q, BC(P))$ in additional $O(\log n + |Vis(C_q, BC(P))|)$ time. Then, $Vis(q)$ is obtained and is represented as a cyclically ordered list of visible edges and vertices. The correctness of the algorithm follows from our discussion in Section 4.1.

To analyze the query time, let $k = |Vis(q)|$. First, $|Vis(q, \mathcal{M})|$ plus the sum of all $|Vis(C_q, BC(P))|$'s is $O(k)$. Second, the number of cones in Step (3) is $O(h')$ because only h' obstacles of $\mathcal{P}'$ are visible to q. Therefore, the overall time of the query algorithm is $O(k + h' \log n)$. Clearly, $h' \leq h$ and $h' \leq k$.

Next, we discuss the case when q is in a bay, say $bay(\overline{cd})$. In Step (1), we compute the visibility polygon $Vis(q, bay(\overline{cd}))$ in $bay(\overline{cd})$, in $O(\log^2 n + |Vis(q, bay(\overline{cd}))|)$ time using the data structure in [2]. If $\overline{cd}$ has a sub-segment $\overline{c'd'}$ on the boundary of $Vis(q, bay(\overline{cd}))$, then in Step (2), we compute the visibility polygon of q outside $bay(\overline{cd})$ seeing through the cone with apex q and delimited by $\overline{c'd'}$. This step and the rest of the algorithm are basically the same as the former case of $q \in \mathcal{M}$. One difference is that we use Corollary 1 instead of Lemma 3 to compute $Vis(q, \mathcal{M})$. Similarly to the analysis above, the overall query time is $O(\log^2 n + k + \min\{k, h\} \log n)$. Note that we have an additive $O(\log^2 n)$ time due to using the data structure for simple polygons [2].

The remaining case when q is in a canal is the same as the bay case except that we process the two canal gates separately. The time of each query is also $O(\log^2 n + k + \min\{k, h\} \log n)$. In summary, we have the following result.

Theorem 2. *For a polygonal domain $\mathcal{P}$, we can build a data structure of size $O(n^2)$ in $O(n^2 \log n)$ preprocessing time that can answer each visibility query in $O(\log^2 n + k + \min\{k, h\} \log n)$ time.*

4.4 The Second Data Structure

The main difference between our second data structure and the first one is that for bays and canals, we do not preprocess them using the data structures in [2] and Lemma 4. Instead, we build a ray-shooting data structure in simple polygons [5,10] for each bay and canal, which takes totally $O(n)$ preprocessing time and space. But, we still keep the data structures for Lemma 3 and Corollary 1. The overall preprocessing time and space then become $O(n + h^2 \log h)$ and $O(n + h^2)$, respectively. Below, we discuss the query algorithm.

Consider a query point q. We first discuss the case of $q \in \mathcal{M}$. In the first step, we still compute $Vis(q, \mathcal{M})$ by Lemma 3. In the second step, we check the boundary of $Vis(q, \mathcal{M})$. If a sub-segment of a bay/canal gate appears on $\partial Vis(q, \mathcal{M})$, then we use the ray-shooting approach [2] to compute the visibility polygon of q in the bay/canal through that sub-segment, which takes $O(k' \log n)$ time, where k' is the output size of this visibility polygon. For the query time, the first step takes $O(|Vis(q, \mathcal{M})| + h' \log n)$ time. Again, $h' = O(k)$. For the second step, clearly, the sum of all such k' terms is $O(k)$. Therefore, the query time is $O(k \log n)$. The other cases when q is in a bay or canal are very similar and we omit the discussions of them. In summary, we have the following result.

Theorem 3. *For a polygonal domain $\mathcal{P}$, we can build a data structure of size $O(n + h^2)$ in $O(n + h^2 \log h)$ preprocessing time that can answer each visibility query in $O(k \log n)$ time.*

5 Conclusions

In this paper we propose new data structures for ray-shooting queries and computing visibility polygons for query points in polygonal domains, which benefit in a large part from the extended corridor structure [8]. It would be interesting to see whether further improvements are possible. In addition, the current best visibility query data structures on simply polygons have complexities $O(n^3 \log n, n^3, k + \log n)$ and $O(n^2 \log n, n^2, k + \log^2 n)$, respectively; improving these results would also be interesting, and in particular, an open question is whether $O(n^2 \log n, n^2, k + \log n)$ complexity data structures exist.

Acknowledgments. The authors would like to thank Tiancong Chen for helpful discussions in early phases of this work.

References

1. Agarwal, P., Sharir, M.: Ray shooting amidst convex polygons in 2D. Journal of Algorithms 21(3), 508–519 (1996)
2. Aronov, B., Guibas, L., Teichmann, M., Zhang, L.: Visibility queries and maintenance in simple polygons. Discrete and Computational Geometry 27(4), 461–483 (2002)
3. Bar-Yehuda, R., Chazelle, B.: Triangulating disjoint Jordan chains. International Journal of Computational Geometry and Applications 4(4), 475–481 (1994)
4. Bose, P., Lubiw, A., Munro, J.: Efficient visibility queries in simple polygons. Computational Geometry: Theory and Applications 23(3), 313–335 (2002)
5. Chazelle, B., Edelsbrunner, H., Grigni, M., Gribas, L., Hershberger, J., Sharir, M., Snoeyink, J.: Ray shooting in polygons using geodesic triangulations. Algorithmica 12(1), 54–68 (1994)
6. Chazelle, B., Guibas, L.: Visibility and intersection problems in plane geometry. Discrete and Computational Geometry 4, 551–589 (1989)
7. Chen, D.Z., Wang, H.: A nearly optimal algorithm for finding L_1 shortest paths among polygonal obstacles in the plane. In: Demetrescu, C., Halldórsson, M.M. (eds.) ESA 2011. LNCS, vol. 6942, pp. 481–492. Springer, Heidelberg (2011)
8. Chen, D.Z., Wang, H.: Computing the visibility polygon of an island in a polygonal domain. In: Czumaj, A., Mehlhorn, K., Pitts, A., Wattenhofer, R. (eds.) ICALP 2012, Part I. LNCS, vol. 7391, pp. 218–229. Springer, Heidelberg (2012)
9. Guibas, L., Hershberger, J., Leven, D., Sharir, M., Tarjan, R.: Linear-time algorithms for visibility and shortest path problems inside triangulated simple polygons. Algorithmica 2(1-4), 209–233 (1987)
10. Hershberger, J., Suri, S.: A pedestrian approach to ray shooting: Shoot a ray, take a walk. Journal of Algorithms 18(3), 403–431 (1995)
11. Inkulu, R., Kapoor, S.: Planar rectilinear shortest path computation using corridors. Computational Geometry: Theory and Applications 42(9), 873–884 (2009)
12. Inkulu, R., Kapoor, S.: Visibility queries in a polygonal region. Computational Geometry: Theory and Applications 42(9), 852–864 (2009)
13. Kapoor, S., Maheshwari, S., Mitchell, J.: An efficient algorithm for Euclidean shortest paths among polygonal obstacles in the plane. Discrete and Computational Geometry 18(4), 377–383 (1997)
14. Lee, D., Preparata, F.: Euclidean shortest paths in the presence of rectilinear barriers. Networks 14(3), 393–410 (1984)
15. Lu, L., Yang, C., Wang, J.: Point visibility computing in polygons with holes. Journal of Information and Computational Science 8(16), 4165–4173 (2011)
16. Nouri, M., Ghodsi, M.: Space–query-time tradeoff for computing the visibility polygon. In: Deng, X., Hopcroft, J.E., Xue, J. (eds.) FAW 2009. LNCS, vol. 5598, pp. 120–131. Springer, Heidelberg (2009)
17. Pocchiola, M.: Graphics in flatland revisited. In: Gilbert, J.R., Karlsson, R. (eds.) SWAT 1990. LNCS, vol. 447, pp. 85–96. Springer, Heidelberg (1990)
18. Pocchiola, M., Vegter, G.: Pseudo-triangulations: Theory and applications. In: Proc. of the 12th Annual Symposium on Computational Geometry, pp. 291–300 (1996)
19. Pocchiola, M., Vegter, G.: Topologically sweeping visibility complexes via pseudo-triangulations. Discrete and Computational Geometry 16(4), 419–453 (1996)
20. Pocchiola, M., Vegter, G.: The visibility complex. International Journal of Computational Geometry and Applications 6(3), 279–308 (1996)
21. Zarei, A., Ghodsi, M.: Query point visibility computation in polygons with holes. Computational Geometry: Theory and Applications 39(2), 78–90 (2008)

Lift-and-Project Methods for Set Cover and Knapsack[*]

Eden Chlamtáč[1], Zachary Friggstad[2], and Konstantinos Georgiou[2]

[1] Ben Gurion University, Department of Computer Science, Beer Sheva, Israel
chlamtac@cs.bgu.ac.il
[2] University of Waterloo, Department of Combinatorics and Optimization,
Waterloo, ON, Canada
{zfriggstad,k2georgiou}@math.uwaterloo.ca

Abstract. We study the applicability of lift-and-project methods to the
SET COVER and KNAPSACK problems. Inspired by recent work of Karlin, Mathieu, and Nguyen [IPCO 2011], who examined this connection for
KNAPSACK, we consider the applicability and limitations of these methods for SET COVER, as well as extending extending the existing results
for KNAPSACK.

For the SET COVER problem, Cygan, Kowalik, and Wykurz [IPL 2009]
gave sub-exponential-time approximation algorithms with approximation ratios better than $\ln n$. We present a very simple combinatorial algorithm which has nearly the same time-approximation tradeoff as the
algorithm of Cygan et al. We then adapt this to an LP-based algorithm
using the LP hierarchy of Lovász and Schrijver. However, our approach
involves the trick of "lifting the objective function". We show that this
trick is essential, by demonstrating an integrality gap of $(1 - \varepsilon) \ln n$ at
level $\Omega(n)$ of the stronger LP hierarchy of Sherali and Adams (when the
objective function is not lifted).

Finally, we show that the SDP hierarchy of Lovász and Schrijver (LS_+)
reduces the integrality gap for KNAPSACK to $(1 + \varepsilon)$ at level $O(1)$. This
stands in contrast to SET COVER (where the work of Aleknovich, Arora,
and Tourlakis [STOC 2005] rules out any improvement using LS_+), and
extends the work of Karlin et al., who demonstrated such an improvement only for the more powerful SDP hierarchy of Lasserre. Our LS_+
based rounding and analysis are quite different from theirs (in particular, not relying on the decomposition theorem they prove for the Lasserre
hierarchy), and to the best of our knowledge represents the first explicit
demonstration of such a reduction in the integrality gap of LS_+ relaxations after a constant number of rounds.

Keywords: Set Cover, Sub-exponential Algorithms, Approximation
Algorithms, Lift-and-Project Methods, Knapsack.

---

[*] Full version available as arXiv:1204.5489. The first author's work is partially supported by the Lynn and William Frankel Center for Computer Science.

F. Dehne, R. Solis-Oba, and J.-R. Sack (Eds.): WADS 2013, LNCS 8037, pp. 256–267, 2013.
© Springer-Verlag Berlin Heidelberg 2013

1   Introduction

The SET COVER problem and the KNAPSACK problem are two of the most fundamental and well-studied problems in approximation algorithms, and both appeared on Karp's original list of 21 NP-complete problems [12]. We consider the minimum cost (or weighted) version of SET COVER: given a set X of n items and a collection $\mathcal{S} \subseteq 2^X$ of m subsets of X called "cover-sets", where each cover set has an associated non-negative cost $c(S)$, the SET COVER problem on instance $(X, \mathcal{S})$ is the problem of finding a collection $\mathcal{C}$ of cover-sets in $\mathcal{S}$ of minimum total cost, such that $X = \bigcup_{S \in \mathcal{C}} S$. In the KNAPSACK problem, we are given n items which we identify with the integers $[n]$, and each item $i \in [n]$ has some associated (nonnegative) reward r_i and cost (or size) c_i, and the goal is to choose a set of items which fit in the knapsack, i.e. whose total cost does not exceed some bound C, so as to maximize the total reward.

As is well known, the SET COVER problem can be approximated within a logarithmic factor. Specifically, Johnson [10] showed that for uniform costs, a simple greedy algorithm gives an H_n-approximation (where $H_n = \ln n + O(1)$ is the n'th harmonic number $\sum_{k=1}^{\lfloor n \rfloor} 1/k$), and this was extended to the weighted case by Chvátal [4]. Moreover, the cost of the solution found by the greedy algorithm is at most an H_n-factor larger than the optimum value of the natural linear programming (LP) relaxation [14,4]. As shown by Feige [6], this approximation is tight in the sense that a $(1 - \varepsilon) \ln n$-approximation in polynomial time would imply that all problems in NP can be solved deterministically in time $n^{O(\log \log n)}$.

Recently, Cygan, Kowalik, and Wykurz [5] showed that SET COVER can be approximated within $(1 - \epsilon) \cdot \ln n + O(1)$ in time $2^{n^\epsilon + O(\log m)}$. Note that this time-approximation tradeoff is essentially optimal assuming Moshkovitz's Projection Games Conjecture [16] and the Exponential Time Hypothesis (ETH) [9].

The KNAPSACK problem, on the other hand, is famously easy to approximate. In particular, it has a well-known FPTAS [8].

1.1   Hierarchies of Convex Relaxations

One of the most powerful and ubiquitous tools in approximation algorithms has been the use of mathematical programming relaxations, such as linear programming (LP) and semidefinite programming (SDP). The common approach is as follows: solve a convex (LP or SDP) relaxation for the 0-1 program, and "round" the relaxed solution to give a (possibly suboptimal) feasible 0-1 solution. Since the approximation ratio is usually analyzed by comparing the value of the relaxed solution to the value of the output (note that the 0-1 optimum is always sandwiched between these two), a natural obstacle is the worst case ratio between the relaxed optimum and the 0-1 optimum, known as the *integrality gap*.

While for many problems, this approach gives optimal approximations (see, e.g., Raghavendra [17]), there are still many cases where natural LP and SDP relaxations have large integrality gaps not matched by any hardness of approximation results. This limitation can be circumvented by considering more powerful relaxations. In particular, Sherali and Adams [19] Lovász and Schrijver [15], and Lasserre [13] each have devised different systems, collectively known as *hierarchies* or *lift-and-project methods*, by which a simple relaxation can be strengthened until the polytope (or the convex body) it defines converges to the convex hull of feasible 0-1 solutions. It is known that, for each of these hierarchies, if the original relaxation has n variables and $n^{O(1)}$ constraints, then the relaxation at level t of the hierarchy can be solved optimally in time $n^{O(t)}$. Thus, to achieve improved approximations for a problem in polynomial (resp. sub-exponential time), we would like to know if we can beat the integrality gap of the natural relaxation by using a relaxation at level $O(1)$ (resp. $o(n/\log n)$) of some hierarchy. For a survey on both algorithms using hierarchies and integrality gap results, see [3].

While we often apply lift-and-project to the constraints, and optimise the original objective function over the lifted feasible region, another approach is "lifting the objective function". In this variant, we guess some bound on the objective function, add this bound as a constraint, and then apply lift-and-project to the amended set of constraints. In this approach, we want the optimum bound (on the objective function) for which the lifted relaxation is feasible.

One important way to evaluate the usefulness of LP and SDP hierarchies for approximation algorithms is to study how their application affects the integrality gap of natural relaxations for well-understood problems. This was done recently by Karlin et al. [11] for KNAPSACK. They showed that, while the Sherali-Adams LP hierarchy requires $\Omega(n)$ levels to bring the integrality gap below $2 - o(1)$, level k of the Lasserre SDP hierarchy brings the integrality gap down to $1 + O(1/k)$. While we would like to emulate the success of their lift-and-project-based approach (and give an alternative sub-exponential algorithm for SET COVER), as we shall see, in the case of SET COVER, this requires "lifting the objective function".

1.2   Our Results

To facilitate our lift-and-project based approach, we start by giving in Section 3.1 a simple new sub-exponential time combinatorial algorithm for SET COVER which nearly matches the time-approximation tradeoff guarantee in [5].

Theorem 1. *For any instance of* SET COVER *with n items and m cover-sets and for any (not necessarily constant) $1 \leq d \leq n$, there is an $H_{n/d}$-approximation algorithm running in time* $\text{poly}(n,m) \cdot m^{O(d)}$.

The algorithm is combinatorial and does not rely on linear programming techniques. By choosing $d = n^{\varepsilon}$, we get a sub-exponential time algorithm whose approximation guarantee is better than $\ln n$ by a constant factor. While this theorem is slightly weaker than the previous best known guarantee, our algorithm is remarkably simple, and will be instrumental in designing a similar lift-and-project based SET COVER approximation, which is summarised in the following theorem, which we prove in Section 3.2.

Theorem 2. *For any $1 \leq d \leq n$, the integrality gap of LP relaxation obtained by taking the standard LP, and applying d rounds of the LS hierarchy while lifting the objective function, is at most $H_{n/d}$.*

On the other hand, without the trick of "lifting the objective function", we show in Section 4 that even the stronger LP hierarchy of Sherali-Adams [19] has an integrality gap of at least $(1 - \varepsilon) \ln n$ at level $\Omega(n)$. Specifically, we show the following

Theorem 3. *For every $0 < \varepsilon, \gamma \leq \frac{1}{2}$, and for sufficiently large values of n, there are instances of* SET COVER *on n cover-sets (over a universe of n items) for which the integrality gap of the level-$\lfloor \frac{\gamma(\varepsilon - \varepsilon^2)}{1 + \gamma} n \rfloor$ Sherali-Adams LP relaxation is at least $\frac{1-\varepsilon}{1+\gamma} \ln n$.*

For SDPs, it seems that lifting the objective function is necessary as well, due to the work of Alekhnovich et al. [1] which gives a similar integrality gap for LS_+ (note that the LS_+ hierarchy and the Sherali-Adams hierarchy are incomparable). As we will show, this stands in contrast to the KNAPSACK problem. For this problem, we show that even without lifting the objective function, we can obtain a PTAS using LS_+ (reducing the integrality gap from 2 for the standard LP relaxation).

Theorem 4. *The integrality gap of level k of the LS_+ relaxation for* KNAPSACK *is at most $1 + O(k^{-1/3})$.*

This extends the work of Karlin et al. [11], who showed a similar result, but only for the much stronger Lasserre hierarchy. Both our rounding algorithm (for LS_+) and analysis deviate significantly from theirs. Indeed, the algorithm of Karlin et al. [11] relied a powerful decomposition theorem for the Lasserre hierarchy which does not seem to be applicable to LS_+. Instead, our analysis introduces a novel approach of bounding the number of rounds of LS_+ needed via an upper bound on the integrality gap of the standard LP relaxation.

In what follows, and before we start the exposition of our results, we present in Section 2 the Lovász-Schrijver system along with some well-known facts that we will need later on. We end with a discussion of future directions in Section 6.

2   Preliminaries on the **Lovász-Schrijver** System

For any polytope P, this system begins by introducing a nonnegative auxiliary variable x_0, so that in every constraint of P, constants are multiplied by x_0. This yields the cone $K_0(P) := \{(x_0, x_0\mathbf{x}) \mid x_0 \geq 0 \ \& \ \mathbf{x} \in P\}$. For an n-dimensional polytope P, the Lovász-Schrijver system finds a hierarchy of nested cones $K_0(P) \supseteq K_1(P) \supseteq \ldots \supseteq K_n(P)$ (in the SDP variant, we will write $K_t^+(P)$), defined recursively, and which enjoy remarkable algorithmic properties. In what follows, let $\mathcal{P}_k$ denote the space of vectors indexed by subsets of $[n]$ of size at most k, and for any $\mathbf{y} \in \mathcal{P}_k$, define the moment matrix $Y^{[\mathbf{y}]}$ to be the square matrix with rows and columns indexed by sets of size at most $\lfloor k/2 \rfloor$, where the entry at row A and column B is $y_{A \cup B}$. Also we denote by $\mathbf{e}_0, \mathbf{e}_1, \ldots, \mathbf{e}_n$ the standard orthonormal basis of dimension $n + 1$, such that $Y^{[\mathbf{y}]}\mathbf{e}_i$ is the i-th column of the moment matrix.

Definition 1 (The Lovász-Schrijver (LS) and Lovász-Schrijver SDP (LS$_+$) systems). *Consider the conified polytope $K_0(P)$ defined earlier (let us also write $K_0^+(P) = K_0(P)$). The level-t Lovász-Schrijver cone (relaxation or tightening) $K_t(P)$ (resp. $K_t^+(P)$) of LS (resp. LS$_+$) is recursively defined as all $n + 1$ dimensional vectors $(x_0, x_0\mathbf{x})$ for which there exist $\mathbf{y} \in \mathcal{P}_2$ such that $Y^{[\mathbf{y}]}\mathbf{e}_i, Y^{[\mathbf{y}]}(\mathbf{e}_0 - \mathbf{e}_i) \in K_{t-1}(P)$ (resp. $K_{t-1}^+(P)$) and $(x_0, x_0\mathbf{x}) = Y^{[\mathbf{y}]}\mathbf{e}_0$. The level-$t$ Lovász-Schrijver SDP tightening of LS$_+$ asks in addition that $Y^{[\mathbf{y}]}$ is a positive-semidefinite matrix.*

In the original work of Lovász and Schrijver [15] it is shown that the cone $K_n(P)$ (even in the LS system) projected on $x_0 = 1$ is exactly the integral hull of the original LP relaxation, while one can optimize over $K_t(P)$ in time $n^{O(t)}$, given that the original relaxation admits a (weak) polytime separation oracle. The algorithm in this section, as well as the one in Section 5, both rely heavily on the following facts, which follow easily from the above definition:

Fact 5. *For any vector $\mathbf{x} \in [0,1]^n$ such that $(1, \mathbf{x}) \in K_t(P)$, and corresponding moment vector $\mathbf{y} \in \mathcal{P}_2$, and for any $i \in [n]$ such that $x_i > 0$, the rescaled column vector $\frac{1}{x_i}Y^{[\mathbf{y}]}\mathbf{e}_i$ is in $K_{t-1}(P) \cap \{(1, \mathbf{x}') \mid \mathbf{x}' \in [0,1]^n\}$.*

Fact 6. *For any vector $\mathbf{x} \in [0,1]^n$ such that $(1, \mathbf{x}) \in K_t(P)$, and any coordinate j such that x_j in integral, and any vector $\mathbf{x}'$ such that $(1, \mathbf{x}') \in K_{t-1}(P)$ derived from $\mathbf{x}$ as in Fact 5, we have $x'_j = x_j$.*

3 Sub-Exponential Algorithms for Set Cover

In this section, we present a simple combinatorial sub-exponential algorithm for SET COVER, and then adapt it to a lift-and-project based algorithm. In what follows, we let $(X, \mathcal{S})$ denote a SET COVER instance with items X and cover-sets $\mathcal{S}$ where each $S \in \mathcal{S}$ has cost $c(S)$. We use n to denote the number of items in X and m to denote the number of cover-sets in $\mathcal{S}$.

3.1 Sketch of our Combinatorial Set Cover Algorithm

Recall that the standard greedy algorithm for approximating SET COVER iteratively selects the cover-set S of minimum density $c(S)/|S \setminus \bigcup_{T \in \mathcal{C}} T|$ where $\mathcal{C}$ is the collection of cover-sets already chosen. The approximation guarantee of this algorithm is H_b, where b is the size of the largest cover-set. Our algorithm builds on this result simply by guessing up to d cover-sets in the optimal solution before running the greedy algorithm.

However, some of the cover-sets in $\mathcal{S}$ that were not guessed (and might still contain uncovered items) are discarded before running the greedy algorithm.

Specifically, we discard the cover-sets that contain more than $\frac{n}{d}$ uncovered items after initially guessing the d sets. While it may seem counter-intuitive to discard sets that cover many items, we do this to exploit the stronger approximation guarantee of the greedy algorithm when the sizes of the cover-sets are bounded.

To that end, we show that for some choice of d sets in the optimal solution, no remaining set in the optimum solution covers more than $\frac{n}{d}$ uncovered items. Thus, running the greedy algorithm on the remaining sets is actually an $H_{n/d}$-approximation. The full description of the algorithm along with all details of the proof of Theorem 1 are in the full version of the paper.

3.2 Proof Based on the Lovász-Schrijver System

In this section we prove Theorem 2, giving an alternative LP-based approximation algorithm for SET COVER with the same performance as in Section 3.1. Specifically, we adapt our combinatorial algorithm to give a rounding algorithm for an LP obtained by applying lift-and-project and "lifting the objective function". Consider the standard LP relaxation for SET COVER on instance $(X, \mathcal{S})$:

$$\text{minimize} \quad \sum_{S \in \mathcal{S}} c(s) x_S$$

$$\text{subject to} \quad \sum_{S \ni i} x_S \geq 1 \qquad \forall\, i \in X \qquad (1)$$

$$0 \leq x_S \leq 1 \qquad \forall\, S \in \mathcal{S} \qquad (2)$$

Now, consider the corresponding feasibility LP where instead of explicitly minimizing the objective function, we add the following bound on the objective function as a constraint (we will later guess the optimal value q by binary search):

$$\sum_{S \in \mathcal{S}} c(S) x_S \leq q. \qquad (3)$$

We will work with the feasibility LP consisting only of constraints (1), (2), and (3) (and no objective function). Denote the corresponding polytope of feasible solutions by P_q.

In what follows we strengthen polytope P_q using the Lovász-Schrijver lift-and-project system. Next we show that the level-d Lovász-Schrijver relaxation $K_d(P_q)$ can give a $H_{\frac{n}{d}}$-factor approximation algorithm. We note here that applying the Lovász-Schrijver system to the feasibility P_q (which includes the objective function as a constraint) and not on the standard LP relaxation of SET COVER is crucial, since by Alekhnovich et al. [1] the latter LP has a very bad integrality gap even when strengthened by $\Omega(n)$ rounds of LS_+ (which is even stronger than LS).

To that end, let q be the smallest value such that $K_d(P_q)$ is not empty (note that $q \leq \text{OPT}$). The value q can be found through binary search (note that in each stage of the binary search we check $K_d(P_{q'})$ for emptyness for some q', which takes time $m^{O(d)}$). Our goal is to show that for this q we can find a SET COVER of cost at most $q \cdot H_{\frac{n}{d}}$.

Let $\mathbf{x}^{(0)}$ be such that $(1, \mathbf{x}^{(0)}) \in K_d(P_q)$. For any coordinate i in the support of $\mathbf{x}^{(0)}$ we can invoke Fact 5 and get a vector $\mathbf{x}^{(1)}$ such that $(1, \mathbf{x}^{(1)}) \in K_{d-1}(P_q)$ and $\mathbf{x}_i^{(1)} = 1$. By Fact 6, by iterating this step, we eventually obtain a vector $\mathbf{x}^{(d)} \in P_q$ which is integral in at least d coordinates. Note that by constraint (3), this solution has cost at most q. We refer to this subroutine as the *Conditioning Phase*, which is realised in d inductive steps.

If at some step $0 \leq i \leq d$, the sets whose coordinates in $\mathbf{x}^{(i)}$ are set to 1 cover all universe elements X, we have solved the SET COVER instance with cost $\sum_{S:\mathbf{x}_S^{(i)}=1} c(S) \leq q \leq \text{OPT}$. Otherwise, we need to solve a smaller instance of SET COVER defined by all elements $Y \subseteq X$ not already covered, using cover-sets $\mathcal{T} = \{S \cap Y \mid \mathbf{x}_S^{(d)} > 0\}$. We introduce some structure in the resulting instance $(Y, \mathcal{T})$ of SET COVER by choosing the indices we condition on greedily (the reader can see more details in the full version of the paper).

Lemma 1. *If at each step of the Conditioning Phase we condition on a set S in the support of the current solution $\mathbf{x}^{(d')}$ containing the most uncovered elements in X, then for all $T \in \mathcal{T}$ we have $|T| \leq \frac{n}{d}$.*

Proof. For $1 \leq i \leq d$ let S_i denote the cover-set chosen at step i of the Conditioning Phase. Note that, by Fact 6, the support of $\mathbf{x}^{(i')}$ contains the support of $\mathbf{x}^{(i)}$ for all $i' < i$, and all the coordinates corresponding to the sets S_i have value 1 in $\mathbf{x}^{(d)}$. Therefore, all d sets $S_1, \ldots, S_d$ are in the support of all vectors $\mathbf{x}^{(0)}, \ldots, \mathbf{x}^{(d)}$.

Since at step i we chose the largest (with respect to the uncovered items) cover-set S_i in the support of $\mathbf{x}^{(i-1)}$ we have $|S_j \setminus \bigcup_{i' < j} S_{i'}| \leq |S_j \setminus \bigcup_{i' < i} S_{i'}| \leq |S_i \setminus \bigcup_{i' < i} S_{i'}|$ for every $j > i$. Thus, letting $\alpha_i := |S_i \setminus \bigcup_{i' < i} S_{i'}|$ for $1 \leq i \leq d$, we have $\alpha_1 \geq \alpha_2 \geq \ldots \geq \alpha_d$. Since these represent cardinalities of disjoint sets, we have $d \cdot \alpha_d \leq \sum_{i=1}^d \alpha_i = |\bigcup_{i=1}^d S_i| \leq n$, and so $\alpha_d \leq \frac{n}{d}$.

Again, by our choice of S_d and the fact that the support of $\mathbf{x}^{(d)}$ is contained in the support of $\mathbf{x}^{(d-1)}$, it follows that for every set $S \in \mathcal{S} \setminus \{S_1, \ldots, S_d\}$ which is in the support of $\mathbf{x}^{(d)}$, we have $|S \setminus \bigcup_{i=1}^d S_i| \leq |S \setminus \bigcup_{i=1}^{d-1} S_i| \leq \alpha_d \leq \frac{n}{d}$. Thus, the instance $(Y, \mathcal{T})$ has $|T| \leq \frac{n}{d}$ for any $T \in \mathcal{T}$.

Let $\mathcal{D}$ be the collection of cover-sets chosen as in Lemma 1. Observe that the vector $\mathbf{x}^{(d)}$ projected on the cover-sets $\mathcal{S} \setminus \mathcal{D}$ that were *not* chosen in the Conditioning Phase is feasible for the LP relaxation of the instance $(Y, \mathcal{T})$. In particular, the cost of the LP is at most $q - \sum_{S \in \mathcal{D}} c(S)$, and by Lemma 1 all cover-sets have size at most $\frac{n}{d}$. The greedy algorithm will then find a solution for $(Y, \mathcal{T})$ of cost at most $H_{\frac{n}{d}} \cdot (q - \sum_{S \in \mathcal{D}} c(S))$. Altogether, this gives a feasible solution for $(X, \mathcal{S})$ of cost $H_{\frac{n}{d}} \cdot \left(q - \sum_{S \in \mathcal{D}} c(S)\right) + \sum_{S \in \mathcal{D}} c(S) \leq H_{\frac{n}{d}} \cdot q \leq H_{\frac{n}{d}} \cdot \text{OPT}$.

4   Linear **Sherali-Adams** Integrality Gap for Set Cover

The level-ℓ Sherali-Adams relaxation is a tightened LP that can be derived systematically starting with any 0-1 LP relaxation. While in this work we are interested in tightening the SET COVER polytope, the process we describe below is applicable to any other relaxation.

Definition 2 (The Sherali-Adams system). *Consider a polytope over the variables $y_1, \ldots, y_n$ defined by finitely many constraints (including the box-constraints $0 \leq y_i \leq 1$). The level-ℓ Sherali-Adams relaxation is an LP over the variables $\{y_A\}$ where A is any subset of $\{1, 2, \ldots, n\}$ of size at most $\ell + 1$, and where $y_\emptyset = 1$. For every constraint $\sum_{i=1}^n a_i y_i \geq b$ of the original polytope and for every disjoint $P, E \subseteq \{1, \ldots, n\}$ with $|P| + |E| \leq \ell$, the following is a constraint of the level-ℓ Sherali-Adams relaxation $\sum_{i=1}^n a_i \sum_{\emptyset \subseteq T \subseteq E} (-1)^{|T|} y_{P \cup T \cup \{i\}} \geq b \sum_{\emptyset \subseteq T \subseteq E} (-1)^{|T|} y_{P \cup T}$.*

We will prove Theorem 3 in this section. For this we will need two ingredients: (a) appropriate instances, and (b) a solution of the Sherali-Adams LP as described in Definition 2. Our hard instances are described in the following lemma, which is due to Alekhnovich et al. [1].

Lemma 2 (Set Cover instances with no small feasible solutions).
For every $\varepsilon > \eta > 0$, and for all sufficiently large n, there exist SET COVER instances over a universe of n elements and n cover-sets, such that:
(i) Every element of the universe appears in exactly $(\varepsilon - \eta)n$ cover-sets, and
(ii) There is no feasible solution that uses less than $\log_{1+\varepsilon} n$ cover-sets.

In order to prove Theorem 3 we will invoke Lemma 2 with appropriate parameters. Then we will define a vector solution for the level-ℓ Sherali-Adams relaxation as described. The proof of Lemma 3 below involves a number of extensive calculations which we give in the full version of the paper.

Lemma 3. *Consider a SET COVER instance on n cover-sets as described in Lemma 2. Let f denote the number of cover-sets covering every element of the universe. For $f \geq 3\ell$, the vector $\mathbf{y}$ indexed by subsets of $\{1, \ldots, n\}$ of size at most $\ell + 1$ defined as $y_A := \frac{(f-\ell-1)!}{(f-\ell-1+|A|)!}$, $\forall A \subseteq \{1, \ldots, n\}$, $|A| \leq \ell + 1$, satisfies the level-ℓ Sherali-Adams LP relaxation of the SET COVER polytope.*

Assuming the lemma, we are ready to prove Theorem 3.

Proof (Proof of Theorem 3). Fix $\varepsilon > 0$ and invoke Lemma 2 with $\eta = \varepsilon^2$ to obtain a SET COVER instance on n universe elements and n cover-sets for which (i) every universe element is covered by exactly $(\varepsilon - \varepsilon^2)n$ cover-sets, and (ii) no feasible solution exists of cost less than $\log_{1+\varepsilon} n$. Note that in particular (i) implies that in the SET COVER LP relaxation, every constraint has support exactly $f = (\varepsilon - \varepsilon^2)n$.

Set $\ell = \frac{\gamma(\varepsilon - \varepsilon^2)}{1+\gamma} n$ and note that $f/\ell \geq 3$, since $\gamma \leq \frac{1}{2}$. This means we can define a feasible level-ℓ Sherali-Adams solution as described in Lemma 3. The values of the singleton variables are set to $y_{\{i\}} = \frac{1}{(\varepsilon - \varepsilon^2)n - \ell} = \frac{1+\gamma}{(\varepsilon - \varepsilon^2)n}$. But then, the integrality gap is at least $\frac{\text{OPT}}{\sum_{i=1}^n y_{\{i\}}} \geq \frac{\varepsilon - \varepsilon^2}{1+\gamma} \cdot \log_{1+\varepsilon} n = \frac{\varepsilon - \varepsilon^2}{(1+\gamma)\ln(1+\varepsilon)} \ln n$. The lemma follows once we observe that $\ln(1 + \varepsilon) = \varepsilon - \frac{1}{2}\varepsilon^2 + \Theta(\varepsilon^3)$.

5 An LS$_+$-Based PTAS for Knapsack

Consider an instance of the KNAPSACK problem with rewards $r_1 \ldots, r_n$ and costs (or sizes) $c_1, \ldots, c_n$ and total capacity C. In what follows we will use the natural LP relaxation for KNAPSACK:

$$
\begin{aligned}
&\text{maximize} && \sum_{i=1}^{n} r_i x_i \\
&\text{subject to} && \sum_{i=1}^{n} c_i x_i \leq C \\
& && 0 \leq x_i \leq 1 && \forall\, i \in [n]
\end{aligned}
$$

Denote the polytope corresponding to the above constraints by P. We will consider the SDP derived by applying sufficiently many levels of LS$_+$ (as defined in Section 3.2) to the above LP. That is, for some $\ell > 0$, we consider the SDP

$$
\begin{aligned}
&\text{maximize} && \sum_{i=1}^{n} r_i x_i \\
&\text{subject to} && (1, \mathbf{x}) \in K_\ell^+(P).
\end{aligned}
$$

(note that we do *not* lift the objective function).

There is a well-known simple greedy algorithm for KNAPSACK: Sort the items by decreasing order of r_i/c_i, and add them to the knapsack one at a time until the current item does not fit. The following lemma (which is folklore) relates the performance of the greedy algorithm to the value of the LP relaxation P:

Lemma 4. *Let $\mathbf{x}$ be a solution to P, and R_G be the reward given by the greedy algorithm. Then $\sum_i r_i x_i \leq R_G + \max_i r_i$.*

This gives a trivial bound of 2 on the integrality gap, assuming that $c_i \leq C$ for all i (that is, that each item can be a solution on its own), since we then have $R_G + \max_i r_i \leq 2\mathsf{OPT}$.[1] We note that the above assumption is not needed when using lift-and-project methods since they will place a weight of 0 on any item i with $c_i > C$. Lemma 4 has the following easy corollary: Consider the above greedy algorithm, with the modification that we first add all items which have $x_i = 1$ and discard all items which have $x_i = 0$. Then the following holds:

Corollary 1. *Let $\mathbf{x}$ be a solution to P, and let R_G' be the total reward given by the above modified greedy algorithm. Then $\sum_i r_i x_i \leq R_G' + \max_{i:0<x_i<1} r_i$.*

We will show that, for any constant $\varepsilon > 0$, there is a constant L_ε such that the SDP relaxation for KNAPSACK arising from level L_ε of LS$_+$ has integrality gap at most $1 + O(\varepsilon)$. For the Lasserre hierarchy, this has been shown for level $1/\varepsilon$ [11]. We will show this for level $L_\varepsilon = 1/\varepsilon^3$ in the case of LS$_+$.

Our rounding algorithm will take as input the values of the KNAPSACK instance $(r_i)_i$, $(c_i)_i$, and C, an optimal solution $\mathbf{x}$ s.t. $(1, \mathbf{x}) \in K_\ell^+(P)$ (for some level $\ell > 0$, initially $\ell = L_\varepsilon$), and parameters ε and ρ. The parameter ρ is

---

[1] Here, as before, OPT denotes the optimal 0-1 solution.

intended to be the threshold $\varepsilon \cdot \mathsf{OPT}$ in the set $S_{\varepsilon\mathsf{OPT}} = \{i \mid r_i > \varepsilon \cdot \mathsf{OPT}\}$. Rather than guessing a value for OPT, though, we will simply try all values of $\rho \in \{r_i \mid i \in [n]\} \cup \{0\}$ and note that for exactly one of those values, the set $S_{\varepsilon\mathsf{OPT}}$ coincides with the set $\{i \mid r_i > \rho\}$ (also note that ρ is a parameter of the rounding, and not involved at all in the SDP relaxation).

The intuition behind our rounding algorithm is as follows: As we did for SET COVER, we would like to repeatedly "condition" on setting some variable to 1, by using Fact 5. If we condition only on (variables corresponding to) items in $S_{\varepsilon\mathsf{OPT}}$, then after at most $1/\varepsilon$ iterations, the SDP solution will be integral on that set, and then by Corollary 1 the modified greedy algorithm will give a $1 + O(\varepsilon\mathsf{OPT})$ approximation relative to the value of the objective function (since items outside $S_{\varepsilon\mathsf{OPT}}$ have reward at most $\varepsilon\mathsf{OPT}$). The problem with this approach (and the reason why LP hierarchies do not work), is the same problem as for SET COVER: the conditioning step does not preserve the value of the objective function. While the optimum value of any relaxation is at least OPT by definition, after conditioning, the value of the new solution may be much smaller than OPT, which then makes the use of Corollary 1 meaningless. The key observation is that we can avoid this scenario by using SDPs:

Lemma 5. *Let* $(1, \mathbf{x})$ *be a solution to* $K_\ell^+(P)$ *for some* $\ell \geq 1$, *with the corresponding moment vector* $\mathbf{y}$. *Then the solution satisfies*

$$\sum_{i=1}^n r_i \sum_{j=1}^n r_j y_{\{i,j\}} \geq \left(\sum_{i=1}^n r_i x_i \right)^2 \left(= \sum_{i=1}^n r_i \sum_{j=1}^n x_i r_j x_j \right).$$

Indeed, this ensures that we can choose some item i to condition on without any decrease in the new objective function $\sum_j r_j(y_{i,j}/x_i)$.[2] Unfortunately, there may not necessarily be such an item *specifically in* $S_{\varepsilon\mathsf{OPT}}$. However, if all items in $S_{\varepsilon\mathsf{OPT}}$ cause the objective function to decrease after conditioning, then the above lemma guarantees that some item outside $S_{\varepsilon\mathsf{OPT}}$ will cause the objective function to *increase* after conditioning. Counter-intuitively, we will actually condition on such an item in this case. We bound the number of times this can occur via the following evidently new idea: if $\mathbf{x}^{(0)}$ is an optimal solution to $K_\ell^+(P)$ (for some ℓ), then it has objective value at least OPT; If by a series of conditioning steps we obtain a sequence of solutions $\mathbf{x}^{(1)}, \ldots, \mathbf{x}^{(d)}$ with sufficient increase in the objective value such that the value of $\mathbf{x}^{(d)}$ is more than twice the value of $\mathbf{x}^{(0)}$, then this contradicts the fact that the integrality gap (even of the standard LP) is always bounded by 2. Our rounding algorithm **KS-Round** is described in Algorithm 1. We defer the performance analysis to the full version of the paper.

---

[2] Note that this is crucial for a maximization problem like KNAPSACK, while for a minimization problem like SET COVER it does not seem helpful (and indeed, by the integrality gap of Alekhnovich et al. [1], we know it does not help).

Algorithm 1. KS-Round$((r_i)_i, (c_i)_i, C, \mathbf{x}, \varepsilon, \rho)$

1: Let $\mathbf{y} \in \mathcal{P}_2$ be the moment vector associated with $(1, \mathbf{x})$.
2: Let $S_\rho \leftarrow \{i \mid r_i > \rho\}$, and let $S^b \leftarrow \{i \mid x_i = b\}$ for $b = 0, 1$.
3: **if** $S_\rho \subseteq S^0 \cup S^1$ **then**
4:     Run the modified greedy algorithm.
5: **else if** $\displaystyle\sum_{i \in S_\rho \setminus S^1} r_i x_i < \varepsilon \cdot \sum_{i=1}^{n} r_i x_i$ **then**
6:     Run the modified greedy algorithm on items in $([n] \setminus S_\rho) \cup S^1$.
7: **else if** there is some $i \in S_\rho \setminus (S^0 \cup S^1)$ s.t. $\displaystyle\sum_{j=1}^{n} r_j y_{\{i,j\}} \geq (1 - \varepsilon^2) x_i \cdot \sum_{j=1}^{n} r_j x_j$ **then**
8:     Run **KS-Round**$((r_i)_i, (c_i)_i, C, \frac{1}{x_i} Y^{[\mathbf{y}]} \mathbf{e}_i, \varepsilon, \rho)$. $\triangleright$ See Fact 5
9: **else**
10:     Choose $i \in [n] \setminus (S_\rho \cup S^0)$ s.t. $\displaystyle\sum_{j=1}^{n} r_j y_{\{i,j\}} > (1 + \varepsilon^3) x_i \cdot \sum_{j=1}^{n} r_j x_j$
11:     Run **KS-Round**$((r_i)_i, (c_i)_i, C, \frac{1}{x_i} Y^{[\mathbf{y}]} \mathbf{e}_i, \varepsilon, \rho)$. $\triangleright$ See Fact 5
12: **end if**

---

6 Conclusion

As we have seen, lift-and-project methods can give rise to LP and SDP relaxations which match the guarantee of combinatorial algorithms in cases where the standard relaxations have a large integrality gap. For packing problems, such as KNAPSACK, it seems like (even relatively weak) SDP hierarchies can accomplish this by keeping the value of the objective relatively stable, while for covering problems, such as SET COVER, lifting the objective function explicitly becomes necessary.

Note that our LS_+-based algorithm for KNAPSACK shows that in some instances reduced integrality gaps which rely heavily on properties of the Lasserre hierarchy can be achieved using the weaker LS_+ hierarchy. This raises the question of whether the problems discussed in the recent series of Lasserre-based approximation algorithms [2,7,18] also admit similar results using LS_+. On the flip side, it would also be interesting to see whether any such problems have strong integrality gap lower bounds for LS_+, which would show a separation between the two hierarchies.

Acknowledgements. We would like to thank Dana Moshkovitz for pointing out the blowup in her SET COVER reduction. We would also like to thank Mohammad R. Salavatipour for preliminary discussions on sub-exponential time approximation algorithms in general. Finally, we would like to thank Claire Mathieu for insightful past discussions of the KNAPSACK-related results in [11].

References

1. Alekhnovich, M., Arora, S., Tourlakis, I.: Towards strong nonapproximability results in the Lovász-Schrijver hierarchy. In: Proceedings of ACM Symposium on Theory of Computing, pp. 294–303 (2005)

2. Barak, B., Raghavendra, P., Steurer, D.: Rounding semidefinite programming hierarchies via global correlation. In: Proceedings of IEEE Symposium on Foundations of Computer Science, pp. 472–481 (2011)
3. Chlamtáč, E., Tulsiani, M.: Convex relaxations and integrality gaps. In: Anjos, M.F., Lasserre, J.B. (eds.) Handbook on Semidefinite, Conic and Polynomial Optimization. International Series in Operations Research & Management Science, vol. 166, pp. 139–169. Springer, Heidelberg (2012)
4. Chvátal, V.: A greedy heuristic for the set-covering problem. Mathematics of Operations Research 4(3), 233–235 (1979), doi:10.2307/3689577
5. Cygan, M., Kowalik, L., Wykurz, M.: Exponential-time approximation of weighted set cover. Inf. Process. Lett. 109(16), 957–961 (2009)
6. Feige, U.: A threshold of $\ln n$ for approximating set cover. J. ACM 45(4), 634–652 (1998)
7. Guruswami, V., Sinop, A.K.: Lasserre hierarchy, higher eigenvalues, and approximation schemes for graph partitioning and quadratic integer programming with PSD objectives. In: Proceedings of IEEE Symposium on Foundations of Computer Science, pp. 482–491 (October 2011)
8. Ibarra, O.H., Kim, C.E.: Fast approximation algorithms for the knapsack and sum of subset problems. J. ACM 22(4), 463–468 (1975)
9. Impagliazzo, R., Paturi, R.: On the complexity of k-SAT. J. Comput. Syst. Sci. 62(2), 367–375 (2001)
10. Johnson, D.S.: Approximation algorithms for combinatorial problems. J. Comput. Syst. Sci. 9(3), 256–278 (1974)
11. Karlin, A.R., Mathieu, C., Nguyen, C.T.: Integrality gaps of linear and semidefinite programming relaxations for knapsack. In: Günlük, O., Woeginger, G.J. (eds.) IPCO 2011. LNCS, vol. 6655, pp. 301–314. Springer, Heidelberg (2011)
12. Karp, R.M.: Reducibility among combinatorial problems. In: Complexity of Computer Computations, pp. 85–103 (1972)
13. Lasserre, J.B.: An explicit equivalent positive semidefinite program for nonlinear 0-1 programs. SIAM Journal on Optimization 12(3), 756–769 (2002)
14. Lovász, L.: On the ratio of optimal integral and fractional covers. Discrete Mathematics 13(4), 383–390 (1975)
15. Lovász, L., Schrijver, A.: Cones of matrices and set-functions and 0-1 optimization. SIAM J. Optim. 1(2), 166–190 (1991)
16. Moshkovitz, D.: The projection games conjecture and the NP-hardness of $\ln n$-approximating set-cover. In: Gupta, A., Jansen, K., Rolim, J., Servedio, R. (eds.) APPROX 2012 and RANDOM 2012. LNCS, vol. 7408, pp. 276–287. Springer, Heidelberg (2012)
17. Raghavendra, P.: Optimal algorithms and inapproximability results for every csp? In: Proceedings of ACM Symposium on Theory of Computing, pp. 245–254 (2008)
18. Raghavendra, P., Tan, N.: Approximating csps with global cardinality constraints using sdp hierarchies. In: Proceedings of ACM-SIAM Symposium on Discrete Algorithms, pp. 373–387. SIAM (2012)
19. Sherali, H.D., Adams, W.P.: A hierarchy of relaxations between the continuous and convex hull representations for zero-one programming problems. SIAM Journal on Discrete Mathematics 3(3), 411–430 (1990)

Optimal Time-Convex Hull under the L_p Metrics[*]

Bang-Sin Dai[3], Mong-Jen Kao[1], and D.T. Lee[1,2,3]

[1] Research Center for Infor. Tech. Innovation, Academia Sinica, Taiwan
[2] Dep. of Computer Sci. and Engineering, National Chung-Hsing Uni., Taiwan
[3] Dep. of Computer Sci. and Infor. Engineering, National Taiwan Uni., Taiwan
f94922074@ntu.edu.tw, mong@citi.sinica.edu.tw, dtlee@ieee.org

Abstract. We consider the problem of computing the time-convex hull of a point set under the general L_p metric in the presence of a straight-line highway in the plane. The traveling speed along the highway is assumed to be faster than that off the highway, and the shortest time-path between a distant pair may involve traveling along the highway. The time-convex hull TCH(P) of a point set P is the smallest set containing both P and *all* shortest time-paths between any two points in TCH(P). In this paper we give an algorithm that computes the time-convex hull under the L_p metric in optimal $\mathcal{O}(n \log n)$ time for a given set of n points and a real number p with $1 \leq p \leq \infty$.

1 Introduction

Path planning, in particular, shortest time-path planning, in complex transportation networks has become an important yet challenging issue in recent years. With the usage of heterogeneous moving speeds provided by different means of transportation, the *time-distance* between two points, i.e., the amount of time it takes to go from one point to the other, is often more important than their straight-line distance. With the reinterpretation of distances by the time-based concept, fundamental geometric problems such as convex hull, Voronoi diagrams, facility location, etc. have been reconsidered recently in depth and with insights [1,4,6].

From the theoretical point of view, straight-line highways which provide faster moving speed and which we can enter and exit at any point is one of the simplest transportation models to explore. The speed at which one can move along the highway is assumed to be $v > 1$, while the speed off the highway is 1. Generalization of convex hulls in the presence of highways was introduced by Hurtado et al. [8], who suggested that the notion of convexity be defined by the inclusion of shortest time paths, instead of straight-line segments, i.e., a set S is said to be **convex** if it contains the shortest time-path between any two points of S. Using

---

[*] This work was supported in part by National Science Council (NSC), Taiwan, under grants NSC-101-2221-E-005-026 and NSC-101-2221-E-005-019.

F. Dehne, R. Solis-Oba, and J.-R. Sack (Eds.): WADS 2013, LNCS 8037, pp. 268–279, 2013.

this new definition, the time-convex hull $\text{TCH}(P)$ for a set P is the closure of P with respect to the inclusion of shortest time-paths.

In following work, Palop [10] studied the structure of $\text{TCH}(P)$ in the presence of a highway and showed that it is composed of convex clusters possibly together with segments of the highway connecting all the clusters. A particularly interesting fact implied by the hull-structure is that, the shortest time-path between each pair of inter-cluster points must contain a piece of traversal along the highway, while similar assertions do not hold for intra-cluster pairs of points: A distant pair of points (p, q) whose shortest time-path contains a segment of the highway could still belong to the same cluster, for there may exist other points from the same cluster whose shortest time-path to either p or q does not use the highway at all. This suggests that, the structure of $\text{TCH}(P)$ in some sense indicates the degree of convenience provided by the underlying transportation network. We are content with clusters of higher densities, i.e., any cluster with a large ratio between the number of points of P it contains and the area of that cluster. For sparse clusters, we may want to break them and benefit distant pairs they contain by enhancing the transportation infrastructure.

The approach suggested by Palop [10] for the presence of a highway involves enumeration of shortest time-paths between all pairs of points and hence requires $\Theta(n^2)$ time, where n is the number of points. This problem was later studied by Yu and Lee [11], who proposed an approach based on incremental point insertions in a highway-parallel monotonic order. However, the proposed algorithm does not return the correct hull in all circumstances as particular cases were overlooked. The first sub-quadratic algorithm was given by Aloupis et al. [3], who proposed an $\mathcal{O}(n \log^2 n)$ algorithm for the L_2 metric and an $\mathcal{O}(n \log n)$ algorithm for the L_1 metric, following the incremental approach suggested by [11] with careful case analysis. To the best of our knowledge, no previous results regarding metrics other than L_1 and L_2 were presented.

Our Focus and Contribution. In this paper we address the problem of computing the time-convex hull of a point set in the presence of a straight-line highway under the L_p metric for a given real number p with $1 \le p \le \infty$. First, we adopt the concept of *wavefront propagation*, a notion commonly used for path planning [2, 6], and derive basic properties required for depicting the hull structure under the general L_p metric. When the shortest path between two points is not uniquely defined, e.g., in L_1 and L_∞ metrics, we propose a re-evaluation on the existing definition of convexity. Previous works concerning convex hulls under metrics other than L_2, e.g., Ottmann et al. [9] and Aloupis et al. [3], assume a particular path to be taken when multiple choices are available. However, this assumption allows the boundary of a convex set to contain reflex angles, which in some sense deviates from the intuition of a set being convex.

In this work we adopt the definition that requires a convex set to include *every shortest path* between any two points it contains. Although this definition fundamentally simplifies the shapes of convex sets for L_1 and L_∞ metrics, we show that the nature of the problem is not altered when time-based concepts are considered. In particular, the problem of deciding whether any pair of the given points belong to the same cluster under the L_p metric requires $\Omega(n \log n)$ time under the algebraic computation model [5], for all $1 \le p \le \infty$.

Second, we provide an optimal $\mathcal{O}(n \log n)$ algorithm for computing the time-convex hull for a given set of points. The known algorithm due to Aloupis et al. [3] stems from a scenario in the cluster-merging step where we have to check for the existence of intersections between a line segment and a set of convex curves composed of parabolae and line segments, which leads to their $\mathcal{O}(n \log^2 n)$ algorithm. In our paper, we tackle this situation by making an observation on the duality of cluster-merging conditions and reduce the problem to the geometric query of deciding if any of the given points lies above a line segment of an arbitrary slope. This approach greatly simplifies the algorithm structure and can be easily generalized to other L_p-metrics for $1 \le p \le \infty$. For this particular geometric problem, we use a data structure due to Guibas et al [7] to answer this query in logarithmic time. All together this yields our $\mathcal{O}(n \log n)$ algorithm. We remark that, although our adopted definition of convexity simplifies the shape of convex sets under the L_1 and the L_∞ metrics, the algorithm we propose does not take advantage of this specific property and also works for the original notion for which only a particular path is to be included.

2   Preliminaries

In this section, we give precise definitions of the notions as well as sketches of previously known properties that are essential to present our work. We begin with the general L_p distance metric and basic time-based concepts.

Definition 1 (Distance in the L_p-metrics). *For any real number $p \ge 1$ and any two points $q_i, q_j \in \mathbb{R}^n$ with coordinates $(i_1, i_2, \ldots, i_n)$ and $(j_1, j_2, \ldots, j_n)$, the distance between q_i and q_j under the L_p-metric is defined to be $d_p(q_i, q_j) = \left(\sum_{k=1}^{n} |i_k - j_k|^p \right)^{\frac{1}{p}}$.*

Note that when p tends to infinity, $d_p(q_i, q_j)$ converges to $\max_{1 \le k \le n} |i_k - j_k|$. This gives the definition of the distance function in the L_∞-metric, which is $d_\infty(q_i, q_j) = \max_{1 \le k \le n} |i_k - j_k|$. For the rest of this paper, we use the subscript p to indicate the specific L_p-metric, and the subscript p will be omitted when there is no ambiguity.

A *transportation highway* $\mathcal{H}$ in $\mathbb{R}^n$ is a hyperplane in which the moving speed in $\mathcal{H}$ is $v_{\mathcal{H}}$, where $1 < v_{\mathcal{H}} \le \infty$, while the moving speed off $\mathcal{H}$ is assumed to be *unit*. Given the moving speed in the space, we can define the *time-distance* between any two points in $\mathbb{R}^n$.

Definition 2. *For any $q_i, q_j \in \mathbb{R}^n$, a continuous curve $\mathcal{C}$ connecting q_i and q_j is said to be a* shortest time-path *if the traveling time required along $\mathcal{C}$ is minimum among all possible curves connecting q_i and q_j. The traveling time required along $\mathcal{C}$ is referred to as the time-distance between q_i and q_j, denoted $\hat{d}(q_i, q_j)$.*

For any two points q_i and q_j, let $\text{STP}(q_i, q_j)$ denote the set of shortest time-paths between q_i and q_j. For any $\mathcal{C} \in \text{STP}(q_i, q_j)$, we say that $\mathcal{C}$ *enters the highway* $\mathcal{H}$ if $\mathcal{C} \cap \mathcal{H} \neq \emptyset$. The *walking-region* of a point $q \in \mathbb{R}^n$, denoted $\text{WR}(q)$, is defined to be the set of points whose set of shortest time-paths to q contains a time-path that does not enter the highway $\mathcal{H}$. For any $\mathcal{C} \in \text{STP}(q_i, q_j)$, we say that $\mathcal{C}$ *uses the highway* $\mathcal{H}$ if $\mathcal{C} \cap \mathcal{H}$ contains a piece with non-zero length, i.e., at some point $\mathcal{C}$ enters the highway $\mathcal{H}$ and walks along it.

Convexity and Time-Convex Hulls. In classical definitions, a set of points is said to be *convex* if it contains every line segment joining each pair of points in the set, and the convex hull of a set of points $Q \subseteq \mathbb{R}^n$ is the minimal convex set containing Q. When time-distance is considered, the concept of convexity as well as convex hulls with respect to time-paths is defined analogously. A set of points is said to be *convex with respect to time*, or, *time-convex*, if it contains every shortest time-path joining each pair of points in the set.

Definition 3 (Time-convex hull). *The time-convex hull, of a set of points $Q \subseteq \mathbb{R}^n$, denoted $\text{TCH}(Q)$, is the minimal time-convex set containing Q.*

Although the aforementioned concepts are defined in $\mathbb{R}^n$ space, in this paper we work in $\mathbb{R}^2$ plane with an axis-parallel highway placed on the x-axis as higher dimensional space does not give further insights: When considering the shortest time-paths between two points in higher dimensional space, it suffices to consider the specific plane that is orthogonal to $\mathcal{H}$ and that contains the two points.

Time-Convex Hull under the L_1 and the L_2 Metrics. The structure of time-convex hulls under the L_1 and the L_2 metrics has been studied in a series of work [3, 10, 11]. Below we review important properties. See also Fig. 1 for an illustration.

Proposition 1 ([10,11]). *For the L_2-metric and any point $q = (x_q, y_q)$ with $y_q \geq 0$, we have the following properties.*
1. *If a shortest time-path starting from q uses the highway $\mathcal{H}$, then it must enter the highway with an incidence angle $\alpha = \arcsin 1/v_{\mathcal{H}}$ toward the direction of the destination.*
2. *The walking region of a point $q \in \mathbb{R}^2$ is characterized by the following two parabolae: (a)* right discriminating parabola, *which is the curve satisfying*

$$\begin{cases} x \geq x_q + y_q \tan \alpha, \quad and \\ \sqrt{(x - x_q)^2 + (y - y_q)^2} = y_q \sec \alpha + y \sec \alpha + \frac{1}{v_{\mathcal{H}}}((x - y \tan \alpha) - (x_q + y_q \tan \alpha)). \end{cases}$$

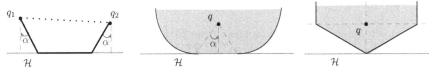

Fig. 1. (a) The only two possible paths for being a shortest time-path between q_1 and q_2 in L_2. (b)(c) The walking regions of a point $q \in \mathbb{R}^2$ under L_2 and L_1, respectively.

(b) The left discriminating parabola *is symmetric to the right discriminating parabola with respect to the line* $x = x_q$.

Proposition 2 ([3]). *For the L_1-metric, the walking region of a point $q = (x_q, y_q)$ with $y_q \geq 0$ is formed by the intersection of the following regions: (a) the vertical strip $x_q - y_q/\beta \leq x \leq x_q - y_q/\beta$, and (b) $y \geq \pm\beta(x - x_q)$, where $\beta = \frac{1}{2}\left(1 - \frac{1}{v_{\mathcal{H}}}\right)$.*

3 Hull-Structure under the General L_p-Metrics

In this section, we derive necessary properties to describe the structure of time-convex hulls under the general L_p metrics. First, we adopt the notion of *wavefront propagation* [2, 6], which is a well-established model used in path planning, and derive the behavior of a shortest time-path between any two points. Then we show how the corresponding walking regions are formed, followed by a description of the desired structural properties.

Wavefronts and Shortest Time-Paths. For any $q \in \mathbb{R}^2$, $t \geq 0$, and $p \geq 1$, the wavefront with *source* q and *radius* t under the L_p-metric is defined as

$$W_p(q, t) = \left\{ s \colon s \in \mathbb{R}^2, \hat{d}_p(q, s) = t \right\}.$$

Literally, $W_p(q, t)$ is the set of points whose time-distances to q are exactly t. Fig. 2 (a) shows the wavefronts, i.e., the "unit-circles" under the L_p metric, or, the p-circles, for different p with $0 < p \leq \infty$ when the highway is not used. The shortest time-path between q and any point $q' \in W_p(q, t)$ is the trace on which q' moves as t changes smoothly to zero, which is a straight-line joining q and q'.

When the highway $\mathcal{H}$ is present and the time-distance changes, deriving the behavior of a shortest time-path that uses $\mathcal{H}$ becomes tricky. Let $q_1, q_2 \in \mathbb{R}^2$, $q_1 \neq q_2$, be two points in the plane, and let $\hat{t}_{1/2}(q_1, q_2) \geq 0$ be the smallest real number such that

$$\text{Bisect}(q_1, q_2) \equiv W_p\big(q_1, \hat{t}_{1/2}(q_1, q_2)\big) \cap W_p\big(q_2, \hat{t}_{1/2}(q_1, q_2)\big) \neq \emptyset.$$

In other words, $\text{Bisect}(q_1, q_2)$ is the set of points at which $W_p(q_1, t)$ and $W_p(q_2, t)$ meet for the first time. The following lemma shows that $\text{Bisect}(q_1, q_2)$ characterizes the set of "*middle points*" of all shortest time-paths between q_1 and q_2.

Lemma 1. *For each $\mathcal{C} \in \text{STP}(q_1, q_2)$, we have $\mathcal{C} \cap \text{Bisect}(q_1, q_2) \neq \emptyset$. Moreover, for each $q \in \text{Bisect}(q_1, q_2)$, there exists $\mathcal{C}' \in \text{STP}(q_1, q_2)$ such that $q \in \mathcal{C}'$.*

Given the set $\text{Bisect}(q_1, q_2)$, a shortest time-path between q_1 and q_2 can be obtained by joining $\mathcal{C}_1$ and $\mathcal{C}_2$, where $\mathcal{C}_1 \in \text{STP}(q_1, q)$ and $\mathcal{C}_2 \in \text{STP}(q, q_2)$ for some $q \in \text{Bisect}(q_1, q_2)$. By expanding the process in a recursive manner we get a set of middle points. Although the cardinality of the set we identified is countable while any continuous curve in the plane contains uncountably infinite

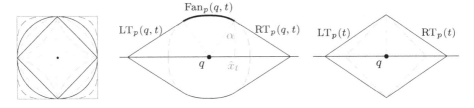

Fig. 2. (a) p-circles for different values of p: bold rhombus for $p = 1$, bold circle for $p = 2$, and bold square for $p = \infty$. (b) $W_p(q, t)$ for a point $q \in \mathcal{H}$, $v_{\mathcal{H}} < \infty$, and $p > 1$, where the angle α satisfies $\sin \alpha = \frac{\hat{x}_t}{\sqrt{\hat{x}_t^2 + y_p(\hat{x}_t)^2}}$. (c) $W_p(q, t)$ for $v_{\mathcal{H}} < \infty$ and $p = 1$.

points, it is not difficult to see that, the set of points we locate is *dense*[1] in the underlying curve, and therefore can serve as a representative.

To describe the shape of a wavefront when the highway may be used, we need the following lemma regarding the propagation of wavefronts.

Lemma 2. *Let* $\mathcal{C}_p^\circ(q, t)$ *denote the p-circle with center* q *and radius* t. *Then* $W_p(q, t)$ *is formed by the boundary of*

$$\mathcal{C}_p^\circ(q, t) \cup \bigcup_{s:\ s \in \mathbb{R}^2, \hat{d}_p(q,s) < d_p(q,s) \le t} W_p\left(s, t - \hat{d}_p(q, s)\right).$$

In the following we discuss the case when $1 < p < \infty$ and leave the discussion of shortest time-paths in L_∞ to the appendix for further reference. Let $\mathcal{H}$ be the highway placed on the x-axis with moving speed $v_{\mathcal{H}} > 1$. For any $t \ge 0$, any $0 \le x \le t$, and any $1 < p < \infty$, we use $y_p(x, t) = (t^p - |x|^p)^{1/p}$ to denote the y-coordinate of the specific point on the p-circle with x-coordinate x. Let $\hat{x}_t = t \cdot v_{\mathcal{H}}^{1/(1-p)}$. We have the following lemma regarding $W_p(q, t)$. Also refer to Fig. 2 (b) for an illustration.

Lemma 3. *For* $1 < p < \infty$, $v_{\mathcal{H}} < \infty$, *and a point* $q \in \mathcal{H}$ *which we assume to be* $(0, 0)$ *for the ease of presentation, the upper-part of* $W_p(q, t)$ *that lies above* $\mathcal{H}$ *consists of the following three pieces:*

– Fan$_p(q, t)$: *the circular-sector of the p-circle with radius* t, *ranging from* $(-\hat{x}_t, y_p(-\hat{x}_t, t))$ *to* $(\hat{x}_t, y_p(\hat{x}_t, t))$.
– LT$_p(q, t)$, RT$_p(q, t)$: *two line segments joining* $(-v_{\mathcal{H}} \cdot t, 0)$, $(-\hat{x}_t, y_p(-\hat{x}_t, t))$, *and* $(\hat{x}_t, y_p(\hat{x}_t, t))$, $(v_{\mathcal{H}} \cdot t, 0)$, *respectively. Moreover,* LT$_p(q, t)$ *and* RT$_p(q, t)$ *are tangent to* Fan$_p(q, t)$.

The lower-part that lies below $\mathcal{H}$ *follows symmetrically. For* $v_{\mathcal{H}} = \infty$, *the upper-part of* $W_p(q, t)$ *consists of a horizontal line* $y = t$.

---

[1] *Dense* is a concept used in classical analysis to indicate that any element of one set can be approximated to any degree by elements of a subset being dense within.

For each $1 \leq p < \infty$ and $1 < v_{\mathcal{H}} \leq \infty$, we define the real number $\alpha(p, v_{\mathcal{H}})$ as follows. If $p = 1$ or $v_{\mathcal{H}} = \infty$, then $\alpha(p, v_{\mathcal{H}})$ is defined to be zero. Otherwise, $\alpha(p, v_{\mathcal{H}})$ is defined to be

$$
\arcsin \frac{v_{\mathcal{H}}^{1/(1-p)}}{\sqrt{v_{\mathcal{H}}^{2/(1-p)} + \left(1 - v_{\mathcal{H}}^{p/(1-p)}\right)^{2/p}}}.
$$

Note that, when $p = 2$, this is exactly $\arcsin(1/v_{\mathcal{H}})$. For brevity, we simply use α when there is no ambiguity. The behavior of a shortest time-path that takes the advantage of traversal along the highway is characterized by the following lemma.

Lemma 4. *For any point $q = (x_q, y_q)$, $1 \leq p < \infty$, and $1 < v_{\mathcal{H}} \leq \infty$, if a shortest time-path starting from q uses the highway $\mathcal{H}$, then it must enter the highway with an incidence angle α.*

Walking Regions. For any point $q = (x_q, y_q) \in \mathbb{R}^2$ with $y_q \geq 0$, let $q_{\mathcal{H}}^+$ and $q_{\mathcal{H}}^-$ be two points located at $(x_q \pm y_q \tan \alpha, 0)$, respectively. By Lemma 4, we know that, $q_{\mathcal{H}}^+$ and $q_{\mathcal{H}}^-$ are exactly the points at which any shortest time-path from q will enter the highway if needed. This gives the walking region for any point. Let $\alpha = \pi/4$ when $p = \infty$. The following lemma is an updated version of Proposition 1 for general p with $1 \leq p \leq \infty$.

Lemma 5. *For any p with $1 \leq p \leq \infty$ and any point $q = (x_q, y_q)$ with $y_q \geq 0$, $\mathrm{WR}_p(q)$ is characterized by the following two curves: (a) right discriminating curve, which is the curve $q' = (x', y')$ satisfying $x' \geq x_q + y_q \tan \alpha$ and*

$$
\left|\overline{qq'}\right|_p = \left|\overline{qq_{\mathcal{H}}^+}\right|_p + \left|\overline{q'q_{\mathcal{H}}^-}\right|_p + \frac{1}{v_{\mathcal{H}}}\left|\overline{q_{\mathcal{H}}^+q_{\mathcal{H}}^-}\right|_p.
$$

(b) The left discriminating curve is symmetric with respect to the line $x = x_q$.

For any point q, let $\mathrm{WR}_\ell(q)$ and $\mathrm{WR}_r(q)$ denote the left- and right- discriminating curves of $\mathrm{WR}(q)$, respectively. We have the following *dominance property* of the walking regions.

Lemma 6. *Let $q_1 = (x_1, y_1)$ and $q_2 = (x_2, y_2)$ be two points such that $x_1 \leq x_2$. If $y_1 \geq y_2$, then $\mathrm{WR}_\ell(q_2)$ lies to the right of $\mathrm{WR}_\ell(q_1)$. Similarly, if $y_1 \leq y_2$, then $\mathrm{WR}_r(q_1)$ lies to the left of $\mathrm{WR}_r(q_2)$.*

Lemma 6 suggests that, to describe the leftmost and the rightmost boundaries of the walking-regions for a set of points, it suffices to consider the *extreme points*. Let $e = \overline{q_1 q_2}$ be a line seg-

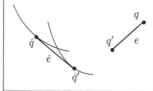

Fig. 3. The left boundary of the walking region for the edge $e = (q', q)$

ment between two points q_1 and q_2, where q_1 lies to the left of q_2. If e has non-positive slope, then the left-boundary of the walking region for e is dominated by $\mathrm{WR}_\ell(q_1)$. Otherwise, we have to consider $\bigcup_{q \in e} \mathrm{WR}_\ell(q)$. By parameterizing each point of e, it is not difficult to see that the left-boundary consists of $\mathrm{WR}_\ell(q_1)$, $\mathrm{WR}_\ell(q_2)$, and their common tangent line. See also Fig. 3 for an illustration.

Closure and Time-Convex Hull of a Point Set. By Lemma 1, to obtain the union of possible shortest time-paths, it suffices to consider the set of all possible bisecting sets that arise inside the recursion. We begin with the closure between pairs of points.

Lemma 7. *Let $q_1, q_2 \in \mathbb{R}^2$ be two points. When the highway is not used, the set of all shortest time-paths between q_1 and q_2 is:*
- *The smallest bounding rectangle of $\{q_1, q_2\}$, when $p = 1$.*
- *The straight line segment $\overline{q_1 q_2}$ joining q_1 and q_2, when $1 < p < \infty$.*
- *The smallest bounding parallelogram whose slopes of the four sides are ± 1, i.e., a rectangle rotated by $45°$, that contains q_1 and q_2.*

Lemma 7 suggests that when the highway is not used and when $1 < p < \infty$, the closure, or, convex hull, of a point set **S** with respect to the L_p-metric is identical to that in L_2, while in L_1 and L_∞ the convex hulls are given by the bounding rectangles and bounding square-parallelograms.

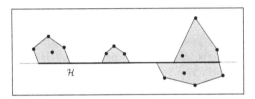

Fig. 4. Time-convex hull for a set of points under the L_p metric where $1 < p < \infty$

When the highway may be used, the structure of the time-convex hull under the general L_p-metric consists of a set of clusters arranged in a way such that the following holds: (1) Any shortest time-path between intra-cluster pair of points must use the highway. (2) If any shortest time-path between two points does not use the highway, then the two points must belong to the same cluster. Fig. 4 and Fig. 5 illustrate examples of the time-convex hull for the L_p metrics with $1 < p < \infty$ and $p = \infty$, respectively. Note that, the shape of the closure for each cluster does depend on p and $v_\mathcal{H}$, as they determine the incidence angle α.

4   Constructing the Time-Convex Hull

In this section, we present our algorithmic results for this problem. First, we show that, although our definition of convexity simplifies the structures of the resulting convex hulls, e.g., in L_1 and L_∞, the problem of deciding if any given pair of points belongs to the same cluster already requires $\Omega(n \log n)$ time. Then we present our optimal $\mathcal{O}(n \log n)$ algorithm.

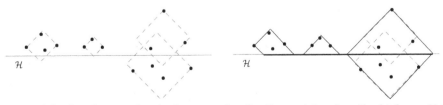

Fig. 5. (a) The closure of each cluster under the L_∞ metric when the highway $\mathcal{H}$ is not considered. (b) The closure, i.e., the time-convex hull, for the L_∞ metric.

4.1 Problem Complexity

We make a reduction from the *minimum gap problem*, which is a classical problem known to have the problem complexity of $\Theta(n \log n)$. Given n real numbers $a_1, a_2, \ldots, a_n$ and a target gap $\epsilon > 0$, the minimum gap problem is to decide if there exist some i, j, $1 \leq i, j \leq n$, such that $|a_i - a_j| \leq \epsilon$.

For any $y \geq 0$, consider the point $q(y) = (0, y)$. Let $\mathcal{C}_{q(y)} \colon \left(x, f_{q(y)}(x)\right)$ denote the right discriminating curve of $\mathrm{WR}_p(q(y))$. For any $\epsilon > 0$, let $y_0(p, \epsilon)$ denote the specific real number such that $f_{q(y_0(p,\epsilon))}(\epsilon) = y_0(p, \epsilon)$. Our reduction is done as follows. Given a real number $p \geq 1$ and an instance $\mathcal{I}$ of minimum gap, we create a set $\mathbf{S}$ consisting of n points $q_1, q_2, \ldots, q_n$, where $q_i = (a_i, y_0(p, \epsilon))$ for $1 \leq i \leq n$. The following lemma shows the correctness of this reduction and establishes the $\Omega(n \log n)$ lower bound.

Lemma 8. $y_0(p, \epsilon)$ *is well-defined for all p with $1 \leq p \leq \infty$ and all $\epsilon > 0$. Furthermore, the answer to the minimum gap problem on $\mathcal{I}$ is "yes" if and only if the number of clusters in the time-convex hull of $\mathbf{S}$ is less than n.*

Corollary 1. *Given a set of points $\mathbf{S}$ in the plane, a real number p with $1 \leq p \leq \infty$, and a highway $\mathcal{H}$ placed on the x-axis, the problem of deciding if any given pair of points belongs to the same cluster requires $\Omega(n \log n)$ time.*

4.2 An Optimal Algorithm

In this section, we present our algorithm for constructing the time-convex hull for a given point set $\mathbf{S}$ under a given metric L_p with $p \geq 1$. The main approach is to insert the points incrementally into the partially-constructed clusters in ascending order of their x-coordinates. In order to prevent a situation that leads to an undesirably complicated query encountered in the previous work by Aloupis et al. [3], we exploit the symmetric property of cluster-merging conditions and reduce the sub-problem to the following geometric query.

Definition 4 (One-Sided Segment Sweeping Query). *Given a set of points $\mathbf{S}$ in the plane, for any line segment L of finite slope, the one-sided segment sweeping query, denoted $\mathcal{Q}(\mathrm{L})$, asks if $\mathbf{S} \cap \mathrm{L}^+$ is empty, where L^+ is the intersection of the half-plane above L and the vertical strip defined by the end-points of L. That is, we ask if there exists any point $p \in \mathbf{S}$ such that p lies above L.*

In the following, we first describe the algorithm and our idea in more detail, assuming the one-sided segment sweeping query is available. Then we show how this query can be answered efficiently.

The given set $\mathbf{S}$ of points is partitioned into two subsets, one containing those points lying above $\mathcal{H}$ and the other containing the remaining. We compute the time-convex hull for the two subsets separately, followed by using a linear scan on the clusters created on both sides to obtain the closure for the entire point set. Below we describe how the time-convex hull for each of the two subsets can be computed.

Let $q_1, q_2, \ldots, q_n$ be the set of points sorted in ascending order of their x-coordinates with ties broken by their y-coordinates. During the execution of the algorithm, we maintain the set of clusters the algorithm has created so far, which we further denote by $\mathcal{C} = \{C_1, C_2, \ldots, C_k\}$. For ease of presentation, we denote the left- and right-boundary of the walking region of C_i by $\mathrm{WR}_\ell(C_i)$ and $\mathrm{WR}_r(C_i)$, respectively. Furthermore, we use $q \in \mathrm{WR}_\ell(C_i)$ or $q \in \mathrm{WR}_r(C_i)$ to indicate that point q lies to the right of $\mathrm{WR}_\ell(C_i)$ or to the left of $\mathrm{WR}_r(C_i)$, respectively.

In iteration i, $1 \leq i \leq n$, the algorithm inserts q_i into $\mathcal{C}$ and checks if a new cluster has to be created or if existing clusters have to be merged. This is done in the following two steps.

(a) *Point inclusion test.* In this step, we check if there exists any j, $1 \leq j \leq k$, such that $q_i \in \mathrm{WR}_r(C_j)$. If not, then a new cluster C_{k+1} consisting of the point q_i is created and we enter the next iteration. Otherwise, the smallest index j such that $q_i \in \mathrm{WR}_r(C_j)$ is located. The clusters $C_j, C_{j+1}, \ldots, C_k$ and the point q_i are merged into one cluster, which will in turn replace $C_j, C_{j+1}, \ldots, C_k$. Let $\mathbf{E}$ be the set of newly created edges on the upper-hull of this cluster whose slopes are positive. Then we proceed to step (b).

(b) *Edge inclusion test.* Let k be the number of clusters, and x_0 be the x-coordinate of the leftmost point in C_k. Pick an arbitrary edge $e \in \mathbf{E}$, let $\hat{e}$ denote the line segment appeared on $\mathrm{WR}_\ell(e)$ to which e corresponds, and let $\hat{e}(x_0)$ be the intersection of $\hat{e}$ with the half-plane $x \leq x_0$. Then we invoke the one-sided segment sweeping query $\mathcal{Q}(\hat{e}(x_0))$. If no point lies above $\hat{e}(x_0)$, then e is removed from $\mathbf{E}$. Otherwise, C_k is merged with C_{k-1}. Let e' be the newly created bridge edge between C_k and C_{k-1}. If e' has positive slope, then it is added to the set $\mathbf{E}$. This procedure is repeated until the set $\mathbf{E}$ becomes empty.

An approach has been proposed to resolve the *point inclusion test* efficiently, e.g., Yu and Lee [11], and Aloupis et al. [3]. Below we state the lemma directly and leave the technical details to the appendix for further reference.

Lemma 9 ([3,11]). *For each iteration, say i, the smallest index j, $1 \leq j \leq k$, such that $q_i \in \mathrm{WR}_r(C_j)$ can be located in amortized constant time.*

To see that our algorithm gives the correct clustering, it suffices to argue the following two conditions: (1) Each cluster-merge our algorithm performs is valid. (2) At the end of each iteration, no more clusters have to be merged.

Apparently these conditions hold at the end of the first iteration, when q_1 is processed. For each of the succeeding iterations, say i, if no clusters are merged in step (a), then the conditions hold trivially. Otherwise, the validity of the cluster-merging operations is guaranteed by Lemma 9 and the fact that if any point lies above $\hat{e}$, then it belongs to the walking region of e, meaning that the last cluster, C_k, has to be merged again. See also Fig. 6 for an illustration.

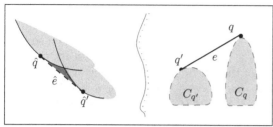

Fig. 6. When two clusters $C_{q'}$ and C_q are merged and new hull edge e is created, it suffices to check the new walking region e corresponds to, i.e., the dark-gray area

To see that the second condition holds, let $e = (q', q) \in \mathbf{E}$ be a newly created hull edge, and let $C_{q'}$ and C_q be the two corresponding clusters that were merged. By our assumption that the clusters are correctly created before q_i arrives, we know that the walking-regions of $C_{q'}$ and C_q contain only points that do belong to them, i.e., the light-gray area in the left-hand side of Fig. 6 contains only points from $C_{q'}$ or C_q. Therefore, when $C_{q'}$ and C_q are merged and e is created, it suffices to check for the existence of points other than C_k inside the new walking region e corresponds to, which is exactly the dark-gray area in Fig. 6. Furthermore, by the dominance property stated in Lemma 6, it suffices to check those edges with positive slopes. This shows that at the end of each iteration when $\mathbf{E}$ becomes empty, no more clusters need to be merged. We have the following theorem.

Theorem 1. *Provided that the one-sided segment sweeping query can be answered in $Q(n)$ time using $P(n)$ preprocessing time and $S(n)$ storage, the time-convex hull for a given set $\mathbf{S}$ of n points under the given L_p-metric can be computed in $\mathcal{O}(n \log n + nQ(n) + P(n))$ time using $\mathcal{O}(n + S(n))$ space.*

Regarding the One-Sided Segment Sweeping Query. Below we sketch how this query can be answered efficiently in logarithmic time. Let $\mathbf{S}$ be the set of points, L be the line segment of interest, and $\mathcal{I}_L$ be the vertical strip defined by the two end-points of L. We have the following observation, which relates the query $\mathcal{Q}(L)$ to the problem of computing the upper-hull of $\mathbf{S} \cap \mathcal{I}_L$.

Lemma 10. *Let $\mathcal{I}$ be an interval, $\mathcal{C} \colon \mathcal{I} \to \mathbb{R}$ be a convex function, i.e., we have $\mathcal{C}\left(\frac{1}{2}(x_1 + x_2)\right) \geq \frac{1}{2}(\mathcal{C}(x_1) + \mathcal{C}(x_2)) \; \forall x_1, x_2 \in \mathcal{I}$, that is differentiable almost everywhere, L be a segment with slope θ_L, $-\infty < \theta_L < \infty$, and $q = (x_q, \mathcal{C}(x_q))$ be a point on the curve $\mathcal{C}$ such that*

$$\lim_{x \to x_q^-} \frac{\mathrm{d}\mathcal{C}(x)}{\mathrm{d}x} \geq \theta_L \geq \lim_{x \to x_q^+} \frac{\mathrm{d}\mathcal{C}(x)}{\mathrm{d}x}.$$

If q lies under $\overleftrightarrow{L}$, then the curve $\mathcal{C}$ never intersects L.

To help compute the upper-hull of $\mathbf{S} \cap \mathcal{I}_L$, we use a data structure due to Guibas et al [7]. For a given simple path $\mathcal{P}$ of n points with an x-sorted ordering of the points, with $\mathcal{O}(n \log \log n)$ preprocessing time and space, the upper-hull of any subpath $p \in \mathcal{P}$ can be assembled efficiently in $\mathcal{O}(\log n)$ time, represented by a balanced search tree that allows binary search on the hull edges. Note that, $q_1, q_2, \ldots, q_n$ is exactly a simple path by definition. The subpath to which $\mathcal{I}_L$

corresponds can be located in $\mathcal{O}(\log n)$ time. In $\mathcal{O}(\log n)$ time we can obtain the corresponding upper-hull and test the condition specified in Lemma 10. We conclude with the following lemma.

Lemma 11. *The* one-sided segment sweeping query *can be answered in $\mathcal{O}(\log n)$ time, where n is the number of points, using $\mathcal{O}(n \log n)$ preprocessing time and $\mathcal{O}(n \log \log n)$ space.*

5 Conclusion

We conclude with a brief discussion as well as an overview on future work. In this paper, we give an optimal algorithm for the time-convex hull in the presence of a straight-line highway under the general L_p-metric where $1 \le p \le \infty$. The structural properties we provide involve non-trivial geometric arguments. We believe that our algorithm and the approach we use can serve as a base to the scenarios for which we have a more complicated transportation infrastructure, e.g., modern city-metros represented by line-segments of different moving speeds. Furthermore, we believe that approaches supporting dynamic settings to a certain degree, e.g., point insertions/deletions, or, dynamic speed transitions, are also a nice direction to explore.

References

1. Abellanas, M., Hurtado, F., Sacristán, V., Icking, C., Ma, L., Klein, R., Langetepe, E., Palop, B.: Voronoi diagram for services neighboring a highway. Inf. Process. Lett. 86(5), 283–288 (2003)
2. Aichholzer, O., Aurenhammer, F., Palop, B.: Quickest paths, straight skeletons, and the city Voronoi diagram. In: Proceedings of SCG 2002, pp. 151–159 (2002)
3. Aloupis, G., Cardinal, J., Collette, S., Hurtado, F., Langerman, S., O'Rourke, J., Palop, B.: Highway hull revisited. Comput. Geom. Theo. Appl. 43(2), 115–130 (2010)
4. Bae, S.W., Kim, J.-H., Chwa, K.-Y.: Optimal construction of the city voronoi diagram. In: Asano, T. (ed.) ISAAC 2006. LNCS, vol. 4288, pp. 183–192. Springer, Heidelberg (2006)
5. Ben-Or, M.: Lower bounds for algebraic computation trees. In: Proceedings of STOC 1983, pp. 80–86 (1983)
6. Gemsa, A., Lee, D.T., Liu, C.-H., Wagner, D.: Higher order city voronoi diagrams. In: Fomin, F.V., Kaski, P. (eds.) SWAT 2012. LNCS, vol. 7357, pp. 59–70. Springer, Heidelberg (2012)
7. Guibas, L., Hershberger, J., Snoeyink, J.: Compact interval trees: a data structure for convex hulls. In: Proceedings of SODA 1990, pp. 169–178 (1990)
8. Hurtado, F., Palop, B., Sacristán, V.: Diagramas de Voronoi con distancias temporales. In: Actas de los VIII Encuentros de Geometra Computacional, pp. 279–288 (1999) (in Spanish)
9. Ottmann, T., Soisalon-Soininen, E., Wood, D.: On the definition and computation of rectilinear convex hulls. Information Sciences 33(3), 157–171 (1984)
10. Palop, B.: Algorithmic Problems on Proximity and Location under Metric Constraints. Ph.D thesis, Universitat Politécnica de Catalunya (2003)
11. Yu, T.-K., Lee, D.T.: Time convex hull with a highway. In: Proceedings of ISVD 2007, pp. 240–250 (2007)

Blame Trees

Erik D. Demaine[1], Pavel Panchekha[1], David A. Wilson[1], and Edward Z. Yang[2]

[1] Massachusetts Institute of Technology, Cambridge, Massachusetts
{edemaine,pavpan,dwilson}@mit.edu
[2] Stanford University, Stanford, California
ezyang@cs.stanford.edu

Abstract. We consider the problem of merging individual text documents, motivated by the single-file merge algorithms of document-based version control systems. Abstracting away the merging of conflicting edits to an external conflict resolution function (possibly implemented by a human), we consider the efficient identification of conflicting regions. We show how to implement tree-based document representation to quickly answer a data structure inspired by the "blame" query of some version control systems. A "blame" query associates every line of a document with the revision in which it was last edited. Our tree uses this idea to quickly identify conflicting edits. We show how to perform a merge operation in time proportional to the sum of the logarithms of the shared regions of the documents, plus the cost of conflict resolution. Our data structure is functional and therefore confluently persistent, allowing arbitrary version DAGs as in real version-control systems. Our results rely on concurrent traversal of two trees with short circuiting when shared subtrees are encountered.

1 Introduction

The document-level merge operation is a fundamental primitive in version control systems. However, most current implementations of this operation take linear time in the size of the document, and rely on the ability to identify the least common ancestor of the two revisions to be merged. For large documents, we can improve on this naïve bound by not spending time on non-conflicting portions of the document. More abstractly, the lowest common ancestor may not be unique, or even exist, in a fully confluent setting. In this paper, we describe the practical motivation for this problem, our model of the theoretical problem, and our solution.

Document Merge. Single-document merging forms the core of many modern version control systems, as it is critical for reconciling multiple, concurrent branches of development.

Single-document merging has been implemented in a variety of different ways by different version control systems. The most basic merge strategy is the three-way merge, as implemented by Git, Mercurial, and many other version control systems. In a three-way merge, three revisions of the file are specified: the

F. Dehne, R. Solis-Oba, and J.-R. Sack (Eds.): WADS 2013, LNCS 8037, pp. 280–290, 2013.
© Springer-Verlag Berlin Heidelberg 2013

"source" version, the "target" version, and the "base" version, which is the least common ancestor of the source and target. Any intermediate history is thrown out, and the result of the merge relies on the diffs between the base and source and between the base and target. These diffs are split into changed and unchanged segments, which are then used to build the new document.

There are a few elaborations on the basic three-way merge. In the case of multiple least common ancestors, Git will recursively merge the common ancestors together, and use those as the base version.[1] Additionally, there is some question of whether two chunks that have applied an identical change (as opposed to conflicting ones) should silently merge together: most systems opt for not reporting a conflict. This behavior has lead to some highly publicized edge cases in the merge algorithm [3].

A more sophisticated merging algorithm is implemented by Darcs. It uses whether two patches commute as the test for whether a merge conflict should be generated; and it performs the merge patch by patch. Because Darcs uses the intermediate history, this often results in a higher quality merge, but requires time at least linear in the number of patches, and can result in exponential behavior in some cases.

An interesting but largely obsolete representation for an entire history which was implemented by the SCCS and BitKeeper systems is the "Weave" [9], where every document in the repository contains all lines of text present in any revision of the document, with metadata indicating what revisions they correspond to. Merging on this representation takes time proportional to the size of the entire history.

Prior Work. Demaine, Langerman, and Price [4] considered the problem of efficient merge at the directory/file level, using confluently persistent data structures. They cite the document-level merge problem as "relatively easy to handle", assuming that the merge may take linear time. The goal of this work is to beat this bound.

In the algorithms community, merging usually refers to the combination of two sorted arrays into one sorted array. This problem can be viewed as similar, particularly if we imagine that equal-key items get combined by some auxiliary function. The standard solution to this problem (as in, e.g., mergesort) takes linear time in the input arrays. Adaptive merging algorithms [2,5,10,11] achieve the optimal bound of $\Theta(\sum_{i=1}^{k} \lg g_i)$ if the solution consists of g_1 items from the first set, then g_2 items from the second set, then g_3 items from the first set, etc. Our result is essentially a confluent data structure built around a dynamic form of this one-shot algorithm.

In this paper, we will refer repeatedly to the well-known results of the functional programming community [12], in particular confluence via purely functional data structures with path copying as the primary technique. The idea of

---

[1] This situation is rare enough in practice that very few other VCSes implement this behavior, although this strategy is reported to reduce conflicts in merges on the Linux kernel.

parametrizing an algorithm on a human-driven component is a basic technique of human-based computation; however, we do not refer to any of the results in that literature.

Theoretical Model. A natural model of a text document in a version control system is a confluently persistent sequence of characters. Persistent data structures [7, 8] preserve old versions of a document as modifications are made to a document. While a fully persistent data structure permits both queries and modifications to old versions of document, a confluently persistent structure additionally supports a merge operation. Any two documents can be merged, which means that the version dependencies can form a directed acyclic graph (DAG). Our paper presents an implementation of this data structure which admits an efficient implementation of this merge operation.

For the purposes of our treatment, we consider a more expressive model of text documents as confluently persistent sorted associative maps (dictionaries), whose values are arrays of characters, and whose keys are an ordered data type supporting constant-time comparison and a split operation with the property $a < \text{split}(a, b) < b$. (Data structures for maintaining order in this way are well known [1, 6], and also common in maintaining full persistence [7].) The user can decide to divide the document into one entry (key/value pair) per character, or one entry per line, or some other level of granularity. The supported operations are then insertion, deletion, and modification of entries in the map, which correspond to equivalent operations on characters, lines, etc., of the document.[2] These operations take one (unchanged) version as input and produce a new version (with the requested change) as output. In addition, the merge operation takes two maps (versions) and a conflict-resolution function, which takes two conflicting submaps ranging between keys i and j and combines them into a single submap ranging between keys i and j. In order to amortize some costs of our approach, we assume that conflict resolution requires at least $\Theta(|M| + |N|)$ time, where $|M|$ and $|N|$ are the number of entries in the input submaps (our bounds do not hold for a $\Theta(1)$ resolution function); in practice, the cost of conflict resolution is likely to be $\Theta(k(|M| + |N|))$, where k is the average length of the sequences of characters inside the submaps.

Entries of these maps have one extra piece of metadata: a unique ID identifying the revision that this entry was last updated. We also assume that we have a source of unique IDs (occupying $O(1)$ space each), which can be used to allocate new revision numbers for marking nodes. In practice, cryptographically secure hash functions are used to generate these IDs.

Sharing. The performance of our document merge operation depends on the underlying structure of our documents; if common and conflicting regions are interleaved $\Theta(n)$ times, then we cannot hope to do any better than a linear-time merge. Thus, we define the *disjoint shared regions* S of our sorted maps to be

---

[2] It is worth emphasizing that the keys do *not* correspond to character-indexes or line-numbers; aside from order, they are completely arbitrary.

the set of maximal disjoint ranges which have the same contents (matching keys, values, and revision ID): these can be thought as the non-conflicting regions of two documents. Note that, if two independent editors make the same revision to a document, the resulting two entries are considered in conflict, as the revision IDs will differ.

It will also be useful to refer to the *non-shared entries* $\mathcal{N}$ of two documents, which is defined to be the set of all entries in either document that are contained in no $S \in \mathcal{S}$.

Let $f(|\mathcal{N}|)$ denote the cost of conflict resolution among the $|\mathcal{N}|$ non-shared entries, which we assume to be $\Omega(|\mathcal{N}|)$. In our analyses, we will often charge the cost of traversing conflicting entries to the execution of the conflict resolution function, when considering the overall cost of a merge.

Main Result. Our main result is a functional (and thus confluently persistent) data structure supporting insert, delete, modifying, and indexing on a map of size n in $O(\log n)$ time per operation; and merging two (versions of) maps in $O(f(|\mathcal{N}|)+\sum_{S \in \mathcal{S}} \log |S|)$ time. For example, for a constant number of non-shared nodes, the merge cost is logarithmic; or more generally, for a constant number of shared regions, the merge cost is logarithmic plus the conflict resolution cost. We expect that this adaptive running time will be substantially smaller than the standard linear-time merge in most practical scenarios.

2   Blame Trees, Version 1

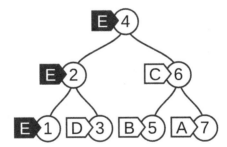

Fig. 1. An example blame tree, where the latest update was revision E made to key 1. The values on the leaves are omitted for clarity.

Blame trees represent a text document as a balanced binary search tree containing strings, augmented with length annotations, to facilitate efficient indexing into arbitrary locations of the document. (Because these annotations are not relevant for merges, we omit them from our presentation.) Blame trees are further augmented with a revision **rev** annotation, which tracks the latest revision

Tree is either:

1: NODE(rev, key, val, left, right)
2: LEAF

Fig. 2. Our trees are classic binary search trees annotated with an extra revision field, indicating the last edit which affected a node or any node in its subtree

to the data structure which affected this node. Updates generate a fresh revision for the edit and record it on all nodes they touch.

For binary trees, disjoint shared regions correspond directly to disjoint shared subtrees, e.g., the set of maximum-sized disjoint subtrees which exist identically (matching keys, values and annotations) in both trees. Any shared region S can be represented by $O(\log |S|)$ shared subtrees. To simplify analysis, we will instead consider disjoint shared subtrees $\tilde{\mathcal{S}}$, and then translate our bounds back into disjoint shared regions. A useful fact which we will refer to repeatedly is that $|\tilde{\mathcal{S}}| \leq 2 \sum_{S \in \mathcal{S}} \log |S|$.

As any modifications to a node must modify all of its parent nodes, the definition of two shared subtrees in Figure 3 is equivalent (e.g., we only need to check roots of shared subtrees for equality).

1: **function** SHARED(a, b)
2:     **return** a.key $= b$.key $\wedge$ a.rev $= b$.rev
3: **end function**

Fig. 3. The definition of two shared subtrees

When the trees in question have identical structure, merging two blame trees is trivial: traverse both structures in-order and simultaneously. Because the structures are identical, traversals will be in lock-step, and we can immediately identify a shared subtree when we first encounter it, and skipping it entirely. We pay only the cost of visiting the root of every shared subtree, so the cost of traversal is $O(|\mathcal{N}| + |\tilde{\mathcal{S}}|)$, which in particular is $O(|\mathcal{N}| + \sum_{S \in \mathcal{S}} \log |S|)$. The overall cost is $O(f(|\mathcal{N}|) + \sum_{S \in \mathcal{S}} \log |S|)$, with the cost of traversing conflicting nodes charged to the conflict resolution function f.

This naïve traversal doesn't work, however, when the two tree have differing structures. Additionally, the conflict resolution may return a new subtree to be spliced in, and thus we need to manage rebalancing the resulting trees. Consequently, our general strategy for merging balanced search trees of different shapes will be to identify the disjoint shared subtrees of the two trees, split the trees into conflicting and shared regions, resolve the conflicting regions, and concatenate the trees back together. The core of our algorithm is this:

Lemma 1. *It is possible to determine the disjoint shared subtrees $\tilde{\mathcal{S}}$ of two balanced blame trees in time $O(|\mathcal{N}| + \sum_{S \in \tilde{\mathcal{S}}} \log |S|) \subseteq O(|\mathcal{N}| + \sum_{S \in \mathcal{S}} \log^2 |S|)$.*

```
1: function INORDER(a)
2:     if a is NODE then
3:         INORDER(a.left)
4:         skip ← yield a                      ▷ Suspend the coroutine to visit the node
5:         if skip ≠ SKIP then
6:             INORDER(a.right)
7:         end if
8:     end if
9: end function
```

Fig. 4. In-order traversal as a coroutine. Execution of this function proceeds normally until the **yield** a statement is reached; at this point, execution of the function is suspended and the value a is returned to the caller of the coroutine. When the coroutine is initially invoked, it returns a resumption continuation, which the caller can use to resume the execution of the coroutine. In our case, the resumption continuation requires the caller to provide a value $skip$, which indicates whether or not to skip traversal of the right subtree.

```
 1: function TRAVERSE(a, b)
 2:     n_a, k_a ← INORDER(a)
 3:     n_b, k_b ← INORDER(b)
 4:     while n_a, n_b not NULL do
 5:         if SHARED(n_a, n_b) then                    ▷ n_a = n_b
 6:             yield n_a                    ▷ Add n_a to list of shared subtrees
 7:             n_a ← k_a(SKIP)                    ▷ Skip the right subtree
 8:             n_b ← k_b(SKIP)
 9:         else if n_a.key ≤ n_b.key then
10:             n_a ← k_a(NOSKIP)
11:         else if n_a.key > n_b.key then
12:             n_b ← k_b(NOSKIP)
13:         end if
14:     end while
15: end function
```

Fig. 5. Concurrent in-order traversal of two trees which reports shared subtrees. The algorithm begins by initiating in-order traversal on a and b, retrieving the left-most nodes n_a and n_b and the resumption continuations of the traversals k_a and k_b. The algorithm then repeatedly checks for shared subtrees, advancing the traversal with the lowest key, skipping right subtrees when a shared subtree is found.

Proof. Perform an in-order traversal concurrently on both trees, advancing the traversal on the tree with the lower key. This traversal can easily be expressed as a pair of coroutines, as seen in Figure 4 and Figure 5 and illustrated in Figure 6. If the two nodes being traversed are roots of shared subtrees, record the node as a shared subtree and skip traversal of the right child of both trees; continue traversal from the parent.

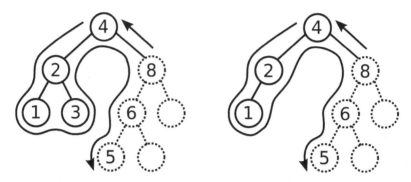

Fig. 6. An illustration of concurrent in-order traversal, where the dashed portion of the tree is shared. Both traversals start at 1 and progressively increase (the left tree stepping twice at key 3) until node 5 is reached: at this point, we discover the tree is shared. At this point, we walk back up the tree (5, 6, 8); each node is shared, so the right subtrees are skipped.

The resulting list r of shared subtrees will contain "runs" of shared nodes where $r[i] = r[i + 1]$.left (that is, the parent of the shared subtree was also shared). Discard all shared subtrees except the final subtree of each run (as any such tree i is strictly contained in $i + 1$), and return the resulting list.

Because in-order traversal returns nodes with monotonically increasing keys, it is easy to see that, without short circuiting, this traversal will discover all shared subtrees. Furthermore, because we only skip subtrees of shared subtrees (which must also be shared), it is easy to see that no shared subtrees are skipped.

An ordinary in-order traversal will take $O(n)$ time, so we need to show that with our short-circuiting we spend only $O(\log |S|)$ per shared subtree S. Suppose that we have accessed the leftmost node of a shared subtree, with cost $O(\log |S|)$; this must occur before any other nodes of the shared subtree are traversed, as we are doing in-order traversal. This node is itself a shared subtree (though not the maximal node), and we will short circuit to the parent. This will occur repeatedly for the entire path contained within the maximal shared subtree, so after another $O(\log |S|)$ steps, we reach the shared node rooting the maximal shared subtree, and continue traversal of the rest of the tree. Clearly only $O(\log |S|)$ total is spent through a shared subtree, for $O(\sum_{S \in \mathcal{S}} \log S)$.

To translate this bound from shared subtrees into shared regions, we observe that for any shared region S, the shared subtrees $\tilde{S}$ have the following property $O(\sum_{s \in \tilde{S}} \log |s|) \subseteq O(\log^2 |S|)$ (recalling that only $O(\log |S|)$ subtrees are necessary to encode a region S, each with maximum size $|S|$). The bound $O(|\mathcal{N}| + \sum_{S \in \mathcal{S}} \log^2 |S|)$ follows.   □

From here, it is easy to implement merge in general:

Theorem 1. *Given the ability to split and concatenate a sequence M of blame trees in $O(\sum_M \log |M|)$ time, it is possible to merge two balanced blame trees with shared disjoint regions $\mathcal{S}$ in time $O(f(|\mathcal{N}|) + \sum_{S \in \mathcal{S}} \log^2 |S|)$.*

Proof. We need to show that the cost of traversal and conflict resolution dominates the cost of splitting and concatenating trees; that is, that $O(\sum_M \log |M|) \subseteq O(f(|\mathcal{N}|) + \sum_{S \in \tilde{S}} \log |S|)$. We split this bound into shared regions $\mathcal{S}$ and unshared regions $\mathcal{U}$. For $\mathcal{S}$, the cost contributed by each shared subtree $|S|$ is a $\log |S|$ factor better than the cost of traversal as stated in the lemma. For $\mathcal{U}$, we observe $\sum_{U \in \mathcal{U}} \log |U| \leq |\mathcal{N}|$, so the cost of splitting and concatenating unshared nodes can be charged to conflict resolution.   □

3   Faster Traversal

We now show how to achieve a logarithmic speedup when considering a specific type of balanced binary tree, namely red-black trees. Our traversal time improves from $O(|\mathcal{N}| + \sum_{S \in \mathcal{S}} \log^2 |S|)$ to $O(|\mathcal{N}| + \sum_{S \in \mathcal{S}} \log |S|)$. This method relies on level-order traversal. We first describe the algorithm for perfect binary trees, and then sketch how to apply this to red-black trees, which are not perfectly balanced, but are perfectly balanced on black nodes.

Lemma 2. *It is possible to determine the shared subtrees $\tilde{S}$ of two perfect binary trees annotated with precise max-depth in time $O(|\mathcal{N}| + |\tilde{S}|)$, e.g. $O(|\mathcal{N}| + \sum_{S \in \mathcal{S}} \log |S|)$.*

Proof. We maintain two queues, one for tree A and one for tree B. Each queue starts containing the root node of its respective tree. Without loss of generality, assume the max-depth of the two trees are equal (if they are not, recursively deconstruct the tree until you have a forest of correct depth; the nodes removed by the deconstruction are guaranteed not to be shared, because they have the wrong height.)

Claim. The list of elements extracted from each queue consists of the nodes of depth d whose parents were non-shared and had depth $d + 1$, sorted in order of their key.

   If the claim is true, we can perform a merge of sorted lists in linear time, and for any matching keys we check if the nodes are shared subtrees. (This is sufficient, as nodes with different max-depths or non-equal keys cannot be shared subtrees.) Finally, for all non-shared nodes add their children to their corresponding queue in order of the lists and repeat. The full algorithm is presented in Figure 8 and illustrated in Figure 7.

   We first show that our claim is true by induction on d. The base case is trivial. Suppose that the algorithm has fulfilled the claim up until the current round d. We consider the possible sources of nodes of depth d; by the properties of a perfect binary tree, the parent of a node with max-depth d must have max-depth $d + 1$. Furthermore, the node would not have been added if the parent were shared, as desired.

   Observing that only non-shared nodes, or the roots of maximal shared subtrees ever enter the queue, if queue pop and push operations take $O(1)$, then time

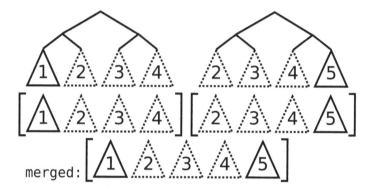

merged:

Fig. 7. An illustration of level-order traversal. Given the two trees shown above, this diagram shows the state of the two queues (in brackets) after finishing traversing the second level of the tree, and the resulting merged list of trees. Subsequent traversals will only traverse over subtrees 1 and 5. (As an optimization, we can note when subtrees are disjoint and immediately mark them as conflicts.)

```
 1: function PUSHCHILDRENPOP(Q)
 2:     x ← POP(Q)
 3:     PUSH(Q, x.left)
 4:     PUSH(Q, x.right)
 5: end function
 6: function LEVELTRAVERSE(a, b)
 7:     Q_a, Q_b ← SINGLETON(a), SINGLETON(b)
 8:     while ¬(EMPTY(Q_a) ∨ EMPTY(Q_b)) do
 9:         n_a, n_b ← PEEK(Q_a), PEEK(Q_b)
10:         if SHARED(n_a, n_b) then
11:             yield n_a
12:             POP(Q_a)
13:             POP(Q_b)
14:         else if n_a.depth > n_b.depth then
15:             PUSHCHILDRENPOP(Q_a)
16:         else if n_a.depth < n_b.depth then
17:             PUSHCHILDRENPOP(Q_b)
18:         else if n_a.key ≤ n_b.key then
19:             PUSHCHILDRENPOP(Q_a)
20:         else if n_a.key > n_b.key then
21:             PUSHCHILDRENPOP(Q_b)
22:         end if
23:     end while
24: end function
```

Fig. 8. Level-order traversal of two perfect trees

bound $O(|\mathcal{N}| + |\tilde{\mathcal{S}}|)$ follows easily; the alternate formulation of the bound falls easily out of the fact that any $S \in \mathcal{S}$ can only contribute $\log |S|$ roots of shared subtrees. □

In general, most practical self-balancing binary search trees will not be perfect. However, in the case of red-black trees, the number of black nodes down any path of the tree is constant. So we can adapt the algorithm for perfect trees to only count black nodes towards depth: when we would add a red node into the queue, we instead push its two black children. This means that we do not ever check red nodes to see if they are shared subtrees, but this only adds a constant factor extra time on the analysis. Note that this technique does not work if the tree is not perfectly balanced in some fashion: without perfect balance, we will encounter nodes whose max-depths are much lower than the current max-depth which still must be queued. Switching our queue to a priority queue to accommodate these nodes would result in a logarithmic slowdown, destroying our bound.

Our final result follows easily:

Theorem 2. *It is possible to merge two red-black blame trees with shared disjoint regions $\mathcal{S}$ in time $O(f(|\mathcal{N}|) + \sum_{S \in \mathcal{S}} \log |S|)$.*

Proof. The analysis proceeds identically as our previous Theorem 1, except that the cost of splitting and concatenating the red-black trees is the same as the traversal in the case of shared regions $\mathcal{S}$. □

Acknowledgments. This work began during an open-problem solving session for MIT's Advanced Data Structures class (6.851) in Spring 2012. Thanks to Anders Kaseorg, Andrea Lincoln, and any unnamed participants of the session for contributing to this solution. This research is partially funded by the DARPA Clean-Slate Design of Resilient, Adaptive, Secure Hosts (CRASH) program, BAA-10-70, under contract #N66001-10-2-4088 (*Bridging the Security Gap with Decentralized Information Flow Control*), and by MADALGO — Center for Massive Data Algorithmics — a Center of the Danish National Research Foundation.

References

1. Bender, M.A., Cole, R., Demaine, E.D., Farach-Colton, M., Zito, J.: Two simplified algorithms for maintaining order in a list. In: Möhring, R.H., Raman, R. (eds.) ESA 2002. LNCS, vol. 2461, pp. 152–164. Springer, Heidelberg (2002)
2. Carlsson, S., Levcopoulos, C., Petersson, O.: Sublinear merging and natural mergesort. Algorithmica 9, 629–648 (1993)
3. Cohen, B.: Git can't be made consistent (April 2011), `http://bramcohen.livejournal.com/74462.html`
4. Demaine, E.D., Langerman, S., Price, E.: Confluently persistent tries for efficient version control. Algorithmica 57(3), 462–483 (2010)
5. Demaine, E.D., López-Ortiz, A., Munro, J.I.: Adaptive set intersections, unions, and differences. In: Proceedings of the 11th Annual ACM-SIAM Symposium on Discrete Algorithms, San Francisco, California, pp. 743–752 (January 2000)

6. Dietz, P.F., Sleator, D.D.: Two algorithms for maintaining order in a list. In: Proceedings of the 19th Annual ACM Symposium on Theory of Computing, New York City, pp. 365–372 (May 1987)

7. Driscoll, J.R., Sarnak, N., Sleator, D.D., Tarjan, R.E.: Making data structures persistent. Journal of Computer and System Sciences 38(1), 86–124 (1989)

8. Fiat, A., Kaplan, H.: Making data structures confluently persistent. In: Proceedings of the 12th Annual Symposium on Discrete Algorithms, Washington, DC, pp. 537–546 (January 2001)

9. Hudson, G.: Notes on keeping version histories of files (October 2002), http://web.mit.edu/ghudson/thoughts/file-versioning

10. Mehlhorn, K.: Data Structures and Algorithms. Sorting and Searching, vol. 1, pp. 240–241. Springer (1984)

11. Moffat, A., Petersson, O., Wormald, N.C.: A tree-based Mergesort. Acta Informatica 35(9), 775–793 (1998)

12. Okasaki, C.: Purely functional data structures. Cambridge University Press, New York (1998)

Plane 3-trees: Embeddability and Approximation
(Extended Abstract)

Stephane Durocher* and Debajyoti Mondal**

Department of Computer Science, University of Manitoba
{durocher,jyoti}@cs.umanitoba.ca

Abstract. We give an $O(n \log^3 n)$-time linear-space algorithm that, given a plane 3-tree G with n vertices and a set S of n points in the plane, determines whether G has a point-set embedding on S (i.e., a planar straight-line drawing of G where each vertex is mapped to a distinct point of S), improving the $O(n^{4/3+\varepsilon})$-time $O(n^{4/3})$-space algorithm of Moosa and Rahman. Given an arbitrary plane graph G and a point set S, Di Giacomo and Liotta gave an algorithm to compute 2-bend point-set embeddings of G on S using $O(W^3)$ area, where W is the length of the longest edge of the bounding box of S. Their algorithm uses $O(W^3)$ area even when the input graphs are restricted to plane 3-trees. We introduce new techniques for computing 2-bend point-set embeddings of plane 3-trees that takes only $O(W^2)$ area. We also give approximation algorithms for point-set embeddings of plane 3-trees. Our results on 2-bend point-set embeddings and approximate point-set embeddings hold for partial plane 3-trees (e.g., series-parallel graphs and Halin graphs).

1   Introduction

A *planar drawing* of a graph G is an embedding (i.e., a mapping) of G onto the Euclidean plane $\mathbb{R}^2$, where each vertex in G is assigned a unique point in $\mathbb{R}^2$ and each edge in G is a simple curve in $\mathbb{R}^2$ joining the points corresponding to its endvertices such that no two curves intersect except possibly at their endpoints. A graph is *planar* if it has a planar drawing. A *straight-line drawing* of a planar graph is a planar drawing, where each edge is drawn as a straight line segment. The straight-line drawing style is popular since it naturally produces drawings that are easier to read and to display on smaller screens [1,2]. To meet the requirements of different practical applications, researchers have examined the straight-line drawing problem under various constraints, e.g., when the vertices are constrained to be placed on a set of pre-specified locations [3,4]. If the pre-specified locations for placing the vertices of the input graph are points on the Euclidean plane, then we call the problem a point-set embedding problem. Such

---

* Work of the author is supported in part by the Natural Sciences and Engineering Research Council of Canada (NSERC).
** Work of the author is supported in part by a University of Manitoba Graduate Fellowship.

F. Dehne, R. Solis-Oba, and J.-R. Sack (Eds.): WADS 2013, LNCS 8037, pp. 291–303, 2013.

problems have applications in VLSI circuit layout, where different circuits need to be mapped onto a fixed printed circuit board, simultaneous display of different social and biological networks, and construction of a desired network among a set of fixed locations. Formally, a *point-set embedding* of a *plane graph G* (i.e., a fixed combinatorial planar embedding of G) with n vertices on a set S of n points is a straight-line drawing of G, where the vertices are placed on distinct points of S.

Point-Set Embeddings. In 1994, Ikebe et al. [5] gave an $O(n^2)$-time algorithm to embed any tree with n vertices on any set of n points in *general position*, i.e., no three points are collinear. Later, Bose et al. [6] devised a divide and conquer algorithm that runs in $O(n \log n)$ time. In 1996, Castañeda and Urrutia [7] gave an $O(n^2)$-time algorithm to construct point-set embeddings of maximal outerplanar graphs. Later, Bose [3] improved the running time of their algorithm to $O(n \log^3 n)$ using a dynamic convex hull data structure. In the same paper Bose posed an open problem that asks to determine the time complexity of testing the point-set embeddability for planar graphs. In 2006, Cabello [4] proved the problem to be NP-complete for graphs that are 2-connected and 2-outerplanar. The problem remains NP-complete for 3-connected planar graphs [8], even when the treewidth is constant [9].

In the last few years researchers have examined the point-set embeddability problem restricted to *plane 3-trees* (also known as stacked polytopes, Apollonian networks, and maximal planar graphs with treewidth three) because of their wide range of applications in many theoretical and applied fields [10]. Nishat et al. [11] first gave an $O(n^2)$-time algorithm for deciding point-set embeddability of plane 3-trees, and proved an $\Omega(n \log n)$-time lower bound. Later, Durocher et al. [12] and Moosa and Rahman [13] independently improved the running time to $O(n^{4/3+\varepsilon})$, for any $\varepsilon > 0$. Since $\Omega(n^{4/3})$ is a lower bound on the worst-case time complexity for solving various geometric problems [14], it may be natural to accept the possibility that the $O(n^{4/3+\varepsilon})$-time algorithm could be asymptotically optimal. In fact, Moosa and Rahman mention that an $o(n^{4/3})$-time algorithm seems unlikely using currently known techniques. However, in this paper we prove that the $\Omega(n \log n)$ lower bound is nearly tight, giving an $O(n \log^3 n)$-time algorithm for deciding point-set embeddability of plane 3-trees.

Universal Point Set. Observe that a planar graph may not always admit point-set embedding on a given point set. Attempts have been made at constructing a set S of $k \geq n$ points such that every planar graph with n vertices admits a point-set embedding on a subset of S [15,16]. Such a point set that *supports* all planar graphs with n vertices is called a *universal point set for n*. A long standing open question in graph drawing asks to design a set of $O(n)$ points that is universal for all planar graphs with n vertices [15]. Recently, Everett et al. [16] have designed a 1-*bend universal point set* S_n for planar graphs with n vertices, i.e., every planar graph with n vertices admits a straight-line drawing on S_n such that each vertex is mapped to a distinct point and each edge is drawn as a chain of at most two straight line segments.

The point-set embeddability problem seems to have close relation with the universal point set problem. Castañeda and Urrutia [7] proved that any set of n points in general position is universal for all outerplanar graphs with n vertices. Later, Kaufmann and Wiese [17] proved that any set S of n points is 2-*bend universal* for n (i.e., every planar graph with n vertices admits a straight-line drawing on S such that each vertex is mapped to a distinct point and each edge is drawn as a chain of at most three straight line segments). However, the area required for the drawing could be exponential in W, where W is the length of the side of the smallest axis-parallel square that encloses S. Di Giacomo and Liotta [18, Theorem 7] showed that using the concept of monotone topological book embeddingone can reduce the area requirement to $O(W^3)$. Even when restricted to simpler classes of graphs (e.g., series parallel graphs or plane 3-trees), the technique of Di Giacomo and Liotta is the best known, which still requires $O(W^3)$ area. In this paper, we contribute a new technique that uses only $O(W^2)$ area to compute 2-bend point set embeddings of plane 3-trees, and hence also for partial plane 3-trees (e.g., series-parallel graphs and Halin graphs).

Approximate Point-Set Embeddings. Although any set of n points in general position is universal for n-vertex outerplanar graphs [7], a plane 3-tree with n vertices may not admit a point-set embedding on a given set of n points [11]. On the other hand, while allowing two bends per edge, any set of n points in general position is 2-bend universal for plane 3-trees. Due to this apparent difficulty of defining algorithms that simultaneously minimize area, the number of bends, and running time, we consider algorithms that provide approximate solutions, that is, at least a fraction ρ of the vertices of the input graph are mapped to distinct points of the given point set. Specifically, if the input points are in general position, then we prove that the point-set embeddability of plane 3-trees is approximable with factor $\Omega(1/\sqrt{n})$.

2 Faster Point-Set Embeddings of Plane 3-Trees

In this section we give an $O(n \log^3 n)$-time algorithm for deciding point-set embeddability of plane 3-trees. Before going into details, we review a few definitions.

A *plane 3-tree* G with $n \geq 3$ vertices is a triangulated plane graph such that if $n > 3$, then G contains a vertex whose deletion yields a plane 3-tree with $n-1$ vertices. Let r, s, t be a cycle of three vertices in G. By G_{rst} we denote the subgraph induced by r, s, t and the vertices that lie interior to the cycle. Every plane 3-tree G with $n > 3$ vertices contains a vertex that is the common neighbor of all the three outer vertices of G. We call this vertex the *representative vertex* of G. Let p be the representative vertex of G and let a, b, c be the three outer vertices of G in clockwise order. Then each of the subgraphs G_{abp}, G_{bcp} and G_{cap} is a plane 3-tree. Let S be a set of n points in the plane. Let p, q and r be three points that do not necessarily belong to S. Then $S(pqr)$ consists of the points of S that lie either on the boundary or in the interior of the triangle pqr.

Overview of Known Algorithms. Let G be a plane 3-tree with n vertices, and let a, b, c and p be the three outer vertices and the representative vertex of G, respectively. Nishat et al. [11]'s algorithm is as follows.

Step 1. If the number of points on the boundary of the convex hull C of S is not exactly three, then G does not admit a point-set embedding on S. Otherwise, let x, y, z be the points on C.

Step 2. For each of the possible six different mappings of the outer vertices a, b, c to the points x, y, z, execute Step 3.

Step 3. Let n_1, n_2 and n_3 be the number of vertices of G_{abp}, G_{bcp} and G_{cap}, respectively. Without loss of generality assume that the current mapping of a, b and c is to x, y and z, respectively. Find the unique mapping of the representative vertex p of G to a point $w \in S$ such that the triangles xyw, yzw and zxw properly contain exactly n_1, n_2 and n_3 points, respectively. If no such mapping of p exists, then G does not admit a point-set embedding on S for the current mapping of a, b, c; hence go to Step 2 for the next mapping. Otherwise, recursively compute point-set embeddings of G_{abp}, G_{bcp} and G_{cap} on $S(xyw), S(yzw)$ and $S(zxw)$, respectively. See Figures 1(a)–(d).

Observe that the recurrence relation for the time taken in Step 3 is $T(n) = T(n_1) + T(n_2) + T(n_3) + \mathcal{T}$, where $\mathcal{T}$ denotes the time required to find the mapping of the representative vertex. The algorithm of Nishat et al. [11] preprocesses the set S in $O(n^2)$ time so that the computation for the mapping of a representative vertex takes $O(n)$ time. Hence $\mathcal{T} = O(n)$ and the overall time complexity becomes $O(n^2)$. Moosa and Rahman [13] used a binary search technique with the help of a triangular range search data structure of Chazelle et al. [19] to obtain $\mathcal{T} = \min\{n_1, n_2, n_3\} \cdot n^{1/3+\varepsilon}$ and $T(n) = O(n^{4/3+\varepsilon})$. Durocher et al. [12] use the same idea, but instead of a binary search they use a randomized search.

Embedding Plane 3-Trees in $O(n \log^3 n)$ time. We speed up the mapping of the representative vertex as follows. We first select $O(\min\{n_1, n_2, n_3\})$ points interior to the triangle xyz in $O(\min\{n_1 + n_2, n_2 + n_3, n_1 + n_3\} \log^2 n)$ time using a dynamic convex hull data structure. We prove that these are the only candidates for the mapping of the representative vertex. We then make some non-trivial observations to test and compute a mapping for the representative

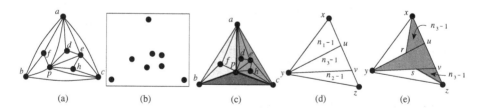

(a)   (b)   (c)   (d)   (e)

Fig. 1. (a) A plane 3-tree G. (b) A point set S. (c) A valid mapping of the representative vertex of G, and the recursive computation of the three subproblems. (d)–(e) Illustration for the lines uy, vy, xr and zs. The region of interest is shown in gray.

vertex in $O(\min\{n_1, n_2, n_3\})$ time. Hence we obtain $\mathcal{T} = O(\min\{n_1 + n_2, n_2 + n_3, n_1 + n_3\} \log^2 n)$ and a running time of $T(n) = O(n \log^3 n)$, which dominates the $O(n \log^2 n)$ time for building the initial dynamic convex hull data structure.

In the following we use three lemmas to obtain our main result. Lemma 1 selects a region R containing the candidate points inside the triangle xyz. Lemma 2 reduces the problem of finding a mapping inside the triangle xyz to the problem of finding a point satisfying specific criteria inside R. Lemma 3 gives an efficient technique to find such a point. Finally, we use these lemmas to obtain a mapping for the representative vertex in $O(\min\{n_1 + n_2, n_2 + n_3, n_1 + n_3\} \log^2 n)$ time.

Without loss of generality assume that $n_3 \leq n_2 \leq n_1$. Observe that $n_1 + n_2 + n_3 - 5 = n$. Let S be a set of n points in general position such that the convex hull of S contains exactly three points x, y, z on its boundary. Without loss of generality assume that the vertices outer vertices a, b, c are mapped to the points x, y, z, respectively.

Let u and v be two points on the straight line segment xz such that $|S(uxy)| = n_1 - 1$ and $|S(vzy)| = n_2 - 1$, as shown in Figure 1(d). It is straightforward to verify that if a *valid mapping* for the representative vertex exists (i.e, there exists a point $w \in S$ such that $|S(wxy)| = n_1$, $|S(wyz)| = n_2$ and $|S(wzx)| = n_3$), then the corresponding point (i.e., the point w) must lie inside $S(uvy)$. Let r and s be two points on the straight line segments uy and vy, respectively, such that $|S(rux)| = |S(svz)| = n_3 - 1$. We call the region defined by the simple polygon x, u, v, z, s, y, r, x the *region of interest*. An example is shown in Figures 1(e). We will use the following lemma whose proof is omitted due to space constraints.

Lemma 1. *If there exists a point $w \in S$ that corresponds to a valid mapping for the representative vertex of G, then the straight line segments wx, wy and wz lie inside the region of interest R. Moreover, the number of points in R that belong to S is $O(n_3)$, and the following properties hold.*
(a) If the points s, y, z (respectively, points r, x, y) are distinct, then $|S(syz)| = n_2 - n_3 + 2$ (respectively, $|S(rxy)| = n_1 - n_3 + 2$).
(b) Otherwise, point s (respectively, point r) coincides with y (respectively, y) and $|S(syz)| = 2$ (respectively, $|S(rxy)| = 2$).

Let $S' \subseteq S$ be the set that consists of the points lying on the boundary of R and the points lying in the proper interior of R. We call S' the *set of interest*. By Lemma 1, $|S'| = O(n_3)$. We reduce the problem of finding a valid mapping in S to the problem of finding a point with certain properties in S', as shown in the following lemma. We omit its proof due to space constraints.

Lemma 2. *There exists a valid mapping for the representative vertex of G in S if and only if there exists a point $w' \in S'$ such that $|S'(w'yz)| = n_2 - |S(yzs)| + 3$, $|S'(w'xy)| = n_1 - |S(xyr)| + 3$ and $|S'(w'xz)| = n_3$.*

Since a valid mapping for the representative vertex is unique, w' must be unique. We call the point w' the *principal point* of S'. Observe that this principal point corresponds to the valid mapping of the representative vertex of G in S.

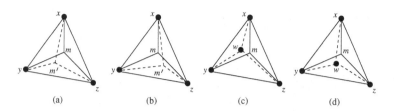

Fig. 2. Illustration for the proof of Lemma 3, where $\{m, m'\} \cap S = \varnothing$ and $\{x, y, z, w\} \subset S$

Lemma 3. *Let S be a set of $t \geq 4$ points in general position such that the convex hull of S is a triangle xyz. Let i, j, k be three non-negative integers, where $i \geq 3, j \geq 3$ and $k = t + 5 - i - j$. Then we can decide in $O(t)$ time whether there exists a point $w \in S$ such that $|S(wxy)| = i, |S(wyz)| = j$ and $|S(wxz)| = k$, and compute such a point if it exists.*

Proof. Consider first a variation of the problem, where we want to construct a point $m \notin S$ interior to xyz such that $|S(mxy)| = i + 1, |S(myz)| = j - 1$ and $|S(mxz)| = k - 1$. Steiger and Streinu [20] proved the existence of m and gave an $O(t)$-time algorithm to find m. If there exists a point $w \in S$ such that $|S(wxy)| = i, |S(wyz)| = j$ and $|S(wxz)| = k$, then it is straightforward to observe that there exists a point $m \notin S$ interior to xyz such that $|S(mxy)| = i+1, |S(myz)| = j-1$ and $|S(mxz)| = k - 1$. We now prove that the existence of m implies a unique partition of S. Hence we can efficiently test whether w exists.

We claim that if there exists a point $m' \neq m$, where $m' \notin S$, such that $|S(m'xy)| = i + 1, |S(m'yz)| = j - 1$ and $|S(m'xz)| = k - 1$, then the sets $S(m'xy), S(m'yz)$ and $S(m'xz)$ must coincide with the sets $S(mxy), S(myz)$ and $S(mxz)$. To verify the claim assume without loss of generality that $m' \in S(myz)$. Since the triangle $m'yz$ lies interior to the triangle myz, the sets $S(m'yz)$ and $S(myz)$ must be identical. On the other hand, either the triangle mxz lies interior to the triangle $m'xz$, or the triangle mxy lies interior to the triangle $m'xy$, as shown in Figures 2(a)–(b). Therefore, either the sets $S(mxz)$ and $S(m'xz)$, or the sets $S(mxy)$ and $S(m'xy)$ must be identical. Consequently, the remaining pair of sets must also be identical.

Observe that if the point $w \in S$ we are looking for exists, then w must lie interior to $S(mxy)$, as shown in Figure 2(c). Otherwise, if $w \in S(myz)$ (respectively, $w \in S(mxz)$), then $|S(myz)| \geq |S(wyz)| = j$ (respectively, $|S(mxz)| \geq |S(wxz)| = k$), which contradicts our initial assumption that $|S(myz)| = j - 1$ (respectively, $|S(mxz)| = k - 1$). Figure 2(d) depicts such a scenario. If w exists, then the convex hull of $S(mxy)$ must be a triangle xym'', where $m'' \in S(mxy)$. If $|S(m''xy)| = i, |S(m''yz)| = j$ and $|S(m''xz)| = k$, then m'' is the required point w. Otherwise, no such w exists.

We can test whether the convex hull of $S(mxy)$ is a triangle in $O(t)$ time (e.g., find the leftmost point a, the rightmost point b and the point c with the largest perpendicular distance to the line determined by the line segment ab, and then test whether triangle abc contains all the points). It is also straightforward to compute the values $|S(m''xy)|, |S(m''yz)|$ and $|S(m''xz)|$ in $O(t)$ time.   □

Given the set of interest $S' \subseteq S$, we use Lemmas 2 and 3 to find the principal point $w' \in S'$ in $O(n_3)$ time. Observe that this principal point corresponds to the valid mapping of the representative vertex of G in S. We now show how to compute the set S' in $O((n_2 + n_3) \log^2 n)$ time using the dynamic planar convex hull data structure of Overmars and van Leeuwen [21], which supports a single update (i.e., a single insertion or deletion) in $O(\log^2 n)$ time.

Step A. Assume that the points of S are placed in a dynamic convex hull data structure $\mathcal{D}$. We recursively delete the neighbor of y on the boundary of the convex hull of S starting from z in anticlockwise order. After deleting $n_2 - 2$ points, we insert all the deleted points into a new dynamic convex hull data structure $\mathcal{D}'$. We then insert a copy of y into $\mathcal{D}'$. Recall u and v from Figure 1(e). Observe that all the points of $S(vyz)$ are placed in $\mathcal{D}'$. In a similar way we construct another dynamic convex hull data structure $\mathcal{D}''$ that maintains all the points of $S(uvy)$. Consequently, $\mathcal{D}$ now only maintains the points of $S(uxy)$. Since a single insertion or deletion takes $O(\log^2 n)$ time, all the above $O(n_2 + n_3)$ insertions and deletions take $O((n_2 + n_3) \log^2 n)$ time in total.

Step B. We now construct two other dynamic convex hull data structures $\mathcal{D}_1$ and $\mathcal{D}_2$ using $\mathcal{D}$ and $\mathcal{D}'$ such that they maintain the points of $S(rux)$ and $S(svz)$, respectively. Since $|S(rux)| + |S(svz)| = O(n_3)$, this takes $O(n_3 \log^2 n)$ time.

Step C. We construct the point set S' using the points maintained in $\mathcal{D}'', \mathcal{D}_1$ and $\mathcal{D}_2$, which also takes $O(n_3 \log^2 n)$ time. In similar way we can restore the original point set S and the initial data structure $\mathcal{D}$ in $O((n_2 + n_3) \log^2 n)$ time.

The time for the construction of S' using Steps A–C is $O((n_2 + n_3) \log^2 n)$, which dominates the time required for the computation of the valid mapping of the representative vertex p. Let w be the point that corresponds to the valid mapping. We now need to construct the point sets $S(wxy), S(wyz)$ and $S(wzx)$ for recursively testing the point-set embeddability of G_{abp}, G_{bcp} and G_{cap}, respectively. We can construct $S(wxy), S(wyz)$ and $S(wzx)$ and their corresponding dynamic convex hull data structures in $O((n_2 + n_3) \log^2 n)$ time as follows. Let l be the point of intersection of the straight lines determined by the line segments wy and xz. First construct the set $S(lyz)$ and then modify it to obtain the sets $S(wyz)$ and $S(lwz)$, which takes $O((n_2 + n_3) \log^2 n)$ time. Now modify the set $S(lxy)$ to construct the set $S(lwx)$, and then use the sets $S(lwx)$ and $S(lwz)$ to construct $S(wzx)$, which takes $O(n_3 \log^2 n)$ time. Observe that after the modification of the set $S(lxy)$, we are left with the set $S(wxy)$.

We now show that the total time taken is $T(n) \leq dn \log^3 n$, for some constant d, as follows. There exists $c > 0$ such that for all $d \geq c$,

$$T(n) = T(n_1) + T(n_2) + T(n_3) + O((n_2 + n_3) \log^2 n)$$
$$\leq dn_1 \log^3 n_1 + dn_2 \log^3 n_2 + dn_3 \log^3 n_3 + c(n_2 + n_3) \log^2 n$$
$$\leq dn_1 \log^3 n + dn_2 \log^2 n \log \tfrac{n}{2} + dn_3 \log^2 n \log \tfrac{n}{2} + c(n_2 + n_3) \log^2 n$$
$$= dn_1 \log^3 n + dn_2 \log^2 n(\log n - 1) + dn_3 \log^2 n(\log n - 1) + c(n_2 + n_3) \log^2 n$$
$$= d(n_1 + n_2 + n_3) \log^3 n - (d - c)(n_2 + n_3) \log^2 n$$
$$\leq dn \log^3 n.$$

Observe that the construction of the initial data structure $\mathcal{D}$ takes $O(n \log^2 n)$ time, which is dominated by $T(n)$. The dynamic planar convex hull of Brodal and Jacob [22] takes amortized $O(\log n)$ time per update. Therefore, using their data structure instead of Overmars and van Leeuwen's data structure [21] we can improve the expected running time of our algorithm. Since the algorithms of [21,20] take linear space, the space complexity of our algorithm is $O(n)$.

Theorem 1. *Given a plane 3-tree G with n vertices and a set S of n points in general position in $\mathbb{R}^2$, we can decide the point-set embeddability of G on S in $O(n \log^3 n)$ time and $O(n)$ space, and compute such an embedding if it exists.*

Under the assumption that the algorithms of Overmars and van Leeuwen [21] and Steiger and Streinu [20] can handle degenerate cases, it is straightforward to modify our algorithm for the case when the input points are not necessarily in general position.

3 Universal Point Set for Plane 3-Trees

In this section we give an algorithm to compute 2-bend point-set embeddings of plane 3-trees on a set of n points in general position in $O(W^2)$ area, where W is the length of the side of the smallest axis-parallel square that encloses S.

We describe an outline of the algorithm. Given a plane 3-tree G and a set of points S in general position, we first construct a straight-line drawing Γ of G such that every point of S other than a pair of points on the convex hull of S lies in the proper interior of some distinct inner face in Γ, as shown in Figure 3(b). While constructing Γ, we compute a bijective function ϕ from the vertices of Γ to the points of S. We then extend each edge (u, v) in Γ using two bends to place the vertices u and v onto the points $\phi(u)$ and $\phi(v)$, respectively, as shown in Figure 3(c). We prove that Γ and ϕ maintain certain properties so that the resulting drawing Γ' remains planar.

In the following we describe the algorithm in detail. Let H be the convex hull of S. Construct a triangle xyz with $O(W^2)$ area such that xyz encloses H and the side yz passes through a pair of consecutive points y', z' on the boundary of H. Assume that y' is closer to y than z'. Set $\phi(y) = y'$ and $\phi(z) = z'$. Set $\phi(x) = x'$, where x' is the point on the convex hull of $S(xyz)$ for which the angle

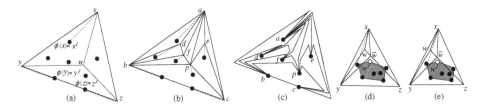

Fig. 3. (a) Illustration for the triangle xyz. (b) Γ and ϕ, where ϕ is illustrated with dashed lines. (c) A 2-bend point-set embedding of G on S. (d)–(e) Construction of w and $\phi(w)$, where $\phi(w) = w'$ is shown in white and the convex hull of $S(xyz)$ in gray.

$\angle xyx'$ is smallest. Figure 3(a) illustrates the triangle xyz and the function ϕ. We call the straight line segments $x\phi(x), y\phi(y), z\phi(z)$ the *wings* of xyz. Observe that only $x\phi(x)$ among the three wings of xyz lie in the proper interior of xyz. We use this invariant throughout the algorithm, i.e., every face f in the drawing will contain at most one wing that is in the proper interior of f. We call such a wing the *major wing* of f.

Let a, b, c be the outer vertices of G in anticlockwise order and let p be the representative vertex of G. Map the vertices a, b, c to the points x, y, z. Let $S \setminus \{x', y', z'\}$ be the point set S'. Let n_1, n_2 and n_3 be the number of inner vertices of G_{abp}, G_{bcp} and G_{cap}, respectively. Since the major wing of xyz is incident to x, we construct a point $w \notin S$ such that $S'(wxy) = n_1, S'(wyz) = n_2 + 1$ and $S'(wxz) = n_3$, as shown in Figure 3(a). Steiger and Streinu [20] proved that such a point always exists and gave an $O(|S'|)$-time algorithm to find w. Since the angle $\angle xy\phi(x)$ is the smallest, if wy or wz intersects $x\phi(x)$, then by continuity there must exist another point $\bar{w}$ on the line wz such that $S'(\bar{w}xy) = n_1, S'(\bar{w}yz) = n_2 + 1, S'(\bar{w}xz) = n_3$ holds, and we choose $\bar{w}$ as the point w. Figures 3(d)–(e) depict such scenarios. Set $\phi(w) = w'$, where w' is the point on the convex hull of $S'(wyz)$ for which the angle $\angle wyw'$ is smallest. Since wyz does not contain $x\phi(x)$, the mapping we compute maintains the invariant that every face contains at most one major wing.

We now recursively construct the drawings of G_{abp}, G_{bcp} and G_{cap} with the point sets $S'(xyw), S'(yzw) \setminus w'$ and $S'(zxw)$, respectively. Note that while recursively constructing a point w for the representative vertex inside some triangle xyz, then the triangle may not have any major wing. Also in this case, it suffices to compute w such that $S'(wxy) = n_1, S'(wyz) = n_2 + 1$ and $S'(wxz) = n_3$ holds. Once we complete the recursive computation, we obtain a straight-line drawing Γ of G, and a bijective function ϕ from the vertices of Γ to the points of S. We now extend each edge (u, v) in Γ using two bends to place the vertices u and v onto the points $\phi(u)$ and $\phi(v)$, respectively. We use ϕ and the property that every face in Γ contains at most one major wing, to maintain planarity. We omit the details due to space constraints.

Theorem 2. *Given a plane 3-tree G with n vertices and a point set S of $n points in general position, we can compute a 2-bend point-set embedding of G*

in $O(n \log^3 n)$ time with $O(W^2)$ area, where W is the length of the side of the smallest axis-parallel square that encloses S.

4   Approximate Point-Set Embeddings

Let Γ be a straight-line drawing of G. Then $S(\Gamma)$ denotes the number of vertices in Γ that are mapped to distinct points of S. The *optimal point-set embedding* of G is a straight-line drawing Γ^* such that $S(\Gamma^*) \geq S(\Gamma')$ for any straight-line drawing Γ' of G. A ρ-approximation point-set embedding algorithm computes a straight-line drawing Γ of G such that $S(\Gamma)/S(\Gamma^*) \geq \rho$. In this section we show that given a plane 3-tree G with n vertices, we can construct a straight-line drawing Γ of G such that $S(\Gamma) = \Omega(\sqrt{n})$, and hence point-set embeddability is approximable with factor $\Omega(1/\sqrt{n})$ for plane 3-trees.

Let G be a plane 3-tree with the outer vertices a, b, c and representative vertex p, and let the number of vertices of G be n. Then the *representative tree T_{n-3}* of G satisfies the following conditions [11].

(a) If $n = 3$, then T_{n-3} is empty.
(b) If $n = 4$, then T_{n-3} consists of a single vertex.
(c) If $n > 4$, then the root p of T_{n-3} is the representative vertex of G and the subtrees rooted at the three counter-clockwise ordered children p_1, p_2 and p_3 of p in T_{n-3} are the representative trees of G_{abp}, G_{bcp} and G_{cap}, respectively.

Since a rooted tree with n nodes is a partially ordered set under the 'successor' relation, by Dilworth's theorem [23], either the height or the number of leaves in the tree is at least $\sqrt{n}$. Let G be the input plane 3-tree with n vertices and let T be its representative tree with $n - 3$ vertices [11].

If T has $\Omega(\sqrt{n})$ leaves, then it is straightforward to construct a drawing Γ of G using the algorithm of Steiger and Streinu [20] such that exactly the leaves are mapped to the points of S, i.e., $S(\Gamma) = \Omega(\sqrt{n})$. Otherwise, the height of T is $\Omega(\sqrt{n})$. In this case we prove that G has a 'canonical ordering tree'(also, called Schnyder's realizer [2]) with height $\Omega(\sqrt{n})$, as shown in Lemma 4. There exists a simple algorithm (one can also modify de Fraysseix et al.'s algorithm [1]) to obtain a straight-line drawing Γ of G such that $S(\Gamma) = \Omega(\sqrt{n})$. We omit the details due to space constraints.

Lemma 4. *Let G be a plane 3-tree and let T be its representative tree. If the height of the representative tree is $\Omega(\sqrt{n})$, then G has a canonical ordering tree with height $\Omega(\sqrt{n})$.*

Proof. Let $P = (v_1, v_2, \ldots, v_k)$, $k = \Omega(\sqrt{n})$, be the longest path from the root v_1 of T to some leaf v_k. Without loss of generality assume that k is even. By G_i we denote the plane 3-tree induced by the outer vertices of G and the vertices $v_1, v_2, \ldots, v_k$. We now incrementally construct G_k. First construct a triangle xyz, place the vertex v_1 interior to xyz and add the segments $v_1 x, v_1 y, v_1 z$. Since v_2 is a child of v_1, v_2 must be placed interior to one of the triangles incident to

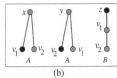 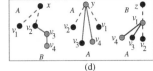

(a)   (b)   (c)   (d)

Fig. 4. (a) Illustration for G_2, where T_x, T_y and T_z are shown in red, green and blue. (b) Illustration for the connectors, shown in gray. (c)–(d) Example of a connection of A, A, B with B, A, A, respectively.

v_1. Since v_{i+1}, where $i+1 \le k$, is a child of v_i, this condition holds throughout the construction. Let T_x, T_y, T_z be the trees of the Schnyder's realizer rooted at x, y, z, respectively. Figure 4(a) illustrates the realizer of G_2, where the height of T_x, T_y and T_z is two, two and three, respectively. By A and B we denote the rooted trees isomorphic to T_x and T_z of G_2, respectively. The nodes of $T_w, w \in \{x, y, z\}$, where the realizer grows while adding v_{i+1} to G_i, $i \ge 2$, are called the *connectors* of T_w in G_i. See Figure 4(b).

Consider now the steps when we obtain the graphs $G_2, G_4, \dots, G_k$. Observe that each time some tree of the form A (or B) gets connected with some $T_w, w \in \{x, y, z\}$, of G_i, the connectors of A (or B) become the only connectors of T_w in G_{i+2}. Figures 4(c)–(d) illustrate such a scenario. Consequently, each time some tree of the form B gets connected with some $T_w, w \in \{x, y, z\}$, of G_i, the height of T_w increases by one in G_{i+2}. Since we need to encounter $k/2$ steps before we obtain G_k, one of T_x, T_y or T_z must have height at least $k/6 = \Omega(\sqrt{n})$. Since each tree of the Schnyder's realizer of G_k is a subtree of a distinct tree of the Schnyder's realizer of G, the proof is complete.   □

Theorem 3. *Given a plane 3-tree G with n vertices and a point set S of n points in general position in $\mathbb{R}^2$, we can compute a straight-line drawing Γ of G in polynomial time such that the number of vertices in Γ that are mapped to distinct points of S is $\Omega(1/\sqrt{n})$ times to the optimal. Hence the point-set embeddability of plane 3-trees is approximable with factor $\Omega(1/\sqrt{n})$.*

5   Conclusion

Using techniques that are completely different from those used in the previously best known approaches for testing point-set embeddability of plane 3-trees (achieving $O(n^{4/3+\varepsilon})$ time and $O(n^{4/3})$ space), in Section 2 we described an algorithm that solves the problem for a given plane 3-tree in $O(n \log^3 n)$ time using $O(n)$ space. As suggested by an anonymous reviewer, one possibility for potentially reducing the running time further might be to apply the algorithm of Moosa and Rahman [13], where an orthogonal range search would be used instead of a triangular range search. Specifically, given points x and y and an integer k, a triangle wxy that contains k points can be found by encoding each point w using two values: the slopes of wx and wy. The triangle wxy is then mapped to a two-sided axis-aligned orthogonal range query. It is not obvious, however, how this technique would be applied in recursive levels. One possibility might be to use a dynamic orthogonal range counting data structure. Another interesting issue is to examine the amount of scale up required to ensure that the vertices and bend points of the drawings produced in Section 3 lie on integer coordinates, i.e., the area requirement under minimum resolution assumption. In Section 4 we gave an $\Omega(1/\sqrt{n})$-approximation algorithm for plane 3-trees. Hence a natural question is to examine whether a constant factor approximation algorithm exists.

Acknowledgement. We thank Valentin Polishchuk and the other anonymous reviewers for many constructive and helpful comments.

References

1. de Fraysseix, H., Pach, J., Pollack, R.: How to draw a planar graph on a grid. Combinatorica 10(1), 41–51 (1990)
2. Schnyder, W.: Embedding planar graphs on the grid. In: SODA, pp. 138–148. ACM (1990)
3. Bose, P.: On embedding an outer-planar graph in a point set. Computational Geometry: Theory and Applications 23(3), 303–312 (2002)
4. Cabello, S.: Planar embeddability of the vertices of a graph using a fixed point set is NP-hard. Journal of Graph Algorithms and Applications 10(2), 353–363 (2006)
5. Ikebe, Y., Perles, M.A., Tamura, A., Tokunaga, S.: The rooted tree embedding problem into points in the plane. Discrete & Comp. Geometry 11(1), 51–63 (1994)
6. Bose, P., McAllister, M., Snoeyink, J.: Optimal algorithms to embed trees in a point set. Journal of Graph Algorithms and Applications 1(2), 1–15 (1997)
7. Castañeda, N., Urrutia, J.: Straight line embeddings of planar graphs on point sets. In: CCCG, pp. 312–318 (1996)
8. Durocher, S., Mondal, D.: On the hardness of point-set embeddability. In: Rahman, M.S., Nakano, S.-I. (eds.) WALCOM 2012. LNCS, vol. 7157, pp. 148–159. Springer, Heidelberg (2012)
9. Biedl, T., Vatshelle, M.: The point-set embeddability problem for plane graphs. In: SoCG, pp. 41–50. ACM (2012)
10. Andrade Jr., J.S., Herrmann, H.J., Andrade, R.F.S., da Silva, L.R.: Apollonian networks: Simultaneously scale-free, small world, euclidean, space filling, and with matching graphs. Physical Review Letters 94 (2005)
11. Nishat, R.I., Mondal, D., Rahman, M.S.: Point-set embeddings of plane 3-trees. Computational Geometry: Theory and Applications 45(3), 88–98 (2012)
12. Durocher, S., Mondal, D., Nishat, R.I., Rahman, M.S., Whitesides, S.: Embedding plane 3-trees in $\mathbb{R}^2$ and $\mathbb{R}^3$. In: Speckmann, B. (ed.) GD 2011. LNCS, vol. 7034, pp. 39–51. Springer, Heidelberg (2011)
13. Moosa, T.M., Sohel Rahman, M.: Improved algorithms for the point-set embeddability problem for plane 3-trees. In: Fu, B., Du, D.-Z. (eds.) COCOON 2011. LNCS, vol. 6842, pp. 204–212. Springer, Heidelberg (2011)
14. Erickson, J.: On the relative complexities of some geometric problems. In: CCCG, pp. 85–90 (1995)

15. Brandenburg, F.J., Eppstein, D., Goodrich, M.T., Kobourov, S.G., Liotta, G., Mutzel, P.: Selected open problems in graph drawing. In: Liotta, G. (ed.) GD 2003. LNCS, vol. 2912, pp. 515–539. Springer, Heidelberg (2004)
16. Everett, H., Lazard, S., Liotta, G., Wismath, S.K.: Universal sets of n points for one-bend drawings of planar graphs with n vertices. Discrete & Computational Geometry 43(2), 272–288 (2010)
17. Kaufmann, M., Wiese, R.: Embedding vertices at points: Few bends suffice for planar graphs. J. of Graph Algorithms and Applications 6(1), 115–129 (2002)
18. Di Giacomo, E., Liotta, G.: The Hamiltonian augmentation problem and its applications to graph drawing. In: Rahman, M. S., Fujita, S. (eds.) WALCOM 2010. LNCS, vol. 5942, pp. 35–46. Springer, Heidelberg (2010)
19. Chazelle, B., Sharir, M., Welzl, E.: Quasi-optimal upper bounds for simplex range searching and new zone theorems. Algorithmica 8(5&6), 407–429 (1992)
20. Steiger, W.L., Streinu, I.: Illumination by floodlights. Computational Geometry: Theory and Applications 10(1), 57–70 (1998)
21. Overmars, M.H., van Leeuwen, J.: Maintenance of configurations in the plane. Journal of Computer and System Sciences 23(2), 166–204 (1981)
22. Brodal, G.S., Jacob, R.: Dynamic planar convex hull. In: FOCS, pp. 617–626. IEEE (2002)
23. Dilworth, R.P.: A decomposition theorem for partially ordered sets. Annals of Mathematics 51(1), 161–166 (1950)

A Dynamic Data Structure for Counting Subgraphs in Sparse Graphs

Zdeněk Dvořák and Vojtěch Tůma*

Computer Science Institute, Charles University
Prague, Czech Republic
{rakdver,voyta}@iuuk.mff.cuni.cz

Abstract. We present a dynamic data structure representing a graph G, which allows addition and removal of edges from G and can determine the number of appearances of a graph of a bounded size as an induced subgraph of G. The queries are answered in constant time. When the data structure is used to represent graphs from a class with bounded expansion (which includes planar graphs and more generally all proper classes closed on topological minors, as well as many other natural classes of graphs with bounded average degree), the amortized time complexity of updates is polylogarithmic.

Keywords: sparse graphs, subgraphs, data structure.

1 Introduction

In this paper, we deal with the problem of determining whether a fixed graph H is an induced subgraph of another graph G. We consider a dynamic setting, that is, we construct a data structure representing a graph, supporting efficient edge additions and removals, and keeping track of whether H appears in the current graph as an induced subgraph.

An exemplar application of such a data structure is an algorithm for finding 5-coloring of a graph on torus, based on the result of Thomassen [21]. Here, the algorithm performs various reductions of the considered graph, and after each reduction, it needs to test whether the reduced graph contains some of four specific subgraphs. Rather than running a subgraph testing algorithm each time, it makes more sense to update the information about subgraphs dynamically.

We actually deal with the counting version of the problem, i.e., determining how many times does a fixed graph H appear as an induced subgraph in the represented graph. This generalisation has applications in Bioinformatics and Social Networking research – for instance, see [13] and [20].

* The work leading to this invention has received funding from the European Research Council under the European Union's Seventh Framework Programme (FP7/2007-2013)/ERC grant agreement no. 259385. The second author received support under project GAUK/592412 of Grant agency of Charles University, KONTAKT II LH12095 and SVV 267313.

F. Dehne, R. Solis-Oba, and J.-R. Sack (Eds.): WADS 2013, LNCS 8037, pp. 304–315, 2013.

The problem of determining whether a graph H is an (induced) subgraph of another graph G is W[1]-complete when parameterized by H, see [3]. Furthermore, as observed in [6], this result still holds when G is restricted to belong to a hereditary class of graphs, as long as this class is not nowhere-dense. Consequently, unless W[1] = FPT, the discussed data structure cannot have both subpolynomial update and query times when used to represent general graphs (or even graphs from a hereditary class that is not nowhere-dense).

The time complexity of queries in our data structure is constant depending only on H. The time complexity of the updates is subpolynomial when the represented graph is restricted to belong to a nowhere-dense class of graphs. Furthermore, if the class has bounded expansion, the complexity becomes polylogarithmic. Let us recall that the concepts of *nowhere-denseness* and *bounded expansion* were introduced by Nešetřil and Ossona de Mendez [15,17]; we give the definitions below. Here, let us just note that every class with bounded expansion is nowhere-dense, and that many natural classes of graphs with bounded average degree (e.g., proper minor-closed classes of graphs, classes of graphs with bounded maximum degree, classes of graphs excluding a subdivision of a fixed graph, classes of graphs that can be embedded in a fixed surface with bounded number of crossings per each edge, see [18]) have bounded expansion.

Let us now state the main result of this paper precisely.

Theorem 1. *Let H be a fixed graph and let $\mathcal{G}$ be a class of graphs.*

There exists a data structure $\mathrm{ISub}_H(G)$ representing a graph $G \in \mathcal{G}$ which supports the following operations.

- *Determine the number of induced subgraphs of G isomorphic to H.*
- *Add an edge e, i.e., transform $\mathrm{ISub}_H(G)$ to $\mathrm{ISub}_H(G + e)$, under the assumption that $G + e$ is in $\mathcal{G}$.*
- *Delete an edge e, i.e., transform $\mathrm{ISub}_H(G)$ to $\mathrm{ISub}_H(G - e)$, under the assumption that $G - e$ is in $\mathcal{G}$.*

If $\mathcal{G}$ has bounded expansion, then the time complexity of query and edge removal is $O(1)$, while the amortized time complexity of edge addition is $O(\log^h |V(G)|)$, where $h = \binom{|V(H)|}{2} - 1$. The structure can be initialized in $O(|V(G)|)$ and the space complexity for the structure is $O(|V(G)|)$. If $\mathcal{G}$ is nowhere-dense, then the time complexity of query is $O(1)$, the amortized time complexity of edge addition or removal is $O(|V(G)|^\varepsilon)$, the time complexity of the initialization is $O(|V(G)|^{1+\varepsilon})$ and the space complexity is $O(|V(G)|^{1+\varepsilon})$, for every $\varepsilon > 0$.

Using this data structure, we can also count graph inclusions other than induced subgraphs (e.g., subgraphs and homomorphisms), as these counts only depend on which and how many small induced subgraphs appear in G. Furthermore, it is easy to modify the data structure to apply to objects other than undirected graphs, e.g., to directed graphs with colors on vertices and edges.

The problem of dynamic subgraph counting was introduced by Eppstein et al. in [10] and later extended in [9]. Using a different approach, they obtain a data structure parametrized by the number of vertices of G of large degree (which is usually worse than parametrization by expansion that we use) , and determines only the number of all induced subgraphs with *at most four vertices*. As a trade-off, it has substantially better time complexity per operation.

Our result extends, in a certain sense, the result of Dvořák, Král' and Thomas [5], who provided a data structure for first-order testing. For a fixed first-order formula ϕ and a class of graphs C with bounded expansion, this data structure represents a graph $G \in C$ and can be initialized in time $O(|V(G)|)$. The data structure enables testing whether the graph satisfies ϕ in constant time. However, the data structure is only semi-dynamic – the graph can be modified by adding and removing edges in constant time, but the edge additions are restricted: we can only add edges that were removed before. On the other hand, the data structure from [5] handles general FO properties, whereas the data structure developed in this paper deals only with the case of induced subgraph testing – formally, this amounts to deciding existential FO properties for classes of relational structures with bounded expansion (or nowhere-dense).

Finally, let us briefly mention some results for the non-dynamic setting, that is, when the graph G is only tested once. Eppstein [8] gave a linear-time algorithm for this problem for planar graphs. This was subsequently generalized by Frick and Grohe [11] to FO property testing in graphs with locally bounded tree-width, by Nešetřil and Ossona de Mendez [14] to subgraph testing in nowhere-dense graphs, and by Dvořák, Král' and Thomas [5] to FO property testing in graphs with locally bounded expansion (the question whether FO property testing is fixed-parameter tractable in nowhere-dense classes of graphs is still open).

When there is no restriction on G, the situation is far worse. As we mentioned before, the subgraph problem is $W[1]$-hard when parameterized by H [3], and consequently it is unlikely to admit an algorithm with time complexity $f(|H|)|V(G)|^{O(1)}$ for any function f. The best known general algorithms are based on matrix multiplication; Nešetřil and Poljak [19] gave an $O(|V(G)|^{\omega|V(H)|/3})$-time algorithm, where ω is the exponent in the complexity of matrix multiplication. This was subsequently refined in [12,7].

The rest of the paper is organized as follows. In Section 2, we describe a basic idea of the data structure for induced subgraphs. In Section 3, we give some definitions and auxiliary results needed in the rest of the paper. Section 4 contains a more detailed description of the key parts of the data structure.

2   Basic Idea

For concreteness, in this section we consider the class of planar graphs, rather than an arbitrary class with bounded expansion. Suppose that we want to keep track of triangles in a planar graph G. A simple way to do this is as follows. Orient the edges of G so that every vertex has in-degree at most 6, which is possible by 5-degeneracy of planar graphs. For an edge xy of G, we write $x \to y$ if the edge is oriented towards y. For each vertex $u \in V(G)$, we maintain

- the number $n_1(u)$ of pairs of vertices $v, w \in V(G)$ such that $u \to v$, $v \to w$ and $u \to w$;

- the number $n_2(u)$ of pairs of vertices $v, w \in V(G)$ such that $u \to v$, $v \to w$ and $w \to u$.

We also maintain the sums $N_1 = \sum_{u \in V(G)} n_1(u)$ and $N_2 = \sum_{u \in V(G)} n_2(u)$. Consider a triangle $T \subseteq G$ with vertex set $\{x, y, z\}$. By symmetry, we can assume that $x \to y$ and $y \to z$. If $x \to z$, then T contributes 1 to $n_1(x)$. If $z \to x$, then T contributes 1 to each of $n_2(x)$, $n_2(y)$ and $n_2(z)$. Therefore, the number of triangles in G is $N_1 + N_2/3$.

Let us add an edge xy to G and choose its orientation, say $x \to y$. Assume for now that in the resulting orientation, the in-degree of y is still at most 6. Which of the numbers that we maintain are affected? Clearly, if $n_1(u)$ or $n_2(u)$ changes, then either u is incident with the edge xy, or u is an in-neighbor of x. Thus, we only need to update information for at most 8 vertices of G.

Updating an in-neighbor u of x in constant time is easy, as we just need to check whether the path $u \to x \to y$ contributes to $n_1(u)$ and $n_2(u)$ or not. For y, the number $n_1(y)$ is unchanged, while the number $n_2(y)$ increases by the number of vertices v such that $y \to v \to x$. We can enumerate such vertices in a constant time, as they are in-neighbors of x. Similarly, we can update $n_2(x)$ in a constant time, as the path $x \to y \to z$ only contributes to $n_2(x)$ if z is an in-neighbor of x.

There are two ways $n_1(x)$ can be affected by the addition of $x \to y$. It could be that there exists $v \in V(G)$ with $x \to v \to y$. All these vertices are in-neighbors of y, and they can be enumerated in constant time. The most complicated case is that there exists a vertex z with $x \to z$ and $y \to z$. Here, we cannot easily enumerate all possibilities for z, as we do not have any bound on out-degrees of vertices. Therefore, we need one more piece of information. For each pair of distinct vertices $u, v \in V(G)$, let $n_3(u, v)$ be the number of vertices $w \in V(G)$ with $u \to w$ and $v \to w$. In a hash table, we store

- the number $n_3(u, v)$ for all pairs $u, v \in V(G)$ such that $n_3(u, v) \neq 0$.

Hence, in the last case of the update of $n_1(x)$, we just need to add $n_3(x, y)$.

Let us note that since each vertex has at most 6 in-neighbors, the number $n_3(u, v)$ is non-zero for at most $\binom{6}{2}|V(G)|$ pairs $u, v \in V(G)$, and thus n_3 can be stored in linear space. We need to consider how the addition of $x \to y$ affects n_3. If $n_3(u, v)$ changes, then by symmetry we can assume that $u = x$ and $v \to y$. Consequently, v is an in-neighbor of y, and thus we can enumerate in constant time all (at most 5) pairs $u, v \in V(G)$ such that $n_3(u, v)$ increases.

This finishes the description of the update in the case of edge addition. Edge removal is handled similarly. One problem that we skipped is what to do when the addition of an edge would violate the constraint on the maximum in-degree. However, Brodal and Fagerberg [1] provided an algorithm for maintaining an orientation with bounded maximum in-degree, which only needs to change orientation of $O(\log |V(G)|)$ edges per update (amortized). In our data structure, the edge reorientations can be handled similarly to edge additions.

Therefore, we have described a data structure for counting the number of triangles in a planar graph (or indeed, any graph with bounded degeneracy), with logarithmic time complexity per update. For a general subgraph H, there appear additional complications. The idea of maintaining for each vertex v the number of copies in H in the subgraph reachable by short paths from v (a similar idea appears already in Chrobak and Eppstein [2]) only works for the orientations of H that contain a directed Hamiltonian path. As a first step, we extend this to the case that H contains a spanning outbranching without cross edges, at the expense of only counting homomorphisms from H to G instead of subgraphs. This is not a big problem, as counting subgraphs is equivalent to counting homomorphisms through a standard inclusion-exclusion argument.

However, how to deal with the orientations that do not admit such an outbranching? For this, we use the idea of fraternal augmentations of Nešetřil and Ossona de Mendez [15]. Essentially, we add new edges to H and G, obtaining new graphs H' and G', in such a way that we can recover the original number of homomorphisms, but H' contains a spanning outbranching without cross edges. The results of [15] and the assumption of bounded expansion or nowhere-dense class of graphs ensure that G' has bounded degeneracy. This is the most technically complicated part of the argument, formalized in Lemmas 1 and 2.

3   Definitions and Auxiliary Results

The graphs considered in this paper are simple, without loops or parallel edges, unless specified otherwise. For directed graphs, we also do not allow edges joining a single pair of vertices in opposite directions.

A graph H is said to be a *minor of depth r* of a graph G, if it can be obtained from a subgraph of G by contracting vertex-disjoint subgraphs of radius at most r into single vertices, with arising parallel edges and loops suppressed. The *Greatest Reduced Average Density* at depth r of graph G then denotes the value

$$\nabla_r(G) = \max\{|E(H)|/|V(H)| : H \text{ is a minor of depth } r \text{ of } G\}.$$

A graph G has *expansion bounded by f*, if f is a function from $\mathbf{N}$ to $\mathbf{R}^+$ and $\nabla_r(G) \leq f(r)$ for every r. A class of graphs $\mathcal{G}$ has *bounded expansion*, if there is a function f such that every graph in $\mathcal{G}$ has expansion bounded by f. Let us note that the average degree of a graph G is at most $2\nabla_0(G)$; hence, graphs in any class of graphs with expansion bounded by f have average degree bounded by a constant $2f(0)$. Similarly, we conclude that every $G \in \mathcal{G}$ has an orientation (even acyclic one) with in-degree at most $D = 2f(0)$.

The nowhere dense classes introduced in [17] that generalize classes with bounded expansion can be defined in a similar manner – here, let us just remark that unlike the case of bounded expansion, the average degree does not have to be bounded by a constant, but is $n^{o(1)}$, i.e., for any nowhere-dense class $\mathcal{G}$ and for every $\varepsilon > 0$ there exists a function $g(n) = O(n^\varepsilon)$ such that every graph $G \in \mathcal{G}$ has average degree at most $g(|V(G)|)$. These two concepts of sparsity turned out to be very powerful. We refer the reader to surveys [4,16] for more information.

Suppose that G is a directed graph. Vertices $u, v \in V(G)$ form a *fork* if u and v are distinct and non-adjacent and there exists $w \in V(G)$ with $u \to w, v \to w \in E(G)$. Let G' be a graph obtained from G by adding the edge $u \to v$ or $v \to u$ for every pair of vertices u and v forming a fork. Then G' is called a *fraternal augmentation* of G. Let us remark that a directed graph can have several different fraternal augmentations, depending on the choices of directions of newly added edges. If G has no fork, then G is called *elder graph*. For an undirected graph G, a *k-th augmentation of G* is a directed graph G' obtained from an orientation of G by iterating fraternal augmentation (for all forks) k times. Note that $\left(\binom{|V(G)|}{2} - 2\right)$-th augmentation of G is an elder graph, because any graph with at most 1 edge is already elder and fraternal augmentation of a non-elder graph adds at least one edge.

The following result of Nešetřil and Ossona de Mendez [15] shows that fraternal augmentation preserve bounded expansion and nowhere-denseness.

Theorem 2. *There exist polynomials f_0, f_1, f_2, ... with the following property. Let G be a graph with expansion bounded by a function g and let G_1 be an orientation of G with in-degree at most D. If G' is the underlying undirected graph of a fraternal augmentation of G_1, then G' has expansion bounded by the function $g'(r) = f_r(g(2r + 1), D)$.*

The fraternal augmentations are a basic tool for deriving properties of graphs with bounded expansion, e.g., existence of low tree-depth colorings (see [14] for a definition). Once such a coloring is found, the subgraph problem can be reduced to graphs with bounded tree-width, where it can be easily solved in linear time by dynamic programing. However, we do not know how to maintain a low tree-depth coloring dynamically (indeed, not even an efficient data structure for maintaining say a proper 1000-coloring of a planar graph during edge additions and deletions is known). The main contribution of this paper is showing that we can count subgraphs using just the fraternal augmentations, which are easier to update.

To maintain orientations of a graph, we use the following result by Brodal and Fagerberg [1]:

Theorem 3. *There exists a data structure that, for a graph G with $\nabla_0(G) \le d$, maintains an orientation with maximum in-degree at most $4d$ within the following bounds:*

- *an edge can be added to G (provided that the resulting graph G' still satisfies $\nabla_0(G') \le d$) in amortized $O(\log n)$ time, and*
- *an edge can be removed in $O(1)$ time, without affecting the orientation of any other edges.*

The data structure can be initialized in time $O(|V(G)| + |E(G)|)$. During the updates, the edges whose orientation has changed can be reported in the same time bounds. The orientation is maintained explicitly, i.e., each vertex stores a list of in- and out-neighbors.

Let us remark that the multiplicative constants of the O-notation in Theorem 3 do not depend on d, although the implementation of the data structure as described in the paper of Brodal and Fagerberg requires the knowledge of d.

Using Theorem 2, we obtain the following modification of the Theorem 3.

Theorem 4. *For every $k \geq 0$, there exists an integer k' and a polynomial g with the following property. Let $\mathcal{G}$ be a class of graphs and $h(n, r)$ a computable function such that the expansion of every graph $G \in \mathcal{G}$ is bounded by $f(r) = h(|V(G)|, r)$. There exists a data structure representing a k-th augmentation $\tilde{G}_k$ of a graph $G \in G$ with n vertices within the following bounds, where $D = g(h(n, k'))$:*

- *the maximum in-degree of $\tilde{G}_k$ is at most D,*
- *an edge can be added to G (provided that the resulting graph still belongs to $\mathcal{G}$) in an amortized $O(D \log^{k+1} n)$ time, and*
- *an edge can be removed in $O(D)$ time, without affecting the orientation of any other edges.*

The data structure can be initialized in time $O(Dn + t)$, where t is the time necessary to compute D. The orientation is maintained explicitly, i.e., each vertex stores a list of in- and out-neighbors.

Let G and H be undirected graphs, a mapping $\phi \colon V(H) \to V(G)$ is a *homomorphism* if for every edge $uv \in E(H)$, we have that $\phi(u)\phi(v)$ is an edge of G (in particular, $\phi(u) \neq \phi(v)$). A homomorphism is a *subgraph* if it is injective. It is an *induced subgraph* if it is injective and $\phi(u)\phi(v) \in E(G)$ implies $uv \in E(H)$, for every $u, v \in V(H)$. Let $\hom(H, G)$, $\mathrm{sub}(H, G)$ and $\mathrm{isub}(H, G)$ denote the number of homomorphisms, subgraphs and induced subgraphs, respectively, of H in G. Let us note that the definitions of subgraph and induced subgraph distinguish the vertices, i.e., $\mathrm{sub}(H, H) = \mathrm{isub}(H, H)$ is equal to the number of automorphisms of H.

Similarly, if H and G are directed graphs, a mapping $\phi \colon V(H) \to V(G)$ is a *homomorphism* if $u \to v \in E(H)$ implies that $\phi(u) \to \phi(v)$ is an edge of G, and $\hom(H, G)$ denotes the number of homomorphisms from H to G.

4   Dynamic Data Structure for Induced Subgraphs

In this section, we aim to design the data structure ISub as described in the introduction.

To implement the data structure $\mathrm{ISub}_H(G)$, we first perform several standard transformations using principle of inclusion and exclusion, reducing the problem to counting homomorphisms from connected graphs (for the straightforward argument, see the full version of the paper). That is, we only need to design a data structure $\mathrm{Hom}_{H'}(G)$ for a connected graph H', which counts the number $\hom(H', G)$ of homomorphisms from H' to G, and allows additions and removals of edges in G.

4.1   Reduction to Elder Augmentations

In order to implement the data structure $\mathrm{Hom}_H(G)$, we use fraternal augmentations. Essentially, we would like to find a bijection between homomorphisms from H to G and between homomorphisms from all possible h-th augmentations of H to an h-th augmentation of G, where $h = \binom{|V(H)|}{2} - 2$. These augmentations of H are elder graphs, whose structure we exploit in the design of the data structure. The results are presented in a slightly simplifed form and without proofs, for the details we refer to the appendix.

However, the situation is more complicated, as taking augmentations of H does not suffice. The problem is that a homomorphism of H can map two non-adjacent vertices u, v of H to a single vertex of G, but in all augmentations of H there is an edge between u and v, thus there is no corresponding homomorphism. Therefore we consider the set $\mathcal{H}^e$, which contains augmentations of H and some other elder graphs, which are formed basically by identifying some vertices of H and whose number is bounded by a function of size of H. Additionally, one has to work with colored edges.

Lemma 1. *Let H and G be graphs and let $h = \binom{|V(H)|}{2} - 2$. If G' is an h-th augmentation of G, then*

$$\text{hom}(H, G) = \sum_{H' \in \mathcal{H}^e} \text{hom}(H', G').$$

All graphs in the set $\mathcal{H}^e$ are elder, which implies good connectivity properties, as formalized bellow. A directed tree T with all edges directed away from the root is called an *outbranching*. The root of T is denoted by $r(T)$. Let H be a supergraph of an outbranching T with $V(H) = V(T)$, such that for every edge $t_1 \to t_2 \in E(H)$, there exists a directed path in T either from t_1 to t_2 or from t_2 to t_1. We call such a pair (H, T) a *vineyard*.

Lemma 2. *If H is a connected elder graph, then there exists an outbranching $T \subseteq H$ such that (H, T) is a vineyard.*

In the following subsection, we design a data structure $\text{AHom}_{(H',T'),D}(G')$ for an elder vineyard (H', T') and a directed graph G' of maximum in-degree at most D. The data structure $\text{AHom}_{(H',T'),D}$ counts the number $\text{hom}(H', G')$ of homomorphisms from H' to G' and allows additions, removals and reorientations of edges in G'.

The data structure $\text{Hom}_H(G)$ then basically consists of a collection of such AHom structures – one for every $H' \in \mathcal{H}^e$, with D derived from the Theorem 4 and T' obtained from Lemma 2. Edge additions and removals propagate to all these AHom structures, and a query is answered by summing respective queries – as hinted at by the Lemma 1. Similarly, the data structure $\text{ISub}_H(G)$ is realised via $\text{Hom}_H(G)$ structures, whose number is bounded by a function of the size H.

4.2   Homomorphisms of Elder Graphs

Consider a directed graph and a set S of its vertices. Let $N_d^+(S)$ denote the set of vertices that are reachable from S by a directed path of length at most d, and let $N_\infty^+(S)$ we denote the set of vertices reachable from S by a directed path of any length. Similarly, $N_d^-(S)$ and $N_\infty^-(S)$ denote the sets of vertices from that S can be reached by a directed path of length at most d and by a directed path of any length, respectively. We also use $N_d^+(v)$, $N_\infty^+(v)$, $N_d^-(v)$, $N_\infty^-(v)$ as shorthands for $N_d^+(\{v\})$, $N_\infty^+(\{v\})$, $N_d^-(\{v\})$, $N_\infty^-(\{v\})$, respectively.

Let (H, T) be an elder vineyard. A *clan* is a subset C of vertices of H such that $N_1^+(C) = C$ and the subgraph T' of T induced by C is an outbranching. Let $r(C)$ denote the root of this outbranching T'. The *ghosts* of a clan C are the vertices $N_1^-(C) \setminus C$.

Lemma 3. *Let (H, T) be an elder vineyard.*

1. *For every $v \in V(H)$, the set $N_\infty^+(v)$ is a clan.*
2. *The ghosts of a clan C are exactly the vertices in $N_1^-(r(C)) \setminus C$.*
3. *All ghosts of a clan C are on the path from $r(H)$ to $r(C)$ in T.*

Proof. Let us prove the claims separately:

1. Let $C = N_\infty^+(v)$. Clearly, $N_1^+(C) = C$. Note that the subgraph $H[C]$ of H induced by C is connected. If the subgraph of T induced by C is not an outbranching, then it contains two components T_1 and T_2 joined by an edge of $H[C]$. Observe that no directed path in T contains a vertex both in T_1 and T_2. This contradicts the assumption that (H, T) is a vineyard.
2. Suppose that v is a ghost of C, i.e., $v \notin C$ and there exists an edge $v \to w \in E(H)$ for some $w \in C$. Let w be such a vertex whose distance from $r(C)$ in T is minimal. If $w \neq r(C)$, then consider the in-neighbor z of w in T. Since H is an elder graph, v and z are adjacent in H. Since v does not belong to C, we have $z \to v \notin E(H)$, and thus $v \to z \in E(H)$. However, the distance from $r(C)$ to z in T is smaller than the distance from $r(C)$ to w, which is a contradiction. Therefore, we have $w = r(C)$ as required.
3. This follows from the definition of vineyard.

The *extended clan* C^* for a clan C is obtained from the subgraph of H induced by C and its ghosts by removing the edges joining pairs of ghosts. We want to count homomorphisms from clans of H to G, but we need to control the behaviour of ghosts of the clan – let be ghosts $g_1, \ldots, g_m$ of a clan C listed in the increasing order by their distance from $r(C)$ in T and let v and $w_1, \ldots, w_m$ be (not necessarily distinct) vertices of G. Note that g_1 is the in-neighbor of $r(C)$ in T. By $\hom_{(H,T)}(C, v, w_1, \ldots, w_m, G)$ we denote the number of homomorphisms from C^* to G such that $r(C)$ maps to v and $g_1, \ldots, g_m$ map to $w_1, \ldots, w_m$ in order, and by $\hom((H, T), G, v)$ we denote the number of homomorphisms from H to G such that $r(T)$ maps to v.

Theorem 5. *Let (H, T) be an elder vineyard and let D be an integer. There exists a data structure $\mathrm{AHom}_{(H,T),D}(G)$ representing a directed graph G and maximum in-degree at most D supporting the following operations in $O(D^{|V(H)|^2})$ time.*

1. *Addition of an edge e to G such that the maximum in-degree of $G + e$ is at most D.*

2. *Reorientation of an edge in G such that the maximum in-degree of the resulting graph is at most D.*
3. *Removal of an edge.*

The data structure can be used to determine $\hom((H,T),G,v)$ for a vertex $v \in V(G)$, as well as $\hom(H,G)$, in $O(1)$. The data structure can be built in time $O(D^{|V(H)|^2+1}|V(G)|)$ and has space complexity $O(D^{|V(H)|}|V(G)|)$.

Proof. We store the following information:

- For each clan $C \neq V(H)$ with m ghosts and each m-tuple of vertices $w_1, \ldots, w_m$ of G we record the number

$$S(C, w_1, \ldots, w_m) = \sum_{v \in N_1^+(w_1)} \hom_{(H,T)}(C, v, w_1, \ldots, w_m),$$

  that is the number of homomorphisms of C^* to G such that the ghosts of C map to $w_1, \ldots, w_m$ and $r(C)$ maps to some outneighbor v of w_1.
- For each $v \in V(G)$, the number $\hom((H,T),G,v)$.
- The sum $\hom(H,G)$ of these numbers over all vertices of G.

The number $S(C, w_1, \ldots, w_m)$ is only stored for those combinations of C and $w_1, \ldots, w_m$ for that it is non-zero. The values are stored in a hash table, so that they can be accessed in constant time. By Lemma 3, if $\hom_{(H,T)}(C, v, w_1, \ldots, w_m)$ is non-zero, then $w_1, \ldots, w_m$ are in-neighbors of v in G. Since the maximum in-degree of G is at most D, each vertex v contributes at most $D^{|V(H)|}$ non-zero values (and each of the numbers is smaller or equal to $|V(G)|^{|V(H)|}$), thus the space necessary for the storage is $O(D^{|V(H)|}|V(G)|)$. Queries can be performed in constant time by returning the stored information.

The addition of an edge $x \to y$ to G is implemented as follows. We process the clans of (H,T) in the decreasing order of size, i.e., when we use the information stored for the smaller clans, it still refers to the graph G without the new edge. Let us consider a clan $C \neq V(H)$ with ghosts $g_1, \ldots, g_m$. For each non-empty set X of edges of C^*, we are going to find all vertices v and $w_1, \ldots, w_m$ such that there exists a homomorphism of C^* mapping $r(C)$ to v and the ghosts of C to $w_1, \ldots, w_m$ which maps precisely the edges of X to $x \to y$. We will also determine the numbers of such homomorphisms, and decrease the number $S(C, w_1, \ldots, w_m)$ by this amount. Note that the number of choices of X is constant (bounded by a function of H).

Consider now a fixed set X. Let M be the set of vertices $z \in V(C^*)$ such that there exists a directed path in C^* from z to the head of an edge of X. Note that $r(C)$ and all ghosts of C belong to M. Let $C_1, \ldots, C_t$ be the vertex sets of connected components of $C^* - M$, and observe that they are clans. Now, let $\mathcal{F}$ be the set of all homomorphisms from the subgraph of C^* induced by M to $G + (x \to y)$ such that exactly the edges of X are mapped to $x \to y$. Note that if z is an image of a vertex of M in such a homomorphism, then G contains a directed path from z to y of length at most $|V(H)|$, thus there are only $O(D^{|V(H)|})$ vertices of G to that M can map, and consequently only $O(D^{|V(H)|^2})$ choices for the homomorphisms. Each such choice fixes the image of $r(C)$ as well as all the ghosts.

Consider $\phi \in \mathcal{F}$. We need to determine in how many ways ϕ extends to a homomorphism of C^* that maps no further edges to $x \to y$ (this number is then added to the value $S(C, \phi(g_1), \ldots, \phi(g_m)))$. Note that for $1 \le i \le t$, the ghosts of C_i are contained in M, and thus their images are fixed by the choice of ϕ. Therefore, if $g^i_1, \ldots, g^i_{m_i}$ are the ghosts of C_i, then the number of the homomorphisms extending ϕ is

$$\prod_{i=1}^{t} S(C_i, \phi(g^i_1), \ldots, \phi(g^i_{m_i})).$$

Here, we use the fact that the values $S(C_i, \ldots)$ were not updated yet, and thus in the homomorphisms that we count, no other edge maps to $x \to y$. These products can be determined in a constant time.

The values $\hom((H, T), G, v)$ are updated similarly, before the values $S(C, \ldots)$ are updated. The changes in the values of $\hom((H, T), G, v)$ are also propagated to the stored value of $\hom(H, G)$. The complexity of the update is given by the number of choices of partial homomorphisms $\mathcal{F}$, i.e., $O(D^{|V(H)|^2})$.

Edge removal works in the same manner, except that the information is subtracted in the end, and that the clans are processed in the opposite direction, i.e. starting from the inclusion-wise smallest clans, so that the values for the graph without the edge are used in the computations. Change of the orientation of an edge can be implemented as subsequent deletion and addition. The data structure can be initialized by adding edges one by one, starting with the data structure for an empty graph G whose initialization is trivial.

5   Concluding Remarks

A natural question is whether one can design a fully dynamic data structure to decide properties expressible in First Order Logic on graphs with bounded expansion. For this purpose, it would be convenient to be able to maintain low tree-depth colorings of [14], which however appears to be difficult.

Possibly a much easier problem is the following. We have described a dynamic data structure for counting the number of appearances of H as an induced subgraph of G, for graphs from a class with bounded expansion. If this number is non-zero, can we find such an appearance? Getting this from our data structure is not entirely trivial, due to the use of the principle of inclusion and exclusion.

By a famous result of Courcelle, any property expressible in Monadic Second Order Logic can be tested for graphs of bounded tree-width in linear time. Can one design a dynamic data structure for this problem? It is not even clear how to maintain a tree decomposition of bounded width dynamically.

References

1. Brodal, G.S., Fagerberg, R.: Dynamic representations of sparse graphs. In: Dehne, F., Gupta, A., Sack, J.-R., Tamassia, R. (eds.) WADS 1999. LNCS, vol. 1663, pp. 342–351. Springer, Heidelberg (1999)
2. Chrobak, M., Eppstein, D.: Planar orientations with low out-degree and compaction of adjacency matrices. Theoretical Computer Science 86(2), 243–266 (1991)
3. Downey, R., Fellows, M.: Fixed-parameter tractability and completeness. II. On completeness for W[1]. Theoretical Computer Science 141, 109–131 (1995)
4. Dvořák, Z., Král', D.: Algorithms for classes of graphs with bounded expansion. In: Paul, C., Habib, M. (eds.) WG 2009. LNCS, vol. 5911, pp. 17–32. Springer, Heidelberg (2010)
5. Dvořák, Z., Král', D., Thomas, R.: Deciding first-order properties for sparse graphs. In: FOCS, pp. 133–142. IEEE Computer Society (2010)
6. Dvořák, Z., Král', D., Thomas, R.: Testing first-order properties for subclasses of sparse graphs. ArXiv e-prints, 1109.5036 (January 2013)
7. Eisenbrand, F., Grandoni, F.: On the complexity of fixed parameter clique and dominating set. Theoretical Computer Science 326, 57–67 (2004)
8. Eppstein, D.: Subgraph isomorphism in planar graphs and related problems. J. Graph Algorithms Appl. 3, 1–27 (1999)
9. Eppstein, D., Goodrich, M.T., Strash, D., Trott, L.: Extended dynamic subgraph statistics using h-index parameterized data structures. In: Wu, W., Daescu, O. (eds.) COCOA 2010, Part I. LNCS, vol. 6508, pp. 128–141. Springer, Heidelberg (2010)
10. Eppstein, D., Spiro, E.S.: The h-index of a graph and its application to dynamic subgraph statistics. In: Dehne, F., Gavrilova, M., Sack, J.-R., Tóth, C.D. (eds.) WADS 2009. LNCS, vol. 5664, pp. 278–289. Springer, Heidelberg (2009)
11. Frick, M., Grohe, M.: Deciding first-order properties of locally tree-decomposable structures. J. ACM 48, 1184–1206 (2001)
12. Kloks, T., Kratsch, D., Müller, H.: Finding and counting small induced subgraphs efficiently. Information Processing Letters 74, 115–121 (2000)
13. Milenkoviæ, T., Pržulj, N.: Uncovering biological network function via graphlet degree signatures. Cancer Informatics 6, 257 (2008)
14. Nešetřil, J., Ossona de Mendez, P.: Linear time low tree-width partitions and algorithmic consequences. In: Proceedings of the Thirty-Eighth Annual ACM Symposium on Theory of Computing, STOC 2006, pp. 391–400. ACM (2006)
15. Nešetřil, J., Ossona de Mendez, P.: Grad and classes with bounded expansion I. Decomposition. European J. Combin. 29, 760–776 (2008)
16. Nešetřil, J., Ossona de Mendez, P.: Structural properties of sparse graphs. Bolyai Society Mathematical Studies 19, 369–426 (2008)
17. Nešetřil, J., Ossona de Mendez, P.: First order properties on nowhere dense structures. J. Symbolic Logic 75, 868–887 (2010)
18. Nešetřil, J., Ossona de Mendez, P., Wood, D.: Characterisations and examples of graph classes with bounded expansion. Eur. J. Comb. 33, 350–373 (2012)
19. Nešetřil, J., Poljak, S.: Complexity of the subgraph problem. Comment. Math. Univ. Carol. 26, 415–420 (1985)
20. Robins, G., Morris, M.: Advances in exponential random graph (p^*) models. Social Networks 29(2), 169–172 (2007)
21. Thomassen, C.: Five-coloring graphs on the torus. J. Combin. Theory, Ser. B 62, 11–33 (1994)

Combinatorial Pair Testing: Distinguishing Workers from Slackers

David Eppstein, Michael T. Goodrich, and Daniel S. Hirschberg

Dept. of Computer Science, University of California, Irvine, CA 92697 USA

Abstract. We formalize a problem we call *combinatorial pair testing* (CPT), which has applications to the identification of uncooperative or unproductive participants in pair programming, massively distributed computing, and crowdsourcing environments. We give efficient adaptive and nonadaptive CPT algorithms and we show that our methods use an optimal number of testing rounds to within constant factors. We also provide an empirical evaluation of some of our methods.

1 Introduction

Pair programming [19] is a software development paradigm where programmers are teamed in pairs and write software together using a single workstation. This paradigm is said to produce fewer software bugs and shorter programs than when programmers work alone [20]. Consequently, it is often used to teach software design in introductory programming courses [15], including courses at the authors' institution [13], the University of California, Irvine. This design paradigm presents an additional challenge, however, for evaluative purposes. Namely, if programmers are always working in pairs, how can a manager or instructor evaluate the performance of programmers as individuals?

For instance, suppose 100 students enroll in an introductory programming course, among whom 80 are conscientious and 20 are lazy. We will call the conscientious students *workers* and the lazy ones *slackers*. In order to assign final grades to these students, the instructor would like to distinguish the workers from the slackers, but whenever she pairs a worker and a slacker on a project, the worker will do the assignment individually and the project will be completed successfully in spite of the slacker's laziness. Based on their performances, the instructor can only detect slackers when two slackers are paired together. Therefore, it would be useful for her to have systematic and effective strategies for pairing the students in order to distinguish workers from slackers.

Motivated by this evaluation problem, we are interested in this paper in the design of efficient algorithms for generating testing schemes that can distinguish workers from slackers. We formulate such problems in a general framework, which we call *combinatorial pair testing* (CPT), and we consider a number of different assessment settings, such as whether all tests must be specified in advance or whether tests may be determined adaptively. This approach allows us to focus on natural performance characteristics of such problems and provides a general framework that unifies other diagnosis problems under the CPT heading.

F. Dehne, R. Solis-Oba, and J.-R. Sack (Eds.): WADS 2013, LNCS 8037, pp. 316–327, 2013.

Combinatorial Pair Testing. Suppose we are given a set X of n individuals, ϵn of whom are *slackers* and $(1 - \epsilon)n$ of whom are *workers*, where ϵ may or may not be known in advance. A *pairwise test* is a function $T(x, y)$ that takes as its arguments two members x and y of X, and produces as output a Boolean value, the result of a test performed for x and y based solely on the worker/slacker status of x and y. Naturally, although this framework allows for T to be any Boolean function, some Boolean functions will be more interesting than others. In this paper, we are particularly interested in the following type of test:

– *Performance-based testing:* In a performance-based test, we pair two individuals, x and y, and evaluate their output performance as a team. Thus, if both x and y are slackers, then $T(x, y) = $ **false**, indicating that the two slackers, x and y, have been paired together and didn't complete the assigned project. If, on the other hand, x, y, or both, are workers, then $T(x, y) = $ **true**, indicating that the project was completed.

Performance-based testing is symmetric, so $T(x, y) = T(y, x)$, and, indeed, this test is equivalent to a Boolean OR of x and y, where a slacker corresponds to a 0 and a worker corresponds to a 1. Moreover, by De Morgan's laws, any CPT algorithm that uses OR for $T(x, y)$ can be easily modified to produce a CPT algorithm that uses AND for $T(x, y)$.

In *combinatorial pair testing* (CPT), only pairwise tests are allowed. The tests are organized in a sequence of *rounds*, in which each member of X may be tested at most once, so up to $\lfloor n/2 \rfloor$ pairwise tests can be performed in a single round. The choices made by CPT algorithms can be determined adaptively or non-adaptively and may be based on decisions that are either deterministic or randomized. In some cases we will also require some prior knowledge of the relative numbers of slackers and workers; for instance, using only performance-based tests, it is not possible to distinguish the case of there being only one slacker in X from that of there being none. Moreover, the efficiency of a given testing scheme may depend on assumptions about the number of slackers.

Because our intended applications may involve sensitive information about individual misbehavior, we may also desire CPT algorithms to have additional security or privacy guarantees. For instance, we may want our algorithms to be implementable in a way that allows an instructor to outsource the evaluation of the tests without revealing the input data [1]. Such an approach is common in privacy-preserving computations (e.g., see [21]).

One additional security condition that we study in this paper, which appears to be novel, is that of a detection algorithm that is *participant oblivious*. A detection algorithm is participant oblivious if an individual cannot detect whether he has been identified by the evaluator as a worker or slacker based only on the pairings to which he has been assigned (without knowing the status of his or her partners or the outcome of their tests). A nonadaptive algorithm must be participant oblivious, but we show that some adaptive algorithms can also be participant oblivious. The advantage of a participant-oblivious algorithm is that it allows the evaluator to impose penalties to slackers or rewards to workers after the completion of the tests without tipping off a participant during the testing process that the evaluator might already know his or her status.

Prior Related Work. Combinatorial pair testing is related to *combinatorial group testing* [6]. In combinatorial group testing, we are given a set, S, of n items, at most d of which are "defective." A test consists of selecting a subset, T, and determining whether T contains any defective items. Thus, combinatorial pair testing with performance-based testing is a restricted type of combinatorial group test in which every subset is a pair. There are many known results and applications for algorithmic problems in combinatorial group testing (e.g., see [6, 9,11]), but we are not aware of any results for the case where every subset must be a pair and in which tests are issued in groups of $O(n)$ independent tests. The closest previous analysis is by Hwang [12], who analyzes random size-k tests that are issued independently (that is, not in groups). Instead, insisting that every test to be a pair and that the pairs are issued in groups, as is required in combinatorial pair testing, goes against a standard approach in combinatorial group testing, according to which one performs tests to limit the defective items to a subset of size at most $O(d \log n)$ and then tests each such item individually.

Combinatorial pair testing is also a generalization of *processor fault diagnosis*. In this problem, we are given a set of n processors, each of which can be either faulty or good. One processor can check another, but the result of this check can only be trusted if the processor doing the testing is good. Often, in fact, one assumes that faulty processors deliberately misidentify the ones they are testing [5, 17]. Beigel *et al.* [2–4] show that if the number of faulty processors is sufficiently far below $n/2$, then $O(n)$ tests can be organized into a sequence of $O(1)$ parallel testing rounds, where each processor tests at most one other in each round, so as to identify all faulty processors. Thus, processor fault diagnosis forms a type of combinatorial pair testing problem where the tests are based on queries and, in the case when faulty processors deliberately misidentify the ones they are testing, the Boolean function that determines the outcome of a test is the exclusive-or function.

In addition, combinatorial pair testing can be applied to cheater detection in *massively distributed computations* [10], such as SETI@home and distributed.net. These systems break very large computations into independent tasks, which are then sent out to be executed to participants of the system (typically by using the idle time of individual personal computers). The problem is that some participants cheat: instead of performing the requested tasks, they rig their computers to return false or partial results, often merely for the sake of appearing on a leader board of top participants. To deal with this problem, these distributed systems often will send out the same task to two participants at the same time, and if they both return the same answer, then the output is accepted and the participants are labeled as being honest (e.g., see [7]). One challenge is that when two answers don't agree, the system doesn't immediately know which participant(s) cheated. The problem of identifying all the honest participants (and, hence, all the cheaters) in a distributed computing environment can be formulated using the approach of this paper, and solved, using combinatorial pair testing with performance-based tests based on the AND function. Previous work on cheater-detection in distributed computations does not take this approach, however, and is instead based on ad hoc solutions or reductions to processor fault diagnosis (e.g., see [7,8,10]).

Along these same lines, combinatorial pair testing also has applications to *crowdsourcing*, where complex, independent tasks, such as labeling images, is farmed out to a large set of individuals to perform. One challenge in this case is that the group of individuals contains both "experts," who are competent and diligent with their work, and "spammers," whose performance is no better than a random oracle [14]. Combinatorial pair testing can be applied in this context to weed out the spammers, much in the same way as it applies to cheater detection for massively distributed computations.

Our Results. Given a set, X, of n individuals such that ϵn of them are slackers, we formalize the combinatorial pair testing (CPT) problem, and we present and analyze several efficient CPT algorithms for identifying the slackers in X. For the adaptive case, we give an algorithm that uses $O(1/\epsilon)$ testing rounds, and we show this to be optimal to within constant factors. Moreover, we show that our algorithm is participant oblivious and we extend our algorithm to work in $O(1/\epsilon)$ testing rounds even if we don't know the value of ϵ in advance. We also give both deterministic and randomized nonadaptive CPT algorithms, and we show that the performance of these algorithms is optimal to within constant factors. For example, our randomized nonadaptive CPT algorithm uses $O((1/\epsilon)\log n)$ testing rounds and succeeds in identifying all slackers with high probability. Our analysis of this algorithm is based on an extension to the coupon collectors problem, which we call the coupon packet collectors problem. In addition, we give an empirical study of our randomized CPT algorithm that provides experimental bounds for the number of tests needed to identify various percentages of the slackers in X.

2   Adaptive Algorithms

In this section, we describe an adaptive participant-oblivious algorithm for identifying all the slackers in a performance-based testing problem.

The Two-Phase Algorithm. Assume that we know there are ϵn slackers. In phase one, we perform the following computation:

- *Phase One:* We group the individuals into $\lfloor \epsilon n/2 \rfloor$ "bins" of size at most $\lceil 2/\epsilon \rceil$ each. We then do $\lceil 2/\epsilon \rceil$ "round-robin" rounds of testing to compare all pairs of items in the same bin as each other, across all bins in parallel.

This completes phase one, and gives us the following.

Lemma 1. *After phase one completes, we will have identified all the slackers in each bin that has at least 2 slackers.*

Proof: If a bin contains 0 or 1 slackers, then each pairing of two individuals in that bin will contain a worker. Thus, every test for that bin has the same outcome (true). If, on the other hand, a bin contains 2 or more slackers, then each slacker in that bin will eventually be paired with another slacker; hence, we discover each slacker in that bin.   ∎

More importantly, we also have the following.

Lemma 2. *After phase one completes, we will have identified at least $\lceil \epsilon n/2 \rceil$ slackers.*

Proof: By the previous lemma, a slacker can go undiscovered only if he is the sole slacker assigned to a given bin. Since there are $\lfloor \epsilon n/2 \rfloor$ bins, then, by a generalized pigeonhole argument, there has to be at least $\epsilon n - \lfloor \epsilon n/2 \rfloor = \lceil \epsilon n/2 \rceil$ slackers that are assigned to bins that each contain at least two slackers.   ∎

Given that we now have identified at least $\lceil \epsilon n/2 \rceil$ slackers, in phase two we perform the following computation.

– *Phase Two:* We choose $\lceil \epsilon n/2 \rceil$ known slackers and assign one of them to each bin randomly. We assign the remaining individuals to bins, while keeping the bins to be of size at most $\lceil 2/\epsilon \rceil$. Moreover, we choose these assignments uniformly at random, subject to the rule that each bin contains a slacker and that no two individuals who were paired in round one are assigned to the same bin as each other. We then do $\lceil 2/\epsilon \rceil$ "round-robin" rounds of testing to compare all pairs of items in the same bin as each other, across all bins in parallel.

This completes phase two.

From the perspective of any individual, their bin assignment is done at random, with every bin being equally likely, and the people they are paired with are equally likely to come from any other bin from phase one. Moreover, the only nonadaptive step is the assignment of known slackers to bins in phase two, which is done via a random permutation, similar to how elements not known to be slackers are assigned. Thus, so long as individuals in our group do not collude, this algorithm is participant oblivious. Note, in addition, that any bin that now contains a previously undiscovered slacker, will necessarily contain at least two slackers. Thus, by Lemma 1, we will discover this (and all other) remaining slackers in phase two.

Theorem 1. *Given a set, X, of n workers and slackers, such that ϵn of the individuals in X are known to be slackers, we can identify all the slackers in X in $O(1/\epsilon)$ rounds of disjoint pairwise tests, in a participant-oblivious adaptive fashion.*

This bound is optimal, to within constant factors, as the following theorem establishes.

Theorem 2. *Given a set, X, of n workers and slackers, such that ϵn of the individuals in X are slackers, then identifying all the slackers in X (either deterministically in the worst case or randomly with success probability $\geq 1/2$) requires at least $\Omega(1/\epsilon)$ rounds of disjoint pairwise tests.*

Proof: We consider the randomized case first, and we assume a randomized input distribution in which all permutations of workers and slackers are equally likely. Let x be a random variable whose value is one of the slackers in the input, chosen uniformly at random among the slackers. In the first $1/(2\epsilon)$ rounds of testing, at most $n/2$ of the members of X may become identified. In any given round of testing in which x has not already been identified as a slacker, at most $\epsilon n - 1$ of the unidentified members of X can be paired with (identified or unidentified) slackers other than x, and x is equally likely to be any one of the $\geq n/2$ unidentified members, so the probability that x becomes identified by being paired with a slacker is at most $(\epsilon n - 1)/(n/2) < \epsilon/2$. By the union bound, after $1/(4\epsilon) = \Omega(1/\epsilon)$ rounds, x will remain unidentified with probability greater than $1/2$, so the probability that all slackers are identified is less than $1/2$.

Since this randomized input distribution fools even a randomized algorithm with probability at least $1/2$, after $\Omega(1/\epsilon)$ rounds, it follows that for every deterministic algorithm there exists an input in this distribution that is certain to fool the algorithm with the same number of rounds.   ∎

Estimating Epsilon. Suppose now that there are ϵn slackers, but we do not know the value of ϵ. Instead, let us assume we have an estimate, ϵ', and our goal is to use $O(1/\epsilon')$ rounds, and either find all ϵn slackers, with $\epsilon' \leq 2\epsilon$, or determine that $\epsilon' > \epsilon$.

Consider again the above two-phase algorithm, but now assume that it is calibrated for ϵ' instead of ϵ. One possible outcome of phase one, is that we discover at least $\lceil \epsilon' n/2 \rceil$ slackers, which then allows us to discover all the slackers in phase two. In this case,

$$\epsilon n \geq \epsilon' n/2,$$

hence, $\epsilon' \leq 2\epsilon$.

Alternatively, phase one may discover fewer than $\lceil \epsilon' n/2 \rceil$ slackers. Since a bin that appears to hold no slackers can hold at most one, this implies that

$$\epsilon < \epsilon'/2 + \epsilon'/2 = \epsilon'.$$

Thus, our two-phase algorithm achieves our goal.

We can therefore now use our two-phase algorithm in an iterative fashion. We start with $\epsilon' = 1/2$, and use the two-phase algorithm with this estimate for ϵ. If we discover all the slackers, then we are done. Otherwise, we determine that $\epsilon < \epsilon'$. In this case, we set $\epsilon' \leftarrow \epsilon'/2$ and we repeat the process with this estimate. Eventually, we will reach a point where we discover all the slackers, with $\epsilon' \leq 2\epsilon$. Moreover, since the previous iteration, if there is one, would have failed, we also know that $\epsilon < 2\epsilon'$, that is, $\epsilon' > \epsilon/2$. The number of testing rounds is therefore proportional to

$$2 + 4 + 8 + \cdots + 1/\epsilon' \leq 2 + 4 + 8 + \cdots + 2/\epsilon \leq 4/\epsilon.$$

Therefore, even without knowing the value of ϵ, the number of rounds is $O(1/\epsilon)$, which implies the following.

Theorem 3. *Given a set, X, of n workers and slackers, such that ϵn of the individuals in X are slackers, we can identify all the slackers in X in $O(1/\epsilon)$ rounds of $O(n)$ pairwise tests per round, in a participant-oblivious adaptive algorithm, without knowing ϵ in advance.*

In the full version of this paper, we explore optimizations to the constant factors in the above bounds, in adaptive CPT algorithms for the case when $\delta = 1 - \epsilon \leq 1/2$, that is, when at least half of the individuals are slackers. Such instances of the combinatorial pair testing problem arise naturally in massively distributed and crowdsourcing applications, for example, where the roles of slackers and workers are reversed and the testing function, T, is Boolean AND instead of OR.

3   Nonadaptive Pair Testing

In this section, we study nonadaptive algorithms for combinatorial pair testing, to identify ϵn slackers in a group of n individuals. In this case, if we assume that we do not know the value of ϵ, then the only valid algorithm is the trivial brute-force algorithm that compares every pair of individuals, since a nonadaptive algorithm must specify all its tests in advance and it is possible that $\epsilon = 2/n$. Therefore, we assume that we know in advance that there are ϵn slackers.

Deterministic Nonadaptive Pair Testing. Unfortunately, nonadaptive deterministic pair testing is not very interesting, because it requires a linear number of rounds. The argument is simple: suppose a deterministic nonadaptive pair testing algorithm could use at most $(1 - \epsilon)n/2$ rounds. Then, in the graph of pairs that are tested by the algorithm, each vertex would have at most $(1-\epsilon)n/2$ neighbors. An adversary could choose one edge of the graph, make one of its two endpoints a slacker and the other endpoint a worker, set all neighbors of these two vertices to be workers, and fill out the rest of the graph arbitrarily to fit whatever number of slackers and workers is desired. From the set of tests that are performed, there is no way to distinguish which of the two endpoints of the chosen edge is the slacker and which is the worker. Therefore, there must be at least $\Omega((1 - \epsilon)n)$ rounds in a deterministic nonadaptive CPT algorithm, which, for any fixed $\epsilon < 1$, is asymptotically not any better than the brute-force algorithm that tests every pair.

This bound can be achieved as an upper bound, as well, using an algorithm that pairs each individual, x, with at least $(1-\epsilon)n+1$ other distinct individuals, using $O((1 - \epsilon)n)$ rounds. For this algorithm, at least one of the individuals paired with each such x must be a slacker.

Randomized Nonadaptive Pair Testing. Despite the nonexistence of efficient deterministic nonadaptive pair testing algorithms, there is a simple randomized algorithm for nonadaptive randomized testing, which succeeds with high probability using many fewer tests than the deterministic nonadaptive solution. In particular, let us repeatedly choose a random matching of all the members of the set, X, for some value, k, number of rounds. Each matching corresponds to a round of testing. For instance, for $k = (c/\epsilon) \log n$, for a sufficiently large constant, $c \geq 1$, then this scheme uses $O((1/\epsilon) \log n)$ rounds and $O((n/\epsilon) \log n)$ tests in total.

Relation to the Coupon Collector's Problem. The expected performance of the nonadaptive randomized algorithm described above can be analyzed precisely using a variant of the classical *coupon collector's problem*.

In the coupon collector's problem, a collector wishes to collect a set of n trading cards, by randomly acquiring one card at a time, and the problem is to calculate the number of steps that are required until, with high probability, all cards have been collected. Now consider a slight variation, which we call the *coupon packet collector's problem*: instead of buying one card at a time, the collector buys the cards in packets of m cards [18]. Each packet of trading cards is guaranteed to have no duplicates, and is uniformly random among all m-card samples of the whole set of cards. How does this affect the total time required for the collector? If m is much smaller than n, the difference between this problem and the standard coupon collector's problem is very small: a random sample of m cards, each independently and uniformly randomly sampled, is very likely to be duplicate-free. But if m is a constant fraction of n, then the avoidance of duplicates in each packet is very likely to cause the number of packets that the collector needs to collect to be smaller by a constant fraction than the number that a one-at-a-time collector would need. But what is the fraction?

In the coupon packet collector's problem, the probability that a card remains uncollected after k rounds is $(1 - m/n)^k$. So, after k rounds, by the linearity of expectation, the expected number of uncollected cards is $n(1 - m/n)^k$. Thus, for $k = (1 + \alpha) \log_{1/(1-m/n)} n$ rounds, the expected number of uncollected cards is $1/n^\alpha$; hence, by Markov's inequality, with very high probability, $1 - 1/n^\alpha$, all the cards are collected.

In the pair testing problem, observe that a slacker's status is identified whenever the slacker is paired with another slacker, and a student's status is identified whenever that student is paired with a known slacker. If we allow these identifications to be made retroactively (*i.e.*, once we find a known slacker we use that identity to confirm as workers all the other students the slacker has already been paired with) then there is a very simple criterion for whether we have identified everybody: we have done so if and only if all students have been paired at least once with a slacker. More weakly, we have identified all slackers whenever the slackers have all been paired with another slacker in some round of testing. Suppose that there are m slackers and n total students. In each round, exactly m students will be paired with slackers, so it is very much like the coupon packet collector's problem, where the trading cards in a packet correspond to the students that are paired with slackers. There is a small complication, however: in the pair testing problem the sets of students that are identified are not quite uniformly random over all m-element subsets of students. In particular, the slackers are slightly less likely to be paired with other slackers than the workers, because there are fewer other slackers for them to be paired with.

To be precise, in the case that there are an even number of students, a slacker has probability exactly $(n-m)/(n-1)$ of remaining unidentified after one round, because there are $n-1$ students the slacker could be paired with, each of which is equally likely, and $n-m$ of which (the workers) fail to identify the slacker. The probability that a specific student is identified in any one round is independent of the same probability for the same student in a different round, so after

$$k = (1+\alpha) \log_{\frac{n-1}{n-m}} m$$

rounds, the probability that an individual slacker remains unidentified is $1/m^{1+\alpha}$. Similarly, a worker has probability exactly $(n-m-1)/(n-1)$ of not having been paired with a slacker after one round, and probability $1/(n-m)^{1+\alpha}$ of never having been paired with a slacker after

$$k = (1+\alpha) \log_{\frac{n-1}{n-m-1}} (n-m)$$

rounds. Different students have probabilities that are not independent of each other, but by linearity of expectation after

$$k = (1+\alpha) \max \left\{ \log_{\frac{n-1}{n-m}} m, \log_{\frac{n-1}{n-m-1}} (n-m)) \right\}$$

rounds the expected number of students who have not been paired with a slacker is $\min\{1/m^\alpha, 1/(n-m)^\alpha\}$, so by Markov's inequality, with high probability all students will be identified. In the case that there are an odd number of students, there are n alternatives for each student in each round rather than $n-1$, so the number of rounds needed is instead

$$k = (1+\alpha)(\log_{\frac{n}{n-m+1}} m + \log_{\frac{n}{n-m}} (n-m)).$$

In either case, for $m = \epsilon n$ slackers, if we extend the above two bounds so that the number of rounds is increased to

$$k = (1+\alpha) \log_{1/(1-\epsilon)} n,$$

then the expected number of unclassified students is $1/n^\alpha$. Thus, by Markov's inequality, there are no unclassified students with high probability, $1 - 1/n^\alpha$. Choosing $\alpha \geq 1$ to be a fixed constant, and using the inequality, $x < -\ln(1-x)$, for $0 < x < 1$, we get the following result.

Theorem 4. *Given a set, X, of n individuals, such that $\epsilon n \geq 2$ of them are slackers and the rest are workers, we can distinguish the workers and slackers using $O((1/\epsilon) \log n)$ rounds of random performance-based tests, with $O(n)$ tests per round, with high probability, $1 - 1/n^c$, in a nonadaptive fashion, for any fixed constant $c \geq 1$.*

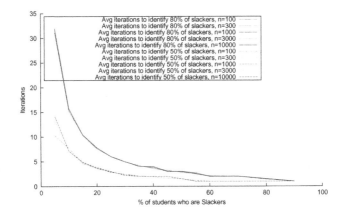

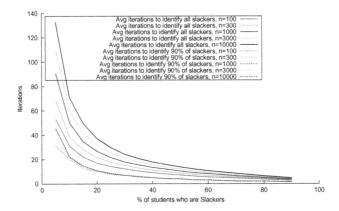

Fig. 1. Results for number of random tests needed to identify various percentages of slackers, for various values of the set size, n, and slacker percentage, ϵ

In a nonadaptive randomized strategy, the most information is gathered by randomly matching of the members of X and testing each matched pair. Thus, for any slacker, s, the probability s is not paired with another slacker is at least $(1 - \epsilon)$. So, after k independent rounds of testing, the probability s has not been discovered to be a slacker is at least $(1 - \epsilon)^k$, which we can bound as

$$(1 - \epsilon)^k \geq \left(\frac{1 - \epsilon}{e}\right)^{\epsilon k},$$

by an inequality due to Niculescu and Vernescu [16]. Thus, we have the following.

Theorem 5. *For $2/n \leq \epsilon \leq 1/2$, we require $\Omega((1/\epsilon) \log n)$ rounds of testing for each slacker to be identified, with probability at least $1 - 1/n$, in a nonadaptive randomized testing scheme for a set of n members having ϵn slackers.*

Therefore, the above analysis is tight to within constant factors.

Experimental Results. To get a better handle on the expected number of tests needed to identify various percentages of slackers, we performed an experimental study of the above nonadaptive randomized CPT algorithm. We performed tests for values of n ranging from 100 to 10000, with percentage of slackers ranging from 5% to 90%. We then performed tests to determine the average number of tests required in order to identify 50%, 80%, 90%, and 100% of the slackers. We show the results in Figure 1.

4   Conclusion

In this paper, we have given efficient algorithms for solving combinatorial pair testing problems, along with lower bounds showing that our algorithms are optimal to within constant factors. All of our algorithms assume we are using performance-based tests. Therefore, one possible direction for future work would be to explore CPT algorithms and applications for other kinds of tests (other than the exclusive-or tests used in processor fault diagnosis [2–5, 17]). Another direction would be to enlarge the size of tested groups beyond two and explore the effect of different group sizes on the numbers of rounds needed for testing.

Acknowledgments. This research was supported in part by the National Science Foundation under grants 1011840, 1217322, and 1228639, and by the Office of Naval Research under MURI grant N00014-08-1-1015.

References

1. Atallah, M.J., Frikken, K.B., Blanton, M., Cho, Y.: Private combinatorial group testing. In: ACM Symp on Information, Computer and Communications Security (ASIACCS), pp. 312–320 (2008)

2. Beigel, R., Hurwood, W., Kahale, N.: Fault diagnosis in a flash. In: Proc. IEEE Foundations of Computer Science (FOCS), pp. 571–580 (October 1995)
3. Beigel, R., Kosaraju, S.R., Sullican, G.F.: Locating faults in a constant number of parallel testing rounds. In: ACM Symp. on Parallel Algorithms and Architectures (SPAA), pp. 189–198 (1989)
4. Beigel, R., Margulis, G., Spielman, D.A.: Fault diagnosis in a small constant number of parallel testing rounds. In: ACM Symp. on Parallel Algorithms and Architectures (SPAA), pp. 21–29 (1993)
5. Blecher, P.M.: On a logical problem. Discrete Mathematics 43(1), 107–110 (1983)
6. Du, D.-Z., Hwang, F.: Combinatorial Group Testing and Its Applications. Series on Applied Mathematics. World Scientific (2000)
7. Du, W., Goodrich, M.T.: Searching for high-value rare events with uncheatable grid computing. In: Ioannidis, J., Keromytis, A., Yung, M. (eds.) ACNS 2005. LNCS, vol. 3531, pp. 122–137. Springer, Heidelberg (2005)
8. Du, W., Jia, J., Mangal, M., Murugesan, M.: Uncheatable grid computing. In: 24th Int. Conf. on Distributed Computing Systems (ICDCS), pp. 4–11 (2004)
9. Eppstein, D., Goodrich, M.T., Hirschberg, D.S.: Improved combinatorial group testing algorithms for real-world problem sizes. SIAM J. Comput. 36(5), 1360–1375 (2006)
10. Goodrich, M.T.: Pipelined algorithms to detect cheating in long-term grid computations. Theoretical Computer Science 408(2/3), 199–207 (2008)
11. Goodrich, M.T., Atallah, M.J., Tamassia, R.: Indexing information for data forensics. In: Ioannidis, J., Keromytis, A., Yung, M. (eds.) ACNS 2005. LNCS, vol. 3531, pp. 206–221. Springer, Heidelberg (2005)
12. Hwang, F.K.: Random k-set pool designs with distinct columns. Probab. Eng. Inf. Sci. 14(1), 49–56 (2000)
13. Jacobson, N., Schaefer, S.K.: Pair programming in CS1: overcoming objections to its adoption. SIGCSE Bull. 40(2), 93–96 (2008)
14. Liu, Q., Peng, J., Ihler, A.: Variational inference for crowdsourcing. In: Bartlett, P., Pereira, F.C.N., Burges, C.J.C., Bottou, L., Weinberger, K.Q. (eds.) Advances in Neural Information Processing Systems (NIPS), pp. 701–709 (2012)
15. Nagappan, N., Williams, L., Ferzli, M., Wiebe, E., Yang, K., Miller, C., Balik, S.: Improving the CS1 experience with pair programming. In: Proc. 34th SIGCSE Technical Symp. on Computer Science Education (SIGCSE 2003). SIGCSE Bulletin, vol. 35(1), pp. 359–362 (2003)
16. Niculescu, C.P., Vernescu, A.: A two-sided estimate of $e^x - (1 + x/n)^n$. Journal of Inequalities in Pure and Applied Mathematics 5(3) (2004)
17. Pelc, A., Upfal, E.: Reliable fault diagnosis with few tests. Comb. Probab. Comput. 7(3), 323–333 (1998)
18. Stadje, W.: The collector's problem with group drawings. Advances in Applied Probability 22(4), 866–882 (1990)
19. Williams, L., Kessler, R.R.: Pair Programming Illuminated. Addison-Wesley (2003)
20. Williams, L., Kessler, R.R., Cunningham, W., Jeffries, R.: Strengthening the case for pair programming. IEEE Software 17(4), 19–25 (2000)
21. Yao, A.C.: How to generate and exchange secrets. In: Proceedings of the 27th Annual Symposium on Foundations of Computer Science, pp. 162–167. IEEE Computer Society, Washington, DC (1986)

Approximation Algorithms for B_1-EPG Graphs

Dror Epstein[1,2], Martin Charles Golumbic[1,2], and Gila Morgenstern[2]

[1] Department of Computer Science, University of Haifa
[2] Caesarea Rothschild Institute (CRI), University of Haifa

Abstract. The edge intersection graphs of paths on a grid (or EPG graphs) are graphs whose vertices can be represented as simple paths on a rectangular grid such that two vertices are adjacent if and only if the corresponding paths share at least one edge of the grid. We consider the case of single-bend paths, namely, the class known as B_1-EPG graphs. The motivation for studying these graphs comes from the context of circuit layout problems. It is known that recognizing B_1-EPG graphs is NP-complete, nevertheless, optimization problems when given a set of paths in the grid are of considerable practical interest.

In this paper, we show that the coloring problem and the maximum independent set problem are both NP-complete for B_1-EPG graphs, even when the EPG representation is given. We then provide efficient 4-approximation algorithms for both of these problems, assuming the EPG representation is given. We conclude by noting that the maximum clique problem can be optimally solved in polynomial time for B_1-EPG graphs, even when the EPG representation is not given.

1 Introduction

Edge intersection graphs of paths on a grid (or for short EPG graphs) were first introduced by Golumbic, Lipshteyn and Stern in [9]. This is the class of graphs whose vertices can be represented as simple paths on a rectangular grid so that two vertices are adjacent if and only if the corresponding paths share at least one edge of the grid.

EPG graphs have a practical use, e.g., in the context of circuit layout setting, which may be modeled as paths (wires) on a grid. In the knock-knee layout model, two wires may either cross or bend (turn) at a common grid point, but are not allowed to share a grid edge; that is, overlap of wires is not allowed. In this context, some of the classical optimization graph problems are relevant, e.g., maximum independent set and coloring. More precisely, the layout of a circuit may have multiple layers, each of which contains no overlapping paths. Referring to a corresponding EPG graph, then each layer is an Independent Set and a valid partitioning into layers corresponds to a proper coloring.

In [9], the authors show that every graph is an EPG graph. That is, for every graph $G = (V, E)$ there exists an EPG representation $\langle \mathcal{P}, \mathcal{G} \rangle$ where $\mathcal{P} = \{P_v : v \in V\}$ is a collection of paths on a grid $\mathcal{G}$, corresponding to the vertices of V and satisfying: paths $P_v, P_u \in \mathcal{P}$ share a grid edge of $\mathcal{G}$ if and only if $(v, u) \in E$.

F. Dehne, R. Solis-Oba, and J.-R. Sack (Eds.): WADS 2013, LNCS 8037, pp. 328–340, 2013.
© Springer-Verlag Berlin Heidelberg 2013

Moreover, they showed that if G has n vertices and m edges, then there exists an EPG representation $\langle \mathcal{P}, \mathcal{G} \rangle$ of G in which $\mathcal{G}$ is a grid of size $n \times (n + m)$ and the paths in $\mathcal{P}$ are monotonic. As such, much of the current research today focuses on subclasses of EPG graphs, and, in particular, limiting the type of paths allowed.

A turn of a path at a grid point is called a *bend* and a graph is called a *k-bend EPG graph* (denoted B_k-EPG) if it has an EPG representation in which each path has at most k bends. It is both interesting mathematically, and justified by the circuit layout application described above, to consider subclasses of graphs, e.g., by bounding the number of bends allowed in each path.

A number of mathematical results on B_k-EPG graphs have been shown recently. In [2], the authors show that for any k, only a small fraction of all labeled graphs on n vertices are B_k-EPG. Improving a result of [3], it was shown in [12] that every planar graph is a B_4-EPG graph. It is still open whether $k = 4$ is best possible. So far it is only known that there are planar graphs that are B_3-EPG graphs and not B_2-EPG graphs. The authors in [12] also show that all outerplanar graphs are B_2-EPG graphs thus proving a conjecture of [3]. For the case of B_1-EPG graphs, Golumbic, Lipshteyn and Stern [9] showed that every tree is a B_1-EPG graph, and Asinowski and Ries [1] showed that every B_1-EPG graph on n vertices contains either a clique or a stable set of size at least $n^{1/3}$. In [1], the authors also give a characterization of the B_1-EPG graphs among some subclasses of chordal graphs, namely, chordal bull-free graphs, chordal claw-free graphs, chordal diamond-free graphs, and special cases of split graphs. In [5], a characterization of the subfamily of cographs that are B_1-EPG graphs is given by a complete family of minimal forbidden induced subgraphs.

The simplest case, B_0-EPG graphs, where all paths a straight line segments, are exactly the well studied class of *interval graphs* (the intersection graphs of intervals on a line), and it is well-known that these can be colored optimally with the exact minimum number of colors $\chi(G)$ in polynomial time (see [8]). This is no longer the case when $k > 0$.

In this paper, we consider approximation algorithms for B_1-EPG which are the edge intersection graphs of (at most) single bend paths on a rectangular grid. Heldt et al. [11] have proved that the recognition problem for B_1-EPG is NP-complete. Moreover, Cameron, Chaplick and Hoang [4] proved that even the recognition of a subclass of B_1-EPG know as ∟-EPG is NP-complete; we define this subclass in Section 2 below. Thus, for all of the algorithms that we will later present, an EPG representation $\langle \mathcal{P}, \mathcal{G} \rangle$ of G is assumed to be given as part of the input.

MAXIMUM INDEPENDENT SET, MINIMUM COLORING, and MAXIMUM CLIQUE are fundamental optimization problems in graph-theory. These problems arises naturally in many scenarios involving resource allocation in the presence of interference. The graph coloring problem deals with assigning colors to the vertices of a graph such that no two adjacent vertices share the same color, and the number of colors used is minimized. A coloring using at most c colors is called a (proper) c-coloring. The smallest number of colors needed to color a graph G is called its *chromatic number*, and is denoted $\chi(G)$. The graph coloring problem is known to be NP-hard. The current best known approximation ratio for the graph coloring problem is $O(n\frac{(\log\log n)^2}{(\log n)^3})$, where n is the number of vertices in the graph; see [10]. In graph theory, an *Independent Set* (*Stable Set*) is a set of vertices in a graph, where no two of which are adjacent. This corresponds to a *Clique* in the graphs complement. The size of a maximum independent set of a graph G is denoted by $\alpha(G)$, and the size of a maximum clique is denoted by $\omega(G)$. The problem of finding a largest independent set for a given graph G is called the *Maximum Independent Set Problem* (MIS) which is NP-hard. Even for graphs whose maximum degree is bounded by b, the current best known approximation ratio for the MIS problem are a fraction of b, see references in [13].

The paper is organized as follows. We begin with preliminary definitions in Section 2. In Section 3, we first prove that coloring B_1-EPG graphs is NP-complete, and then we present a 4-approximation algorithm for coloring B_1-EPG graphs in polynomial time. Similarly, in Section 4, we prove that finding a maximum independent set in B_1-EPG graphs is NP-complete, and then present a 4-approximation algorithm for the problem. Conclusions and open problems are given in Section 5 where we note that the maximum clique problem can be optimally solved in polynomial time for B_1-EPG graphs.

2 Preliminaries

Let $\langle \mathcal{P}, \mathcal{G}\rangle$ be a B_1-EPG representation of a graph $G = (V, E)$. We say that paths P_v and P_u are adjacent paths if v and u are adjacent vertices in G, i.e., P_v and P_u share a common grid edge of $\mathcal{G}$. We also say that $G = EPG(\langle \mathcal{P}, \mathcal{G}\rangle)$. In B_1-EPG graphs, each vertex corresponds to a path of one of the following shapes: ∟, Γ, ⌐ or ⌐, allowing horizontal or vertical segments as well. We refer to a path of shape $\tau \in \{$∟, Γ, ⌐, ⌐, $|$, $-\}$ as an τ-path. We denote by $\mathcal{P}_∟$ the collection of ∟-paths in $\mathcal{P}$, and similarly we use the notations $\mathcal{P}_⌐$, $\mathcal{P}_Γ$ and $\mathcal{P}_⌐$. For no-bend paths we complete the definition by referring them as ∟-paths. Sometimes, it is of interest to consider even finer, more restrictive subclasses of B_1-EPG by limiting the type of bends that are allowed, namely, the subclasses formed by the subsets of the four single bend shapes (i.e., $\{∟\}$, $\{⌐, ∟\}$, $\{∟, ⌐\}$, $\{∟, Γ, ⌐\}$, where all other subsets are isomorphic to these up to 90° rotation), allowing paths with no-bend as well. We denote these classes by ∟-EPG, ⌐∟-EPG, ∟⌐-EPG and ∟ Γ ⌐-EPG respectively.

Let G be a ⌐∟-EPG with grid representation $\langle \mathcal{P}, \mathcal{G}\rangle$. We define the lexicographic (LEX) order $\prec$ on the paths in $\mathcal{P}$ as follows; see Figure 1. For path $P_v \in \mathcal{P}$ we denote by ∂P_v the bottommost-leftmost grid point that is contained in P_v, that is, $\partial P_v = \min_y \{\min_x \{(x, y) \in P_v\}\}$. We say that $P_v \prec P_u$ if ∂P_v lies below ∂P_u or they both lie in the same row and ∂P_v is left of ∂P_u. We complete $\prec$ to a total order by arbitrarily breaking ties.

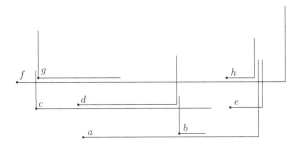

Fig. 1. The LEX ordering of a ⌐L-EPG representation: $a \prec b \prec c \prec d \prec e \prec f \prec g \prec h$

3   Coloring B_1-EPG Graphs

3.1   Hardness Result for Coloring B_1-EPG Graphs

In this section, we prove that coloring problem on B_1-EPG graphs is NP-complete by a reduction from the problem of coloring circle graphs which was shown to be NP-complete in [7].

We start by defining circle graphs. A *circle graph* is the intersection graph of a set of chords of a circle. That is, it is an undirected graph whose vertices can be associated with chords of a circle such that two vertices are adjacent if and only if the corresponding chords cross each other. We may assume without loss of generality that no two chords in the diagram of chords of the circle share a common endpoint. Coloring circle graphs remains NP-complete even if the graph is given by its chord model [7].

Theorem 1. *Let G be a B_1-EPG graph. Coloring G with the exact number of colors $\chi(G)$ is NP-complete.*

Proof. Let G be a circle graph. We construct a B_1-EPG representation for a graph G' so that G is c-colorable if and only if G' is. The construction is as follows; see Figures 2 and 3 for an illustration. We slide all the endpoints of the chords to the upper right quadrant of the circle, while preserving their order on the circle (thus, intersections are not changed under these transformations). Now, we replace each chord by an L-shape bend path, where every vertex v in G corresponds to a path P_v with the same endpoints on the circle. Note that since we assumed that all endpoints are distinct, the horizontal segment of each path lies on a unique horizontal line, and the vertical segment lies on a unique vertical line. Moreover, the intersection points of pairs of paths are in one-to-one correspondence with the edges of the graph.

Consider an intersection point between two paths P_v and P_u in the representation, where the horizontal section of P_v intersects with the vertical segment of P_u. We split P_v at the intersection point into two disjoint parts; the left part is a L-path, and the right one is a −-path. We complete the latter to a L-path by joining it to a vertical segment that overlaps only P_u. We also add $(c-1)$ short

Fig. 2. (a) A circle diagram. (b) Each chord is replaced by a single-bend path on the grid.

--paths overlapping only these two segments of the former path P_v. Perform this transformation for every intersection point, and let G' be the B_1-EPG-graph of this transformed representation. This, of course, may have split P_v into several segments, $P_{v_1}, P_{v_2}, \ldots, P_{v_k}$, with consecutive segments P_{v_i} and $P_{v_{i+1}}$ being joined by such a set of $(c-1)$ short horizontal paths: a $(c-1)$-clique in G' overlapping only P_{v_i} and $P_{v_{i+1}}$. See Figure 3 for an illustration.

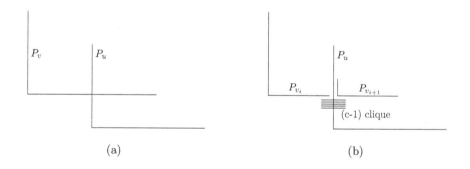

Fig. 3. (a) Intersecting paths. (b) The horizontal is "split" and "glued" using a $(c-1)$-clique.

It is clear from the transformation that the obtained graph G' is indeed a B_1-EPG graph. Moreover, the transformation can be performed in polynomial time and the size of G' is polynomial in the size of G, since $|V(G')| = n + ce \leq n + n^3$, where G has n vertices and e edges.

We now claim that G is c-colorable if and only if G' is c-colorable. Let $\varphi : V \mapsto \{1, \cdots, c\}$ be a valid assignment of colors for G. Then to color G' it suffices to (1) color each vertex from G' that came from an original path P_v (including its vertical segment and all of its horizontal split segments $P_{v_1}, \ldots, P_{v_k}$) with the color used in G, and (2) for each newly added $(c-1)$-clique (the short segments overlapping only P_{v_i} and $P_{v_{i+1}}$ which have the same color in the construction), we can use the $(c-1)$ remaining colors. This clearly colors G' in c colors.

We now show that if G' is c-colorable then G is c-colorable. Assume we have a c-coloring of the graph G'. Since the $(c-1)$-clique connecting any P_{v_i} and $P_{v_{i+1}}$ requires $(c-1)$ colors, consequently, P_{v_i} and $P_{v_{i+1}}$ have the same remaining c^{th} color. Moreover, let P_u be the path that intersects P_v in G and whose intersection point with P_v is the split point between P_{v_i} and $P_{v_{i+1}}$, then P_u and $P_{v_{i+1}}$ are adjacent in G', thus get distinct colors. Since the coloring of G' is proper, it also gives a proper coloring of G: color the path representing v in G with the same (common) color of its split segments $P_{v_1}, \dots, P_{v_k}$ in G'. This concludes the proof of the theorem.   □

Observe that by our construction, the paths in G' are either ⌐-paths or —-paths, we thus conclude:

Corollary 2. *Let G be a* ⌐*-EPG graph. Coloring G with the exact number of colors $\chi(G)$ is NP-complete.*

3.2   A 4-Approximation Algorithm for Coloring B_1-EPG Graphs

We start by presenting a "subroutine" in Algorithm 3.1 that computes an approximation solution for a ⌐⌐-EPG representation. We then apply it more generally to an arbitrary B_1-EPG representation. It is a greedy First-Fit algorithm using the LEX ordering $\prec$, defined in Section 2 so clearly, it produces a proper coloring. Lemma 1 will show that when used for a ⌐⌐-EPG graph, Algorithm 3.1 achieves a 2-approximation. We will use the notation $c(v)$ for the color assigned to vertex v.

Algorithm 3.1 Greedy-⌐⌐-EPG-Coloring (Input: $\mathcal{P} = \mathcal{P}_\lrcorner \cup \mathcal{P}_\llcorner$)

1: **for each** $P_v \in \mathcal{P}$ (in increasing order $\prec$) **do**
2:     $c(v) \leftarrow$ least color not in use among v's neighbors
3: **return** total number k of distinct colors used and the coloring

---

Applying Algorithm 3.1 to the representation in Figure 1 gives the coloring: $c(a) = c(c) = c(f) = 1$; $c(b) = c(e) = c(g) = 2$; $c(d) = c(h) = 3$.

For every path $P_v \in \mathcal{P}$ we denote by $\widetilde{\Gamma}(P_v)$ the collection of paths adjacent to P_v that have been colored by Algorithm 3.1 prior to P_v. When convenient, we refer to $\widetilde{\Gamma}(P_v)$ as a set of vertices. The color assigned to P_v by Algorithm 3.1 is dependent only on the colors assigned to paths in $\widetilde{\Gamma}(P_v)$, thus we have Observation 3.

Observation 3. *Let $\langle \mathcal{P}, \mathcal{G} \rangle$ be a* ⌐⌐*-EPG representation of a graph $G = (V, E)$, and let P_v and P_u be adjacent paths. If $P_u \prec P_v$, then P_v and P_u share at least one of two grid edges e_1 and e_2 as follows:*

- If P_v is a ∟-path, then e_1 and e_2 are respectively the horizontal and vertical grid edges contained in P_v and attached to its bend point.
- If P_v is a ⌐-path, then e_1 is the left-most horizontal grid edge contained in P_v and e_2 is the vertical grid edge attached to its bend point.
- If P_v is a |-path, then e_1 is the bottom-most vertical grid edge contained in P_v (e_2 in this case is undefined).
- If P_v is a −-path, then e_1 is the left-most horizontal grid edge contained in P_v (e_2 in this case is undefined).

Lemma 1. *Let G be a ⌐∟-EPG graph, then Algorithm 3.1 uses at most $2\chi(G)$ colors.*

Proof. Let k be the maximum color used by Algorithm 3.1, we show that $k \leq 2\chi(G)$. Indeed, put $G = (V, E)$ and let $v \in V$ be a vertex for which $c(v) = k$. Notice that whenever Algorithm 3.1 colors a vertex, the assigned color is determined by its previous-colored neighbors $\widetilde{\Gamma}(P_v)$. Notice that if Algorithm 3.1 colored v with color k, then k is the least color that not in use for any vertex $u \in \widetilde{\Gamma}(P_v)$, thus $k \leq \widetilde{\Gamma}(P_v) + 1$. Moreover, by Observation 3, we have that each path in $\widetilde{\Gamma}(P_v)$ shares at least one of two specified grid edges contained in $\mathcal{P}_v$ (denoted e_1 and e_2). We conclude that at least half of the paths in $\widetilde{\Gamma}(P_v)$ contain one of those edges and without loss of generality, we assume it is e_1. Now, observe that any collection of paths containing a common edge corresponds to a clique in G, in particular, those paths in $\widetilde{\Gamma}(P_v)$ that contain e_1 together with v itself, form a clique. We get $\frac{1}{2}\widetilde{\Gamma}(v) + 1 \leq \omega(G) \leq \chi(G)$, thus $k \leq \widetilde{\Gamma}(P_v) + 1 < 2\omega(G) \leq 2\chi(G)$, which completes the proof.   □

Remark 1. Clearly, by rotating a representation by 180°, Algorithm 3.1 can be "turned" from Greedy-⌐∟-EPG-Coloring into Greedy-¬⌐-EPG-Coloring.

We now use Algorithm 3.1 as a building block in Algorithm 3.2 in order to colors B_1-EPG graphs.

Algorithm 3.2 B_1-EPG Coloring 4-Approximation (Input: $G = EPG(\langle \mathcal{P}, \mathcal{G} \rangle)$)

1: Let $\mathcal{P} = \mathcal{P}_∟ \cup \mathcal{P}_⌐ \cup \mathcal{P}_¬ \cup \mathcal{P}_⌐$
2: $k_1 \leftarrow$ **Greedy-⌐∟-EPG-Coloring**$(\mathcal{P}_∟ \cup \mathcal{P}_⌐)$
3: $k_2 \leftarrow$ **Greedy-¬⌐-EPG-Coloring**$(\mathcal{P}_⌐ \cup \mathcal{P}_¬)$ // using different color names //
4: **return** total number of distinct colors used and the coloring

---

Algorithm 3.2 partitions the paths in $\mathcal{P}$ into two subsets $\mathcal{P}_∟ \cup \mathcal{P}_⌐$ and $\mathcal{P}_⌐ \cup \mathcal{P}_¬$, each induces a subgraph of G, which is a ⌐∟-EPG graph (denoted G_1 and G_2 respectively). Then, it colors each of these two graphs G_1 and G_2 using Algorithm 3.1, with distinct "palettes" of colors. Clearly, the coloring produced by

Algorithm 3.2 is proper. Notice that in order to color a graph G, one needs at least the maximum of $\chi(G_1), \chi(G_2)$ colors. By Lemma 1, Algorithm 3.2 uses at most $2\chi(G_1) + 2\chi(G_2) \leq 4\chi(G)$ colors, we thus have Theorem 4 below.

Theorem 4. *Let G be a B_1-EPG graph, then Algorithm 3.2 uses at most $4\chi(G)$ colors.*

4   Maximum Independent Set on B_1-EPG Graphs

4.1   Hardness Result for Finding Maximum Independent Set on B_1-EPG Graphs

In this section, we show that the MAXIMUM INDEPENDENT SET on B_1-EPG graphs is NP-complete. We use a reduction from MAXIMUM INDEPENDENT SET on planar graphs with maximum degree four, which is known to be NP-complete [6]; our proof is inspired by [14].

Theorem 5. MAXIMUM INDEPENDENT SET *on B_1-EPG graphs is NP-complete.*

Proof. Let $G = (V, E)$ be a planar graph with maximum degree four; MAXIMUM INDEPENDENT SET on planar graph with maximum degree four is NP-complete [6]. We construct a B_1-EPG representation of a graph $G' = (V', E')$ so that a maximum independent set in G' corresponds to a maximum independent set in G and vice versa.

Fix an embedding of G in a grid $\mathcal{G}$ such that edges of G are piecewise linear curves following the grid lines (such an embedding in a linear sized grid always exists and is constructible in polynomial time [16]). Each edge $e \in E$ is thus corresponds to a path π_e in the grid $\mathcal{G}$, and denote by k_e the number of segments (*links*) π_e consists of. Note further, that these paths intersect only at their endpoints, namely, in the vertices of G since the embedding is planar.

Let G' be a graph obtained from G by subdividing every edge e with $2 \left\lceil \frac{k_e+1}{2} \right\rceil$ new vertices; we denote the set of new vertices corresponding to an edge e by U_e and by U the set of all such new vertices, we thus have $V' = V \cup U$. Notice that since $|U_e|$ is even for each edge e of G, a maximum independent set in G' contains exactly half of the vertices in U_e, and at most one of the vertices corresponding to the "original" endpoints of e. We thus have

$$\alpha(G') = \alpha(G) + \sum_{e \in E} \left\lceil \frac{k_e + 1}{2} \right\rceil$$

and thus to complete the proof it suffices to show that G' is B_1-EPG graph.

Having the grid embedding of G, we construct a B_1-EPG representation $\langle \mathcal{P}, \mathcal{G} \rangle$ of G' as follows; see Figure 4 for an illustration. We start by placing the vertices in U into $\mathcal{G}$. Let e be an edge of G, by definition π_e has $k_e - 1$ bend points. At each such grid point we place one vertex from U_e, we also place one vertex from U_e in the interiors of the first and last links of π_e. Finally, we place the remaining vertices of U_e arbitrarily along π_e (the order in which the vertices are located along π_e preserves adjacencies). When convenient we may refer to vertices of G' as the grid points they are embedded to. We now associate each vertex v of G' with a path P_v (which is either a single-bend path or a segment) so that P_v and P_u share an edge of $\mathcal{G}$ if and only if v and u are adjacent in G'.

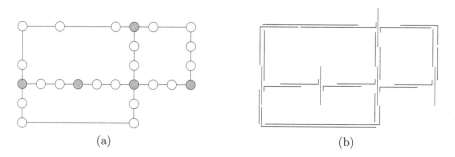

Fig. 4. (a) A rectilinear grid embedding of some graph G'; vertices of V are grayed. (b) A B_1-EPG representation of G'.

For every $v \in V$, set P_v to be a short vertical segment around v. Let $u \in U$, then u has exactly two neighbors, and consider first the case where both are from U. We set P_u to be a path consisting of the two segments connecting u with each of its neighbors. If u is embedded to a bend point of some π_e, then P_u is a single-bend path, otherwise it is just a segment. Finally, let $u \in U$ be a vertex with neighbors $u' \in U$ and $v \in V$ (notice that by construction no vertex in G' has more than one neighbor from V) in this case, u, u', and v are embedded to the same grid row/column and we set P_u as follows, distinguishing between two subcases, according to whether all three vertices are embedded to the same column or row of $\mathcal{G}$. (i) u, u', and v are on the same column: We set P_u to be a vertical segment that begins at u' and almost reaches v (in such a way that it ends close enough to share a grid edge with P_v). (ii) u, u', and v are on the same row: We set P_u to be a $\lrcorner$-path or a $\ulcorner$-path that starts at u' and bends at v, sharing its vertical edge with P_v, avoiding other possible neighbors of v.

It is easy to see that indeed for every $u, v \in V'$ the paths P_u and P_v share a grid edge if and only if u and v are adjacent in G', thus the desired result follows. $\square$

Remark 2. The proof of Theorem 5 can be modified so that it uses only two bend shapes; thus MAXIMUM INDEPENDENT SET is NP-complete already on $\lrcorner\llcorner$-EPG and on $\llcorner\urcorner$-EPG graphs.

4.2   A 4-Approximation Algorithm for Maximum Independent Set on B_1-EPG Graphs

In this section we present a constant-factor approximation algorithm for MAXIMUM INDEPENDENT SET (Algorithm 4.2 below). In a similar way to Section 3.2, we start by presenting a "subroutine" that computes an approximated solution for a subgraph, and then use the subroutine in order to compute an approximated solution for the whole graph. This subroutine is described in Algorithm 4.1 below, which uses a standard greedy Independent Set algorithm (thus clearly, produces an Independent Set). Note that the order in which it examines the vertices is the *reversed order* of that used in Algorithm 3.1, namely, according to the decreasing order of $\prec$. Lemma 2 claims that when used for a ⌐L-EPG graph, Algorithm 4.1 computes a 2-approximation.

Algorithm 4.1 Greedy-⌐L-EPG-Independent-Set (Input: $\mathcal{P} = \mathcal{P}_\lrcorner \cup \mathcal{P}_\llcorner$)

1: $S \leftarrow \emptyset$
2: **for each** $P_u \in \mathcal{P}$ (in decreasing order by $\prec$) **do**
3:     add u to S and remove P_u from $\mathcal{P}$
4:     remove all paths corresponding to u's neighbors from $\mathcal{P}$
5: **return** S

Applying Algorithm 4.1 to the representation in Figure 1 gives the independent set: $\{h, g, d\}$.

Lemma 2. *Let G be a ⌐L-EPG graph, then Algorithm 4.1 finds a maximal independent set of size at least $\frac{1}{2}\alpha(G)$.*

Proof. Let $\langle \mathcal{P}, \mathcal{G} \rangle$ be a ⌐L-EPG representation of a graph $G = (V, E)$. Let OPT be a maximum independent set in G and let S be the maximal Independent Set returned by Algorithm 4.1. We claim that $|OPT| \leq 2|S|$.

Notice that for every $v \in V$ the path P_v is removed from $\mathcal{P}$ at some point (in lines 3 or 4). Moreover, if a path P_v is removed from $\mathcal{P}$ in line 4, then its deletion must occur when the algorithm added to S some vertex u with $v \prec u$. Equivalently, whenever the algorithm adds a vertex u to S, it removes from $\mathcal{P}$ paths P_v adjacent to P_u where $v \prec u$ (in this case, any other vertex v' adjacent to u with $u \prec v'$ has been already removed from S in an earlier stage, necessarily in line 4).

By eliminating vertices in $OPT \cap S$ we may assume that $OPT \cap S = \emptyset$. We therefore assume that the paths corresponding to vertices in OPT were all eliminated from $\mathcal{P}$ in line 4. We define a correspondence $\varphi : OPT \rightarrow S$ as follows:

$\varphi(v) = u$ where P_v was removed from $\mathcal{P}$ in line 4 as a consequence of adding u to S

In particular, if $\varphi(v) = u$ then u and v are adjacent and $v \prec u$. We claim that for every $u \in S$ there exist at most two distinct vertices $v_1, v_2 \in OPT$ with $\varphi(v_1) = \varphi(v_2) = u$ and conclude that $|OPT| \leq 2|S|$. Indeed, assume to the contrary that for some $u \in S$, there exist three vertices $v_1, v_2, v_3 \in OPT$ with $\varphi(v_i) = u$ ($i = 1, 2, 3$). At least two of the three paths share with P_u a grid edge on the same direction; w.l.o.g., assume that P_{v_1} and P_{v_2} share a horizontal edge with P_u. We thus have that P_{v_i} is adjacent to P_u and $v_i \prec u$ ($i = 1, 2$), and in particular P_u, P_{v_1} and P_{v_2} share a common edge (the leftmost-bottommost grid-edge contained in P_u). However, as v_1 and v_2 are both in OPT, they are nonadjacent. – A contradiction. □

We now use Algorithm 4.1 as a building block in Algorithm 4.2 in order to find a maximal Independent Set in B_1-EPG graphs. Here too, as in Remark 1, by rotating a representation by 180°, Algorithm 4.1 can be "turned" from Greedy-⌐ᴸ-EPG-Independent-Set into Greedy-⌐ᴦ-EPG-Independent-Set. Theorem 6 claims that when used on a B_1-EPG graph, Algorithm 4.2 achieves a 4-approximation.

Algorithm 4.2 B_1-EPG Independent Set 4-Approximation($G = \langle \mathcal{P}, \mathcal{G} \rangle$)

1: let $\mathcal{P} = \mathcal{P}_ᴸ \cup \mathcal{P}_ᴶ \cup \mathcal{P}_ᴺ \cup \mathcal{P}_ᴦ$
2: $S_1 \leftarrow$ **Greedy-⌐ᴸ-EPG-Independent-Set**($\mathcal{P}_ᴸ \cup \mathcal{P}_ᴶ$)
3: $S_2 \leftarrow$ **Greedy-⌐ᴦ-EPG-Independent-Set**($\mathcal{P}_ᴦ \cup \mathcal{P}_ᴺ$)
4: **return** the largest amongst S_1, S_2

Theorem 6. *Let G be a B_1-EPG graph, then Algorithm 4.2 finds a maximal Independent Set of size at least $\frac{1}{4}\alpha(G)$.*

Proof. Let $\langle \mathcal{P}, \mathcal{G} \rangle$ be a B_1-EPG representation of G. Put $\mathcal{P} = \mathcal{P}_ᴸ \cup \mathcal{P}_ᴶ \cup \mathcal{P}_ᴺ \cup \mathcal{P}_ᴦ$ and let G_1 and G_2 be the ⌐ᴸ-EPG graphs with representations $\langle \mathcal{P}_ᴸ \cup \mathcal{P}_ᴶ, \mathcal{G} \rangle$ and $\langle \mathcal{P}_ᴦ \cup \mathcal{P}_ᴺ, \mathcal{G} \rangle$, respectively. Clearly, $\alpha(G) \leq \alpha(G_1) + \alpha(G_2)$.

Let S_1 and S_2 be the sets computed in lines 2 and 3 of the algorithm. By Lemma 2, we get

$$\alpha(G) \leq \alpha(G_1) + \alpha(G_2) \leq 2|S_1| + 2|S_2| \leq 4\max\{|S_1|, |S_2|\}$$

which completes the proof. □

5   Concluding Remarks

We observe that MAXIMUM CLIQUE in B_1-EPG graphs can be optimally solved in polynomial time using a brute-force algorithm. In [9] the authors show that each clique in the graph has one of two forms in the B_1-EPG representation, referred to as "edge clique" and "claw clique". An *edge clique* consists of all paths containing a given grid edge; a *claw clique* consists of all paths sharing two-out-of-three edges of a given claw centered at a given grid point (there are 4 different claws at each grid point.) Consequently, given a grid representation of a B_1-EPG graph G, one can simply examine each grid edge and count the number of paths containing that edge, and for each grid point and four corresponding claws, count the number of path containing two out of three edges of that claw. This can be done in time polynomial in the size of $\mathcal{G}$, which may be assumed to be of size at most $2n \times 2n$ for a B_1-EPG representation. This implies an $O(n^3)$ time algorithm for MAXIMUM CLIQUE given a B_1-EPG representation.

A somewhat different approach can solve MAXIMUM CLIQUE for a B_1-EPG graph *without* being given representation based on the fact that the neighborhood of a vertex in B_1-EPG graph is weakly-chordal [1]. It is well known that MAXIMUM CLIQUE in weakly-chordal graphs can be found in $O(n^4)$ time [15]. Since a maximum clique is contained in a closed neighborhood of each of its vertices, then this yields a $O(n^5)$ time algorithm for MAXIMUM CLIQUE given just the B_1-EPG graph and not the representation.

In Algorithms 3.2 and 4.2 we used, respectively, Algorithms 3.1 and 4.1 with subgraphs induced by $\mathcal{P}_\llcorner \cup \mathcal{P}_\lrcorner$ and $\mathcal{P}_\ulcorner \cup \mathcal{P}_\urcorner$. Taking into consideration also the two other options (i.e., $\mathcal{P}_\lrcorner \cup \mathcal{P}_\urcorner$ and $\mathcal{P}_\ulcorner \cup \mathcal{P}_\llcorner$) has no effect on the asymptotic quality of the solutions. However, as a heuristic, one might wish to apply the algorithm to both and take the better of the two.

Algorithm 3.2 and Algorithm 4.2 are greedy. Both have "bad" instances for which the factors mentioned here are tight. It is possible, of course, that a different approach may lead to better approximation factors.

As open problems, we suggest that it would be interesting to find approximation algorithms to find a minimum dominating set or a maximum weighted independent set for B_1-EPG graphs.

References

1. Asinowski, A., Ries, B.: Some properties of edge intersection graphs of single bend paths on a grid. Discrete Mathematics 312, 427–440 (2012)
2. Asinowski, A., Suk, A.: Edge intersection graphs of systems of grid paths with bounded number of bends. Discrete Applied Mathematics 157, 3174–3180 (2009)
3. Biedl, T., Stern, M.: On edge intersection graphs of k-bend paths in grids. Discrete Mathematics & Theoretical Computer Science (DMTCS) 12, 1–12 (2010)
4. Cameron, K., Chaplick, S., Hoang, C.T.: Recognizing Edge Intersection Graphs of ∟-Shaped Grid Paths. In: LAGOS 2013 (to appear, 2013)
5. Cohen, E., Golumbic, M.C., Ries, B.: Characterizations of cographs as intersection graphs of paths on a grid (submitted)
6. Garey, M.R., Johnson, D.S.: Computers and Intractability: a Guide to the Theory of NP-completeness. Freeman, San Francisco (1979)
7. Garey, M.R., Johnson, D.S., Miller, G.L., Papadimitriou, C.: The complexity of coloring circular arcs and chords. SIAM. J. on Algebraic and Discrete Methods 1, 216–227 (1980)
8. Golumbic, M.C.: Algorithmic Graph Theory and Perfect Graphs. Academic Press, New York (1980); Annals of Discrete Mathematics, 2nd edn., vol. 57. Elsevier, Amsterdam (2004)
9. Golumbic, M.C., Lipshteyn, M., Stern, M.: Edge intersection graphs of single bend paths on a grid. Networks 54, 130–138 (2009)
10. Halldórsson, M.M.: A still better performance guarantee for approximate graph coloring. Information Processing Letters 45, 19–23 (1993)
11. Heldt, D., Knauer, K., Ueckerdt, T.: Edge-intersection graphs of grid paths: the bend-number, Arxiv preprint arXiv:1009.2861, arxiv.org (September 2010)
12. Heldt, D., Knauer, K., Ueckerdt, T.: On the bend-number of planar and outerplanar graphs. In: Fernández-Baca, D. (ed.) LATIN 2012. LNCS, vol. 7256, pp. 458–469. Springer, Heidelberg (2012)

13. Kako, A., Ono, T., Hirata, T., Halldórsson, M.M.: Approximation algorithms for the weighted independent set problem in sparse graphs. Discrete Applied Mathematics 157, 617–626 (2009)
14. Kratochvíl, J., Nešetřil, J.: Independent set and clique problems in intersection-defined classes of graphs. Commentationes Mathematicae Universitatis Carolinae 31, 85–93 (1990)
15. Spinrad, J.P., Sritharan, R.: Algorithms for weakly triangulated graphs. Discrete Appl. Math. 59, 181–191 (1995)
16. Valiant, L.G.: Universality considerations in VLSI circuits. IEEE Trans. Comput. 30, 135–140 (1981)

Universal Point Sets for Planar Three-Trees

Radoslav Fulek[1,*] and Csaba D. Tóth[2,**]

[1] Charles University, Prague, Czech Republic
radoslav.fulek@gmail.com
[2] California State University, Northridge, CA, USA and University of Calgary, AB, Canada
cdtoth@acm.org

Abstract. For every $n \in \mathbb{N}$, we present a set S_n of $O(n^{5/3})$ points in the plane such that every planar 3-tree with n vertices has a straight-line embedding in the plane in which the vertices are mapped to a subset of S_n. This is the first subquadratic upper bound on the size of universal point sets for planar 3-trees, as well as for the class of 2-trees and serial parallel graphs.

Keywords: planar 3-tree, universal point set, straight-line embedding.

1 Introduction

Every planar graph has a *straight-line embedding* in the plane [17] where the vertices are mapped to distinct points and the edges to pairwise noncrossing straight line segments between the corresponding vertices. A set $S \subset \mathbb{R}^2$ of points in the plane is called *n-universal* if every n-vertex planar graph has a straight-line embedding in $\mathbb{R}^2$ such that the vertices are mapped into a subset of S. Similarly, $S \subset \mathbb{R}^2$ is *n-universal for a family* $\mathcal{G}$ of planar graphs if every n-vertex planar graph in $\mathcal{G}$ has a straight-line embedding in $\mathbb{R}^2$ such that the vertices are mapped into a subset of S. It is a longstanding open problem to determine the minimum size $f(n)$ of an n-universal point set for all $n \in \mathbb{N}$. Our main result is that there is an n-universal point set of size $O(n^{5/3})$ for the class of planar graphs of treewidth at most three.

Theorem 1. *For every* $n \in \mathbb{N}$, *there is an n-universal point set of size* $O(n^{5/3})$ *for planar 3-trees.*

A graph is called a *k-tree*, for some $k \in \mathbb{N}$, if it can be constructed by the following iterative process: start with a k-vertex clique and successively add new vertices such that each new vertex has exactly k neighbors that form a clique in the current graph. For example, 1-trees are the same as trees; 2-trees are maximal series-parallel graphs, and include also all outerplanar graphs. In general, k-trees are the maximal graphs with treewidth k. A planar 3-tree is a 3-tree that is planar. Theorem 1 is the first subquadratic upper bound on the size of n-universal point sets for planar 3-trees, for 2-trees, and for series-parallel graphs.

---

* The author gratefully acknowledge support from the Swiss National Science Foundation Grant No. 200021-125287/1 and ESF Eurogiga project GraDR as GAČR GIG/11/E023.
** Supported in part by NSERC (RGPIN 35586) and NSF (CCF-0830734).

F. Dehne, R. Solis-Oba, and J.-R. Sack (Eds.): WADS 2013, LNCS 8037, pp. 341–352, 2013.

Related Previous Work. In a pivotal paper, de Fraysseix, Pach and Pollack [10] showed that an n-universal set must have at least $n + (1 - o(1))\sqrt{n}$ points. Chrobak and Karloff [8] improved the lower bound to $1.098n$ and later Kurowski [23] to $(1.235 - o(1))n$. This is the currently known best lower bound for n-universal sets in general. De Fraysseix et al. [10] and Schnyder [25] independently showed that there are n-universal sets of size $O(n^2)$. In fact, an $(n - 2) \times (n - 2)$ section of the integer lattice is n-universal [9,25] for every $n \geq 4$. Alternatively, an $\frac{4}{3}n \times \frac{2}{3}n$ section of the integer lattice is also n-universal [5]. The quadratic upper bound is the best possible if the point set is restricted to sections of the integer lattice: Frati and Patrignani [20] showed (based on earlier work by Dolev et al. [12]) that if a rectangular section of the integer lattice is n-universal, then it must contain at least $n^2/9 + \Omega(n)$ points.

Grid drawings have been studied intensively due to their versatile applications. It is known that sections of the integer lattice with $o(n^2)$ points are n-universal for certain classes of graphs. For example, Di Battista and Frati [11] proved that an $O(n^{1.48})$ size integer grid is n-universal for *outerplanar* graphs. Frati [19] showed that 2-trees on n vertices require a grid of size at least $\Omega(n2^{\sqrt{\log n}})$. Biedl [2] observed that the grid embedding of all n-vertex 2-trees requires an $\Omega(n) \times \Omega(n)$ section of the integer lattice *if* the combinatorial embedding (i.e., all vertex-edge and edge-face incidences) is given. On the other hand, Zhou et al. [26] showed recently that every n-vertex series-parallel graph, and thus, every 2-tree, has a straight-line embedding in a $\frac{2}{3}n \times \frac{2}{3}n$ section of the integer lattice and a section of the integer lattice of area $0.3941n^2$. Researchers have studied classes of planar graphs that admit n-universal point sets of size $o(n^2)$. A classical result in this direction, due to Gritzmann et al. [22] (see also [4]), is that every set of n points in general position is n-universal for *outerplanar graphs*. Recently, Angelini et al. [1] generalized this result and showed that there exists an n-universal point set of size $O(n(\log n / \log \log n)^2)$ for so-called *simply nested* planar graphs. A planar graph is simply nested if it can be reduced to an outerplanar graph by successively deleting chordless cycles from the boundary of the outer face. Theorem 1 provides a new broad class of planar graphs that admit subquadratic n-universal sets.

Algorithmic questions pertaining to the straight-line embedding of planar graphs have also been studied. The *point set embeddability* problem asks whether a given planar graph G has a straight-line embedding such that the vertices are mapped to a given point set $S \subset \mathbb{R}^2$. The problem is known to be NP-hard [7], and remains NP-hard even for 3-connected planar graphs [14], triangulations and 2-connected outerplanar graphs [3]. However, it has a polynomial-time solution for 3-trees [15,24]. In a *polyline* embedding of a plane graph, the edges are represented by pairwise noncrossing polygonal paths. Biedl [2] proved that every 2-tree with n vertices has a polyline embedding where the vertices are mapped to an $O(n) \times O(\sqrt{n})$ section of the integer lattice, and each edge is a polyline with at most two bends. Everett et al. [16] showed that there is a set S_n of n points in the plane, for every $n \in \mathbb{N}$, such that every n-vertex planar graph has a polyline embedding with at most one bend per edge on S. Dujmović et al. [13] constructed a point set S_n' of size $O(n^2/\log n)$ for all $n \in \mathbb{N}$ such that every n-vertex planar graph has a polyline embedding with at most one bend per edge in which the vertices as well as all bend points of the edges are mapped to S_n'.

Organization. We briefly review some structural properties of planar 3-trees (Section 2), then construct a point set $S_n \subset \mathbb{R}^2$ for every $n \in \mathbb{N}$ (Section 3), and show that S_n is n-universal for planar 3-trees (Section 4).

2 Basic Properties of Planar Three-Trees

A graph G is a *planar 3-tree* if it can be constructed by the following iterative procedure. Initially, let $G = K_3$, the complete graph with three vertices. Successively augment G by adding one new vertex u and three new edges that join u to three vertices of a triangle such that no two vertices are connected to all the vertices of the same triangle. A planar 3-tree can be embedded in the plane consistently with the iterative process: the initial triangle forms the outer-face and each new vertex u is inserted in the interior of the face corresponding to the triangle it is attached to.

The iterative augmentation process that produces a 3-tree G can be represented by a rooted tree $T = T(G)$ as follows (this is called a *face-representative tree* in [18]). Refer to Fig. 1. The nodes of T correspond to the triangles of G. For convenience we denote a vertex of T by its corresponding triangle in G. The root of T corresponds to the initial triangle of G. When G is augmented by a new vertex u connected to the vertices of the triangle $\Delta = v_1 v_2 v_3$, we attach three new leaves to Δ corresponding to the triangles $v_1 v_2 u$, $v_1 u v_3$ and $u v_2 v_3$.

For a node Δ of T, let T_Δ denote the subtree of T rooted at Δ. Let V_Δ denote the set of vertices of G embedded in the interior of Δ.

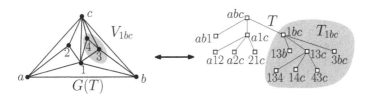

Fig. 1. Left: a 3-tree, constructed from the initial triangle abc by successively adding new vertices $1, \ldots, 4$. Right: The corresponding tree $T = T(G)$. The gray region indicates subtree T_{1bc} rooted at $1bc$, and its corresponding vertex set $V_{1bc} \subseteq V(G)$.

In Section 4, we embed the vertices of a planar 3-tree on a point set by traversing the tree T from the root. The initial triangle abc will be the outer face in the embedding such that edge ab is a horizonal line segment, and vertex c is the top vertex (i.e., it has maximal y-coordinate). We then successively insert the remaining $n - 3$ vertices of G, each of which subdivides a triangular face into three triangles. We label the vertices of each triangle of G as *left*, *right* and *top* vertex, respectively. These labels are assigned (without knowing the specifics of our embedding algorithm) as follows. Label the three vertices of the initial triangle in G arbitrarily as left, right and top, respectively. If G is augmented by a new vertex u and edges uv_1, uv_2, and uv_3, where v_1 is the left, v_2 is the right, and v_3 is the top vertex of an existing triangle $v_1 v_2 v_3$, then v_1, v_2, and v_3 keeps their labels left, right, and top, respectively, in the new triangles $v_1 v_2 u$, $u v_2 v_3$ and $v_1 u v_3$.

Furthermore, the vertex u becomes the top vertex of $v_1 v_2 u$, the left vertex of $u v_2 v_3$, and the right vertex of $v_1 u v_3$. The triangles $v_1 v_2 u$, $v_1 u v_3$ and $u v_2 v_3$, respectively, will be called the *bottom*, *left* and *right* triangles within $v_1 v_2 v_3$. In the tree $T = T(G)$, the three children of a node corresponding to a vertex can be labeled as *bottom*, *left*, and *right* child, analogously.

3   Construction of a Point Set

We construct a point set $S_n \subset \mathbb{R}^2$ of size $O(n^{5/3})$ for every $n \in \mathbb{N}$. Assume in the sequel that $n^{1/3}$ is an integer, otherwise let $S_n = S_{\lceil n^{1/3} \rceil^3}$.

The point set S_n is constructed in two easy steps: we first choose a "sparse" set B_n of $O(n^{5/3})$ points from a $14n \times 14n$ section of the integer lattice, and then "stretch" the points by the transformation $(x, y) \to (x, (28n)^y)$, as described below.

Sparse Grid. Let $A_n = \{(i, j) \in \mathbb{Z}^2 : 0 \le i, j \le 14n\}$ be an $14n \times 14n$ section of the integer lattice. Let $B_n \subset A_n$ be the set of points in A_n with at least one of the following four properties (see Fig. 2, left):

(a) (i, j) such that $i \equiv 0 \mod n^{1/3}$ (*full columns*);
(b) (i, j) such that $j \equiv 0 \mod n^{1/3}$ (*full rows*);
(c) $(i + k, j + k)$ such that $i, j \equiv 0 \mod n^{1/3}$ and $0 \le k < n^{1/3}$ (*forward diagonals*);
(d) $(i + k, j - k)$ such that $i, j \equiv 0 \mod n^{1/3}$ and $0 \le k < n^{1/3}$ (*backward diagonals*).

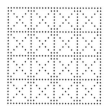 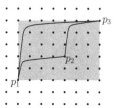

Fig. 2. Left: A schematic picture of a sparse grid: every point is in a full row, a full column, a forward diagonal or a backward diagonal. Right: A grid and three points $p_1 = (a_1, b_1)$, $p_2 = (a_2, b_2)$ and $p_3 = (a_3, b_3)$ with $a_1 < a_2 < a_3$ and $b_1 < b_2 < b_3$. The Jordan arcs between the points represent straight-line segments between the stretched points $\tau(a_1, b_1)$, $\tau(a_2, b_2)$ and $\tau(a_3, b_3)$.

Stretched Grid. We deform the plane by the following transformation.

$$\tau : \mathbb{R}^2 \to \mathbb{R}^2, \qquad\qquad (x, y) \to (x, (28n)^y).$$

For an integer point $(i, j) \in \mathbb{Z}^2$, we use the shorthand notation $\tau(i, j) = \tau((i, j))$. If $A \subset \mathbb{R}^2$ is a rectangular section of the integer lattice (a *grid*), then we call the point set $\tau(A) = \{\tau(p) : p \in A\}$ a *stretched grid*. Note that τ translates every point vertically, and it translates points of the same y-coordinate by the same vector. The purpose of transformation τ is to establish the following property for the stretched grid $\tau(A_n)$.

Observation 2. *Let* $(a_1, b_1), (a_2, b_2), (a_3, b_3) \in A_n$ *such that* (a_2, b_2) *lies in the interior of the axis-aligned rectangle spanned by* (a_1, b_1) *and* (a_3, b_3) *(formally,* $a_1 < a_2 < a_3$ *and either* $b_1 < b_2 < b_3$ *or* $b_3 < b_2 < b_1$*). Then* $\tau(a_2, b_2)$ *lies below the line segment between* $\tau(a_1, b_1)$ *and* $\tau(a_3, b_3)$*. (See Fig. 2, right. See the full version [21] for a proof.)*

Universal Point Set for 3-Trees. We are now in a position to define S_n. Let $S_n = \tau(B_n)$. Intuitively, S_n is a sparse $14n \times 14n$ grid with diagonals inside each "hole," and stretched vertically by τ.

Similarly to [6], our illustrations show the "unstretched" point set $B_n = \tau^{-1}(S_n)$ instead of S_n. The transformation τ^{-1} maps line segments between points in S_n to Jordan arcs between grid points in B_n. In our figures, line segments are drawn as Jordan arcs that correctly represent the above-below relationship between segments and points (Fig. 2, right).

Remark 3. The grid-embedding algorithm by de Fraysseix et al. [10] embeds every n-vertex planar graph on an $(2n - 4) \times (n - 2)$ section of the integer lattice. Their algorithms also works on the stretched grid in place of the integer grid. Specifically, we use their result in the following form. Suppose that G is a planar graph with $n \in \mathbb{N}$ vertices and endowed with a given combinatorial embedding in which u, v and z are the vertices of the outer face. Let $X, Y \subset \mathbb{N}$ be two sets of cardinality $|X| \geq 2n$ and $|Y| \geq n$. Then G has a straight-line embedding such that the vertices are mapped to the stretched cross product $\tau(X \times Y)$ of size at least $2n^2$; the two endpoints of edge uv are mapped to $\tau(\min X, \min Y)$ and $\tau(\max X, \min Y)$, respectively; and z is mapped to a point in the top row $\tau(X \times \max Y)$. By Observation 2, we can shift u or v vertically down to another point of the stretched grid (while keeping all other vertices fixed) without introducing any edge crossings.

4   Embedding Algorithm

Let G be a planar 3-tree with n vertices. We construct a straight-line embedding of G such that the vertices are mapped into S_n. Our embedding algorithm is guided by the tree $T = T(G)$, which represents an incremental process that constructs G from a single triangle. Recall that T_Δ denotes the subtree of T rooted at a node Δ; and V_Δ denotes the set of vertices of G that correspond to nodes in T_Δ.

Let the *weight* of a node Δ of T be weight$(\Delta) = |V_\Delta|$. A node Δ is *heavy* (resp., *light*) in T if its weight is at least (resp., less than) $n^{1/3}$. We say that a node Δ of T is a *big-split* if it is not the root of T, and $n^{1/3} \leq$ weight$(\Delta) \leq$ weight$(\Delta') - n^{1/3}$, where Δ' is the parent of Δ. The tree T is a partition tree. For every node Δ, weight(Δ) equals one plus the total weight of the children of Δ.

We show that T has at most $2n^{2/3}$ big-split nodes. Consider the subtree T' of T induced by the nodes of weight at least $n^{1/3}$. Let T'' denote the tree obtained from T' by adding a leaf to all nonroot vertices of degree two in T'. Observe that every big-split node of T is in T'', and its parent in T'' is either the root or a node of degree at least three in T''. The tree T'' has at most $n^{2/3}$ leaves, since every leaf of T'' accounts for at least $n^{1/3}$ vertices of G. Therefore, T'' has at most $n^{2/3} - 1$ nonroot vertices of degree at least three. Thus, there are at most $2n^{2/3}$ big-split nodes in T.

Overview. We embed the vertices of G while traversing the tree T from its root. For every node Δ with sufficiently large weight, we choose an axis-aligned rectangle $R(\Delta)$ such that the vertices in V_Δ will be mapped to points in $S_n \cap R(\Delta)$. Intuitively, $R(\Delta)$ is a region "allocated" for the vertices in V_Δ. See Fig. 5 for an illustration. For convenience, we describe the dimensions of all rectangles R_Δ in terms of the unstretched grid $B_n = \tau^{-1}(S_n)$.

When the breadth-first traversal of T reaches a node Δ with sufficiently small weight, we use Remark 3 to embed V_Δ into the point set $S_n \cap R(\Delta)$. We can use Remark 3 if $S_n \cap R(\Delta)$ contains a cross product $X \times Y$ where $|X| \geq 2 \cdot \text{weight}(\Delta)$ and $|Y| \geq \text{weight}(\Delta)$. The cross product $X \times Y$ will contain either full rows or full columns in $S_n \cap R(\Delta)$. Since every $n^{1/3}$-th row and every $n^{1/3}$-th column of S_n is full, $R(\Delta)$ must intersect either at least $2\text{weight}(\Delta)$ full columns and $\text{weight}(\Delta)$ arbitrary rows; or at least $2\text{weight}(\Delta)$ arbitrary columns and $\text{weight}(\Delta)$ full rows. Hence, $R(\Delta)$ must intersect either at least $2\text{weight}(\Delta)n^{1/3}$ columns and $\text{weight}(\Delta)$ rows; or at least $2\text{weight}(\Delta)$ columns and $\text{weight}(\Delta)n^{1/3}$ rows of S_n.

Let u denote the vertex of G connected to all three vertices of $\Delta \in V(T)$, if such a vertex exists. Let Δ_1, Δ_2 and Δ_3 denote the children of Δ. The main difficulty of our strategy lies in the fact that at each step of the algorithm we need to allocate three internally disjoint rectangles $R(\Delta_1), R(\Delta_2)$ and $R(\Delta_3)$ such that they intersect in a single point of $S_n \cap R(\Delta)$. Intuitively, we would like to choose rectangles $R(\Delta_1)$, $R(\Delta_2)$ and $R(\Delta_3)$ so that their areas are proportional to their weights. This would be possible (up to integer rounding) if *all* points of $\tau(A_n) \cap R(\Delta)$ were available for embedding u. However, we have to place u at a point of the sparse set $\tau(B_n) \cap R(\Delta)$, and so some distortion is unavoidable. A simple way to achieve that $R(\Delta_1), R(\Delta_2)$ and $R(\Delta_3)$ intersect in a single point of S_n is to "snap" their corners to an intersection point of a full row and a full column of S_n. Each such snapping can "waste" up to $O(n^{1/3})$ units in both horizontal and vertical directions, and hence, we apply it only to $O(n^{2/3})$ big-split nodes of T.

In order to avoid wasting too many points of S_n, we maintain an invariant for heavy nodes Δ that requires the lower-left and lower-right corners of $R(\Delta)$ to be on a forward and, respectively, backward diagonal of S_n (invariant I_3 below). This will allow for allocating the rectangles $R(\Delta_1), R(\Delta_2)$ and $R(\Delta_3)$ economically in the case that Δ_1, Δ_2 or Δ_3 is light.

Snapping. To every big-split node Δ, we would like to assign a rectangle $R(\Delta)$ whose bottom corners are in the intersection of full rows and full columns of S_n. Our algorithm (described below) achieves this property in two steps (refer to Fig. 3a-3c): It first selects a rectangle $R_0(\Delta)$ which may not have this property, and then applies a repair step (called "snapping") to establish the required property. Suppose that Δ is a big-split node, and not all corners of the rectangle $R_0(\Delta)$ are at full rows or full columns of the grid B_n. The repair step increases the width and height of $R_0(\Delta)$ by $2n^{1/3}$ to obtain a larger rectangle $R_0'(\Delta)$; and then snaps the corners of $R_0(\Delta)$ to points lying on full rows and full columns within $R_0'(\Delta)$ (thereby decreasing the width and height of $R_0'(\Delta)$ by at most $n^{1/3}$) as follows.

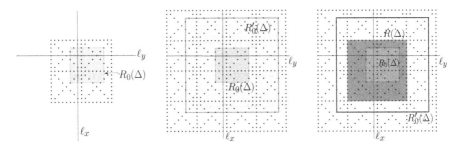

Fig. 3. Left: A rectangle $R_0(\Delta)$ that intersects a vertical line ℓ_x and a horizontal line ℓ_y. Middle: Shift every corner on the left and right of ℓ_x by $n^{1/3}$ to the left and right, resp., and every corner above and below ℓ_y by $n^{1/3}$ up and down, resp.. Right: Let $R(\Delta) \subseteq R_0'(\Delta)$ be the maximal axis-aligned rectangle whose corners are on full rows and full columns.

Let ℓ_x (resp., ℓ_y) be a vertical (resp., horizontal) line passing through $R_0(\Delta)$. For all rectangles $R(\Delta')$, shift every corner lying on the left (right) of ℓ_x to the left (right) by $n^{1/3}$. Similarly, for all the already allocated rectangles $R(\Delta')$, shift every corner above (below) ℓ_y up (down) by $n^{1/3}$. (Shifting is meant in terms of the unstretched grid: for example a point $\tau(i, j)$ right of ℓ_x and above ℓ_y is shifted to $\tau(i + n^{1/3}, j + n^{1/3})$.) This operation maps $R_0(\Delta)$ to a rectangle $R_0'(\Delta)$. Note that the operation does not decrease the width and height of any rectangle. Finally, let $R(\Delta) \subseteq R_0'(\Delta)$ be the maximal axis-aligned rectangle whose corners are on full rows and full columns of B_n.

Since there are at most $2n^{2/3}$ big-split nodes in T, we perform at most $2n^{2/3}$ snapping operations, one for each big-split node. Altogether, the snapping operations increase the width and the height of the bounding box by $2n^{2/3} \cdot 2n^{1/3} = 4n$. The point set B_n is a $14n \times 14n$ section of the sparse grid. If we choose the initial rectangle (assigned to the root of T) as the middle $10n \times 10n$ portion of B_n with margins of $2n$ all around, then all rectangles remain within the point set B_n after snapping.

Each snapping operation changes the width and height of rectangles allocated to several nodes of T. A snapping for a rectangle $R(\Delta)$ affects the dimension of all ancestors of Δ as well as of any other rectangle that intersect ℓ_x or ℓ_y. In the analysis of our algorithm, we do not attempt to maintain the true dimensions of the rectangles. We are satisfied with lower bounds on their widths and heights. Since the snapping operations can only *increase* the dimensions of the rectangles, we can afford to *ignore* their effect completely, and we still retain a lower bound for the true dimensions. We define the *width* (resp., *height*) of an axis-aligned rectangle R with respect to the unstretched grid, and denote them by $w(R)$ and $h(R)$, respectively. Hence a rectangle R intersects at least $w(R)$ (not necessarily full) columns and $h(R)$ (not necessarily full) rows.

Invariants. By traversing the tree T from the root, we assign a rectangle $R(\Delta)$ to every node Δ up to the depth where Remark 3 becomes applicable, that is, $[w(R(\Delta)) \geq 2\text{weight}(\Delta)$ and $h(R(\Delta)) \geq 20n^{1/3}\text{weight}(\Delta)]$ or $[h(R(\Delta)) \geq \text{weight}(\Delta)$ and $w(R(\Delta)) \geq 20n^{1/3}\text{weight}(\Delta)]$. The constant factor of 20 is used merely to simplify the analysis of the algorithm. We call the set of nodes of T where these conditions are first satisfied the *fringe* of T. For a fringe node Δ, we can embed the vertices in V_Δ using Remark 3, and so there is no need to assign rectangles to its descendants.

For all nodes Δ at or above the fringe of T, we maintain the following invariants.

I_1 If $T_\Delta \subseteq T_{\Delta'}$ then $R(\Delta) \subseteq R(\Delta')$; otherwise $R(\Delta)$ and $R(\Delta')$ are interior-disjoint.

I_2 For every node Δ, the the horizontal extent of $R(\Delta)$ lies in the horizontal extent of the triangle Δ; every vertex $u \in V_\Delta$ is embedded in the interior of $R(\Delta)$.

I_3 If weight$(\Delta) \geq n^{1/3}$, the lower-left and lower-right corners of $R(\Delta)$ are in S_n; specifically, the lower-left corner is in a forward diagonal, and the lower-right corner is in a backward diagonal of S_n.

I_4 If weight$(\Delta) \geq n^{1/3}$, then $w(R(\Delta)) \cdot h(R(\Delta)) \geq 100n\text{weight}(\Delta)$.

I_5 If weight$(\Delta) < n^{1/3}$, then
$$[w(R(\Delta)) \geq 2\text{weight}(\Delta) \text{ and } h(R(\Delta)) \geq 20n^{1/3}\text{weight}(\Delta)] \text{ or}$$
$$[h(R(\Delta)) \geq \text{weight}(\Delta) \text{ and } w(R(\Delta)) \geq 20n^{1/3}\text{weight}(\Delta)].$$

Note that invariants I_1, I_2 and I_5 ensure that all light nodes Δ (i.e., nodes with weight$(\Delta) < n^{1/3}$) are on or below the fringe. We now recursively allocate rectangles $R(\Delta)$ for all nodes Δ of T on or above the fringe of T, maintaining invariants I_1–I_5.

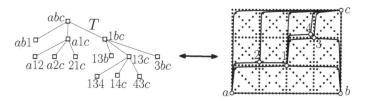

Fig. 4. The embedding of a 3-tree G from Fig. 1 on a sparse grid

Initialization. Denote by abc the initial triangle of G, with a labeled left, b labeled right and c labeled top. Then we have $T = T_{abc}$. Let $R(abc)$ be the bounding box of a $10n \times 10n$ section of S_n. Embed a and b to the lower-left and lower-right corners of $R(abc)$, respectively. Embed c in the upper-right corner of $R(abc)$ (see Fig. 4). It is clear that invariants I_1–I_5 are satisfied for abc.

By construction, every nonleaf node of T has three children: a left, a right and a top child. For a node Δ' the rectangle $R(\Delta')$ is obtained from its parental rectangle $R(\Delta)$ by the following procedure.

Assume that the vertices of triangle Δ have already been embedded and we have a rectangle $R(\Delta)$ satisfying invariants I_1–I_5. If Δ is on the fringe of T, then the embedding of the vertices V_Δ is completed by Remark 3 and invariant I_5. Otherwise, denote the bottom, left and right child of Δ, respectively, by Δ_1, Δ_2 and Δ_3. Suppose that $R(\Delta) = \tau([a, b] \times [c, d])$. We distinguish between two cases depending on the number of heavy children of Δ.

The Node Δ has More Than One Heavy Child. In this case, we partition the area of rectangle $R(\Delta)$ among its three children proportionally to their weights; and establish invariant I_3 by snapping operations for the heavy children. Note that all heavy children of Δ are big-split nodes. Refer to Fig. 5a.

We choose rectangles $R(\Delta_1), R(\Delta_2)$ and $R(\Delta_3) \subset R(\Delta)$ for the children of Δ; and place the vertex of G corresponding to Δ at a point in $R(\Delta) \cap S_n$ such that its x-coordinate corresponds to the right side of $R(\Delta_2)$ and the left side of $R(\Delta_3)$, and its y-coordinate lies (not strictly) below both rectangles and above rectangle $R(\Delta_1)$.

Note that $\text{weight}(\Delta) = \text{weight}(\Delta_1) + \text{weight}(\Delta_2) + \text{weight}(\Delta_3) + 1$. We distribute the height of $R(\Delta)$ between $R_0(\Delta_1)$ and $R_0(\Delta_2) \cup R_0(\Delta_3)$ proportionally to their weights. Then we distribute the width of $R_0(\Delta)$ between $R_0(\Delta_2)$ and $R_0(\Delta_3)$ proportionally to their weights. Finally, the rectangles of heavy children among $R(\Delta_1), R(\Delta_2)$ and $R(\Delta_3)$ are obtained by snapping the corners of $R_0(\Delta_1), R_0(\Delta_2)$ and $R_0(\Delta_3)$, respectively, to full rows and full columns as described above. The rectangles $R(\Delta_i)$ for light children Δ_i are equal to $R_0(\Delta_i)$. Due to snapping, the lower-left (resp., lower-right) corner of $R(\Delta_i)$, for a heavy Δ_i, is on a forward (resp., backward) diagonal of S_n. Hence, we maintain invariant I_3.

The height of the rectangle $R(\Delta_1)$ is at least $\frac{\text{weight}(\Delta_1)}{\text{weight}(\Delta)} \cdot h(R(\Delta))$.

The height of the rectangle $R(\Delta_2)$ ($R(\Delta_3)$ is treated analogously) is at least $\frac{\text{weight}(\Delta_2)+\text{weight}(\Delta_3)}{\text{weight}(\Delta)} \cdot h(R(\Delta))$.

The width of the rectangle $R(\Delta_2)$ ($R(\Delta_3)$ is treated analogously) is as least $\frac{\text{weight}(\Delta_2)}{\text{weight}(\Delta_2)+\text{weight}(\Delta_3)} \cdot w(R(\Delta))$.

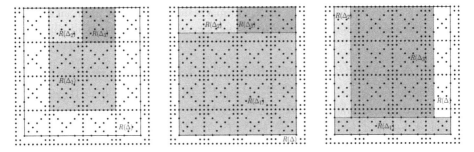

Fig. 5. (a) A step where all three children of Δ are heavy. The corners of all rectangles $R(\Delta_i)$ were snapped to the intersection of full rows and columns. (b) A step where only the bottom child of Δ is heavy. No snapping is necessary. (c) A step where only the right child of Δ is possibly heavy. No snapping is necessary.

The Node u Has at Most One Heavy Child. In this case, we do not use snapping: we choose the height (or the width) of each light child to be at least their weight (or twice their weight); and we establish invariant I_3 using the forward and backward diagonals. Refer to Figs. 5b and 5c.

Let us distinguish between two subcases depending on whether the bottom child is heavy or not.

If the heavy child of Δ happens to be the bottom one, that is Δ_1, then let $r_\Delta = \max\{\text{weight}(\Delta_2), \text{weight}(\Delta_3)\}$ be the weight of a largest top child. We assign

$$R(\Delta_1) := \tau([a, b] \times [c, d - r_\Delta])$$

and we distribute the width of the remaining part of $R(\Delta)$ evenly between $R(\Delta_2)$ and $R(\Delta_3)$ so that we can map the vertex of G corresponding to Δ to a point in $R(\Delta) \cap S_n$ such that its x-coordinate corresponds to the right side of $R(\Delta_2)$ and the left side of $R(\Delta_3)$ and its y-coordinate is $d - 2r_\Delta$.

Otherwise, if, say, the left child Δ_2 is light, let $r_\Delta = \max\{\text{weight}(\Delta_1), \text{weight}(\Delta_2)\}$. We assign

$$R(\Delta_3) := \tau([a + 2r_\Delta, b - 2r_\Delta] \times [c + 2r_\Delta, d])$$

and place the vertex of G corresponding to Δ at the point $\tau([a + 2r_\Delta, c + 2r_\Delta])$ on a forward diagonal. Then we assign $R(\Delta_1) = \tau([a, b] \times [c, c + 2r_\Delta])$ and $R(\Delta_2) = \tau([a, a + 2r_\Delta] \times [c + 2r_\Delta, d])$. The lower-left (resp., lower-right) corner of $R(\Delta_3)$ is on a forward (resp., backward) diagonal of S_n. Thus, we maintain invariant I_3.

If the right child Δ_1 is light, the embedding is done analogously, placing the vertex of G corresponding to Δ on a backward diagonal. This concludes the description of the embedding algorithm.

Maintenance of Invariants. The invariants I_1–I_5 trivially hold when Δ is the root of T. By construction, the invariants I_1, I_2, and I_3 are maintained for subtrees in each step of our algorithm. It remains to verify that invariants I_4 and I_5 are maintained.

To verify invariant I_4, consider a node Δ of T corresponding to a heavy triangle. Let $\Delta_1, \Delta_2, \ldots, \Delta_k$ be the vertices on the path in T from the root Δ_1 to $\Delta_k = \Delta$ corresponding to triangles. Suppose that invariant I_4 holds for Δ_i, $1 \leq i < k$. Assume that the rectangles $R(\Delta_1) \supseteq \ldots \supseteq R(\Delta_k)$ form a nested sequence by invariant I_1. The heights and widths of the rectangles $R(\Delta_i)$ may decrease in three essentially different ways:

(a) $w(R(\Delta_{i+1})) \geq w(R(\Delta_i))$ and $h(R(\Delta_{i+1})) \geq h(R(\Delta_i)) \frac{\text{weight}(\Delta_{i+1})}{\text{weight}(\Delta_i)}$;
(b) $w(R(\Delta_{i+1})) \geq w(R(\Delta_i)) \frac{\text{weight}(\Delta_{i+1})}{c_{i+1}}$ and $h(R(\Delta_{i+1})) \geq h(R(\Delta_i)) \frac{c_{i+1}}{\text{weight}(\Delta_i)}$ where c_{i+1} is a parameter with $c_{i+1} \geq \text{weight}(\Delta_{i+1})$;
(c) $w(R(\Delta_{i+1})) \geq w(R(\Delta_i)) - 4r$ and $h(R(\Delta_{i+1})) \geq h(R(\Delta_i)) - 2r$ where $r = \text{weight}(\Delta_i) - \text{weight}(\Delta_{i+1})$.

Recall that in the analysis we ignore snapping. Case (a) occurs when $R(\Delta_{i+1})$ is a bottom rectangle. Indeed, if Δ_{i+1} is the bottom child of Δ_i, then $w(R(\Delta_{i+1})) = w(R(\Delta_i))$ by construction; and the height of Δ_{i+1} is a proportional fraction of the height of Δ_i, if Δ_i has several heavy children. If Δ_{i+1} is the only heavy child by invariant I_4 we have $h(R(\Delta_i)) \geq 10\text{weight}(\Delta_i)$, since $w(R(\Delta_i)) \leq 10n$. Thus, the height of Δ_{i+1} is more than a proportional fraction in this case.

Case (b) occurs when $R(\Delta_{i+1})$ is a left or right rectangle, and both the left and right child of Δ_i are heavy. Case (c) occurs when $R(\Delta_{i+1})$ is a left or right rectangle, and Δ_{i+1} is the only heavy child of Δ_i.

In cases (a) and (b), the width or the height decreases at most proportionally with the weight, and in case (c) the width and the height decrease by the at most 4 times and 2 times, respectively, the actual decrease in weight. We show that in all three cases, the area of the rectangle decreases proportionally to the weight, that is,

$$\frac{w(R(\Delta_{i+1}))h(R(\Delta_{i+1}))}{w(R(\Delta_i))h(R(\Delta_i))} \geq \frac{\text{weight}(\Delta_{i+1})}{\text{weight}(\Delta_i)}. \tag{1}$$

This is obvious in cases (a) and (b). In case (c), we have:

$$\frac{w(R(\Delta_{i+1}))}{w(R(\Delta_i))} \cdot \frac{h(R(\Delta_{i+1}))}{h(R(\Delta_i))} = \frac{(w(R(\Delta_i)) - 4r)(h(R(\Delta_i)) - 2r)}{w(R(\Delta_i)) \cdot h(R(\Delta_i))}$$

$$\geq \frac{100n\,\text{weight}(\Delta_i) - 6r \cdot 10n}{100n\,\text{weight}(\Delta_i)} \geq \frac{\text{weight}(\Delta_{i+1})}{\text{weight}(\Delta_i)}$$

where we used that $w(R(\Delta_i)) \leq 10n$ and $h(R(\Delta_i)) \leq 10n$ since we ignore snapping, and we have $w(R(\Delta_i)) \cdot h(R(\Delta_i)) \geq 100n\,\text{weight}(\Delta_i)$ by invariant I_4. It follows that (1) holds for $i = 1, \ldots, k - 1$. Therefore we have

$$w(R(\Delta_k)) \cdot h(R(\Delta_k)) \geq 10n \cdot 10\text{weight}(\Delta_1)\frac{\text{weight}(\Delta_2)}{\text{weight}(\Delta_1)} \cdot \ldots \cdot \frac{\text{weight}(\Delta_k)}{\text{weight}(\Delta_{k-1})}$$

$$\geq 100n\,\text{weight}(\Delta_k).$$

This confirms invariant I_4 and similarly we can prove that invariant I_5 is also maintained (see the full version [21] for a complete proof).

5 Conclusion

We have presented a set S_n of $O(n^{5/3})$ points in the plane such that every n-vertex planar 3-tree has a straight-line embedding where the vertices are mapped into S_n. We do not know what is the minimum size of an n-universal point set for planar 3-trees.

The bottleneck of our method is the snapping operation. Recall that snapping is invoked at most $2n^{2/3}$ times, once for each big-split node, and each snapping operation extends the width and the height of the outer face by $2n^{1/3}$. If not for invariant I_3, we could abandon the snapping operations and we could define a sparse grid with resolution $\sqrt{n}$ instead of $n^{1/3}$, yielding a point set of size $O(n^{3/2})$.

The point set S_n, $n \in \mathbb{N}$, defined in Section 3 is n-universal for planar 3-trees. It certainly admits some other n-vertex planar graphs, as well. It remains to be seen whether S_n is n-universal for all n-vertex planar graphs.

Acknowledgements. We are grateful to Vida Dujmović and David Wood for their encouragement and for repeatedly posing the universal point set problem for 2-trees and planar 3-trees.

References

1. Angelini, P., Di Battista, G., Kaufmann, M., Mchedlidze, T., Roselli, V., Squarcella, C.: Small point sets for simply-nested planar graphs. In: Speckmann, B. (ed.) GD 2011. LNCS, vol. 7034, pp. 75–85. Springer, Heidelberg (2011)
2. Biedl, T.: Small drawings of outerplanar graphs, series-parallel graphs, and other planar graphs. Discrete Computational Geometry 45, 141–160 (2011)
3. Biedl, T., Vatshelle, M.: The point-set embeddability problem for plane graphs, in. In: Proc. Symposuim on Computational Geometry, pp. 41–50. ACM Press (2011)
4. Bose, P.: On embedding an outer-planar graph in a point set. Computational Geometry: Theory and Applications 23(3), 303–312 (2002)

5. Brandenburg, F.-J.: Drawing planar graphs on $\frac{8}{9}n^2$ area. Electronic Notes in Discrete Mathematics 31, 37–40 (2008)

6. Bukh, B., Matoušek, J., Nivasch, G.: Lower bounds for weak epsilon-nets and stair-convexity. Israel Journal of Mathematics 182, 199–228 (2011)

7. Cabello, S.: Planar embeddability of the vertices of a graph using a fixed point set is NP-hard. Journal of Graph Algorithms and Applications 10(2), 353–363 (2006)

8. Chrobak, M., Karloff, H.J.: A lower bound on the size of universal sets for planar graphs. SIGACT News 20(4), 83–86 (1989)

9. Chrobak, M., Payne, T.: A linear time algorithm for drawing a planar graph on a grid. Information Processing Letters 54, 241–246 (1995)

10. de Fraysseix, H., Pach, J., Pollack, R.: How to draw a planar graph on a grid. Combinatorica 10(1), 41–51 (1990)

11. Di Battista, G., Frati, F.: Small area drawings of outerplanar graphs. Algorithmica 54(1), 25–53 (2009)

12. Dolev, D., Leighton, F.T., Trickey, H.: Planar embedding of planar graphs. In: Preparata, F. (ed.) Advances in Computing Research, vol. 2. JAI Press Inc., London (1984)

13. Dujmović, V., Evans, W., Lazard, S., Lenhart, W., Liotta, G., Rappaport, D., Wismath, S.: On point-sets that support planar graphs. Computational Geometry: Theory and Applications 46(1), 29–50 (2013)

14. Durocher, S., Mondal, D.: On the hardness of point-set embeddability. In: Rahman, M.S., Nakano, S.-I. (eds.) WALCOM 2012. LNCS, vol. 7157, pp. 148–159. Springer, Heidelberg (2012)

15. Durocher, S., Mondal, D., Nishat, R.I., Rahman, M.S., Whitesides, S.: Embedding plane 3-trees in $\mathbb{R}^2$ and $\mathbb{R}^3$. In: Speckmann, B. (ed.) GD 2011. LNCS, vol. 7034, pp. 39–51. Springer, Heidelberg (2011)

16. Everett, H., Lazard, S., Liotta, G., Wismath, S.: Universal sets of n points for one-bend drawings of planar graphs with n vertices. Discrete and Computational Geometry 43(2), 272–288 (2010)

17. Fáry, I.: On straight lines representation of plane graphs. Acta Scientiarum Mathematicarum (Szeged) 11, 229–233 (1948)

18. Hossain, M. I., Mondal, D., Rahman, M. S., Salma, S.A.: Universal line-sets for drawing planar 3-trees. In: Rahman, M.S., Nakano, S.-I. (eds.) WALCOM 2012. LNCS, vol. 7157, pp. 136–147. Springer, Heidelberg (2012)

19. Frati, F.: Lower bounds on the area requirements of series-parallel graphs. Discrete Mathematics and Theoretical Computer Science 12(5), 139–174 (2010)

20. Frati, F., Patrignani, M.: A note on minimum-area straight-line drawings of planar graphs. In: Hong, S.-H., Nishizeki, T., Quan, W. (eds.) GD 2007. LNCS, vol. 4875, pp. 339–344. Springer, Heidelberg (2008)

21. Fulek, R., Tóth, C.D.: Universal point sets for planar three-tree, http://arxiv.org/abs/1212.6148

22. Gritzmann, P., Mohar, B., Pach, J., Pollack, R.: Embedding a planar triangulation with vertices at specified positions. American Mathematic Monthly 98, 165–166 (1991)

23. Kurowski, M.: A 1.235 lower bound on the number of points needed to draw all n-vertex planar graphs. Information Processing Letters 92, 95–98 (2004)

24. Nishat, R., Mondal, D., Rahman, M.S.: Point-set embeddings of plane 3-trees. Computational Geometry: Theory and Applications 45(3), 88–98 (2012)

25. Schnyder, W.: Embedding planar graphs in the grid, in. In: Proc. 1st Symposium on Discrete Algorithms, pp. 138–147. ACM Press (1990)

26. Zhou, X., Hikino, T., Nishizeki, T.: Small grid drawings of planar graphs with balanced partition. Journal of Combinatorial Optimization 24(2), 99–115 (2012)

Planar Packing of Binary Trees

Markus Geyer[1], Michael Hoffmann[2,*], Michael Kaufmann[1],
Vincent Kusters[2,*], and Csaba D. Tóth[3]

[1] Wilhelm-Schickard-Institut für Informatik, Universität Tübingen, Germany
{geyer,mk}@informatik.uni-tuebingen.de
[2] Institute of Theoretical Computer Science, ETH Zürich, Switzerland
{hoffmann,vincent.kusters}@inf.ethz.ch
[3] California State University Northridge and University of Calgary
cdtoth@acm.org

Abstract. In the graph packing problem we are given several graphs and have to map them into a single host graph G such that each edge of G is used at most once. Much research has been devoted to the packing of trees, especially to the case where the host graph must be planar. More formally, the problem is: Given any two trees T_1 and T_2 on n vertices, we want a simple planar graph G on n vertices such that the edges of G can be colored with two colors and the subgraph induced by the edges colored i is isomorphic to T_i, for $i \in \{1, 2\}$.

A clear exception that must be made is the star tree which cannot be packed together with any other tree. But a popular hypothesis states that this is the only exception, and all other pairs of trees admit a planar packing. Previous proof attempts lead to very limited results only, which include a tree and a spider tree, a tree and a caterpillar, two trees of diameter four and two isomorphic trees.

We make a step forward and prove the hypothesis for any two binary trees. The proof is algorithmic and yields a linear time algorithm to compute a plane packing, that is, a suitable two-edge-colored host graph along with a planar embedding for it. In addition we can also guarantee several nice geometric properties for the embedding: vertices are embedded equidistantly on the x-axis and edges are embedded as semi-circles.

* Partially supported by the ESF EUROCORES programme EuroGIGA, CRP GraDR and the Swiss National Science Foundation, SNF Project 20GG21-134306.

F. Dehne, R. Solis-Oba, and J.-R. Sack (Eds.): WADS 2013, LNCS 8037, pp. 353–364, 2013.

1   Introduction

Finding subgraphs with specific properties within a given graph or more generally determining relationships between a graph and its subgraphs is one of the most studied topics in graph theory. The *subgraph isomorphism* problem [21,11,5] asks to find a subgraph H in a graph G. The *graph thickness* problem [16] asks for the minimum number of planar subgraphs which the edges of a graph can be partitioned into. The *arboricity* problem [4] asks to determine the minimum number of forests which a graph can be partitioned into. Another related classical combinatorial problem is the k edge-disjoint spanning trees problem which dates back at least to Tutte [20] and Nash-Williams [17], who gave necessary and sufficient conditions for the existence of k edge-disjoint spanning trees in a graph. Every (maximal) planar graph can be partitioned into at most three edge-disjoint trees or into three forests, respectively, also known as *Schnyder woods* [19]. Finally, Gonçalves [13] proved that every planar graph can be partitioned in two edge-disjoint outerplanar graphs.

Of course, the study of relationships between a graph and its subgraphs can also be done the other way round. Instead of asking for specific types of subgraphs of a given graph, one can ask for a graph G that encompasses a given set of graphs $G_1, \ldots, G_k$ and satisfies certain properties in addition. This topic occurs with different flavors in the computational geometry and graph drawing literature. It is motivated by visualization aims, such as the display of networks evolving over time and the simultaneous visualization of relationships involving the same entities. In the *simultaneous embedding* problem [2,12,7] the graph $G = \bigcup G_i$ is given and the goal is to draw it so that the drawing of each G_i is plane. The *simultaneous embedding without mapping* problem [2] is to find a graph G on n vertices such that: (i) G contains all the G_i as subgraphs, and (ii) G can be drawn with straight-line edges so that the drawing of each G_i is plane.

The *packing problem* is to find a graph G on n vertices that contains a given collection $G_1, \ldots, G_k$ of graphs on n vertices each as edge-disjoint subgraphs. This problem has been studied in a wide variety of scenarios (see, e.g., [1], [6], [3]). Much attention has been devoted to the packing of trees. Hedetniemi [14] showed that any two non-star trees can be packed into a subgraph of K_n. A *star* is a tree with exactly one vertex of degree greater than one. Maheo et al. [15] gave a characterization of the triples of trees that can be packed into K_n.

In the *planar packing* problem the graph G is required to be planar. García et al. [10] conjectured that there exists a planar packing for any two non-star trees, that is, for any two trees with diameter greater than two. Notice that the hypothesis that none of the trees is a star is necessary, since a star uses all edges incident to one vertex and so there is no edge left to connect that vertex in the other tree. García et al. proved their conjecture for the cases that the trees are isomorphic and that one of the trees is a path. Oda and Ota [18] addressed the case that one of the trees is a caterpillar or that one of the trees is a spider of diameter at most four. A *caterpillar* is a tree that becomes a path when all leaves are deleted and a *spider* is a tree with at most one vertex of degree greater than two. Frati et al. [9] gave an algorithm to construct a planar packing of any spider with any tree. Finally, Frati [8] proved the conjecture for the case that both trees have diameter at most four. In this paper we will prove the following:

Theorem 1. *Any two non-star binary trees admit a planar packing.*

Binary trees are a major step in the study of planar tree packing, because they offer far more variety than the path-like sub-structures in spiders or the sub-trees of constant depth in caterpillars and—more generally—trees of bounded diameter. We believe that the techniques used here shed some new light on the structure and complexity of the problem that might also help to attack the general case.

2   Definitions and Overview

A *binary tree* is a tree in which no vertex has more than three neighbors. All trees considered in the following are binary. A *rooted tree* is a directed tree with exactly one sink (vertex of out-degree zero). In a rooted tree, every vertex v other than the root has exactly one outgoing edge vp. The target $p(v) = p$ is the *parent* of v, and conversely v is a *child* of p. In figures we denote the root of a tree by an outgoing vertical arrow. For a vertex v of a rooted tree T, denote by $t(v)$ the *subtree rooted at* v, that is, the subtree of T induced by the vertices from which v can be reached on a directed path. Furthermore, denote by $|v|$ the *size* (number of vertices) of $t(v)$. The *path* on n vertices is denoted by P_n. A tree in which all vertices—except for at most one that is called the *center*—have degree one is called a *star* and the star on n vertices is denoted by S_n. A *two page book embedding* (2PBE) of a graph $G = (V, E)$ is a plane embedding of G, such that the vertices of G are aligned on a horizontal line and the set of edges can be partitioned into two sets, one of which is embedded in the closed halfplane above the line and the other is embedded in the closed halfplane below the line. Similarly, a *one page book embedding* (1PBE) uses only one of the closed halfplanes. We embed vertices equidistantly along the positive x-axis and refer to them by their x-coordinate, that is, $P = \{1, \ldots, n\}$. An *interval* $[i, j]$ in P is a sequence of the form $i, i + 1, \ldots, j$, for $1 \le i \le j \le n$, or $i, i - 1, \ldots, j$, for $1 \le j \le i \le n$. Observe that we consider an interval $[i, j]$ as oriented and so we can have $i > j$. Denote the *length* of an interval $[i, j]$ by $|[i, j]| = |i - j| + 1$.

Overview. We explicitly construct a plane drawing of two trees $T_1 = (V_1, E_1)$ and $T_2 = (V_2, E_2)$ on the point set $P = \{(i, 0) : 1 \le i \le n\}$. For the most part we will actually work with a 2PBE where the trees give the partition of the edges. In certain situations, however, we will also embed an edge of a tree "on the other side" of the x-axis. So while the final embedding is a 2PBE, in general the partition of the edges is not just according to the trees T_1 and T_2.

Every edge $\{p, q\} \in \binom{P}{2}$ is embedded as an upper or lower semicircle with diameter pq. A joint embedding for T_1 and T_2 is then determined by a map π that assigns to each vertex in $V = V_1 \cup V_2$ a distinct point from P and determines for each edge $e \in E = E_1 \cup E_2$ whether e is embedded above or below the x-axis. If π is a 2PBE, then it is plane, if the following holds: for any two edges $\{u, v\}, \{w, x\}$ that are embedded on the same page, $\pi(u) < \pi(w) < \pi(v)$ implies $\pi(u) \leq \pi(x) \leq \pi(v)$.

As a first step we construct an embedding π_1 for T_1 onto P, using only the halfplane above P to route the edges. The embedding is guided by specific rules which are discussed in Section 3.

Next we recursively construct an embedding for T_2 to pair up with $\pi_1(T_1)$. In principle we follow the same strategy as for T_1, except for the first step, which introduces a "shift". Also, we sometimes have to deviate from this strategy and in some cases even π_1 has to be adjusted locally in order to obtain the final embedding π.

Although neither of the two trees T_1 and T_2 we start with is a star, it is possible—in fact, unavoidable—that stars appear as subtrees during the recursion. Given that we deal with binary trees, these stars can have at most four vertices, though. We have to deal with these stars explicitly whenever they arise, because the general recursive step works for non-stars only. Mostly this can be done simply by gathering all the information on the embedding of T_1 and then computing the proper intervals into which to embed, but at some points subtle changes of the general plan are required. For instance, we occasionally "flip" the embedding of some subtree of T_1: when *flipping* a subtree A of $\pi(T_1)$ that is embedded on an interval $[i, j]$, we reflect the embedded tree at the vertical line $x = \frac{i+j}{2}$ through the midpoint of $[i, j]$. The recursion ends with explicit constructions for subtrees of size at most four.

Each step of the algorithm fixes the embedding for at least one vertex and looks at and works with a constant number of vertices and edges only. As the vertex degree in our graphs is constant, it is straightforward to represent the trees and the host graph so that we can test for the presence of an edge and add or remove an edge in constant time. The sizes of the subtrees in the rooted trees T_1 and T_2 can be precomputed in linear time. Therefore we obtain a linear time algorithm overall.

3   Embedding of T_1

We begin by defining a preliminary 1PBE π_1 for T_1. Throughout the paper, whenever we embed a tree T on an interval $[i, j]$, we assume w.l.o.g. that $i < j$. Since we are free to choose a root r_1 for T_1, let r_1 be any leaf of T_1. We start the recursive procedure by embedding T_1 rooted at r_1 onto $[1, n]$.

In every recursive step, we are given a tree T rooted at a vertex r and an interval $[i, j]$ of length $|r|$. We place r at position i and embed its one or two children according to two rules. The *larger-subtree-first rule* dictates that in the embedding of T on $[i, j]$, the larger subtree of r is embedded on an interval bordering the position of r. The *one-side rule* dictates that all neighbors of a vertex v mapped to $k \in [i, j]$ are mapped to either $[i, k-1]$ or to $[k+1, j]$. Note that these rules imply that every subtree $T \subseteq T_1$ is embedded onto an interval

$[i, j] \subseteq [1, n]$, using the edge $\{i, j\}$. Together with the placement of the root of T_1 at position 1, these two rules completely define the embedding algorithm below. For an example see Fig. 1.

Fig. 1. Example for the embedding of T_1

Algorithm 1. $Embed(T, I)$

Input: A rooted binary tree $T = (V, E)$ and a directed interval I with
$\quad\quad I \subseteq [1, n]$.
Output: A map $\pi_1 : V \to I$.
Let r be the root of T and let $[i, j] = I$ (w.l.o.g. $i < j$).
$\pi_1(r) \leftarrow i$
if r *has degree one in* T **then**
$\quad\big|$ Let r' be the child of r in t.
$\quad\big|$ $Embed(\mathrm{t}(r'), [j, i + 1])$
else if r *has degree two in* T **then**
$\quad\big|$ Let r_1, r_2 be the two children of r, such that $|r_1| \geq |r_2|$.
$\quad\big|$ $Embed(\mathrm{t}(r_1), [i + |r_1|, i + 1])$
$\quad\big|$ $Embed(\mathrm{t}(r_2), [j, i + |r_1| + 1])$

Observe that $\pi_1(T_1)$ satisfies the following property:

Proposition 1. *The edge* $\{n - 1, n\}$ *is not used by* π_1, *for* $n \geq 5$.

This edge will be used to start the embedding of T_2. Consider the embedding π_1 of a tree T_1 onto $[1, n]$. When restricting the focus to some subinterval $[i, j] \subseteq [1, n]$, we see the embedding $\pi_1(F)$ of a forest $F := \pi_1^{-1}([i, j]) \subseteq T_1$. Given that a specific embedding is used for T_1, we should also be able to derive some properties of $\pi_1(F)$. The following lemma describes some embeddings that cannot be produced by Algorithm 1. We will use this during the embedding of T_2.

Lemma 1. *Consider a tree* $T \subseteq T_1$ *such that* $T = \pi_1^{-1}([i, j])$, *for some interval* $[i, j] \subseteq [1, n]$, *and suppose that* $\pi_1(T)$ *uses the edges* $\{i, k\}$ *and* $\{\ell, j\}$, *for some* $i < k < \ell < j$. *Then at least one of the following two conditions holds:*

$$a > b + c \quad \wedge \quad c > b \tag{1}$$

$$c > a + b \quad \wedge \quad a > b, \tag{2}$$

where $a = |[i, k]|$, $b = |[k + 1, \ell - 1]|$, *and* $c = |[\ell, j]|$.

Proof. Note that the second equation is obtained from the first by exchanging the roles of a and c. Hence, we may assume w.l.o.g. that $p(j) = i$.

Suppose that $b = 0$. Since i has children k and j, vertex i has no incoming edges besides $\{i, k\}$ and $\{i, j\}$. The larger-subtree-first rule for i implies $a - 1 \geq b + c$, which together with the trivial $c > 0$ yields (1).

It remains to consider the case $b > 0$. Since i already has indegree two and j already has one child in ℓ, the vertices at $[k + 1, \ell - 1]$ must form a tree B with root r_B and $p(r_B) = j$. The larger-subtree-first rule for j implies $c - 1 \geq b$ and the one for i implies $a - 1 \geq b + c$, which together gives (1).   □

Corollary 1. *Consider a tree $T \subseteq T_1$ such that $T = \pi_1^{-1}([i, j])$, for some interval $[i, j] \subseteq [1, n]$, and suppose that $\pi_1(T)$ uses the edges $\{i, k\}$ and $\{\ell, j\}$, for some $i < k < \ell < j$. Then $k - i \neq j - \ell$.*

Proof. If $k - i = j - \ell$, then we have $a = c$ in Lemma 1. At least one of the two conditions must hold, but both imply $a > b + a$, which is impossible for $a, b \geq 0$.   □

4   Embedding of T_2

In this section we describe how to obtain an embedding for T_2 that is compatible with the already constructed embedding of T_1. We will do this in a recursive way similar to the embedding of T_1. The difference is that we have to take the edges used by T_1 into account and adapt our strategy accordingly. Sometimes it is also necessary to change the embedding of T_1. When doing so, we have to be very careful in order to not destroy the properties of the embedding that were discussed in the preceding section and that play a crucial role in driving the embedding algorithm for T_2.

Small subtrees of size at most four may be stars and so lead to unsolvable subproblems in the recursion. We resolve this by giving explicit solutions for these cases. These therefore serve as base cases for the recursive embedding of T_2. In the recursion we will keep as an invariant—we call it the *placement invariant*—that whenever the recursive embedding for a subtree $t(v)$ on an interval $I = [i, j]$ is invoked, the placement of v on i is valid in the sense that the edge from i to the point where $p(v)$ is embedded is not used by T_1 (and so it is available for T_2). Sometimes this invariant does not only apply to the parent of v but also to another child that has already been embedded. Generally speaking, whenever we call the algorithm recursively for some tree T to be embedded on some interval I, we have to ensure that placing the root r of T on i does not induce any edge with the already embedded neighbors of r outside of I that is already used by T_1. This invariant allows us to work with the current interval locally, without having to care about where vertices are placed outside of this interval.

Whenever we would like to map r to a position different from i, we have to be careful, because there may be edges that we do not see when only considering the situation on I locally: A vertex a of T_1 mapped into I may have edges to vertices outside of I, in particular, to a vertex that is mapped to the same position as the parent of r in T_2. In such a case, we say that r and a are *in conflict*. Obviously we must not map two vertices that are in conflict to the same position. Note that vertices of T other than r do not have conflicts and hence can be placed safely, as long as we ensure that no edge on I is used on both sides.

Stars occur not only in the base cases of the recursion, but they may also occur as subtrees of a large tree. So whenever we are at a subtree $T \subset T_2$ with root r in the recursion and one of the subtrees of r has strictly fewer than five vertices, we cannot use the recursive procedure for that subtree but have to handle that situation explicitly.

The remainder of this section details our recursive embedding algorithm. We start by handling the base cases of the recursion, which consist of certain trees on at most nine vertices ("A small tree."). It follows a description of the general step, which consists of two parts: "the first step" of the recursive procedure selects the starting vertex (root) for the recursion and then the general case handles "a large tree", in which at least one subtree of the root has five or more vertices. Due to space limitations many proofs have to be omitted in this extended abstract.

A small tree. We begin with the case where $|b| \leq 4$. Let w.l.o.g. $I = [1, |b|]$. Note that there are stars of size at most four, so it will not always be possible to find an embedding. We will describe precisely when an embedding is possible. For $|b| = 1$ an embedding is trivially possible. For $|b| = 2$ an embedding is possible if and only if $\{1, 2\}$ is not used (in which case we are embedding against a star). For $|b| = 3$ there are two rooted versions of P_3 that we denote by P_3^1 (rooted at a leaf) and P_3^2 (rooted at the interior vertex of degree two).

Lemma 2. *Given a forest $A \subseteq T_1$ embedded on an interval $I = [i, i+2] \subseteq [1, n]$ using π_1, we can pack any rooted tree B of size 3 together with A onto I, assuming the placement invariant holds (the root b of B can be placed at i), unless:*

- *$A = P_3$; or*
- *$B = P_3^2$, $\pi_1(A)$ uses the edge $\{i, i+1\}$, and b has a conflict at $i+2$.*

Proof. Suppose that $A \neq P_3$. Then $\pi_1(A)$ uses at most one edge on I, which is either $\{i, i+1\}$ or $\{i+1, i+2\}$. If $\pi_1(A)$ uses $\{i, i+1\}$, then $B = P_3^1$ can be embedded by placing b on i and using the edges $\{i, i+2\}$ and $\{i+1, i+2\}$. Using the same edges also $B = P_3^2$ can be embedded by placing b on $i + 2$, unless it has a conflict there. If $\pi_1(A)$ uses $\{i+1, i+2\}$, then $B = P_3^2$ can be embedded by placing b on i and using the edges $\{i, i+1\}$ and $\{i, i+2\}$. Using the same edges also $B = P_3^1$ can be embedded by placing b on $i + 2$. If b has a conflict at $i + 2$, flip the edge $\{i+1, i+2\}$ in $\pi(A)$ to match b with the leaf $i + 1$ of T_1.   □

For $|b| = 4$, there are three possible trees that we want to embed. We will give explicit embeddings for each of these. We distinguish the following cases in Fig. 2, from left to right. If edge $\{1, 4\}$ is used, then we are embedding against one of the three possible trees of size 4. The star never works, which leaves us with two options (case 1 and 2). If $\{1, 4\}$ is not used, then $\{1, 3\}$ might be used. In this case, $\{3,4\}$ is certainly not used, and $[1,3]$ may contain an embedding of P_3^2 (case 3) or P_3^1 (case 4). The case where $\{2,4\}$ is used instead of $\{1,3\}$ is symmetric (we can use symmetry here since in this particular case, we never use the placement invariant). If none of the edges $\{1,4\}$ or $\{1,3\}$ or $\{2,4\}$ are used, then the only remaining possible edges are between consecutive vertices (cases 5 to 8). Finally, if no edges are used, then any embedding will work. Observe that, in particular, all cases where we embed a non-star tree with a tree work, which settles the theorem for $n \leq 4$.

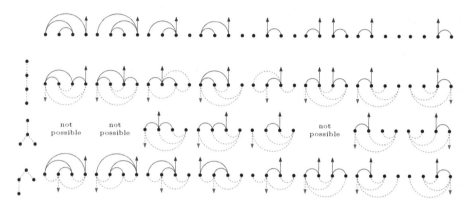

Fig. 2. The embeddings of all trees of size 4 with all possible embeddings of π_1, except the case where a star of size 4 was embedded by π_1

The implications of this table are summarized in the following lemma:

Lemma 3. *Given a forest $A \subseteq T_1$ embedded on an interval $I = [i, i+3] \subseteq [1, n]$ using π_1, we can pack any rooted tree B of size 4 together with A onto I, assuming the placement invariant holds (the root b of B can be placed at i), unless:*

- *a star is embedded at $[i, i+3]$; or*
- *B is a star, and either a non-star tree is embedded at $[i, i+3]$, or $\{i, i+1\}$ and $\{i+2, i+3\}$ are both used.*

The situation changes for $n \geq 5$, because there is no binary star on more than four vertices. The following lemma—whose proof is omitted due to space limitations—shows that an embedding is always possible for certain small trees, in particular, for all trees on five vertices.

Lemma 4. *Consider a forest $A \subseteq T_1$ embedded on an interval $I = [i, i+k] \subseteq [1, n]$ using π_1, for some $k \geq 4$, and a tree B on $k+1$ vertices rooted at a vertex b with $\deg_B(b) \leq 2$ and such that $|c| \leq 4$, for every child c of b. Then we can pack B together with A onto I, assuming the placement invariant holds (b can be placed at i).*

The general step. We have to embed a non-star subtree $B \subset T_2$ with root b, $|b| \geq 6$, onto $I = [i, j]$ with $|I| = |b|$. W.l.o.g. assume that $i < j$. On the other side, there is a forest $A \subset T_1$ that consists of trees $A_1, \ldots, A_k$ that have been embedded in this order onto I using π_1. Each A_i has a single vertex, its root, that is connected to vertices outside of I.

We have to be careful when pairing b with a root of some A_i because there are edges that we do not see when only considering the situation on I locally. More precisely, we must not map two vertices that are in conflict to the same position. Recall that non-root vertices do not have conflicts and hence can be placed safely, as long as we ensure that no edge on I is used on both sides.

From the placement invariant we know that b can be safely mapped to i, that is, b is not in conflict with whichever vertex of A is mapped to i. Whenever we want to map b to a point different from i, it has to be ensured that there is no conflict with the corresponding vertex of A. Moreover, knowing that b is connected to some vertex outside of I, we have to ensure that such edges can be drawn without crossing any of the edges used for the embedding of B. As long as b is mapped to i, which is an endpoint of I, this is clearly true. But when mapping b to a point $y \in I \setminus \{i\}$, there must not be any edge $\{x, z\}$ in B for which $y \in [\pi(x), \pi(z)]$. A similar care has to be taken for the roots of A. In fact, to ensure this invariant, we argue locally only: In the algorithm we map b to a point different from i only if then b is paired with a vertex of A on the other side that does not have any edge to a vertex outside of the current interval.

Also, sometimes we have to change the embedding of T_1, which in principle might destroy the carefully derived properties of the embdding π_1. Indeed, in the final embedding these properties do not necessarily hold. However, we ensure that they do hold for any interval that we invoke the recursive algorithm on. That is, whenever the recursive procedure is invoked on some interval, then the subforest of T_1 embedded on it looks "as-if" it came from an embedding of type π_1 (in particular, it satisfies the one-side rule and the larger-subtree-first rule).

The first step. This paragraph describes the first step of the recursive procedure that serves only to select a suitable starting vertex. Choose a leaf b of T_2 as a root and let b' denote the child of b in T_2. Map b to n and recursively embed $t(b')$ onto $[n - 1, 1]$. For $n = 6$ this can be done by Lemma 4. For $n \geq 7$, we apply the general recursive step, as described in the remainder. Observe that the placement invariant for the recursive step is guaranteed by Proposition 1.

A large tree. The root b may have degree one or two in B. We consider these two cases separately.

A large tree with a degree one root. So suppose that $\deg_B(b) = 1$. Let b_1 denote the child of b in B and let $B_1 = t(b_1)$. Note that $|b_1| \geq 5$ and so, in particular, B_1 is not a star and we can embed it recursively. We would like to embed b on i and B_1 recursively onto $[j, i + 1]$ (Fig. 3a). In case the edge $\{i, j\}$ is used by $\pi_1(A)$ already, we change this plan and embed B_1 onto $[i + 1, j]$ instead (Fig. 3b). If also the edge $\{i, i + 1\}$ is used by $\pi_1(A)$ already, we rearrange the embedding for A by exchanging the mappings to i and $i+1$ (Fig. 3c). (Note that by the one-side rule $i+1$ is a leaf of $\pi_1(A)$ and, therefore, such a rearrangement does not introduce a crossing.) After this change we can resort to the original plan and embed B_1 onto $[j, i+1]$ (Fig. 3d).

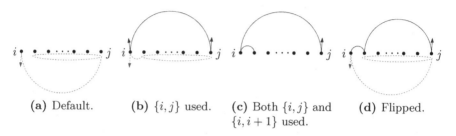

(a) Default.   (b) $\{i,j\}$ used.   (c) Both $\{i,j\}$ and   (d) Flipped.
                                              $\{i, i+1\}$ used.

Fig. 3. Cases for a large tree with a degree one root

A large tree with a degree two root. Consider now the case $\deg_B(b) = 2$. Denote the children of b by b_1 and b_2 so that w.l.o.g. $|b_1| \geq |b_2|$, and let $B_1 = \mathrm{t}(b_1)$ and $B_2 = \mathrm{t}(b_2)$. By Lemma 4 we may assume that $|b_1| \geq 5$. Let B_1^+ be B_1 extended with b as a root of degree one. As B_1^+ has at least six vertices, it can be handled recursively. The general plan is the following: embed B_2 onto $[j, i+|b_1|+1]$ and then embed B_1^+ onto $[i, i+|b_1|]$. We distinguish subcases depending on the size of B_2. Due to space limitations, we discuss the cases $|b_2| \geq 5$ and $|b_2| \leq 2$ only.

Case L1: $|b_2| \geq 5$. Then neither B_1 nor B_2 is a star. If $\pi_1(A)$ does not use $\{i,j\}$, we can just follow the plan mentioned. Similarly, if $\pi_1(A)$ does not use $\{i, i+|b_1|+1\}$, then we can adjust the plan to embed B_2 onto $[i+|b_1|+1, j]$ instead. So suppose that $\pi_1(A)$ uses both $\{i,j\}$ and $\{i, i+|b_1|+1\}$. Due to the presence of $\{i, i+|b_1|+1\}$ we know that there is no conflict for b at $i+|b_1|$ and that $\{i+|b_1|, j\}$ is not used by $\pi_1(A)$. We first embed B_2 onto $[j, i+|b_1|+1]$ recursively, treating $i+|b_1|+1$ as a conflict in case that $\pi_1(A)$ uses $\{i+|b_1|, i+|b_1|+1\}$. Then B_1^+ can be embedded onto $[i+|b_1|, i]$ recursively. Note that b may have a conflict at i, if the embedding for B_2 chose to map b_2 to a neighbor of i in $\pi(A)$ (which is no problem).

Case L2: $|b_2| = 1$. Suppose that $\pi_1(A)$ uses $\{i,j\}$. Then by Corollary 1 $\pi_1(A)$ does not use both $\{i, i+1\}$ and $\{j-1, j\}$. Since $\{i,j\}$ is used, none of the vertices $i+1, \ldots, j-1$ are in conflict with b. Therefore we can recursively embed B_1^+ onto $[i+1, j]$ (if $\{i, i+1\}$ is not used) or $[j-1, i]$ (if $\{j-1, j\}$ is not used) and then map b_2 to i or j, respectively.

Alternatively, suppose that $\pi_1(A)$ does not use $\{i,j\}$. By the placement invariant, it is safe to embed b on i, so we can embed b_2 on j and B_1^+ onto $[i, j-1]$.

Case L3: $|b_2| = 2$. If $\pi_1(A)$ does not use either of $\{i,j\}$ or $\{j-1, j\}$, then we can embed B_2 onto $[j, j-1]$ and B_1^+ onto $[i, j-2]$. So we may suppose that $\pi_1(A)$ uses one of these edges.

Case L3.1: $\pi_1(A)$ uses $\{i,j\}$. Then by Corollary 1 $\pi_1(A)$ does not use both $\{i, i+1\}$ and $\{j-1, j\}$. If $\pi_1(A)$ does not use $\{j-1, j\}$, then we embed B_2 onto

$[j-1,j]$ and B_1^+ onto $[i,j-2]$—unless $\pi_1(A)$ uses $\{i,j-1\}$. In that case we embed B_2 onto $[j,j-1]$ and B_1^+ onto $[j-2,i]$ instead.

Hence we may suppose that $\pi_1(A)$ uses $\{j-1,j\}$ but does not use $\{i,i+1\}$. By the larger-subtree-first rule j cannot be the root of $\pi_1(A)$ (the leaf at $j-1$ is right next to it), and so b does not have a conflict at j and $\pi_1(A)$ does not use $\{i+1,j\}$. Therefore we can embed B_2 onto $[i+1,i]$ and B_1^+ onto $[j,i+2]$.

Case L3.2: $\pi_1(A)$ does not use $\{i,j\}$ but it uses $\{j-1,j\}$. Our plan is to embed B_2 onto $[j-2,j-1]$, as this edge is not used by $\pi_1(A)$ according to the one-side rule. For this we would like to map $b \mapsto j$ and B_1 onto $[i,j-3]$. This works fine, unless $\{j-2,j\}$ is used by $\pi_1(A)$ or b has a conflict at j.

Case L3.2.1 $\pi_1(A)$ uses $\{j-2,j\}$. Consider the subtree S of $\pi_1(A)$ on $[j-2,j]$. We will argue below that regardless of whether the parent of S is at $j-2$ or at j, we can flip S by exchanging the mapping for $j-2$ and j and then embed B_2 onto $[j-1,j]$ and B_1^+ onto $[i,j-2]$. Note that after the flip $\{j-1,j\}$ is not used by $\pi(A)$ anymore and the neither before nor after the flip $\pi(A)$ uses $\{i,j-1\}$.

It remains to argue about the validity of the flip. If the parent of S lies to the right, it is in a different interval and so it does not know about the details of S, anyway. If the parent lies to the left and within $[i,j-3]$, then the (already embedded) subtree B_2 is outside of the interval $[i,j-2]$ for B_1^+, effectively shrinking a subtree on three vertices to a subtree on one vertex. But as this subtree is certainly the one furthest from its parent in the current interval of interest, decreasing its size does not affect the larger-subtree-first rule.

Case L3.2.2 $\pi_1(A)$ does not use $\{j-2,j\}$ but b has a conflict at j. Due to the conflict, the root of $\{j-1,j\}$ is at j and so b does not have a conflict at $j-1$. Just as in the previous case, we can flip the edge by exchanging the mapping for $j-1$ and j and then embed B_2 onto $[j-2,j-1]$ (not used by $\pi(A)$, because it corresponds to $\{j-2,j\}$ before the flip), map b on j (no conflict, because it corresponds to $j-1$ before the flip), and finally embed B_1 onto $[i,j-3]$ recursively ($\{i,j\}$ is not used by $\pi(A)$, because it corresponds to $\{i,j-1\}$ before the flip and $j-1$ was a leaf connected to j in $\pi_1(A)$).

5   Conclusion

In this paper, we gave a proof of the well-known planar packing conjecture for the case of binary non-star trees. The major open problem is the proof of the hypothesis for general non-star trees. We strongly suspect this to be true and we hope that the techniques developed for the binary case here can also be used to attack the general problem.

Of course, the inclusion of more than two trees should be considered. Since a tight packing into a planar graph of n vertices is not possible, the question there would be to minimize the size of the planar graph which comprises the union of the trees. For example a planar packing of three n-vertex paths on $n+1$ vertices can be obtained, while appropriate generalizations to more general graph classes and/or a larger number of subgraphs have not been obtained as far as we know.

References

1. Akiyama, J., Chvátal, V.: Packing paths perfectly. Discrete Mathematics 85(3), 247–255 (1990)
2. Braß, P., Cenek, E., Duncan, C.A., Efrat, A., Erten, C., Ismailescu, D., Kobourov, S.G., Lubiw, A., Mitchell, J.S.B.: On simultaneous planar graph embeddings. Comput. Geom. 36(2), 117–130 (2007)
3. Caro, Y., Yuster, R.: Packing graphs: The packing problem solved. Electr. J. Comb. 4(1) (1997)
4. Eppstein, D.: Arboricity and bipartite subgraph listing algorithms. Information Processing Letters 51(4), 207–211 (1994)
5. Eppstein, D.: Subgraph isomorphism in planar graphs and related problems. J. Graph Algorithms & Applications 3(3), 1–27 (1999)
6. Frank, A., Szigeti, Z.: A note on packing paths in planar graphs. Math. Program. 70(2), 201–209 (1995)
7. Frati, F.: Embedding graphs simultaneously with fixed edges. In: Kaufmann, M., Wagner, D. (eds.) GD 2006. LNCS, vol. 4372, pp. 108–113. Springer, Heidelberg (2007)
8. Frati, F.: Planar packing of diameter-four trees. In: 21st Canadian Conference on Computational Geometry (CCCG 2009), pp. 95–98 (2009)
9. Frati, F., Geyer, M., Kaufmann, M.: Planar packings of trees and spider trees. Information Processing Letters 109(6), 301–307 (2009)
10. García, A., Hernando, C., Hurtado, F., Noy, M., Tejel, J.: Packing trees into planar graphs. J. Graph Theory, 172–181 (2002)
11. M. R. Garey and D. S. Johnson. *Computers and Intractability, A Guide to the Theory of NP-Completeness.* W.H. Freeman and Company, New York, 1979.
12. Geyer, M., Kaufmann, M., Vrt'o, I.: Two trees which are self–intersecting when drawn simultaneously. In: Healy, P., Nikolov, N.S. (eds.) GD 2005. LNCS, vol. 3843, pp. 201–210. Springer, Heidelberg (2006)
13. Gonçalves, D.: Edge partition of planar graphs into two outerplanar graphs. In: Proc. 37th Annu. ACM Sympos. Theory Comput., pp. 504–512 (2005)
14. Hedetniemi, S.M., Hedetniemi, S.T., Slater, P.J.: A note on packing two trees into K_N. Ars Combin. 11, 149–153 (1981)
15. Maheo, M., Saclé, J.-F., Woźniak, M.: Edge-disjoint placement of three trees. European J. Combin. 17(6), 543–563 (1996)
16. Mutzel, P., Odenthal, T., Scharbrodt, M.: The thickness of graphs: A survey. Graphs and Combinatorics 14(1), 59–73 (1998)
17. Nash-Williams, C.S.J.A.: Edge-Disjoint Spanning Trees of Finite Graphs. Journal of the London Mathematical Society-second Series s1-36, 445–450 (1961)
18. Oda, Y., Ota, K.: Tight planar packings of two trees. In: European Workshop on Computational Geometry, pp. 215–216 (2006)
19. Schnyder, W.: Planar graphs and poset dimension. Order 5, 323–343 (1989)
20. Tutte, W.T.: On the problem of decomposing a graph into n connected factors. Journal of the London Mathematical Society s1-36(1), 221–230 (1961)
21. Ullmann, J.R.: An algorithm for subgraph isomorphism. J. ACM 23(1), 31–42 (1976)

Hierarchies of Predominantly Connected Communities

Michael Hamann, Tanja Hartmann, and Dorothea Wagner

Department of Informatics, Karlsruhe Institute of Technology (KIT)
michael@content-space.de, {t.hartmann,dorothea.wagner}@kit.edu

Abstract. We consider communities whose vertices are predominantly connected, i.e., the vertices in each community are stronger connected to other community members of the same community than to vertices outside the community. Flake et al. introduced a hierarchical clustering algorithm that finds predominantly connected communities of different coarseness depending on an input parameter. We present a simple and efficient method for constructing a clustering hierarchy according to Flake et al. that supersedes the necessity of choosing feasible parameter values and guarantees the completeness of the resulting hierarchy, i.e., the hierarchy contains all clusterings that can be constructed by the original algorithm for any parameter value. However, predominantly connected communities are not organized in a single hierarchy. Thus, we further develop a framework that, after precomputing at most $2(n-1)$ maximum flows, admits a linear time construction of a clustering $\Omega(S)$ of predominantly connected communities that contains a given community S and is maximum in the sense that any further clustering of predominantly connected communities that also contains S is hierarchically nested in $\Omega(S)$. We further generalize this construction yielding a clustering with similar properties for k given communities in $O(kn)$ time. This admits the analysis of a network's structure with respect to various communities in different hierarchies.

1 Introduction

There exist many different approaches to find communities in networks, many of which are inspired by graph clustering techniques originally developed for special applications in fields like physics and biology. Graph clustering is based on the assumption that the given network is a compound of dense subgraphs, so called *clusters* or *communities*, that are only sparsely connected among each other, and aims at finding a *clustering* that represents these subgraphs. However, evaluating the quality of a found clustering is often difficult, since there are no generally applicable criteria for good clusterings and clustering properties that are well interpretable in the network's context are rarely guaranteed. In this work we thus focus on predominantly connected communities in undirected edge-weighted graphs. Predominant connectivity is easy to interpret and guarantees that only vertices whose membership to a community is clearly indicated by the networks's structure are assigned to a community. The latter is in particular desired if the analysis of the community structure is meant to support costly or risky decisions.

Contribution and Outline. We discuss different types of predominantly connected communities (cp. Table 1 for an overview) in Section 2 and argue that considering

F. Dehne, R. Solis-Oba, and J.-R. Sack (Eds.): WADS 2013, LNCS 8037, pp. 365–377, 2013.
© Springer-Verlag Berlin Heidelberg 2013

Table 1. Overview of different types of predominantly connected communities. The columns to the right describe the relations between the types in terms of inclusion.

A subgraph $S \subseteq V$ is a			WC	ES	SC
WC	$\forall u \in S$	$c(\{u\}, S \setminus \{u\}) \gneqq c(\{u\}, V \setminus S)$	x	x	
ES	$\forall U \subset S$	$c(U, S \setminus U) \gneqq c(U, V \setminus S)$		x	
SC	$\exists s \in S : \forall U \subset S, s \notin U$	$c(U, S \setminus U) \gneqq c(U, V \setminus S)$		x	x

source communities (SCs) in networks is reasonable. We further give a characterization of SCs and introduce basic nesting properties. In Section 3, we review the cut clustering algorithm by Flake et al. [2], which takes an input parameter α and decomposes a given network into SCs, each of which providing an intra-cluster density of at least α and an inter-cluster sparsity of at most α. At the same time, α controls the coarseness of the resulting clustering such that for varying values the algorithm returns a clustering hierarchy. Flake at al. refer to Gallo et al. [3] for the question how to choose α such that all possible hierarchy levels are found. However, they give no further description how to extend the approach of Gallo et al., which finds all breakpoints of α for a single parametric flow, to a fast construction of a complete hierarchy. They just propose a binary-search approach to find good values for α. We introduce a parametric-search approach that guarantees the completeness of the resulting hierarchy and exceeds the running time of a binary search-based approach, whose running time strongly depends on the discretization of the parameter range.

Experimental evaluations further showed that the cut clustering algorithm finds meaningful clusters in real-world instances [2], but yet, it often happens that even in a complete hierarchy non-singleton clusters are only found for a subgraph of the initial network, while the remaining vertices stay unclustered even on the coarsest non-trivial hierarchy level [2,6]. Motivated by this observation, in Section 4, we develop a framework that is based on a set $M(G)$ of $n \leq |M(G)| \leq 2(n-1)$ maximal SCs in the graph G, i.e., each further SC is nested in a SC in $M(G)$, and is represented by a special cut tree, which can be constructed by at most $2(n-1)$ max-flow computations. After computing $M(G)$ in a preprocessing step, the framework efficiently answers the following queries: (i) Given an arbitrary SC S, what does a clustering $\Omega(S)$ look like that consists of S and further SCs such that any SC not intersecting with S is nested in a cluster of $\Omega(S)$? In particular, $\Omega(S)$ is maximum in the sense that any clustering of SCs that contains S is hierarchically nested in $\Omega(S)$. We show that $\Omega(S)$ can be determined in linear time. (ii) Given k disjoint SCs, which is the maximal clustering $\Omega(S_1, \ldots, S_k)$ that contains the given SCs, is nested in each $\Omega(S_i)$, $i = 1, \ldots, k$, and guarantees that any clustering of SCs that also contains the given ones is nested in $\Omega(S_1, \ldots, S_k)$? Computing $\Omega(S_1, \ldots, S_k)$ takes $O(kn)$ time. These queries allow to further examine the community structure of a given network, beyond the complete clustering hierarchy according to Flake et al. We exemplarily apply both queries to a small real world network, thereby finding a new clustering beyond the hierarchy that contains all non-singleton clusters of the best clustering in the hierarchy but far less singletons. Proofs omitted due to space constraints can be found in the full version [7].

Preliminaries. Throughout this work we consider an undirected, weighted graph $G = (V, E, c)$ with vertices V, edges E and a positive edge cost function c, writing $c(u, v)$ as a shorthand for $c(\{u, v\})$ with $\{u, v\} \in E$. Whenever we consider the degree $\deg(v)$ of $v \in V$, we implicitly mean the sum of all edge costs incident to v. A *cut* in G is a partition of V into two *cut sides* S and $V \setminus S$. The cost $c(S, V \setminus S)$ of a cut is the sum of the costs of all edges *crossing* the cut, i.e., edges $\{u, v\}$ with $u \in S$, $v \in V \setminus S$. For two disjoint sets $A, B \subseteq V$ we define the cost $c(A, B)$ analogously. Two cuts are *non-crossing* if their cut sides are pairwise nested or disjoint. Two sets $S, T \subset V$ are *separated* by a cut if they lie on different cut sides. A minimum S-T-cut is a cut that separates S and T and is the cheapest cut among all cuts separating these sets. We call a cut a *minimum separating cut* if there exists an arbitrary pair $\{S, T\}$ for which it is a minimum S-T-cut. We identify singleton sets with the contained vertex without further notice. We further denote the *connectivity* of $\{S, T\} \subseteq 2^V$ by $\lambda(S, T)$, describing the cost of a minimum S-T-cut. A *clustering* Ω of G is a partition of V into subsets $C^1, \ldots, C^k$, which define vertex-induced subgraphs, called *clusters*. A cluster is *trivial* if it corresponds to a connected component. A vertex that forms a singleton cluster although it is no singleton in G, is *unclustered*. A clustering is *trivial* if it consists of trivial clusters or if $k = n$. A *hierarchy of clusterings* is a sequence $\Omega_1 \leq \cdots \leq \Omega_r$ such that $\Omega_i \leq \Omega_j$ implies that each cluster in Ω_i is a subset of a cluster in Ω_j. We say $\Omega_i \leq \Omega_j$ are *hierarchically nested*. A clustering Ω is *maximal* with respect to a property $\mathcal{P}$ if there is no other clustering Ω' with property $\mathcal{P}$ and $\Omega \leq \Omega'$.

2   Predominantly Connected Communities

In the context of large web-based graphs, Flake et al. [1] introduce *web communities* (WCs) in terms of predominant connectivity of single vertices: A set $S \subseteq V$ is a web community if $c(\{u\}, S \setminus \{u\}) \gneq c(\{u\}, V \setminus S)$ for all $u \in S$. Web communities are not necessarily connected (cp. Fig. 1) and decomposing a graph into k web communities is NP-complete [2]. Extending the predominant connectivity from vertices to arbitrary subsets yields *extreme sets* (ESs), which satisfy a stricter prop-erty that guarantees connectivity and gives a good intuition why the vertices in ESs belong together: A set $S \subseteq V$ is an extreme set if $c(U, S \setminus U) \gneq c(U, V \setminus S)$ for all $U \subsetneq S$. The extreme sets in a graph can be computed in $O(nm + n^2 \log n)$ time with the help of maximum adjacency orderings [10]. They form a subset of the *maximal components* of a graph, which subsume vertices that are not separated

Fig. 1.   Unconnected web community (left)

by cuts cheaper than a certain lower bound. Maximal components are either nested or disjoint and can be deduced from a cut tree, whose construction needs $n - 1$ maximum flow computations [4]. They are used in the context of image segmentation by Wu and Leahy [12].

In, for example, social networks, we are also interested in communities that sur-round a designated vertex, for instance a central person. Complying with this view, *source communities* (SCs) describe vertex sets where each subset that does not contain a designated vertex is predominantly connected to the remainder of the group: A set $S \subseteq V$ is a SC with *source* $s \in S$ if $c(U, S \setminus U) \gneq c(U, V \setminus S)$ for all $U \subsetneq S \setminus \{s\}$.

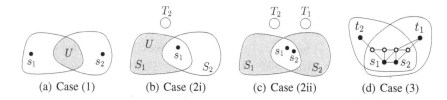

(a) Case (1)　　　(b) Case (2i)　　　(c) Case (2ii)　　　(d) Case (3)

Fig. 3. Situation in Lemma 3

The members of a SC can be interpreted as *followers* of the source in that sense that each subgroup feels more attracted by the source (and other group members) than by the vertices outside the group. The predominant connectivity of SCs implements a close relation to minimum separating cuts. In fact, SCs are characterized as follows.

Lemma 1. *A set $S \subset V$ is a SC of $s \in S$ iff there is $T \subseteq V \setminus S$ such that $(S, V \setminus S)$ is the minimum s-T-cut in G that minimizes the number of vertices on the side containing s.*

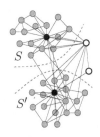

Based on this characterization, we introduce some further notations and two basic lemmas on nesting properties of SCs, which we will mainly use in Section 4. Note that a minimum s-T-cut in G must not be unique, however, the minimum s-T-cut that minimizes the number of vertices on the side containing s is unique. We call such a cut, which induces a SC S, a *community cut*, S the SC of s *with respect to* T and T the *opponent* of s. Hence, $\mathcal{SC}: V \times 2^V \to 2^V$, $\mathcal{SC}(s, T) \mapsto \{$the SC of s with respect to $T\}$ is well defined providing $\mathcal{SC}(s, T)$ as future notation. The corresponding maximum flow between s and T also induces an *opposite*

Fig. 2. Indecisive vertices (white)

SC $S' := \mathcal{SC}(T, s)$, if we consider T as a compound node. If the community cut is the only minimum s-T-cut, it is $S' = V \setminus S$. Otherwise, $X := V \setminus (S \cup S') \neq \emptyset$ and the vertices in X are neither predominantly connected within $S \cup X$ nor within $S' \cup X$, i.e., $c(U, S \cup X) \leq c(U, V \setminus (S \cup X))$ for all $U \subseteq X$ (analogously for S'). In, for example, a social network this can be interpreted as follows. Whenever s and the group T become rivals, the network decomposes into followers of s (in S), followers of T (in S') and possibly some *indecisive* individuals in $V \setminus (S \cup S')$. Figure 2 exemplarily shows two indecisive vertices in the (unweighted[1]) karate club network gathered by Zachary [13]. Note that a SC can have several sources, and a vertex can have different SCs w.r.t. different opponents. The SCs of a vertex are partially nested as stated in Lemma 2, which is a special case of (2i) of Lemma 3 summarizing the intersection behavior of arbitrary SCs. See Figure 3 for illustration and an example of neither nested nor disjoint SCs.

Lemma 2. *Let S denote a SC of s and $T \cap S = \emptyset$. Then $S \subseteq \mathcal{SC}(s, T)$.*

As a consequence, each SC $S \neq V$ is nested in a SC S' that is a SC w.r.t. a single vertex t, while any SC $\bar{S}$ with $S' \subsetneq \bar{S}$ contains t. In this sense, SCs w.r.t. single vertices are *maximal*. We denote the set of maximal SCs in G by $M(G)$.

---

[1] Zachary considers the weighted network and therein the minimum cut that separates the two central vertices of highest degree (black). In the weighted network this cut is unique.

Lemma 3. *Consider $S_1 := \mathcal{SC}(s_1, T_1)$ and $S_2 := \mathcal{SC}(s_2, T_2)$.*
(1) If $\{s_1, s_2\} \cap (S_1 \cap S_2) = \emptyset$, then $S_1 \cap S_2 = \emptyset$.
(2) If $T_2 \cap S_1 = \emptyset$ and $s_1 \in S_2$, then $S_1 \subseteq S_2$ (i). If further $T_1 \cap S_2 = \emptyset$ and $s_2 \in S_1$,
then $S_1 = S_2$ (ii).
(3) Otherwise, S_1 and S_2 are neither nested nor disjoint.

3   Complete Hierarchical Cut Clustering

The clustering algorithm of Flake et al. [2] exploits the properties of minimum separating cuts together with a parameter α in order to get clusterings where the clusters are SCs with the following additional property: For each cluster $C \in \Omega$ and each $U \subsetneq C$ it holds

$$\frac{c(C, V \setminus C)}{|V \setminus C|} \leq \alpha \leq \frac{c(U, C \setminus U)}{\min\{|U|, |C \setminus U|\}} \tag{1}$$

According to the left side of this inequality separating a cluster C from the rest of the graph costs at most $\alpha|V \setminus C|$ which guarantees a certain inter-cluster sparsity. The right side further guarantees a good intra-cluster density in terms of expansion, a measure introduced by [8], saying that splitting a cluster C into U and $C \setminus U$ costs at least $\alpha \min\{|U|, |C \setminus U|\}$. Hence, the vertex sets representing valid candidates for clusters must be very tight—in addition to the predominant connectivity they must also provide an expansion that exceeds a given bound.

Flake et al. develop their parametric cut clustering algorithm step by step starting from an idea involving cut trees [4]. The final approach, however, just uses community-cuts in a modified graph in order to identify clusters that satisfy condition (1). We refer to this approach by CutC. Here we give a more direct description of this method. Given a graph $G = (V, E, c)$ and a parameter $\alpha > 0$, as a preprocessing step, augment G by inserting an artificial vertex t and connect-ing t to each vertex in G by an edge of

---

Algorithm 1. CUTC

Input: Graph $G_\alpha = (V_\alpha, E_\alpha, c_\alpha)$
1   $\Omega \leftarrow \emptyset$
2   **while** $\exists\, u \in V_\alpha \setminus \{t\}$ **do**
3      $C^u \leftarrow \mathcal{SC}(u, t)$ in G_α
4      $r(C^u) \leftarrow u$
5   **forall the** $C^i \in \Omega$ **do**
6   **if** $r(C^i) \in C^u$ **then**
           $\Omega \leftarrow \Omega \setminus \{C^i\}$
7      $\Omega \leftarrow \Omega \cup \{C^u\}$;
      $V_\alpha \leftarrow V_\alpha \setminus C^u$
8   **return** $\mathcal{C}$

---

cost α. Denote the resulting graph by $G_\alpha = (V_\alpha, E_\alpha, c_\alpha)$. Then apply CutC (Alg. 1) by iterating V and computing $\mathcal{SC}(u, t)$ for each vertex u not yet contained in a previously computed community. The source u becomes the representative of the newly computed SC (line 4). Since SCs with respect to a common vertex t are either disjoint or nested (Lemma 3(1),(2i)), we finally get a set Ω of SCs in G_α, which together decompose V. Since the vertices in G_α are additionally connected to t, each SC in G_α with respect to t is also a SC in G. However it is not necessarily a maximal SC in $M(G)$.

Applying CutC iteratively with decreasing α yields a hierarchy of at most n different clusterings (cp. Figure 4). This is due to a special nesting property for different param-eter values. Let C_1 denote the SC of u in G_{α_1} and C_2 the SC of u in G_{α_2}. Then it is

$C_1 \subseteq C_2$ if $\alpha_1 \geq \alpha_2$. The hierarchy is bounded by two trivial clusterings, which we already know in advance. The clustering at the top consists of the connected components of G and is returned by CutC for $\alpha_{\max} = 0$, the clustering at the bottom consists of singletons and comes up if we choose α_0 equal to the maximum edge cost in G.

Simple Parametric Search Approach. The crucial point with the construction of such a hierarchy, however, is the choice of α. If we choose the next value too close to a previous one, we get a clustering we already know, which implies unnecessary effort. If we choose the next value too far from any previous, we possibly miss a clustering. Flake et al. propose a binary search for the choice of α. However, this necessitates a discretization of the parameter range—an issue where again limiting the risk of missing interesting values by small steps is opposed to improving the running time by wide steps. In practice the choice of a good coarseness of the discretization requires previous knowledge on the graph structure, which we usually do not have. Thus, we introduce a simple parametric search approach for constructing a complete[2] hierarchy that does not require any previous knowledge.

For two consecutive hierarchy levels $\Omega_i < \Omega_{i+1}$ we call α' the *breakpoint* if CutC returns Ω_i for α' and Ω_{i+1} for $\alpha' - \varepsilon$ with $\varepsilon \to 0$. The simple idea of our approach is to compute good candidates for breakpoints during a recursive search with the help of cut-cost functions of the clusters, such that each candidate that is no breakpoint yields a new clustering instead. In this way, we apply CutC at most twice

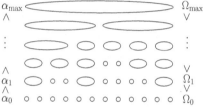

Fig. 4. Clustering hierarchy by CutC. Note, $\alpha_{\max} < \alpha_0$ whereas $\Omega_{\max} > \Omega_0$.

per level in the final hierarchy. Beginning with the trivial clusterings $\Omega_0 < \Omega_{\max}$ ($\alpha_0 > \alpha_{\max}$), the following theorem directly implies an efficient algorithm.

Theorem 1. *Let $\Omega_i < \Omega_j$ denote two different clusterings with parameter values $\alpha_i > \alpha_j$. In time $O(|\Omega_i|)$ a parameter value α_m with 1) $\alpha_j < \alpha_m \leq \alpha_i$ can be computed such that 2) $\Omega_i \leq \Omega_m < \Omega_j$, and 3) $\Omega_m = \Omega_i$ implies that α_m is the breakpoint between Ω_i and Ω_j.*

Sketch of proof. We use *cut-cost functions* that represent, depending on α, the cost $\omega_S(\alpha)$ of a cut $(S, V_\alpha \setminus S)$ in G_α based on the cost of the cut $(S, V \setminus S)$ in G and the size of S.

$$\omega_S : \mathbb{R}_0^+ \longrightarrow [c(S, V \setminus S), \infty) \subset \mathbb{R}_0^+$$
$$\omega_S(\alpha) := c(S, V \setminus S) + |S|\,\alpha$$

The main idea is the following. Let $\Omega_i < \Omega_j$ denote two hierarchically nested clusterings. We call a cluster $C' \in \Omega_i$ that is nested in $C \in \Omega_j$ a *child* of C and C the *parent* of C'. If there exists another level Ω' between Ω_i and Ω_j, at least two clusters in Ω_i must be merged yielding a larger cluster in Ω'. The maximal parameter value where this happens is a value α^* where a child C' in Ω_i becomes more expensive than its parent C

---

[2] The completeness refers to all clusterings that can be obtained by CutC for a value α.

in Ω_j, and thus, is dominated by C in the sense that it will not become a cluster in any hierarchy level above α^* (i.e., where $\alpha < \alpha^*$). For two nested clusters $C' \subseteq C$ this point is marked by the intersection point of the cut-cost functions $\omega_{C'}$ and ω_C (Figure 5). Thus, this intersection point is a good candidate for a breakpoint between Ω_i and Ω'. We choose $\alpha_m := \min_{C \in \Omega_j} \lambda_C$ with $\lambda_C := \max_{C' \in \Omega_i : C' \subset C} \{\alpha \mid \omega_C(\alpha) = \omega_{C'}(\alpha)\}$ and prove that Claim 1) to 3) as stated in Theorem 1 hold with this choice of α_m. The proofs are rather technical, which is why we only provide them in the full version [7].

For the running time, observe that α_m is well-defined as each parent function intersects with at least one child function. In practice we construct α_m by iterating the list of representatives stored for Ω_i. These representatives are assigned to a cluster in Ω_j, thus, matching children to their parents can be done in time $O(|\Omega_i|)$. The computation of the intersection points takes only constant time, given that the sizes and costs of the clusters are stored with the representatives by CutC. In total, the time for computing α_m is thus in $O(|\Omega_i|)$.   □

Running Time. The parametric search approach calls CutC twice per level in the final hierarchy, once when computing a level the first time and again right before detecting that the level already exists and a breakpoint is reached. The trivial levels $\Omega_{\max}$ and Ω_0 are calculated in advance without using CutC. Nevertheless, Ω_0 is recalculated once when the breakpoint to the lowest non-trivial level is found. This yields $2(h-2)+1$ applications of CutC, with h the number of levels. We denote

Fig. 5. Intersecting cut-cost functions

the running time of CutC by $T(n)$ without further analysis. For a more detailed discussion on the running time of CutC see [2]. Since common min-cut algorithms run in $O(n^2\sqrt{m})$ time, a single min-cut computation already dominates the costs for determining α_m and further linear overhead. The running time of our simple parametric approach thus is in $O(2h\,T(n))$, where $h \leq n-1$. This obviously improves the running time of a binary search, which is in $O(h\,\log(d)\,T(n))$, with d the number of discretization steps—in particular since we may assume $d \gg n$ in order to minimize the risk of missing levels. We also tested the practicability of our simple approach by a brief experiment. The results confirm the improved theoretical running time. We provide them in the full version [7] as bonus.

4   Framework for Analyzing SC Structures

In general, clusterings in which all clusters are SCs are only partially hierarchically ordered. Hence, hierarchical algorithms like the cut clustering algorithm of Flake et al. [2] provide only a limited view on the whole SC structure of a network. In this section we develop a framework for efficiently analyzing different hierarchies in the SC structure after precomputing at most $2(n-1)$ maximum flows. The basis of our framework is the set $M(G)$ of maximal SCs in G. This can be represented by a cut tree of special community cuts, together with some additionally stored SCs, as we will show in the following.

A (general) *cut tree* is a weighted tree $\mathcal{T}(G) = (V, E_{\mathcal{T}}, c_{\mathcal{T}})$ on the vertices of an undirected, weighted graph $G = (V, E, c)$ (with edges not necessarily in G) such that each $\{s, t\} \in E_{\mathcal{T}}$ induces a minimum s-t-cut in G (by decomposing $\mathcal{T}(G)$ into two connected components) and such that $c_{\mathcal{T}}(\{s, t\})$ is equal to the cost of the induced cut. The cut tree algorithm, which was first introduced by Gomory and Hu [4] in their pioneering work on cut trees and later simplified by Gusfield [5], applies $n - 1$ cut computations. For a detailed description of this algorithm see [4,5] or the full version [7]. The main idea of the cut tree algorithm is to iteratively choose vertices s and t that are not yet separated by a previous cut, and separating them by a minimum s-t-cut, which is represented by a new tree edge $\{s, t\}$. Depending on the shape of the found cut it might be necessary to reconnect previous edges in the intermediate tree. Gomory and Hu showed that a reconnected edge also represents a minimum s'-t'-cut for the new vertices s' and t' incident to the edge after the reconnection. Furthermore, the constructed cuts need to be non-crossing in order to be representable by a tree. While Gomory and Hu prevent crossings with the help of contractions, Gusfield shows that a crossing of an arbitrary minimum s-t-cut with another minimum separating cut can be easily resolved, if the latter does not separate s and t. Hence, the cut tree algorithm basically admits the use of arbitrary minimum cuts.

For our special cut tree we choose the following community cuts: for a vertex pair $\{s, t\}$ let $(S, V \setminus S)$ denote the community cut inducing $S := \mathcal{SC}(s, t)$ and let $(T, V \setminus T)$ denote the community cut inducing $T := \mathcal{SC}(t, s)$. If $|S| \leq |T|$, we choose $(S, V \setminus S)$, and $(T, V \setminus T)$ otherwise. Furthermore, we direct the corresponding tree edge to the chosen SC, and we associate the opposite SC, which was not chosen, also with the edge, storing it elsewhere for further use. In the full version [7] we show that the so chosen "smallest" community cuts are already non-crossing, hence a transformation according to Gusfield is not necessary. This guarantees that the cuts represented in the final tree are the same community cuts as chosen for the construction. We further show that after reconnecting an edge, the corresponding cut still induces a "smallest" SC for the vertex the edge points to. Altogether, this proves the following.

Theorem 2. *For an undirected, weighted graph $G = (V, E, c)$ there exists a rooted cut tree $\mathcal{T}(G) = (V, E_{\mathcal{T}}, c_{\mathcal{T}})$ with edges directed to the leaves such that each edge $(t, s) \in E_{\mathcal{T}}$ represents $\mathcal{SC}(s, t)$, and $|\mathcal{SC}(s, t)| \leq |\mathcal{SC}(t, s)|$. Such a tree can be constructed by $n - 1$ maximum flow[3] computations.*

At the price of $O(n^2)$ additional space, the opposite SCs resulting from the cut tree construction can be naively stored in an $(n - 1) \times n$ matrix, which admits to check the membership of a vertex to an opposite SC in constant time. In many cases we even need only $k \leq (n - 1)$ rows in the matrix, since some edges share the same SC, and we can deduce these edges during the cut tree construction. However, for few edges the determined opposite SC might become invalid again, due to a special situation while reconnecting the edge. For these edges we need to recalculate the opposite SCs in a second step. Hence, the construction of $\mathcal{T}(G)$ together with the opposite SCs associated with the edges in $\mathcal{T}(G)$ can be done by at most $2(n - 1)$ max-flow computations.

---

[3] Max-flows are necessary in order to determine a smallest SC. For general cut trees preflows (after the first phase of common max-flow-push-relabel algorithms) suffice.

We now show that each SC in $M(G)$ is either given by an edge or is an opposite SC associated with an edge in $\mathcal{T}(G)$.

Theorem 3. *For an undirected weighted graph $G = (V, E, c)$ it is $n \leq |M(G)| \leq 2(n-1)$. Constructing $M(G)$ needs at most $2(n-1)$ max-flow computations.*

Sketch of proof. For the full proof see [7]. Recall that the maximal SCs in $M(G)$ are the SCs with respect to single vertices. We consider a path $\pi(u, v)$ from u to v in $\mathcal{T}(G)$ as the set of edges or the set of vertices on it, as convenient, ignoring the direction. The cut tree structure of $\mathcal{T}(G)$ induces that for two (also non-adjacent) vertices u and v in $\mathcal{T}(G)$ a minimum u-v-cut is given by the cheapest edge on $\pi(u, v)$. Together with the direction of the edges and the fact that $\mathcal{T}(G)$ represents "smallest" community cuts, it follows that if q is a successor of p, $\mathcal{SC}(q, p)$ is given by the cheapest edge on $\pi(p, q)$ that is closest to q. In the full proof we further show that $\mathcal{SC}(p, q)$ is the opposite SC associated with the cheapest edge on $\pi(p, q)$ that is closest to p. This implies at least n different SCs in $M(G)$—one per successor in $\mathcal{T}(G)$ and an additional one for the root in $\mathcal{T}(G)$. If u and v are vertices in disjoint subtrees with r the nearest common predecessor, we prove that $\mathcal{SC}(u, v)$ equals $\mathcal{SC}(u, r)$ if no edge on $\pi(r, v)$ is cheaper than the cheapest edge on $\pi(r, u)$, and that it equals $\mathcal{SC}(r, v)$, otherwise. Hence, each SC in $M(G)$ either corresponds to an edge or an opposite SC associated with an edge in $\mathcal{T}(G)$, which yields the upper bound. $\qquad\square$

After precomputing $M(G)$, which includes the construction of $\mathcal{T}(G)$ (we denote this by $M(G) \supset \mathcal{T}(G)$), the following tools allow to efficiently analyze the SC structure of G with respect to different SCs that are already known, for example, from the cut clustering algorithm of Flake et al. or the set $M(G)$. The key is Lemma 4. It limits the shape of arbitrary SCs to subtrees in $\mathcal{T}(G)$, which admits an efficient enumeration of disjoint SCs by a depth-first search (DFS), as we will see in the following.

Lemma 4. *The subgraph $\mathcal{T}[T]$ induced by a SC T in $\mathcal{T}(G)$ is connected.*

Maximal SC Clustering for one SC. Given an arbitrary SC S, the first tool returns a clustering $\Omega(S)$ of G that contains S, consists of SCs and is maximum in the sense that each clustering that also consists of S and further SCs is hierarchically nested in $\Omega(S)$. This implies that $\Omega(S)$ is the unique maximal clustering among all clusterings consisting of S and further SCs. We call $\Omega(S)$ the *maximal SC clustering* for S.

Theorem 4. *Let S denote a SC in G. The unique maximal SC clustering for S can be determined in $O(n)$ time after preprocessing $M(G) \supset \mathcal{T}(G)$.*

Proof. The maximal SC clustering for $S =: S_0$ can be determined by the following construction, which directly implies a simple algorithm. Let r denote the root of $\mathcal{T}(G) =: \mathcal{T}_0$ and $\mathcal{T}[S_0]$ the subtree induced by S_0 in $\mathcal{T}_0$ (Lemma 4). Deleting $\mathcal{T}[S_0]$ decomposes $\mathcal{T}_0$ into connected components, each of which representing a SC, apart from the one containing r if $r \notin S_0$. If $r \in S_0$, we are done. Otherwise, let $\mathcal{T}_1$ denote the component containing r and r_0 the root of $\mathcal{T}[S_0]$. Obviously is $p_0 \in \mathcal{T}_1$ for $(p_0, r_0) \in E_{\mathcal{T}}$ and $\mathcal{SC}(p_0, r_0) =: S_1$ induces a subtree $\mathcal{T}[S_1]$ in $\mathcal{T}_1$. Thus, S_1 and $\mathcal{T}_1$ adopt the roles of S_0 and $\mathcal{T}_0$.

Continuing in this way, we finally end up with a SC S_k containing r, such that deleting $\mathcal{T}[S_k]$ yields only SCs. The resulting clustering $\Omega(S)$ consists of $S = S_0, S_i$, $i = 1, \ldots, k$, and the remaining SCs resulting from the decompositions of $\mathcal{T}_0, \ldots, \mathcal{T}_k$. The proof of the maximality of $\Omega(S)$ is based on the following lemma.

Lemma 5. *Each SC in $\Omega(S) \setminus \{S\}$ is a SC with respect to the source of S.*

Let Q denote an arbitrary SC with source q that does not intersect S, let s denote the source of S, and let C denote the SC in $\Omega(S) \setminus \{S\}$ with $q \in C$. Since C is a SC with respect to $s \notin Q$ (Lemma 5) and $q \in Q \cap C$, it is $Q \subseteq C$, according to Lemma 3(2i). Thus, each SC not intersecting S is nested in a cluster in $\Omega(S)$.

For the running time we assume that S is given in a structure that allows to check the membership of a vertex in time $O(1)$. Then identifying all clusters in $\Omega(S)$ (which are subtrees) by applying a DFS[4] starting from the first vertex found in each cluster can be done in $O(n)$ time, since checking if a visited vertex is still in S_i takes constant time for $i = 1, \ldots, k$ (recall, that we store the opposite SCs in a matrix). The remaining subtrees share their leaves with $\mathcal{T}(G)$.   □

Overlay Clustering for k Disjoint SCs. Given k disjoint arbitrary SCs $S_1, \ldots, S_k$, the second tool returns a clustering $\Omega(S_1, \ldots, S_k)$ of G that contains $S_1, \ldots S_k$, is nested in each maximal SC clustering $\Omega(S_1), \ldots, \Omega(S_k)$ and is maximum in the sense that each clustering that consists of SCs and also contains $S_1, \ldots, S_k$ is hierarchically nested in $\Omega(S_1, \ldots, S_k)$. Basically, according to the construction described below, $\Omega(S_1, \ldots, S_k)$ is the unique maximal clustering among all clusterings that are nested in the maximal SC clusterings $\Omega(S_1), \ldots, \Omega(S_k)$. The further properties result from the maximality of the SC clusterings, as for each $\Omega(S_j)$ and each arbitrary SC S that does not intersect $S_1, \ldots, S_k$ (or equals a given SC) there exists a cluster $C \in \Omega(S_j)$ with $S \subseteq C$. Note that the clusters in $\Omega(S_1, \ldots S_k) \setminus \{S_1, \ldots, S_k\}$ are not necessarily SCs. We call $\Omega(S_1, \ldots, S_k)$ the *overlay clustering* for $S_1, \ldots S_k$.

Theorem 5. *Let $S_1, \ldots, S_k$ denote disjoint SCs in G. The unique overlay clustering for $S_1, \ldots, S_k$ can be determined in $O(kn)$ time after preprocessing $M \supset \mathcal{T}(G)$.*

Proof. The overlay clustering for $S_1, \ldots, S_k$ can be determined by the following inductive construction, which directly implies a simple algorithm. We first compute the maximal SC clustering $\Omega(S_1)$ and color the vertices in each cluster, using different colors for different clusters. Now consider the overlay clustering $\Omega(S_1, \ldots, S_i)$ for the first i maximal SC clusterings and color the vertices in S_{i+1}, which is nested in a cluster of $\Omega(S_1, \ldots, S_i)$, with a new color. During the computation of $\Omega(S_{i+1})$, we then construct the intersections of each newly found cluster C with the clusters in $\Omega(S_1, \ldots, S_i)$. To this end we exploit that the intersection of two subtrees in a tree is again a subtree. Hence, the clusters in $\Omega(S_1, \ldots, S_i, S_{i+1})$ will be subtrees in $\mathcal{T}(G)$, since the clusters in $\Omega(S_1), \ldots, \Omega(S_i)$ and $\Omega(S_{i+1})$ are subtrees in $\mathcal{T}(G)$ by Lemma 4.

Let r' denote the first vertex found in C during the computation of $\Omega(S_{i+1})$. We mark r' as root of a new cluster in $\Omega(S_1, \ldots, S_i, S_{i+1})$ and choose a new color x for r',

---

[4] This induces a rooted subtree independent from the orientation in $\mathcal{T}(G)$.

besides the color it already has in $\Omega(S_1, \ldots, S_i)$. When constructing C (by applying a DFS), we assign the current color x to all vertices visited by the DFS as long as the underlying color in $\Omega(S_1, \ldots, S_i)$ does not change. Whenever the DFS visits a vertex r'' (still in C) with a new underlying color, we chose a new color y for r'' and mark r'' as root of a subtree of a new cluster in $\Omega(S_1, \ldots, S_i, S_{i+1})$. When the DFS passes r'' on the way back to the parent[5] p of r'', the color of p in $\Omega(S_1, \ldots, S_i, S_{i+1})$ becomes the current color again. Continuing in this way yields a coloring that indicates the intersections of C with $\Omega(S_1, \ldots, S_i)$. Repeating this procedure for all clusters in $\Omega(S_{i+1})$ finally yields $\Omega(S_1, \ldots, S_{i+1})$. The running time is in $O(kn)$, since we just apply k computations of maximal SC clusterings. □

Example. We extract two of the many faces of the SC structure of the weighted co-appearance network (called "lesmis") of the characters in the novel Les Miserables [9]. Figure 6(a) shows the cut tree $\mathcal{T}$("lesmis"), the root r is depicted as filled square. Figure 6(b) shows the maximal SC clustering $\Omega(R_1)$ for the SC R_1 (filled vertices in squared box). The subtree $\mathcal{T}[R_1]$ induced by R_1 in $\mathcal{T}$("lesmis") is indicated by filled vertices in Figure 6(a). Since $r \in R_1$, deleting $\mathcal{T}[R_1]$ immediately decomposes $\mathcal{T}$("lesmis") into the unframed singleton SCs and the round framed SCs shown in Figure 6(b). The SC R_1 is the larger of the only two non-singleton clusters in the best cut clustering (with respect to *modularity* [11]) found by the cut clustering algorithm of Flake et al. On the other hand, R_1 is the smallest reasonable SC that was found by the cut clustering algorithm containing r. The next smaller SC in the hierarchy that contains r consists of only three vertices. The second non-singleton cluster besides R_1 in the best cut clustering is also in $\Omega(R_1)$, namely A. Nevertheless, $\Omega(R_1)$ is not nested in any clustering of the hierarchy. This is, we found a new clustering that contains all non-singleton clusters of the best cut clustering but far less unclustered vertices. Due to the maximality of $\Omega(R_1)$, there is also no SC clustering with less singletons containing R_1.

Figure 6(c) shows the overlay clustering $\Omega(S_1, \ldots, S_6, R_2)$ with $S_1, \ldots, S_6$ defined by the non-singleton subtrees of r in $\mathcal{T}$("lesmis"). The SC R_2 (filled vertices in squared box) has been computed additionally. It equals $\mathcal{SC}(r, T)$ with $T := \bigcup_{i=1}^6 S_i$. If we consider the filled vertices in Figure 6(c) as one cluster $F := V \setminus T$, then $S_1, \ldots, S_6$ together with F represent the overlay clustering $\Omega(S_1, \ldots, S_6)$. However, $\Omega(S_1, \ldots, S_6)$ does not only consist of SCs since F is no SC: Observe that for the two vertices $v_1, v_2 \in F \setminus R_2$ there exists a vertex $u \in T$ (unfilled square) such that $\mathcal{SC}(v_i, u) \subseteq F$ ($i = 1, 2$) is a singleton. Hence, according to Lemma 3(2i), any SC in F, apart from $\{v_1\}$ and $\{v_2\}$, must be in R_2. This is, in contrast to $\Omega(S_1, \ldots, S_6)$, the overlay clustering $\Omega(S_1, \ldots, S_6, R_2)$ consists of SCs and any clustering that also consists of SCs and contains $S_1, \ldots, S_6$ is nested in $\Omega(S_1, \ldots, S_6, R_2)$.

5   Conclusion

Based on minimum separating cuts and maximum flows, respectively, we characterized SCs, a special type of predominantly connected communities. We introduced a method

---

[5] The predecessor adjacent to r'' in the rooted subtree induced by the DFS.

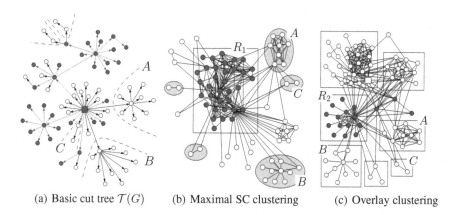

(a) Basic cut tree $\mathcal{T}(G)$ (b) Maximal SC clustering   (c) Overlay clustering

Fig. 6. Exemplary clusterings of the lesmis-network; A, B, C appear in both clusterings

for efficiently computing a complete hierarchy of clusterings consisting of SCs according to Flake et al. [2]. Furthermore, we exploited the structure of cut trees [4] in order to develop a framework that admits the efficient construction of maximal SC clusterings and overlay clusterings for given SCs, after precomputing at most $2(n-1)$ maximum flows. In most cases, however, we expect only around $n-1$ maximum flows for the preprocessing, since the cases that cause the additional flow computations (when the opposite SC becomes invalid during the construction of the cut tree) are rare in practice. For the "lesmis" network in the previous example we needed only $n+3$ maximum flows with $n = 77$. We remark that a single maximal SC clustering for S can be also constructed directly by iteratively computing maximal SCs of the vertices not in S with respect to the source of S. However, in the worst case, this needs $|V \setminus S|$ flow computations, if the SCs are singletons or if they are considered in an order that causes many unnecessary computations of nested SCs. In contrast, due to its short query times, our framework efficiently supports the detailed analysis of a networks's SC structure with respect to many different maximal SC clusterings and overlay clusterings.

References

1. Flake, G.W., Lawrence, S., Giles, C.L̃., Coetzee, F.M.: Self-Organization and Identification of Web Communities. IEEE Computer 35(3), 66–71 (2002)
2. Flake, G.W., Tarjan, R.E., Tsioutsiouliklis, K.: Graph Clustering and Minimum Cut Trees. Internet Mathematics 1(4), 385–408 (2004)
3. Gallo, G., Grigoriadis, M.D., Tarjan, R.E.: A fast parametric maximum flow algorithm and applications. SIAM Journal on Computing 18(1), 30–55 (1989)
4. Gomory, R.E., Hu, T.C.: Multi-terminal network flows. Journal of the Society for Industrial and Applied Mathematics 9(4), 551–570 (1961)
5. Gusfield, D.: Very simple methods for all pairs network flow analysis. SIAM Journal on Computing 19(1), 143–155 (1990)
6. Hamann, M., Hartmann, T., Wagner, D.: Complete Hierarchical Cut-Clustering: A Case Study on Expansion and Modularity. In: Bader, D.A., Meyerhenke, H., Sanders, P., Wagner, D. (eds.) Graph Partitioning and Graph Clustering: Tenth DIMACS Implementation Challenge. DIMACS Book, vol. 588, American Mathematical Society (to appear, 2013)

7. Hamann, M., Hartmann, T., Wagner, D.: Hierarchies of predominantly connected communities. arXiv e-print (2013), http://arxiv.org/abs/1305.0757
8. Kannan, R., Vempala, S., Vetta, A.: On Clusterings - Good, Bad and Spectral. In: Proceedings of the 41st Annual IEEE Symposium on Foundations of Computer Science (FOCS 2000), pp. 367–378 (2000)
9. Knuth, D.E.: The Stanford GraphBase: a platform for combinatorial computing. Addison-Wesley (1993)
10. Nagamochi, H.: Graph Algorithms for Network Connectivity Problems. Journal of the Operations Research Society of Japan 47(4), 199–223 (2004)
11. Newman, M.E.J., Girvan, M.: Finding and evaluating community structure in networks. Physical Review E 69(026113), 1–16 (2004)
12. Wu, Z., Leahy, R.: An Optimal Graph Theoretic Approach to Data Clustering: Theory and its Application to Image Segmentation. IEEE Transactions on Pattern Analysis and Machine Intelligence 15(11), 1101–1113 (1993)
13. Zachary, W.W.: An Information Flow Model for Conflict and Fission in Small Groups. Journal of Anthropological Research 33, 452–473 (1977)

Joint Cache Partition and Job Assignment on Multi-core Processors

Avinatan Hassidim[1,*], Haim Kaplan[2,**], and Omry Tuval[2,**]

[1] Dept. of Computer Science, Bar Ilan University
avinatan@cs.biu.ac.il
[2] Dept. of Computer Science, Tel-Aviv University
{haimk,omrytuva}@tau.ac.il

Abstract. Multicore shared cache processors pose a challenge for designers of embedded systems who try to achieve minimal and predictable execution time of workloads consisting of several jobs. To address this challenge the cache is statically partitioned among the cores and the jobs are assigned to the cores so as to minimize the makespan. Several heuristic algorithms have been proposed that jointly decide how to partition the cache among the cores and assign the jobs. We initiate a theoretical study of this problem which we call the joint cache partition and job assignment problem.

By a careful analysis of the possible cache partitions we obtain a constant approximation algorithm for this problem. For some practical special cases we obtain a 2-approximation algorithm, and show how to improve the approximation factor even further by allowing the algorithm to use additional cache. We also study possible improvements that can be obtained by allowing dynamic cache partitions and dynamic job assignments.

We define a natural restriction of the well known scheduling problem on unrelated machines in which machines are ordered by "strength". We call this restriction the *ordered unrelated machines scheduling problem.* We show that our joint cache partition and job assignment problem is harder than this scheduling problem. The ordered unrelated machines scheduling problem is of independent interest and we give a polynomial time algorithm for certain natural workloads.

Keywords: cache partition, job assignment.

1 Introduction

We study the problem of assigning n jobs to c cores on a multi-core processor, and simultanously partitioning a shared cache of size K among the cores. Each

* This work was partially supported under ISF grant 1241/12 and under GIF young.
** This work was partially supported by the ISF grant no. 822-10. Israeli Centers of Research Excellence (I-CORE) program (Center No. 4/11). Google Inter-university center for Electronic Markets and Auctions.

F. Dehne, R. Solis-Oba, and J.-R. Sack (Eds.): WADS 2013, LNCS 8037, pp. 378–389, 2013.
© Springer-Verlag Berlin Heidelberg 2013

job j is given by a non-increasing function $T_j(x)$ indicating the running time of job j on a core with cache of size x. A solution is a cache partition p, assigning $p(i)$ cache to each core i, and a job assignment S assigning each job j to core $S(j)$. The total cache allocated to the cores in the solution is K, that is $\sum_{i=1}^{c} p(i) = K$. The *makespan* of a cache partition p and a job assignment S is $\max_i \sum_{j|S(j)=i} T_j(p(i))$. Our goal is to find a cache partition and a job assignment that minimize the makespan.

Multi-core processors are the prevalent computational architecture used today in PCs, mobile devices and high performance computing. Having multiple cores running jobs concurrently, while sharing the same level 2 and/or level 3 cache, results in complex interactions between the jobs, thereby posing a significant challenge in determining the makespan of a set of jobs. Cache partitioning has emerged as a technique to increase run time predictability and increase performance on multi-core processors [1,2]. Theoretic research on online multi-core caching shows that the cache partition (which may be dynamic) has more influence on the performance than the eviction policy [3,4]. To obtain effective cache partitions, methods have been developed to estimate the running time of a job as a function of allocated cache, that is the function $T_j(x)$ (see for example the cache locking technique of [5]).

Recent empirical research [6] suggests that jointly solving for the cache partition and for the job assignment leads to significant improvements over combining separate algorithms for the two problems. Liu et al. [6] suggest and test heuristic algorithms for the joint cache partition and job assignment problem. Our work initiates the theoretic study of this problem.

We study this problem in the context of multi-core caching, but our formulation and results are applicable in a more general setting, where the running time of a job depends on the availability of some shared resource (cache, CPU, RAM, budget, etc.) that is allocated to the machines. This setting is applicable, for example, for users of a public cloud infrastructure like Amazon's Elastic Cloud. When a user decides on her public cloud setup, there is usually a limited resource (e.g. budget), that can be spent on different machines in the cloud. The more budget is spent on a machine, it runs jobs faster and the user is interested in minimizing the makespan of her set of jobs, while staying within the given budget.

Our Results

We show that the joint cache partition and job assignment problem is related to an interesting special case of scheduling on unrelated machines that we call the *ordered unrelated machines scheduling problem*. In this problem there is a total order on the machines which captures their relative strength. Each job has a different running time on each machine and these running times are non-increasing with the strength of the machine. In Section 2 we give an approximation preserving reduction from scheduling on ordered unrelated machines to the joint cache partition and job assignment problem. In the full version [7] we consider a special case of the ordered unrelated machines problem in which each job j runs in time l_j on machines $1, \ldots x_{j-1}$ and in time $h_j > l_j$ on machines $x_j, \ldots, c$ and there is a constant number of different l_j's and h_j's. We present a polynomial time dynamic program that finds an assignment of minimal makespan.

We present, in Section 3, for any constant $0 < \epsilon < \frac{1}{2}$, an 18-approximation algorithm for the joint cache partition and job assignment problem that uses $(1 + \frac{5}{2}\epsilon)K$ cache by showing that it suffices to consider a subset of the cache partitions of size polynomial in c. We obtain a 36-approximation algorithm that uses at most K cache, by considering a subset of the cache partitions that is of size K times a polynomial in c.

We obtain better approximation guarantees for special cases of the joint cache partition and job assignment problem. When each job has a fixed running time a_j and a minimal cache demand x_j, we present, in Section 4, a 2-approximation algorithm, a $\frac{3}{2}$-approximation algorithm that uses $2K$ cache and a $\frac{4}{3}$-approximation algorithm that uses $3K$ cache. We call this special case the *single load minimal cache demand* problem. Our $\frac{4}{3}$-approximation algorithm is based on an algorithm that finds a dominant perfect matching in a threshold graph that has a perfect matching, presented in Section 4.4. This algorithm and the existence of such a matching in such a threshold graph are of independent interest. We present a polynomial time approximation scheme for the single load minimal cache demand problem, in the case where the jobs' loads and cache demands are correlative, that is $a_j \le a_{j'}$ iff $x_j \le x_{j'}$ (Section 4.5).

In Section 5 we generalize the joint cache partition and job assignment problem and consider dynamic cache partitions and dynamic job schedules. We show upper and lower bounds on the makespan improvement that can be gained by using dynamic partitions and dynamic assignments.

All omitted proofs in this paper can be found in the full version [7].

2   The Ordered Unrelated Machines Problem

The ordered unrelated machines scheduling problem is defined as follows. There are c machines and a set J of jobs. The input is a matrix $T(i, j)$ giving the running time of job j on machine i, such that for any two machines $i_1 < i_2$ and any job j, $T(i_1, j) \ge T(i_2, j)$. The goal is to assign the jobs to the machines such that the makespan is minimized.

The ordered unrelated machines scheduling problem is a special case of scheduling on unrelated machines in which there is a total order on the machines that captures their relative strengths. This special case is natural since in many practical scenarios the machines have some underlying notion of strength and jobs run faster on a stronger machine. For example a newer computer typically dominates an older one in all parameters, or a more experienced employee does any job faster than a new recruit.

Lenstra et al [8] gave a 2 approximation algorithm for scheduling on unrelated machines based on rounding an optimal fractional solution of a linear program, and proved that it is NP-hard to approximate the problem to within a factor better than $\frac{3}{2}$. It is currently an open question if there are better approximation algorithms for ordered unrelated machines than the more general algorithms that approximate unrelated machines.

Another well-studied scheduling problem is scheduling on uniformly related machines. In this problem, the time it takes for machine i to run job j is $\frac{l_j}{s_i}$ where l_j is the load of job j and s_i is the speed of machine i. A polynomial time approximation scheme for related machines is described in [9]. It is easy to see that the problem of scheduling on related machines is a special case of the problem of scheduling on ordered unrelated machines, and therefore the ordered unrelated machines problem is NP-hard.

The ordered unrelated machines problem is closely related to the joint cache partition and job assignment problem. Consider an instance of the joint cache partition and job assignment problem with c cores, K cache and a set of jobs J such that $T_j(x)$ is the load function of job j. If we fix the cache partition to be some arbitrary partition p, and we index the cores in non-decreasing order of their cache allocation, then we get an instance of the ordered unrelated machines problem, where $T(i, j) = T_j(p(i))$. Our constant approximation algorithm for the joint cache partition and job assignment problem, described in Section 3, uses this observation as well as Lenstra's 2-approximation algorithm for unrelated machines. In the rest of this section we prove that the joint cache partition and job assignment problem is at least as hard as the ordered unrelated machines scheduling problem.

We reduce the ordered unrelated machines problem to the joint cache partition and job assignment problem. Consider the decision version of the ordered unrelated machines scheduling problem, with c machines and $n = |J|$ jobs, where job j takes $T(i, j)$ time to run on machine i. We want to decide if it is possible to schedule the jobs on the machines with makespan at most M.

Define the following instance of the joint cache partition and job assignment problem. This instance has c cores, a total cache $K = c(c+1)/2$ and $n' = n + c$ jobs. The first n jobs ($1 \le j \le n$) correspond to the jobs in the original ordered unrelated machines problem, and c jobs are new jobs ($n + 1 \le j \le n + c$). The load function $T_j(x)$ of job j, where $1 \le j \le n$, equals $T(x, j)$ if $x \le c$ and equals $T(c, j)$ if $x > c$. The load function $T_j(x)$ of job j, where $n + 1 \le j \le n + c$, equals $M + \delta$ if $x \ge j - n$ for some $\delta > 0$ and equals ∞ if $x < j - n$. Our load functions $T_j(x)$ are non-increasing because the original $T(i, j)$'s are non-increasing in the machine index i.

Lemma 1. *The makespan of the joint cache partition and job assignment instance defined above is at most $2M + \delta$ if and only if the makespan of the original ordered unrelated machines scheduling problem is at most M.*

Proof. Assume there is an assignment S' of the jobs in the original ordered unrelated machines instance of makespan at most M. We show a cache partition p and job assignment S for the joint cache partition and job assignment instance with makespan at most $2M + \delta$.

The cache partition p is defined such that $p(i) = i$ for each core i. The partition p uses exactly $K = c(c+1)/2$ cache. The job assignment S is defined such that for a job $j > n$, $S(j) = j - n$ and for a job $j \le n$, $S(j) = S'(j)$. The partition p

assigns i cache to core i, which is exactly enough for job $n + i$, which is assigned to core i by S, to run in time $M + \delta$. It is easy to verify that p, S is a solution to the joint cache partition and job assignment instance with makespan at most $2M + \delta$.

Assume there is a solution p, S for the joint cache partition and job assignment instance, with makespan at most $2M + \delta$. Job j, such that $n < j \leq n + c$, must run on a core with cache at least $j - n$, or else the makespan would be infinite. Moreover, no two jobs $j_1 > n$ and $j_2 > n$ are assigned by S the same core, as this would give a makespan of at least $2M + 2\delta$. Combining these observations with the fact that the total available cache is $K = c(c + 1)/2$, we get that the cache partition must be $p(i) = i$ for each core i. Furthermore, each job $j > n$ is assigned by S to core $j - n$ and all the other jobs assigned by S to core $j - n$ are jobs corresponding to original jobs in the ordered unrelated machines instance. Therefore, the total load of original jobs assigned by S to core i is at most M.

We define S', a job assignment for the original ordered unrelated machines instance, by setting $S'(j) = S(j)$ for each $j \leq n$. Since S assigns original jobs of total load at most M on each core, it follows that the makespaen of S' is at most M.   □

The following theorem follows immediately from Lemma 1.

Theorem 1. *There is a polynomial-time reduction from the ordered unrelated machines scheduling problem to the joint cache partition and job assignment problem.*

The reduction in the proof of Lemma 1 does not preserve approximation guarantees. However by choosing δ carefully we can get the following result.

Theorem 2. *Given an algorithm A for the joint cache partition and job assignment problem that approximates the optimal makespan up to a factor of $1 + \epsilon$, for $0 < \epsilon < 1$, we can construct an algorithm for the ordered unrelated machines scheduling problem that approximates the optimal makespan up to a factor of $1 + 2\epsilon + \frac{2\epsilon^2}{1 - \epsilon - \chi}$ for any $\chi > 0$.*

This approximation preserving reduction is given not as a practical means to obtain approximation algorithms for the ordered unrelated machines scheduling problem but rather as a testament to the hardness of the joint cache partition and job assignment problem.

3   Joint Cache Partition and Job Assignment

We first obtain, for any constant $0 < \epsilon < \frac{1}{2}$, an 18-approximation algorithm for the joint cache partition and job assignment problem that uses $(1 + \frac{5}{2}\epsilon)K$ cache by showing that it suffices to consider only a subset of the cache partitions of size polynomial in c. We then show another algorithm that uses K cache and approximates the makespan up to a factor of 36 and considers a subset of the cache partitions that is of size K times a polynomial in c.

Our first algorithm, denoted by A, enumerates over a subset of cache partitions, denoted by $P(K, c, \epsilon)$. For each partition in this set A approximates the makespan of the corresponding scheduling problem, using Lenstra's algorithm, and returns the partition and associated job assignment with the smallest makespan.

Let $K' = (1 + \epsilon)^{\lceil \log_{1+\epsilon}(K) \rceil}$, the smallest integral power of $(1 + \epsilon)$ which is at least K. The set $P(K, c, \epsilon)$ contains cache partitions in which the cache allocated to each core is an integral power of $(1 + \epsilon)$ and the number of different integral powers used by the partition is at most $\log_2(c)$. We denote by b the number of different cache sizes in a partition. Each core is allocated $\frac{K'}{(1+\epsilon)^{l_j}}$ cache, where $l_j \in \mathbb{N}$ and $1 \leq j \leq b$. The smallest possible cache allocated to any core is the smallest integral power of $(1 + \epsilon)$ which is at least $\frac{K\epsilon}{c}$ and the largest possible cache allocated to a core is K'. We denote by $\hat{\sigma}_j$ the number of cores with cache at least $\frac{K'}{(1+\epsilon)^{l_j}}$. It follows that there are $(\hat{\sigma}_j - \hat{\sigma}_{j-1})$ cores with $\frac{K'}{(1+\epsilon)^{l_j}}$ cache. We require that $\hat{\sigma}_j$ is an integral power of 2 and that the total cache used is at most $\left(1 + \frac{5}{2}\epsilon\right) K$. Formally,

$$P(K, c, \epsilon) = \{(l = < l_1, \ldots, l_b >, \hat{\sigma} = < \hat{\sigma}_0, \ldots, \hat{\sigma}_b >) \mid b \in \mathbb{N}, 1 \leq b \leq \log_2 c \quad (1)$$

$$\forall j, \; l_j \in \mathbb{N}, 0 \leq l_j \leq \log_{1+\epsilon}\left(\frac{c}{\epsilon}\right) + 1, \quad \forall j, \; l_{j+1} > l_j \quad (2)$$

$$\forall j \; \exists u_j \in \mathbb{N} \quad s.t. \quad \hat{\sigma}_j = 2^{u_j}, \hat{\sigma}_0 = 0, \hat{\sigma}_b \leq c, \quad \forall j \; \hat{\sigma}_{j+1} > \hat{\sigma}_j \quad (3)$$

$$\sum_{j=1}^{b} (\hat{\sigma}_j - \hat{\sigma}_{j-1}) \frac{K'}{(1 + \epsilon)^{l_j}} \leq \left(1 + \frac{5}{2}\epsilon\right) K\} \quad (4)$$

When the parameters are clear from the context, we use P to denote $P(K, c, \epsilon)$. Let $M(p, S)$ denote the makespan of cache partition p and job assignment S. The following theorem specifies the main property of P.

Theorem 3. *Let p, S be any cache partition and job assignment. A cache partition $\hat{p}$ and a job assignment $\hat{S}$ exist such that $\hat{p} \in P$ and $M(\hat{p}, \hat{S}) \leq 9M(p, S)$.*

An immediate corollary of Theorem 3 is that algorithm A described above finds a cache partition and job assignment with makespan at most 18 times the optimal makespan.

Lemma 2 shows that A is a polynomial time algorithm.

Lemma 2. *The size of P is polynomial in c.*

In the remainder of this section we give the outline of the proof of Theorem 3. Let (p, S) be a cache partition and a job assignment that use c cores, K cache and have a makespan $M(p, S)$. Define a cache partition p_1 such that for each core i, if $p(i) < \frac{K\epsilon}{c}$ then $p_1(i) = \frac{K\epsilon}{c}$ and if $p(i) \geq \frac{K\epsilon}{c}$ then $p_1(i) = p(i)$. For each core i, $p_1(i) \leq p(i) + \frac{K\epsilon}{c}$ and hence the total amount of cache allocated by p_1 is bounded by $(1 + \epsilon)K$. For each core i, $p_1(i) \geq p(i)$ and therefore $M(p_1, S) \leq M(p, S)$.

Let p_2 be a cache partition such that for each core i, $p_2(i) = (1+\epsilon)^{\lceil \log_{1+\epsilon}(p_1(i)) \rceil}$, the smallest integral power of $(1+\epsilon)$ that is at least $p_1(i)$. For each i, $p_2(i) \geq p_1(i)$

and thus $M(p_2, S) \leq M(p_1, S) \leq M(p, S)$. We increased the total cache allocated by at most a multiplicative factor of $(1 + \epsilon)$ and therefore the total cache used by p_2 is at most $(1 + \epsilon)^2 K \leq (1 + \frac{5}{2}\epsilon)K$ since $\epsilon < \frac{1}{2}$.

Let φ be any cache partition that allocates to each core an integral power of $(1 + \epsilon)$ cache. We define the notion of *cache levels*. We say that core i is *of cache level l in φ* if $\varphi(i) = \frac{K'}{(1+\epsilon)^l}$. Let $c_l(\varphi)$ denote the number of cores in cache level l in φ. The vector of c_l's, which we call the *cache levels vector* of φ, defines the partition φ completely since any two partitions that have the same cache level vector are identical up to a renaming of the cores.

Let $\sigma(\varphi)$ be the vector of prefix sums of the cache levels vector of φ. Formally,
$$\sigma_l(\varphi) = \sum_{i=0}^{l} c_i(\varphi).$$
Note that $\sigma_l(\varphi)$ is the number of cores in cache partition φ with at least $\frac{K'}{(1+\epsilon)^l}$ cache and that for each l, $\sigma_l(\varphi) \leq c$.

For each such cache partition φ, we define the *significant cache levels $l_i(\varphi)$* recursively as follows. The first significant cache level $l_1(\varphi)$ is the first cache level l such that $c_l(\varphi) > 0$. Assume we already defined the $i-1$ first significant cache levels and let $l' = l_{i-1}(\varphi)$ then $l_i(\varphi)$ is the smallest cache level $l > l'$ such that $\sigma_l(\varphi) \geq 2\sigma_{l'}(\varphi)$.

Lemma 3. *Let l_j and l_{j+1} be two consecutive significant cache levels of φ, then the total number of cores in cache levels in between l_j and l_{j+1} is at most $\sigma_{l_j}(\varphi)$. Let l_b be the last significant cache level of φ then the total number of cores in cache levels larger than l_b is at most $\sigma_{l_b}(\varphi)$.*

For each core i, $\frac{K\epsilon}{c} \leq p_2(i) \leq K'$, so we get that if l is a cache level in p_2 such that $c_l(p_2) \neq 0$ then $0 \leq l \leq \log_{1+\epsilon}(\frac{c}{\epsilon}) + 1$.

Let b be the number of significant cache levels in p_2. We adjust p_2 and S to create a new cache partition p_3 and a new job assignment S_3. Cache partition p_3 has cores only in the significant cache levels $l_1(p_2), \ldots, l_b(p_2)$. We obtain p_3 from p_2 as follows. Let f be a non-significant cache level in p_2. If there is a j such that $l_{j-1}(p_2) < f < l_j(p_2)$ then we take the $c_f(p_2)$ cores in cache level f in p_2 and reduce their cache so they are now in cache level $l_j(p_2)$ in p_3. If $f > l_b(p_2)$ then we remove the $c_f(p_2)$ cores at level f from our solution. It is easy to check that the significant cache levels of p_3 are the same as of p_2, that is $l_j(p_2) = l_j(p_3)$ for $1 \leq j \leq b$. To simplify notation, we denote $l_j = l_j(p_2) = l_j(p_3)$ for any $1 \leq j \leq b$. Since we only reduce the cache allocated to some cores, the new cache partition p_3 uses no more cache than p_2 which is at most $(1 + \frac{5}{2}\epsilon)K$.

We construct S_3 by changing the assignment of the jobs assigned by S to cores in non-significant cache levels in p_2. As before, let f be a nonsignificant cache level and let l_{j-1} be the maximal significant cache level such that $l_{j-1} < f$. For each core i in cache level f in p_2 we move all the jobs assigned by S to core i, to a target core in cache level l_{j-1} in p_3. Lemma 4 specifies the key property of this job-reassignment.

Lemma 4. *We can construct S_3 such that each core in a significant level of p_3 is the target of the jobs from at most two cores in a nonsignificant level of p_2.*

Corollary 1. $M(p_3, S_3) \leq 3M(p, S)$

We now define another cache partition $\hat{p}$ based on p_3. Let $u_j = \lfloor \log_2(\sigma_{l_j}(p_3)) \rfloor$. The partition $\hat{p}$ has 2^{u_1} cores in cache level l_1, and $2^{u_j} - 2^{u_j-1}$ cores in cache level l_j for $1 < j \leq b$. The significant cache levels of $\hat{p}$ and p_3 are the same, that is $l_j(\hat{p}) = l_j(p_3) = l_j$ for $1 \leq j \leq b$. Furthermore, $\hat{p}$ has cores only in its significant cache levels.

Lemma 5. $3c_{l_j}(\hat{p}) \geq c_{l_j}(p_3)$

Lemma 5 shows that cache partition $\hat{p}$ has in each cache level l_j at least a third of the cores that p_3 has at cache level l_j. Therefore, there exists a job assignment $\hat{S}$ that assigns to each core of cache level l_j the jobs that S_3 assigns to at most 3 cores in cache level l_j. We only moved jobs within the same cache level and thus their load remains the same and the makespan $M(\hat{p}, \hat{s}) \leq 3M(p_3, S_3) \leq 9M(p, s)$.

Lemma 6. *Cache partition $\hat{p}$ is in the set $P(K, c, \epsilon)$.*

This concludes the proof of Theorem 3, and establishes that our algorithm A is an 18-approximation algorithm for the problem, using $(1 + \frac{5}{2}\epsilon)K$ cache.

We provide a variation of algorithm A that uses at most K cache, finds a 36-approximation for the optimal makespan and considers a subset of all cache partitions of size that is K times a polynomial in c. Algorithm B enumerates on r, $1 \leq r \leq K$, the amount of cache allocated to the first core. It then enumerates over the set of partitions $P = P(\frac{K-r}{2}, c - 1, \frac{2}{5})$. For each partition in P it adds another core with r cache and applies Lenstra's approximation algorithm on the resulting instance of the unrelated machines scheduling problem, to assign all the jobs in J to the c cores. Algorithm B returns the partition and assignment with the minimal makespan it encounters.

Theorem 4. *If there is a solution of makespan M that uses at most K cache and at most c cores then algorithm B returns a solution of makespan $36M$ that uses at most K cache and at most c cores.*

4   Single Load and Minimal Cache Demand

We consider a special case of the general joint cache partition and job assignment problem where each job has a minimal cache demand x_j and single load value a_j. Job j must run on a core with at least x_j cache and it contributes a load of a_j to the core. We want to decide if the jobs can be assigned to c cores, using K cache, such that the makespan is at most M? W.l.o.g. we assume $M = 1$.

In Section 4.1 we describe a 2-*approximate decision* algorithm that if the given instance has a solution of makespan at most 1, returns a solution with makespan at most 2 and otherwise may fail. In Sections 4.2 and 4.3 we improve the approximation guarantee to $\frac{3}{2}$ and $\frac{4}{3}$ at the expense of using $2K$ and $3K$ cache, respectively. These approximate decision algorithms can be transformed into approximate optimization algorithms by using a standard binary search technique [8,7].

4.1   2-Approximation

We present a 2-approximate decision algorithm, denoted by A_2. Algorithm A_2 sorts the jobs in a non-increasing order of their cache demand. It then assigns the jobs to the cores in this order. It keeps assigning jobs to a core until the load on the core exceeds 1. Then, A_2 starts assigning jobs to the next core. Note that among the jobs assigned to a specific core the first one is the most cache demanding and it determines the cache allocated to this core by A_2. Algorithm A_2 fails if the generated solution uses more than c cores or more than K cache. Otherwise, A_2 returns the generated cache partition and job assignment.

Theorem 5. *If there is a cache partition and job assignment of makespan at most 1 that use c cores and K cache then algorithm A_2 finds a cache partition and job assignment of makespan at most 2 that use at most c cores and at most K cache.*

4.2 $\frac{3}{2}$-Approximation with $2K$ Cache

We define a job to be *large* if $a_j > \frac{1}{2}$ and *small* otherwise. Our algorithm $A_{\frac{3}{2}}$ assigns one large job to each core. Let s_i be the load on core i after the large jobs are assigned. Let $r_i = 1 - s_i$. We process the small jobs by non-increasing order of their cache demand x_j, and assign them to the cores in non-increasing order of the cores' r_i's. We stop assigning jobs to a core when its load exceeds 1 and start loading the next core. Algorithm $A_{\frac{3}{2}}$ allocates to each core the cache demand of its most demanding job. Algorithm $A_{\frac{3}{2}}$ fails if the resulting solution uses more than c cores or more than $2K$ cache.

Theorem 6. *If there is a cache partition and job assignment of makespan at most 1 that use c cores and K cache then $A_{\frac{3}{2}}$ finds a cache partition and job assignment that use at most $2K$ cache, at most c cores and have a makespan of at most $\frac{3}{2}$.*

4.3 $\frac{4}{3}$-Approximation with $3K$ Cache, Using Dominant Matching

We present a $\frac{4}{3}$ approximate decision algorithm, $A_{\frac{4}{3}}$, that uses at most $3K$ cache. The main challenge is assigning the *large jobs*, which here are defined as jobs of load greater than $\frac{1}{3}$.

There are at most $2c$ large jobs in our instance, because we assume there is a solution of makespan at most 1 that uses c cores. Algorithm $A_{\frac{4}{3}}$ matches these large jobs into pairs, and assigns each pair to a different core. In order to perform the matching, we construct a graph G where each vertex represents a large job j of weight $a_j > \frac{1}{3}$. If needed, we add artificial vertices of weight zero to have a total of exactly $2c$ vertices in the graph. Each two vertices have an edge between them if the sum of their weights is at most 1. The weight of an edge is the sum of the weights of its endpoints. A perfect matching in a graph is a subset of edges such that every vertex in the graph is incident to exactly one edge in the subset.

We note that there is a natural bijection between perfect matchings in the graph G and assignments of makespan at most 1 of the large jobs to the cores.

A *dominant perfect matching* in G is a perfect matching Q such that for every i, the i heaviest edges in Q are a maximum weight matching in G of i edges. The graph G is a threshold graph [10], and in Section 4.4 we provide a polynomial time algorithm that finds a dominant perfect matching in any threshold graph that has a perfect matching. If there is a solution for the given instance of makespan at most 1 then the assignment of the large jobs in that solution correspond to a perfect matching in G and thus algorithm $A_{\frac{4}{3}}$ can find a dominant perfect matching, Q, in G.

Algorithm $A_{\frac{4}{3}}$ then assigns the small jobs (load $\leq \frac{1}{3}$) similarly to algorithms A_2 and $A_{\frac{3}{2}}$ described in Sections 4.1 and 4.2, respectively. It greedily assigns jobs to a core, until the core's load exceeds 1. Jobs are assigned in a non-increasing order of their cache demand and the algorithm goes through the cores in a non-decreasing order of the sum of loads of the large jobs on each core. Once all the jobs are assigned, the algorithm allocates cache to the cores according to the cache demand of the most demanding job on each core. Algorithm $A_{\frac{4}{3}}$ fails if it does not find a dominant perfect matching in G or if the resulting solution uses more than c cores or more than $3K$ cache.

Theorem 7. *If there is a solution that assigns the jobs to c cores with makespan at most 1 and uses K cache then algorithm $A_{\frac{4}{3}}$ assigns the jobs to c cores with makespan at most $\frac{4}{3}$ and uses at most $3K$ cache.*

4.4 Dominant Perfect Matching in Threshold Graphs

Let $G = (V, E)$ be an undirected graph with $2c$ vertices where each vertex $x \in V$ has a weight $w(x) \geq 0$. The edges in the graph are defined by a threshold $t > 0$ to be $E = \{(x, y) \mid w(x) + w(y) \leq t, x \neq y\}$. Such a graph G is known as a threshold graph [11,10]. We say that the *weight* of an edge (x, y) is $w(x, y) = w(x) + w(y)$.

A perfect matching A in G is a subset of the edges such that every vertex in V is incident to exactly one edge in A. Let A_i denote the i-th heaviest edge in A. We assume, w.l.o.g, that there is some arbitrary predefined order of the edges in E that is used, as a secondary sort criteria, to break ties in case several edges have the same weight. In particular, this implies that A_i is uniquely defined. A perfect matching A *dominates* a perfect matching B if for every $x \in \{1, \ldots, c\}$ $\sum_{i=1}^{x} w(A_i) \geq \sum_{i=1}^{x} w(B_i)$. A perfect matching A is a *dominant perfect matching* if A dominates any other perfect matching B.

Let A and B be two perfect matchings in G. We say that A and B *share a prefix of length l* if $A_i = B_i$ for $i \in \{1, \ldots, l\}$. The following greedy algorithm finds a dominant perfect matching in a threshold graph G that has a perfect matching. We start with $G_0 = G$. At step i, the algorithm selects the edge (x, y) with maximum weight in the graph G_i. If there are several edges of maximum weight, then (x, y) is the first by the predefined order on E. The graph G_{i+1} is obtained from G_i by removing vertices x, y and all edges incident to x or y.

Theorem 8. *If G is a threshold graph with $2c$ vertices that has a perfect matching, then the greedy algorithm described above finds a dominant perfect matching.*

4.5 PTAS for Jobs with Correlative Single Load and Minimal Cache Demand

We define an instance of the single load minimal cache demand problem to be *correlative* if for any two jobs j, j' such that x_j and $x_{j'}$ are non-zero, $a_j \leq a_{j'} \iff x_j \leq x_{j'}$.

Theorem 9. *There is a polynomial time approximation scheme for the joint cache partition and job assignment problem for jobs with correlative single load and minimal cache demands.*

5   Joint Dynamic Cache Partition and Job Scheduling

We consider a generalization of the joint cache partition and job assignment problem in which the cache partition and the job assignment are dynamic. We define the generalized problem as follows. As before, J denotes the set of jobs, there are c cores and a total cache of size K. Each job $j \in J$ is described by a non-increasing function $T_j(x)$.

A dynamic cache partition $p = p(t, i)$ indicates the amount of cache allocated to core i at time unit t.[1] For each time unit t, $\sum_{i=1}^{c} p(t, i) \leq K$. A dynamic assignment $S = S(t, i)$ indicates for each core i and time unit t, the index of the job that runs on core i at time t. If no job runs on core i at time t then $S(t, i) = -1$. If $S(t, i) = j \neq -1$ then for any other core $i_2 \neq i$, $S(t, i_2) \neq j$. Each job has to perform 1 *work unit*. If job j runs for α time units on a core with x cache, then it completes $\frac{\alpha}{T_j(x)}$ work. A partition and schedule p, S are *valid* if all jobs complete their work. Formally, p, S are valid if for each job j, $\sum_{<t,i> \in S^{-1}(j)} \frac{1}{T_j(p(t,i))} = 1$. The *load* of core i is defined as the maximum t such that $S(t, i) \neq -1$. The makespan of (p, S) is defined as the maximum load on any core. The goal is to find a valid dynamic cache partition and dynamic job assignment with a minimal makespan.

It is easy to verify that dynamic cache partition and dynamic job assignment, as defined above, generalize the static partition and static job assignment. The partition is static if for every fixed core i, $p(t, i)$ is constant with respect to t. The schedule is a static assignment if for every job j, there are times $t_1 < t_2$ and a core i such that $S^{-1}(j) = \{< t, i > \mid t_1 \leq t \leq t_2\}$.

We consider four variants of the joint cache partition and job assignment problem. The static partition and static assignment variant studied so far, the variant in which the cache partition is dynamic and the job assignment is static, the variant in which the job assignment is dynamic and the cache partition is static and the variant in which both are dynamic.

---

[1] To simplify the presentation we assume that time is discrete.

Theorem 10. *Allowing a dynamic partition and a dynamic assignment can improve the makespan by a factor of at most c, the number of cores, and there is an instance where by using a dynamic partition and a static assignment we achieve an improvement factor arbitrarily close to c. Allowing a dynamic assignment of the jobs, while keeping the cache partition static, improves the makespan by at most a factor of 2, and there is an instance where an improvement of $2 - \frac{2}{c}$ is achieved, for $c \geq 2$.*

References

1. Lin, J., Lu, Q., Ding, X., Zhang, Z., Zhang, X., Sadayappan, P.: Gaining insights into multicore cache partitioning: Bridging the gap between simulation and real systems. In: HPCA, pp. 367–378 (2008)
2. Molnos, A.M., Cotofana, S.D., Heijligers, M.J.M., van Eijndhoven, J.T.J.: Throughput optimization via cache partitioning for embedded multiprocessors. In: ICSAMOS, pp. 185–192 (2006)
3. Hassidim, A.: Cache replacement policies for multicore processors. In: ICS, pp. 501–509 (2010)
4. López-Ortiz, A., Salinger, A.: Paging for multi-core shared caches. In: ITCS, pp. 113–127. ACM (2012)
5. Liu, T., Li, M., Xue, C.J.: Instruction cache locking for multi-task real-time embedded systems. Real-Time Systems 48, 166–197 (2012)
6. Liu, T., Zhao, Y., Li, M., Xue, C.J.: Joint task assignment and cache partitioning with cache locking for WCET minimization on MPSoC. J. Parallel Distrib. Comput. 71, 1473–1483 (2011)
7. Hassidim, A., Kaplan, H., Tuval, O.: Joint cache partition and job assignment on multi-core processors. CoRR abs/1210.4053 (2012)
8. Lenstra, J.K., Shmoys, D.B., Tardos, É.: Approximation algorithms for scheduling unrelated parallel machines. Math. Program. 46, 259–271 (1990)
9. Hochbaum, D.S., Shmoys, D.B.: A polynomial approximation scheme for scheduling on uniform processors: Using the dual approximation approach. SIAM J. Comput. 17, 539–551 (1988)
10. Mahadev, N.V.R., Peled, U.N.: Threshold graphs and related topics. Annals of Discrete Mathematics, vol. 56. Elsevier (1995)
11. Chvátal, V., Hammer, P.L.: Set-packing problems and threshold graphs. Technical Report CORR 73-21, Dep. of Combinatorics and Optimization, Waterloo, Ontario (1973)

Finding the Minimum-Weight k-Path

Avinatan Hassidim[*], Orgad Keller, Moshe Lewenstein[**], and Liam Roditty

Department of Computer Science, Bar-Ilan University, Ramat-Gan, Israel
{avinatan,kellero,moshe,liamr}@cs.biu.ac.il

Abstract. Given a weighted n-vertex graph G with integer edge-weights taken from a range $[-M, M]$, we show that the minimum-weight simple path visiting k vertices can be found in time $\tilde{O}(2^k \text{poly}(k) M n^\omega) = O^*(2^k M)$. If the weights are reals in $[1, M]$, we provide a $(1 + \varepsilon)$-approximation which has a running time of $\tilde{O}(2^k \text{poly}(k) n^\omega (\log \log M + 1/\varepsilon))$. For the more general problem of k-tree, in which we wish to find a minimum-weight copy of a k-node tree T in a given weighted graph G, under the same restrictions on edge weights respectively, we give an exact solution of running time $\tilde{O}(2^k \text{poly}(k) M n^3)$ and a $(1+\varepsilon)$-approximate solution of running time $\tilde{O}(2^k \text{poly}(k) n^3 (\log \log M + 1/\varepsilon))$. All of the above algorithms are randomized with a polynomially-small error probability.

1 Introduction

Given an n-vertex graph $G = (V, E)$ and a parameter k, in the k-*path* problem we wish to find a path in G consisting of k vertices, if such exists. The k-path problem can be easily shown to be NP-complete: when $k = n$, it is exactly the Hamiltonian path problem. While a trivial $O^*(n^k)$ solution[1] is to try all $\binom{n}{k}$ combinations of k vertices, better can be obtained; Monien was the first to show an improvement [11], with an $O^*(k!)$ algorithm. In their seminal result, Alon, Yuster, and Zwick [2] introduced the *color-coding* technique. They used it to present a randomized $O^*((2e)^k)$ algorithm for this problem, which can be derandomized, replacing the $2e$ term with a large constant. Their result thus shows that the LOGPATH problem of determining whether a graph contains a path of length $\log n$ can be solved in polynomial time. Later, two independent results [8,5] presented randomized $O^*(4^k)$ algorithms, again with larger constants when derandomized, having running times of $O^*(16^k)$ [8] and $O^*(12.5^k)$ [5].

While these results were combinatorial in nature, the next improvements used algebraic techniques: Koutis [9] presented an algorithm solving the problem in $O^*(2.83^k)$ time. His method was perfected by Williams [12], reducing the running time to $O^*(2^k)$. This had somewhat closed the gap between the k-path problem

---

[*] Research supported by ISF grant 1241/12 and by GIF Young.

[**] Research supported by BSF grant 2010437, a Google Research Award and GIF grant 1147/2011.

[1] Here and throughout, the O^* notation discards all factors that are polynomial in n, k, and $\log M$ from the running time. Similarly, the $\tilde{O}$ expressions discard polylogarithmic factors.

F. Dehne, R. Solis-Oba, and J.-R. Sack (Eds.): WADS 2013, LNCS 8037, pp. 390–401, 2013.
© Springer-Verlag Berlin Heidelberg 2013

and the best method known for the specific case of finding a Hamiltonian path in a directed graph, which is $O^*(2^n)$ (though the latter is combinatorial in nature). For undirected graphs, recent results presented $O^*(1.657^n)$ [3] and later $O^*(1.657^k)$ [4,1] running times for Hamiltonian path and k-path, respectively.

It is worthwhile to focus on Koutis' and Williams' techniques, as they are the basis to this paper. They reduce k-path and other problems to the problem of determining whether a given n-variable polynomial contains a k-multilinear-monomial (that is, a term which is the multiplication of k distinct variables) in its sum-product expansion. The problem is then solved by (roughly) evaluating this polynomial over random values taken from an adequate choice of an algebraic structure. In a later result [10] they both show that, in the evaluation framework they use, their technique for finding a k-multilinear-monomial is essentially optimal, as any choice of an algebraic structure for the polynomial evaluation would require that the elements in this structure have an $\Omega(2^k/k)$-sized representation.

One of the most natural generalizations coming to mind, is the *minimum-weight k-path* problem: in this scenario, the graph edges are weighted and we wish to find a k-path having minimum weight in the graph. In [12] this was referred to as the *short cheap tour* problem and mentioned that while the $O^*(4^k)$ methods can be easily extended to accommodate for this version, the algebraic methods do not seem to support such extension, and left this as an open problem. We solve this problem for the specific case in which the edge weights are integers in the range $[-M, M]$, incurring a running time which also has a superlinear dependency on M. If the weights are reals in $[1, M]$ (or can be normalized to this range, as is the case if they are in the range $[\ell, h]$ for $0 < \ell < h$), we provide a $(1+\varepsilon)$-approximation which reduces this dependency to $\log \log M$. Notice that by this we conform to the important line of research in recent years, of discussing variants of distance problems in which edge-weights are integers taken from a bounded range, see e.g., [14,6].

Another problem that generalizes k-path is presented in [10]: in the k-tree problem, given an n-vertex graph G and a k-node tree T, find a copy of T in G. For a similar generalization of this problem to *minimum-weight k-tree*, and under similar restrictions on the edge weights, we show similar exact and approximate results.

Paper Organization. In Section 3, we first present an $\tilde{O}(2^k \text{poly}(k) M n^\omega)$ algorithm for computing the weight of the minimum-weight k-path when edge weights are integers in $[-M, M]$, where $\omega < 2.3727$ stands for the matrix multiplication exponent [13]. In Section 4, we show how to find the path itself, incurring an $O(k \cdot \text{poly} \log n)$ multiplicative overhead for the above algorithm. Finally, in Section 5, for the case of real edge-weights in $[1, M]$, we provide a $(1+\varepsilon)$-approximation algorithm that reduces the dependency on M to $\log \log M$, by using a technique of careful adaptive scaling of the edge weights. The overall running time of this algorithm is $\tilde{O}(2^k \text{poly}(k) n^\omega (\log \log M + 1/\varepsilon))$.

In Section 6 we turn to the k-tree problem, and show similar results: we present an $\tilde{O}(2^k \text{poly}(k) M n^3)$ algorithm for finding the minimum-weight k-tree when edge weights are integers in $[-M, M]$, and for the case the edge-weights are reals in $[1, M]$, provide a $(1 + \varepsilon)$-approximation algorithm having running time $\tilde{O}(2^k \text{poly}(k) n^3 (\log \log M + 1/\varepsilon))$.

2   Preliminaries

We follow Williams' notation [12]. Let $\mathbb{F}$ be a field and G be a multiplicative group. The group algebra $\mathbb{F}[G]$ is defined over the set of elements of the form

$$\sum_{g \in G} a_g g \tag{1}$$

where $a_g \in \mathbb{F}$ for all $g \in G$, i.e., on the set of sums of elements from G with coefficients from $\mathbb{F}$. Addition is computed component-wise as

$$\sum_{g \in G} a_g g + \sum_{g \in G} b_g g = \sum_{g \in G} (a_g + b_g) g \ , \tag{2}$$

multiplication is defined in the form of a convolution:

$$\left(\sum_{g \in G} a_g g \right) \left(\sum_{g \in G} b_g g \right) = \sum_{g, h \in G} a_g b_h gh = \sum_{g \in G} \left(\sum_{h \in G} a_h b_{h^{-1} g} \right) g \ , \tag{3}$$

(since G is a multiplicative group, the expression $h^{-1} g$ here replaces the expression of the type $g - h$ which is usually found in a convolution definition) and multiplication by a scalar $c \in \mathbb{F}$ as

$$c \left(\sum_{g \in G} a_g g \right) = \sum_{g \in G} c a_g g \ . \tag{4}$$

Let $0_{\mathbb{F}}, 1_{\mathbb{F}}$ be the addition and multiplication identities of $\mathbb{F}$, respectively. Let 1_G be the identity of G. It is easy to verify that $\mathbb{F}[G]$ is a ring where the addition identity element $0_{\mathbb{F}[G]} = \sum_{g \in G} 0_{\mathbb{F}} \cdot g$ is the element having all coefficients taken as $0_{\mathbb{F}}$, and the multiplication identity element $1_{\mathbb{F}[G]} = 1_{\mathbb{F}} \cdot 1_G = 1_G$. For ease of notation, hereafter 0 and 1 will denote $0_{\mathbb{F}[G]}$ and $1_{\mathbb{F}[G]}$, respectively.

Let z be a symbolic variable. Our computations are done on the set $(\mathbb{F}[G])[z]$ of univariate polynomials on z with coefficients in $\mathbb{F}[G]$. Notice that the set of polynomials with coefficients in a ring is a ring by itself.

For our algorithm, we follow Williams and choose G to be $\mathbb{Z}_2^k$ (i.e., the set of binary vectors of dimension k) with multiplication between elements of $\mathbb{Z}_2^k$ defined as entry-wise addition modulo 2. It follows that 1_G is the k-dimensional all-zeros vector. Notice that for all $u, v \in \mathbb{Z}_2^k$, $u \cdot v = 1_G$ iff $u = v$. We also choose $\mathbb{F} = \text{GF}(2^\ell)$ for $\ell = \log k + 3$. Notice that since $\mathbb{F} = \text{GF}(2^\ell)$ has characteristic 2, it holds that for all $c \in \mathbb{F}$, $c + c = 0_{\mathbb{F}}$, and therefore that for all $v \in \mathbb{F}[G]$, $v + v = 0$.

3   Method

Given a weighted, directed or undirected graph $H = (V, E, w)$ on the vertex-set $V = \{1, \ldots, n\}$, with integer edge-weights in $[-M, M]$, we first show how to compute the weight of the minimum-weight k-path with high probability. We can assume the edge weights are actually in $[0, M]$, otherwise we re-define $w(i, j) \leftarrow w(i, j) + M$ for each $(i, j) \in E$ and then $M \leftarrow 2M$: as this process incurs a penalty of $(k - 1)M$ for each k-path, it maintains the order relation on k-path weights. Define a k-walk to be a walk in the graph comprised of k (not necessarily distinct) vertices, and let $I = \langle i_1, \ldots, i_k \rangle$ be some arbitrary k-walk in H. With a slight abuse of notation, we will also use I to denote the set of edges participating in the walk.

We define a collection $\{B_c\}_{c=1}^{k-1}$ of polynomial matrices B_c as follows:

$$B_c[i, j] = \begin{cases} y_{i,j,c} \cdot x_i \cdot z^{w(i,j)} & \text{if } (i, j) \in E, \\ 0 & \text{if } (i, j) \notin E; \end{cases} \tag{5}$$

where each variable $y_{i,j,c}$ shall be assigned with a randomly selected value from $\mathbb{F}$ and each x_i will be assigned with a value chosen from $\mathbb{F}[G]$ by a method to be described shortly. Notice that each x_i corresponds to vertex i. Assume the values $\{y_{i,j,c}\}_{i,j,c}$ have already been chosen. Recall that z is a symbolic variable. We define the polynomial P as follows: $P(x_1, \ldots, x_n, z) = \mathbf{1} \cdot B_1 \cdots B_{k-1} \cdot \boldsymbol{x}$, where $\mathbf{1}$ is the n-dimensional all-ones vector and $\boldsymbol{x}$ is the vector $(x_1, \ldots, x_n)$. Re-writing P as its sum-product expansion we get:

$$P(x_1, \ldots, x_n, z) = \sum_{\substack{I \\ I = \langle i_1, \ldots, i_k \rangle \text{ is a walk in } H}} \left(\prod_{c=1}^{k-1} B_c[i_c, i_{c+1}] \right) x_{i_k}, \tag{6}$$

that is, P is an aggregate sum over all k-walks in H, where each walk $I = \langle i_1, \ldots, i_k \rangle$ is represented by the product of its corresponding components in $B_1, \ldots, B_{k-1}$, finally multiplied by x_{i_k} which corresponds to the final vertex of the walk. By substituting the $B_c[i_c, i_{c+1}]$'s for their values, and re-arranging the walk's product such that the $y_{i,j,c}$ terms appear first, then the x_i terms, and finally the z term, it follows that

$$P(x_1, \ldots, x_n, z) = \sum_{\substack{I \\ I = \langle i_1, \ldots, i_k \rangle \text{ is a walk in } H}} y^I \cdot x^I \cdot z^{w(I)}, \tag{7}$$

where $y^I = \prod_{c=1}^{k-1} y_{i_c, i_{c+1}, c}$, $x^I = x_{i_1} \cdots x_{i_k}$, and $w(I) = \sum_{e \in I} w(e)$ is the weight of walk I.

3.1   Algorithm

Given H, randomly choose all values $y_{i,j,c} \in \mathbb{F}$, and randomly pick n vectors $v_1, \ldots, v_n$ from $G = \mathbb{Z}_2^k$. Now compute the polynomial $P'(z) = P(1_G + $

$v_1, \ldots, 1_G + v_n, z)$. Let $\text{coeff}_z^d P'(z)$ be the d-th degree term coefficient of $P'(z)$, and let $d' = \min\{d \mid \text{coeff}_z^d P'(z) \text{ is not } 0\}$ (if such exists). If d' exists, return it. Otherwise output "no k-path exists in H".

3.2 Proof of Correctness

Let I be the minimum-weight k-simple-path in H, and notice that $w(I)$ is represented in P by the term $z^{w(I)}$ in the product corresponding to I. Notice that while no degrees $d < w(I)$ occur in P, it might be that the $w(I)$-th degree term of P was eliminated when (partially) evaluating P. Our goal is to show that this happens with low probability. For a walk I, notice that if I is simple, i.e., it visits every node at most once, then x^I is multilinear, or equivalently, square-free, since each variable x_i appears in it at most once. On the other hand, if I is non-simple, then x^I must contain some square x_j^2. Therefore, in order to prove the algorithm correct, we need to show that w.h.p., (a) products corresponding to non-simple paths vanish, (b) products corresponding to simple-k-paths do not vanish by their evaluation, and that (c) products corresponding to simple-k-paths are not eliminated when they are summed with other (same-degree) products.

These notions are captured by the following propositions, which are similar to the ones in [12]. Due to lack of space we defer the proofs to the full version of the paper.

Proposition 1. *If x^I is non-multilinear, it vanishes.*

Let $J = \sum_{v \in G} v$ be the sum of all vectors from $G = \mathbb{Z}_2^k$ (addition here is the addition of $\mathbb{F}[G]$).

Proposition 2. *Let $I = \langle i_1, \ldots, i_k \rangle$ be a k-walk. If x^I is multilinear (i.e., I is a k-path), then if the vectors $v_{i_1}, \ldots, v_{i_k} \in \mathbb{Z}_2^k$ are linearly independent w.r.t. entry-wise addition modulo 2, then $x^I = J$.*

Corollary 1. *Let $I = \langle i_1, \ldots, i_k \rangle$ be a k-walk. If x^I is multilinear (i.e., I is a k-path), then with probability at least 0.28 it does not vanish.*

We have shown that with at least constant probability, multilinear terms do not vanish when they are assigned values as described. However, it still might happen that such multilinear terms will get eliminated when they are summed up with other multilinear terms. The next two propositions show that this can happen with at most constant probability.

Proposition 3. *Let $I = \langle i_1, \ldots, i_k \rangle$ be a k-walk. If the variables $v_{i_1}, \ldots, v_{i_k} \in \mathbb{Z}_2^k$ are linearly dependent w.r.t. entry-wise addition modulo 2, then x^I vanishes.*

Recall that $P(x_1, \ldots, x_n, z)$ is a polynomial in z and therefore can be viewed as

$$P(x_1, \ldots, x_n, z) = \sum_{d=0}^{kM} \sum_{\substack{I \\ I = \langle i_1, \ldots, i_k \rangle \text{ is a walk in } H \\ w(I) = d}} y^I \cdot x^I \cdot z^d . \tag{8}$$

It is therefore easy to see that the minimum-degree term in P corresponds to minimum-weight k-paths in H. Let d' be the minimum degree of P and let

$$\mathrm{coeff}_z^{d'} P(x_1, \ldots, x_n, z) = \sum_{\substack{I \\ I \text{ is a walk in } H \\ w(I) = d'}} y^I \cdot x^I \qquad (9)$$

be its corresponding coefficient. Our goal is to show that with at least constant probability, $\mathrm{coeff}_z^{d'} P$ does not vanish when it is evaluated.

Proposition 4. $\mathrm{coeff}_z^{d'} P'(z)$ *does not vanish with probability at least* $1/5$.

3.3   Running Time Analysis

The running time of the algorithm is dominated by k matrix multiplications, where the basic arithmetic operations are done over the polynomial ring $(\mathbb{F}[G])[z]$. Therefore, we need to account for the the cost of each such operation. Notice that for any arithmetic operation in $(\mathbb{F}[G])[z]$ performed by our algorithm, the maximum degree of the operand polynomials and resulting polynomial, is at most kM. We can therefore focus on the set R of polynomials in $(\mathbb{F}[G])[z]$ with degree at most kM. By treating the polynomials in R as periodic with period kM (since there will be no carry or overflow to greater degrees), R continues to be a ring. Let T be the upper-bound on the time required for an arithmetic operation in R; trivially, $T = \Omega(2^k \cdot kM \log|\mathbb{F}|)$. It follows that the algorithm requires $O(kn^\omega T)$ time, and it remains to compute T.

Addition. Addition of two polynomials can be easily done component-wise in time $O(kM \cdot 2^k \cdot \log|\mathbb{F}|) = O(2^k \mathrm{poly}(k)M)$.

Multiplication. Multiplication is trickier and is done by employing a multidimensional fast Fourier transform-type approach.[2] We now describe the multiplication process in more detail.

The multiplication process will be easier to describe on the ring $\mathbb{F}[\mathbb{Z}_2^k \times [kM]]$ which is isomorphic to R, as will be shown immediately. Given a vector $v = (v_1, \ldots, v_k) \in \mathbb{Z}_2^k$ and an integer $d \in [kM]$, let $(v; d)$ denote the vector $(v_1, \ldots, v_k, d) \in \mathbb{Z}_2^k \times [kM]$. A polynomial $p \in R$ can be uniquely described as a sum $\sum_{v,d} a_{(v;d)} \cdot (v; d)$ of at most $N = 2^k kM$ summands, where each $a_{(v;d)} \in \mathbb{F}$ is the coefficient of v appearing in $\mathrm{coeff}_z^d p$ (i.e., if $\mathrm{coeff}_z^d p = \sum_{v \in G} b_v v$, then $a_{(v;d)} = b_v$). Our definition of multiplication over $G = \mathbb{Z}_2^k$ can be naturally extended to $\mathbb{Z}_2^k \times [kM]$: multiplication still corresponds to entry-wise addition, only that now addition is done modulo 2 for dimensions $1, \ldots, k$ and modulo kM for dimension $k + 1$. With that in mind, our definitions of addition, multiplication,

---

[2] Here, as opposed to Williams [12], the Walsh-Hadamard transform is not an adequate choice anymore due to the existence of the variable z which can have a degree up to kM.

and identity elements for R are extended appropriately, thus forming the ring $\mathbb{F}[\mathbb{Z}_2^k \times [kM]]$. The bottom line is that now any $p \in R$ can be viewed as a sum of elements with coefficients taken from a multidimensional array indexed by values from $\mathbb{Z}_2^k \times [kM]$ and that multiplication is still a convolution, an important fact to be used later.

Moving to $\mathbb{F} = \mathrm{GF}(2^\ell)$, being a finite field, all elements in $\mathbb{F}$ can be represented in the usual manner as a degree-ℓ polynomials with coefficients in $\mathbb{Z}_2 = \mathrm{GF}(2)$ and operations that are done modulo some predefined irreducible polynomial of degree ℓ (this irreducible polynomial can even be found naïvely as $\ell = \log k + 3$). For the purpose of using FFT, we treat polynomials in $\mathbb{Z}_2[x]$ as if they were actually in $\mathbb{C}[x]$, i.e., the set of univariate polynomials over the complex numbers. At the end of the multiplication process, we will appropriately convert polynomials in $\mathbb{C}[x]$ back to $\mathrm{GF}(2^\ell)$ as will be described shortly.

By the above arguments, given two polynomials $p, q \in R$ to be multiplied, they can be taken as the sums $\sum_{v,d} p_{(v;d)} \cdot (v; d)$ and $\sum_{v,d} q_{(v;d)} \cdot (v; d)$, respectively, where $p_{(v;d)}, q_{(v;d)} \in \mathbb{C}[x]$ for each $v \in \mathbb{Z}_2^k$ and $d \in [kM]$. As the multiplication corresponds to a convolution, by the convolution theorem, it holds that $p*q = \mathrm{DFT}^{-1}(\mathrm{DFT}(p) \cdot \mathrm{DFT}(q))$, where $*$ denotes a convolution, $\cdot$ denotes pointwise multiplication, and DFT denotes the $(k+1)$-dimensional discrete Fourier transform for values indexed by vectors of type $(v_1, \ldots, v_k, d) \in \mathbb{Z}_2^k \times [kM]$. Let $D(\ell)$ denote the time required for an arithmetic operation on degree-ℓ polynomials in $\mathbb{C}[x]$—including converting them back to $\mathrm{GF}(2^\ell)$ by division by an irreducible polynomial—and notice that $D(\ell) = O(\ell^2) = O(\mathrm{poly} \log k)$ as multiplication and division here are quadratic by nature. Then the above DFT operations can be computed efficiently in time $O(N \log N \cdot D(\ell)) = \tilde{O}(2^k k^2 M)$ by using the multidimensional FFT algorithm. Once we have computed $\mathrm{DFT}(p)$ and $\mathrm{DFT}(q)$, thus obtaining for each of them N values in $\mathbb{C}[x]$ (indexed as well by vectors in $\mathbb{Z}_2^k \times [kM]$), we point-wise multiply them, obtaining a sum $w = \mathrm{DFT}(p) \cdot \mathrm{DFT}(q)$, and compute $\mathrm{DFT}^{-1}(w)$, again by using FFT on multidimensional coefficients in $\mathbb{C}[x]$. Finally, we reduce $\mathbb{C}[x]$ terms (which are actually in $\mathbb{Z}[x]$, as convolution over integer values returns integer values) by dividing them by the irreducible polynomial used before and the appropriate modulo operations.

We conclude that multiplication of polynomials in R can be performed in time $\tilde{O}(2^k \mathrm{poly}(k) M)$, and therefore $T = \tilde{O}(2^k \mathrm{poly}(k) M)$.

4   Finding the Actual Path

Let $G = (V, E, w)$ be a weighted graph with integer edge-weights in $[-M, M]$. Given the algorithm from the previous section, we show that it is possible to find the minimum-weight k-path itself with only $O(k \mathrm{poly} \log n)$ multiplicative overhead w.r.t. the previous algorithm and with a polynomially small error probability. We denote by $\mathcal{A}$ the algorithm from the previous section, amplified by running $O(\log n)$ iterations of it and choosing the minimal result, such that its error probability is bounded by $1/n^{c'}$ for some constant c'. The algorithm for finding the actual path uses $\mathcal{A}$ as a sub-routine. Its pseudo-code is provided as Algorithm 1. Full analysis is deferred to the full version of the paper.

Algorithm 1. Finding the minimum-weight k-path.

```
1  d ← A(G, k)
2  while |V(G)| > 10k do
3       for Θ(log n) times do
4            G' ← a copy of G in which each vertex is removed with probability 1/k
5            if at least Ω(|V(G)|/k) were removed and A(G', k) = d then
6                 G ← G'
7                 Go to the while loop
8       return "Fail"
9  foreach remaining vertex v ∈ V(G) and until |V(G)| = k do
10      G' ← G \ v   /* G \ v is G with v and its incident edges removed */
11      if A(G', k) = d then G ← G'
12 return E(G)
```

5   Approximation

The main drawback of the previous algorithm is that its running time has a superlinear dependency in M, the bound on an edge weight. If the weights are in $[1, M]$ (or can be normalized to this range), we show that if we settle for a $(1+\varepsilon)$-approximation algorithm to the problem, this dependency can be brought down to $\log \log M$, by using a technique of careful adaptive scaling of the edge weights, thus bringing the overall running time to $\tilde{O}(2^k \mathrm{poly}(k) n^\omega (\log \log M + 1/\varepsilon))$. Our techniques are in the spirit of the FPTAS of Ergün et al. [7] for the restricted shortest path problem. We start with the following proposition:

Proposition 5. *Given a graph G with integer edge-weights in $[0, M]$, a parameter k, and a value B, it is possible to find an exact solution to the minimum-weight k-path problem of weight at most B, if such exists, or to return that no such solution exists, in time $\tilde{O}(2^k \mathrm{poly}(k) B n^\omega) = O^*(2^k B)$ and polynomially-small error probability.*[3]

Proof. The algorithm is identical to the previous one, except that as a first step, edges of weight greater than B are deleted from the graph, and that when multiplying two polynomials in $(\mathbb{F}[G])[z]$ of degree at most B, we truncate from the resulting polynomial any term of degree greater than B, thus keeping all polynomials throughout the algorithm at degree of at most B. As every polynomial multiplication now takes $\tilde{O}(2^k \mathrm{poly}(k) B)$ time, the running time analysis follows.   □

We denote with $\mathcal{B}$ the algorithm that finds an exact solution to the k-path problem of weight at most B, if such exists, or to returns that no such solution exists. We will use it as a sub-routine in our approximation algorithm.

Define $k' = k - 1$ (the number of edges in a k-path), and let OPT be the minimum-weight k-path. Our approximation algorithm starts by defining an

---

[3] B does not have to be an integer, but the effect in this case is as if $\lfloor B \rfloor$ is used.

upper and a lower bound, U and L, respectively, to the weight of OPT. At first, $U = k'M$ and $L = k'$. It then iteratively fine-tunes U and L to the point where the ratio U/L is less than or equal to 2, while maintaining the invariant that $L \leq w(OPT) \leq U$. This fine tuning is done as follows.

At each iteration we let the value $X = \sqrt{LU}$ be the geometric mean of L and U, and define the value $\delta = (L/U)^{1/3} - \sqrt{L/U}$ which will serve as a scaling coefficient. Notice that $\delta > 0$ as $U > L$. We then scale-down the edge weights by a factor of $\delta U/k'$, thus defining a new weight $w'(i,j) = \left\lfloor \frac{w(i,j)}{\delta U/k'} \right\rfloor$ for each edge (i,j), and let $G' = (V, E, w')$ be the graph with the new weights. Ideally, we would like to test whether the weight of the optimal solution is less than or greater than X by calling $\mathcal{B}(G', k, \frac{X}{\delta U/k'})$; here notice that the value $\frac{X}{\delta U/k'}$ is the scaled-down equivalent of X in G'. However, while the scaling guarantees that this test can be done without incurring a high running time cost, it also introduces a loss of precision due to the floor function in the scaling: define $w_{\text{eff}}(i,j) = (\delta U/k')w'(i,j)$ as the effective weight $w'(i,j)$ simulates, then we have that $w_{\text{eff}}(i,j) \leq w(i,j) \leq w_{\text{eff}}(i,j) + \delta U/k'$, and therefore for a k-path P, we have that $w_{\text{eff}}(P) \leq w(P) \leq w_{\text{eff}}(P) + \delta U$. Therefore, in the case $w'(OPT) > \frac{X}{\delta U/k'}$ we have that $w(OPT) \geq w_{\text{eff}}(OPT) > X$, but if $w'(OPT) \leq \frac{X}{\delta U/k'}$ (and therefore $w_{\text{eff}}(OPT) \leq X$) then all we can assert is that $w(OPT) \leq X + \delta U$. Therefore, a k-path returned by a call to $\mathcal{B}(G', k, \frac{X}{\delta U/k'})$ has weight at most $X + \delta U$ (and not X) w.r.t. the original graph. According to the outcome of the call to $\mathcal{B}(G', k, \frac{X}{\delta U/k'})$, we redefine U and L: if $\mathcal{B}(G', k, \frac{X}{\delta U/k'})$ returned a result, we set $U \leftarrow X + \delta U$; otherwise we set $L \leftarrow X$.

When the main loop is done (convergence is shown to exist below), we again redefine a new weight function: $w'(i,j) = \left\lfloor \frac{w(i,j)}{\varepsilon L/k'} \right\rfloor$ for each edge (i,j), the graph $G' = (V, E, w')$, and return the result of a call to $\mathcal{B}(G', k, \frac{U}{\varepsilon L/k'})$. The full algorithm pseudo-code is given as Algorithm 2.

Running-Time. We first show that the main loop performs $O(\log\log M)$ iterations. Let L_i, U_i be the respective values of L, U at the start of iteration i; we will show that $U_{i+1}/L_{i+1} \leq (U_i/L_i)^{2/3}$. At the end of each iteration i, we have that either $L_{i+1} \leftarrow L_i$ and $U_{i+1} \leftarrow X + \delta U_i$, or that $L_{i+1} \leftarrow X$ and $U_{i+1} \leftarrow U_i$, where $X = \sqrt{L_i U_i}$ and $\delta = (L_i/U_i)^{1/3} - \sqrt{L_i/U_i}$. In the former case we have that

$$\frac{U_{i+1}}{L_{i+1}} = \frac{X + \delta U_i}{L_i} = \frac{\sqrt{L_i U_i} + \left(\left(\frac{L_i}{U_i}\right)^{1/3} - \sqrt{\frac{L_i}{U_i}}\right)U_i}{L_i} = \frac{\left(\frac{L_i}{U_i}\right)^{1/3}U_i}{L_i} = \left(\frac{U_i}{L_i}\right)^{2/3}, \tag{10}$$

and in the latter

$$\frac{U_{i+1}}{L_{i+1}} = \frac{U_i}{X} = \frac{U_i}{\sqrt{L_i U_i}} = \sqrt{\frac{U_i}{L_i}} \leq \left(\frac{U_i}{L_i}\right)^{2/3}. \tag{11}$$

Algorithm 2. Approximation algorithm.

```
1  k' ← k − 1
2  L ← k'
3  U ← k'M
4  while U > 2L do
5  │   X ← √LU
6  │   δ ← (L/U)^(1/3) − √(L/U)
7  │   Define w': E → ℕ such that w'(i,j) = ⌊w(i,j)/(δU/k')⌋
8  │   G' ← (V, E, w')
9  │   if B(G', k, X/(δU/k')) returns a result then
10 │   │   U ← X + δU
11 │   else
12 │   │   L ← X
13 Define w': E → ℕ such that w'(i,j) = ⌊w(i,j)/(εL/k')⌋
14 G' ← (V, E, w')
15 return B(G', k, U/(εL/k'))
```

In both cases we have that $U_{i+1}/L_{i+1} \leq (U_i/L_i)^{2/3}$. Therefore it converges to a constant after $O(\log \log M)$ iterations. Notice that an invocation of $\mathcal{B}(G', k, \frac{X}{\delta U/k'})$ costs $\tilde{O}(2^k \text{poly}(k)n^\omega)$ by Proposition 5, with the bound $B = \frac{X}{\delta U/k'}$ which is $O(k)$, as $\delta U = \Omega(X)$. We conclude that the overall cost of the main loop is $\tilde{O}(2^k \text{poly}(k)n^\omega \log \log M)$.

As for the final call to $\mathcal{B}(G', k, \frac{U}{\varepsilon L/k'})$, we have that its running time is $\tilde{O}(2^k \text{poly}(k)n^\omega/\varepsilon)$ by Proposition 5, with the bound $B = \frac{U}{\varepsilon L/k'}$ which is $O(k/\varepsilon)$ since at this stage $U \leq 2L$. We conclude that the overall running time of the approximation algorithm is $\tilde{O}(2^k \text{poly}(k)n^\omega(\log \log M + 1/\varepsilon))$.

Correctness. Throughout the execution, the algorithm maintains the invariant that $L < X < X + \delta U < U$. That can be easily seen by substituting X and δ for their values and observing that $L < \sqrt{LU} < L^{1/3}U^{2/3} < U$. Assume there exist a k-path in G, and let OPT be the minimum-weight k-path. By the scaling arguments, and the fact that we have brought the loss of precision due to scaling into consideration when redefining U and L, we have that the invariant $L \leq w(OPT) \leq U$ always holds. Due to the running-time argument, when the main loop is done we have $U/L \leq 2$. Let P^* be the result of the call to $\mathcal{B}(G', k, \frac{U}{\varepsilon L/k'})$ at line 15 of the pseudo-code, and notice the the weights defined at line 13 incur an $\varepsilon L/k'$ loss of precision per edge, or equivalently εL per k-path. By the call to the exact algorithm, we have that $w'(P^*) \leq w'(OPT)$ and therefore also $w_{\text{eff}}(P^*) \leq w_{\text{eff}}(OPT)$. Accounting for the loss of precision, we have that $w(P^*) \leq w_{\text{eff}}(P^*) + \varepsilon L \leq w_{\text{eff}}(OPT) + \varepsilon L \leq (1 + \varepsilon)w(OPT)$.

6 k-Tree

In [10], they provide a solution to the k-tree problem: given an n-vertex graph G and a k-node tree T, is there a (not necessarily induced) copy of T in G. Again their solution is based on a reduction to the question of is there a k-multilinear-monomial in the sum-product expansion of a given polynomial. We show how to handle the *minimum-weight k-tree* problem—in which we are given a weighted graph G, and wish to find a minimum-weight copy of T in it, across all copies of T in it—again, when the weights are integers in a given range $[-M, M]$.

Theorem 1. *Given a graph G, if the edge-weights are integers in $[-M, M]$, the minimum-weight k-tree can be found in $\tilde{O}(2^k \mathrm{poly}(k) M n^3)$ time. If the edge-weights are reals in $[1, M]$, the problem can be approximated within $(1 + \varepsilon)$ in $\tilde{O}(2^k \mathrm{poly}(k) n^3 (\log \log M + 1/\varepsilon))$ time.*

Let $N_G(i)$ be the neighbor-set of vertex i in G, and let $X = \{x_1, \ldots, x_n\}$ be a variable-set corresponding to $V(G)$. We use the following polynomial on X, implemented as an arithmetic circuit:

Let $V(G) = [n]$ and $V(T) = [k]$. The polynomial $C_{T,i,j}(x_1, \ldots, x_n)$ is defined as follows. If $|V(T)| = 1$, then $C_{T,i,j} = x_j$. Otherwise, $C_{T,i,j}$ is defined recursively: let $\{T_{i,\ell} \mid \ell \in N_T(i)\}$ be the subtrees of T created by removing node i from T, where $T_{i,\ell}$ is the subtree containing ℓ. Then

$$C_{T,i,j} = \prod_{\ell \in N_T(i)} \left(\sum_{j' \in N_G(j)} y_{(i,\ell),(j,j')} \cdot z^{w(j,j')} C_{T_{i,\ell},\ell,j'} \right), \tag{12}$$

where as before, z is a symbolic variable, and the values $\{y_{e,e'} \mid e \in E(T), e' \in E(G)\}$ are random values drawn from $\mathbb{F}$.[4] Finally, define the polynomial $Q = \sum_{j \in V(G)} C_{T,1,j}$. Each $C_{T,1,j}$ is a circuit containing at most $|E(T)| \cdot |E(G)|$ addition and multiplication gates and therefore Q contains $n \cdot |E(T)| \cdot |E(G)| = O(n^3 k)$ such gates. Q is a sum over all homomorphisms from T to subgraphs of G of size at most k: specifically $C_{T,i,j}$ aggregates over all homomorphisms that map $i \in V(T)$ to $j \in V(G)$ (proof can be found in [10][5]). Therefore, a monomial $x_{j_1} \cdots x_{j_k}$ appears in the sum-product expansion of Q if an only if there is a homomorphism mapping $V(T)$ to $\{j_1, \ldots, j_k\}$ such that if $(i, \ell) \in E(T)$, then $(j_i, j_\ell) \in E(G)$. If such a monomial is multilinear, it corresponds to such a homomorphism in which $j_1, \ldots, j_k$ are distinct vertices, i.e., a vertex in G was not used more than once for the sake of a single mapping. From this point, the same algorithms given before follow (only this time, evaluating Q over $(\mathbb{F}[G])[z]$), and propositions similar to Propositions 1–4 apply. Full proofs are deferred to the full version of the paper. We obtain that the minimum-weight k-tree problem

---

[4] In [10], the y-values are implicit and come from the multiplication of the output of each multiplication gate with a random value taken from $\mathbb{F}$.

[5] Their arithmetic circuit is defined as $Q = \sum_{i \in V(T), j \in V(G)} C_{T,i,j}$, however, it seems to contain redundancy.

with integer edge-weights in $[-M, M]$ can be solved in $\tilde{O}(2^k \text{poly}(k) M n^3)$ time and that if the edge-weights are reals in $[1, M]$, it can be approximated within $(1 + \varepsilon)$ in $\tilde{O}(2^k \text{poly}(k) n^3 (\log \log M + 1/\varepsilon))$ time.

Acknowledgments. We would like to thank Ryan Williams and Danny Raz for helpful comments.

References

1. Abasi, H., Bshouty, N.H.: A simple algorithm for undirected hamiltonicity. Electronic Colloquium on Computational Complexity (ECCC) 20, 12 (2013)
2. Alon, N., Yuster, R., Zwick, U.: Color-coding. J. ACM 42(4), 844–856 (1995)
3. Björklund, A.: Determinant sums for undirected hamiltonicity. In: FOCS, pp. 173–182. IEEE Computer Society (2010)
4. Björklund, A., Husfeldt, T., Kaski, P., Koivisto, M.: Narrow sieves for parameterized paths and packings. CoRR, abs/1007.1161 (2010)
5. Chen, J., Lu, S., Sze, S.-H., Zhang, F.: Improved algorithms for path, matching, and packing problems. In: Bansal, N., Pruhs, K., Stein, C. (eds.) SODA, pp. 298–307. SIAM (2007)
6. Cygan, M., Gabow, H.N., Sankowski, P.: Algorithmic applications of baur-strassen's theorem: Shortest cycles, diameter and matchings. In: FOCS, pp. 531–540. IEEE Computer Society (2012)
7. Ergün, F., Sinha, R.K., Zhang, L.: An improved fptas for restricted shortest path. Inf. Process. Lett. 83(5), 287–291 (2002)
8. Kneis, J., Mölle, D., Richter, S., Rossmanith, P.: Divide-and-color. In: Fomin, F.V. (ed.) WG 2006. LNCS, vol. 4271, pp. 58–67. Springer, Heidelberg (2006)
9. Koutis, I.: Faster algebraic algorithms for path and packing problems. In: Aceto, L., Damgård, I., Goldberg, L.A., Halldórsson, M.M., Ingólfsdóttir, A., Walukiewicz, I. (eds.) ICALP 2008, Part I. LNCS, vol. 5125, pp. 575–586. Springer, Heidelberg (2008)
10. Koutis, I., Williams, R.: Limits and applications of group algebras for parameterized problems. In: Albers, S., Marchetti-Spaccamela, A., Matias, Y., Nikoletseas, S., Thomas, W. (eds.) ICALP 2009, Part I. LNCS, vol. 5555, pp. 653–664. Springer, Heidelberg (2009)
11. Monien, B.: How to find long paths efficiently. Annals of Discrete Mathematics 25, 239–254 (1985)
12. Williams, R.: Finding paths of length k in $o^*(2^k)$ time. Inf. Process. Lett. 109(6), 315–318 (2009)
13. Williams, V.V.: Multiplying matrices faster than coppersmith-winograd. In: Karloff, H.J., Pitassi, T. (eds.) STOC, pp. 887–898. ACM (2012)
14. Zwick, U.: All pairs shortest paths using bridging sets and rectangular matrix multiplication. J. ACM 49(3), 289–317 (2002)

Compressed Persistent Index for Efficient Rank/Select Queries

Wing-Kai Hon[1,*], Lap-Kei Lee[2,**], Kunihiko Sadakane[3,***],
and Konstantinos Tsakalidis[4]

[1] Department of Computer Science, National Tsing Hua University, Taiwan
[2] HKU-BGI Bioinformatics Algorithms & Core Technology Research Laboratory,
University of Hong Kong, Hong Kong
[3] National Institute of Informatics, 2-1-2 Hitotsubashi, Tokyo 101-8430, Japan
[4] Department of Computer Science & Engineering, HKUST, Hong Kong

Abstract. We design compressed persistent indices that store a bit vector of size n and support a sequence of k bit-flip update operations, such that rank and select queries at any version can be supported efficiently. In particular, we present partially and fully persistent compressed indices for offline and online updates that support all operations in time polylogarithmic in n and k. This improves upon the space or time complexities of straightforward approaches, when $k = O(\frac{n}{\log n})$, which is common in biological applications. We also prove that any partially persistent index that occupies $O((n + k) \log(nk))$ bits requires $\omega(1)$ time to support the rank query at a given version.

1 Introduction

In this paper we consider the problem of maintaining persistently a compressed bit vector under (online and offline) bit-flip updates, such that *rank* and *select* queries (and even updates) can be supported at any version of the bit vector. We consider the word-RAM model of computation. Although many persistent implementations have been devised for specific data structures, such as deques, dictionaries, etc. [8], this is the first study of making a compressed data structure persistent. A potential application of our data structures can be found in *temporal* indexing of similar DNA sequences. Many existing index implementations are for a single DNA sequence and rely on rank/select queries over compressed bit vectors to support pattern searching queries, e.g., FM-index [5], wavelet tree [6]. By interpreting differences between sequences as offline updates and temporal modifications of the sequences as online updates, our structures provide the extra capability of temporal rank/select queries over any version of the sequences.

Specifically, let $B[1..n]$ be a bit vector of length n. For a bit $c \in \{0,1\}$ and an integer $i \in [1, n]$, the query operation $rank_c(B, i)$ returns the number of occurrences of c in the prefix $B[1..i]$ of B, and the query operation $select_c(B, i)$

---

* W.K. Hon was supported by Taiwan NSC Grant 99-2221-E-007-123.
** L.K. Lee was supported by Hong Kong Research Grant Council HKU 713512E.
*** K. Sadakane was supported by JSPS KAKENHI 23240002.

F. Dehne, R. Solis-Oba, and J.-R. Sack (Eds.): WADS 2013, LNCS 8037, pp. 402–414, 2013.

returns the position of the i-th occurrence of c in B. In the dynamic case, the index also supports the *bit-flip*(B, i) update operation that flips the bit $B[i]$ from 1 to 0, or from 0 to 1. Yet such an index is *ephemeral*, meaning that an update operation creates a new version of B *without* maintaining previous versions.

In this paper, we are interested in maintaining a *persistent* index that more-over remembers all versions of B when updates are performed to it. In particular, we consider two notions of persistence: A *partially persistent* index allows only updates to the latest version of B and the other versions are read-only; the versions of B form a list called *version list*. A *fully persistent* index allows updates and queries to any version of B; the versions form a tree called *version tree*.

The ephemeral static and dynamic data structures proposed for this problem are all *succinct* (see [13,14] and references therein), namely their space usage is as close as possible to the information-theoretic lower bound. Following the literature, for a bit vector B of length n that stores m occurrences of 1-bits, this is n times the empirical zero-order entropy $H_0(B) = \frac{m}{n} \log \frac{n}{m} + \frac{n-m}{m} \log \frac{n}{n-m}$. However, a complication arises when independent update operations are maintained persistently, since after k updates at least $\log k + \log n$ bits are required in order to store respectively both the version number and the position of the bit flip of each update. Therefore, we define a persistent index to be *compressed* when it uses $nH_0(B_0) + o(n) + O(k \cdot \log(kn))$ bits of space, where B_0 is the initial version of the bit vector of length n. In other words, the initial bit vector is to be represented by a succinct data structure using space close to the information-theoretic lower bound, while we simultaneously maintain the information of each update using only $O(\log(kn)) = O(\log k + \log n)$ bits. Notice that after $k = \omega(\frac{n}{\log n})$ updates, the $O(k \log(kn))$ term dominates the space complexity and thus the structure occupies $\omega(n)$ bits. Then, we can straightforwardly modify a regular persistent binary tree [4] to support the operations in $O(\log n)$ time, using $O(n)$ *words*.

Therefore we focus on "small" sequences of $k = O(\frac{n}{\log n})$ updates wherein the structure occupies $O(n)$ *bits*. This is a typical scenario in biological applications: We want to store a set of related DNA strings together, so that pattern searching queries can be supported efficiently. We may think of one string as a modification of the other. Here, the number k of DNA mutations is much smaller than the length of a DNA string. For example, for a human genome k is in the order of millions, while its length is around 3 billion nucleotides [15]. We study the problem under two types of updates, namely offline and online updates. For *offline updates*, all the k updates (and thus all the k versions of B) are known in advance. For *online updates*, the updates to B arrive in an online fashion such that an update must be performed before the next update arrives.

Previous Results. In the word-RAM model, Raman et al. [13] present a static succinct data structure that supports rank and select queries in $O(1)$ time. Sadakane and Navarro [14] present the *range min-max tree*, a dynamic succinct data structure that supports all operations in $O(\frac{\log n}{\log \log n})$ time. If we utilize this structure and store every version explicitly, the space usage will degrade to $O(n)$ words after only $k = O(\log n)$ updates. On the other hand, if we maintain only the information relevant to an update operation and reproduce a queried

Table 1. Asymptotic time bounds for persistent rank, select and bit-flip operations, where n is the size of bit vector B, k is the number of updates/versions and ϵ is any positive constant. The fully persistent index for online updates occupies $nH_0(B_0) + o(n) + O(k \log n \log(kn))$ bits, while the other indices are compressed. † is amortized.

	Offline updates	Online updates
Partially persistent	$\frac{\log k}{\log \log k}$, $\log n(\frac{\log k}{\log \log k})$, $-$	$(\frac{\log k}{\log \log k})^2$, $\log n(\frac{\log k}{\log \log k})^2$, $\log^{4+\epsilon} k$
Fully persistent	$\frac{\log^2 k}{\log \log k}$, $\log n(\frac{\log^2 k}{\log \log k})$, $-$	$\log^3 n$, $\log^3 n$, $\log^2 n \log \log n$†

version by the sequence of updates that created it, then the query time has a linear dependence on k in the worst case.

There exist generic techniques to render a data structure persistent in the pointer machine [4], word-RAM [3,9] and external memory [1] models. It is natural to consider applying these techniques to the range min-max tree [14]. The node splitting technique of Driscoll et al. [4] is applicable to pointer-based structures of constant-size nodes, which is not the case for the range min-max tree. Alternatively, we can store the tree in arrays and make them persistent using techniques in [3,9]. However, the arrays are not succinct and the update time is only efficient in expectation.

Our Contributions. This paper presents partially and fully persistent compressed indices for bit vectors that support efficient rank and select queries under sequences of offline and online bit-flip updates (see Table 1). They improve the space usage of straightforward approaches, as long as the number of bit flips k is $O(\frac{n}{\log n})$, where n is the bit vector size. These are the first compressed persistent indices that support all operations in time polylogarithmic in n and k.

In Section 2 we present the partially persistent indices for offline and online updates. They are obtained by storing the initial bit vector in a static structure for rank and select queries [13], and maintaining the information relevant to every update operation in a static (respectively dynamic) structure that supports planar orthogonal range counting queries [7,11]. Then we show how to obtain the answers of rank and select queries to a particular version without reconstructing the queried version, but instead by interpreting them as range counting queries appropriately. We follow a similar approach in the case of the fully persistent index for offline updates (Section 3), where we moreover apply centroid path decomposition (see, e.g., [2]) to the version tree in order to efficiently determine the updates that have created a queried version. To obtain the fully persistent index for online updates (Section 4), we first present the *range sum tree*, a simplification of the range min-max tree [14] that is succinct and supports rank, select and bit-flip in $O(\log n)$ time. Then we parametrize the I/O-efficient technique for full persistence of [1] such that it can handle nodes of non-constant size in the word-RAM model, and we apply it to the range sum tree.

Finally, in Section 5 we prove a superconstant lower bound on the rank query time of any partially persistent index that supports offline bit-flip updates and uses $O((n+k)\log(nk))$ bits of space. This is in contrast to the non-persistent setting, where there exist succinct representations of the bit vector that support rank queries in $O(1)$ time [13].

2   Compressed Partially Persistent Index

In this section, we present two compressed partially persistent indices for offline and online updates, respectively. Let k be the number of updates and let n be the size of the bit vector B. The main results are stated below.

Theorem 1. *There is a compressed partially persistent index for offline updates that occupies $nH_0(B_0)+o(n)+O(k\log(kn))$ bits, and supports at any version, rank queries in $O(\frac{\log k}{\log\log k})$ time and select queries in $O(\frac{\log n\log k}{\log\log k})$ time.*

Theorem 2. *There is a compressed partially persistent index for online updates that occupies $nH_0(B_0)+o(n)+O(k\log(kn))$ bits, and supports at any version, rank queries in $O((\frac{\log k}{\log\log k})^2)$ time, select queries in $O(\log n(\frac{\log k}{\log\log k})^2)$ time and accessing a bit in $O((\frac{\log k}{\log\log k})^2)$ time. An update at the latest version takes $O(\log^{4+\epsilon} k)$ time, for any constant $\epsilon>0$.*

2.1   Data Structure and Algorithm for Offline Updates

We now show Theorem 1. The compressed partially persistent index consists of two components. The first component is a succinct data structure for the initial bit vector B_0. We use the data structure of Raman et al. [13] that occupies $nH_0(B_0)+o(n)$ bits and supports rank and select queries on B_0 in $O(1)$ time.

Lemma 1. *[13] A bit vector $B_0[1..n]$ can be stored using $nH_0(B_0)+O(\frac{n\lg\lg n}{\lg n})$ bits to support in $O(1)$ time the queries $rank_c(B_0,i)$ and $select_c(B_0,i)$, for any $1\leq i\leq n$ and $c\in\{0,1\}$.*

The second component stores the information of the update for each version. We reduce the rank and select query to the problem of *planar range counting* which, given Z points on a $N\times N$ grid, asks for the number of points in a given range $[x_1,x_2]\times[y_1,y_2]$. We employ the data structure of JáJá et al. [7].

Lemma 2. *[7] Let Z points lie on an $N\times N$ grid. Planar range counting queries can be supported in $O(\frac{\log Z}{\log\log Z})$ time, using $O(Z\log N)$ bits.*

We define two grids G_0 and G_1 of size $\max(k,n)\times\max(k,n)$, such that an update on bit i of version B_{t-1} from 1 to 0, which creates the new version B_t, corresponds to the point (t,i) on grid G_0 (similarly, on G_1 for bit-flips from 0 to 1).[1] We maintain two data structures of Lemma 2 for the grids G_0 and G_1, respectively. They occupy in total $2\cdot O(k\log(\max(k,n))) = O(k\log(kn))$ bits. Thus, the total space of both components is the space stated in Theorem 1.

---

[1]   For offline updates, we can determine if the bit is flipped from 0 or 1 at no cost.

Query Algorithm. We can answer the queries $rank_c(B_t, i)$ and $select_c(B_t, i)$, for any version t, position $1 \leq i \leq n$ and bit $c \in \{0, 1\}$, as follows. Let $count_c(t, i)$ be the number of points in the range $[0, t] \times [0, i]$ of G_c.

- $rank_c(B_t, i)$: First, we obtain $rank_c(B_0, i)$ from the succinct data structure for B_0. Then, we make two range counting queries on the data structures for grids G_c and G_{1-c} to obtain $count_c(t, i)$ and $count_{1-c}(t, i)$. Then $rank_c(B_t, i) = rank_c(B_0, i) + count_c(t, i) - count_{1-c}(t, i)$.
- $select_c(B_t, i)$: $select_c(B_t, i)$ is the smallest $j \in [1, n]$ such that $rank_c(B_t, j) = i$. We find such a j, by a binary search on $rank_c(B_t, j)$.

Lemma 3. *For any version t, bit position $1 \leq i \leq n$ and bit $c \in \{0, 1\}$, the above query algorithms correctly answer $rank_c(B_t, i)$ and $select_c(B_t, i)$ in $O(\frac{\log k}{\log \log k})$ time and $O(\frac{\log n \log k}{\log \log k})$ time, respectively.*

Proof. We prove the correctness of answering $rank_c(B_t, i)$ by induction on the version t. When $t=0$, since $count_c(0, i) = count_{1-c}(0, i) = 0$, we have $rank_c(B_t, i) = rank_c(B_0, i) + count_c(0, i) - count_{1-c}(0, i) = rank_c(B_0, i)$. Assume that for some version $t \geq 1$, $rank_c(B_{t-1}, i)$ can be correctly answered, i.e., $rank_c(B_{t-1}, i) = rank_c(B_0, i) + count_c(t-1, i) - count_{1-c}(t-1, i)$. Recall that for a partially persistent index, an update on B_{t-1} (i.e., version $t-1$ of B) is a single bit-flip on B_{t-1}, which creates the bit vector B_t. There are three cases: **(1)** If a bit in position $[i+1, n]$ is flipped, $rank_c$ does not change. Since $count_c$ and $count_{1-c}$ remain the same, $rank_c(B_t, i) = rank_c(B_{t-1}, i)$. **(2)** If a bit in position $[1, i]$ is flipped from $1-c$ to c, then $rank_c$ is increased by 1. We have the point (t, i) in grid G_c, so $count_c(t, i) = count_c(t-1, i) + 1$, while $count_{1-c}$ is unchanged. Thus, $rank_c(B_t, i) = rank_c(B_{t-1}, i) + 1$. **(3)** If a bit in position $[1, i]$ is flipped from c to $1-c$, then $rank_c$ is decreased by 1. The point (t, i) is in grid G_{1-c}, so $count_{1-c}(t, i) = count_{1-c}(t-1, i) + 1$ while $count_c$ is unchanged. Thus, $rank_c(B_t, i) = rank_c(B_{t-1}, i) - 1$. Therefore, $rank_c(B_t, i)$ is correctly answered for all version t. It takes $O(1)$ time to obtain $rank_c(B_0, i)$ and $O(\frac{\log k}{\log \log k})$ time to obtain both $count_c(t, i)$ and $count_{1-c}(t, i)$. The total time is $O(\frac{\log k}{\log \log k})$.

The correctness of $select_c(B_t, i)$ follows from its definition. The binary search makes at most $O(\log n)$ queries on $rank_c(B_t, j)$ for $j \in [1, n]$, and each takes $O(\frac{\log k}{\log \log k})$ time, which implies the stated time complexity.   □

2.2   Data Structure and Algorithm for Online Updates

We now consider online updates and show Theorem 2. For online updates, we will define a new query $access(B_t, i)$ that returns bit i in B_t. Similarly to Section 2.1, we divide the compressed partially persistent index into two components. The first component is the succinct data structure for the initial bit vector B_0 given in Lemma 1. The second component stores the update for each version, utilizing a dynamic data structure for the planar range counting problem.

Specifically, for an online update on bit i of B_{t-1} that creates B_t, we need to add the point (t, i) to one of the grids G_0 and G_1 in an online fashion. Nekrich [11] has presented data structures for the *dynamic* planar range counting problem, where points can be added to or removed from the grid dynamically.

Lemma 4. *[11] Let Z points lie on an $N \times N$ grid. Planar range counting queries can be supported in $O((\frac{\log Z}{\log \log Z})^2)$ time, and updates in $O(\log^{4+\epsilon} Z)$ time, for any constant $\epsilon > 0$, using $O(Z \log N)$ bits.*

In the case of online updates, the maximum version number, denoted by K, is not fixed. We can set K to some constant and double it, whenever the current version number k is equal to K. In this way, K is always at most $2k$, and thus a version number can be represented in $O(\log k)$ bits. Similarly to Section 2.1, we define two grids G_0 and G_1 of size $\max(K, n) \times \max(K, n)$, such that an update on bit i of B_{t-1} from 1 to 0 (that gives B_t) corresponds to the point (t, i) on grid G_0 (similarly, on G_1 for 0-to-1 bit-flips). Here, to determine if an update is a bit-flip from 0 to 1 or vice versa, we need to call $access(B_{t-1}, i)$.

We maintain two data structures of Lemma 4 for the grids G_0 and G_1, respectively, which occupy $2 \cdot O(k \log(\max(K, n))) = O(k \log(kn))$ bits of space in total. Thus, the total space of both components is the space stated in Theorem 2.

Query Algorithm. For any version t, position $1 \le i \le n$ and bit $c \in \{0, 1\}$, we answer $rank_c(B_t, i)$ and $select_c(B_t, i)$ in the same way as in Section 2.1. We answer $access(B_t, i)$, as follows. First, we obtain $access(B_0, i)$, which is the value of $B_0[i]$, from the succinct data structure for B_0. We make four planar range counting queries on grids G_0 and G_1 to obtain $count_0(t, i-1)$, $count_0(t, i)$, $count_1(t, i-1)$ and $count_1(t, i)$. Then we report $access(B_t, i)$ to be

$$access(B_0, i) + (count_1(t, i) - count_1(t, i-1)) - (count_0(t, i) - count_0(t, i-1)).$$

The correctness of the rank and select queries follows directly from Lemma 3. Their time complexities are blown up by a factor of $O(\frac{\log k}{\log \log k})$, because we use the data structure of Lemma 4, instead of that of Lemma 2. Thus, the following lemma suffices to complete the proof of Theorem 2.

Lemma 5. *For any version t and bit position $1 \le i \le n$, the above query algorithm correctly answers $access(B_t, i)$ in $O((\frac{\log k}{\log \log k})^2)$ time. Furthermore, an update at the latest version takes $O(\log^{4+\epsilon} k)$ time, for any constant $\epsilon > 0$.*

Proof. Note that $count_1(t, i) - count_1(t, i-1)$ is the number of times bit i is flipped from 0 to 1 up to version t, while $count_0(t, i) - count_0(t, i-1)$ is the number of times bit i is flipped from 1 to 0 up to version t. Therefore, their difference is equal to the change of bit i from B_0 to B_t, and the correctness of $access(B_t, i)$ follows. The access query involves a call to $access(B_0, i)$ that takes $O(1)$ time, and four planar range counting queries that take $O((\frac{\log k}{\log \log k})^2)$ time, which implies the time complexity stated in Theorem 2.

For an online update on bit i of B_{t-1} that creates B_t, we need a query on $access(B_{t-1}, i)$ to determine which of grid G_0 or G_1 to add the point (t, i) to. This takes $O((\frac{\log k}{\log \log k})^2)$ time. By Lemma 4, adding the point (t, i) to a grid takes $O(\log^{4+\epsilon} k)$ time. Thus, each update takes $O(\log^{4+\epsilon} k)$ time in total.   □

3   Compressed Fully Persistent Index for Offline Updates

This section considers offline updates and presents a compressed fully persistent index. Let k be the number of updates and let n be the size of the bit vector B.

Theorem 3. *There is a compressed fully persistent index for offline updates that occupies $nH_0(B_0)+o(n)+O(k\log(kn))$ bits, and supports at any version, rank queries in $O(\frac{\log^2 k}{\log\log k})$ time and select queries in $O(\frac{\log n\log^2 k}{\log\log k})$ time.*

The fully persistent index allows updates to any version. A version B_t is created by flipping a single bit in B_p for some $p<t$. Let T be the version tree.

Centroid Path Decomposition. We decompose the version tree T using centroid path decomposition (see, e.g., [2]), as follows. For any internal node u, let v be the child of u with the largest number of leaves in its subtree (ties are broken arbitrarily). We refer to edge uv as a *core edge*, and to non-core edges as *side edges*. A *centroid path* C is a maximal path connecting consecutive core edges. The root of C, denoted by $r(C)$, is the top-most node of C. We denote by $\Delta(T)$, the set of all centroid paths in T. The following property is well-known.

Property 1. Let T be a tree of k nodes with a centroid path decomposition. The path from the root of T to any node v traverses at most $\log k$ centroid paths.

Data Structure and Algorithm. The compressed fully persistent index consists of three components. The first component is the succinct data structure for the initial bit vector B_0 given in Lemma 1. The second component stores the version tree T and three pieces of auxiliary information for each node in T. In particular, for each version v, we maintain the version number $p(v)$ of its parent. We also assign a *node label* $\ell(v)$ from 1 to k to each node v in ascending order of their depth, such that the node labels are strictly increasing along the path from the root of T to any node v of T. Finally, every node v is in some centroid path C, and we define $f(v)$ to be the root $r(C)$ of C. We can store $p(v)$, $\ell(v)$ and $f(v)$ in three arrays of size k, which allows $O(1)$ time access, given the version number v. In total, the second component takes $O(k\log(kn))$ bits of space.

The third component stores the information of the update for each version, using the data structure for the planar range counting problem of Lemma 2, as follows. For each centroid path C, we define two grids $G_0(C)$ and $G_1(C)$ of size $\max(k,n)\times\max(k,n)$. Consider each update on a version p that creates a version $t>p$, where $t\in C$. If the update flips bit i of B_p from 1 to 0, there is a point $(\ell(t),i)$ on grid $G_0(C)$ (similarly, on $G_1(C)$ for 0-to-1 bit-flips). For each centroid path C, we maintain two structures of Lemma 2 for the grids $G_0(C)$ and $G_1(C)$, respectively. These data structures are associated with the node $r(C)$. For all centroid paths, this takes $2\cdot\sum_{C\in\Delta(T)}O(|C|\log(\max(k,n)))=2\cdot O(k\log(\max(k,n)))=O(k\log(kn))$ bits of space. Thus, the total space of all components is the space stated in Theorem 3.

Query Algorithm. Consider any version t, position $1\leq i\leq n$ and bit $c\in\{0,1\}$. We answer the query on $select_c(B_t,i)$ by using $rank_c$ in the same way as in

Section 2.1. We now give the query algorithm for answering $rank_c(B_t, i)$. Let $count_c(C, t, i)$ be the number of points in the range $[0, t] \times [0, i]$ of $G_c(C)$.

- $rank_c(B_t, i)$: First, we obtain $rank_c(B_0, i)$ from the succinct data structure for B_0. Then, we consider all updates along the path from the root of T to version t. Let $U = (u_0 = 0, u_1, u_2, \ldots, u_{x-1}, u_x = t)$ be the path that contains x versions. Suppose U traverses y centroid paths in the order of $C_1, C_2, \ldots, C_y$. The roots of all these y centroid paths must be in U; we denote them by $u_{z(1)}, u_{z(2)}, \ldots, u_{z(y)}$. Note that $u_{z(1)} = u_0 = 0$.
  Let $count'_c = \sum_{j=1}^{y-1} count_c(C_j, \ell(u_{z(j+1)-1}), i) + count_c(C_y, \ell(t), i)$, and let $count'_{1-c} = \sum_{j=1}^{y-1} count_{1-c}(C_j, \ell(u_{z(j+1)-1}), i) + count_{1-c}(C_y, \ell(t), i)$. We compute them as follows. Since $t = u_x$ is in C_y, the root of C_y is $u_{z(y)} = f(t)$. Since $u_{z(y)-1}$ is in C_{y-1}, the root of C_{y-1} is $u_{z(y-1)} = f(u_{z(y)-1})$. We repeat the above to identify the roots of all the y centroid paths, and make $2y$ range counting queries on grids $G_c(C_j)$ and $G_{1-c}(C_j)$ for $1 \le j \le y$, respectively, to compute the counts. Finally, $rank_c(B_t, i) = rank_c(B_0, i) + count'_c - count'_{1-c}$.

To establish Theorem 3, it suffices to prove the correctness and time complexity for the rank query, since for select they follow similarly to Lemma 3.

Lemma 6. *For any version t, position $1 \le i \le n$ and $c \in \{0, 1\}$, the above query algorithm correctly answers $rank_c(B_t, i)$ in $O(\frac{\log^2 k}{\log \log k})$ time.*

Proof. It suffices to prove that counter $count'_c$ (resp. $count'_{1-c}$) counts correctly the updates along the path U that flip the bits in position $[1, i]$ from $1 - c$ to c (resp. from c to $1 - c$). Since each such flip contributes 1 (resp. -1) to $rank_c(B_t, i)$, $rank_c(B_t, i) = rank_c(B_0, i) + count'_c - count'_{1-c}$ will follow.

Recall that $count'_c = \sum_{j=1}^{y-1} count_c(C_j, \ell(u_{z(j+1)-1}), i) + count_c(C_y, \ell(t), i)$. For convenience, we set $z(y+1) = t+1$. We focus on the path $U_j = (u_{z(j)}, u_{z(j)+1}, \ldots, u_{z(j+1)-1})$ for some $1 \le j \le y$. Then $U_j \subseteq C_j$. By the definition of node labels, we have that $\ell(u_{z(j)}) < \ell(u_{z(j)+1}) < \cdots < \ell(u_{z(j+1)-1})$ and all other nodes in C_j have a node label larger than $\ell(u_{z(j+1)-1})$. Thus, on the grid $G_c(C_j)$, $count_c(C_j, \ell(u_{z(j+1)-1}), i)$ correctly counts the updates along the path U_j that flip bits in position $[1, i]$ from $1 - c$ to c. Summing over all j, it follows that $count'_c = \sum_{j=1}^{y} count_c(C_j, \ell(u_{z(j+1)-1}), i)$ correctly counts such bit flips made by the updates along the path U. The correctness of $count'_{1-c}$ follows similarly.

Regarding time complexity, it takes $y \cdot O(1) = O(y)$ time to identify the roots of the y centroid paths. By Lemma 2, it takes $2y \cdot O(\frac{\log k}{\log \log k}) = O(y \cdot \frac{\log k}{\log \log k})$ time for the $2y$ range counting queries. By Property 1, the number of centroid paths in U is $y = O(\log k)$. Therefore, the total time complexity is $O(\frac{\log^2 k}{\log \log k})$. □

4   Compressed Fully Persistent Index for Online Updates

This section considers online updates and presents a compressed fully persistent index. Let k be the number of updates and let n be the size of the bit vector B.

Theorem 4. *There is a compressed fully persistent index for online updates that occupies $nH_0(B_0)+o(n)+O(k\log n\log(kn))$ bits, and supports at any version, rank, select and access queries in $O(\log^3 n)$ worst case time. An update at any version takes $O(\log^2 n\log\log n)$ amortized time.*

Overview. To show Theorem 4, we first present the *range sum tree*, a simplification of the *range min-max tree* of Sadakane and Navarro [14], that supports rank, select and bit-flip in $O(\log n)$ time. Then we make it fully persistent using a generic method from [1], which is designed for the I/O model and can be applied to the word-RAM model with a modest blow-up on time.

Range Sum Tree. The range sum tree is a balanced binary tree T, where each node corresponds to a range $[i, j]$ of B and it stores i, j and a value $e(i, j)$ that represents the number of 1's in $B[i, j]$. We divide the bit vector B into segments of length $L=\log^2 n$ and each leaf of T corresponds to the range of a segment. Let $[i_z, j_z]$ be the range of a node z. An internal node z with left child u and right child v has the range $[i_u, j_v]$ and $e(i_z, j_z) = e(i_u, j_u)+e(i_v, j_v)$. Therefore, the number of nodes in T is $O(\frac{n}{\log^2 n})$ and each node needs $O(\log n)$ bits, which sums up to $O(\frac{n}{\log n})$ bits of space.

Each leaf node also stores the bits in $B[i, j]$ for the query, as follows. We further divide the length-L segment into $2\log n$ sub-segments of length $t=\frac{\log n}{2}$. A leaf node has $2\log n$ extra fields, each representing a sub-segment succinctly [13]: Each sub-segment with x bits belongs to a class x of t-bitmaps. E.g., if $t=2$, class 0 is $\{00\}$, class 1 is $\{01, 10\}$ and class 2 is $\{11\}$. As class x contains $\binom{t}{x}$ elements, we can use $\lceil\log\binom{t}{x}\rceil$ bits and $\lceil\log(t+1)\rceil$ bits respectively to represent its element index within the class and the class identifier. As shown in [13], all sub-segments take at most $nH_0(B)+O(\frac{n}{\log n})$ bits of space.

Let $P_{x,y}$ be the length-t sub-segment represented by element y of class x. We maintain three universal tables U_{rank}, $U_{select,0}$ and $U_{select,1}$ for each class, such that given class x, element index y and an integer $0\leq i\leq t$, $U_{rank}(P_{x,y}, i)$ returns the number of 1's in $P_{x,y}[1, i]$; and $U_{select,0}(P_{x,y}, i)$ (resp. $U_{select,1}(P_{x,y}, i)$) returns the smallest index j such that $P_{x,y}[1, j]$ contains i 1's (resp. i 0's). These tables need $3\cdot O(2^t\cdot t\cdot\log t) = O(\sqrt{n}\log n\log\log n) = o(n)$ bits. Thus, the range sum tree takes $nH_0(B)+2\cdot O(\frac{n}{\log n})+o(n) = nH_0(B)+o(n)$ bits in total.

Query Algorithm. We traverse T to answer a query on any bit position $1\leq i\leq n$. Initially, we set z to be the root of T. Note that $[i_z, j_z]=[1, n]$. We traverse T depending on whether z is an internal node or leaf node as follows.

- *$rank_1(B, i)$*: We count the number of 1's in $[1, i]$ using a counter $count_1$ initiated to 0. **(1)** z is an internal node with left child u and right child v: If $i\in[i_v, j_v]$, we add $e(i_u, j_u)$ (i.e., the number of 1's in $[i_u, j_u]$) to $count_1$, and set $z=v$. If $i\in[i_u, j_u]$, we set $z=u$. Then we repeat this procedure. **(2)** z is a leaf node: Let $(S_1, S_2, ..., S_{2\log n})$ be the sub-segments of z. Suppose position i is in S_j. We make j queries to the universal tables U_{rank} to determine the number of 1's in $S_1, S_2, ..., S_{j-1}$ and S_j up to position i and add them to $count_1$. Finally, we return $count_1$ that is clearly the number of 1's in $[1, i]$.

- $select_1(B, i)$: We find the i-th 1-bit using a variable j (initiated to i) as follows. **(1)** z is an internal node with left child u and right child v: If $j > e(i_u, j_u)$, the i-th 1-bit is not in $[i_u, j_u]$. We decrease j by $e(i_u, j_u)$ and set $z = v$. If $j \leq e(i_u, j_u)$, the i-th 1-bit is in $[i_u, j_u]$ and we set $z = u$. Then we repeat this procedure. **(2)** z is a leaf node: If $j > e(i_z, j_z)$, the i-th 1-bit does not exist and we simply return $select_1(B, i) = 0$. Otherwise, let $(S_1, S_2, \ldots, S_{2\log n})$ be the sub-segments of z. We make at most $2\log n$ queries to the universal table $U_{rank}(S_\ell, t)$, where $t = \frac{\log n}{2}$, from $\ell = 1, 2, \ldots$, until we find an x, such that $\sum_{\ell=1}^{x} U_{rank}(S_\ell, t) \geq j$.[2] Then the i-th 1-bit is in S_x. We make a query on $U_{select,1}(S_x, j - \sum_{\ell=1}^{x-1} U_{rank}(S_\ell, t))$ to obtain the position of B's i-th 1-bit in S_x. We return $select_1(B, i) = i_z + (x-1) \cdot (\frac{\log n}{2}) - 1 + U_{select,1}(S_x, j - \sum_{\ell=1}^{x-1} U_{rank}(S_\ell, t))$.

Note that the number of 0's in a range $[i, j]$ is equal to $(j - i + 1) - e(i, j)$. Thus, we can answer $rank_0(B, i)$ and $select_0(B, i)$ in a similar way, where for $select_0(B, i)$ we query $U_{select,0}$ instead of $U_{select,1}$. To answer $access(B, i)$, we traverse the path from the root to the leaf z containing $B[i]$, and identify the sub-segment S_x in z that contains $B[i]$, as described for $rank_1(B, i)$. Let $B[i]$ be bit j of S_x. We obtain $B[i] = U_{rank}(S_x, j) - U_{rank}(S_x, j-1)$ with two queries on U_{rank}.

Regarding query time, each query takes $O(1)$ time for an internal node, and $O(\log n)$ time for a leaf node, since we make $O(\log n)$ queries on universal tables, where each takes $O(1)$ time. Since a path contains $O(\log(\frac{n}{\log^2 n})) = O(\log n)$ internal nodes and a leaf node, a query takes $O(\log n \cdot 1 + \log n) = O(\log n)$ time.

Updating Bit i. To update bit i, we first make a query on $access(B, i)$ to locate the leaf node z that contains $B[i]$. We update the sub-segment with $B[i]$ to a new sub-segment in $O(\log n)$ time, since the sub-segment is of length $t = \frac{\log n}{2}$. Then, we update each node u on the path from the root to z, as follows. If the update flips $B[i]$ from 0 to 1, we increase $e(i_u, j_u)$ by 1; otherwise, we decrease it by 1. The update time is $O(\log(\frac{n}{\log^2 n})) = O(\log n)$.

Lemma 7. *The range sum tree for a length-n bit vector B is a balanced search tree, where each internal node contains $O(1)$ fields and each leaf node contains $O(\log n)$ fields. It occupies $nH_0(B) + o(n)$ bits and supports access, rank and select queries and updating a bit of B in $O(\log n)$ time by accessing $O(\log n)$ internal nodes and a leaf node (and for update, modifying a field in each of them).*

Fully Persistent Range Sum Tree. We apply on the range sum tree T the following result of [1] for the I/O model with disk block size of $\mathcal{B}$ words.

Lemma 8. *[1] Let T be a pointer-based ephemeral data structure that supports queries in $O(q)$ worst case I/Os and where updates make $O(u)$ modifications to T in the worst case. Given that every node of T occupies at most $O(1)$ blocks and has $O(1)$ maximum in-degree, T can be made fully persistent such that a query to a particular version is supported in $O(q)$ worst case I/Os, and an update to any*

---

[2] Such an x exists, because $j \leq e(i_z, j_z) = \sum_{\ell=1}^{2\log n} U_{rank}(S_\ell, t)$.

version is supported in $O(u \log \mathcal{B})$ amortized I/Os. After performing a sequence of k updates, the fully persistent structure occupies $O(u \cdot \frac{k}{\mathcal{B}})$ blocks of space.

In the scheme of [1] for the above lemma, each I/O can be simulated by $O(\mathcal{B})$ RAM operations, so that the time complexity in the word-RAM model is $O(\mathcal{B})$ times that in the I/O model. We set the block size $\mathcal{B} = \log n$, such that a block contains $\log n$ words and each (internal or leaf) node of T occupies $O(1)$ blocks. Since all algorithms are implemented only by top-down traversals of T, the tree can be implemented such that each node has in-degree 1. We set $q = O(\log n)$, since accessing an internal node takes $O(1)$ time and a leaf node takes $O(\log n)$ time. By Lemma 7, the rank, select and access queries access $O(\log n)$ nodes and thus take $O(\log n \cdot q \cdot \mathcal{B}) = O(\log^3 n)$ time. We set $u = O(\log n)$, since by Lemma 7, an update makes $O(\log n)$ modifications to T. Thus, the update time is $O(u \cdot \mathcal{B} \cdot \log \mathcal{B}) = O(\log^2 n \log \log n)$ amortized. The fully persistent structure occupies $O(\log n \cdot \frac{k}{\mathcal{B}})$ blocks $= O(\log n \cdot k)$ words $= O(\log n \cdot k \cdot \log(kn))$ bits, since the word size is $O(\log(kn))$. This gives Theorem 4.

5   Lower Bound

In this section, we show that even for offline updates, a partially persistent index for a length-n bit vector B that occupies $O((n+k) \log(kn))$ bits, where k is the number of updates, must answer the rank query at any version in $\omega(1)$ time.

Our proof is based on a reduction of the problem of *planar dominance counting*, which is defined as follows: on a grid $[1, N] \times [1, N]$ with N points, a dominance counting query (x, y) asks for the number of points in a given range $[0, x] \times [0, y]$. Pătraşcu [12] has shown that any static data structure of size $O(N)$ words must take $\Omega(\frac{\lg N}{\lg \lg N})$ time to answer a dominance counting query.

Theorem 5. *Let B be a length-n bit vector, where k offline bit-flip updates have been performed. A partially persistent index for B that occupies $O((n+k) \log(kn))$ bits of space must answer the rank query at any version in $\omega(1)$ time.*

Proof. Suppose for the sake of contradiction, the partially persistent index, denoted by I, can answer the rank query at any version in $O(1)$ time. We show how to use I in combination with y-fast tries [16] to answer a dominance counting query on a grid $G = [1, n] \times [1, n]$ with n points in $O(\log \log n)$ time, using only $O(n)$ words of space. This contradicts the lower bound of $\Omega(\frac{\lg n}{\lg \lg n})$ time for dominance counting queries [12] and thus proves the theorem.

Based on the n points on G, we construct a bit vector $B[1..n]$ and n offline bit-flip updates, as follows. All n bits in B are initially 0, which is the initial version B_0 of B. For each point (i, j) on G, suppose that among all the n points, i is the p-th smallest x-coordinate and j is the q-th smallest y-coordinate, where ties are broken arbitrarily. We construct an update operation that flips the p-bit of B_{q-1} from 0 to 1 to create version B_q.

We maintain the partially persistent index I for B that uses $O(2n \log n^2) = O(n \log n)$ bits, i.e., $O(n)$ words. In addition, we maintain a y-fast trie for all the distinct x-coordinates X and another y-fast trie for all the distinct y-coordinates Y. These two y-fast tries occupy $O(|X|+|Y|)=O(2n)=O(n)$ words and allow us, given a dominance counting query (a, b), to determine in $O(\log \log n)$ time the predecessor $pred_X(a)$ of a in X (i.e., the largest element $c \in X$ such that $c \leq a$) and the predecessor $pred_Y(b)$ of b in Y. For each element $c \in X$ (resp. $c \in Y$), we also store the number $r_X(c)$ (resp. $r_Y(c)$) of points on G whose x- (resp. y-) coordinates are at most c. This requires $O(n)$ words.

To answer the dominance counting query (a, b), it is not hard to see that we can ask I for the rank of $r_X(pred_X(a))$ in version $r_Y(pred_Y(b))$ of B. The query time is $O(\log \log n)$ and the space is $O(n)$ words, completing the proof.   □

6   Conclusion

In this paper we presented the first efficient compressed persistent indices for *bit* vectors that support temporal rank/select queries and *independent* bit-flip updates. Extending our results to handle general alphabets and/or correlated updates (that exhibit a smaller information-theoretic space lower bound) may find important applications in computational biology [10] and other fields. We leave as open the problem of designing a compressed fully persistent index for online updates. The rest of our structures can be improved by use of *succinct* (static or dynamic) data structures for planar range counting.

References

1. Brodal, G.S., Sioutas, S., Tsakalidis, K., Tsichlas, K.: Fully persistent B-trees. In: Proc. SODA, pp. 602–614 (2012)
2. Cole, R., Gottlieb, L.A., Lewenstein, M.: Dictionary matching and indexing with errors and don't cares. In: Proc. STOC, pp. 91–100 (2004)
3. Dietz, P.F.: Fully Persistent arrays. In: Dehne, F., Santoro, N., Sack, J.-R. (eds.) WADS 1989. LNCS, vol. 382, pp. 67–74. Springer, Heidelberg (1989)
4. Driscoll, J.R., Sarnak, N., Sleator, D.D., Tarjan, R.E.: Making data structures persistent. J. Comput. Syst. Sci. 38(1), 86–124 (1989)
5. Ferragina, P., Manzini, G.: Opportunistic data structures with applications. In: Proc. FOCS, pp. 390–398 (2000)
6. Grossi, R., Gupta, A., Vitter, J.S.: High-order entropy-compressed text indexes. In: Proc. SODA, pp. 841–850 (2003)
7. JáJá, J., Mortensen, C.W., Shi, Q.: Space-efficient and fast algorithms for multidimensional dominance reporting and counting. In: Fleischer, R., Trippen, G. (eds.) ISAAC 2004. LNCS, vol. 3341, pp. 558–568. Springer, Heidelberg (2004)
8. Kaplan, H.: Persistent data structures. In: Handbook on Data Structures and Applications, ch. 31, pp. 31-1–31-26. CRC Press (2004)
9. Kopelowitz, T.: On-line indexing for general alphabets via predecessor queries on subsets of an ordered list. In: Proc. FOCS, pp. 283–292 (2012)
10. Mäkinen, V., Navarro, G., Sirén, J., Välimäki, N.: Storage and retrieval of highly repetitive sequence collections. J. Comp. Biology 17(3), 281–308 (2010)
11. Nekrich, Y.: Orthogonal range searching in linear and almost-linear space. Comput. Geom. 42(4), 342–351 (2009)

12. Pătraşcu, M.: Lower bounds for 2-dimensional range counting. In: Proc. STOC, pp. 40–46 (2007)
13. Raman, R., Raman, V., Satti, S.R.: Succinct indexable dictionaries with applications to encoding k-ary trees, prefix sums and multisets. ACM Transactions on Algorithms 3(4), 43 (2007)
14. Sadakane, K., Navarro, G.: Fully-functional succinct trees. In: Proc. SODA, pp. 134–149 (2010)
15. The 1000 Genomes Project Consortium. A map of human genome variation from population-scale sequencing. Nature 467(7319), 1061–1073 (2010)
16. Willard, D.E.: Log-logarithmic worst-case range queries are possible in space $\Theta(n)$. Information Processing Letters 17(2), 81–84 (1983)

Tight Bounds for Low Dimensional Star Stencils in the External Memory Model

Philipp Hupp and Riko Jacob

Institute of Theoretical Computer Science, ETH Zürich, Zürich, Switzerland
philipp.hupp@inf.ethz.ch, rjacob@inf.ethz.ch

Abstract. Stencil computations on low dimensional grids are kernels of many scientific applications including finite difference methods used to solve partial differential equations. On typical modern computer architectures such stencil computations are limited by the performance of the memory subsystem, namely by the bandwidth between main memory and the cache. This work considers the computation of star stencils, like the 5-point and 7-point stencil, in the external memory model. The analysis focuses on the constant of the leading term of the non-compulsory I/Os. Optimizing stencil computations is an active field of research, but so far, there has been a significant gap between the lower bounds and the performance of the algorithms. In two dimensions, matching constants for lower and upper bounds are provided closing a gap of 4. In three dimensions, the bounds match up to a factor of $\sqrt{2}$ improving the known results by a factor of $2\sqrt{3}\sqrt{B}$, where B is the block (cache line) size of the external memory model. For higher dimensions n, the presented lower bounds improve the previously known by a factor between 4 and 6 leaving a gap of $\sqrt[n-1]{n!} \approx \frac{n}{e}$.

Keywords: Hierarchical Memories, Lower Bounds, High Performance Computing, Isoperimetric Inequalities, Non-compulsory I/Os, Capacity Cache Misses.

1 Introduction

Solving Partial Differential Equations (PDEs) is one of the most common tasks in scientific computing. A standard way to discretize low dimensional Euclidean spaces for these computations are regular grids. Applying a finite difference method on this discretization turns a differential operator into a linear function of a grid point and its neighbors. Such a linear function is also called stencil and results in a very regular sparse system of linear equations. To make use of the sparsity, such systems are typically solved with iterative solvers like the Jacobi or Gauss-Seidel method. The kernel of these methods is the evaluation of the underlying stencil, making it the most performance critical component of the computation.

Stencil operations are performed on grids for which each vertex possesses a value. The task is to recompute the values at the vertices of the grid according to

F. Dehne, R. Solis-Oba, and J.-R. Sack (Eds.): WADS 2013, LNCS 8037, pp. 415–426, 2013.

the stencil. Hereby the stencil states which neighboring vertices of a grid point are necessary to update the grid point. To clarify how stencils are used to solve PDEs we give a simple example. Consider the one-dimensional heat equation which describes the variation of temperature on a pole over time. For a function $u(t, x)$ describing the temperature of the pole at time t and position x, this problem can formally be written as the PDE $\frac{\partial u}{\partial t} = \frac{\partial^2 u}{\partial x^2}$. We approximate the pole by a one-dimensional grid and use an explicit finite difference method to calculate the temperature of the grid points at time $t + 1$ given the temperature at time t. In this setting, the PDE can be approximated by $\frac{u(t+1, x)-u(t,x)}{\Delta t} = \frac{u(t, x-1)-2u(t, x)+u(t, x+1)}{(\Delta x)^2}$. Abbreviating $c := \frac{\Delta t}{(\Delta x)^2}$, this solves to $u(t + 1, x) = c \cdot u(t, x - 1) + (1 - 2c) \cdot u(t, x) + c \cdot u(t, x + 1)$ which in turn gives rise to the one-dimensional 1-star stencil. Another well known example is the linear approximation of the Laplacian on a regular two dimensional grid as given by $\Delta u(x, y) \doteq \frac{1}{h^2}[u((x-h), y)+u((x+h), y)+u(x, y-h)+u(x, y+h)-4u(x, y)]$. This defines the so called 5-point or 1-star stencil.

Stencil operations are not only easy in the sense that the floating point operations are predetermined. The majority of the I/O operations is already needed for reading the input and writing the output. In fact, many simple algorithms for the 5-point stencil are within a factor of 5 of this lower bound and the classical asymptotic analysis is too coarse to give interesting insights. I/O operations related to the initial read of the input and the final write of the output are called *compulsory* I/Os or *cold* cache misses. All other I/Os are called *non-compulsory* I/Os or *capacity* misses (because they are unnecessary for sufficiently large main memory M). The analysis of stencil operations carried out in this paper gives almost matching bounds in the sense that it focuses on the constant of the leading term in the asymptotics of the non-compulsory I/Os.

1.1   Problem Definition

The *computational model* we consider is an I/O Model similar to [11] and [1]. There are two levels of memory, an *external memory* of infinite size on which all data is stored initially, and an *internal memory* of size M to which the data has to be loaded to perform computations. The external memory is organized in *blocks* of size B. An I/O operation is the transfer of one block of data from external to internal memory (read) or from internal to external memory (write). We classify the I/Os into *compulsory I/Os* (cold misses), which account for the first access to a block and writing the final output, and *non-compulsory I/Os* (capacity misses). Non-compulsory I/Os are due to the limited size of the internal memory.

We further assume that all I/Os are simple, i.e. data elements are moved instead of copied between internal and external memory. While this facilitates the derivation of our bounds, this assumption is not crucial and matching bounds assuming simple I/Os translate to matching bounds using non-simple I/Os as we discuss in Sect. 1.4.

Let $[k]$ abbreviate $\{0, \ldots, k-1\}$ and $[k_1] \times \ldots \times [k_n]$ denote the n-dimensional *grid* and $\mathbb{Z}_{k_1} \times \ldots \times \mathbb{Z}_{k_n}$ the n-dimensional *torus* of side lengths k_i. Denote by $\|\cdot\|_1$ the ℓ^1-norm which is defined as usual for the grid and for an element $v \in \mathbb{Z}_{k_1} \times \ldots \times \mathbb{Z}_{k_n}$ of the torus it is given by $\|v\|_1 = \sum_{i=1}^{n} \min\{(-v_i \mod k_i), (v_i \mod k_i)\}$ (assuming $(v_i \mod k_i) \in \{0, \ldots, k_i - 1\}$).

The problem of evaluating a stencil on the grid (torus) is formally defined as follows: Denote by V the vertices of the grid (torus), by $V_{in} := V \times \{in\}$ the input layer and by $V_{out} := V \times \{out\}$ the output layer of the stencil computation. The input of the computation are values $f(v_{in})$ for all vertices $v_{in} \in V_{in}$ of the grid (torus). The function which maps the values of V_{in} to V_{out} is described by a stencil. The task is to evaluate the stencil for all points of the output layer, i.e. to compute all values of the output grid (torus) and to write these results to external memory. We consider so called *star stencils*. Denote by $v_{in} \in V_{in}$ and $v_{out} \in V_{out}$ corresponding vertices of the input and output layer, i.e. the first n coordinates of these vertices of the n-dimensional grid (torus) are identical. The s-star stencil S_s for a vertex $v_{out} \in V_{out}$ is defined as all vertices within distance s from v_{in}, $S_s(v_{out}) := \{w \in V_{in} : \|w - v_{in}\|_1 \le s\}$. We connect the input layer to the output layer by adding edges (w, v_{out}) for all vertices $w \in S_s(v_{out})$. Doing this for all vertices in $v_{out} \in V_{out}$ gives the edge set of the computation graph, $E := \{(w, v_{out}) \in V_{in} \times V_{out} : \|w - v_{in}\|_1 \le s\}$. The computation graph is then $(V_{in} \dot\cup V_{out}, E)$.

We consider computing the value for one grid point $v_{out} \in V_{out}$ as an *atomic* operation, i.e. all input required to compute $f(v_{out})$, namely $S_s(v_{out})$, needs to reside in internal memory to do the calculation and partial computations are not allowed. When we later argue about the stencil computations, the distinction between input and output layer is less strict. We say to evaluate a vertex v of the grid (torus) V when we compute the stencil for v_{out} and have the input $S_s(v_{out}) \subset V_{in}$ in internal memory.

The 1-star stencils are the most common stencils. Since upper (lower) complexity bounds for the s-star stencil in the I/O model induce upper (lower) bounds for all stencils which are subsets (supersets) of the s-star stencil meaningful choices also include $s = 2$ and $s = 3$. So, for the asymptotic notation, we assume throughout the paper that s is a small constant.

1.2 Results

This work examines the leading term of the non-compulsory I/Os of the s-point stencil. In two dimensions, matching lower and upper bounds are given closing a multiplicative gap of 4. In three dimensions the provided bounds match up to a factor of $\sqrt{2}$ improving the known results by a factor of $2\sqrt{3}\sqrt{B}$. For dimensions n bigger than three, the lower bounds are improved between a factor of 4 and 6 leaving a gap of $\sqrt[n-1]{n!} \approx \frac{n}{e}$ for higher dimensions n.

We assume that the grid sides are ordered, $k_1 \ge \ldots \ge k_n$, and significantly larger than the internal memory, namely $k_1, k_2 \ge 2nM + M + 1$ and $k_i \ge 2nM + 1$ for $i \in \{3, \ldots, n\}$. The asymptotics are considered for $k_i \to \infty$, $M \to \infty$ and $B \to \infty$ while assuming $\frac{k_n}{M} \to \infty$ and $\frac{M}{B} \to \infty$. Denote by $C_s(k_1, \ldots, k_n)$ the number of simple I/Os to evaluate the s-point stencil on $[k_1] \times \ldots \times [k_n]$. Then the following holds (assuming n and s are constant):

$$C_s(k_1, k_2) = \left(2 + 4\frac{s^2}{M} \cdot \left\{ \begin{array}{l} 1 + \mathcal{O}\left(\frac{B}{M} + \frac{M}{k_1}\right) \\ 1 - \mathcal{O}\left(\frac{1}{M} + \frac{M}{k_2}\right) \end{array} \right\} \right) \cdot \frac{k_1 k_2}{B}$$

$$C_s(k_1, k_2, k_3) = \left(2 + \frac{8}{\sqrt{3}} \cdot \frac{s^{3/2}}{\sqrt{M}} \cdot \left\{ \begin{array}{l} \sqrt{2} + \mathcal{O}\left(\sqrt{\frac{B}{M}}\right) \\ 1 - \mathcal{O}\left(\frac{1}{\sqrt{M}} + \frac{\sqrt{M}}{k_3}\right) \end{array} \right\} \right) \cdot \frac{k_1 k_2 k_3}{B}$$

$$C_s(k_1, \ldots, k_n) =$$

$$= \left(2 + 4 \cdot 2^{1/(n-1)} \cdot (n-1) \cdot \sqrt[n-1]{\frac{s^n}{M}} \cdot \left\{ \begin{array}{l} 1 \quad + \mathcal{O}\left(\sqrt[n-1]{\frac{B}{M}}\right) \\ \frac{1}{\sqrt[n-1]{n!}} - \mathcal{O}\left(\frac{1}{\sqrt[n-1]{M}} + \frac{\sqrt[n-1]{M}}{k_n}\right) \end{array} \right\} \right) \cdot \frac{\prod_{i=1}^{n} k_i}{B} \ .$$

The bounds consist of three parts. The first part is the constant 2 accounting for the compulsory I/Os. The second part is the leading term of the non-compulsory I/Os on which this work focuses. The third part characterizes lower order terms that we do not explore further.

Both lower bounds and algorithms can be transferred to parallel external memory (PEM) as introduced in [2], as long as the number P of processors is smaller than $\frac{1}{M} \prod_{i=1}^{n-1} k_i$. In this case, the complexities are reduced by a factor of P. Unlike with classical computational complexity (i.e. on a PRAM), there cannot be a general simulation of a parallel algorithm on a single processor that is only P times slower (it is possible to make use of the combined internal memory of size $P \cdot M$). Still, the lower bounds in this paper work in the parallel setting just as well: One round of the parallel computations, as defined by a certain number of non-compulsory I/Os, cannot evaluate more stencils than its serial counterpart. Regarding the algorithms, as the external memory is assumed to be large enough so that we do not need to work in-place, all evaluations of stencils are independent from each other and could in principle be done in parallel. For moderate P we use the proposed serial algorithms and merely split the computation into P contiguous parts. The only additional non-compulsory I/Os are used to initially fill the local memory. Assuming $P \le \frac{1}{M} \prod_{i=1}^{n-1} k_i$, this is a lower order term, namely the one that we analyze as the difference between the torus and the grid.

1.3 Related Work

The external memory or I/O model was introduced by Hong and Kung [11] for $B = 1$. Using essentially an isoperimetric argument they apply it to problems like the Fast-Fourier-Transform (FFT), matrix-matrix multiplication and products of graphs. The latter yields the first bounds for the number of I/Os for directed grid graphs: $\Theta\left(\frac{1}{\sqrt[n]{M}} \cdot \prod_{i=1}^{n} k_i\right)$. Being directed, these graphs and the notion of

Table 1. Comparison of the bounds for the leading term of the non-compulsory I/Os for the 1-star stencil. All to be multiplied with the number of grid points $\prod_{i=1}^{n} k_i$.

	Presented Result	Frumkin and Wijngaart	Leopold
Lower Bound 2D	$\frac{4}{BM}$	$\frac{8}{9}\frac{1}{BM}$	$\frac{2}{BM}$
Lower Bound 3D	$\frac{8}{\sqrt{3}}\frac{1}{B\sqrt{M}}$	$\frac{2}{\sqrt{3}}\frac{1}{B\sqrt{M}}$	$\frac{2}{B\sqrt{M}}$
Low. Bnd. Arbitrary D	$\frac{4\cdot 2^{1/(n-1)}\cdot(n-1)}{\sqrt[n]{n!}}\frac{1}{B\,^{n-1}\!\sqrt{M}}$	$\left(\frac{2}{3}\right)^{\frac{n}{n-1}}\frac{n}{\sqrt[n-1]{(n-1)!}}\frac{1}{B\,^{n-1}\!\sqrt{M}}$	n.a.
Upper Bound 2D	$\frac{4}{BM}$	$\mathcal{O}\left(\frac{1}{M}\right)$	$\frac{8}{BM}$
Upper Bound 3D	$\frac{8\sqrt{2}}{\sqrt{3}}\frac{1}{B\sqrt{M}}$	$\mathcal{O}\left(\frac{1}{\sqrt{M}}\right)$	$\frac{4\sqrt{6}}{\sqrt{B}\sqrt{M}}$
Upp. Bnd. Arbitrary D	$4\cdot 2^{\frac{1}{n-1}}(n-1)\frac{1}{B\,^{n-1}\!\sqrt{M}}$	$\mathcal{O}\left(\frac{1}{\sqrt[n-1]{M}}\right)$	n.a.

boundary on them differs significantly from our setup. Aggarwal and Vitter [1] generalized Hong and Kungs model to arbitrary B. Irony et al. [13] extend Hong and Kungs lower bound for matrix-matrix multiplication to a distributed memory setup and Ballard et al. [3] generalize these results to various linear algebra algorithms like factorization and eigenvalue algorithms. They also derive lower bounds for Strassen like algorithms [4] by relating the I/O complexity to expansion properties of the computation graphs.

The I/O complexity of the 1-star stencil has been discussed further independently by Frumkin and Wijngaart [10] and Leopold [15,14,16]. The different results for the leading term of the non-compulsory I/Os are given in Table 1 and have to be multiplied by the number of vertices $\prod_{i=1}^{n} k_i$. Frumkin and Wijngaart consider arbitrary dimensions but focus on the asymptotic behavior of the non-compulsory I/Os. The lower bound uses an isoperimetric argument similar to the one presented in this article but does not exploit its full strength. We improve these results by a factor between 4 and 6. The upper bound focuses on the asymptotic behavior and is an existence results. Leopold focuses on the two and three dimensional cases. Her lower bounds exploit a weak isoperimetric result [15,16] which we improve by a factor of 2 and $\frac{4}{\sqrt{3}}$ for two respective three dimensions. The upper bounds discuss row and column layouts. By using a data layout suited for our algorithms we decrease the upper bounds by $\frac{1}{2}$ and $\frac{2}{3\sqrt{B}}$ for two and three dimensions. Leopold also discusses two spatial and one temporal dimension [14], which is out of the scope of this paper (also see Sect. 1.4).

There is vast ongoing research about optimizing stencil computations, mostly in two and three dimensions, on modern computer architectures. This research focuses on improving the I/O behavior of the algorithms. When implementing one may need to be careful about the tradeoff between a more complicated data layout making a sophisticated padding scheme necessary and (theoretically) optimal I/O behavior. Addressing these problems is out of the scope of this work. However, diagonal hyperspace cuts, similar to the ones proven optimal in this work, are often employed in empirical work to select suitable substructures for computation. The literature includes work on compiler optimization [17]. In the cache oblivious model asymptotic upper bounds are derived [8] which are then shown to be achieved [9,18]. A recent survey of the field is [7].

1.4 Discussion: Upper Bounds and Real World Programs

This paper focuses on the lower bounds and their derivation. The details of the upper bounds can be found in the full version of this article [12], and are here merely stated to show that the bounds are tight in the two dimensional case, differ by a factor of $\sqrt{2}$ for three dimensions and $\sqrt[n-1]{n!} \approx \frac{n}{e}$ for higher dimensions.

All the upper bounds have in common that a *sweep shape* is moved through the grid in unit shifts in a *sweep order* resulting in *working bands* (see [12] for the details). To achieve good results two things are important: First, the data layout has to reproduce the shape of the sweep shapes. Second, to evaluate all vertices of the grid, the working bands have to overlap, dividing the working band into *core* and *wing bands*. For optimal asymptotic behavior, vertices in different bands have to be saved in separate blocks. In two dimensions, for example, the lower bound suggests to work in adjacent ℓ^1 balls where neighboring balls form a diagonal working band. For $B = 1$ or a data layout supporting such a data access, the diagonal sweep through the data provides a matching upper bound, but it is not optimal for many other data layouts.

Although memory access is very important for high performance code, it is not the only factor influencing the runtime. It needs to be determined in a progress of algorithms engineering if and to which extent the benefits from on optimized data layout and access lead to faster code. Other options that may influence runtime include the more complicated index computations, optimizing for several layers of the memory hierarchy, vectorization and loop unrolling enabling scalar replacement. In particular in the parallel setting modifications to the algorithm may prove useful to optimize the communication and synchronization required between different processes.

As we want to give precise bounds we need a theoretical model that provides enough details to prove these bounds. The theoretical model chosen restricts the results to atomic stencil operations, simple I/Os and allows to work not-in-place.

Some implementations, for example, evaluate the stencil partially which requires a more general lower bound while the upper bounds still apply. Although we do not present a proof, we think the assumption that stencil operations are atomic can be dropped without weakening the lower bounds. Given a set of vertices which we want to evaluate in one round of the algorithm, the isoperimetric inequalities yield how many grid points need to be transferred to (or have already been transferred from) other rounds. This does not assume that the stencil is indivisible but only states that neighboring values are needed to evaluate the stencil. Reducing the number of vertices that need to be transferred from one round to another would mean to compress the data which has to be disallowed for the I/O model to make sense.

Also, the assumption that all I/Os are simple, i.e. data elements are moved instead of copied between external and internal memory, is not crucial and matching bounds assuming simple I/Os translate to matching bounds using non-simple I/Os. The key observation is that for simple I/Os a non-compulsory read of an item corresponds to exactly one non-compulsory write (writing the same item back to external memory beforehand). Hence, dropping the assumption that I/Os are simple reduces the number of non-compulsory I/Os by a factor of 2 for both the lower and the upper bounds.

Having to work in in-place, however, requires a modification of the upper bounds in the parallel and serial case. Vertices belonging to wing bands shared by several core bands need to be buffered to be available at a later time. This buffering requires additional memory such that we would not work completely in-place. As working in-place is more restrictive, the lower bounds carry over.

In this paper we do not consider a time step which introduces a directed dimension and hence changes the structure of the computation graph, the stencil defining the neighborhood of a set and hence the isoperimetric sets and inequalities. However, in a setting with time step the number of computations is multiplied by the number of time steps whereas the number of compulsory I/Os solely depends on the spatial dimensions. This implies that an isoperimetric argument, as presented in this paper, would analyze the constant of the leading instead of the second order term. Whereas it seems difficult to transfer the lower bounds to a time step setting this should be easier for the upper bounds. The structure of the two and three dimensional algorithms is compatible with the setting of one temporal and one respectively two spatial dimensions. For higher dimension the algorithms need to be altered to allow for parallelism.

For three and higher dimensions both lower and upper bounds do not seem optimal and it remains open if the complexity can be pinpointed. It would also be interesting to examine the I/O complexity of stencils different from the star stencils given by ℓ^1 balls. Canonical candidates are stencils described by ℓ^∞ balls and mixtures between ℓ^1 and ℓ^∞ stencils appearing in finite element methods. Finally, the lower bounds have not been tuned to account for different data layouts as this would also change the isoperimetric inequalities. However, while this further restricts the theoretical model it may be a key aspect to get matching lower and upper bounds for different layouts.

2 The Lower Bounds

The lower bound is derived by splitting an arbitrary algorithm into rounds of an equal number of non-compulsory I/Os. The work which can be done in each of these rounds is then bounded by an isoperimetric inequality. This yields the minimum number of rounds which have to be performed by any algorithm. Multiplying this with the number of non-compulsory I/Os that define a round yields the lower bound. The lower bound is first deduced assuming that an I/O operation accesses one element ($B = 1$) and is then generalized for arbitrary B.

2.1   The Isoperimetric Inequality

The isoperimetric inequality states how many vertices can be enclosed by a fixed number of boundary vertices. The optimal sets in this sense are called isoperimetric sets and, as proven by Bollobás and Leader [6], the isoperimetric sets in $\mathbb{Z}_k^n$ are (fractional) ℓ^1 balls.[1] To state this result precisely we introduce some notation, mainly from [6]:

A fractional system or simply system f is a function from $\mathbb{Z}_k^n$ or $\mathbb{Z}^n$ to the unit interval $[0, 1]$. For $f : \mathbb{Z}^n \to [0, 1]$ the function can take non-zero values only for a finite number of grid points. The weight w of a system f is $w(f) = \sum_{x \in \mathbb{Z}_k^n} f(x)$ or $w(f) = \sum_{x \in \mathbb{Z}^n} f(x)$ according to the domain of f. A fractional system f on $\mathbb{Z}_k^n$ or $\mathbb{Z}^n$ is therefore a generalization of a subset S of $\mathbb{Z}_k^n$ or $\mathbb{Z}^n$ respectively. If a fractional systems f takes just the values 0 and 1, then f is naturally identified with the set $S = f^{-1}(1)$ and the weight $w(f)$ is the cardinality of S. We define the *inner core* Δf of f by

$$\Delta f(x) = \begin{cases} 0, & f(x) < 1 \\ \min_{||x-y||_1 = 1}\{f(y)\}, & f(x) = 1 \end{cases}$$

and the *inner-s-core* by applying the operator repeatedly, $\Delta_s f = \underbrace{\Delta \ldots \Delta}_{s \text{ times}} f$.

This is now used to define the *inner-s-boundary* by $\Gamma_s f(x) = f(x) - \Delta_s f(x)$. The fractional ℓ^1 ball $b_y^{(r,\,\alpha)}$ of radius $r \in \mathbb{N}_0$, $0 \leq r \leq \frac{k}{2}$, surplus $\alpha \in (0, 1)$ and center $y \in \mathbb{Z}_k^n$ is defined as

$$b_y^{(r,\,\alpha)}(x) := \begin{cases} 1, & ||x-y||_1 \leq r \\ \alpha, & ||x-y||_1 = r+1 \\ 0, & ||x-y||_1 > r+1 \end{cases}$$

For $0 \leq v \leq k^n$ we also use the notation b_y^v which describes the unique ball of weight v and center y. For the isoperimetric inequalities the centers of the balls are irrelevant and hence we omit the subscript y when it is not needed. Bollobás and Leader [6] proved that balls have the smallest closure of all systems of the same weight. We need a version of this result which allows us to bound the number of interior vertices given the number of inner-boundary vertices.

Theorem 1 (The boundary bounds the core on $\mathbb{Z}^n$ – [12]).
Let $s \in \mathbb{N}$ and f be a fractional system on $\mathbb{Z}^n$. For $v \in \mathbb{R}_0^+$ the following holds:

$$(w(\Gamma_{2s}f) \leq w(\Gamma_{2s}b^v)) \Rightarrow (w(\Delta_s f) \leq w(\Delta_s b^v)) . \tag{1}$$

We conclude this section by giving the asymptotic expansion for the number of vertices of a ball and its inner-boundary with respect to the radius r in $\mathbb{Z}^n$. As long as the sides of the torus or grid are big enough, $k \geq 2(r+1)$, the formulas

---

[1] It is known that the isoperimetric sets in the continuous domains $\mathbb{R}^n$ are ℓ^2 balls.

apply there also. Note that all lower order terms have positive coefficients. The formulas are derived in [12].

$$w\left(b^{(r,0)}\right) = \frac{2^n}{n!} \cdot r^n + \mathcal{O}\left(r^{n-1}\right) \text{ and } w\left(\Gamma_1 b^{(r,0)}\right) = \frac{2^n}{(n-1)!} \cdot r^{n-1} + \mathcal{O}\left(r^{n-2}\right) . \quad (2)$$

2.2 Pathwidth

We employ pathwidth [5] to ensure that we are working on the "inside" of the torus and can treat it like the infinite grid which allows to use the isoperimetric results. Please refer to [12] for the details as only the results used are stated.

- An algorithm evaluating the s-star stencil on G without non-compulsory I/Os and internal memory of size M implies that $pathwidth(G) \leq M - 1$.
- The two dimensional grid $[k_1] \times [k_2]$ has pathwidth $\min\{k_1, k_2\}$. Hence there have to be non-compulsory I/Os if we want to evaluate the s-star stencil on a two dimensional grid or torus with $\min\{k_1, k_2\} \geq M$.
- If the subgraph H of a two dimensional grid or torus consists of $p+1$ complete rows and complete columns, then $pathwidth(H) \geq p$.

2.3 Splitting into Rounds and Deducing the Lower Bound

To derive the lower bounds assume an arbitrary algorithm evaluating the s-star stencil on $\mathbb{Z}_{k_1} \times \ldots \times \mathbb{Z}_{k_n}$ is given. When $\min\{k_1, k_2\} \geq M$ the pathwidth of $\mathbb{Z}_{k_1} \times \ldots \times \mathbb{Z}_{k_n}$ is at least M and hence the algorithm causes non-compulsory I/O operations. We can count these operations and split the algorithm into rounds of c non-compulsory I/Os. c denotes the round length and hence all rounds except the last one cause c non-compulsory I/Os. This approach is similiar to the idea presented by Hong and Kung [11] and therefore we call the rounds Hong-Kung rounds.

To apply the isoperimetric inequality we need to establish a link between the inner-core, the inner-boundary and the rounds. Choose one of the Hong-Kung rounds and denote with S the set of vertices which are in internal memory at some point of this round.

Let *Transfer(S)* be the *transfer vertices* of S, i.e. vertices which are also present in internal memory during other rounds. Precisely, a vertex is a transfer vertex if at least one of four cases applies:

- The vertex is transferred from the previous to the current round by residing in internal memory at the beginning of the current round.
- The vertex has been written back to external memory in a preceding round and is read again in the current round.
- The vertex is written from internal to external memory in the current round to be read again in a subsequent round.
- The vertex is transferred from the current to the proceeding round by residing in internal memory at the end of the current round.

We denote further $Eval(S)$ the *evaluated vertices* which are all vertices of S for which the s-point stencil is evaluated in the current round. The following two observations relate these sets to the inner-core and the inner-boundary:

$$\Gamma_{2s}(S) \subset Transfer(S) \qquad \text{and} \qquad Eval(S) \subset \Delta_s(S) \ . \tag{3}$$

A vertex can only be evaluated in a round if all its neighbors within distance s are in S as well. $\Delta_s(S)$ consists of exactly these vertices. Equivalently $\Gamma_s(S)$ are the vertices which cannot be evaluated in round S. Take any $x \in \Gamma_s(S)$. All vertices which are within distance s from x need to be in the round in which x is evaluated. Hence they need to be transferred. The set of all vertices of S within distance s from any of the vertices of $\Gamma_s(S)$ is $\Gamma_{2s}(S)$. Therefore these vertices are a subset of the transfer vertices.

Furthermore, we can give an upper bound for the number of transfer vertices of a round. At the beginning and at the end of a round there are at most M vertices in internal memory. Together these account for at most $2M$ transfer vertices. The only other way a vertex can be a transfer vertex is that it has been rewritten to external memory in a previous round and is reloaded in the current round or rewritten to external memory in the current round to be reloaded in a subsequent round. So either the reload or write of the vertex causes a non-compulsory I/O. Since there are at most c non-compulsory I/Os per round, the total number of transfer vertices is at most $2M + c$,

$$w(Transfer(S)) \leq 2M + c \ . \tag{4}$$

By definition, the vertices of $S \setminus Transfer(S)$ do not cause non-compulsory I/Os and they are in internal memory only in one single round. Hence their s-star stencil has to be computed in the current round and their pathwidth is limited by

$$pathwidth\left(S \setminus Transfer(S)\right) \leq M - 1 \ . \tag{5}$$

Assuming $k_1, k_2 \geq 2M + c + (M + 1)$ and $k_i \geq 2M + c + 1$ for $i \in \{3, \ldots, n\}$, we know by (4) that the vertices of (at least) $M + 1$ hyperplanes of normal x_1, $M + 1$ hyperplanes of normal x_2 and one hyperplane of normal x_i ($3 \leq i \leq n$) do not belong to $Transfer(S)$. These hyperplanes form a connected component in $\mathbb{Z}_{k_1} \times \cdots \times \mathbb{Z}_{k_n}$. So they could either be a subset of $S \setminus Transfer(S)$ or disjoint from S. Taking the union of all hyperplanes of normal x_1 and normal x_2 and intersecting them with all other hyperplanes results in a subset of a two dimensional torus of at least $M + 1$ complete rows and columns which has $pathwidth(S \setminus Transfer(S)) \geq M$ by Sect. 2.2. By (5) such a set cannot be a subset of $S \setminus Transfer(S)$. Therefore, at least one hyperplane for each normal direction x_i ($1 \leq i \leq n$) is disjoint from S. Deleting these hyperplanes allows to embed S in the infinite grid $\mathbb{Z}^n$.

Combining (3) and (4) we get $w(\Gamma_{2s}S) \leq 2M + c$. Denote with v_0 the weight such that $w(\Gamma_{2s}b^{v_0}) = 2M + c$. Using the assumption that s is small and constant we simplify this equation before solving. Denote (r_0, α_0) the radius and surplus such that $b^{v_0} = b^{(r_0, \alpha_0)}$. Using (2), the asymptotic expansion of $w(\Gamma_{2s}b^{v_0})$ is given by

$$w(\Gamma_{2s}b^{v_0}) = \sum_{i=0}^{2s-1} \frac{2^n \cdot (r_0 - i)^{n-1}}{(n-1)!} + \mathcal{O}\left(r_0^{n-2}\right) = \frac{2s \cdot 2^n}{(n-1)!}(r_0 - 2s)^{n-1} + \mathcal{O}\left(r_0^{n-2}\right). \quad (6)$$

Since all coefficients in the lower order terms are non-negative, dropping the lower order terms before solving (6) increases r_0 and v_0, increases $w(\Delta_s b^{v_0})$ and hence weakens the lower bound as we will see in the lower bound formula (8). Solving (6) without the lower order terms yields

$$r_0 = \sqrt[n-1]{(n-1)! \frac{2M+c}{2s \cdot 2^n}} + 2s \ . \quad (7)$$

To determine r_0 we fix the round length c such that it gives the best lower bound. Combining a version of the isoperimetric inequality [12] and (3) bounds the number of evaluable vertices of S by $w(Eval(S)) \le w(\Delta_s b^{v_0})$. Therefore, a lower bound is given by

$$\frac{c}{w(\Delta_s b^{v_0})} \cdot \prod_{i=1}^{n} k_i \ . \quad (8)$$

The best round length c is hence chosen by plugging (7) into (8) and maximizing over c (setting the derivative to 0 and checking that the solution is a maximum). Disregarding lower order terms it is approximately $c = 2(n-1) \cdot M$. Using this round length in (7), we determine the upper bound $r_0 = \sqrt[n-1]{\frac{n!}{2^n} \frac{M}{s}} + 2s$ for the radius of a ball to be handled in one round. Finally, by plugging this radius into (8), the lower bound reads

$$\frac{2(n-1)M}{w\left(\Delta_s b^{(r_0,0)}\right)} \cdot \prod_{i=1}^{n} k_i = \left(4(n-1) \sqrt[n-1]{\frac{2}{n!}} s^n \cdot \frac{1}{\sqrt[n-1]{M}} - \mathcal{O}\left(\frac{1}{\sqrt[n-1]{M^2}}\right)\right) \cdot \prod_{i=1}^{n} k_i \ .$$

This bound was derived on the torus $\mathbb{Z}_{k_1} \times \cdots \times \mathbb{Z}_{k_n}$ and we can apply it to the grid $[k_1] \times \cdots \times [k_n]$ using a reduction.

Lemma 1. *Any algorithm using internal memory of size M and evaluating the s-point stencil on the grid $[k_1] \times \cdots \times [k_n]$ induces an algorithm, using internal memory M and evaluating the s-point stencil, on the torus $\mathbb{Z}_{k_1} \times \cdots \times \mathbb{Z}_{k_n}$ causing at most $\mathcal{O}\left(\prod_{i=1}^{n-1} k_i\right)$ additional I/Os.*

Proof. When the algorithm for the grid is evaluated on the torus, only the vertices close the boundary of the grid have to be treated differently. If a vertex is within ℓ^1 distance $s-1$ in a unit direction from a bounding hyperplane, at most half of the points of the s-point stencil, corresponding to that unit direction, have to be read and written additionally for this vertex. Altogether these are at most $\frac{b^{(s,0)}}{2} \cdot 2n \cdot s \cdot \prod_{i=1}^{n-1} k_i = \mathcal{O}\left(\prod_{i=1}^{n-1} k_i\right)$ I/Os. $\square$

Furthermore, the lower bound can be generalized to arbitrary B by the simple observation that one I/O operation affects at most B elements. Hence for the grid the total number of I/Os (including compulsory ones) is

$$\left(2 + \frac{4(n-1)\sqrt[n-1]{\frac{2}{n!}s^n}}{\sqrt[n-1]{M}} - \mathcal{O}\left(\frac{1}{\sqrt[n-1]{M^2}} + \frac{1}{k_n} \right) \right) \frac{\prod_{i=1}^{n} k_i}{B} .$$

References

1. Aggarwal, A., Vitter, J.S.: The input/output complexity of sorting and related problems. Commun. ACM 31(9), 1116–1127 (1988)
2. Arge, L., Goodrich, M.T., Nelson, M., Sitchinava, N.: Fundamental parallel algorithms for private-cache chip multiprocessors. In: Proc. of SPAA 2008. ACM (2008)
3. Ballard, G., Demmel, J., Holtz, O., Schwartz, O.: Minimizing communication in numerical linear algebra. SIAM J. Matrix Analysis Appl. 32(3), 866–901 (2011)
4. Ballard, G., Demmel, J., Holtz, O., Schwartz, O.: Graph expansion and communication costs of fast matrix multiplication. J. ACM 59(6), 32 (2012)
5. Bodlaender, H.L.: A partial k-arboretum of graphs with bounded treewidth. J. Algorithms, 1–16 (1998)
6. Bollobás, B., Leader, I.: An isoperimetric inequality on the discrete torus. SIAM J. Discret. Math. 3, 32–37 (1990)
7. Datta, K., Kamil, S., Williams, S., Oliker, L., Shalf, J., Yelick, K.: Optimization and performance modeling of stencil computations on modern microprocessors. SIAM Rev. 51(1), 129–159 (2009)
8. Frigo, M., Strumpen, V.: Cache oblivious stencil computations. In: Proc. of 19th Annual ICS 2005, ICS 2005, pp. 361–366. ACM (2005)
9. Frigo, M., Strumpen, V.: The memory behavior of cache oblivious stencil computations. J. Supercomput. 39(2), 93–112 (2007)
10. Frumkin, M.A., Van der Wijngaart, R.F.: Tight bounds on cache use for stencil operations on rectangular grids. J. ACM 49, 434–453 (2002)
11. Hong, J.-W., Kung, H.T.: I/O complexity: The red-blue pebble game. In: Proceedings of STOC 1981, pp. 326–333. ACM, New York (1981)
12. Hupp, P., Jacob, R.: Tight bounds for low dimensional star stencils in the external memory model. CoRR, abs/1205.0606 (2012)
13. Irony, D., Toledo, S., Tiskin, A.: Communication lower bounds for distributed-memory matrix multiplication. J. Parallel Distrib. Comput. 64(9), 1017–1026 (2004)
14. Leopold, C.: An analytical evaluation of tiling for stencil codes with time loop. In: Proc. of the 16th IPDPS. IEEE Computer Society (2002)
15. Leopold, C.: On optimal locality of linear relaxation. In: Proc. Int. Symp. on Parallel and Distributed Computing and Network, IASTED, pp. 201–206 (2002)
16. Leopold, C.: Tight bounds on capacity misses for 3D stencil codes. In: Sloot, P.M.A., Tan, C.J.K., Dongarra, J., Hoekstra, A.G. (eds.) ICCS-ComputSci 2002, Part I. LNCS, vol. 2329, pp. 843–852. Springer, Heidelberg (2002)
17. Tang, Y., Chowdhury, R.A., Kuszmaul, B.C., Luk, C.-K., Leiserson, C.E.: The pochoir stencil compiler. In: Proceedings of SPAA 2011, pp. 117–128. ACM (2011)
18. Zeiser, T., Wellein, G., Nitsure, A., Iglberger, K., Rüde, U., Hager, G.: Introducing a parallel cache oblivious blocking approach for the lattice Boltzmann method. Progress in Computational Fluid Dynamics 8(1-4), 179–188 (2008)

Neighborhood-Preserving Mapping between Trees*

Jan Baumbach[1,3], Jiong Guo[2], and Rashid Ibragimov[1]

[1] Max Planck Institute für Informatik, Saarbrücken 66123, Germany
ribragim@mpi-inf.mpg.de
[2] Universität des Saarlandes, Campus E 1.7, Saarbrücken 66123, Germany
jguo@mmci.uni-saarland.de
[3] University of Southern Denmark, Campusvej 5, 5230 Odense M, Denmark
jan.baumbach@imada.sdu.dk

Abstract. We introduce a variation of the graph isomorphism problem, where, given two graphs $G_1 = (V_1, E_1)$ and $G_2 = (V_2, E_2)$ and three integers l, d, and k, we seek for a set $D \subseteq V_1$ and a one-to-one mapping $f : V_1 \to V_2$ such that $|D| \leq k$ and for every vertex $v \in V_1 \setminus D$ and every vertex $u \in N_{G_1}^l(v) \setminus D$ we have $f(u) \in N_{G_2}^d(f(v))$. Here, for a graph G and a vertex v, we use $N_G^i(v)$ to denote the set of vertices which have distance at most i to v in G. We call this problem NEIGHBORHOOD-PRESERVING MAPPING (NPM). The main result of this paper is a complete dichotomy of the classical complexity of NPM on trees with respect to different values of l, d, k. Additionally, we present two dynamic programming algorithms for the case that one of the input trees is a path.

Keywords: tree edit distance, graph algorithms, complexity, graph matching.

1 Introduction

Applications of the graph isomorphism problem, which seeks for a one-to-one mapping between the vertices of two graphs, given certain constrains, can be found in many fields, for example bioinformatics, pattern recognition, computer vision [1, 2, 3].

A class of graph isomorphism problems was formulated and studied on trees [4, 5, 6, 7]. The quality of the mapping between trees and general graphs usually can be expressed by the minimal cost of edit operations (like deletions and insertions of vertices or edges, needed to make one graph equal to the other).

Motivated by Protein-Protein Interaction Networks (PPINs) [1], we introduce a new isomorphism problem, called NEIGHBORHOOD-PRESERVING MAPPING (NPM). Here, we ask if there is a mapping between two graphs, such that the neighborhoods of the vertices of the first graph, except for few vertices, are preserved in the second graph. More precisely, given two graphs $G_1 = (V_1, E_1)$

---

* Partially supported by the DFG Cluster of Excellence MMCI and the International Max Planck Research School.

F. Dehne, R. Solis-Oba, and J.-R. Sack (Eds.): WADS 2013, LNCS 8037, pp. 427–438, 2013.
© Springer-Verlag Berlin Heidelberg 2013

Table 1. Summary of the cases for NPM on trees with $k = 0$

	$d = 1$	$d = 2$	$d \geq 3$
$l = 1$	P (Thm. 4)	NPC (Thm. 1)	NPC (Thm. 1)
$l = 2$	P (Thm. 5)	P (Thm. 6)	NPC (Thm. 1)
$l = 3$	P (Thm. 5)	P (Thm. 7)	NPC (Thm. 2)
$l \geq 4$	P (Thm. 5)	P (Thm. 7)	NPC (Thm. 3)

and $G_2 = (V_2, E_2)$ and three integers l, d, k, NPM asks for a set $D \subseteq V_1$ and an one-to-one mapping $f : V_1 \to V_2$ such that $|D| \leq k$ and for every vertex $v \in V_1 \setminus D$ and every vertex $u \in N_{G_1}^l(v) \setminus D$ it holds $f(u) \in N_{G_2}^d(f(v))$. Hereby, $N_G^i(v)$ denotes the set of vertices which have distance at most i to v in the graph G. The set D is called the *isolation set*.

Similar problem can be formulated for mappings between Protein-Protein Interaction Networks (PPI *network alignment*). Building a neighborhood-preserving mapping, in contrast to the classic subisomorphism problem, provides more freedom by setting closeness constraints on the sought mapping. This freedom may help to deal with data incompleteness (missing edge or nodes) as well as noise (erroneous edges or nodes), respecting at the same time topological distance. Then, a mapping with the larger number of more important mapped nodes is thought to be more biologically meaningful.

In the paper we focus on the classical complexity of NPM on trees, that is both input graphs are trees. We first briefly introduce main definitions used in the paper. Then we study NPM on trees with $k = 0$ and provide proofs for NP-hard and polynomial-time solvable cases. Table 1 summarizes our findings. Next we investigate the problem when $k > 0$ and prove that NPM with $k > 0$ is NP-hard for all values of l and d. We complete the paper with presenting two algorithms for NPM on trees when one of the input trees is restricted to be a path.

Preliminaries. Throughout this paper, we consider only simple, undirected graphs without self-loop. Given a graph G, we use $V(G)$ and $E(G)$ to denote the vertex and edge sets of G, respectively. The direct neighborhood of a vertex v in a graph G, denoted by $N_G(v)$, is the set of vertices which are adjacent to v. The *degree* of v is $|N_G(v)|$. We use $N_G[v]$ to denote $N_G(v) \cup \{v\}$. We call the set of vertices, which have distance at most i to v, the i-neighborhood of v for integer $i > 1$, denoted by $N_G^i(v)$.

2 NPM on Trees with $k = 0$

In this section, we provide a dichotomy of the classical complexity of NEIGHBORHOOD-PRESERVING MAPPING on trees with $k = 0$, see Table 1 for an overview.

2.1 NP-Hardness Results

First, we show that if $k = 0$ and $1 \leq l < d$, then NPM on trees is NP-hard.

Theorem 1. NEIGHBORHOOD-PRESERVING MAPPING *on trees is NP-complete with $k = 0$, $d > 1$, and $l < d$.*

Proof. Clearly, NPM is in NP. To show the hardness of this case, we reduce from the NP-hard 3-PARTITION problem [8]:

> **Input**: An integer B and a set of $3n$ non-negative integers $S = \{a_1, \ldots, a_{3n}\}$
> **Question**: Can S be partitioned into n disjoint sets $S_1, S_2, \ldots, S_n$ such that for every $1 \leq i \leq n$ we have $|S_i| = 3$ and $\sum_{a \in S_i} a = B$?

3-PARTITION remains NP-hard even if $B/4 < a_i < B/2$ for every $1 \leq i \leq 3n$ [8]. We first describe the reduction for NPM on trees with $k = 0$, $d = 2$, and $l = 1$ and then indicate how to extend the reduction for other cases with $1 \leq l < d$.

Given an instance (B, S) of 3-PARTITION, we construct an instance of NPM on trees: the first tree T_1 consists of $3n$ paths and a star of $2(n+1)B+n+2$ vertices (with the center c). Each path P_i one-to-one corresponds to an element a_i of S and contains a_i vertices. One end-vertex of each path is made adjacent to a leaf of the star, denoted by r, by adding an edge. In order to construct the second tree T_2, we first create a big star with $2(n+1)B+1$ vertices, n small stars, each having $B + 1$ vertices, and a special vertex r'. Then, we add edges to connect the centers of all small stars to r'. Finally, we add an edge between r' and an arbitrary leaf, denoted by c', of the big star.

To prove the equivalence between the instances, observe that the vertex c in T_1 has to be mapped to c' in T_2, since this is the only possible way to preserve the $2(n+1)B+n+1$ neighbors of c in T_2. Thus, all vertices of the big star in T_2, except for c', are mapped to the neighbors of c. The remaining $n + 1$ neighbors of c, including r, have to be mapped to r' and the centers of n small stars in T_2. Moreover, the vertex r has to be mapped to r', since r is adjacent to the end-vertices of all $3n$ paths in T_1. This implies that the vertices of the $3n$ paths have to be mapped to the leaves of the small stars in T_2. Finally, to preserve the neighborhoods of the vertices on these paths, we have to group the paths of T_1 into n groups, each of which contains 3 paths with exactly B vertices in total, corresponding to 3-partition of the integers.

By subdividing the edges of T_1 and T_2, we can extend the reduction for other cases of d and l with $l < d$. □

For the case of $k = 0$ and $l \geq d \geq 3$ we proved the following theorems:

Theorem 2. NEIGHBORHOOD-PRESERVING MAPPING *on trees is NP-complete with $k = 0$ and $l = d \geq 3$.*

Theorem 3. NEIGHBORHOOD-PRESERVING MAPPING *is NP-complete with $k = 0$ and $l > d \geq 3$.*

2.2 Polynomial-Time Solvable Cases

In this subsection we present polynomial-time solvable cases of NPM on trees with $k = 0$. Obviously, with $k = 0$ and $l = d = 1$, NEIGHBORHOOD-PRESERVING

MAPPING (NPM) on trees is equivalent to the classical tree isomorphism problem and it is well-known that the isomorphism problem of trees can be solved in polynomial time [9].

Theorem 4. NEIGHBORHOOD-PRESERVING MAPPING *on trees can be solved in polynomial time for the case $k = 0$ and $l = d = 1$.*

Theorem 5. NEIGHBORHOOD-PRESERVING MAPPING *on trees can be solved in polynomial time for the case $k = 0$ and $l > d = 1$.*

Next we present an algorithm for the case of NPM on trees when $l = 2$, $d = 2$, $k = 0$. We first present some conditions for vertices, under which a neighborhood-preserving mapping can exist. Assume that $|V_1| = |V_2| \geq 3$. We let leaves(T) to denote the set of leaves of the tree T.

In the following we consider only the case where the diameter of T_2 is at least 4. If T_2 has a diameter of 2, then T_2 is a star and thus, an arbitrary mapping from T_1 to T_2 is a solution. For the case that T_2 has a diameter 3, there is a path in T_2 with 4 vertices a, b, c, d. Clearly, all other vertices are leaves adjacent to b or c. Observe that the diameter of T_1 should be at least 3, since otherwise, the given instance has no solution. Moreover, we cannot map two non-adjacent vertices u and v to b and c, since, otherwise, the neighborhoods of the vertices on the path between u and v cannot be preserved. Further, we cannot map a leaf u of T_1 to b or c, since otherwise, say mapping u to b, the whole T_1 has to be mapped to the star centered at b. Thus, if there exists a neighborhood-preserving mapping f, then we have two adjacent, non-leaf vertices v, u with $f(v) = b$ and $f(u) = c$. Clearly, there cannot be two neighbors of v such that one is mapped to a leaf adjacent to b and one to a leaf adjacent to c. So they are either all mapped to the leaves adjacent to b or all mapped to the leaves adjacent to c. Then, we can simply compare the numbers of leaves adjacent to b and to c with $|T_1(u)|$ and $|T_1(v)|$ to decide whether a neighborhood-preserving mapping exists. Here, $T_1(u)$ and $T_1(v)$ denote the subtrees, that result by deleting (u, v) from T_1 and contain u and v, respectively. This is clearly doable in polynomial time.

Some observations. In the following we present some observations which are crucial for the correctness of the algorithm.

Lemma 1. *Let $u, v \in V_1$ with $(u, v) \in E_1$. Suppose that there is a neighborhood-preserving mapping f with $(f(u), f(v)) \notin E_2$. Let a be the vertex in T_2 with $(f(u), a) \in E_2$ and $(f(v), a) \in E_2$. Then, it holds that for every vertex $z \in N_{T_1}(u) \cup N_{T_1}(v)$, $f(z) \in N_{T_2}[a]$.*

Lemma 2. *Let $u, v \in V_1$ with $(u, v) \in E_1$. Suppose that the diameter of T_2 is at least 4. Then, a neighborhood-preserving mapping f with $(f(u), f(v)) \notin E_2$ exists, if and only if both $f(u)$ and $f(v)$ are in leaves(T_2).*

Lemma 3. *Let v be a leaf of T_1 with u being its only neighbor. Suppose that the diameter of T_2 is at least 4. If there is a neighborhood-preserving mapping f with $f(u) \notin$ leaves(T_2), then $f(v) \in$ leaves(T_2).*

Lemma 4. *Suppose that the diameter of T_2 is at least 4. If a neighborhood-preserving mapping f exists with $f(u) \notin$ leaves(T_2) for $u \in V_1$, then for every $x \in N_{T_1}(u)$, we have $f(x) \in N_{T_2}(f(u))$.*

Lemma 5. *For $u, v \in V_1$ with $(u, v) \in E_1$, let $T_1(v)$ denote the subtree which results by deleting (u, v) from T_1 and contains v. Suppose that the diameter of T_2 is at least 4. If a neighborhood-preserving matching f exists with $f(u) \notin$ leaves(T_2) and $f(v) \in$ leaves(T_2), then for every vertex $x \in V(T_1(v))$, we have $f(x) \in$ leaves(T_2) and $(f(u), f(x)) \in E_2$.*

If there is a mapping f which fulfills the condition of Lemma 5, that is, $f(u) \notin$ leaves(T_2) and $f(v) \in$ leaves(T_2) for two vertices $u, v \in V_1$ with $(u, v) \in E_1$, then we say that the subtree $T_1(v)$ is *absorbed* at $f(u)$. Clearly, subtree $T_1(v)$ cannot be absorbed, if the number of leaves adjacent to $f(u)$ is smaller than the number of vertices in $T_1(v)$. The following lemma summarizes the above observations.

Lemma 6. *Suppose T_2 has a diameter at least 4. Let $u, v \in T_1$ with $(u, v) \in E_1$. Let $T_1(u)$ and $T_1(v)$ be the subtrees resulting by deleting (u, v) from T_1 and containing u and v, respectively. Suppose that there exists a neighborhood-preserving mapping f with $(f(u), f(v)) \in E_2$. Let $T_2(f(u))$ and $T_2(f(v))$ be the subtrees resulting by deleting $(f(u), f(v))$ from T_2 and containing $f(u)$ and $f(v)$, respectively. Then, it holds that*

1. *either one of $f(u)$ and $f(v)$ (say $f(u)$) is a leaf of T_2 and for every vertices $x \in V(T_1(u))$, we have $f(x)$ being a leaf adjacent to $f(v)$,*
2. *or $|V(T_1(u))| = |V(T_2(f(u)))|$, $|V(T_1(v))| = |V(T_2(f(v)))|$, $|$leaves$(T_1(u))| \leq |$leaves$(T_2(f(u)))|$, and $|$leaves$(T_1(v))| \leq |$leaves$(T_2(f(v)))|$.*

The Algorithm. To ease the presentation, we assume that both trees T_1 and T_2 are rooted at root(T_1) and root(T_2), respectively, and the mapping f sought for satisfies $f(\text{root}(T_1)) = \text{root}(T_2)$. Further we assume that root(T_2) is not a leaf. For a vertex v of a rooted tree T, we use $T(v)$ to denote the subtree rooted at v. The *labels* of a vertex $v \in V_2$, denoted by labels(v), is a set of vertices from V_1 that can potentially be mapped to v. Clearly, labels$(\text{root}(T_2)) = \{\text{root}(T_1)\}$. For $U' \subseteq V_1$, we define labels$(v, U') :=$ labels$(v) \cap U'$. *Discarding* a label $u \in$ labels(v) is denoted by labels$(v) :=$ labels$(v) \setminus \{u\}$. By labels(V') with $V' \subseteq V_2$ we denote the set $\bigcup_{v \in V'}$ labels(v). The algorithm consists of two phases, the first phase top-down preparing the labels of all vertices of T_2 and the second phase constructing the mapping from the labels in a bottom-up manner.

Phase 1. Starting at root(T_2) with labels$(\text{root}(T_2)) = \{\text{root}(T_2)\}$, the algorithm iterates over all non-leaf vertices in T_2 according the breath-first order, and builds label sets for the children of a vertex $v \in V_2$ from the label set of v. Let ch(u) denote the set of children of a vertex u in a rooted tree. For a vertex $v \in V_2$ and one of its labels $u \in$ labels(v), we process the children of u and v depending on their degrees as follows.

(Leaf children of u) We first consider the leaf children of u. By Lemma 3, if v can be mapped to u, then all leaf children of u have to be mapped to the leaf children of v. Let l_u and l_v be the numbers of the leaf children of u and v, respectively. Thus, if $l_u > l_v$, then we discard u from labels(v); otherwise, we select l_u many v's leaf children and store u's leaf children one-to-one in the label sets of the corresponding v's leaf children. We denote these l_u leaf children of v by $\mathcal{M}_{v,u}$.

(Non-leaf children of v). For each non-leaf child v' of v, we iterate over all non-leaf children of u. If there is one non-leaf child u' of u satisfying $|V(T_1(u'))| = |V(T_2(v'))|$ and $|\operatorname{leaves}(T_1(u'))| \leq |\operatorname{leaves}(T_2(v'))|$, then we add u' to labels(v'); otherwise, we discard u from labels(v). This is correct due to Lemma 6.

Now, labels(v') $\neq \emptyset$ for all non-leaf children $v' \in \operatorname{ch}(v)$ and all leaf children of u are in the label sets of the leaf children in $\mathcal{M}_{v,u}$. The algorithm moves to the next vertex according the breath-first order.

Phase 2. In this phase, the algorithm processes the non-leaf vertices of T_2, in a reversed order of the first phase. For a vertex $v \in V_2$ and a label u from labels(v), it computes a maximum matching on the bipartite graph consisting of the non-leaf children of v and the non-leaf children of u. There is an edge between a non-leaf child v' of v and a non-leaf child u' of u, if and only if $u' \in$ labels(v'). If the matching is not perfect for the non-leaf children of v, then discard u from labels(v); otherwise, consider the non-leaf children of u which are not in the matching. By Lemma 5, all subtrees rooted at these non-leaf children of u have to be absorbed, that is, mapped to the leaf children of v, excluding the leaf children in $\mathcal{M}_{v,u}$. Then, the algorithm compares the total size of the subtrees rooted at these non-leaf children of u and the number of the leaf children of v that are not in $\mathcal{M}_{v,u}$. If they are not equal, then discard u from labels(v). If afterwards labels(v) $= \emptyset$, then return "no"; otherwise move to the next vertex.

Finally, at the root of T_2, if we have labels(root(T_2)) $= \{$root(T_1)$\}$, then we can answer "yes".

Theorem 6. NEIGHBORHOOD-PRESERVING MAPPING *on trees can be solved in $O(n^{4+\omega})$ time for the case $k = 0$ and $l = d = 2$, where $n = |V_1| = |V_2|$ and $\omega = 2.38$.*

By combining the ideas for proving Theorems 5 and 6, we can show that a neighborhood-preserving mapping between trees exists for $k = 0$ and $l > d = 2$, if and only if $|V_1| \leq 3$ or T_2 is a star.

Theorem 7. NEIGHBORHOOD-PRESERVING MAPPING *on trees can be solved in polynomial time for the case $k = 0$ and $l > d = 2$.*

3   NPM on Trees with $k > 0$

We show next that NEIGHBORHOOD-PRESERVING MAPPING on trees with $k > 0$ is NP-complete for all values of l and d. Then, we give two polynomial-time algorithms solving the special case of NPM that $k > 0$, $l = d = 1$, and one input tree is a path.

3.1   Two Input Trees

The NP-hardness results for NPM on trees with $k = 0$ can be easily generalized for the case $k > 0$. In the following we focus on the cases where NPM on trees with $k = 0$ can be solved in polynomial time.

Theorem 8. NEIGHBORHOOD-PRESERVING MAPPING *on trees is NP-complete, even if* $k = 1$ *and* $l = d = 1$.

Theorem 9. NEIGHBORHOOD-PRESERVING MAPPING *on trees is NP-hard for the case* $k > 0$ *and* $l > d = 1$.

Proof. We reduce again from 3-PARTITION. Suppose that the given 3-PARTITION instance (B, S) satisfies that all elements of S are even.

The tree T_1 contains only one vertex c with degree greater than 2; the degree of c is equal to $3n + 2$. Each neighbor of c is an end-vertex of a path with $\lfloor l/2 \rfloor$ vertices. In one special path, to the other end-vertex t, if l is even, we add two leaves as neighbors; if l is odd, we add one leaf as a neighbor to the only neighbor of t. We call the so far resulting tree a "spider".

To the end-vertices of the remaining paths, we attach the following $3n + 1$ paths. One path is a long path consisting of $4B + 2B(l - 1) + 1$ vertices. The others correspond to the elements in S, i.e., the i-th path consisting of $a_i + (l - 1) \cdot (a_i/2 - 1)$ vertices.

The tree T_2 has only one vertex c' with degree greater than 2; the degree of c' is equal to $(3n+2) \cdot \lfloor l/2 \rfloor + 2B(l-1) + \sum_{a_i \in S}(l-1) \cdot (a_i/2-1) + p + 1$, where $p = 1$ if l is even, and $p = 0$, otherwise. Among these neighbors of c', n of them are the end-vertices of n paths, each of length $B - 1$; one neighbor is an end-vertex of a path of length 1; two of the neighbors are the end-vertices of 2 paths, each with $2B$ vertices. Finally, we set $k := (3n+2) \cdot \lfloor l/2 \rfloor + 2B(l-1) + \sum_{a_i \in S}(l-1) \cdot (a_i/2-1) + p$ with $p = 1$ if l even, and $p = 0$, otherwise.

For the equivalence, observe that, from a set of vertices in T_1, that have pairwise distance at most l, at most two vertices can be in $V(T_1) \setminus D$; otherwise, we would need cycles in T_2. Thus, at most two vertices of the spider can be "kept", i.e., not in the isolation set. Further, for a path with $x + (l - 1) \cdot x/2$ vertices with x being even, we need to delete at least $(l - 1) \cdot x/2$ vertices to get a set of vertices such that no three vertices in this set have pairwise distance at most l. Then, we can conclude that, with k isolations allowed, if a mapping f exists, then, after deleting the isolation vertices, the remaining vertices must "induce with their l-neighborhoods" a set $\mathcal{P}$ of paths. Given a tree T and a set V' of vertices in T, the graph induced by V' with their l-neighborhoods has V' as its vertex set. There is an edge between $u, v \in V'$, if and only if the distance between u and v in T is at most l. In $\mathcal{P}$, there is a path with $4B + 1$ vertices (remaining vertices of the long path), a length-1 path (two vertices kept in the spider), and $3n$ element paths, where the i-th path contains a_i vertices. Clearly, the length-$4B$ path has to be mapped to the two length-$2B$ paths. The two vertices kept in the spider are mapped the length-1 path attached to a neighbor of c'. The element paths can be mapped to the n paths of length $B - 1$, if and only if the given 3-PARTITION instance is a yes-instance.   □

Theorem 10. NEIGHBORHOOD-PRESERVING MAPPING on trees is NP-complete, if $k > 0$ and $l \geq d = 2$.

3.2 $l = d = 1$, $k > 0$, and a Tree and a Path as Input

In contrast to the NP-hardness result of NPM on trees with $l = d = 1$ (Theorem 8), we show that if one of the input trees is a path, then NPM is polynomial-time solvable with $l = d = 1$.

The second tree is a path. To simplify the presentation, we reformulate NPM on trees with $l = d = 1$ and the second tree being a path as the following problem:

> CUTTING TREE INTO PATHS (CTP)
> **Input**: A tree T, an integer k
> **Question**: Can we transform T to a set $\mathcal{P}$ of paths by deleting at most k vertices?

Lemma 7. *Given the second tree being a path, NPM on trees with $l = d = 1$ is equivalent to CTP.*

Next, we give a dynamic programming based algorithm solving CTP. Assume that T is rooted at an arbitrary vertex r, and let $T(v)$ denote the subtree rooted at a vertex v. Hereby, we distinguish at every vertex $v \in V(T)$ the following four cases:

1. v is deleted,
2. v is an end-vertex of a path in $\mathcal{P}$ and all children of v are deleted,
3. v is on a path in $\mathcal{P}$ and exactly one end-vertex of this path is in $V(T(v)) \backslash \{v\}$,
4. v is on a path in $\mathcal{P}$ and both end-vertices of this path are in $V(T(v)) \backslash \{v\}$.

We define further a function $d_v(c)$ for v with $c \in \{1, 2, 3, 4\}$, denoting one of the previously defined cases. This function $d_v(c)$ stores the minimal number of deletions needed in $T(v)$ to derive a set of paths, where v follows Case c. We recursively compute $d_v(c)$ for all vertices $v \in V(T)$ and all four cases in a bottom-up way. Clearly, at the root r, if $\min_{c=1...4} d_r(c) \leq k$, then we return "yes"; otherwise, return "no".

At a leaf vertex v, Cases 3 and 4 clearly cannot be applied. We can easily set $d_v(1) := 1$, $d_v(2) := 0$, and $d_v(3) = d_4(v) := \infty$. The computation of $d_v(c)$ for a non-leaf vertex v distinguishes again four cases:

1. $d_v(1) := \sum_{u \in \mathrm{ch}(v)} \min_{c=1...4} d_u(c) + 1$,
2. $d_v(2) := \sum_{u \in \mathrm{ch}(v)} d_u(1)$,
3. $d_v(3) := \min_{u \in \mathrm{ch}(v)} (\sum_{u' \in (\mathrm{ch}(v) \backslash \{u\})} d_{u'}(1) + \min_{c=2,3} d_u(c))$,
4. $d_v(4) := \min_{u_1, u_2 \in \mathrm{ch}(v)} (\sum_{u \in (\mathrm{ch}(v) \backslash \{u_1, u_2\})} d_u(1) + \min_{c=2,3} d_{u_1}(c) + \min_{c=2,3} d_{u_2}(c))$.

Theorem 11. *CTP can be solved in $O(|V(T)|^3)$ time.*

Corollary 1. *NPM on trees with the second tree being a path can be solve in $O(|V(P)|^3)$ time for $l = d = 1$.*

The First Tree Is a Path. Again, NPM on trees with $l = d = 1$ and the first tree being a path can be reformulated as the following problem:

FITTING PATH TO TREE BY DELETIONS (FPTD)
Input: A path P and a tree T with $|V(P)| = |V(T)|$, an integer k
Question: Can we delete at most k vertices from P such that there exists a subgraph T' of T isomorphic to the resulting set $\mathcal{P}$ of paths?

With $l = d = 1$, the following lemma is easy to prove.

Lemma 8. *If the first tree is a path, then NPM on trees with $l = d = 1$ is equivalent to FPTD.*

In the following, we give a polynomial-time algorithm solving FPTD. Again we assume that T is rooted at an arbitrary vertex r and denote by $T(v)$ the subtree rooted at a vertex v. Let D denote the set of the vertices whose deletion from P results in a set $\mathcal{P}$ of paths. We extend the isomorphic mapping f from $V(\mathcal{P})$ to $V(T')$ to a mapping from $V(P)$ to $V(T)$, by assigning an arbitrary one-to-one correspondence between D and $V(T) \setminus V(T')$. This is doable since $|V(P)| = |V(T)|$. The algorithm processes the vertices in T in a bottom-up manner. At each vertex v, it distinguishes the following 6 cases concerning the way how v is mapped by the mapping f:

1. v is mapped to a vertex in D;
2. v is mapped to a path $p \in \mathcal{P}$ with $|V(p)| = 1$;
3. v is mapped to an end-vertex of a path in $\mathcal{P}$, whose other end-vertex is mapped to a vertex not in $V(T(v))$;
4. v is mapped to an end-vertex of a path in $\mathcal{P}$, whose other end-vertex is mapped to a vertex in $V(T(v)) \setminus \{v\}$;
5. v is mapped to a non-end vertex of a path in $\mathcal{P}$, which has one end-vertex mapped to a vertex in $V(T(v)) \setminus \{v\}$ and another one mapped to a vertex not in $V(T(v))$;
6. v is mapped to a non-end vertex of a path in $\mathcal{P}$, whose both end-vertices are mapped to vertices in $V(T(v)) \setminus \{v\}$.

For each of the cases, the algorithm checks whether it is possible to delete some vertices to create a set of paths, which can be mapped to $T(v)$, given the mapping of v following this case. If so, it stores the minimum possible number of deletions. Notice that Case 1 causes additional caution in this check. On the one hand, the subtree $T(v)$ could be mapped to some set of paths, which however need to delete a lot of vertices from P. These deleted vertices might be mapped to vertices of Case 1, which are outside of $T(v)$. On the other hand, we might have a lot of vertices in $T(v)$ with Case 1. However, the paths mapped to $T(v)$ do not cause so many vertex deletions. Thus, we introduce an additional parameter s to record this information with $-k \le s \le k$. If we say that there are s "mappable" vertices in $T(v)$, we mean the following: If $s < 0$, there are $|s|$ vertices which are deleted to create the paths mapped to $T(v)$ but are not mapped to the vertices with Case 1 in $T(v)$; otherwise, there are s vertices with Case 1 in $T(v)$, which can be mapped to vertices deleted to create paths mapped to vertices outside of $T(v)$. Thereupon, we define the dynamic programming table F_v at vertex v with two parameters, one representing the 6 cases and the other being s. The entry $F_v(c, s)$ contains the minimal number of vertex deletions needed to create a set of paths in P, which are mapped to $T(v)$, under the conditions that v follows Case c and there are s mappable vertices in $T(v)$.

To ease the presentation, we say to "open" a path at v, if v is in Cases 2 and 3. Notice that, once we open a path, we increase the number of vertex deletions by one. However, since by deleting i vertices from P we can create $i + 1$ paths, we check whether $F_r(c, -1) \leq k + 1$ for some Case c at the root r. If so, we return "yes"; otherwise, we return "no".

It remains to describe the computation of $F_v(c, s)$. At a leaf v, it is clear that only Cases 1, 2, and 3 can be applied. Thus, all entries of F_v are set to ∞, except for three entries, where we set $F_v(1, 1) := 0$, $F_v(2, -1) := 1$, and $F_v(3, -1) := 1$. The correctness here is obvious.

At a non-leaf vertex v, let $\mathrm{ch}(v) = \{u_1, \ldots, u_d\}$, where d is the number of children of v. We define three additional tables:

- For $-k \leq s \leq k$ and $1 \leq i \leq d$, $A_v(s, i)$ stores the minimal number of deletions needed to create a set of paths in P, which are mapped to the subtrees rooted at $u_1, \ldots, u_i$, under the conditions that $u_1, \ldots, u_i$ are of Case 1, 2, 4 or 6, and there are s mappable vertices in these subtrees;
- For $-k \leq s \leq k$, $1 \leq i \leq d$, and $1 \leq j \leq i$, $B_v(s, i, j)$ stores the minimal number of deletions needed to create a set of paths in P, which are mapped to the subtrees rooted at the vertices in $\{u_1, \ldots, u_i\} \setminus \{u_j\}$, under the conditions that all vertices in $\{u_1, \ldots, u_i\} \setminus \{u_j\}$ are of Case 1, 2, 4 or 6, and there are s mappable vertices in these subtrees;
- For $-k \leq s \leq k$, $1 \leq i \leq d$, and $1 \leq j_1, < j_2 \leq i$, $C_v(s, i, j_1, j_2)$ stores the minimal number of deletions needed to create a set of paths in P, which are mapped to the subtrees rooted at $\{u_1, \ldots, u_i\} \setminus \{u_{j_1}, u_{j_2}\}$, under the conditions that all vertices in $\{u_1, \ldots, u_i\} \setminus \{u_{j_1}, u_{j_2}\}$ are of Case 1, 2, 4 or 6, and there are s mappable vertices in these subtrees;
- For $-k \leq s \leq k$, $1 \leq i \leq d$, and $i < j \leq d$, $D_v(s, i, j)$ stores the minimal number of deletions needed to create a set of paths in P, which are mapped to the subtrees rooted at u_i and u_j, under the conditions that u_i and u_j are of Case 3 or 5, and there are s mappable vertices in these subtrees.

To compute the three tables, we apply the following recursions: initialize $A_v(s, 0) := 0$ for $-k \leq s \leq k$. For each $i = 1, \ldots, d$ and each $s = -k, \ldots, k$, set

$$A_v(s, i) := \min_{-k \leq q \leq k} \left(A_v(s - q, i - 1) + \min_{c \in \{1, 2, 4, 6\}} F_{u_i}(c, q) \right) .$$

In order to fill in $B_v(s, i, j)$, initialize $B_v(s, i, i) := A_v(s, i - 1)$ for every $-k \leq s \leq k$ and $2 \leq i \leq d$. For $1 \leq j < i$, the recursion for B_v is as follows:

$$B_v(s, i, j) := \min_{-k \leq q \leq k} \left(B_v(s - q, i - 1, j) + \min_{c \in \{1, 2, 4, 6\}} F_{u_i}(c, q) \right) .$$

Then, initialize $C_v(s, i, j, i) := B_v(s, i, j)$ for every $-k \leq s \leq k$, $2 \leq i \leq d$, and $1 \leq j < i$. For $1 \leq j_1 < j_2 < i$, the recursion for C_v is as follows:

$$C_v(s, i, j_1, j_2) := \min_{-k \leq q \leq k} (C_v(s - q, i - 1, j_1, j_2) + \min_{c \in \{1,2,4,6\}} F_{u_i}(c, q)) \ .$$

Finally, for each $1 \leq i < j \leq d$ and $-k \leq s \leq k$, we compute D_v as follows:

$$D_v(s, i, j) := \min_{-k \leq q \leq k} (\min_{c \in \{3,5\}} F_{u_i}(c, q) + \min_{c \in \{3,5\}} F_{u_j}(c, s - q)) \ .$$

The correctness of the computation of the tables follows from the recursions. We compute F_v for each of the 6 cases as follows:

Case 1. Here, v should be mapped to a deleted vertex. Then, the children of v should be of Cases 1, 2, 4, and 6. We need only to sum up the deletions needed to create the paths mapped to the subtrees rooted at the children. Notice that we have one additional vertex v of Case 1 which is not mapped. We set $F_v(1, s) := A_v(s - 1, d)$.

Case 2. We have to open a new path p with $|V(p)| = 1$ mapped to v. This implies that we need one more vertex deleted in $T(v)$ than in the forest consisting of the subtrees rooted at the children of v. The cases for the children are the same as in Case 1. Thus, we set $F_v(2, s) := A_v(s + 1, d) + 1$.

Case 3. As in Case 2, we open a new path p at v. Therefore, $F_v(3, s) := F_v(2, s)$.

Case 4. One path should end at v and has its other end-vertex in $V(T(v)) \setminus \{v\}$. Thus, at least one of v's children has to be of Case 3 or 5, while the other are of Cases 1, 2, 4, and 6. We set

$$F_v(4, s) := \min_{u_i \in ch(v), -k \leq q \leq k} (B_v(s - q, d, i) + \min_{c \in \{3,5\}} F_{u_i}(c, q)) \ .$$

Case 5. The vertex v is mapped to a non-end vertex of a path with one end-vertex mapped inside of $T(v)$ and the other outside of $T(v)$. Therefore, one child of v must be of Case 3 or 5, while the others are of Cases 1, 2, 4, and 6. We have the same situation as Case 4: $F_v(5, s) := F_v(4, s)$.

Case 6. With both end-vertices mapped inside of $T(v)$, two children of v must be of Cases 3 and 5. Note that with two paths "merging" at v, we have in $T(v)$ one path less than in the forest consisting of the subtrees rooted at the children of v. With C_v and D_v, we compute $F_v(6, s)$ as follows:

$$F_v(6, s) := \min_{u_i, u_j \in ch(v), -k \leq q \leq k} (C_v(s + 1 - q, d, i, j) + D_v(q, i, j)) - 1 \ .$$

At the root r, if $\min_{c \in \{1,2,4,6\}} F_r(c, -1) \leq k + 1$, then we return "yes"; otherwise, "no".

Theorem 12. *FPTD can be solved in $O(|V(T)|^4 \cdot k^2)$ time.*

Corollary 2. *NPM on trees with the first tree being a path can be solved in $O(|V(T)|^4 \cdot k^2)$ time.*

4   Conclusion

A variation of the graph isomorphism problem, called NEIGHBORHOOD-PRESERVING MAPPING (NPM), has been introduced. We studied the computational complexity of NPM on trees and presented a complete dichotomy with respect to l, d, and k. The result is that NPM on tress is polynomial-time solvable only for the cases when $k = 0$, $l \geq d$, and $d \leq 2$. Additionally, we considered NPM on trees with one of the input trees restricted to be a path. For this two polynomial-time algorithms were developed.

Future research directions could include development of effective heuristics for input graph/trees with certain properties (tree-like graphs for example). The next natural step is to apply the proposed or similar model to real (biological) network data, and study the correlation with known graph measure. From the theoretical point of view, it would be interestingly to study NPM on graphs with bounded treewidth and to examine the connection between NPM and the graph homomorphism problems.

References

1. Heath, A.P., Kavraki, L.E.: Computational challenges in systems biology. Computer Science Review 3, 1–17 (2009)
2. Bunke, H., Riesen, K.: Recent advances in graph-based pattern recognition with applications in document analysis. Pattern Recognition 44, 1057–1067 (2011)
3. Riesen, K., Bunke, H.: Approximate graph edit distance computation by means of bipartite graph matching. Image and Vision Computing 27, 950–959 (2009)
4. Akutsu, T., Fukagawa, D., Halldórsson, M.M., Takasu, A., Tanaka, K.: Approximation and parameterized algorithms for common subtrees and edit distance between unordered trees. Theor. Comput. Sci. 470, 10–22 (2013)
5. Akutsu, T., Fukagawa, D., Takasu, A., Tamura, T.: Exact algorithms for computing the tree edit distance between unordered trees. Theor. Comput. Sci. 412, 352–364 (2011)
6. Bille, P.: A survey on tree edit distance and related problems. Theor. Comput. Sci. 337, 217–239 (2005)
7. Lozano, A., Pinter, R.Y., Rokhlenko, O., Valiente, G., Ziv-Ukelson, M.: Seeded tree alignment. IEEE/ACM Transactions on Computational Biology and Bioinformatics 5, 503–513 (2008)
8. Garey, M.R., Johnson, D.S.: Computers and Intractability: A Guide to the Theory of NP-Completeness. W. H. Freeman (1979)
9. Aho, A.V., Hopcroft, J.E., Ullman, J.D.: The Design and Analysis of Computer Algorithms, 1st edn. Addison-Wesley Longman Publishing Co., Inc., Boston (1974)

Bounding the Running Time of Algorithms for Scheduling and Packing Problems*

Klaus Jansen, Felix Land, and Kati Land

Institute of Computer Science, University of Kiel, 24118 Kiel, Germany
{kj,fku,kla}@informatik.uni-kiel.de

Abstract. We investigate the implications of the exponential time hypothesis on algorithms for scheduling and packing problems. Our main focus is to show tight lower bounds on the running time of these algorithms. For exact algorithms we investigate the dependence of the running time on the number n of items (for packing) or jobs (for scheduling). We show that many of these problems, including SUBSETSUM, KNAPSACK, BINPACKING, $\langle P2 \mid \mid C_{\max} \rangle$, and $\langle P2 \mid \mid \sum w_j C_j \rangle$, have a lower bound of $2^{o(n)} \times \|I\|^{O(1)}$. We also develop an algorithmic framework that is able to solve a large number of scheduling and packing problems in time $2^{O(n)} \times \|I\|^{O(1)}$. Finally, we show that there is no PTAS for MULTIPLEKNAPSACK and 2D-KNAPSACK with running time $2^{o\left(\frac{1}{\varepsilon}\right)} \times \|I\|^{O(1)}$ and $n^{o\left(\frac{1}{\varepsilon}\right)} \times \|I\|^{O(1)}$.

Keywords: scheduling, packing, exponential time hypothesis, exact algorithms, lower bounds.

1 Introduction

The usual assumption $P \neq NP$ allows us to rule out polynomial time algorithms for many decision and optimization problems. Often the preferred way for dealing with such NP-hard problems are heuristics and approximate algorithms. In recent years however, the interest in super-polynomial exact algorithms has increased. A big problem is that, under the assumption $P \neq NP$, we cannot know what super-polynomial running times are possible for these problems.

A stronger assumption was introduced by Impagliazzo and Paturi, the *Exponential Time Hypothesis* (ETH). The subject of the ETH is the satisfiability problem 3-SAT. In contrast to classical complexity theory the running time assumed in the ETH not only depends on the length $\|\varphi\|$ of the instance, but on a special parameter of the instance, the number n of variables.

Conjecture 1 (Exponential Time Hypothesis [15]). There is positive real δ such that 3-SAT cannot be decided in time $2^{\delta n} \times \|\varphi\|^{O(1)}$.

* A full version of this work is available as technical report [18]. Research supported by German Research Foundation (DFG) projects JA 612/16-1 and JA 612/12-1.

F. Dehne, R. Solis-Oba, and J.-R. Sack (Eds.): WADS 2013, LNCS 8037, pp. 439–450, 2013.

Another way to formulate the conjecture is that 3-SAT with parameter n has no *sub-exponential* algorithm. Here, we follow the notation of Flum and Grohe [7]: a function f is called sub-exponential if $f(n) = o(n)$, where $f = o(g)$ if there is a non-decreasing, unbounded function μ such that $g(n) \leq \frac{f(n)}{\mu(n)}$.

The ETH can be used to show lower bounds on the running time of algorithms for other problems by the use of *strong reductions*, i.e. reductions which increase the parameter at most linearly [15]. Another important result is implied by the Sparsification Lemma due to Impagliazzo, Paturi, and Zane [16]: Under assumption of the ETH there is no algorithm that decides 3-SAT with m clauses in time $2^{o(m)} \times \|\varphi\|^{O(1)}$. This allows us to parametrize by the number of clauses.

Our main focus in this paper are consequences of the ETH for scheduling and packing problems. We investigate the dependence of the running time on the number of jobs respectively the number of items, which we will denote by n. We also develop algorithms that are able to solve a broad class of scheduling and packing problems and whose running time matches the lower bound for many problems. We will first concentrate on SUBSETSUM and related problems. These will then be used to show bounds for other problems.

Notation. We use the notation $f(S) = \sum_{s \in s} f(s)$ for any function f and any subset S of the domain of f throughout the paper. For a minimization or maximization problem and $\alpha > 1$, an algorithm A is called α-approximate if $A(I) \leq \alpha \operatorname{OPT}(I)$ or $A(I) \geq \frac{1}{\alpha} \operatorname{OPT}(I)$ holds for each instance I, respectively.

Known Results. There is a large number of lower bounds based on the ETH, mostly in the area of graph problems. For example it is known that CLIQUE (and the equivalent INDEPENDENTSET) cannot be decided in time $2^{o(n)} \times \|I\|^{O(1)}$ [16]. For a good survey of these results and useful techniques we refer to the work of Lokshtanov, Marx, and Saurabh [24]. Only few lower bounds have been obtained for scheduling and packing problems: Chen et al. [3] showed that precedence constrained scheduling on m machines cannot be decided in time $f(m)\|I\|^{o(m)}$ and set packing cannot be decided in time $f(k)\|I\|^{o(k)}$, where k is the size of the packing. Kulik and Shachnai [20] observed that sized subset sum, where k is the size of set to be found, cannot be decided in time $f(k)\|I\|^{o(\sqrt{k})}$ and used this result to show that there is no PTAS for the 2-dimensional vectorial knapsack problem with running time $f(\varepsilon)\|I\|^{o(\sqrt{\frac{1}{\varepsilon}})}$. These results are actually based on the assumption that not all problems in SNP are solvable in sub-exponential time. Since 3-SAT $\in$ SNP, this assumption in weaker than the ETH [28]. Pǎtraşcu and Williams [29] showed a lower bound of $n^{o(k)}$ for sized subset sum, even when the encoding length of the item sizes is bounded by $O(d \log n)$. Finally, Jansen et al. [19] proved that bin packing into m bins cannot be solved in time $f(m)\|I\|^{o(m/\log m)}$ when the item sizes are encoded in unary.

Exact algorithms for $\langle Pm \mid |C_{\max}\rangle$ with $2 \leq m \leq 4$ that have running times $\sqrt{2}^n$, $\sqrt{3}^n$, and $(1 + \sqrt{2})^n$ were developed by Lenté et al [23]. BINPACKING can be solved in time $nB2^n$ [8] or $n^{O(m)}2^{O(m\sqrt{\|I\|})}$ [26], where m is the number of bins. SUBSETSUM, PARTITION and KNAPSACK can all be solved in time $2^{\frac{n}{2}} \times \|I\|^{O(1)}$ [14,8] and $2^{O(\sqrt{\|I\|})}$ [27,26].

New Results and Organization. In Sect. 2 we investigate exact algorithms for SUBSETSUM and related problems, including PARTITION, BINPACKING, and MULTIPROCESSORSCHEDULING. We prove the lower bounds $2^{o(n)} \times \|I\|^{O(1)}$ and $2^{o(\sqrt{\|I\|})}$ for these problems. In Sect. 3 we give a lower bound of $2^{o(n)} \times \|I\|^{O(1)}$ for different types of scheduling problems. We present an algorithmic framework in Sect. 4 that is able to solve nearly all problems mentioned in Sects. 2 and 3 in time $2^{O(n)} \times \|I\|^{O(1)}$, showing that the corresponding bounds are tight. Finally, in Sect. 5 we consider approximation schemes for knapsack problems. We prove that there are no PTAS for MULTIPLEKNAPSACK and 2D-KNAPSACK with running times $2^{o(\frac{1}{\varepsilon})} \times \|I\|^{O(1)}$ and $n^{o(\frac{1}{\varepsilon})} \times \|I\|^{O(1)}$, respectively.

2   The Subset Sum Family

In this section we will prove tight lower bounds on the running time of algorithms for several problems related to SUBSETSUM and PARTITION, when parametrized by the number n of items or the input size $\|I\|$.

2.1   Lower Bounds for Subset Sum and Partition

Wegener [31] presented a chain of reductions from 3-SAT to PARTITION via the subset sum problem. We will omit the proofs of correctness and only give a brief description of the construction.

From 3-SAT to SUBSETSUM. Denote the variables by $x_1, \ldots, x_n$ and the clauses by $C_1, \ldots, C_m$. For each variable x_i we create two items a_i and b_i with sizes

$$s(a_i) = \sum_{\substack{j \in [m] \\ x_i \in C_j}} 10^{n+j-1} + 10^{i-1} \quad \text{and} \quad s(b_i) = \sum_{\substack{j \in [m] \\ \bar{x}_i \in C_j}} 10^{n+j-1} + 10^{i-1}. \quad (1)$$

These numbers have at most $n+m$ digits when encoded in base 10. Additionally we create two dummy items c_j and d_j for each clause C_j with $s(c_j) = s(d_j) = 10^{n+j-1}$. The item set is $A = \{a_i, b_i \mid i \in [n]\} \cup \{c_j, d_j \mid j \in [m]\}$ The target value is

$$B = \sum_{j=1}^{m} 3 \times 10^{n+j-1} + \sum_{i=1}^{n} 10^{i-1}. \quad (2)$$

In total the instance $I = (A, B)$ has $2n+2m$ items, hence the reduction is strong.

From SUBSETSUM to PARTITION. Let (A, B) be an instance of SUBSETSUM. First assume that $s(A) \geq B$, otherwise we can output a trivial no-instance. We introduce two new items p and q with $s(p) = 2\,s(A) - B$ and $s(q) = s(A) + B$. The instance of PARTITION is $A' = A \cup \{p, q\}$. Note that $s(p) \in \mathbb{N}$ because $s(A) \geq B$.

Theorem 2. *The problems* PARTITION *and* SUBSETSUM *cannot be decided in time* $2^{o(n)} \times \|I\|^{O(1)}$*, unless the ETH fails.*

The above bounds are asymptotically tight: A naïve enumeration algorithm solves both problems by testing all 2^n subsets of A in time $2^n \times \|I\|^{O(1)}$. The fastest known algorithms have asymptotic running time $2^{\frac{n}{2}} \times \|I\|^{O(1)}$ [14].

2.2   Implications for Scheduling and Packing

Packing in One Bin. A generalization of SUBSETSUM is the well-known knapsack problem.

Theorem 3. *There is no algorithm deciding* 0-1-INTEGERPROGRAMMING *(even for one constraint and only positive coefficients) or* KNAPSACK *in time* $2^{o(n)} \times \|I\|^{O(1)}$*, unless the ETH fails.*

These results are again asymptotically tight.

Bin Packing and Multiprocessor Scheduling. Another fundamental packing problem is BINPACKING. The decision problem asks if the given items fit into a given number of bins and is known to be strongly NP-hard [10]. Even the case where the number m of bins is a fixed constant, called m-BINPACKING, remains weakly NP-hard [22]. This result originates from the hardness of PARTITION, which is equivalent to 2-BINPACKING with $B = \frac{1}{2}\,\mathrm{s}(A)$. Marx [25] observed that the gap creation technique that is commonly used to show inapproximability can be used in context of the ETH. In combination with Theorem 2 we obtain:

Theorem 4. *For* $\alpha < 2$ *there is no* α-*approximate algorithm for* BINPACKING *and no exact algorithm for* 2-BINPACKING *with running time* $2^{o(n)} \times \|I\|^{O(1)}$*, unless the ETH fails.*

The simplest variant of scheduling is the multiprocessor scheduling problem MPS. It asks if there is a schedule of the given jobs on m machines that finishes within a given deadline D and is equivalent to BINPACKING.

Theorem 5. *There is no algorithm deciding* MPS *in time* $2^{o(n)} \times \|I\|^{O(1)}$*, unless the ETH fails. This also holds for a fixed number* $m \geq 2$ *of machines.*

We present algorithms for MPS and BINPACKING with running time $2^{O(n)} \times \|I\|^{O(1)}$ in Sect. 4, which closes the gap between upper and lower bounds.

2.3   Input Length as Complexity Measure

When the running time is measured in the encoding length of the input the fastest known algorithms for SUBSETSUM, PARTITION, KNAPSACK and m-BINPACKING have running time $2^{O(\sqrt{\|I\|})}$ [27,26].

Theorem 6. SUBSETSUM, PARTITION, KNAPSACK *and* m-BINPACKING *with* $m \geq 2$ *cannot be decided in time* $2^{o(\sqrt{\|I\|})}$*, unless the ETH fails.*

Proof. Consider an instance of SUBSETSUM as constructed by the reduction for Theorem 2. It contains $2n + 2m$ numbers, and each can be encoded (in base 10) with at most $n + m$ digits. Because we can assume that $n = O(m)$ we know that $\|I\| = O(m^2)$. If an algorithm for SUBSETSUM with running time $2^{o(\sqrt{\|I\|})}$ existed, one could use it to solve 3-SAT in time $2^{o(m)} \times \|\varphi\|^{O(1)}$. The reductions to the other problems do not increase the encoding length of the instance significantly.

2.4   Special Cases with Size Restrictions

If φ is some predicate on the instances of SUBSETSUM or PARTITION, we denote the problem restricted to instances for which the predicate is TRUE by SUBSETSUM-φ or PARTITION-φ, respectively.

We first restrict SUBSETSUM to instances (A, B) with the following property: If a subset $S \subseteq A$ with $s(S) = B$ exists, then it contains exactly half of the elements, or more formally the following predicate φ holds:

$$\varphi((A,B)) \iff \forall S \subseteq A\colon \left(s(S) = B \implies |S| = \tfrac{|A|}{2} \right). \tag{3}$$

Lemma 7. *There is no algorithm that decides* SUBSETSUM-φ *in time* $2^{o(n)} \times \|I\|^{O(1)}$, *unless the ETH fails.*

Proof (Sketch). We give a strong reduction from SUBSETSUM. Let (A, B) be an instance of SUBSETSUM. For each item $a \in A$ we construct two items a_1 and a_2 with $s(a_1) = 2n\,s(a) + 1$ and $s(a_2) = 1$, and let $A' = \{a_1, a_2 \mid a \in A\}$ and $B' = 2nB + n$. It remains to prove that (A', B') satisfies φ and (A, B) is a yes-instance iff (A', B') is a yes-instance. For this, partition the elements of a solution of (A', B') into the elements of form a_1 and a_2. The items $a \in A$ for which a_1 is in the solution correspond to a solution of (A, B) and vice-versa.

We can transform the instances of SUBSETSUM-φ to PARTITION using the same construction as for Theorem 2. Recall that we added two items p and q. In every feasible partition the added items are in different sets of the partition. Thus the constructed instance A' has a property similar to the instances of SUBSETSUM-φ: If a partition $A = A_1 \mathbin{\dot\cup} A_2$ with $s(A_1) = s(A_2)$ exists, then $|A_1| = |A_2|$, or more formally they fulfill the predicate φ' defined by

$$\varphi'(A) \iff \forall A_1, A_2 \subseteq A\colon \left(A = A_1 \mathbin{\dot\cup} A_2 \wedge s(A_1) = s(A_2) \implies |A_1| = |A_2| \right). \tag{4}$$

Lemma 8. *There is no algorithm that solves* PARTITION-φ' *in time* $2^{o(n)} \times \|A\|^{O(1)}$, *unless the ETH fails.*

Interestingly, the restriction SUBSETSUM-φ is a special case of the so called SIZEDSUBSETSUM, for which the cardinality of the set to be found is given as part of the instance, and PARTITION-φ' is a special case of BALANCEDPARTITION, for which only partitions $A = A_1 \mathbin{\dot\cup} A_2$ are feasible that satisfy $|A_1| = |A_2|$.

Corollary 9. SIZEDSUBSETSUM *and* BALANCEDPARTITION *cannot be solved in time* $2^{o(n)} \times \|I\|^{O(1)}$, *unless the ETH fails.*

Table 1. Summary of obtained bounds. Parenthesis around job characteristics denote that the bound holds with and without these. An asterisk (*) after the citation shows that the reduction was modified. The polynomial terms $\|I\|^{O(1)}$ in the bounds are omitted. A value in the column *Approx.* denotes that the bound also holds for approximate algorithms with a strictly better approximation ratio than the given number.

Problem	Reduced from	Source	Bound	Approx.
$\langle 1 \mid r(j), d(j) \mid \text{any} \rangle$	PARTITION	[10]	$2^{o(n)}$	
$\langle 1 \mid r(j) \mid \sum w_j C_j \rangle$	SUBSETSUM	[30]	$2^{o(n)}$	
$\langle P2 \mid\mid \sum w_j C_j \rangle$	PARTITION	[22]	$2^{o(n)}$	
$\langle P \mid \text{prec}, t(j) = 1 \mid C_{\max} \rangle$	CLIQUE	[21]	$2^{o(\sqrt{n})}$	3/2
$\langle R \mid t(j,k) \in \{t(j), \infty\}, (\text{pmtn}) \mid C_{\max} \rangle$	3-SAT	[6]	$2^{o(n)}$	3/2
$\langle P2 \mid \text{para}, (\text{pmtn}) \mid C_{\max} \rangle$	PARTITION	[10]	$2^{o(n)}$	
$\langle P \mid \text{para}, (\text{pmtn}\mid\text{migr}) \mid C_{\max} \rangle$	PARTITION	[4]	$2^{o(n)}$	3/2
$\langle P2 \mid \text{mall}, (\text{pmtn}) \mid C_{\max} \rangle$	PARTITION	[10]*	$2^{o(n)}$	
$\langle P \mid \text{mall}, (\text{pmtn}\mid\text{migr}) \mid C_{\max} \rangle$	PARTITION	[4]*	$2^{o(n)}$	3/2
$\langle O3 \mid\mid C_{\max} \rangle$	MONOTONE-NAE-SAT	[32]	$2^{o(n)}$	5/4
$\langle O2 \mid\mid \sum w_j C_j \rangle$	4-PARTITION	[1]*	$2^{o(n)}$	
$\langle O2 \mid \text{pmtn} \mid \sum C_j \rangle$	BALANCEDPARTITION	[5]	$2^{o(n)}$	
$\langle O \mid (\text{pmtn}) \mid \sum C_j \rangle$	3-SAT	[13]*	$2^{o(n)}$	
$\langle F3 \mid (\text{pmtn}) \mid C_{\max} \rangle$	PARTITION	[11]	$2^{o(n)}$	
$\langle F2 \mid\mid \sum w_j C_j \rangle$	4-PARTITION	[9]	$2^{o(n)}$	
$\langle F \mid (\text{pmtn}) \mid \sum C_j \rangle$	3-SAT	[13]*	$2^{o(n)}$	
$\langle J2 \mid (\text{pmtn}) \mid C_{\max} \rangle$	PARTITION	[11]	$2^{o(n)}$	

3   More Scheduling Problems

We conducted a review of existing reductions in the scheduling area. Our findings are summarized in Table 1. We had to modify some of the existing reductions, in particular those starting from 3-PARTITION, for which no strong reduction is known. We have been able to tweak the reduction to 3D-MATCHING by Garey and Johnson [10], and utilized it to obtain a lower bound of $2^{o(n)} \times \|I\|^{O(1)}$ for 4-PARTITION on $4n$ numbers. Most reductions from 3-PARTITION can be altered to start from 4-PARTITION instead.

4   Exact Solution in $2^{O(n)}$

We now present an algorithmic framework that can optimally solve many scheduling and packing problems in time $2^{O(n)} \times \|I\|^{O(1)}$. The algorithms optimize general classes of objective functions that include the popular choices $C_{\max}$ and $\sum w_j C_j$. Here, a schedule σ is a pair of functions $\sigma_m \colon J \to [m]$, $\sigma_s \colon J \to \mathbb{N}_0$ that assign to each job its machine and starting time, respectively.

4.1   Sequencing on a Constant Number of Machines

We start with an algorithm that can solve problems that involve precedence or exclusion constraints (e.g. for open shop). We require that the objective function is of the form $f(\sigma) = \mathrm{Op}_{j \in J} g_j(\sigma_m(j), \sigma_s(j))$, where Op is one of $\sum$, min, and max. Assume that we want to minimize or maximize f and all functions $g_j(k, \cdot)$ are non-decreasing or non-increasing, respectively. Then for any feasible schedule there is an equivalent *compact schedule*, i.e a schedule in which all jobs start as early as possible.

Our algorithm is loosely based on the dynamic programming approach of Held and Karp [12] for sequencing jobs on one machine. In contrast to their setting, we must allow idle time, because it may be beneficial (or even required) to wait for a job to finish on another machine. For this, we create a set T containing all possible starting and finishing times of jobs. A small addition also allows our algorithm to deal with job-specific release times.

Lemma 10. *We can compute a set T that contains the starting and finishing times of jobs in all compact schedules in time $2^{O(n)} \times \|I\|^{O(1)}$ and $|T| = 2^{O(n)}$.*

The basic idea of the algorithm is to examine possible *outlines* of schedules. Consider a schedule σ for a subset $S \subseteq J$ of jobs. For each machine $k \in [m]$ there is a job ℓ_k that is scheduled last, unless it has no jobs. The outline of σ is the restriction $\sigma_{|L}$ of σ to the jobs $L(\sigma) = \{\ell_k \mid k \in [m], \text{machine } k \text{ has jobs}\}$. We denote by $\ell(\sigma)$ the job in $L(\sigma)$ that starts last with respect to σ (ties may be broken arbitrarily). An S-outline is a schedule τ for $L \subseteq S$ that is its own outline such that the placement of $\ell(\tau)$ is feasible and S contains no successor of $\ell(\tau)$. Note that there may be S-outlines that are not the outline of any feasible schedule for S. We denote by $O_S(\tau)$ the set of $(S \setminus \{\ell(\tau)\})$-outlines τ' such that τ' agrees with τ on $L(\sigma) \setminus \{\ell(\tau)\}$, the jobs in $L(\tau') \setminus L(\tau)$ finish before $\tau_s(\ell(\tau))$, and τ' only uses machines that are used by τ.

We use a dynamic program to calculate, for each set $S \subseteq J$ and S-outline τ, the best objective value $\mathrm{B}[S, \tau]$ of a feasible schedule for S with outline τ. This is possible because of the following theorem. For simplicity we assume that $f = C_{\max}$. For a description of the general case we refer to the full version of the paper.

Lemma 11. *Let $S \subseteq J$ be a nonempty set of jobs and τ be an S-outline. Then the following recurrence equation holds:*

$$\mathrm{B}[S, \tau] = \min_{\tau' \in O_S(\tau)} \max\{\mathrm{B}[S \setminus \{\ell(\tau)\}, \tau'], C_\tau\}, \tag{5}$$

where C_τ is the completion time of $\ell(\tau)$ according to τ.

There are at most $|T|^m \times (|S| + 1)^m = 2^{O(n)}$ S-outlines and 2^n subsets $S \subseteq J$, so our dynamic program runs in $2^{O(n)}$ iterations. The makespan of an optimal schedule for all jobs then is $\min_{\tau \ J\text{-outline}} \mathrm{B}[J, \tau]$.

Our algorithm can solve a broad class of problems, including $\langle Om \mid\mid f \rangle$, $\langle Jm \mid\mid f \rangle$, $\langle Fm \mid\mid f \rangle$, and $\langle Rm \mid \mathrm{prec}, \mathrm{r}(j), \mathrm{d}(j) \mid f \rangle$, in time $2^{O(n)} \times \|I\|^{O(1)}$.

It can also be extended for parallel and malleable tasks. For $f \in \{C_{\max}, \sum w_j C_j\}$, the problems $\langle O3 \,|\, |\, f \rangle$, $\langle J3 \,|\, |\, f \rangle$, $\langle F3 \,|\, |\, f \rangle$, and $\langle P2 \,|\, |\, f \rangle$ cannot be solved asymptotically faster, unless the ETH fails (see Sect. 3).

4.2 Scheduling on an Arbitrary Number of Machines

We now describe an exact algorithm for scheduling on arbitrary many machines. For a schedule σ and $k \in [m]$ we denote by $J_{\sigma,k}$ the set $\sigma_m^{-1}(k)$ of jobs to be processed on machine k.

The main idea is again to use dynamic programming over subsets of jobs. For each $S \subseteq J$ and $k \in [m]$ we denote by $B[S, k]$ the best possible objective value when scheduling the jobs S on the first k machines. For each machine k and set S of jobs the algorithm finds and sequences the jobs $S' \subseteq S$ that should be processed on machine k. It does not look back and modify the schedule on the previously filled machines $1, \ldots, k-1$. Thus we demand that there are no constraints on the starting or finishing times of jobs on different machines (e.g. precedence constraints). We must further assume that the objective function of the whole schedule can be calculated iteratively when adding a new machine with jobs to the current schedule, i.e. the objective function is of the form $f(\sigma) = \mathrm{Op}_{k \in [m]} g_k(J_{\sigma,k}, \sigma^{(k)})$, where Op is one of $\sum$, min, and max, and the functions g_k can be computed in time $2^{O(n)} \times \|I\|^{O(1)}$. If the functions g_k are of the form as in Sect. 4.1 we can also use the algorithm presented there to sequence the jobs on each machine.

Again, we restrict ourselves to $f = C_{\max}$ for the explanation. We use dynamic programming to calculate the values $B[S, k]$ by utilizing the recurrence equation

$$B[k, S] = \begin{cases} \sum_{j \in S} t(j, 1) \\ \min_{S' \subseteq S} \max\{B[k-1, S \setminus S'], \sum_{j \in S'} t(j, k)\} & \text{otherwise,} \end{cases} \tag{6}$$

where $t(j, k)$ denotes the processing time of job j on machine k. After computing all values the makespan of an optimal schedule can be read from $B[m, J]$. The dynamic program needs at most $4^n \times m$ iterations.

We have to be careful with the dependence of the running time on m. On identical machines, i.e. when $g_1 = \cdots = g_m$ we can assume $m \leq n$, because an optimal schedule uses at most n machines. For different machines (e.g. scheduling on uniform or unrelated machines) this does not work. However, the m functions (or some parameters to distinguish them) then have to be encoded in the input, so we have $\|I\| = \Omega(m)$. Thus, our algorithm has a total running time of $2^{O(n)} \times \|I\|^{O(1)}$.

Our algorithm is able to solve the general problem $\langle R \,|\, r_j, d_j \,|\, f \rangle$. This contains $\langle 1 \,|\, r_j \,|\, \sum w_j C_j \rangle$, $\langle P2 \,|\, |\, \sum w_j C_j \rangle$, $\langle 1 \,|\, r_j, d_j \,|\, f \rangle$, and $\langle P2 \,|\, |\, C_{\max} \rangle$ as special cases. In Sect. 3 we have shown that none of them can be solved asymptotically faster under assumption of the ETH. The algorithm can also be adapted to packing problems with multiple containers, e.g. BinPacking and MultipleKnapsack.

5   Approximation Schemes for Knapsack Problems

5.1   The Multiple Knapsack Problem

In contrast to the regular knapsack problem instances of MULTIPLEKNAPSACK (MKS) may contain multiple knapsacks with individual capacities.

Theorem 12. *There is no approximation scheme for* MULTIPLEKNAPSACK *with running time* $2^{o\left(\frac{1}{\varepsilon}\right)} \times \|I\|^{O(1)}$, *unless ETH fails. This bound even holds for* $m = 2$ *knapsacks of equal capacity and when either*

(i) *all items have the same profit, or*
(ii) *the profit of each item equals its size.*

The case of condition (i) is a natural one: by scaling, we can assume that the profit of each item is 1, i.e. we are maximizing the number of packed items. With condition (ii) we maximize the size of the packed items, which is known as the multiple subset sum problem. The fastest known PTAS for the general case has a running time of $2^{O\left(\frac{1}{\varepsilon} \log^4 \frac{1}{\varepsilon}\right)} + \|I\|^{O(1)}$ [17].

Also note that both problem restrictions contain PARTITION as special case. Thus the lower bound $2^{o(n)} \times \|I\|^{O(1)}$ applies to exact algorithms. The running time of the algorithm described in Sect. 4.2 matches this bound.

Instances with a Special Profit Structure. To prove Theorem 12 we embed PARTITION into MKS. We then show that an approximation scheme for MKS can be used to decide PARTITION.

Note that for an instance $I = (A, B)$ of MKS, we can regard A as an instance of PARTITION by ignoring the profits. For each set $\mathcal{I}$ of instances of MKS we define $\mathcal{I}_P = \{A \mid (A, B) \in \mathcal{I}\}$ as the set of corresponding instances of PARTITION.

Lemma 13. *Let $\mathcal{I}$ be a set of instances of* MKS, *and $\alpha \geq 1$ such that for every instance $I = (A, B) \in \mathcal{I}$ there is a $C \in \mathbb{N}$ with*

(i) *I has $m = 2$ knapsacks of capacity $\frac{1}{2}\operatorname{s}(A)$ (note that $\operatorname{s}(A)$ must be even)*
(ii) *$|C| = \|A\|^{O(1)}$,*
(iii) *$\operatorname{p}(A) \leq n\alpha C$, and*
(iv) *$\operatorname{p}(a) \geq C$ for each item $a \in A$.*

Unless each instance $A \in \mathcal{I}_P$ can be decided in time $2^{o(n)} \times \|A\|^{O(1)}$, there is no approximation scheme that approximates all instances $I \in \mathcal{I}$ within $(1 + \varepsilon)$ of the optimum in time $2^{o\left(\frac{1}{\varepsilon}\right)} \times \|I\|^{O(1)}$.

Proof. Assume there is an approximation scheme P that finds an $(1 + \varepsilon)$-approximate solution for every instance $I \in \mathcal{I}$ in time $2^{o\left(\frac{1}{\varepsilon}\right)} \times \|I\|^{O(1)}$. Let an arbitrary instance $I = (A, B) \in \mathcal{I}$ be given. First we point out that a packing that packs all items into the two knapsacks exists if and only if A is a yes-instance of PARTITION. Now let $\varepsilon = \frac{1}{n\alpha}$ and solve I approximately using P_ε.

Recall that $\operatorname{p}(A) \leq n\alpha C$, thus $\frac{1}{n\alpha}\operatorname{p}(A) \leq C$. A short calculation shows that, if all items can be packed, the packing found by P_ε has profit at least $\operatorname{p}(A) - C$. Since the profit of all items is at least C, there is no unpacked item.

Therefore one can decide whether A admits a partition by testing if a $(1 + \varepsilon)$-approximate packing packs all items. Because condition (ii) implies $\|I\| = \|A\|^{O(1)}$, the required running time is $2^{o\left(\frac{1}{\varepsilon}\right)} \times \|I\|^{O(1)} = 2^{o(n)} \times \|A\|^{O(1)}$. A contradiction, since not all instances in $\mathcal{I}_P$ can be decided in this running time.

We can now prove the first part of Theorem 12. Let $\mathcal{I}$ be the set of all instances of MULTIPLEKNAPSACK that satisfy condition (i) and have items of the same profit. Let $I = (A, B) \in \mathcal{I}$ and $p \in \mathbb{N}$ such that the profit $\mathrm{p}(a) = p$ for each item $a \in A$. Then conditions (ii) to (iv) hold for $C = p$ and $\alpha = 1$. Furthermore, the set $\mathcal{I}_P$ actually contains every instance of the PARTITION with even $\mathrm{s}(A)$. However, this restriction does not simplify the problem because instances with odd $\mathrm{s}(A)$ must always be no-instances. By Theorem 2 we can apply Theorem 13 to get the desired result.

The Multiple Subset Sum Problem. We have to find a set $\mathcal{I}$ of instances of the multiple subset sum problem that satisfies the preconditions of Theorem 13. First, we can restrict ourselves to instances that satisfy the knapsack condition (i). Any set $\mathcal{I}$ of such instances is unambiguously determined by $\mathcal{I}_P$. Therefore we only need to give the set $\mathcal{I}_P$ and α. The conditions (ii) to (iv) can be equivalently expressed as: For each instance $A \in \mathcal{I}_P$ there is a $C \in \mathbb{N}$ with

(ii) $|C| = \|A\|^{O(1)}$,
(iii) $\mathrm{s}(A) \leq n\alpha C$, and
(iv) $\mathrm{s}(a) \geq C$ for each item $a \in A$.

By a linear reduction from PARTITION-φ' (see Sect. 2.4), we will show that there is such a set $\mathcal{I}_P$ and not every instance $A \in \mathcal{I}_P$ can be solved in time $2^{o(n)} \times \|A\|^{O(1)}$ if the ETH holds true. For this, transform the instances of PARTITION-φ' such that the sizes of all items are similar, i.e. every instance A fulfills the predicate $\psi(A)$:

$$\psi(A) \iff \exists\, C \in \mathbb{N}\, \forall\, a \in A \colon C \leq \mathrm{s}(a) \leq 3C. \tag{7}$$

Lemma 14. *There is no algorithm that decides* PARTITION-ψ *in time* $2^{o(n)} \times \|A\|^{O(1)}$, *unless the ETH fails.*

Proof (Sketch). Add a suitably large value C to the size of all items. Since a solution contains exactly $\frac{n}{2}$ elements the target B must be increased by $\frac{n}{2}C$.

We are now able to prove the second part of Theorem 12. Let $\mathcal{I}_P$ be the set of instances of PARTITION-ψ for which $\mathrm{s}(A)$ is even. Observe that $\psi(A)$ implies $\mathrm{s}(A) \leq n3C$ for any instance $A \in \mathcal{I}_P$. The set $\mathcal{I} = \{(A, B_A) \mid A \in \mathcal{I}_P\}$ with $B_A = \left(\frac{1}{2}\mathrm{s}(A), \frac{1}{2}\mathrm{s}(A)\right)$ will therefore satisfy the preconditions of Theorem 13 for $\alpha = 3$. Combining Theorem 14 with Theorem 13 yields the desired result.

5.2   Multi-dimensional Knapsack

Theorem 15. *There is no PTAS for* 2D-KNAPSACK *with running time* $n^{o\left(\frac{1}{\varepsilon}\right)} \times \|I\|^{O(1)}$, *unless the ETH fails.*

Proof (Sketch). Pătraşcu and Williams [29] showed that, under assumption of the ETH, SIZEDSUBSETSUM with n items and solution size k cannot be decided in time $n^{o(k)}$. Combined with the reduction to 2D-KNAPSACK by Kulik and Shachnai [20] this yields the proposed bound on the running time.

This bound asymptotically matches the running time $n^{O\left(\frac{1}{\varepsilon}\right)} \times \|I\|^{O(1)}$ of known approximation schemes [2].

6   Open Questions

Some questions regarding exact algorithms remain open, for example no strong lower bound is known for $\langle Om \mid \mid \sum C_j \rangle$ and $\langle Fm \mid (\text{pmtn}) \mid \sum C_j \rangle$. More importantly no non-trivial upper bound for many problems with arbitrary many machines, e.g. $\langle O \mid \mid C_{\max} \rangle$, is known. Another open question is whether $\langle P \mid \mid C_{\max} \rangle$ admits an approximation scheme with running time $2^{o(1/\varepsilon)} \times \|I\|^{O(1)}$.

References

1. Achugbue, J.O., Chin, F.Y.: Scheduling the open shop to minimize mean flow time. SIAM Journal on Computing 11(4), 709–720 (1982)
2. Caprara, A., Kellerer, H., Pferschy, U., Pisinger, D.: Approximation algorithms for knapsack problems with cardinality constraints. European Journal of Operational Research 123(2), 333–345 (2000)
3. Chen, J., Huang, X., Kanj, I., Xia, G.: On the computational hardness based on linear FPT-reductions. Journal of Combinatorial Optimization 11, 231–247 (2006)
4. Drozdowski, M.: On The Complexity of Multiprocessor Task Scheduling. Bulletin of the Polish Academy of Sciences. Technical Sciences 43(3), 381–392 (1995)
5. Du, J., Leung, J.Y.-T.: Minimizing Mean Flow Time in Two-Machine Open Shops and Flow Shops. Journal of Algorithms 14(1), 24–44 (1993)
6. Ebenlendr, T., Krčál, M., Sgall, J.: Graph balancing: a special case of scheduling unrelated parallel machines. In: Teng, S.-H. (ed.) Proceedings of the Nineteenth Annual ACM-SIAM Symposium on Discrete Algorithms, pp. 483–490. SIAM, Philadelphia (2008)
7. Flum, J., Grohe, M.: Parameterized Complexity Theory. Springer (2006)
8. Fomin, F.V., Kratsch, D.: Exact Exponential Algorithms. Springer (2010)
9. Garey, M.R., Johnson, D.S., Sethi, R.: The complexity of flowshop and jobshop scheduling. Mathematical Operations Research 1, 117–129 (1976)
10. Garey, M.R., Johnson, D.S.: Computers and Intractability: A Guide to the Theory of NP-Completeness. W. H. Freeman & Co., New York (1979)
11. Gonzalez, T., Sahni, S.: Flowshop and jobshop schedules: complexity and approximation. Operations Research 26(1), 36–52 (1978)
12. Held, M., Karp, R.: A Dynamic Programming Approach to Sequencing Problems. Journal of the Society for Industrial and Applied Mathematics 10(1), 196–210 (1962)
13. Hoogeveen, H., Schuurman, P., Woeginger, G.J.: Non-approximability results for scheduling problems with minsum criteria. Journal on Computing 13(2), 157–168 (2001)

14. Horowitz, E., Sahni, S.: Computing Partitions with Applications to the Knapsack Problem. Journal of the ACM 21(2), 277–292 (1974)
15. Impagliazzo, R., Paturi, R.: On the Complexity of k-SAT. Journal of Computer and System Sciences 62(2), 367–375 (2001)
16. Impagliazzo, R., Paturi, R., Zane, F.: Which Problems Have Strongly Exponential Complexity? Journal of Computer and System Sciences 63(4), 512–530 (2001)
17. Jansen, K.: A Fast Approximation Scheme for the Multiple Knapsack Problem. In: Bieliková, M., Friedrich, G., Gottlob, G., Katzenbeisser, S., Turán, G. (eds.) SOFSEM 2012. LNCS, vol. 7147, pp. 313–324. Springer, Heidelberg (2012)
18. Jansen, K., Land, K., Land, F.: Bounding the Running Time of Algorithms for Scheduling and Packing Problems. Technical Report 1302. Institute of Computer Science, University of Kiel, Germany (2013)
19. Jansen, K., Kratsch, S., Marx, D., Schlotter, I.: Bin packing with fixed number of bins revisited. Journal of Computer and System Sciences 79(1), 39–49 (2013)
20. Kulik, A., Shachnai, H.: There is no EPTAS for two-dimensional knapsack. Information Processing Letters 110 16, 707–710 (2010)
21. Lenstra, J.K., Rinnooy Kan, A.H.G.: Complexity of Scheduling under Precedence Constraints. Operations Research 26(1), 22–35 (1978)
22. Lenstra, J.K., Rinnooy Kan, A.H.G., Brucker, P.: Complexity of Machine Scheduling Problems. In: Hammer, P., Johnson, E., Korte, B., Nemhauser, G. (eds.) Studies in Integer Programming. Annals of Discrete Mathematics, vol. 1, pp. 343–362. Elsevier (1977)
23. Lenté, C., Liedloff, M., Soukhal, A., T'kindt, V.: Exponential-time algorithms for scheduling problems. In: 10th Workshop on Models and Algorithms for Planning and Scheduling Problems (MAPSP 2011), Nymburk, Czech Republic (2011)
24. Lokshtanov, D., Marx, D., Saurabh, S.: Lower bounds based on the Exponential Time Hypothesis. Bulletin of the EATCS 105, 41–72 (2011)
25. Marx, D.: Parameterized complexity and approximation algorithms. The Computer Journal 51(1), 60–78 (2008)
26. O'Neil, T.E.: Sub-Exponential Algorithms for 0/1-Knapsack and Bin Packing. In: Arabnia, H.R., Gravvanis, G.A., Solo, A.M.G. (eds.) Proceedings of the 2011 International Conference on Foundations of Computer Science, pp. 209–214. CSREA Press (2011)
27. O'Neil, T.E., Kerlin, S.: A simple $2^{O(\sqrt{x})}$-algorithm for Partition and Subset Sum. In: Arabnia, H.R., Gravvanis, G.A., Solo, A.M.G. (eds.) Proceedings of the 2010 International Conference on Foundations of Computer Science, pp. 55–58. CSREA Press (2010)
28. Papadimitriou, C.H., Yannakakis, M.: Optimization, approximation, and complexity classes. Journal of Computer and System Sciences 43(3), 425–440 (1991)
29. Pătraşcu, M., Williams, R.: On the possibility of faster SAT algorithms. In: Charikar, M. (ed.) Proceedings of the Twenty-First Annual ACM-SIAM Symposium on Discrete Algorithms, pp. 1065–1075. SIAM, Philadelphia (2010)
30. Rinnooy Kan, A.H.G.: Machine scheduling problems: classification, complexity and computations. Stenfert Kroese (1976)
31. Wegener, I.: Complexity Theory: Exploring the Limits of Efficient Algorithms. Trans. from the German by R. J. Pruim. Springer (2003)
32. Williamson, D.P., Hall, L.A., Hoogeveen, J.A., Hurkens, C.A.J., Lenstra, J.K., Sevast'janov, S.V., Shmoys, D.B.: Short shop schedules. Operations Research 45(2), 288–294 (1997)

When Is Weighted Satisfiability FPT?

Iyad A. Kanj[1] and Ge Xia[2]

[1] School of Computing, DePaul University, Chicago, IL
ikanj@cs.depaul.edu
[2] Dept. of Computer Science, Lafayette College, Easton, PA
xiag@lafayette.edu

Abstract. The weighted monotone and antimonotone satisfiability problems on normalized circuits, abbreviated WSAT$^+$[t] and WSAT$^-$[t], are canonical problems in the parameterized complexity theory. We study the parameterized complexity of WSAT$^-$[t] and WSAT$^+$[t], where $t \geq 2$, with respect to the genus of the circuit. For WSAT$^-$[t], we give a fixed-parameter tractable (FPT) algorithm when the genus of the circuit is $n^{o(1)}$, where n is the number of the variables in the circuit. For WSAT$^+$[2] (*i.e.*, weighted monotone CNF-SAT) and WSAT$^+$[3], which are both W[2]-complete, we also give FPT algorithms when the genus is $n^{o(1)}$. For WSAT$^+$[t] where $t > 3$, we give FPT algorithms when the genus is $O(\sqrt{\log n})$. We also show that both WSAT$^-$[t] and WSAT$^+$[t] on circuits of genus $n^{\Omega(1)}$ have the same W-hardness as the general WSAT$^+$[t] and WSAT$^-$[t] problem (*i.e.*, with no restriction on the genus), thus drawing a precise map of the parameterized complexity of WSAT$^-$[t], and of WSAT$^+$[t], for $t = 2, 3$, with respect to the genus of the underlying circuit.

As a byproduct of our results, we obtain, via standard parameterized reductions, tight results on the parameterized complexity of several problems with respect to the genus of the underlying graph.

1 Introduction

We consider the *weighted satisfiability* problems on monotone and antimonotone normalized circuits of depth at most $t \geq 2$. In the ANTIMONOTONE WEIGHTED SATISFIABILITY problem on normalized circuits of depth at most $t \geq 2$, abbreviated WSAT$^-$[t], we are given a circuit C of depth t in the *normalized* form [9,10] (*i.e.*, the output gate is an AND-gate, and the gates alternate between AND-gates and OR-gates) whose input literals are all negative, and an integer parameter $k \geq 0$, and we need to decide if C has a satisfying assignment of weight k (*i.e.*, assigning k variables in C the value 1). In the MONOTONE WEIGHTED SATISFIABILITY on normalized circuits of depth at most $t \geq 2$, abbreviated WSAT$^+$[t], we are given a circuit C of depth t in the normalized form whose input literals are positive, and an integer parameter $k \geq 0$, and we need to decide if C has a satisfying assignment of weight k. Our goal in this paper is to study the parameterized complexity of WSAT$^-$[t] and WSAT$^+$[t] with respect to the genus of the circuit. We define the genus of the circuit to be the genus of the underlying undirected graph after the output gate is removed. The reason we exclude the output gate

F. Dehne, R. Solis-Oba, and J.-R. Sack (Eds.): WADS 2013, LNCS 8037, pp. 451–462, 2013.

of the circuit in the definition of the genus is two-fold. First, excluding the output gate allows us to use standard fpt-reductions to model problems on graphs satisfying a certain genus upper bound as WSAT$^-[t]$ and WSAT$^+[t]$ problems on circuits that satisfy the same genus upper bound, whereas such modeling would not be possible if the genus is defined to be that of the whole circuit. Second, as it turns out, one obtains the same results obtained in the current paper if the genus is defined to be that of the whole circuit. To see this, observe that all positive results (*i.e.*, FPT results) obtained in this paper carry over because an upper bound on the genus of the whole circuit implies the same upper bound on the genus of the circuit with the output gate removed; on the other hand, it is straightforward to show that all the W-hardness results obtained in this paper hold if the genus is defined to be that of the whole circuit. We mention that the WEIGHTED CIRCUIT SATISFIABILITY problem on depth-t planar circuits with the output gate included is solvable in polynomial time [4], whereas it can be easily shown that WSAT$^-[t]$ and WSAT$^+[t]$ are $\mathcal{NP}$-complete on planar circuits (and hence on circuits of any genus) with the output gate removed.

1.1 Motivation and Related Work

The problems under consideration are of prime interest both theoretically and practically. From the theoretical perspective, they naturally represent the weighted satisfiability of (montone/antimontone) t-normalized propositional formulas, *i.e.*, products-of-sums-of-products... (see [9,10]), including the canonical problems weighted antimonotone/monotone CNF-SAT. Recently, Marx [16] proved that weighted monotone/antimonotone circuit satisfiability has no FPT approximation algorithm with any approximation ratio function ρ, unless FPT = W[1]. Moreover, the WSAT$^-[t]$ and the WSAT$^+[t]$ problems are the canonical complete problems for the different levels of the parameterized complexity hierarchy — the W-hierarchy, and the W-hierarchy can be defined based on them [9,10]. Therefore, determining the underlying structure that makes these problems (parameterized) tractable is important from the perspective of complexity theory. Recently, Marx [16] proved that weighted monotone/antimonotone circuit satisfiability has no FPT approximation algorithm with any approximation ratio function ρ, unless FPT = W[1]. From a more practical perspective, WSAT$^-[t]$ and WSAT$^+[t]$ can model several natural problems. Therefore, parameterized algorithms for WSAT$^-[t]$ and WSAT$^+[t]$ can be used to obtain parameterized algorithms for some natural problems via reductions to/from WSAT$^-[t]$ and WSAT$^+[t]$, as we shall see in Section 6.

   Algorithms for many natural problems on planar graphs and, more generally, on graphs whose genus meets certain upper bounds were extensively researched (see [2,5,6,7,11,12], among others). Moreover, research results on planar circuits, and on satisfiability problems defined on certain structures that are planar or that satisfy certain structural properties, are abundant. Planar Boolean circuits were researched because they can be used to study VLSI chips (for example, see [18]). Khanna and Motwani [15] studied the approximation of instances of satisfiability problems (weighted and unweighted) whose underlying structure is planar. Cai *et al.* [3] studied the parameterized complexity of the satisfiability problems introduced by Khanna and Motwani [15], and showed that these problems are $W[1]$-hard even when the underlying incidence graph is planar. Researchers have also studied the parameterized complexity of CNF-SAT with respect to the treewidth of graphs defined based on the circuit (see the survey [19]).

1.2   Our Results

We obtain the following results regarding the WSAT$^-[t]$ ($t \geq 2$), which is $W[t]$-complete for odd t and $W[t-1]$-complete for even t [9,10] (below n is the number of the variables in the circuit):

(i) **Tight results**: We give an FPT algorithm for WSAT$^-[t]$ when the genus is $n^{o(1)}$, and show that WSAT$^-[t]$ has the same W-hardness as the general WSAT$^-[t]$ problem when the genus is $n^{\Omega(1)}$.

(ii) **Applications**: We show that INDEPENDENT SET ON HYPERGRAPHS and the RED-BLUE NONBLOCKER problems are FPT on (hyper)graphs of genus $N^{o(1)}$ and $W[1]$-complete on (hyper)graphs of genus $N^{\Omega(1)}$ (N is the number of red vertices in RED-BLUE NONBLOCKER and the total number of vertices in INDEPENDENT SET ON HYPERGRAPHS).

We obtain the following results regarding the WSAT$^+[t]$ ($t \geq 2$), which is known to be $W[t]$-complete for even t and $W[t-1]$-complete for odd t [9,10]:

(1) **Tight results for $t = 2, 3$**: We give FPT algorithms for WSAT$^+[2]$ (*i.e.*, weighted monotone CNF-SAT) and WSAT$^+[3]$ when the genus is $n^{o(1)}$, and show that they are $W[2]$-complete when the genus is $n^{\Omega(1)}$.

(2) **Results for $t > 3$**: We give an FPT algorithm for WSAT$^+[t]$ when the genus is $O(\sqrt{\log n})$, and show that WSAT$^+[t]$ has the same W-hardness as the general WSAT$^+[t]$ when the genus is $n^{\Omega(1)}$.

(3) **Applications**: We show that RED-BLUE DOMINATING SET, HITTING SET, and SET COVER are FPT if the underlying graph/hypergraph has genus $N^{o(1)}$, and $W[2]$-complete if the underlying graph/hypergraph has genus $N^{\Omega(1)}$ (N is the number of red vertices in RED-BLUE DOMINATING SET, the cardinality of the vertex-set in HITTING SET, and the number of sets in SET COVER).

Remark. None of the algorithms presented in the current paper needs to know in advance, nor needs to decide, whether or not the minimum genus of the input circuit satisfies the required upper bounds.

2   Preliminaries

We assume familiarity with the basic terminology and definitions in graph theory and parameterized complexity, and refer the reader to [9,10,20].

A *hypergraph* $\mathcal{H} = (V, E)$ consists of a *vertex set* $V = V(\mathcal{H})$ and an *edge set* $E = E(\mathcal{H})$ so that $e \subseteq V$ for every $e \in E$. If E is allowed to be a multiset we call $\mathcal{H}$ a *multihypergraph*. We also call the edges in a hypergraph *hyperedges*.

A graph has *genus* g if it can be drawn on a surface of genus g (a sphere with g handles) without intersections. We say a (multi)hypergraph $\mathcal{H}$ is *embeddable in a surface* if the bipartite incidence graph obtained from $\mathcal{H}$ by replacing each of its hyperedges by a vertex adjacent to all the vertices in the hyperedge is embeddable in that surface. In particular, this definition allows us to speak of *(multi)hypergraph of genus* g. We refer the reader to [13] for more information on the genus of a graph.

A *circuit* is a directed acyclic graph. The vertices of indegree 0 are called the (input) *variables*, and are labeled either by *positive literals* x_i or by *negative literals* $\overline{x}_i$. The vertices of indegree larger than 0 are called the *gates* and are labeled with Boolean operators AND or OR. A special gate of outdegree 0 is designated as the *output* gate. We do not allow NOT gates in the above circuit model, since by De Morgan's laws, a general circuit can be effectively converted into the above circuit model. A circuit is said to be *monotone* (resp. *antimonotone*) if all its input literals are positive (resp. negative). The *depth* of a circuit is the maximum distance from an input variable to the output gate of the circuit. A circuit represents a Boolean function in a natural way. The size of a circuit C, denoted $|C|$, is the size of the underlying graph (number of vertices and edges). An *occurrence* of a literal in C is an edge from the literal to a gate in C. Therefore, the total number of occurrences of the literals in C is the number of outgoing edges from the literals in C to its gates. The *genus of a circuit* is the genus of the underlying undirected graph after the output gate has been removed.

We consider circuits whose output gate is an AND-gate and that are in the *normalized* form (see [9,10]). In the normalized form every (nonvariable) gate has outdegree at most 1, and above the output AND-gate, the gates are structured into alternating levels of ORs-of-ANDs-of-ORs... We denote a circuit that is in the normalized form and that is of depth at most $t \geq 2$ by a Π_t circuit. We write Π_t^+ to denote a monotone Π_t circuit, and Π_t^- to denote an antimonotone Π_t circuit. We do not assume that the literals appear at the same level of the circuit.

Throughout the paper, we implicitly assume that the following hold after simplifications: every gate with outdegree 0 except the output gate is removed, every gate has indegree at least 2, and no two gates of the same type such that one is incoming to the other exist.

We say that a truth assignment τ to the variables of a circuit C *satisfies* a gate g in C if τ makes the gate g have value 1, and that τ *satisfies the circuit* C if τ satisfies the output gate of C. A circuit C is *satisfiable* if there is a truth assignment to the input variables of C that satisfies C. The *weight* of an assignment τ is the number of variables assigned value 1 by τ. An indegree-2 gate is called a *2-literal gate* if both its incoming edges are from literals. A *critical gate* in a Π_t circuit C is an OR-gate that is connected to the output AND-gate of the circuit; clearly, any satisfying assignment to C must satisfy all critical gates in C. If we remove the literals from C, we obtain a directed graph whose underlying undirected graph is a tree T_C. If we root T_C at the output gate of C, we can now use the terms *child(ren), parent, grandparent* of a gate in T_C in a natural way. Note that every literal in C is connected to some gates in T_C. For a gate g in T_C, we denote by T_g the subtree of T_C rooted at g. We may regard an edge in T_C between a child g' of a gate g and g, or between a literal and gate g, as an incoming edge to g.

A *parameterized problem* Q is a set of pairs (x, k), where x is the instance and the non-negative integer k is the *parameter*. A parameterized problem Q is *fixed-parameter tractable* [9], shortly *FPT*, if there is an algorithm that decides whether or not an input (x, k) is a member of Q in time $f(k)N^{O(1)}$, where $f(k)$ is a computable function independent of $N = |x|$. Let FPT denote the class of all fixed-parameter tractable parameterized problems. A parameterized problem Q is *fpt-reducible* to a parameterized problem Q' if there is an algorithm that transforms each instance (x, k) of Q into an instance $(x', g(k))$ (g is a function of k only) of Q' in time $f(k)N^{O(1)}$, where f and g are computable functions of k and $N = |x|$, such that $(x, k) \in Q$ if and only if $(x', g(k)) \in Q'$. By *fpt-time*, we denote time complexity of the form $f(k)N^{O(1)}$, where N is the input length, k is the parameter, and f is a computable function of k. Based on the notion of fpt-reducibility [9], a hierarchy of parameterized complexity, *the W-hierarchy* $\bigcup_{t \geq 0} W[t]$, where $W[t] \subseteq W[t+1]$ for all $t \geq 0$, has been introduced, in which the 0-th level $W[0]$ is the class FPT.

For $t \geq 2$, the WEIGHTED Π_t-CIRCUIT SATISFIABILITY problem, abbreviated WSAT$[t]$ is for a given Π_t-circuit C and a given parameter k, to decide if C has a satisfying assignment of weight k. The WEIGHTED MONOTONE Π_t-CIRCUIT SATISFIABILITY problem, abbreviated WSAT$^+[t]$, and the WEIGHTED ANTIMONOTONE Π_t-CIRCUIT SATISFIABILITY problem, abbreviated WSAT$^-[t]$ are the WSAT$[t]$ problems on monotone circuits and antimonotone circuits, respectively. We denote by WSAT$^-$ the WSAT$^-[2]$ problem, and by WSAT$^+$ the WSAT$^+[2]$ problem (*i.e.*, the weighted antimonotone/monotone CNF-SAT problem). It is known that for each integer $t \geq 2$: WSAT$^+[t]$ is $W[t]$-complete for even t and $W[t-1]$-complete for odd t, and WSAT$^-[t]$ is $W[t]$-complete for odd t and $W[t-1]$-complete for even t [9,10].

3 A Structural Result

The following result shows that any Π_t circuits whose genus is at most linear can be reduced to an equivalent one whose size is linear and in which the number of occurrences of the literals is linear:

Proposition 1. *Let C be a Π_t circuit on n variables of genus $g(n) \leq n$. In polynomial time we can reduce C to an equivalent Π_t circuit C' of genus $g(n)$ on the same set of variables such that the number of occurrences of the literals in C' is $O(n)$, and such that the size of C' is $O(n)$.*

Proof. (Sketch) We start by applying simplification and reduction rules to remove logically-equivalent gates from the circuit. We then use counting arguments based on amortized analysis and Euler-type combinatorial results for graphs and (multi)hypergraphs (see Lemmas 4.4–4.6 in [14]) to upper bound the number of occurrences of the literals and subsequently the size by $O(n)$. □

4   The Antimonotone Case

In this section we give tight results on the parameterized complexity of the WSAT$^-[t]$ problem, where $t \geq 2$ is an integer, with respect to the circuit genus.

Definition 1. Let C be a Π_t^- circuit, and let x_i be a variable in C. We say that x_i is a *zero-variable* for C if assigning $x_i = 1$ causes C to evaluate to 0. Therefore, any zero-variable for C must be assigned the Boolean value 0 in a satisfying truth assignment for C. A *nonzero-variable* for C is a variable that is not a zero-variable for C. A Π_t^- circuit C has *no zero-variables* if all the variables in C are nonzero-variables.

Proposition 2. *Let (C, k) be an instance of WSAT$^-[t]$ ($t \geq 2$) such that the genus of C is $g(n) = n^{o(1)}$. In fpt-time, we can reduce (C, k) to an equivalent instance (C', k) where C' has genus at most $g(n)$ and no zero-variables, and such that the number of variables n' in C' satisfies $g(n) \leq n' \leq n$.*

The following theorem shows that a Π_t^- circuit with no zero-variables and with a linear number of (literal) occurrences can always be satisfied with an (increasing) function of n variables assigned 1.

Theorem 1. *Let C be Π_t^- circuit with n variables such that C has no zero-variables and the number of occurrences of the literals in C is $O(n)$. C has a satisfying assignment in which at least $f(n) = \log^{(d^t)} n$ variables are assigned 1, where $\log^{(i)}$ indicates the logarithm (base 2) applied i times, and $d > 0$ is an integer constant.*

Proof. (Sketch) The proof is by induction on t, the depth of the circuit. The base case when $t = 2$ can be easily handled by reducing it to the INDEPENDENT SET problem on multigraphs of bounded degree, which has a solution of size $\Omega(n)$. When $t \geq 3$, we define an intricate recursive procedure in which each step either assigns a variable 1, or reduces the degree of the variables by 1. The procedure will end when either enough variables are assigned 1 (we are done), or when the degree of the variables is at most 1. By a careful analysis, we can prove that in the case when the degree of the variables is at most 1 there will be enough variables left that can be assigned 1 in a satisfying assignment. □

Theorem 2. *The WSAT$^-[t]$ ($t \geq 2$) problem on circuits of genus $g(n) = n^{o(1)}$ (n is the number of variables) is FPT, and is $W[t]$-complete for odd t and $W[t-1]$-complete for even t if $g(n) = n^{\Omega(1)}$.*

Proof. Let $g(n) = n^{o(1)} = n^{1/\mu(n)}$, where $\mu(n)$ is a complexity function[1], and let (C, k) be an instance of the WSAT$^-[t]$ ($t \geq 2$) problem on circuits of genus

---

[1] In this paper, complexity functions are assumed to be unbounded and nondecreasing.

$g(n)$. By Proposition 2, we can assume that C has no zero-variables, and that the number of variables n in C is least $g(n)$. By Proposition 1, we may assume that the number of occurrences of the literals in C is $O(n)$; if this is not the case then the genus of the circuit is not upper bounded by $g(n)$, and we reject the instance. By Theorem 1, C has a satisfying assignment in which at least $f(n)$ variables are assigned the value 1, where $f(n)$ is the function given in the lemma. Therefore, if $k \leq f(n)$ then we accept the instance (C, k); otherwise, $k > f(n)$ and in fpt-time we can decide the instance by a brute-force algorithm that enumerates every weight-k assignment. The hardness result follows by a simple padding argument. □

5   The Monotone Case

In this section we give tight results on the parameterized complexity of the WSAT$^+[t]$ problem, where $t \geq 2$ is an integer, with respect to the circuit genus.

Proposition 3. *Let (C, k) be an instance of WSAT$^+[t]$ ($t \geq 2$) such that C has genus $g(n) = n^{o(1)}$. There is an fpt-time algorithm that reduces (C, k) to $h(k)n^{O(1)}$ many instances (C', k') of WSAT$^+[t]$, where h is a complexity function and $k' \leq k$, such that (C, k) is a yes-instance if and only if at least one of the instances (C', k') is, and such that each instance (C', k') satisfies that: (1) the number of critical gates in C' is at most $2k'$, (2) every variable in C' is incoming to gates in at most two subtrees T_p, T_q of $T_{C'}$ rooted at critical gates p, q in C', and (3) the genus of C' is at most $g(n)$.*

Proof. Let $g(n)$ be a complexity function such that $g(n) = n^{o(1)}$. Since $g(n) = n^{o(1)}$, $g(n) \leq n^{1/\mu(n)}$ for some complexity function $\mu(n)$.

Let (C, k) be an instance of WSAT$^+[t]$, where C is a Π_t^+ circuit with set of variables $X = \{x_1, \ldots, x_n\}$, and k is the parameter. If more than k variables are incoming to the output gate of C, then clearly C has no satisfying assignment of weight k, and we reject the instance (C, k). Otherwise, we can assign the value 1 to the variables incoming to the output-gate of C, remove these variables, and update C and k accordingly. So we may assume, without loss of generality, that C has no variables incoming to its output gates, and that all gates incoming to the output gates are OR-gates (by the simplification rules discussed in Section 2), and hence are critical gates.

For each critical gate p in C, consider the subtree T_p of T_C. In the case when $t = 2$, this subtree is trivial, and consists of gate p. We form an auxiliary graph $\mathcal{B}$ as follows. Starting at each critical gate p, we contract the edges in T_p to form a single vertex p' whose incoming variables are the variables that are incoming to at least one gate in T_p. Note that if a variable is incoming to several gates in T_p, then there will be multiple edges between p' and this variable. Let $\mathcal{G}$ be the set of vertices resulting from contracting each tree T_p corresponding to a critical gate p in C. Let $\mathcal{B} = (\mathcal{G}, X)$ be the underlying undirected bipartite graph resulting from this contraction with the multiple edges removed. That is, there is an (undirected) edge in $\mathcal{B}$ between a variable $x_i \in X$ and a gate p' in

$\mathcal{G}$ if and only if x_i is incoming to some gate in T_p. Clearly, the genus of $\mathcal{B}$ is at most $g(n)$. Observe that since each critical gate p must be satisfied by every assignment that satisfies C, for any vertex p' in $\mathcal{G}$, at least one variable incident to p' in $\mathcal{B}$ must be assigned 1 in any truth assignment satisfying C. Let $n_g = |\mathcal{G}|$.

We partition the variables in X into two sets: $X_{\geq 3}$ that consists of each variable in X whose degree in $\mathcal{B}$ is at least 3, and $X_{\leq 2}$ consisting of each variable in X whose degree in $\mathcal{B}$ is at most 2. Let $n_3 = |X_{\geq 3}|$ and $n_2 = |X_{\leq 2}|$. By defining a multihypergraph whose vertex-set is $\mathcal{G}$, and whose hyperedges correspond to the neighborhoods of the variables in $X_{\geq 3}$, we obtain from Lemma 4.4 in [14] that $n_3 \leq 2n_g + 4g(n)$; if the preceding upper bound on n_3 does not hold, then we reject the instance (this means that the genus of the circuit is not at most $g(n)$). We use the following search-tree algorithm $\mathcal{A}$ that distinguishes two cases:

Case 1. $n_g \leq n^{1/\mu(n)}$. In this case we have $n_3 \leq 2n_g + 4g(n) \leq 6n^{1/\mu(n)}$. The number of subsets of $X_{\geq 3}$ of size at most k is at most $\Sigma_{i=0}^{k} \binom{n_3}{i} \leq kn_3^k \leq k \cdot (6n^{1/\mu(n)})^k$. We try each such subset of $X_{\geq 3}$ as a candidate subset of variables that will be assigned value 1 by a satisfying assignment of weight k. For each such candidate subset S, we update the gates in C in a natural way according to the partial assignment assigning the variables in S the value 1, and those in $X_{\geq 3} \setminus S$ the value 0. We remove all variables in $X_{\geq 3}$ from C, and update C and k appropriately. Since each remaining variable is in $X_{\leq 2}$, each variable can satisfy at most 2 critical gates, and hence if the number of critical gates in C is more than $2k$, then we can reject the resulting instance (C, k). Therefore, for each instance resulting from the enumeration of such a subset S of $X_{\geq 3}$, either the number of remaining critical gates in C is more than $2k$ and we reject the instance since k variables in $X_{\leq 2}$ cannot satisfy all the critical gates of C, or the number of critical gates in C is at most $2k$. Since the number of candidate subsets of $X_{\geq 3}$ is at most $k \cdot (6n^{1/\mu(n)})^k$ which can be enumerated in fpt-time, the statement of the theorem follows.

Case 2. $n_g > n^{1/\mu(n)}$. Let G be the subgraph of $\mathcal{B}$ induced by the set of vertices in $\mathcal{G}$ plus those in $X_{\geq 3}$. Since $n_3 \leq 2n_g + 4g(n) \leq 6n_g$, the number of vertices in G is at most $7n_g$. Since the genus of G is at most $g(n)$, by Euler's formula the number of edges in G is at most $21n_g + 6g(n) \leq 27n_g$. Let $Y_{\geq 3}$ be the set of variables in $X_{\geq 3}$ of degree at least $27n_g/\log n$ in G. Since the number of edges in G is at most $27n_g$, it follows that $|Y_{\geq 3}| \leq \log n$. In time $(\log n)^k$, which is fpt-time, we can enumerate each subset of $Y_{\geq 3}$ of size at most k as a candidate subset of variables that are assigned value 1 by a satisfying assignment of weight k. For each such *nonempty* candidate subset, C is updated appropriately (as in **Case 1** above) and k is decreased by at least the size of the subset, which is nonzero, and we can repeat the execution of the whole algorithm $\mathcal{A}$; this algorithm will be repeated at most k times. If the candidate subset is empty, then along this branch we reject the instance (C, k) since C cannot be satisfied by an assignment of weight k. The preceding statement can be justified as follows. In any satisfying assignment, the critical gates, whose number is $n_g > n^{1/\mu(n)}$, must be satisfied. Since the chosen subset of $Y_{\geq 3}$ is empty, we are working under the assumption that no variable in $Y_{\geq 3}$ is assigned 1 by any satisfying assignment. Therefore, the variables assigned 1 by any satisfying assignment must be chosen from $X_{\geq 3} - Y_{\geq 3}$ or from $X_{\leq 2}$. Each variable in $X_{\geq 3} - Y_{\geq 3}$ can satisfy at most $27n_g/\log n$ critical gates in C, and each variable in $X_{\leq 2}$ can satisfy at most 2 critical gates. Therefore, k variables from $(X_{\geq 3} - Y_{\geq 3}) \cup X_{\leq 2}$ can satisfy at most $27kn_g/\log n < n_g$ critical gates in C, and hence cannot satisfy C. We assumed here that $k < \log n/27$; otherwise, we can decide the instance in fpt-time.

It follows that the algorithm $\mathcal{A}$ outlined above runs in fpt-time, and either solves the instance (C, k), or reduces it to $h(k)n^{O(1)}$ many instances (C', k') ($k' < k$), such that (C, k) is a yes-instance if and only if at least one of the instances (C', k') is, and such that each of the instances (C', k') satisfies conditions (1), (2), and (3) in the statement of the theorem. □

Theorem 3. *The* WSAT$^+$ *problem on circuits of genus $g(n)$ is FPT if $g(n) = n^{o(1)}$, and is $W[2]$-complete if $g(n) = n^{\Omega(1)}$.*

Proof. By Proposition 3, in fpt-time we can reduce an instance (C, k) of WSAT$^+$ on circuits of genus $g(n) = n^{o(1)}$ to $h(k)n^{O(1)}$ many instances (C', k') of WSAT$^+$, such that each instance (C', k') satisfies the properties described in Proposition 3. It suffices to show that we can decide each such instance (C', k') in fpt-time. First, observe that since each subtree T_p rooted at a critical gate p consists of a single critical gate of C', each variable in C' has outdegree at most 2; that is, each variable in C' is incoming to at most two gates in C'. For two variables x_i and x_j in C', if the set of gates that x_i is incoming to is a subset of that of x_j, then we say that x_j *dominates* x_i. We perform the following reductions. If more than k' variables are incoming to the output gate of C', then C' has no satisfying assignment of weight k', and we reject (C', k'). Otherwise, we assign the value 1 to the variables incoming to the output gate of C', remove them, and update C' and k' accordingly. For any two 2-literal gates that have the same pair of variables incoming to them, we remove one of the two gates from C'. So assume, without loss of generality, that in the instance (C', k') the circuit C' contains no variables incoming to its output gate, and that there are no two 2-literal gates in C' with the same pair of variables incoming to them. For every two variables x_i and x_j in C, if x_i dominates x_j then remove x_j. After applying the previous reductions, it is easy to see that the number of degree-1 variables is at most $2k'$, and the number of degree-2 variables is at most $\binom{2k'}{2}$. Therefore, the resulting circuit has size $O(k'^2)$, and in fpt-time we can decide (C', k'). The proof of hardness result is a simple padding argument. □

The rest of this section handles the cases when $t > 2$. We follow the terminology of [8]. Let G be a graph, and let $V' \subseteq V(G)$ and $E' \subseteq E(G)$ be such that every vertex in V' is an endpoint of some edge in E'. Let G^- be the graph obtained from G by removing the vertices in V' and the edges in E'. G is said to be (V', E')-*embeddable* (in the plane) if G^- is embeddable in the plane. The vertices in V' and the edges in E' are called *flying*. The flying edges are partitioned into: (1) *bridges*, those are the edges whose both endpoints are in G^-; (2) *pillars*, those are the edges with exactly one endpoint in G^-; and (3) *clouds*, those are the edges whose both endpoints are not in G^-. A *partially triangulated* $(r \times r)$-*grid* is a graph that contains the $(r \times r)$-grid as a subgraph, and is itself a subgraph of a triangulation of the $(r \times r)$-grid. A graph G is an (r, ℓ)-*gridoid* if it is (V', E')-embeddable for some V', E' such that G^- is a partially triangulated $(r' \times r')$-grid for some $r' \geq r$, and E' contains at most ℓ edges and no clouds. The following result was proved in [8]:

Theorem 4 ([8]). *If a graph G of genus g excludes all $(\lambda - 12g, g)$-gridoids as contractions, for some $\lambda \geq 12g$, then the branchwidth of G is at most $4\lambda(g+1)$.*

Lemma 1. *Let (C, k) be an instance of* WSAT$^+[t]$ *such that C has genus $g(n)$ and at most $2k$ critical gates. Let C^- be the circuit resulting from C after removing the output gate. The branchwidth of the underlying graph of C^- is $O(g^2(n))$.*

Proof. We show that the underlying graph of C^- excludes all $(\lceil \sqrt{kg(n)} \rceil, g(n))$-gridoids as contractions. By setting $\lambda = 12g(n) + \lceil \sqrt{kg(n)} \rceil$, the result follows from Theorem 4. (We assume that $k < g(n)$; otherwise, the problem is FPT.)

Suppose, to get a contradiction, that the underlying graph of C^- contains an $(r, g(n))$-gridoid G as a contraction, for some integer $r \geq \lceil \sqrt{kg(n)} \rceil$. Since the depth of C is at most t, every literal and gate in C^- is within distance (*i.e.*, length of a shortest path) at most t from some critical gate of C. Let S be the set of vertices in G, each of which either corresponds to a critical gate of C or to a contraction of a critical gate of C, and note that $|S| \leq 2k$. Clearly, every vertex in G must be within distance at most t from one of the vertices in S. Embed G in the plane, and let G^-, E' and V' be as in the definition of the gridoid. Note that E' contains at most $g(n)$ edges, and each edge of E' must be incident to at least one vertex in G^-. Call an endpoint of an edge in E' that is in G^- an *anchor vertex*. Since $|E'| \leq g(n)$, it follows that the number of anchor vertices is at most $2g(n)$. There is a path of length at most t from every vertex v in the partially-triangulated grid G^- to some vertex in S; fix such a path for every vertex v in G^-, and denote it by P_v. Since the number of grid vertices at distance at most t from some grid vertex is $O(t^2)$, the number of paths P_v that pass through a fixed anchor vertex is $O(t^2)$. Therefore, the number of grid vertices v whose paths P_v go through anchor vertices is $O(t^2) \cdot 2g(n) = O(g(n))$. For any other vertex v, its path P_v lies completely within G^-, and hence the number of such vertices v is $O(t^2) \cdot |S| = O(k)$. Since for every vertex v in G^-, P_v either goes through an anchor vertex or lies completely within the grid, the number of grid vertices is at most $O(g(n)) + O(k) = O(g(n))$ (we assumed that $k < g(n)$, otherwise, we solve the problem in fpt-time). Since the number of vertices in G^- is at least $r^2 = \Omega(kg)$, this is a contradiction since k can be chosen to be larger than any prespecified constant, and in such case there would be grid vertices that are not within distance t from any vertex in S.   □

Using the above lemma, and an intricate dynamic programming based on tree decomposition, we can show the following:

Proposition 4. *Let C be a Π_t^+ circuit, and let $G = (V, E)$ be the undirected underlying graph of C with the output gate removed. If a tree decomposition* *for G of N nodes and treewidth ω is given, then a minimum weight satisfying assignment of C can be computed in time $2^{O(\omega)}N^{O(1)}$.*

Theorem 5. *The* WSAT$^+[t]$ *problem* $(t > 2)$ *on circuits of genus* $g(n) = O(\sqrt{\log n})$ *is FPT, and is* $W[t]$-*complete for even* t *and* $W[t-1]$-*complete for odd* t *if* $g(n) = n^{\Omega(1)}$.

Proof. Let (C, k) be an instance of WSAT$^+[t]$ on circuits of genus $g(n) \leq c\sqrt{\log n}$, for some fixed (known) constant $c > 0$. By Proposition 3, in fpt-time we can reduce the instance (C, k) to $h(k)n^{O(1)}$ many instances (C', k') of WSAT$^+[t]$, where h is a complexity function of k and $k' \leq k$, such that (C, k) is a yes-instance if and only if at least one of the instances (C', k') is, and such that C' has at most $2k'$ critical gates. Therefore, we may assume that C has at most $2k$ critical gates. By Lemma 1, the branchwidth of C is at most $c_1 \log n$, for some fixed constant $c_1 > 0$, and hence, by the results of Robertson and Seymour [17], the treewidth of C is at most $c_2 \log n$ for some fixed constant $c_2 > 0$. Using the algorithm of Amir [1], we can decide if the treewidth of C is at most $c_3 \log n$ for some fixed constant $c_3 > 0$ (if not, the genus does not satisfy the given upper bound and we reject the instance), and if so, the algorithm in [1] returns a tree decomposition of C of width $c_4 \log n$, for some constant $c_4 > 0$, in time $2^{O(\log n)}|C|^{O(1)} = |C|^{O(1)}$. By Proposition 4, we can decide (C, k) in fpt-time. The proof of hardness result is a simple padding argument. $\qquad\square$

The above approach can be extended to WSAT$^+[3]$:

Theorem 6. *The* WSAT$^+[3]$ *problem on circuits of genus* $g(n)$ *is FPT if* $g(n) = n^{o(1)}$, *and* $W[2]$-*complete if* $g(n) = n^{\Omega(1)}$.

6   Applications

We show applications of the above results to natural problems. The RED-BLUE NONBLOCKER problem is: Given a bipartite graph with one partition colored red and the other blue, decide whether or not there exists a set S of k red vertices such that every blue vertex has a red neighbor not in S. The other problems under consideration are RED-BLUE DOMINATING SET, HITTING SET, SET COVER, and INDEPENDENT SET ON HYPERGRAPHS; those are well-known problems, and we refer the reader to [9,10] for their definition, and for some of the standard fpt-reductions showing their W-hardness. We note that HITTING SET is the same as the VERTEX COVER problem on hypergraphs, and SET COVER is the same as the EDGE COVER problem on hypergraphs. Therefore, the underlying hypergraph is naturally defined for HITTING SET and SET COVER, and the genus of the hypergraph is by definition the genus of its bipartite incidence graph whose (in this case) first partition corresponds to the set of elements (vertex-set), and its second partition corresponds to the family of subsets (hyperedges).

Theorem 7. *The parameterized* RED-BLUE DOMINATING SET, HITTING SET, *and* SET COVER *are FPT on graphs/hypergraphs of genus* $N^{o(1)}$ *and* $W[2]$-*complete on graphs/hypergraphs of genus* $N^{\Omega(1)}$, *where* N *is the number of red* *vertices in* RED-BLUE DOMINATING SET, *the cardinality of the vertex-set in* HITTING SET, *and the number of (sets) hyperedges in* SET COVER.

Theorem 8. *The parameterized* RED-BLUE NONBLOCKER *and* INDEPENDENT SET ON HYPERGRAPHS *problems are FPT on graphs/hypergraphs of genus* $N^{o(1)}$ *and* $W[1]$-*complete on (hyper)graphs of genus* $N^{\Omega(1)}$ *(N is the number of red vertices in* RED-BLUE NONBLOCKER, *and the total number of vertices in* INDEPENDENT SET ON HYPERGRAPHS*).*

References

1. Amir, E.: Efficient approximation for triangulation of minimum treewidth. In: Proceedings of UAI, pp. 7–15. Morgan Kaufmann (2001)
2. Bodlaender, H., Fomin, F., Lokshtanov, D., Penninkx, E., Saurabh, S., Thilikos, D.: (Meta) kernelization. In: Proceedings of FOCS, pp. 629–638 (2009)
3. Cai, L., Fellows, M., Juedes, D., Rosamond, F.: The complexity of polynomial-time approximation. Theory of Computing Systems 41(3), 459–477 (2007)
4. Chen, J., Huang, X., Kanj, I., Xia, G.: Polynomial time approximation schemes and parameterized complexity. Discrete Appl. Math. 155(2), 180–193 (2007)
5. Chen, J., Kanj, I., Perkovic, L., Sedgwick, E., Xia, G.: Genus characterizes the complexity of certain graph problems: Some tight results. Journal of Computer and System Sciences 73(6), 892–907 (2007)
6. Demaine, E., Fomin, F., Hajiaghayi, M., Thilikos, D.: Subexponential parameterized algorithms on bounded-genus graphs and H-minor-free graphs. J. ACM 52, 866–893 (2005)
7. Demaine, E., Hajiaghayi, M.: Bidimensionality: new connections between FPT algorithms and PTASs. In: Proceedings of SODA, pp. 590–601 (2005)
8. Demaine, E., Hajiaghayi, M., Thilikos, D.: The bidimensional theory of bounded-genus graphs. SIAM J. Discrete Math. 20(2), 357–371 (2006)
9. Downey, R., Fellows, M.: Parameterized Complexity. Springer, New York (1999)
10. Flum, J., Grohe, M.: Parameterized Complexity Theory. Springer, Berlin (2010)
11. Fomin, F., Lokshtanov, D., Raman, V., Saurabh, S.: Bidimensionality and EPTAS. In: Proceedings of SODA, pp. 748–759 (2011)
12. Fomin, F., Lokshtanov, D., Saurabh, S., Thilikos, D.: Bidimensionality and kernels. In: Proceedings of SODA, pp. 503–510 (2010)
13. Gross, J., Tucker, T.: Topological graph theory. Wiley-Interscience, NY (1987)
14. Kanj, I., Pelsmajer, M., Schaefer, M., Xia, G.: On the induced matching problem. Journal of Computers and System Sciences 77(6), 1058–1070 (2011)
15. Khanna, S., Motwani, R.: Towards a syntactic characterization of PTAS. In: Proceedings of STOC, pp. 468–477 (1996)
16. Marx, D.: Completely inapproximable monotone and antimonotone parameterized problems. J. Comput. Syst. Sci. 79(1), 144–151 (2013)
17. Robertson, N., Seymour, P.D.: Graph minors X: Obstructions to tree-decomposition. J. Comb. Theory, Ser. B 52(2), 153–190 (1991)
18. Savage, J.: Planar circuit complexity and the performance of VLSI algorithms. Technical Report RR-0077, INRIA (May 1981)
19. Szeider, S.: On fixed-parameter tractable parameterizations of sat. In: Giunchiglia, E., Tacchella, A. (eds.) SAT 2003. LNCS, vol. 2919, pp. 188–202. Springer, Heidelberg (2004)
20. West, D.: Introduction to graph theory. Prentice Hall Inc., NJ (1996)

Two-Sided Boundary Labeling with Adjacent Sides*

Philipp Kindermann[1], Benjamin Niedermann[2], Ignaz Rutter[2], Marcus Schaefer[3], André Schulz[4], and Alexander Wolff[1]

[1] Lehrstuhl für Informatik I, Universität Würzburg, Germany
http://www1.informatik.uni-wuerzburg.de/en/staff
[2] Fakultät für Informatik, Karlsruher Institut für Technologie (KIT), Germany
{benjamin.niedermann,rutter}@kit.edu
[3] College of Computing and Digital Media, DePaul University, Chicago, IL, USA
mschaefer@cs.depaul.edu
[4] Institut für Mathematische Logik und Grundlagenforschung, Universität Münster, Germany
andre.schulz@uni-muenster.de

Abstract. In the *Boundary Labeling* problem, we are given a set of n points, referred to as *sites*, inside an axis-parallel rectangle R, and a set of n pairwise disjoint rectangular labels that are attached to R from the outside. The task is to connect the sites to the labels by non-intersecting rectilinear paths, so-called *leaders*, with at most one bend.

In this paper, we study the problem *Two-Sided Boundary Labeling with Adjacent Sides*, where labels lie on two adjacent sides of the enclosing rectangle. We present a polynomial-time algorithm that computes a crossing-free leader layout if one exists. So far, such an algorithm has only been known for the cases that labels lie on one side or on two opposite sides of R (where a crossing-free solution always exists). For the more difficult case where labels lie on adjacent sides, we show how to compute crossing-free leader layouts that maximize the number of labeled points or minimize the total leader length.

1 Introduction

Label placement is an important problem in cartography and, more generally, information visualization. Features such as points, lines, and regions in maps, diagrams, and technical drawings often have to be labeled so that users understand better what they see. Even very restricted versions of the label-placement problem are NP-hard [14], which explains why labeling a map manually is a tedious task that has been estimated to take 50% of total map production time [15]. The ACM Computational Geometry Impact Task Force report [6] identified label placement as an important research area. The point-labeling problem in particular has received considerable attention, from practitioners and theoreticians alike. The latter have proposed approximation algorithms for various objectives (label number versus label size), label shapes (such as axis-parallel rectangles or disks), and label-placement models (so-called fixed-position models versus slider models).

* This research was initiated during the GraDr Midterm meeting at the TU Berlin, which was supported by an ESF networking grant. Ph. Kindermann acknowledges support by the ESF EuroGIGA project GraDR (DFG grant Wo 758/5-1).

F. Dehne, R. Solis-Oba, and J.-R. Sack (Eds.): WADS 2013, LNCS 8037, pp. 463–474, 2013.
© Springer-Verlag Berlin Heidelberg 2013

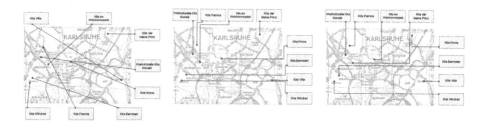

(a) original labeling of kinder- (b) *opo*-labeling computed by (c) *po*-labeling using the same
gartens in Karlsruhe, Germany   the algorithm of Bekos et al. [4]   ports as (b)

Fig. 1. A real-world example of boundary labeling with adjacent sides (taken from [4]). For better
readability, we have simplified the label texts.

The traditional label-placement models for point labeling insist that a label is placed
such that a point on its boundary coincides with the point to be labeled, the *site*. This
can make it impossible to label all sites with labels of sufficient size if some sites are
very close together. For this reason, Freeman et al. [8] and Zoraster [19] advocated the
use of *leaders*, (usually short) line segments that connect sites to labels. In order to
make sure that the background image or map remains visible even in the presence of
large labels, Bekos et al. [4] took a more radical approach. They introduced models and
algorithms for *boundary labeling*, where all labels are placed beyond the boundary of
the map and are connected to the sites by straight-line or rectilinear leaders (see Fig. 1).

Problem Statement. Following Bekos et al. [4], we define the BOUNDARY LABELING
problem as follows. We are given an axis-parallel rectangle $R = [0, W] \times [0, H]$, which
is called the *enclosing rectangle*, a set $P \subset R$ of n points $p_1, \ldots, p_n$, called *sites*, within
the rectangle R, and a set L of $m \leq n$ axis-parallel rectangles $\ell_1, \ldots, \ell_m$, called *labels*,
that lie in the complement of R and touch the boundary of R. No two labels overlap. We
denote an instance of the problem by the triplet (R, L, P). A *solution* to the problem
is a set of m curves $c_1, \ldots, c_m$, called *leaders*, that connect sites to labels such that
the leaders a) produce a matching between the labels and (a subset of) the sites, b) are
contained inside R, and c) touch the associated labels on the boundary of R.

A solution is *planar* if the leaders do not intersect. We call an instance *solvable* if a
planar solution exists. Note that we do not prescribe which site connects to which label.
The endpoint of a curve at a label is called a *port*. We distinguish two versions of the
BOUNDARY LABELING problem: either the position of the ports on the boundary of R
is fixed and part of the input, or the ports *slide*, i.e., their exact location is not prescribed.

We restrict our solutions to *po-leaders*, that is, starting at a site, the first line segment
of a leader is parallel (p) to the side of R containing the label it leads to, and the second
line segment is orthogonal (o) to that side; see Fig. 1c. (Fig. 1b shows a labeling with
so-called *opo-leaders*, which were investigated by Bekos et al. [4]). Bekos et al. [3,
Fig. 12] observed that not every instance (with $m = n$) admits a planar solution with
po-leaders where all sites are labeled.

Previous and related work. For *po*-labeling, Bekos et al. [4] gave a simple quadratic-time algorithm for the one-sided case that, in a first pass, produces a labeling of minimum total leader length by matching sites and ports from bottom to top. In a second pass, their algorithm removes all intersections without increasing the total leader length. This result was improved by Benkert et al. [5] who gave an $O(n \log n)$-time algorithm for the same objective function and an $O(n^3)$-time algorithm for a very general class of objective functions, including, for example, bend minimization. They extend the latter result to the two-sided case (with labels on opposite sides of R), resulting in an $O(n^8)$-time algorithm. For the special two-sided case of leader-length minimization, Bekos et al. [4] gave a simple dynamic program running in $O(n^2)$ time. All these algorithms work both for fixed and sliding ports.

Leaders that contain a diagonal part have been studied by Benkert et al. [5] and by Bekos et al. [2]. Recently, Nöllenburg et al. [16] have investigated a dynamic scenario for the one-sided case, Gemsa et al. [9] have used multi-layer boundary labeling to label panorama images, and Fink et al. [7] have boundary labeled focus regions, for example, in interactive on-line maps.

At its core, the boundary label problem asks for a non-intersecting perfect (or maximum) matching on a bipartite graph. Note that an instance may have a planar solution, although all of its leader-length minimal matchings have crossings. In fact, the ratio between a length-minimal solution and a length-minimal crossing-free matching can be arbitrarily bad; see Fig. 2. When connecting points and sites with straight-line segments, the minimum Euclidean matching is necessarily crossing-free. For this case an $O(n^{2+\varepsilon})$-time $O(n^{1+\varepsilon})$-space algorithm exists [1]. The minimum-length solution using rectilinear paths with an unbounded number of bends in the presence of obstacles is NP-hard, but there is a 2-approximation [18].

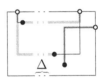

Fig. 2. Length-minimal solutions may have crossings

Boundary labeling can also be seen as a graph-drawing problem where the class of graphs to be drawn is restricted to matchings. The restriction concerning the positions of the graph vertices (that is, sites and ports) has been studied for less restricted graph classes under the name *point-set embeddability (PSE)*, usually following the straight-line drawing convention for edges [10]. More recently, PSE has also been combined with the ortho-geodesic drawing convention [12], which generalizes *po*-labeling by allowing edges to make more than one bend. The case where the mapping between ports and sites is given has been studied in VLSI layout [17].

Our Contribution. We investigate the problem TWO-SIDED BOUNDARY LABELING WITH ADJACENT SIDES where all labels lie on two *adjacent* sides of R, for example, on the top and right side. Note that point data often comes in a coordinate system; then it is natural to have labels on adjacent sides (for example, opposite the coordinate axes). We argue that this problem is more difficult than the case where labels lie on opposite sides, which has been studied before: with labels on opposite sides, (a) there is always a solution where all sites are labeled (if $m = n$) and (b) a feasible solution can be obtained by considering two instances of the one-sided case.

Our main result is an algorithm that, given an instance with n labels and n sites, decides whether a planar solution exists where all sites are labeled and, if yes, computes a layout of the leaders (see Section 3). Our algorithm uses dynamic programming to "guess" a partition of the sites into the two sets that are connected to the leaders on the top side and on the right side. The algorithm runs in $O(n^2)$ time and uses $O(n)$ space.

Notation. We call the labels that lie on the right (top) side of R *right (top) labels.* The *type* of a label refers to the side of R on which it is located. The *type* of a leader (or a site) is simply the type of its label. We assume that no two sites lie on the same horizontal or vertical line, and no site lies on a horizontal or vertical line through a port or an edge of a label.

For a solution $\mathcal{L}$ of a boundary labeling problem, we define several measures that will be used to compare different solutions. We denote the total length of all leaders in $\mathcal{L}$ by $\mathrm{length}(\mathcal{L})$. Moreover, we denote by $|\mathcal{L}|_x$ the total length of all horizontal segments of leaders that connect a right label to a site. Similarly, we denote by $|\mathcal{L}|_y$ the total length of the vertical segments of leaders that connect top labels to sites. Note that in general, it is *not* true that $|\mathcal{L}|_x + |\mathcal{L}|_y = \mathrm{length}(\mathcal{L})$.

We denote the (uniquely defined) leader connecting a site p to a port t of a label ℓ by $\lambda(p, t)$. We denote the bend of the leader $\lambda(p, t)$ by $\mathrm{bend}(p, t)$. In the case of fixed ports, we identify ports with labels and simply write $\lambda(p, \ell)$ and $\mathrm{bend}(p, \ell)$, resp.

2   Structure of Planar Solutions

In this section, we attack our problem presenting a series of structural results of increasing strength. For simplicity, we assume fixed ports. For sliding ports, we can simply fix all ports to the bottom-left corner of their corresponding labels (see the full version of this paper [13]). First we show that we can split a planar two-sided solution into two one-sided solutions by constructing an xy-monotone, rectilinear curve from the top-right to the bottom-left corner of R; see Fig. 4. Afterwards, we provide a necessary and sufficient criterion to decide whether for a given separation there exists a planar solution. This will form the basis of our dynamic programming algorithm, which we present in Section 3.

Lemma 1. *Consider a solution $\mathcal{L}$ for (R, L, P) and let $P' \subseteq P$ be sites of the same type. Let $L' \subseteq L$ be the set of labels of the sites in P'. Let $K \subseteq R$ be a rectangle that contains all bends of the leaders of P'. If the leaders of $P \setminus P'$ do not intersect K, then we can rewire P' and L' such that the resulting solution $\mathcal{L}'$ has the following properties: (i) all intersections in K are removed, (ii) there are no new intersections of leaders outside of K, (iii) $|\mathcal{L}'|_x = |\mathcal{L}|_x$, $|\mathcal{L}'|_y = |\mathcal{L}|_y$, and (iv) $\mathrm{length}(\mathcal{L}') \leq \mathrm{length}(\mathcal{L})$.*

Proof. Without loss of generality, we assume that P' contains top sites; the other cases are symmetric. We first prove that, no matter how we change the assignment between P' and L', new intersection points can arise only in K. Then we show how to establish the claimed solution.

Claim. Let $\ell, \ell' \in L'$ and $p, p' \in P'$ such that ℓ labels p and ℓ' labels p'. Changing the matching by rerouting p to ℓ' and p' to ℓ does not introduce new intersections outside of K.

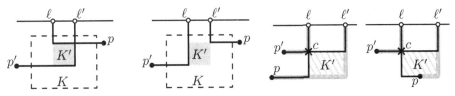

(a) rerouting $\lambda(p, \ell)$ and $\lambda(p', \ell')$ to $\lambda(p, \ell')$ and $\lambda(p', \ell)$ changes leaders only on the boundary of K'

(b) removing the highest crossing c does not increase the total leader length

Fig. 3. Illustration of the proof of Lemma 1

Let $K' \subseteq K$ be the rectangle spanned by $\mathrm{bend}(p, \ell)$ and $\mathrm{bend}(p', \ell')$. When rerouting, we replace $\lambda(p, \ell) \cup \lambda(p', \ell')$ restricted to the boundary of K' by its complement with respect to the boundary of K'; see Fig. 3a for an example. Thus, any changes concerning the leaders occur only in K'. The statement of the claim follows.

Since any rewiring can be seen as a sequence of *pairwise* reroutings, the above claim shows that we can rewire L' and P' arbitrarily without running the risk of creating new conflicts outside of K. In order to resolve the conflicts inside K, we use the length-minimization algorithm for one-sided boundary labeling by Benkert et al. [5], with the sites and ports outside K projected onto the boundary of K. Thus, after finitely many such steps, we find a solution $\mathcal{L}'$ that satisfies properties (i)–(iv) in the statement of the lemma. □

Definition 1. *We call an xy-monotone, rectilinear curve connecting the top-right to the bottom-left corner of R an xy-separating curve; see Fig. 4. We say that a planar solution to* TWO-SIDED BOUNDARY LABELING WITH ADJACENT SIDES *is xy-separated if and only if there exists an xy-separating curve C such that*

a) the top sites and their leaders lie on or above C, and
b) the right sites and their leaders lie below C.

It is not hard to see that a planar solution is not xy-separated if there exists a site p that is labeled to the right side and a site q that is labeled to the top side with $x(p) < x(q)$ and $y(p) > y(q)$. There are exactly four patterns in a possible planar solution that satisfy this condition; see Fig. 5. We claim that these patterns are the only ones that violate xy-separability (for the proof, refer to the full version of the paper [13]).

Lemma 2. *A planar solution is xy-separated if and only if it does not contain any of the patterns P1–P4 in Fig. 5.*

Observe that patterns P1 and P2 can be transformed into patterns P4 and P3, respectively, by mirroring the instance diagonally. Next, we prove constructively that, by rerouting pairs of leaders, any planar solution can be transformed into an xy-separated planar solution.

Proposition 1. *If there exists a planar solution $\mathcal{L}$ to* TWO-SIDED BOUNDARY LABELING WITH ADJACENT SIDES, *then there exists an xy-separated planar solution $\mathcal{L}'$ with $\mathrm{length}(\mathcal{L}') \leq \mathrm{length}(\mathcal{L})$, $|\mathcal{L}'|_x \leq |\mathcal{L}|_x$, and $|\mathcal{L}'|_y \leq |\mathcal{L}|_y$.*

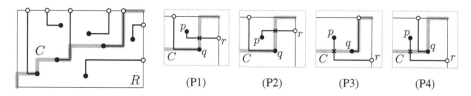

Fig. 4. An xy-separating curve of a planar solution   **Fig. 5.** A planar solution that contains any of the above four patterns P1–P4 is not xy-separated

Proof. Let $\mathcal{L}$ be a planar solution of minimum total leader length. We show that $\mathcal{L}$ is xy-separated. Assume, for the sake of contradiction, that $\mathcal{L}$ is not xy-separated. Then, by Lemma 2, $\mathcal{L}$ contains one of the patterns P1–P4. Without loss of generality, we can assume that the pattern is of type P3 or P4. Otherwise, we mirror the instance diagonally.

Let p be a right site (with port r) and let q be a top site (with port t) such that (p, q) forms a pattern of type P3 or P4. Among all such patterns, pick one where p is rightmost. Among all these patterns, pick one where q is bottommost. Let A be the rectangle spanned by p and t; see Fig. 6. Let A' be the rectangle spanned by $\mathrm{bend}(q, t)$ and p. Let B be the rectangle spanned by q and r. Let B' be the rectangle spanned by q and $\mathrm{bend}(p, r)$. Then we claim the following:

(i) Sites in the interiors of A and A' are connected to the top.
(ii) Sites in the interiors of B and B' are connected to the right.

Property (i) is due to the choice of p as the rightmost site involved in such a pattern. Similarly, property (ii) is due to the choice of q as the bottommost site that forms a pattern with p. This settles our claim.

Our goal is to change the labeling by rerouting p to t and q to r, which decreases the total leader length, but may introduce crossings. We then use Lemma 1 to remove the crossings without increasing the total leader length. Let $\mathcal{L}''$ be the labeling obtained from $\mathcal{L}$ by rerouting p to t and q to r. We have $|\mathcal{L}''|_y \leq |\mathcal{L}|_y - (y(p) - y(q))$ and $|\mathcal{L}''|_x = |\mathcal{L}|_x - (x(q) - x(p))$. Moreover, $\mathrm{length}(\mathcal{L}'') \leq \mathrm{length}(\mathcal{L}) - 2(y(p) - y(q))$, as at least twice the vertical distance between p and q is saved; see Fig. 6. Since the original labeling was planar, crossings may only arise on the horizontal segment of $\lambda(p, t)$ and on the vertical segment of $\lambda(q, r)$.

By properties (i) and (ii), all leaders that cross the new leader $\lambda(p, t)$ have their bends inside A', and all leaders that cross the new leader $\lambda(q, r)$ have their bends inside B'. Thus, we can apply Lemma 1 to the rectangles A' and B' to resolve all new crossings. The resulting solution $\mathcal{L}'$ is planar and has length less than $\mathrm{length}(\mathcal{L})$. This is a contradiction to the choice of $\mathcal{L}$.   □

Since every solvable instance of TWO-SIDED BOUNDARY LABELING WITH ADJACENT SIDES admits an xy-separated planar solution, it suffices to search for such a solution. Moreover, an xy-separated planar solution that minimizes the total leader length has minimum leader length among *all* planar solutions. In Lemma 3 we provide a necessary and sufficient criterion to decide whether, for a given xy-monotone curve C, there is a planar solution that is separated by C. We denote the region of R above C

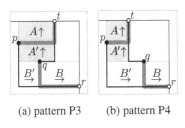

(a) pattern P3   (b) pattern P4

Fig. 6. Types (top = ↑ / right = →) of the sites inside rectangles A, A', B, and B'. Fat edges: result after rerouting.

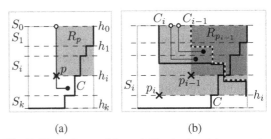

(a)   (b)

Fig. 7. The strip condition. a) The horizontal segments of C partition R_T into the strips $S_0, S_1, \ldots, S_k$. b) Constructing a planar labeling from a sequence of valid rectangles.

by R_T and the region of R below C by R_R. These regions are relatively open at C. For each horizontal segment of C consider the horizontal line through the segment. We denote the part of these lines within R by $h_1, \ldots, h_k$, respectively. Further, let h_0 be the top edge of R. The line segments $h_1, \ldots, h_k$ partition R_T into k strips, which we denote by $S_1, \ldots, S_k$ from top to bottom, such that strip S_i is bounded by h_i from below for $i = 1, \ldots, k$; see Fig. 6a. Additionally, we define S_0 to be the empty strip that coincides with h_0. Note that this strip cannot contain any site of P. For any point p on one of the horizontal lines h_i, we define the rectangle R_p, spanned by p and the top-right corner of R. We define R_p such that it is closed but does *not* contain its top-left corner. In particular, we consider the port of a top label as contained in R_p, except if it is the upper left corner of R_p.

A rectangle R_p is *valid* if the number of sites of P above C that belong to R_p is at least as large as the number of ports on the top side of R_p. The central idea is that the sites of P inside a valid rectangle R_p can be connected to labels on the top side of the valid rectangle by leaders that are completely contained inside the rectangle.

We now prove that, for a given xy-separating curve C, there exists a planar solution in R_T for the top labels if and only if C satisfies the following strip condition for each strip $S_0, \ldots, S_k$ in R_T. The *strip condition* of strip S_i is satisfied if there exists a point $p \in h_i \cap R_T$ such that R_p is valid. We call a region $S \subseteq R$ *balanced* if it contains the same numbers of sites and ports.

Lemma 3. *Let C be an xy-separating curve and let $P_T = P \cap R_T$. There is a planar solution that uses all top labels of R to label the sites in P_T such that all leaders are in R_T if and only if $S_0, \ldots, S_k$ satisfy the strip condition.*

Proof. To show that the conditions are necessary, let $\mathcal{L}$ be a planar solution for which all top leaders are above C. Consider strip S_i, which is bounded from below by line $h_i, 0 \leq i \leq k$. If there is no site of P_T below h_i, rectangle R_p is clearly valid, where p is the intersection of h_i with the left side of R, and thus the strip condition is satisfied. Hence, assume that there is a site $p \in P_T$ that is labeled by a top label, and is in strip S_j with $j > i$; see Fig. 6a. Then, the vertical segment of this leader crosses h_i in R_T. Let p' denote the rightmost such crossing of a leader of a site in P_T with h_i. We claim that $R_{p'}$ is valid. To see this, observe that all sites of P_T top-right of p' are contained in $R_{p'}$. Since no leader may cross the vertical segments defining p', the number of sites in $R_{p'} \cap R_T$ is balanced, i.e., $R_{p'}$ is valid.

Conversely, we show that if the conditions are satisfied, then a corresponding planar solution exists. Let S_k be the last strip that contains sites of P_T. For $i = 0, \ldots, k$, let p'_i denote the rightmost point of $h_i \cap R_T$ such that $R_{p'_i}$ is valid. We define p_i to be the point on $h_i \cap R_T$ with x-coordinate $\min_{j \leq i}\{x(p'_j)\}$. Note that R_{p_i} is a valid rectangle, as, by definition, R_{p_i} contains some valid rectangle $R_{p'_j}$ with $x(p'_j) = x(p_i)$. Also by definition, the sequence $p_0, p_1, \ldots, p_k$ has decreasing x-coordinates, i.e., $R_{p_k} \subseteq \cdots \subseteq R_{p_1} \subseteq R_{p_0}$; see Fig. 6b.

We prove inductively that, for $i = 0, \ldots, k$, there is a planar labeling $\mathcal{L}_i$ that matches the labels on the top side of R_{p_i} to points contained in R_{p_i} such that there exists an xy-monotone curve C_i from the top-left to the bottom-right corner of R_{p_i} that separates the labeled sites from the unlabeled sites without intersecting any leaders. Then $\mathcal{L}_k$ is the claimed labeling.

For $i = 0$, $\mathcal{L}_0 = \emptyset$ is a planar solution. Consider a strip S_i with $0 < i \leq k$; see Fig. 6b. By the induction hypothesis, we have a curve C_{i-1} and a planar labeling $\mathcal{L}_{i-1}$, which matches the labels on the top side of $R_{p_{i-1}}$ to the sites in $R_{p_{i-1}}$ above C_{i-1}. In order to extend $\mathcal{L}_{i-1}$ to a planar solution $\mathcal{L}_i$, we additionally need to match the remaining labels on the top side of R_{p_i} and construct a corresponding curve C_i. Let P_i denote the set of unlabeled sites in R_{p_i}. By the validity of R_{p_i}, this number is at least as large as the number of unused ports at the top side of R_{p_i}. We match these ports from top to bottom to the topmost sites of P_i; the result is the claimed planar labeling $\mathcal{L}_i$. The ordering ensures that no two of the new leaders cross. Moreover, no leader crosses the curve C_{i-1}, and hence such leaders cannot cross leaders in $\mathcal{L}_{i-1}$. It remains to construct the curve C_i. For this, we start at the top-left corner of R_{p_i} and move down vertically, until we have passed all labeled sites. We then move right until we either hit C_{i-1} or the right side of R. In the former case, we follow C_{i-1} until we arrive at the right side of R. Finally, we move down until we arrive at the bottom-right corner of R_{p_i}. Note that all labeled sites are above C_i, unlabeled sites are below C_i, and no leader is crossed by C_i. This is true since we first move below the new leaders and then follow the previous curve C_{i-1}.   □

A symmetric strip condition (with vertical strips) can be obtained for the right region R_R of a partitioned instance. The characterization is completely symmetric.

3   The Algorithm

Now we describe how to find an xy-separating curve C that satisfies the strip conditions. For that purpose we only consider xy-separating curves that lie on the dual of the grid induced by the sites and ports of the given instance. When traversing this grid from grid point to grid point, we either pass a site (*site event*) or a port (*port event*). By passing a site, we decide if the site is connected to the top or to the right side. Clearly, there is an exponential number of possible xy-monotone traversals through the grid. In the following, we describe a dynamic program that finds an xy-separating curve in $O(n^3)$ time.

Let there be m_R ports on the right side of R and m_T ports on the top side of R, then the grid has size $[n + m_T + 1] \times [n + m_R + 1]$. We define the grid points as $G(x, y)$, $0 \le x \le n + m_T + 1$, $0 \le y \le n + m_R + 1$ with $G(0, 0)$ being the bottom-left and $r := G(n + m_T + 1, n + m_R + 1)$ being the top-right corner of R. Further, we define $G_x(s) := x(G(s, 0))$ and $G_y(t) := y(G(0, t))$.

An entry in the table of our dynamic program is described by three values. The first two values are s and t, which give the position of the current search for the curve C. The interpretation is that the entry encodes the possible xy-monotone curves from r to $p_C := G(s, t)$; see Fig. 8. The remaining value u denotes the number of sites above C in the rectangle spanned by r and p_C. Note that it suffices to store u, as the number of sites below the curve C can directly be derived from u and all sites that are contained in the rectangle spanned by r and p_C. We denote the first values describing the positions of the curves by the vector $\mathbf{c} = (s, t)$. Our goal is to compute a table $T[\mathbf{c}, u]$ such that $T[\mathbf{c}, u] = \texttt{true}$ if and only if there exists an xy-separating curve C such that the fol-

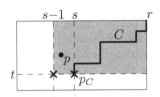

Fig. 8. Possible step of the dynamic program, where p enters the rectangle spanned by r and $G(s - 1, t)$

lowing conditions hold. (i) Curve C starts at r and ends at p_C. (ii) Inside the rectangle spanned by r and p_C, there are u sites of P above C. (iii) For each strip in the two regions R_T and R_R defined by C the strip condition holds.

It follows from these conditions, Proposition 1 and Lemma 3 that the instance admits a planar solution if and only if $T[(0, 0), u] = \texttt{true}$, for some u. Let us now proceed to describe how to compute the table. Initially, we set $\mathbf{c} = (n + m_T + 1, n + m_R + 1)$. We initialize the first entry $T[\mathbf{c}, 0] = \texttt{true}$. The remaining entries are initialized with $\texttt{false}$.

Let $\mathbf{c} := (s, t)$ be the current grid point we checked as endpoint for C. Based on the table $T[\mathbf{c}, \cdot]$ we then compute the entries $T[\mathbf{c} - \Delta\mathbf{c}, \cdot]$ where the vector $\Delta\mathbf{c} = (\Delta s, \Delta t)$ is either $(0, 1)$ or $(1, 0)$. We classify such steps, depending on whether we cross a site or a port. We give a full description for $\Delta\mathbf{c} = (1, 0)$, i.e, we decrease s by 1. The other case is completely symmetric. Assume $T[\mathbf{c}, u] = \texttt{true}$. We distinguish two cases, based on whether we cross a site or a port.

Case 1: Going from s to $s - 1$ is a site event, i.e., there is a site p with $G_x(s) > x(p) > G_x(s - 1)$. Note that by our assumption of general position and the definition of the coordinates, the site p is unique. If $y(p) > G_y(t)$, then p enters the rectangle spanned by $G(s-1, t)$ and r, and it is located above C; see Fig. 8. We thus set $T[\mathbf{c} - \Delta\mathbf{c}, u+1] = \texttt{true}$. Otherwise we set $T[\mathbf{c} - \Delta\mathbf{c}, u] = \texttt{true}$. Note that the strip conditions remain satisfied since we do not decrease the number of sites in any region.

Case 2: Going from s to $s - 1$ is a port event, i.e., there is a label ℓ on the top side, whose port is between $G_x(s - 1)$ and $G_x(s)$. Thus, the region above C contains one more label. We therefore check the strip condition for the strip above the horizontal line through $G(s - 1, t)$. If it is satisfied, we set $T[\mathbf{c} - \Delta\mathbf{c}, u] = \texttt{true}$.

If $T[\mathbf{c}, u] = \texttt{false}$, there is no xy-separating curve that satisfies the conditions given above, so the it suffices to only look at the $\texttt{true}$ table entries. This immediately gives us a polynomial-time algorithm for TWO-SIDED BOUNDARY LABELING WITH ADJACENT SIDES. The running time crucially relies on the number of strip conditions that need to be checked. We show that after a $O(n^2)$ preprocessing phase, such queries can be answered in $O(1)$ time.

To implement the test of the strip conditions, we use a table B_T, which stores in the position $B_T[s, t]$ how large a deficit of top sites to the right can be compensated by sites above and to the left of $G(s, t)$. That is, $B_T[s, t]$ is the maximum value k such that there exists a rectangle $K_{B_T[s,t]}$ with lower right corner $G(s, t)$ whose top side is bounded by the top side of R, and that contains k more sites in its interior, than it has ports on its top side. To compute this matrix, we use a simple dynamic program, which calculates the entries of B_T by going from the left to the right side. Once we have computed this matrix, it is possible to query the strip condition in the dynamic program that computes T in $O(1)$ time. The table can be clearly filled out in $O(n^2)$ time. A similar matrix B_R can be computed for the vertical strips. Altogether, this yields an algorithm for TWO-SIDED BOUNDARY LABELING WITH ADJACENT SIDES that runs in $O(n^3)$ time and uses $O(n^3)$ space. However, the entries of each row and column of T depend only on the previous row and column, which allows us to reduce the storage requirement to $O(n^2)$. Using Hirschberg's algorithm [11], we can still backtrack the dynamic program and find a solution corresponding to an entry in the last cell in the same running time. The detailed approach on how to calculate and use the tables B_T and B_R is given in the full version of the paper [13].

Theorem 1. TWO-SIDED BOUNDARY LABELING WITH ADJACENT SIDES *can be solved in* $O(n^3)$ *time using* $O(n^2)$ *space.*

In order to increase the performance of our algorithm, we can reduce the number of dimensions of the table T by 1. As a first step, we show that for any search position $\mathbf{c}$, the possible values of u, for which $T[\mathbf{c}, u] =\texttt{true}$ form an interval.

Lemma 4. *Let* $T[\mathbf{c}, u] = T[\mathbf{c}, u'] = \texttt{true}$ *with* $u \le u'$. *Then* $T[\mathbf{c}, u''] = \texttt{true}$ *for* $u \le u'' \le u'$.

Proof. Let C be the curve corresponding to the entry $T[\mathbf{c}, u]$. That is C connects r to p_C such that u sites in the rectangle spanned by p_C and r are above C, and the strip conditions (both above and below C) are satisfied. Similarly, let C' be the curve corresponding to $T[\mathbf{c}, u']$.

Since u and u' differ, there is a rightmost site p, such that p is below C and above C'. Let v and v' be the grid points of C and C' that are immediately to the left of p. Note that v is above v' since C is above p and C' is below it. Consider the C'', which starts at r and follows C until v, then it moves down vertically to v', and from their follows C' to p. Obviously C'' is an xy-separating curve, and it has above it the same sites as C', except for p, which is below it. Thus there are $u'' = u' - 1$ sites above C'' in the rectangle spanned by p and r. If all strips defined by C'' satisfy the strip condition, then this implies $T[\mathbf{c}, u''] = \texttt{true}$.

To see that the strip conditions are indeed satisfied, consider a horizontal strip S'' defined by C''. Let S be the lowest horizontal strip defined by C that is not below the lower boundary of S''. We know that S fulfills the strip condition, which is witnessed by some valid rectangle K. We can enlarge this rectangle vertically such that it touches the lower boundary of S''. The enlarged rectangle contains at least as many sites above C'' as there were above C in K. Hence it is a valid rectangle and the strip condition for S'' holds. An analogous statement holds for the vertical strips since C'' is above C' at every x-coordinate. □

Thus, we only need to store the boundaries of the u-interval. Further, we can compute the tables B_T and B_R backwards, i.e., in the direction of the dynamic program, by precomputing the entries of B_T and B_R on the top and right side, respectively. Using Hirschberg's algorithm, this reduces the running time to $O(n^2)$ and the space to $O(n)$. The detailed description is given in the full version of the paper [13].

Theorem 2. TWO-SIDED BOUNDARY LABELING WITH ADJACENT SIDES *can be solved in $O(n^2)$ time using $O(n)$ space.*

4   Conclusion

In this paper, we have studied the problem of testing whether an instance of TWO-SIDED BOUNDARY LABELING WITH ADJACENT SIDES admits a planar solution. We have given the first efficient algorithm for this problem, running in $O(n^2)$ time.

The presented algorithm can also be used to solve a variety of different extensions of the problem. In the full version of the paper [13], we show how to generalize from fixed to sliding ports without increasing the asymptotic running time. Further, we show how to maximize the number of labeled sites such that the solution is planar in $O(n^3 \log n)$ time and we give an extension to the algorithm that minimizes the total leader length in $O(n^8 \log n)$ time.

With some additional work, the presented approach can also be used to solve THREE-SIDED and FOUR-SIDED BOUNDARY LABELING in polynomial time. Namely, it can be shown that if a solution to the four-sided problem exists, there exists one that has a central point z such that xy-monotone curves from z to the four corners of the rectangle R partition the solution without intersecting any leaders. To compute such a partitioned solution, assume we are given, for each side s of the rectangle R, the leader whose segment orthogonal to s is maximum among all leaders of side s. These *extremal leaders* essentially partition the instance into four smaller instances of ADJACENT TWO-SIDED BOUNDARY LABELING, one for each corner. These instances can be processed independently. There are $O(n^8)$ choices for these extremal leaders, trying all of them thus yields a running time of $O(n^{10})$ and space consumption $O(n)$. For THREE-SIDED BOUNDARY LABELING, the running time is $O(n^8)$, but can be improved to $O(n^4)$ by guessing only the extremal leader of the middle side of the rectangle. Also, except for the length minimization, all presented extensions carry over. A proof is given in the full version of the paper [13]. It remains open whether a minimum length solution of THREE- and FOUR-SIDED BOUNDARY LABELING can be computed in polynomial time.

References

1. Agarwal, P.K., Efrat, A., Sharir, M.: Vertical decomposition of shallow levels in 3-dimensional arrangements and its applications. SIAM J. Comput. 29(3), 912–953 (1999)
2. Bekos, M.A., Kaufmann, M., Nöllenburg, M., Symvonis, A.: Boundary labeling with octi-linear leaders. Algorithmica 57(3), 436–461 (2010)
3. Bekos, M.A., Kaufmann, M., Potika, K., Symvonis, A.: Area-feature boundary labeling. Comput. J. 53(6), 827–841 (2010)
4. Bekos, M.A., Kaufmann, M., Symvonis, A., Wolff, A.: Boundary labeling: Models and efficient algorithms for rectangular maps. Comput. Geom. Theory Appl. 36(3), 215–236 (2007), http://dx.doi.org/10.1016/j.comgeo.2006.05.003
5. Benkert, M., Haverkort, H.J., Kroll, M., Nöllenburg, M.: Algorithms for multi-criteria boundary labeling. J. Graph. Algorithms Appl. 13(3), 289–317 (2009)
6. Chazelle, B.: 36 co-authors: The computational geometry impact task force report. In: Chazelle, B., Goodman, J.E., Pollack, R. (eds.) Advances in Discrete and Computational Geometry, vol. 223, pp. 407–463. American Mathematical Society, Providence (1999)
7. Fink, M., Haunert, J.H., Schulz, A., Spoerhase, J., Wolff, A.: Algorithms for labeling focus regions. IEEE Trans. Visual. Comput. Graphics 18(12), 2583–2592 (2012), http://dx.doi.org/10.1109/TVCG.2012.193
8. Freeman, H., Marrinan, S., Chitalia, H.: Automated labeling of soil survey maps. In: ASPRS-ACSM Annual Convention, Baltimore, vol. 1, pp. 51–59 (1996)
9. Gemsa, A., Haunert, J.H., Nöllenburg, M.: Boundary-labeling algorithms for panorama images. In: 19th ACM SIGSPATIAL Int. Conf. Adv. Geogr. Inform. Syst., pp. 289–298 (2011)
10. Gritzmann, P., Mohar, B., Pach, J., Pollack, R.: Embedding a planar triangulation with vertices at specified positions. Amer. Math. Mon. 98, 165–166 (1991)
11. Hirschberg, D.S.: A linear space algorithm for computing maximal common subsequences. Comm. ACM 18(6), 341–343 (1975)
12. Katz, B., Krug, M., Rutter, I., Wolff, A.: Manhattan-geodesic embedding of planar graphs. In: Eppstein, D., Gansner, E.R. (eds.) GD 2009. LNCS, vol. 5849, pp. 207–218. Springer, Heidelberg (2010)
13. Kindermann, P., Niedermann, B., Rutter, I., Schaefer, M., Schulz, A., Wolff, A.: Two-sided boundary labeling with adjacent sides. Arxiv report (May 2013), http://arxiv.org/abs/1305.0750
14. van Kreveld, M., Strijk, T., Wolff, A.: Point labeling with sliding labels. Comput. Geom. Theory Appl. 13, 21–47 (1999), http://dx.doi.org/10.1016/S0925-7721(99)00005-X
15. Morrison, J.L.: Computer technology and cartographic change. In: Taylor, D. (ed.) The Computer in Contemporary Cartography. Johns Hopkins University Press (1980)
16. Nöllenburg, M., Polishchuk, V., Sysikaski, M.: Dynamic one-sided boundary labeling. In: 18th ACM SIGSPATIAL Int. Symp. Adv. Geogr. Inform. Syst., pp. 310–319 (2010)
17. Raghavan, R., Cohoon, J., Sahni, S.: Single bend wiring. J. Algorithms 7(2), 232–257 (1986)
18. Speckmann, B., Verbeek, K.: Homotopic rectilinear routing with few links and thick edges. In: López-Ortiz, A. (ed.) LATIN 2010. LNCS, vol. 6034, pp. 468–479. Springer, Heidelberg (2010)
19. Zoraster, S.: Practical results using simulated annealing for point feature label placement. Cartography and GIS 24(4), 228–238 (1997)

Optimal Batch Schedules for Parallel Machines

Frederic Koehler[1,*] and Samir Khuller[2]

[1] Princeton Univ., Princeton NJ 08544, USA
f.koehler427@gmail.com
[2] Dept. of Computer Science, Univ. of Maryland, College Park, MD 20742, USA
samir@cs.umd.edu

Abstract. We consider the problem of batch scheduling on parallel machines where jobs have release times, deadlines, and identical processing times. The goal is to schedule these jobs in batches of size at most B on m identical machines. Previous work on this problem primarily focused on finding feasible schedules. Motivated by the problem of minimizing energy, we consider problems where the number of batches is significant. Minimizing the number of batches on a single processor previously required an impractical $O(n^8)$ dynamic programming algorithm. We present a $O(n^3)$ algorithm for simultaneously minimizing the number of batches and maximum completion time, and give improved guarantees for variants with infinite size batches, agreeable release times, and batch "budgets". Finally, we give a pseudo-polynomial algorithm for general batch-count-sensitive objective functions and correct errors in previous results.

Keywords: Scheduling, Batching, Optimal Algorithms.

1 Introduction

Batch Scheduling refers to the scheduling of jobs when jobs can be processed in batches of size at most B. The notion of parallel batch scheduling of jobs was initially proposed to model deliveries by trucks of bounded capacity [9]. It has, among other applications, been used to model the management of large multimedia-on-demand systems [2] and "burn-in" operations in an oven where a number of chips can be baked together at once [10]. We focus on the version where all of the jobs in a batch are processed together and start at the same time. In addition, for each job (J_α) the schedule must respect release times (r_α) and deadlines (d_α), times at which jobs become available to process and must be processed by, respectively. This can, for example, model the delivery of people flying into an airport for a conference, where each person must be transported by a given deadline using a fleet of limited capacity vehicles.

Many results in deterministic batch scheduling focus on the version where all jobs have release times, deadlines, and uniform length of p [5,10,9,3,1], where

---

* The first author's work was done as part of his high school research project at the Univ. of Maryland, and later an NSF REU supplement to CCF 0937865. The work of the second author is supported by NSF grants CCF 0728839 and CCF 0937865.

F. Dehne, R. Solis-Oba, and J.-R. Sack (Eds.): WADS 2013, LNCS 8037, pp. 475–486, 2013.

the objective is to find a feasible schedule of batches each containing at most B jobs. The start and end time of the batch must respect the release time and deadline of each job in the batch. Using standard techniques these feasibility algorithms can be used to minimize objectives such as maximum lateness (L_{max}) and maximum completion time (C_{max}). Note that when jobs have different lengths, deciding feasibility becomes an $\mathcal{NP}$-complete problem, although approximation algorithms exist, e.g. [2].

Motivated by issues of savings energy, recently Chang et al. [4] consider the problem of minimizing the time the machine is being used, referred to as *activation time*. In this case of identical job lengths, this translates to scheduling all of the jobs using the fewest number of batches. Chang et al. [4] develop a $O(n^8)$ algorithm for this problem based on the work of Baptiste [1]; the space complexity is also very high, and it is also designed for the single processor case only. The paper also considers a variety of other cases of the activation problem — e.g. when the release times and deadlines are integral and $p = 1$ they present a linear time algorithm. Even though batch scheduling has been studied for over twenty-five years, these are the first algorithms which explicitly aim to minimize batch count. However, we expect the number of batches used almost always affects the energy cost (and thus profit) of the system.

The basic problem dealt with is that of creating a schedule of batches for m identical machines. A batch (or batch instance) B_α in a schedule is associated with three properties:

- The set of jobs contained in the batch. We let $|B_\alpha|$ denote the number of jobs in a batch. In a feasible schedule, $|B_\alpha| \leq B$, where B is the given batch size constraint.
- The start time $s(B_\alpha)$. $\forall J_j \in B_\alpha$, the completion time $C_j = s(B_\alpha) + p$. In a feasible schedule, $r_j + p \leq C_j \leq d_j$.
- The machine $m(B_\alpha)$ that the batch is scheduled upon, occupying the time interval $[s(B_\alpha), s(B_\alpha) + p]$ on that machine. In a feasible schedule, the time intervals of batches scheduled on the same machine must be disjoint.

Our results, briefly: we can find a batch-count and C_{max} minimal schedule in $O(n^3)$ time and improve that in variants of the problem; we produce a $O(n)$ algorithm for the agreeable problem ($r_a < r_b \rightarrow d_a \leq d_b$). We give a pseudo-polynomial algorithm for a notion of batch-count-sensitive objective functions.

1.1   Related Work

Ikura and Gimple originally gave a $O(n^2)$ algorithm for scheduling agreeable jobs on a single processsor with the objective of minimizing C_{max}. Lee et al. found a $O(nB)$ algorithm for this same problem using dynamic programming [10]. Baptiste[1] finally showed that the problem with arbitrary release times was polynomial-time solvable for a broad class of sum-function objectives, such as $\sum C_j$. However his algorithms have extremely high (polynomial) complexity.

Recently, Condotta et al. [5] developed improved algorithms for the feasibility problem for general release times and deadlines: for the single machine case they provide an $O(n^2)$ time algorithm. They also study the previously ignored multiple identical machine case and provide an $O(n^3 \log n)$ time algorithm. These algorithms are generalized forms of algorithms for the non-batching problem $(B = 1)$: the $O(n^2)$ algorithm is based on the "forbidden regions" method of Garey et al. [7], and the $O(n^3 \log n)$ algorithm for the multiprocessor case is based on the "barriers" method of Simons [14].

The barriers and forbidden regions methods for a single processor are both notable for choosing schedules with the property that each job, numbered from the left (or the right), starts as soon as possible. Formally, the start time of the ith job from the left in the generated schedule is a lower bound on the start time of the ith job in *any* feasible schedule. These schedules are thus optimal for the objectives $\sum C_j$ and C_{max}. We shall say that these schedules have UNIT-OPTIMAL structure. Recent algorithms using graph-theoretic techniques find schedules with identical structure [6,12].

The Condotta et al. paper claims that each batch, counting from the left, for their barriers algorithm has minimal start time (Lemma 4). This claim is incorrect: consider a problem instance with large deadlines, two machines, $B = 2$ (or any even number), B jobs released at time 0 and B jobs released at time p. The barriers algorithm will produce a schedule with two full batches, the second at time p. However a feasible schedule exists where the first two batches are started at time zero, each containing $B/2$ jobs. This disproves their claim and invalidates their proof of correctness. However by correcting this claim it is still possible to show their algorithm's correctness. (Here is a sketch: Consider only the classes of schedules where each batch, from the left, greedily takes as many jobs as possible. The schedules generated by the barriers algorithm are those which for any k, both process the *minimum* number of jobs in the first k batches numbered from the left and starts each batch B_k no earlier than any nonempty B'_k in any other schedule of this class. The processing of the minimum number of jobs is crucial to proving the batch start times are minimized.)

They also claim that their algorithms immediately minimize $\sum C_j$ and C_{max} in the batching problem. We do not believe this to be the case. Consider the single machine case: by delaying a job slightly, it may be possible to overlap it in a batch with other jobs, drastically reducing its completion time by not blocking on the processing of the first job. The barriers algorithm only creates barriers when it encounters infeasibility, so if it never encounters infeasibility, no attempt is made to delay jobs to batch them together with later released jobs. Similarly, the forbidden regions algorithm will find no forbidden regions.

The following simple example will demonstrate our claims. Run the barriers algorithm on jobs with r_j, d_j pairs $\{(1, 16), (2, 20), (6, 24))\}$, with the processing length for jobs $p = 8$, with batch size $B = 3$, and one machine: a batch will be created at time $r_1 = 1$ and at $r_1 + p = 9$. An optimal schedule for C_{max} uses only a single batch starting at $r_3 = 6$. Interestingly, an optimal schedule for $\sum C_j$ uses one batch starting at $r_2 = 2$ and another at $r_2 + p = 10$.

Theorem 1. *In the batching problem, there exist instances where minimizing $\sum C_j$ and C_{max} simultaneously is impossible.*

On a different note, when scheduling unit jobs on multiple processors, Simons [14] showed that w.l.o.g. one can only consider the *cyclic* schedules. We will make exactly this assumption in our paper. The original proof of the following claim comes from Simons for the non-batching case [14].

Lemma 1. *For any feasible schedule, a solution identical except in machine assignment exists which is cyclic, i.e. where $\forall x, (B_x, B_{x+m}, \ldots)$ are scheduled on the same machine.*

1.2   Our Approach

We generalize the notion of UNIT-OPTIMALITY. We shall call our structure RIGHT-HEAVY BATCH-OPTIMALITY (RHBO). It comprises the following properties (note the *descending* batch numbering scheme):

(1) Consider any feasible schedule S' composed of batches $B_1' \ldots B_u'$ where B_u' is the *earliest* starting batch (and B_1' the *latest*) containing a job in schedule S'. $\forall i \leq u, \sum_{b=1}^{i} |B_b| \geq \sum_{b=1}^{i} |B_b'|$; i.e. the number of jobs in $\bigcup_{b=1}^{i} B_i$ is an upper bound for feasible schedules.

(2) For any B_i, the start time of batch B_i is a lower bound for feasible schedules; i.e. for any B_i' in any feasible schedule S', $s(B_i) \leq s(B_i')$.

In the case that $B = 1$, the first property is trivial and the second property makes the structure identical to UNIT-OPTIMALITY. Note that unlike in the corrected version of the barriers algorithm, our bounds hold for *all* feasible schedules. Any schedule with these properties is optimal for many objectives:

1. C_{max} because the start time of B_1 is a lower bound. In fact the makespan (availability time) $m_x = B_x$ of *all* of the machines is minimized; so e.g. $\sum m_x$ (average makespan) and a variety of other norms are also minimized.
2. K, the number of batches, by the first property.
3. $\sum_{B_x \in S} s(B_x)$, the sum of batch start times, because a minimal number of batches is used and the start time of each batch is a lower bound.

The first section of our paper gives a low polynomial time complexity algorithm witnessing the existence of these structures. We also use this existence result to produce an optimal recursive algorithm for the *agreeable* batch scheduling problem [9]. The property of simultaneous makespan minimization on multiple machines is crucial to the decomposition.

In general Condotta et al.'s algorithms [5] will use batches efficiently only if that part of the schedule is highly constrained or many jobs share a release time. When batch sizes are larger than e.g. $B = 2$, this becomes evident. By using fewer batches, we also can improve our time complexity bound in the case that a feasible schedule exists with K_* batches ($n/B \leq K_* \leq n$) as excess batches increase algorithmic overhead. In the case of agreeable release times, we produce an elegant algorithm which searches for RHBO schedules. It is both more general and lower complexity than previous algorithms for this problem. This completes our study of RHBO schedules.

Finally we design a pseudo-polynomial algorithm for optimizing a broad array of batch-count-sensitive objectives, generalizing [1]. The lack of structure in this general setting leads to very high complexity. *This result, proofs of auxillary lemmas, and pseudocode versions of the algorithms are omitted for space reasons; the full version is at* http://www.cs.umd.edu/~samir/grant/BatchScheduling.pdf

2   Scheduling Jobs on Multiple Batch Machines

For this section, we study the problem of scheduling *all* of the jobs in a given instance. Thus when we refer to a feasible schedule, this schedule must successfully process all n jobs. We will work through a series of tentative (infeasible) schedules in our algorithms. Each tentative schedule will obey a RHBO structure: we refer to the two properties of a RHBO schedule as Invariant (1) and Invariant (2), matching the numbering in the definition. We say a job J_j is *deadline-available* in a batch B_b if $d_j \geq s(B_b) + p$. Using this notion we will define a third invariant which determines job selection within batches:

(3) $\forall B_x \forall J_j \in B_y$ such that $x < y$, if J_j is deadline-available in B_x then B_x is full of jobs with no less strict release times ($|B_x| = B, r_j \leq min_{J_i \in B_x} r_i$).

This invariant can be viewed directly as expressing a relationship between a single batch B_x, and a set of *preceding jobs* in higher-numbered (earlier) batches. It equivalently states that each batch must prefer to pick latest-released jobs from the set of jobs preceding the next-earliest batch. Note that increasing start times can only reduce the set of deadline-available jobs, and thus only make this invariant easier to satisfy.

We assume w.l.o.g. that $\forall J_j$, $r_j + p \leq d_j$: jobs violating this constraint are impossible to process. Initially let $s(B_b) \leftarrow -\infty$ for all B_b (including those earlier than B_n which cannot actually be used), since $-\infty$ is a trivial lower bound on the start time of any batch.

Lemma 2. *Invariants (2) and (3) imply Invariant (1).*

Proof. Assume Invariant (1) is violated while the other two invariants hold. Let B'_x be the latest batch in a feasible schedule S' such that $\sum_{b=1}^{x} |B'_b| > \sum_{b=1}^{x} |B_b|$. Because we chose the latest batch where the invariant is violated, the invariant holds for $B_{x-1} \ldots B_1$, and so B'_x must contain at least one additional job J_j which is not in B_x. As B_x cannot be full, Invariant (3) implies that $d_j < s(B_x) + p$. By Invariant (2), $s(B_x) \leq s(B'_x)$ and so $d_j < s(B'_x) + p$. The deadline for job J_j is violated, so schedule S' cannot be feasible.

Lemma 3. *If the OPTIMALITY invariants holds for a partial schedule $B_{x-1} \ldots B_1$ then $\forall J_l \notin \bigcup_{b=1}^{x-1} B_b$, for any feasible schedule S' composed of $B'_1, B'_2, \ldots$ it must be true that $r_l \leq s(B'_x)$.*

Proof. Let $J_l \in B'_y$. If $y \geq x$, then we have that $r_l \leq s(B'_y) \leq s(B'_x)$. Otherwise $(y < x)$: because $J_l \notin \bigcup_{b=1}^{x-1} B_b$ and $J_l \in \bigcup_{b=1}^{x-1} B'_b$, by Invariant (1) there exists some job $J_k \in \bigcup_{b=1}^{x-1} B_b$ such that $J_k \notin \bigcup_{b=1}^{x-1} B_b'$. Also, by Invariant (2), $d_l \geq s(B_y') + p \geq s(B_y) + p$, so by Invariant (3), $r_l \leq r_k$. Because $J_k \in B_z'$ with $z \geq x$, as above $s(B_x') \geq r_k \geq r_l$.

2.1   Scheduling with an Unbounded Number of Machines

Theorem 2. *A feasible schedule obeying the* OPTIMALITY *invariants can be computed in $O(n^2)$ time if $m = \infty$.*

Proof. For the first (latest) batch, r_{max} is a lower bound on the start time — thus setting $s(B_1) = r_{max}$ obeys Invariant (2). Invariant (3) determines that this batch should be filled with the maximal number (up to B) of the latest-released deadline-available jobs. For any other batch B_i, we can inductively assume that the partial schedule of $B_{i-1}, \ldots, B_1$ obeys the invariants. Let U be the set of unscheduled jobs. All jobs in U can only be scheduled in B_i and earlier batches. Set $s(B_i)$ to be the latest release time in U; Lemma 3 guarantees that this satisfies Invariant (2). Once again, Invariant (3) dictates that the maximal number of the latest-released deadline-available jobs are chosen to fill the batch.

Every batch created contains at least one job. Thus there are at most n batches and this construction takes $O(n^2)$ time.[1]

An example tentative schedule is shown in Figure 1, based on the input from Table 1 with batches taking three units of time ($p = 3$) to process up to two ($B = 2$) jobs at a time.

Table 1. Jobs for Example 2

j	1	2	3	4	5	6	7	8	9	10	11	12	13	14	15	16
r_j	0	2	1	2	0	4	3	4	1	5	4	3	8	9	10	7
d_j	5	11	6	12	8	13	8	10	7	8	9	9	16	13	14	12

2.2   Scheduling with a Bounded Number of Machines

Theorem 3. *Given a tentative schedule containing all jobs with no more than B jobs in any batch, and obeying the* OPTIMALITY *invariants, in $O(n^3)$ time it is possible to either show no feasible schedule exists or to find a feasible schedule obeying the* OPTIMALITY *invariants.*

Proof. We show how to use invariant-preserving transformations to make this schedule into a feasible one. We use two cooperative alternating passes: PushForward, which increases start times, and MoveBack, which moves jobs which are provably in the wrong batch backward.[2]

---

[1] Though $O(n \log n)$ is possible.

[2] MoveBack tightens the bounds of Invariant (1) while PushForward tightens Invariant (2).

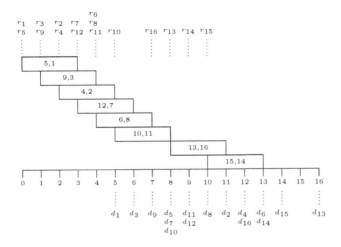

Fig. 1. Schedule constructed by Theorem 2 on Example 2

PushForward is the pass that starts first. It processes batches left-to-right (earliest-to-latest) consecutively; initially, it starts from B_i, the earliest non-empty batch in the schedule. We describe its action. Let B_c be the current batch. Let P be the set of batches which are earlier than B_c (higher indexed): inductively assume that the batches in P (1) are in non-decreasing order of start time (counting higher indexed batches first), (2) can be scheduled without overlap on m machines, and (3) contain jobs whose release times and (4) deadlines are satisfied. To process B_c, let

$$s(B_c) \leftarrow \max\left[\{s(B_{c+m}) + p, s(B_c), s(B_{c+1})\} \cup \{r_l | J_l \in B_c\}\right].$$

As noted before, increasing start times always preserves Invariant (3); we now show Invariant (2) is maintained as well. For any of the first three terms, it is possible that they may reference an empty batch past the end of the schedule: for all such batches their start time has been set to $-\infty$, so the term reduces to $-\infty$ which is a correct lower bound. Otherwise (the normal case), the first term is a valid lower bound because we have restricted ourselves to the class of cyclic schedules, and there can be no overlap between batches run on the same machine. This also satisfies inductive hypothesis (2). For the third term, by definition a lower bound for the start time of B_{c+1} extends to B_c. This satisfies inductive hypothesis (1). The final set of release times are valid lower bounds by Lemma 3, satisfying the inductive hypothesis (3).

After updating the start time, if there are any jobs in B_c which are no longer deadline-available, pick one such J_i arbitrarily and move on to the next phase MoveBack. If there are no such jobs, then the final hypothesis (4) is satisfied. If $B_c = B_1$ then terminate: supposing that all batches in our schedule obey the batch size constraint (which we have not shown yet), then using our inductive hypotheses the requirements for a feasible schedule are satisfied. Otherwise ($B_c \neq B_1$), continue on to the next batch (B_{c-1}).

We now describe MoveBack. This phase will not adjust start times so Invariant (2) is preserved. We will study Invariant (3) separately for each batch and its set of preceding jobs to show that it holds for all batches (when obvious, we will leave implicit which batch the invariant is preserved with respect to). We now describe the action of this phase. The first action this phase takes is to remove J_i from B_c. If there do not exist preceding deadline-available jobs to B_c, this does not affect Invariant (3) with respect to B_c. If there does exist at least one such job, pick the one with latest release time and move it from its current batch B_a into B_c. We say in this case that job $J_{a'}$ was *brought forward* from B_a. This may violate Invariant (3) with respect to B_a; if so, we will show that invariant is restored before the end of this phase. The removal of a job guarantees that $|B_a| < B$. The rest of this phase moves right-to-left over consecutive batches, starting with B_{c+1}. Call the current batch being processed B_z; also let the current job, initially J_i, be called J_j. We now describe the action performed for B_z; remember that when we say this phase *continues*, that means the next batch examined is the preceding batch B_{z+1}.

Case 1 ($|B_z| = B$). Let $J_{j'} = \text{argmin}_{J_y \in B_z} r_y$.

Case 1.a ($r_{j'} \leq r_j$). Swap J_j into B_z, removing $J_{j'}$. Continue MoveBack with $J_{j'}$.

Case 1.b ($r_{j'} > r_j$). Continue MoveBack with J_j.

In either case a new, possibly deadline-available, job will now precede B_z (either J_j or $J_{j'}$). Even if the job is deadline-available, its release time is no bigger than the smallest in B_z so Invariant (3) is preserved.

Case 2 ($|B_z| < B$). Place J_j into B_z.

Case 2.a (Job $J_{a'}$ was brought forward from B_a) Suppose $B_z \neq B_a$. Since $J_{a'}$ preceded B_z before its move (after the execution of the previous phase) and $|B_z| < B$, by Invariant (3) $J_{a'}$ cannot have been deadline-available in B_z. However, it is deadline-available in B_c, and by the action of PushForward we know that this implies $J_{a'}$ is deadline-available in all earlier (higher-numbered) batches. By contradiction $B_z = B_a$.

Since J_j came from some batch later than B_z but not later than B_c, and $J_{a'}$ was deadline-available in this origin-batch, $r_j \geq r_{a'}$ by Invariant (3). Therefore the replacement of J_a by J_j cannot violate Invariant (3) with respect to B_a.

Case 2.b (No job was brought forward) Adding an additional job to a nonfull batch cannot violate Invariant (3), so it is preserved.

In either case, the transformations of this phase are complete. Only batches between B_a and B_c inclusive have been modified. With respect to batches B_f with $f > a$ of $f < c$, this implies that Invariant (3) has been maintained. Thus we have shown that for every batch, Invariant (3) holds with respect to it at the end of this phase. Recall that Invariant (1) holds now by Lemma 2. If $z > n$, declare the scheduling instance infeasible: by Invariant (1), only at most $n - 1$ jobs can be scheduled in $B_1, \ldots, B_n$. Otherwise, continue on to PushForward at Bz: because we have not modified any batches earlier than B_z, the required inductive hypotheses hold for them.

This completes the description of the algorithm itself. As noted before, we still must show that the batch size restriction is obeyed to show that the algorithm is *partially correct*: if it terminates, it gives a correct answer. Recall we required our initial schedule to obey the restriction. Only the MoveBack pass modifies the assignment of jobs to batches, but it only adds a net job to a batch which has at most $B - 1$ jobs. Therefore the batch size restriction is always obeyed.

We must now show that our algorithm terminates. We claim that there can be at most $O(n^2)$ passes: for every MoveBack pass, J_j can never be placed in B_c again, because start times only increase and jobs are brought forward only if they are deadline-available; there are n jobs and at most n batches, so this makes $O(n^2)$ possible passes. Both passes run in $O(n)$ time, so a $O(n^3)$ time bound follows.

A complete algorithm is formed by composing the previous two theorems: feasible schedules for finite m are a subclass of those for unbounded m so the precondition for Theorem 3 holds. See Figure 2 and Figure 3 where $m = 2$. However, as Theorem 3 requires little from its initial schedule, far less intelligent schemes would give the same time bounds.

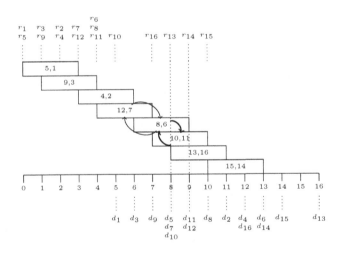

Fig. 2. Operations performed by first (thick arrows) and third (thin) pass of MoveBack

We have not so far discussed how to efficiently represent the contents of a batch. Let each batch's contents be represented by two data structures: a binary min-heap of the jobs ordered by deadlines, and an AVL tree of the jobs' release times maintaining counts in each node for duplicate release times. Our efficiency proofs are omitted for space reasons; they modify the algorithm's internals very slightly to improve its performance. If we are given a fixed batch budget K_* (modifying the algorithm to exit after exceeding its budget of batches rather than n batches), we can call this budget K_* and the improved bounds will hold; alternatively, if a feasible schedule exists with K_* batches this bound also holds.

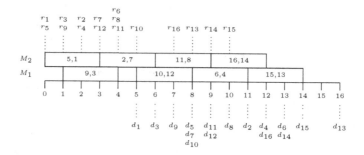

Fig. 3. Final schedule produced for Example 2

Corollary 1. *The batch-budgeted algorithm is* $O(\min(n^2 K_*, nK_*^2 \log B))$.

Corollary 2. *The algorithm terminates in* $O(n^2)$ *time for agreeable release times. The batch-budgeted algorithm is* $O(nK_*)$.

Corollary 3. *The algorithm terminates in* $O(n^2 \log n)$ *time for the unbounded case* $(B = \infty)$. *The batch-budgeted algorithm is* $O(nK_* \log n)$.

Using the enhanced binary search approach outlined in Condotta et al. [5], $L_{max} = \max C_j - d_j$ minimization can be performed with a $O(n^2 \log n)$ preprocessing phase and by calling a feasibility algorithm $O(\log n)$ times. In addition, by following the approach outlined in Condotta et al. [5], we can easily respect start-start precedence constraints. First, the schedule can be passed through their $O(n^2)$ preprocessing phase, which guarantees that if job a precedes job b then $r_a \leq r_b$ and $d_a \leq d_b$. After generating a schedule, swapping will produce a schedule which obeys the precedence constraints.

3   Scheduling Agreeable Jobs

We now design a faster algorithm to solve the problem with agreeable jobs, where $r_i < r_j \Rightarrow d_i \leq d_j$. We assume the jobs are sorted by increasing deadline (giving non-decreasing release time). Let us describe the structure of the solution we search for. By our previous result, if there exists a feasible partial schedule, there exists a RHBO partial schedule. By a simple swapping argument [10], which does not violate our invariants, we also assume w.l.o.g. that each batch consists of consecutively numbered jobs. Finally, we note that these schedules are "left-shifted" (see e.g. [1]). This implies that given an assignment of jobs to batches, the start time of a batch B_x is fully determined: it must be the maximum of r_j for all $J_j \in B_x$ and of the time the previous batch on the machine completes (machine assignments remain determined by cyclic scheduling).

We will need to maintain lists of machine availability times: to do this we use purely functional queues [8,13]. We are given three functions: head(Q) returns the front of the queue Q, tail(Q) returns Q with its front removed, and snoc(X, Q) produces a new queue with X inserted into the back of Q. All of these operations are $O(1)$ and non-destructive. Availability time lists will be maintained sorted ascending order, such that head(A) is the earliest availability time in A. We define a new operation, $U(q, t) =$ snoc(tail$(q), t)$. This will be used to *update* availability times: when a new batch is scheduled ending at time t, by cyclic scheduling it runs on the same machine as B_{m+1} in the resultant schedule, formerly (in the previous partial schedule) B_m.

Now we can easily describe the actual algorithm. Let L_i be defined (see below) such that J_{L_i+1} is the earliest job which can be batched together with J_i in a feasible schedule. Consider the RHBO feasible schedule for i jobs: by Invariant (1), the last batch must consist of jobs $J_{L_i+1}, \ldots, J_i$. Upon removing this final batch, observe that a RHBO feasible schedule is left for the first L_i jobs. Thus we inductively assume we have the RHBO schedules for each of the first $j < i$ ($i \leq n$) jobs (from which we can compute L), and then find the only possible RHBO schedule for i jobs (or fail if none exists). Note that L is a non-decreasing function ($L_{i-1} \leq L_i$): this observation makes the tabulation more efficient. F_i is the availability time list for a RHBO schedule of the first i jobs. $E(j, i) = \max\{r_i, \text{head}(F_j)\} + p$ is the left-shifted end time of the last batch in the schedule for i jobs, where the schedule is composed of a batch of jobs $J_{j+1}, \ldots, J_i$ appended to a RHBO schedule for the first j jobs. Formally:

$$L_0 = 0, \quad L_i = \min\{j \mid \max\{L_{i-1}, i - B\} \leq j < i, E(j, i) \leq d_{j+1}\},$$
$$F_0 = \text{a persistent queue with } m \text{ copies of } 0, \quad F_i = U(F_{L_i}, E(L_i, i)).$$

If at any point L_i is undefined because it minimizes over an empty set, there can exist no RHBO schedule and thus no feasible schedule at all. E and U are not tabulated in the dynamic program. F_n and L_n can be computed in $O(n)$ time. In the case of integer release times and deadlines, the binary search algorithm for L_{max} created by Lee et al. [10] can be combined with our algorithm to solve the multiprocessor problem in $O(n \log(np))$ time.

3.1   Agreeable Processing Times

The relaxation to agreeable processing times was first studied by Li and Lee [11]. Multiprocessor scheduling with no release times and a single deadline ($d_j = d$), which necessarily agrees with the processing times, is unary $\mathcal{NP}$-Hard. However, our algorithm adapts easily to the single processor case.

$$E(j, i) = \max\{r_i, F_j\} + p_i, \quad L_0 = F_0 = 0, \quad F_i = E(L_i, i),$$
$$L_i = \min\{j \mid \max(L_{i-1}, i - B) \leq j < i, E(j, i) \leq d_{j+1}\}.$$

4   Conclusions

The hardness of multi-processor batch scheduling for the objectives not satisfied by RHBO structure remains an open problem: is a pseudo-polynomial algorithm best possible? If so, what are the best approximation algorithms? Most of these problems are open even when $B = 1$; $\sum C_j$ is a notable exception. Because of Theorem 1, it may be difficult to efficiently minimize $\sum C_j$.

References

1. Baptiste, P.: Batching identical jobs. Math. Meth. of O.R. 53, 355–367 (2000)
2. Bar-Noy, A., Guha, S., Katz, Y., Naor, J.(S.), Schieber, B., Shachnai, H.: Through-put Maximization of Real-time Scheduling with Batching. In: Proc. of SODA, pp. 742–751 (2002)
3. Brucker, P.: Scheduling Algorithms. Springer (2007)
4. Chang, J., Gabow, H.N., Khuller, S.: A model for minimizing active processor time. In: Epstein, L., Ferragina, P. (eds.) ESA 2012. LNCS, vol. 7501, pp. 289–300. Springer, Heidelberg (2012), full version at
   http://www.cs.umd.edu/~samir/grant/active.pdf
5. Condotta, A., Knust, S., Shakhlevich, N.V.: Parallel batch scheduling of equal-length jobs with release and due dates. J. of Scheduling 13, 463–477 (2010)
6. Dürr, C., Hurand, M.: Finding total unimodularity in optimization problems solved by linear programs. Algorithmica 59, 256–268 (2011)
7. Garey, M.R., Johnson, D.S., Simons, B., Tarjan, R.E.: Scheduling Unit-Time Tasks with Arbitrary Release Times and Deadlines. SIAM J. on Computing 10(2), 256–269 (1981)
8. Hood, R., Melville, R.: Real-time queue operation in pure LISP. Information Processing Letters 13(2), 50–54 (1981)
9. Ikura, Y., Gimple, M.: Efficient scheduling algorithms for a single batch processing machine. Operations Research Letters 5, 61–65 (1986)
10. Lee, C.-Y., Uzsoy, R., Martin-Vega, L.A.: Efficient algorithms for scheduling semi-conductor burn-in operations. Op. Research 40(4), 764–775 (1992)
11. Li, C.-L., Lee, C.-Y.: Scheduling with agreeable release times and due dates on a batch processing machine. European J. of Operational Research 96(3), 564–569 (1997)
12. López-Ortiz, A., Quimper, C.-G.: A fast algorithm for multi-machine scheduling problems with jobs of equal processing times. In: STACS, pp. 380–391 (2011)
13. Okasaki, C.: Simple and efficient purely functional queues and deques. Journal of Functional Programming 5(04), 583–592 (1995)
14. Simons, B.: Multiprocessor scheduling of unit-time jobs with arbitrary release times and deadlines. SIAM J. Comput. 12(2), 294–299 (1983)

Unions of Onions: Preprocessing Imprecise Points for Fast Onion Layer Decomposition

Maarten Löffler[1] and Wolfgang Mulzer[2]

[1] Department of Information and Computing Sciences, Universiteit Utrecht,
The Netherlands
m.loffler@uu.nl
[2] Institut für Informatik, Freie Universität Berlin, Germany
mulzer@inf.fu-berlin.de

Abstract. Let $\mathcal{D}$ be a set of n pairwise disjoint unit disks in the plane. We describe how to build a data structure for $\mathcal{D}$ so that for any point set P containing exactly one point from each disk, we can quickly find the onion decomposition (convex layers) of P.

Our data structure can be built in $O(n \log n)$ time and has linear size. Given P, we can find its onion decomposition in $O(n \log k)$ time, where k is the number of layers. We also provide a matching lower bound.

Our solution is based on a recursive space decomposition, combined with a fast algorithm to compute the union of two disjoint onion decompositions.

1 Introduction

Let P be a planar n-point set. Take the convex hull of P and remove it; repeat until P becomes empty. This process is called *onion peeling*, and the resulting decomposition of P into convex polygons is the *onion decomposition*, or *onion* for short, of P. It can be computed in $O(n \log n)$ time [6]. Onions provide a natural, more robust, generalization of the convex hull, and they have applications in pattern recognition, statistics, and planar halfspace range searching [7, 14, 22]

Recently, a new paradigm has emerged for modeling data imprecision. Suppose we need to compute some interesting property of a planar point set. Suppose further that we have some advance knowledge about the possible locations of the points, e.g., from an imprecise sensor measurement. We would like to preprocess this information, so that once the precise inputs are available, we can obtain our structure faster. We will study the complexity of computing onions in this framework.

1.1 Related Work

The notion of onion layer decompositions first appears in the computational statistics literature [14], and several rather brute-force algorithms to compute it have been suggested (see [9] and the references therein). In the computational geometry community, Overmars and van Leeuwen [21] presented the first near-linear time algorithm, requiring $O(n \log^2 n)$ time. Chazelle [6] improved this

F. Dehne, R. Solis-Oba, and J.-R. Sack (Eds.): WADS 2013, LNCS 8037, pp. 487–498, 2013.

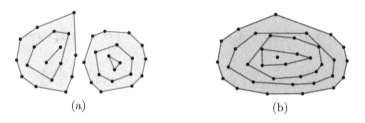

Fig. 1. (a) Two disjoint onions. (b) Their union.

to an optimal $O(n \log n)$ time algorithm. Nielsen [20] gave an output-sensitive algorithm to compute only the outermost k layers in $O(n \log h_k)$ time, where h_k is the number of vertices participating on the outermost k layers. In $\mathbb{R}^3$, Chan [5] described an $O(n \log^6 n)$ expected time algorithm.

The framework for preprocessing regions that represent points was first introduced by Held and Mitchell [12], who show how to store a set of disjoint unit disks in a data structure such that any point set containing one point from each disk can be triangulated in linear time. This result was later extended to arbitrary disjoint regions in the plane by van Kreveld *et al.* [16]. Löffler and Snoeyink first showed that the Delaunay triangulation (or its dual, the Voronoi diagram) can also be computed in linear time after preprocessing a set of disjoint unit disks [17]. This result was later extended by Buchin *et al.* [4], and Devillers gives a practical alternative [8]. Ezra and Mulzer [10] show how to preprocess a set of lines in the plane such that the convex hull of a set of points with one point on each line can be computed faster than $n \log n$ time.

These results also relate to the *update complexity* model. In this paradigm, the input values or points come with some uncertainty, but it is assumed that during the execution of the algorithm, the values or locations can be obtained exactly, or with increased precision, at a certain cost. The goal is then to compute a certain combinatorial property or structure of the precise set of points, while minimising the cost of the updates made by the algorithm [3,11,13,23].

1.2   Results

We begin by showing that the union of two disjoint onions can be computed in $O(n + k^2 \log n)$ time, where k is the number of layers in the resulting onion.

We apply this algorithm to obtain an efficient solution to the onion preprocessing problem mentioned in the introduction. Given n pairwise disjoint unit disks that model an imprecise point set, we build a data structure of size $O(n)$ such that the onion decomposition of an instance can be retrieved in $O(n \log k)$ time, where k is the number of layers in the resulting onion. We present several preprocessing algorithms. The first is very simple and achieves $O(n \log n)$ expected time. The second and third algorithm make this guarantee deterministic, at the cost of worse constants and/or a more involved algorithm.

We also show that the dependence on k is necessary: in the worst case, any comparison-based algorithm can be forced to take $\Omega(n \log k)$ time on some instances.

2   Preliminaries and Definitions

Let P be a set of n points in $\mathbb{R}^2$. The *onion decomposition*, or *onion*, of P, is the sequence $\circledast(P)$ of nested convex polygons with vertices from P, constructed recursively as follows: if $P \neq \emptyset$, we set $\circledast(P) := \{\mathrm{ch}(P)\} \cup \circledast(P \setminus \mathrm{ch}(P))$, where $\mathrm{ch}(P)$ is the convex hull of P; if $P = \emptyset$, then $\circledast(P) := \emptyset$ [6]. An element of $\circledast(P)$ is called a *layer* of P. We represent the layers of $\circledast(P)$ as dynamic balanced binary search trees, so that operations *split* and *join* can be performed in $O(\log n)$ time.

Let $\mathcal{D}$ be a set of disjoint unit disks in $\mathbb{R}^2$. We say a point set P is a *sample* from $\mathcal{D}$ if every disk in $\mathcal{D}$ contains exactly one point from P. We write log for the logarithm with base 2.

3   The Algorithm

Our algorithm requires several pieces, to be described in the following sections.

3.1   Unions of Onions

Suppose we have two point sets P and Q, together with their onions. We show how to find $\circledast(P \cup Q)$ quickly, given that $\circledast(P)$ and $\circledast(Q)$ are disjoint. Deleting points can only decrease the number of layers, so:

Observation 3.1 *Let* $P, Q \subseteq \mathbb{R}^2$. *Then* $\circledast(P)$ *and* $\circledast(Q)$ *cannot have more layers than* $\circledast(P \cup Q)$. $\qquad\square$

The following lemma constitutes the main ingredient of our onion-union algorithm. A *convex chain* is any connected subset of a convex closed curve.

Lemma 3.2. *Let* A *and* B *be two non-crossing convex chains. We can find* $\mathrm{ch}(A \cup B)$ *in* $O(\log n)$ *time.*

Proof. Since A and B do not cross, the pieces of A and B that appear on $\mathrm{ch}(A \cup B)$ are both connected: otherwise, $\mathrm{ch}(A \cup B)$ would contain four points belonging to A, B, A, and B, in that order. However, the points on A must be connected inside $\mathrm{ch}(A \cup B)$; as do the points on B. Thus, the chains A and B cross, which is impossible. Since A and B are convex chains, we can compute $\mathrm{ch}(A), \mathrm{ch}(B)$ in $O(\log n)$ time. Furthermore, since A and B are disjoint, we can also, in $O(\log n)$ time, make sure that $\mathrm{ch}(A) \cap \mathrm{ch}(B) = \emptyset$, by removing parts from A or B, if necessary. Now we can find the bitangents of $\mathrm{ch}(A)$ and $\mathrm{ch}(B)$ in logarithmic time [15]. $\qquad\square$

(a)   (b)

Fig. 2. (a) A half-eaten onion; (b) the restored onion

Lemma 3.3. *Suppose $\circledast(P)$ has k layers. Let A be the outer layer of $\circledast(P)$, and p, q be two vertices of A. Let A_1 be the points on A between p and q, going counter-clockwise. We can find $\circledast(P \setminus A_1)$ in $O(k \log n)$ time.*

Proof. The points p and q partition A into two pieces, A_1 and A_2. Let B be the second layer of $\circledast(P)$. The outer layer of $\circledast(P \setminus A_1)$ is the convex hull of $P \setminus A_1$, i.e., the convex hull of A_2 and B. By Lemma 3.2, we can find it in $O(\log n)$ time. Let $p', q' \in P$ be the points on B where the outer layer of $\circledast(P \setminus A_1)$ connects. We remove the part between p' and q' from B, and use recursion to compute the remaining layers of $\circledast(P \setminus A_1)$ in $O((k-1) \log n)$ time; see Figure 2. □

We conclude with the main theorem of this section:

Theorem 3.4. *Let P and Q be two planar point sets of total size n. Suppose that $\circledast(P)$ and $\circledast(Q)$ are disjoint. We can find the onion $\circledast(P \cup Q)$ in $O(k^2 \log n)$ time, where k is the resulting number of layers.*

Proof. By Observation 3.1, $\circledast(P)$ and $\circledast(Q)$ each have at most k layers. We use Lemma 3.2 to find $\mathrm{ch}(P \cup Q)$ in $O(\log n)$ time. By Lemma 3.3, the remainders of $\circledast(P)$ and $\circledast(Q)$ can be restored to proper onions in $O(k \log n)$ time. The result follows by induction. □

3.2 Space Decomposition Trees

We now describe how to preprocess the disks in $\mathcal{D}$ for fast divide-and-conquer. A *space decomposition tree* (SDT) T is a rooted binary tree where each node v is associated with a planar region R_v. The root corresponds to all of $\mathbb{R}^2$; for each leaf v of T, the region R_v intersects only a constant number of disks in $\mathcal{D}$. Furthermore, each inner node v in T is associated with a directed line ℓ_v, so that if u is the left child and w the right child of v, then $R_u := R_v \cap \ell_v^+$ and $R_w := R_v \cap \ell_v^-$. Here, ℓ_v^+ is the halfplane to the left of ℓ_v and ℓ_v^- the halfplane to the right of ℓ_v; see Figure 3

Let $\alpha, \beta \in (0, 1)$, and let T be an SDT. For a node v of T, let d_v denote the number of disks in $\mathcal{D}$ that intersect R_v. We call T an (α, β)-SDT for $\mathcal{D}$ if for every inner node v we have that (i) the line ℓ_v intersects at most d_v^β disks in R_v; and (ii) $d_u, d_w \leq \alpha d_v$, where u and w are the children of v.

Lemma 3.5. *Let T be an (α, β)-SDT. The tree T has height $O(\log n)$ and $O(n)$ nodes. Furthermore, $\sum_{v \in T} d_v = O(n \log n)$.*

Proof. The fact that T has height $O(\log n)$ is immediate from property (ii) of an (α, β)-SDT. For $i = 0, \ldots, \log n$, let $V_i := \{v \in T \mid d_v \in [2^i, 2^{i+1})\}$, the set of nodes whose regions intersect between 2^i and 2^{i+1} disks. Note that the sets V_i constitute a partition of the nodes. Let $\widetilde{V}_i \subseteq V_i$ be the nodes in V_i whose parent is not in V_i. By property (ii) again, the d_v along any root-leaf path in T are monotonically decreasing, so the nodes in $\widetilde{V}_i$ are unrelated (i.e., no node in $\widetilde{V}_i$ is an ancestor or descendant of another node in $\widetilde{V}_i$). Furthermore, the nodes in V_i induce in T a forest F_i such that each tree in F_i has a root from $\widetilde{V}_i$ and constant height (depending on α).

Let $D_i := \sum_{v \in \widetilde{V}_i} d_v$. We claim that for $i = 0, \ldots, \log n$, we have

$$D_i \leq n \prod_{j=i}^{\log n} \left(1 + c2^{j(\beta-1)}\right), \tag{1}$$

for some large enough constant c. Indeed, consider a node $v \in \widetilde{V}_j$. As noted above, v is the root of a tree F_v of constant height in the forest induced by V_j. By property (i), any node u in this subtree adds at most $d_u^\beta < 2^{(j+1)\beta}$ additional disk intersections (i.e., $d_a + d_b \leq d_u + 2^{(j+1)\beta}$, where a, b are the children of u). Since F_v has constant size, the total increase in disk intersections in F_v is thus at most $c'2^{(j+1)\beta}$, for some constant c'. Since $d_v \geq 2^j$, it follows that the number of disk intersections increases multiplicatively by a factor of at most $1 + c'2^{(j+1)\beta}/2^j \leq 1 + c2^{j(\beta-1)}$, for some constant c. The trees F_v partition T and the root intersects n disks, so for the nodes in $\widetilde{V}_i$, the total number of disk intersections has increased by a factor of at most $\prod_{j=i}^{\log n} \left(1 + c2^{j(\beta-1)}\right)$, giving (1). The product in (1) is easily estimated:

$$D_i \leq n \prod_{j=i}^{\log n} \left(1 + c2^{j(\beta-1)}\right) \leq ne^{\sum_{j=i}^{\log n} c2^{j(\beta-1)}} = ne^{O(1)} = O(n),$$

since $\beta < 1$. Hence, each set $\widetilde{V}_i$ has at most $O(n/2^i)$ nodes for $i = 1, \ldots, \log n$. The total size of all $\widetilde{V}_i$ is $O(n)$. Since each $v \in V_i$ lies in a constant size subtree rooted at a $w \in \widetilde{V}_i$, it follows that T has $O(n)$ nodes. For the same reason, we get that $\sum_{v \in T} d_v = O(n \log n)$. □

Now there are several ways to obtain an (α, β)-SDT for $\mathcal{D}$. A very simple construction is based on the following lemma, which is an algorithmic version of a result by Alon et al. [2, Theorem 1.2]. See Section 4 for alternative approaches.

Lemma 3.6. *There exists a constant $c \geq 0$, so that for any set $\mathcal{D}$ of m congruent nonoverlapping disks in the plane, there is a line ℓ with at least $m/2 - c\sqrt{m \log m}$ disks completely to each side of it. We can find ℓ in $O(m)$ expected time.*

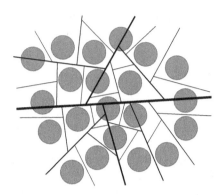

Fig. 3. A space decomposition tree for 21 unit disks

Proof. Our proof closely follows Alon *et al.* [2, Section 2]. Set $r := \lfloor \sqrt{m/\log m} \rfloor$, and pick a random integer z between 1 and $r/2$. Find a line ℓ whose angle with the x-axis is $(z/r)\pi$ and that has $\lfloor m/2 \rfloor$ disk centers on each side. Given z, we can find ℓ in $O(m)$ time by a median computation. The proof by Alon *et al.* implies that with probability at least $1/2$ over the choice of z, the line ℓ intersects at most $c\sqrt{m \log m}$ disks in $\mathcal{D}$, for some constant $c \geq 0$. Thus, we need two tries in expectation to find a good line ℓ. The expected running time is $O(m)$.   □

To obtain a $(1/2 + \varepsilon, 1/2 + \varepsilon)$-SDT T for $\mathcal{D}$, we apply Lemma 3.6 recursively until the region for each node intersects only a constant number of disks. Since the expected running time per node is linear in the number of intersected disks, Lemma 3.5 shows that the total expected running time is $O(n \log n)$.

By Lemma 3.5, the leaves of T induce a planar subdivision G_T with $O(n)$ faces. We add a large enough bounding box to G_T and triangulate the resulting graph. Since G_T is planar, the triangulation has complexity $O(n)$ and can be computed in the same time (no need for heavy machinery—all faces of G_T are convex). With each disk in $\mathcal{D}$, we store the list of triangles that intersect it (recall that each triangle intersects a constant number of disks). This again takes $O(n)$ time and space. We conclude with the main theorem of this section:

Theorem 3.7. *Let $\mathcal{D}$ be a set of n disjoint unit disks in $\mathbb{R}^2$. In $O(n \log n)$ expected time, we can construct an $(1/2 + \varepsilon, 1/2 + \varepsilon)$ space partition tree T for $\mathcal{D}$. Furthermore, for each disk $D \in \mathcal{D}$, we have a list of triangles T_D that cover the leaf regions of T that intersect D.*   □

3.3 Processing a Precise Input

Suppose we have an (α, β)-SDT together with a point location structure as in Theorem 3.7. Let P be a sample from $\mathcal{D}$. Suppose first that we know k, the number of layers in $\circledS(P)$. For each input point p_i, let $D_i \in \mathcal{D}$ be the corresponding disk. We check all triangles in T_{D_i}, until we find the one that contains p_i. Since there are $O(n)$ triangles, this takes $O(n)$ time. Afterwards, we know for each point in P the leaf of T that contains it.

For each node v of T, let n_v be the number of points in the subtree rooted at v. We can compute the n_v's in total time $O(n)$ by a postorder traversal of T. The *upper tree* T_u of T consists of all nodes v with $n_v \geq k^2$. Each leaf of T_u corresponds to a subset of P with $O(k^2)$ points. For each such subset, we use Chazelle's algorithm [6] to find its onion decomposition in $O(k^2 \log k)$ time. Since the subsets are disjoint, this takes $O(n \log k)$ total time. Now, in order to obtain $\circledast(P)$, we perform a postorder traversal of T_u, using Theorem 3.4 in each node to unite the onions of its children. This gives $\circledast(P)$ at the root.

The time for the onion union at a node v is $O(k^2 \log n_v)$. We claim that for $i = 2 \log k, \ldots, \log n$, the upper tree T_u contains at most $O(n/2^i)$ nodes v with $n_v \in [2^i, 2^{i+1})$. Given the claim, the total work is proportional to

$$\sum_{v \in T_u} k^2 \log n_v \leq \sum_{i=2 \log k}^{\log n} \frac{n}{2^i} k^2 (i+1) = nk^2 \sum_{i=2 \log k}^{\log n} \frac{i+1}{2^i} = O(n \log k),$$

since the series $\sum_{i=2 \log k}^{\log n} (i+1)/2^i$ is dominated by the first term $(\log k)/k^2$. It remains to prove the claim. Fix $i \in \{2 \log k, \ldots, \log n\}$ and let V_i be the nodes in T_u with $n_v \in [2^i, 2^{i+1})$, whose parents have $n_v \geq 2^{i+1}$. Since the nodes in V_i represent disjoint subsets of P, we have $|V_i| \leq n/2^i$. Furthermore, by property (i) of an (α, β)-SDT , both children w_1, w_2 for every node $v \in T_u$ have $n_{w_1}, n_{w_2} \leq \alpha n_v$, so that after $O(1)$ levels, all descendants w of $v \in V$ have $n_w < 2^i$. The claim follows.

So far, we have assumed that k is given. Using standard exponential search, this requirement can be removed. More precisely, for $i = 1, \ldots, \log \log n$, set $k_i = 2^{2^i}$. Run the above algorithm for $k = k_0, k_1, \ldots$. If the algorithm succeeds, report the result. If not, abort as soon as it turns out that an intermediate onion has more than k_i layers and try k_{i+1}. The total time is

$$\sum_{i=0}^{\log \log k} O(n2^i) = O(n \log k),$$

as desired. This finally proves our main result.

Theorem 3.8. *Let $\mathcal{D}$ be a set of n disjoint unit disks in $\mathbb{R}^2$. We can build a data structure that stores $\mathcal{D}$, of size $O(n)$, in $O(n \log n)$ expected time, such that given a sample P of $\mathcal{D}$, we can compute $\circledast(P)$ in $O(n \log k)$ time, where k is the number of layers in $\circledast(P)$.* $\square$

Remark. Using the same approach, without the exponential search, we can also compute the outermost k layers of an onion with arbitrarily many layers in $O(n \log k)$ time, for any k. In order to achieve this, we simply abort the union algorithm whenever k layers have been found, and note that by Observation 3.1, the points in P not on the outermost k layers of $\circledast(P)$ will never be part of the outermost k layers of $\circledast(Q)$ for any $Q \supset P$.

4 Deterministic Preprocessing

We now present alternatives to Lemma 3.6. First, we describe a very simple construction that gives a deterministic way to build an $(9/10 + \varepsilon, 1/2 + \varepsilon)$-SDT in $O(n \log n)$ time.

Lemma 4.1. *Let $\mathcal{D}$ be a set of m non-overlapping unit disks. Suppose that the centers of $\mathcal{D}$ have been sorted in horizontal and vertical direction. Then we can find in $O(m)$ time a (vertical or horizontal) line ℓ, such that ℓ intersects $O(\sqrt{m})$ disks and such that ℓ has at least $m/10$ disks from $\mathcal{D}$ completely to each side.*

Proof. Let $\mathcal{D}_l$, $\mathcal{D}_r$, $\mathcal{D}_t$, $\mathcal{D}_b$ be the $m/10$ left-, right-, top-, and bottommost disks in $\mathcal{D}$, respectively. We can find these disks in $O(m)$ time, since we know the horizontal and vertical order of their centers. We call $\mathcal{D}_o := \mathcal{D}_l \cup \mathcal{D}_r \cup \mathcal{D}_t \cup \mathcal{D}_b$ the *outer disks*, and $\mathcal{D}_i := \mathcal{D} \setminus \mathcal{D}_o$ the *inner disks*.

Let R be the smallest axis-aligned rectangle that contains all inner disks. Again, R can be found in linear time. There are $\Omega(m)$ inner disks, and all disks are disjoint, so the area of R must be $\Omega(m)$. Thus, R has width or height $\Omega(\sqrt{m})$; assume wlog that it has width $\Omega(\sqrt{m})$. Let $R' \subseteq R$ be the rectangle obtained by moving the left boundary of R to the right by two units, and the right boundary of R to the left by two units. The rectangle R' still has width $\Omega(\sqrt{m})$, and it intersects no disks from $\mathcal{D}_l \cup \mathcal{D}_r$. There are $\Omega(\sqrt{m})$ vertical lines that intersect R' and that are spaced at least one unit apart. Each such line has at least $m/10$ disks completely to each side, and each disk is intersected by at most one line. Hence, there must be a line that intersects $O(\sqrt{m})$ disks, as claimed. We can find such a line in $O(m)$ time by sweeping the disks from left to right. $\square$

The next lemma improves the constants of the previous construction. It allows us to compute an $(1/2 + \varepsilon, 5/6 + \varepsilon)$-SDT tree in deterministic time $O(n \log^2 n)$, but it requires comparatively heavy machinery.

Lemma 4.2. *Let $\mathcal{D}$ be a set of m congruent non-overlapping disks. In deterministic time $O(m \log m)$, we can find a line ℓ such that there are at least $m/2 - cm^{5/6}$ disks completely to each side of ℓ.*

Proof. Let X be a planar n-point set, and let $1 \leq r \leq n$ be a parameter. A *simplicial r-partition* of X is a sequence $\Delta_1, \dots, \Delta_a$ of $a = \Theta(r)$ triangles and a partition $X = X_1 \dot\cup \cdots \dot\cup X_a$ of X into a pieces such that (i) for $i = 1, \dots, a$, we have $X_i \subseteq \Delta_i$ and $|X_i| \in \{n/r, \dots, 2n/r\}$; and (ii) every line ℓ intersects $O(\sqrt{r})$ triangles Δ_i. Matoušek showed that a simplicial r-partition exists for every planar n-point set and for every r. Furthermore, this partition can be found in $O(n \log r)$ time (provided that $r \leq n^{1-\delta}$, for some $\delta > 0$) [18, Theorem 4.7].

Let $\gamma, \delta \in (0, 1)$ be two constants to be determined later. Set $r := m^\gamma$. Let Q be the set of centers of the disks in $\mathcal{D}$. We compute a simplicial r-partition for Q in $O(m \log m)$ time. Let $\Delta_1, \dots, \Delta_a$ be the resulting triangles and $Q = Q_1 \dot\cup \cdots \dot\cup Q_a$ the partition of Q. Set $s := m^\delta$, and for $i = 1, \dots, s$, let ℓ_i' be the line through the origin that forms an angle $(i/2s)\pi$ with the positive x-axis. Let Y_i be the projection of the triangles $\Delta_1, \ldots, \Delta_a$ onto ℓ'_i. We interpret Y_i as a set of weighted intervals, where the weight of an interval is the size $|Q_j|$ of the associated point set for the corresponding triangle. By the properties of the simplicial partition, the interval set Y_i has *depth* $O(\sqrt{r})$, i.e., every point on ℓ'_i is covered by at most $O(\sqrt{r})$ intervals of Y_i.

Note that the sets Y_i can be determined in $O(sr \log r) = O(m^{\gamma+\delta} \log m) = O(m)$ total time, for γ, δ small enough. Now, for each Y_i, we find a point c_i on ℓ'_i that has intervals of total weight $m/2 - O(\sqrt{r}(m/r)) = m/2 - O(m^{1-\gamma/2})$ completely to each side. Since the depth of Y_i is $O(\sqrt{r})$, we can find such a point in time $O(\log r)$ with binary search, for a total of $O(s \log r) = O(m)$ time (it would even be permissible to spend time $O(r)$ on each Y_i). Let ℓ_i be the line perpendicular to ℓ'_i through c_i.

The analysis of Alon *et al.* shows that for each ℓ_i, there are at most $O(s \log s)$ disks that intersect ℓ_i and at least one other line ℓ_j [2, Section 2]. Thus, it suffices to focus on the disks in $\mathcal{D}$ that intersect at most one line ℓ_i. By simple counting, there is a line ℓ_i that exclusively intersects at most $m/s = m^{1-\delta}$ disks. It remains to find such a line in $O(m)$ time. For this, we compute the arrangement $\mathcal{A}$ of the strips with width 2 centered around each ℓ_i, together with an efficient point location structure. For each cell in the arrangement, we store whether it is covered by 0, 1, or more strips. Using standard techniques, the construction takes $O(s^2) = O(m^{2\delta})$ time. We locate for each triangle Δ_i the cells of $\mathcal{A}$ that contain the vertices of Δ_i. This needs $O(r \log s) = O(m^\gamma \log m)$ steps. Since every line intersects at most $O(\sqrt{r}) = O(m^{\gamma/2})$ triangles, we know that there are at most $O(sm^{\gamma/2}) = O(m^{\delta+\gamma/2})$ triangles that intersect a cell boundary of $\mathcal{A}$. We call these triangles the *bad* triangles.

For all other triangles Δ_i, we know that the associated point set Q_i lies completely in one cell of $\mathcal{A}$. Let $\mathcal{D}_i$ be the corresponding disks. By using the information stored with the cells, we can now determine for each disk $D \in \mathcal{D}_i$ in $O(1)$ time whether D intersects exactly one line ℓ_i. Thus, we can determine in total time $O(m)$ for each line ℓ_i the total number of disks that intersect only ℓ_i and whose center is not associated with a bad triangle. Let ℓ be the line for which this number is minimum.

In total, it has taken us $O(m \log m)$ steps to find ℓ. Let us bound the number of disks that intersect ℓ. First, we know that there are at most $O(m^{\delta+\gamma/2} \cdot m^{1-\gamma}) = O(m^{1+\delta-\gamma/2})$ disks whose centers lie in bad triangles. Then, there are at most $O(m^\delta \log m)$ disks that intersect ℓ and at least one other line. Finally, there are at most $m^{1-\delta}$ disks with a center in a good triangle that intersect only ℓ. Thus, if we choose, say, $\delta = 1/6$ and $\gamma = 2/3$, then ℓ crosses at most $O(m^{5/6})$ disks in $\mathcal{D}$. Furthermore, by construction, ℓ has at least $m/2 - O(m^{2/3})$ disk centers on each side. The result follows. $\square$

Remark. Actually, we can use the approach from Lemma 4.2 to compute an $(1/2+\varepsilon, 5/6+\varepsilon)$-SDT in total deterministic time $O(m \log m)$. The bottleneck lies in finding the simplicial partition for Q. All other steps take $O(m)$ time. However, when applying Lemma 4.2 recursively, we do not need to compute a simplicial partition from scratch. Instead, as in Matoušek's paper, we can recursively refine

Fig. 4. The lower bound construction consists of $n/3$ unit disks centered on a horizontal line (5 in the figure), and two groups of $n/3$ points sufficiently far to the left and to the right of the disks. Distances not to scale.

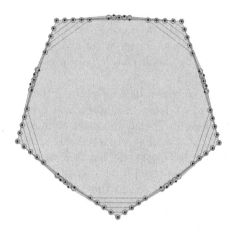

Fig. 5. n/k copies of the construction on a regular n/k-gon

the existing partitions in linear time [18, Corollary 3.5] (while duplicating the triangles for the disks that are intersected by ℓ). Thus, after spending $O(m \log m)$ time on the simplicial partition for the root, we need only linear time per node to find the dividing lines, for a total of $O(m \log m)$, by Lemma 3.5.

5   Lower Bounds

We now show that our algorithm is optimal in the decision tree model. We begin with a lower bound of $\Omega(n \log n)$ for $k = \Omega(n)$. Let n be a multiple of 3, and consider the lines

$$\ell_n^- : y = -1/2 - 6/n - x/n^2; \quad \ell_n^+ : y = -1/2 - 6/n + x/n^2.$$

Let $\mathcal{D}_n$ consist of $n/3$ disks centered on the x-axis at x-coordinates between $-n/6$ and $n/6$; a group of $n/3$ disks centered on ℓ_n^- at x-coordinates between n^2 and $n^2 + n/3$; and a symmetric group of $n/3$ disks centered on ℓ_n^+ at x-coordinates between $-n^2 - n/3$ and $-n^2$. Figure 4 shows $\mathcal{D}_{15}$.

Lemma 5.1. *Let π be a permutation on $n/3$ elements. There is a sample P of $\mathcal{D}_n$ such that p_i (the point for the ith disk from the left in the main group) lies on layer $\pi(i)$ of $\circledcirc(P)$.*

Proof. Take P as the $n/3$ centers of the disks in $\mathcal{D}$ on ℓ_n^-, the $n/3$ centers of the disks in $\mathcal{D}$ on ℓ_n^+, and for each disk $D_i \in \mathcal{D}$ on the x-axis the point $p_i = (i - n/6, \pi(i) \cdot 3/n - 1/2)$. By construction, the outermost layer of $\circledast(P)$ contains at least the leftmost point on ℓ_n^+, the rightmost point on ℓ_n^-, and the highest point (with y-coordinate $1/2$). However, it does not contain any more points: the line segments connecting these three points have slope at most $2/n^2$. The second highest point lies $3/n$ lower, and at most $n/3$ further to the left or the right. The lemma follows by induction. $\qquad\square$

There are $(n/3)! = 2^{\Theta(n \log n)}$ permutations π; so any corresponding decision tree has height $\Omega(n \log n)$. We can strengthen the lower bound to $\Omega(n \log k)$ by taking n/k copies of $\mathcal{D}_k$ and placing them on the sides of a regular (n/k)-gon, see Figure 5. By Lemma 5.1, we can choose independently for each side of the (n/k)-gon one of $(k/3)!$ permutations. The onion depth will be $k/3$, and the number of permutations is $((k/3)!)^{n/k} = 2^{\Theta(n \log k)}$.

Theorem 5.2. *Let $k \in \mathbb{N}$ and $n \geq k$. There is a set $\mathcal{D}$ of n disjoint unit disks in $\mathbb{R}^2$, such that any decision-based algorithm to compute $\circledast(P)$ for a sample P of $\mathcal{D}$, based only on prior knowledge of $\mathcal{D}$, takes $\Omega(n \log k)$ time in the worst case.*

The lower bound still applies if the input points come from an appropriate probability distribution (e.g., [1, Claim 2.2]). Thus, Yao's minimax principle [19, Chapter 2.2] yields a corresponding lower bound for any randomized algorithm.

6   Conclusion and Further Work

It would be interesting how much the parameter k can vary for a set of imprecise bounds and how to estimate k efficiently. Further work includes considering more general regions, such as overlapping disks, disks of different sizes, or fat regions. It would also be interesting to consider the problem in 3D. Three-dimensional onions are not well understood. The best general algorithm is due to Chan and needs $O(n \log^6 n)$ expected time [5], giving more room for improvement.

Acknowledgments. The authors would like to thank an anonymous reviewer for comments that improved the paper. M.L. supported by the Netherlands Organisation for Scientific Research (NWO) under grant 639.021.123. W.M. supported in part by DFG project MU/3501/1.

References

[1] Ailon, N., Chazelle, B., Clarkson, K.L., Liu, D., Mulzer, W., Seshadhri, C.: Self-improving algorithms. SIAM J. Comput. 40(2), 350–375 (2011)
[2] Alon, N., Katchalski, M., Pulleyblank, W.R.: Cutting disjoint disks by straight lines. Discrete Comput. Geom. 4(3), 239–243 (1989)

[3] Bruce, R., Hoffmann, M., Krizanc, D., Raman, R.: Efficient update strategies for geometric computing with uncertainty. Theory of Computing Systems 38(4), 411–423 (2005)

[4] Buchin, K., Löffler, M., Morin, P., Mulzer, W.: Preprocessing imprecise points for Delaunay triangulation: simplified and extended. Algorithmica 61(3), 675–693 (2011)

[5] Chan, T.M.: A dynamic data structure for 3-D convex hulls and 2-D nearest neighbor queries. J. ACM 57(3), Art. 16, 15p. (2010)

[6] Chazelle, B.: On the convex layers of a planar set. IEEE Trans. Inform. Theory 31(4), 509–517 (1985)

[7] Chazelle, B., Guibas, L.J., Lee, D.T.: The power of geometric duality. BIT 25(1), 76–90 (1985)

[8] Devillers, O.: Delaunay triangulation of imprecise points: preprocess and actually get a fast query time. J. Comput. Geom. 2(1), 30–45 (2011)

[9] Eddy, W.F.: Convex hull peeling. In: Proc. 5th Symp. Comp. Statistics (COMPSTAT), pp. 42–47 (1982)

[10] Ezra, E., Mulzer, W.: Convex hull of points lying on lines in $o(n \log n)$ time after preprocessing. Comput. Geom. 46(4), 417–434 (2013)

[11] Franciosa, P.G., Gaibisso, C., Gambosi, G., Talamo, M.: A convex hull algorithm for points with approximately known positions. Internat. J. Comput. Geom. Appl. 4(2), 153–163 (1994)

[12] Held, M., Mitchell, J.S.B.: Triangulating input-constrained planar point sets. Inform. Process. Lett. 109(1), 54–56 (2008)

[13] Hoffmann, M., Erlebach, T., Krizanc, D., Mihalák, M., Raman, R.: Computing minimum spanning trees with uncertainty. In: Proc. 25th Sympos. Theoret. Aspects Comput. Sci. (STACS), pp. 277–288 (2008)

[14] Huber, P.J.: Robust statistics: A review. Ann. Math. Statist. 43, 1041–1067 (1972)

[15] Kirkpatrick, D., Snoeyink, J.: Computing common tangents without a separating line. In: Sack, J.-R., Akl, S.G., Dehne, F., Santoro, N. (eds.) WADS 1995. LNCS, vol. 955, pp. 183–193. Springer, Heidelberg (1995)

[16] van Kreveld, M., Löffler, M., Mitchell, J.S.B.: Preprocessing imprecise points and splitting triangulations. SIAM J. Comput. 39(7), 2990–3000 (2010)

[17] Löffler, M., Snoeyink, J.: Delaunay triangulation of imprecise points in linear time after preprocessing. Comput. Geom. 43(3), 234–242 (2010)

[18] Matoušek, J.: Efficient partition trees. Discrete Comput. Geom. 8(3), 315–334 (1992)

[19] Motwani, R., Raghavan, P.: Randomized algorithms. Cambridge University Press (1995)

[20] Nielsen, F.: Output-sensitive peeling of convex and maximal layers. Inform. Process. Lett. 59, 255–259 (1996)

[21] Overmars, M.H., van Leeuwen, J.: Maintenance of configurations in the plane. J. Comput. System Sci. 23(2), 166–204 (1981)

[22] Suk, T., Flusser, J.: Convex layers: A new tool for recognition of projectively deformed point sets. In: Solina, F., Leonardis, A. (eds.) CAIP 1999. LNCS, vol. 1689, pp. 454–461. Springer, Heidelberg (1999)

[23] Tseng, K.-C.R., Kirkpatrick, D.: Input-thrifty extrema testing. In: Asano, T., Nakano, S.-i., Okamoto, Y., Watanabe, O. (eds.) ISAAC 2011. LNCS, vol. 7074, pp. 554–563. Springer, Heidelberg (2011)

Dynamic Planar Point Location with Sub-logarithmic Local Updates

Maarten Löffler[1], Joseph A. Simons[2], and Darren Strash[2]

[1] Dept. of Information and Computing Sciences, Utrecht University
[2] Dept. of Computer Science, University of California, Irvine

Abstract. We study planar point location in a collection of disjoint fat regions, and investigate the complexity of *local updates*: replacing any region by a different region that is "similar" to the original region. (i.e., the size differs by at most a constant factor, and distance between the two regions is a constant times that size). We show that it is possible to create a linear size data structure that allows for insertions, deletions, and queries in logarithmic time, and allows for local updates in sub-logarithmic time on a pointer machine. We also give results parameterized by the fatness and similarity of the objects considered.

1 Introduction

Planar point location lies at the heart of many geometric problems, and has been a major research topic in computational geometry for the past 40 years. In the static version of the problem, one aims to store a subdivision of the plane such that given a query point q in the plane, the cell of the subdivision containing q can be retrieved quickly [7,14]. In the dynamic version of the problem, one also allows changes to the data set, typically adding or removing line segments to the subdivision [3,10].

The best known dynamic data structures on a real RAM are due to Cheng and Janardan [5], who achieve $O(\log^2 n)$ queries and $O(\log n)$ updates, and Arge et al. [2], who achieve $O(\log n)$ queries, $O(\log^{1+\varepsilon} n)$ insertions, and $O(\log^{2+\varepsilon} n)$ deletions. A central open problem in this area is whether a linear-size data structure exists that can support both queries *and* updates in logarithmic time, although this is known to be possible in more specific settings such as monotone or rectilinear subdivisions [10]. Husfeldt et al. [11] prove that even in the very strong *cell probe model*, there are $\Omega(\log n / \log \log n)$ lower bounds on both queries and updates.

Despite these theoretical results, practical evidence suggests that *updating* a data structure should be fast. Intuitively, an update to a data set should not need to depend on n at all, unless we need to find the place where the update takes place (i.e., we need to do a point location query). Realistic input models are intended for designing algorithms that are provably efficient in practice, and the fat-and-disjoint model is ubiquitous (see e.g. [6]). In this paper, we study point location data structures on a collection of disjoint fat objects in the plane

F. Dehne, R. Solis-Oba, and J.-R. Sack (Eds.): WADS 2013, LNCS 8037, pp. 499–511, 2013.

that support *local updates*: replace any region by a different region that is *similar* to the original. We show that the lower bounds on updates can be broken in this setting, while still allowing $O(\log n)$ queries and using $O(n)$ storage.

The idea of local updates is not new. For example, Nekrich [13] considers (on a word-RAM) the local update operation $\text{insert}_\Delta(x, y)$ which inserts a new element x into a 1-dimensional sorted list, given a pointer to an existing element y that satisfies $|x - y| \leq \Delta$ for some distance parameter Δ. There is also a related concept called *finger updates*, where the position of the update is known; see e.g. Fleischer [9]. However, our results are the first in this area that work in a geometric setting, and they can be implemented on a real-valued pointer machine.

1.1   Problem Description

We define the problem in general dimension d, but restrict our attention to $d \in \{1, 2\}$ in the remainder of this paper. Throughout this paper, we use $|R|$ to denote the diameter of a region $R \subset \mathbb{R}^d$, that is, $|R| = \max_{p,q \in R} |pq|$. We say two *fat*[1] regions $R_1, R_2 \subset \mathbb{R}^d$ are *ρ-similar* if $|R_1 \cup R_2| \leq \rho \min\{|R_1|, |R_2|\}$, see Figure 1.[2]

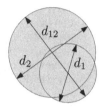

Fig. 1. ρ-similar

Problem 1. Given a set $\mathcal{R}$ of n disjoint fat regions in $\mathbb{R}^d$, store them in a data structure that allows:

- queries: given a point $q \in \mathbb{R}^d$, return the region in $\mathcal{R}$ that contains q (if any) in $Q(n)$ time;
- local updates: given a region $R \in \mathcal{R}$ and a region R' that is ρ-similar to R, replace R by R' in the data structure in $U(n)$ time; and
- global updates: delete an existing region R from the data structure or insert a new region R' into the data structure in $Q(n) + U(n)$ time

such that $Q(n) = O(\log n)$ but $U(n) = o(\log n)$. Note that a local update allows for an arbitrary number of smaller regions to be "between" the old region R and the new region R'.

1.2   Applications

A natural application of our data structure is to keep track of moving objects. One may imagine a number of objects of different sizes moving unpredictably in an environment at different speeds. A popular method for dealing with moving objects is to discretize time and process the new locations of the objects at each time step. The naive way to do this is to simply rebuild an entire data structure

---

[1] We formally define fat regions in Section 4.

[2] This definition captures two ideas at once: firstly, the *sizes* of R_1 and R_2 can differ by at most a factor of ρ, and secondly, the *distance* between R_1 and R_2 can be at most a factor ρ times the smaller of these sizes.

every time step. Our data structure can be used to process such changes more efficiently.

A different reason for studying this problem comes from the desire to cope with *data imprecision*. One way to model an imprecise point is to keep track of a region of possible locations of the point. Although algorithms to deal with static imprecise data are beginning to be well understood, little effort has been devoted to dealing with *dynamic* imprecise points. However, imprecision is often inherently dynamic (e.g. time-dependent or "stale" data), or explicitly made dynamic (e.g. updates from new samples of the same point). Our data structure can be used to store a set of dynamic imprecise points for quickly answering identity queries (i.e., given a query point, is there a point in the data structure that is potentially equal to the query point).

1.3 Results

We show that given constant similarity and fatness parameters:

- A set of n disjoint intervals in $\mathbb{R}^1$ can be maintained in an $O(n)$ size data structure that supports $O(\log n)$ worst-case time insertion, deletion, and point location queries, and $O(1)$ worst-case time local updates (Section 3).
- A set of n disjoint fat regions in $\mathbb{R}^2$ can be maintained in an $O(n)$ size data structure that supports $O(\log n)$ worst-case time insertion, deletion and point location queries, and $O(\log \log n)$ worst-case time local updates (Section 4).
- We also give bounds that can handle arbitrary similarity and fatness parameters in Theorem 1 and Theorem 2 for the $\mathbb{R}^1$ and $\mathbb{R}^2$ case respectively.

Our data structures can be implemented on a real-valued pointer machine. Because of space restrictions, many proofs and details are omitted. We also refer the interested reader to a full, uncompressed version of this text [12].

2 Tools

Quadtrees. Let B be an axis-aligned square.[3] A *quadtree* T on B is a hierarchical decomposition of B into smaller axis-aligned squares called quadtree *cells*. Each node v of T has an associated cell $C_v \subset \mathbb{R}^d$, and v is either a leaf or has 2^d equal-sized children whose cells subdivide C_v [8]. We denote the parent of a node v by $\bar{v}$. A pair of cells are called *neighbors* if they are interior disjoint and meet at an edge or corner. A leaf v is α-*balanced* if $\alpha|C_v| \geq |C_u|$ for every larger neighbor C_u of C_v. We say T is α-*balanced* if every leaf in T is α-balanced. If α is a small constant, then we simply call the quadtree T *balanced*.

Let $P \subset \mathbb{R}^d$ be a set of n points contained in B. We say T is a *valid* quadtree for P if every leaf of T contains at most 1 point of P. T may have unbounded

---

[3] We use the term *square* to mean a d-dimensional hypercube, since our main focus is on $d = 2$.

depth if P has unbounded spread,[4] Given a constant a, an *a-compressed* quadtree replaces some paths in T with *compressed* nodes. A compressed node v has only one child $\tilde{v}$ with $|C_{\tilde{v}}| \leq |C_v|/a$ and such that $C_v \setminus C_{\tilde{v}}$ has no points from P.[5] We assume for convenience that $\tilde{v}$ is *aligned* with v, i.e. if we keep subdividing C_v we will eventually create $C_{\tilde{v}}$.[6]

The compressed nodes of a quadtree T cut the tree into a number of components that correspond to smaller regular (uncompressed) quadtrees. We say T is α-balanced if all these smaller trees are α-balanced. It follows directly from Bern *et al.* [4], that a balanced compressed quadtree of linear complexity exists for any set of points P.

Static edge-oracle trees. Let T be an abstract tree with constant maximum degree d. Suppose that the nodes of T are given unique labels, and suppose that each edge $e \in T$ has an oracle which for any node label x can answer the following question: "If we removed e such that T is split into two components, which component would contain the node labeled x?" The edge-oracle tree is a search structure built over the edges of T which allows us to navigate from any node $u \in T$ to any other node $v \in T$ in $O(\log |T|)$ time and examines only $O(\log |T|)$ edges. We can construct an edge-oracle tree for T by recursively locating an edge which divides T into two components of approximately equal size.

Local updates. For a one-dimensional ordered list, data structures that can handle local (finger) updates are well known. One of the simplest implementations on a pointer machine is due to Fleischer [9].

Marked-ancestor problem. Suppose we have a tree in which nodes may be marked or unmarked. Given a node x, we want to answer the query, "Which is the lowest marked ancestor of x in the tree?". This is known as the *marked-ancestor problem*. We also want to support updates, in which nodes are marked or unmarked, and insertions/deletions of nodes to/from the tree. Alstrup *et al.* [1] gave the following results for the marked-ancestor problem on a word-RAM.

Lemma 1. *We can maintain a data structure over any rooted tree T which supports insertions and deletions of leaves in $O(1)$ amortized time, marking and unmarking nodes in $O(\log \log n)$ worst-case time, and marked ancestor queries in $O(\log n / \log \log n)$ worst-case time.*

---

[4] The *spread* of a point set P is the ratio between the largest and the smallest distance between any two distinct points in P.

[5] Such nodes are also often called *cluster*-nodes in the literature [4].

[6] While this assumption is realistic in practice, on a pure real-valued pointer machine it is not possible to align compressed nodes of arbitrary size difference in constant time. In the full version [12], we show how to adapt the results to unaligned compressed nodes.

2.1   New Tools

Dynamic balanced quadtrees. A *dynamic* quadtree is a data structure that maintains a quadtree Q on a point set P under insertion and deletion of points. In order to maintain a valid quadtree of linear size, we respond with *split* and *merge* operations respectively. A split operation takes a leaf v of Q and adds 2^d children to it; a merge operation takes 2^d leaves with a common parent and removes them. Details are given in the full version [12].

Lemma 2. *We can maintain 4-balance in a dynamic compressed quadtree in $O(1)$ worst-case time per update.*

Dynamic edge-oracle trees. There have been several recent results which generalize classic one-dimensional dynamic structures to a multidimensional setting by combining classic techniques with a quadtree-style space decomposition. However, surprisingly there are no multidimensional data structures which incorporate finger searching techniques, i.e. structures that are able to support both logarithmic queries and worst-case constant time local updates on a quadtree. We show how to build a dynamic edge-oracle tree which combines tree-decomposition and finger searching techniques with a quadtree to support $O(\log n)$ queries and $O(1)$ local updates. Details are given in the full version [12].

Lemma 3. *Let P be a set of n points, and Q be a balanced and compressed quadtree on P. We can maintain P and Q in a data structure that supports $O(\log n)$ point location queries in Q, and local insertions and deletions of points in P (i.e., when given the corresponding cells of Q) in $O(1)$ time.*

Marked-ancestor trees. We show how to answer marked-ancestor queries on a pointer-machine. Details are given in the full version [12].

Lemma 4. *We can maintain a data structure over any rooted tree T which supports insertions and deletions of leaves in $O(1)$ amortized time, marking and unmarking nodes in $O(\log \log n)$ worst-case time, and queries for the lowest marked ancestor in $O(\log n)$ worst-case time. All operations are supported on a pointer machine.*

3   One-Dimensional Case

Our 1D data structure illustrates the key ideas of our approach while being significantly simpler than the 2D version. Note that in $\mathbb{R}^1$, our input set $\mathcal{R}$ of geometric regions is a set of non-overlapping intervals. The difficulty of the problem comes from the fact that a local update may replace any interval by another interval of similar size at a distance related to that size; hence, it may "jump" over an arbitrary number of smaller intervals. Our solution works on a pure Real-valued pointer machine, and achieves constant time updates.

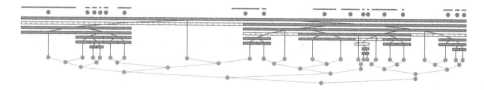

Fig. 2. A set of disjoint intervals and their center points (red); a 4-balanced compressed quadtree on the center points (blue); and a search tree on the leaves (or parts of internal cells not covered by children) of the quadtree (green).

3.1 Definition of the Data Structure

Our data structure consists of two trees. The first is designed to facilitate efficient updates and the second is designed to facilitate efficient queries. The update tree is a compressed quadtree on the center points of the intervals. The quadtree stores a pointer to each interval in the leaf that contains its center point. We also augment the tree with level-links, so that each cell has a pointer to its adjacent cells of the same size (if they exist), and maintain balance in the quadtree as described in Lemma 2. The leaves of the quadtree induce a linear size subdivision of the real line; the query tree is a search tree over this subdivision[7] that allows for fast point location and constant time local updates. We also maintain pointers between the leaves of the two trees, so that when we perform a point location query in the query tree, we also get a pointer to the corresponding cell in the quadtree, and given any leaf in the quadtree, we have a pointer to the corresponding leaf in the query tree. Figure 2 illustrates the data structure.

Details of the following results can be found in the full version [12].

Lemma 5. *Let $I \in \mathcal{R}$ be an interval, and let I' be another interval that is $O(\rho)$-similar to I. Suppose we are given a quadtree storing the midpoints of the intervals in $\mathcal{R}$ and a pointer to the leaf containing the midpoint of I. Then we can find the leaf which contains the midpoint of I' in $O(\log \rho)$ time.*

Theorem 1. *We can maintain a linear size data structure over a set of n non-overlapping intervals such that we can perform point location queries and insertion and deletion of intervals in $O(\log n)$ worst-case time and local updates in $O(\log \rho)$ worst-case time.*

4   Two-Dimensional Case

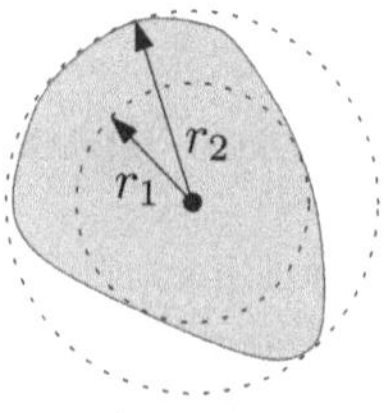

We now focus our attention on disjoint fat regions in the plane. Intuitively, a fat region should not have any long skinny pieces. We consider two types of fat regions which

Fig. 3. β-thick

---
[7] Although we could technically use a search tree directly on the original intervals, we prefer to see it as a tree over the leaves of the quadtree tree in preparation for the situation in $\mathbb{R}^2$.

precisely capture this intuition: *thick* convex regions and *wide* polygons. We say R is β-*thick* if there exists a pair of concentric balls I, O with $I \subseteq R \subseteq O$ and $|O| \leq \beta|I|$, see Figure 3.

Let $\delta \geq 1$. A δ-*corridor* is a isosceles trapezoid whose slanted edges are at most δ times as long as its base. A simple polygon P is δ-*wide* if any isosceles trapezoid $T \subset P$ whose slanted edges lie on the boundary of P is a δ-corridor [15], see Figure 4.[8] Note that any δ-wide polygon R of constant complexity is also β-thick, with $\beta \in \Theta(\delta)$.

We will first solve the problem for convex thick regions, and then extend the result to non-convex wide polygons. Analogously to the 1D case, we will store for each region $R \in \mathcal{R}$ a *representative point* p that lies somehow "in the middle" of R. When the regions are β-thick, we will use the center point of the two concentric disks from the thickness definition as representative point. We denote the set of representative points of the regions in $\mathcal{R}$ by P. Let T be the quadtree built over P. We distinguish between *true* cells, which are necessary in any valid compressed quadtree over P, and *B-cells*, which may further subdivide a true cell and are only added in order to maintain balance. We store each representative point m in T according to the following rule: Let C_v be the smallest quadtree cell containing m. If C_v is a true cell, then m is stored in v. If C_v is a B-cell, then m is stored in u, the lowest (not necessarily proper) ancestor of v in T such that $|C_u| \geq |R|/(4\beta)$.

Several new problems are introduced which were not present in the 1D case. We briefly sketch how to address each of these problems, and then present the complete solution.

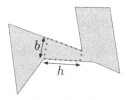

Fig. 4. δ-wide

Linear distance. When performing a query in the one-dimensional case, the location in the quadtree of any intersecting region is at most a constant number of cells away. However, in the two-dimensional case, the location of an intersecting region may be up to a linear number of cells away, as shown in Figure 5(a). We solve this problem with some additional bookkeeping. Given a quadtree cell C_q, we use two different strategies to locate regions intersecting C_q depending on their size. All regions of size at least $2\beta|C_q|$ will be located using a *marked-ancestor* data structure: an additional search structure which we explain in more detail below. All regions of size less than $2\beta|C_q|$ which intersect C_q will register a bidirectional pointer with C_q using the following *tagging* strategy.

Let d be the smallest diameter of a quadtree cell such that $d \geq |R|/(4\beta)$. Let S_R be the set of quadtree cells C which intersect R and are either a leaf or have size $|C| = d$. All cells in S_R will be *tagged* with a pointer to R. Since the quadtree is balanced, given a pointer to any cell in S_R, we can locate all cells in S_R in

---

[8] Many other notions of fatness exist in the literature. We chose to use thickness because it is basic and implied by most other definitions, and wideness because it will be convenient to use Theorem 3.

$O(|S_R|)$ time. By the following lemma, S_R must contain the cell containing the representative point of R.

Lemma 6. *Let R be a β-thick region stored by our data structure. If C is the quadtree cell which stores the representative point of R, then C has side length at least $\frac{|R|}{4\beta}$.*

Proof. If C is a B-cell, then the claim is true by construction. Suppose C is a true cell. Let m be the representative point of R. By the definition of thickness, there exists a disk $I \subseteq R$ centered at m with $|I| \geq |R|/\beta$. I contains no representative points of regions other than R. Let C be the cell containing m. Note that if C contains m and is significantly smaller than $|R|$, then C must be completely contained in I. However, C must be the largest quadtree cell completely contained in I, since if the parent $\bar{C}$ of C in the quadtree is completely contained in R, then $\bar{C}$ would not have been further subdivided because $\bar{C}$ would contain no other points. Therefore, $\bar{C}$ must have some portion outside of I and must have size larger than $|I|/2$. Thus the size of C is at least $|I|/4 \geq |R|/(4\beta)$.   □

Moreover, by the following lemma $|S_R| = O(\beta)$, and therefore, given the cell containing the representative point of R we can tag all cells in S_R in $O(\beta)$ time.

Lemma 7. *Let R be a β-thick region stored in our data structure, and let C be quadtree cell that stores the representative point of R. Then there are at most $O(\beta)$ quadtree cells of size $|C|$ required to cover R.*

Proof. Let I be the largest inscribed disk of R. The boundary of I touches the boundary of R in two or three points. If two points, then these are diametral on I, so R is contained in a strip of width $|I|$. If three points, take the diametral points of these three points and take the strips of width $|I|$ of these three pairs; R is contained in the union of these three strips. Now, if R is beta-thick, the portion of the strips it can be in is at most $\beta|I|$ long. So, R can be covered by $O(\beta)$ disks the size of I. Each such disk can be covered by at most $O(1)$ cells of size $|C|$, by Lemma 6. Thus, $O(\beta)$ cells are required to cover R.   □

Linear overlap. In the one-dimensional case, we store only the center points of our regions, and the number of regions that overlap any quadtree cell is at most three. In two dimensions, it appears that we may have a large number of small regions that intersect a quadtree cell. However, we show in the following lemma that this is not the case.

Lemma 8. *The number of β-thick convex regions intersecting any balanced quadtree leaf is $O(\beta)$.*

Proof. Let R_C be the set of thick convex regions that intersect the boundary of leaf C, and let r be the radius of a large disk D containing all regions in R_C. For each region $R_j \in R_C$ there exists a disk $I_j \subseteq R_j$ with center m_j such that $|I_j| \geq |R_j|/\beta$. Moreover, since each region R_j is convex, it must contain a

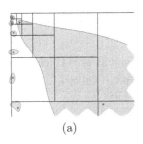
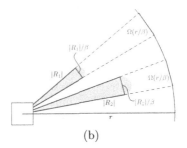

(a)   (b)

Fig. 5. (a) The intersecting region could be stored a linear distance from the query cell (containing the blue point). (b) The number of regions which can intersect quadtree leaf C is at most $O(\beta)$, since each region blocks a $\Omega(1/\beta)$ fraction of a large circle centered at C, by similar triangles.

triangle consisting of the diameter of I_j and some point $p_j \in R_j \cap C$. Each of the four sides of C can "see" at most πr of the perimeter of D. However, by a similar triangles argument each triangle must block the line of sight from one or more sides to at least $\Theta(r/\beta)$ of the perimeter (see Figure 5(b)). Thus, since the regions are convex and disjoint, the number of regions in R_C is at most $O(\beta)$. □

4.1   Definition of the Data Structure

At the core, our data structure is similar to the one-dimensional data structure described above: we have a spacial tree, which allows for efficient updates, and a search tree, which allows for efficient searching over the quadtree. However, our data structure is augmented to address the problems introduced by the two-dimensional case. We maintain a dynamic balanced quadtree Q over P, which we augment to support *mark* and *unmark* operations and marked-ancestor queries, and we maintain a dynamic edge-oracle tree on the edges of Q.

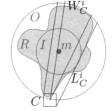

Marked-ancestor tree. Suppose we are given an angle ϕ which divides 2π (i.e., $k\phi = 2\pi$), and consider the set of angular intervals $\Phi_i = [i\phi, (i+1)\phi]$ (modulo 2π), for integers $1 \le i \le k$. For each quadtree cell C of Q with center point c, we define the wedge W_C^i centered at c and with opening angle ϕ to be the union of all halflines from c in a direction in Φ_i. Let $\mathcal{W}_C = \{W_C^i \mid 1 \le i \le k\}$; note that $\mathcal{W}_C$ partitions $\mathbb{R}^2$ into k wedges.

Fig. 6.   Illustrating the claim

For each $1 \le i \le k$, let T_i be a marked-ancestor struc-ture on Q. We mark a cell C in T_i if and only if there is a region $R \in \mathcal{R}$ of size $2\beta|C| \le |R| < 4\beta|C|$ that intersects C, and such that the center point of R lies in W_C^i.

When doing a query, we will only look at the first marked ancestor in each T_i. Lemma 9 captures the essential property of the regions which enables this strategy. First, we need the following claim.

Claim. Let β be given and set $\phi = \frac{2\pi}{\lceil 13\beta \rceil}$. Let C be a cell that is marked in T_i by a β-thick region R. Let L_C^i be the set of lines that start in C, and have a direction in Φ_i. Then every line in L_C^i intersects R.

Proof. Let m be the representative point of R. Since R is β-thick, there exist disks $I \subseteq R \subseteq O$ centered at m with $|O|/|I| \leq \beta$. Since R caused C to be marked, O, must intersect C, and m must lie in W_C^i. See Figure 6.

Now, we need that I intersects all lines in L_C^i. The distance from m to C is at most $\frac{1}{2}|O| \leq \frac{\beta}{2}|I|$. Then, the distance from m to the far edge of W_C^i is at most $\frac{\beta}{2}|I|\sin\phi$, and the distance to the far edge of L_C^i is at most $\frac{\beta}{2}|I|\sin\phi + \frac{1}{2}|C|$. Since $|R| \geq 2\beta|C|$, we know that $|C| \leq \frac{1}{2}|I|$. Using $\phi = \frac{2\pi}{13\beta}$ implies $\beta\sin\phi \leq \frac{2\pi}{13} < \frac{1}{2}$. Combining these, we see that $|I| \geq \beta|I|\sin\phi + |C|$, so, I blocks all lines in L_C^i. $\square$

Lemma 9. *Let C_1 be a cell that is marked in T_i by a convex and β-thick region R_1, and let C_2 be a descendant of C_1 that is marked in T_i by a convex and β-thick region R_2. Then there cannot be a descendant C_3 of C_2 that intersects R_1.*

Proof. Let R_2 and R_1 be convex fat regions which mark cells C_2 and C_1 respectively. Then there is a point $p_2 \in R_2 \cap C_2$. Suppose for contradiction that R_1 intersects C_3; that is, there exists a point $p_1 \in R_1 \cap C_3$. Let r and s be two parallel rays from p_1 and p_2 in some direction $\phi \in \Phi_i$. Note that rays r and s are both in $L_{C_2}^i$. Therefore each ray must intersect both R_1 and R_2 by Claim 4.1. Since each region R_1 and R_2 is convex, their intersection with each ray r (or s) is a single line segment, denoted r_1 and r_2 (s_1 and s_2) respectively. Moreover, since R_1 and R_2 are disjoint, the segments r_1 and r_2 (s_1 and s_2) are also disjoint (see Figure 7).

Since $p_1 \in R_1$, r_1 must come before r_2 on the ray r. Similarly, s_2 must come before s_1 on the ray s. Moreover, R_1 is convex, and thus the convex quadrilateral defined by r_1, s_1 is completely contained in R_1, and likewise $r_2, s_2 \subseteq R_2$. These two quadrilaterals must intersect, which is a contradiction because R_1 and R_2 are disjoint. Therefore there is no point $p_1 \in R_1 \cap C_3$. $\square$

Queries. Given a query point q, we want to find out which region (if any) contains q. We begin by performing a point location query for q in the quadtree Q. By Lemma 3 we can find the leaf cell C in the quadtree which contains q in $O(\log n)$ time using the edge-oracle tree.

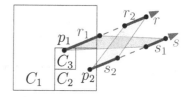

Fig. 7. Illustration of Lemma 9

By Lemma 8, there can only be $O(\beta)$ regions which intersect C. All regions of size at most $2\beta|C|$ will have tagged C with a pointer to themselves, and are immediately available from C. Moreover, we can find all regions of size at least $2\beta|C|$ in $O(\beta \log n)$ time by querying the marked-ancestor structures. We compare each region to our query point, and determine which region (if any) intersects the query point in $O(\beta)$ time. Thus, we can answer the query in total time $O(\beta \log n)$.

Updates. We only store the representative points of the regions in the quadtree. Thus, when performing a local update, it is sufficient to find the new location for the region's representative point, and then update the quadtree, tags, marked-ancestor trees, and edge-oracle trees accordingly.

Given a pointer to a region R, we replace it by another region R' that is ρ-similar to R for any arbitrary parameter $\rho \geq 1$. Let p and p' be the representative points of R and R', respectively. We find the leaf cell of Q containing p' by going up in the quadtreee until the size of the cell we are in is similar to the distance to p', then using level-links to find the ancestor of p' of similar size, and then going back down.

Lemma 10. *The distance in Q between the leaf C containing p and the leaf C' containing p' is at most $O(\log(\rho\beta))$.*

Proof. Recall that by definition, $|R \cup R'| \leq \rho \min\{|R|, |R'|\}$, and by Lemma 6, each region is stored in a quadtree cell proportional to its size, i.e. $|C| \geq \frac{|R|}{4\beta}$. Thus, $|C| \geq \frac{|R \cup R'|}{4\beta\rho}$, and likewise for $|C|'$. Hence, to find C' from C, we move up at most $\log(\beta\rho)$ levels in the quadtree to find a cell of size $\Omega(|R \cup R'|)$, then follow $O(1)$ level-link pointers to find a large cell containing p'. Finally, we move down at most $\log(\beta\rho)$ levels to find C'. □

We must also update the quadtree to reflect the new position of the representative point. By Lemma 2, we can delete p, insert p', and perform the corresponding rebalancing of the quadtree in $O(1)$ worst case time.

A local update replaces an old region R by a new region R' which is ρ-similar to R, but may overlap different quadtree cells than R. Therefore we may require updates to the marked-ancestor structure. Let C be the quadtree cell containing R's representative point. After the update, R' must only intersect $O(\beta)$ quadtree cells which are similar in size to C by Lemma 7. For each of these cells, we test the direction of the representative point of R' and mark it in the corresponding marked-ancestor tree. We also unmark cells which corresponded to the old region R. These updates can be performed in $O(\log \log n)$ time per marked-ancestor structure. We must also remove tags from all cells in S_R and add tags to cells in $S_{R'}$. However, given C and C', this takes $O(\beta)$ time by Lemma 7. By Lemma 3 we can also update the edge-oracle tree in $O(1)$ time.

Theorem 2. *A set of n disjoint convex β-thick objects of constant combinatorial complexity in $\mathbb{R}^2$ can be maintained in a $O(\beta n)$ size data structure that supports insertion, deletion and point location queries in $O(\beta \log n)$ time, and ρ-similar updates in $O(\beta \log \log n + \log(\beta\rho))$ time. All time bounds are worst-case, and the data structure can be implemented on a real-valued pointer machine.*

We can extend the result to non-convex fat regions, by cutting them into convex pieces. This approach only works for polygonal objects, since non-polygonal objects cannot always be partitioned into a finite number of convex pieces. For polygonal objects, we use a theorem by van Kreveld:

Theorem 3 (from [15]). *A δ-wide simple polygon P with n vertices can be partitioned in $O(n \log^2 n)$ time into $O(n)$ β-wide quadrilaterals and triangles, where $\beta = \min\{\delta, 1 - \frac{1}{2}\sqrt{3}\}$.*

We conclude:

Theorem 4. *A set of n disjoint polygonal δ-wide objects of constant combinatorial complexity in $\mathbb{R}^2$ can be maintained in a $O(\delta n)$ size data structure that supports insertion, deletion and point location queries in $O(\delta \log n)$ time, and ρ-similar updates in $O(\delta \log \log n + \log(\delta \rho))$ time. All time bounds are worst-case, and the data structure can be implemented on a real-valued pointer machine.*

5   Discussion

We have shown that given a set of regions in $\mathbb{R}^1$ or $\mathbb{R}^2$ fitting some modest assumptions, we can perform local updates in $\mathbb{R}^1$ in $O(1)$ time and in $\mathbb{R}^2$ in $O(\log \log n)$ time respectively. The following are open problems for future research. Can we also handle local updates in $\mathbb{R}^2$ in $O(1)$ time? Can we relax our assumption that the regions must not intersect each other? Can we adapt our techniques to handle regions in $\mathbb{R}^3$ or higher dimensions?

Acknowledgments. Work on this paper has been partially supported by the Office of Naval Research under MURI grant N00014-08-1-1015. M.L. is further supported by the Netherlands Organisation for Scientific Research (NWO) under grant 639.021.123.

References

1. Alstrup, S., Husfeldt, T., Rauhe, T.: Marked ancestor problems. In: Proc. 39th Symp. on Foundations of Computer Science, pp. 534–543 (1998)
2. Arge, L., Brodal, G.S., Georgiadis, L.: Improved dynamic planar point location. In: Proc. 47th Symp. on Foundations of Computer Science, pp. 305–314 (2006)
3. Bentley, J.L.: Solutions to Klee's rectangle problems. Technical report, Carnegie-Mellon Univ., Pittsburgh, PA (1977)
4. Bern, M., Eppstein, D., Gilbert, J.: Provably good mesh generation. J. Comput. Syst. Sci. 48(3), 384–409 (1994)
5. Cheng, S.W., Janardan, R.: New results on dynamic planar point location. SIAM J. Comput. 21(5), 972–999 (1992)
6. Berg, M.d., Gray, C.: Vertical ray shooting and computing depth orders for fat objects. In: SODA, pp. 494–503. ACM Press (2006)
7. Dobkin, D.P., Lipton, R.J.: Multidimensional searching problems. SIAM J. Comput. 5(2), 181–186 (1976)
8. Finkel, R.A., Bentley, J.L.: Quad trees: A data structure for retrieval on composite keys. Acta Inform. 4, 1–9 (1974)
9. Fleischer, R.: A simple balanced search tree with o(1) worst-case update time. In: Ng, K.W., Balasubramanian, N.V., Raghavan, P., Chin, F.Y.L. (eds.) ISAAC 1993. LNCS, vol. 762, pp. 138–146. Springer, Heidelberg (1993)

10. Giora, Y., Kaplan, H.: Optimal dynamic vertical ray shooting in rectilinear planar subdivisions. ACM Trans. Algorithms 5(3), 28:1–28:51 (2009)
11. Husfeldt, T., Rauhe, T.: Lower bounds for dynamic transitive closure, planar point location, and parentheses matching. Nordic J. Computing 3 (1996)
12. Löffler, M., Simons, J.A., Strash, D.: Dynamic planar point location with sub-logarithmic local updates. Arxiv report, arXiv:1204.4714 [cs.CG] (April 2012)
13. Nekrich, Y.: Data structures with local update operations. In: Gudmundsson, J. (ed.) SWAT 2008. LNCS, vol. 5124, pp. 138–147. Springer, Heidelberg (2008)
14. Sarnak, N., Tarjan, R.E.: Planar point location using persistent search trees. Commun. ACM 29, 669–679 (1986), http://doi.acm.org/10.1145/6138.6151, doi:10.1145/6138.6151
15. van Kreveld, M.: On fat partitioning, fat covering, and the union size of polygons. Comput. Geom. Theory Appl. 9(4), 197–210 (1998)

Parameterized Enumeration of (Locally-) Optimal Aggregations

Naomi Nishimura and Narges Simjour*

Cheriton School of Computer Science, University of Waterloo,
Waterloo, Ontario, Canada
{nishi,nsimjour}@uwaterloo.ca

Abstract. We present a parameterized enumeration algorithm for KE-MENY RANK AGGREGATION, the problem of determining an *optimal aggregation*, a total order that is at minimum total τ-*distance* (k_t) from the input multi-set of m total orders (*votes*) over a set of alternatives (*candidates*), where the τ-distance between two total orders is the number of pairs of candidates ordered differently. Our $O^*(4^{\frac{k_t}{m}})$-time algorithm constitutes a significant improvement over the previous $O^*(36^{\frac{k_t}{m}})$ upper bound.

The analysis of our algorithm relies on the notion of locally-optimal aggregations, total orders whose total τ-distances from the votes do not decrease by any single swap of two candidates adjacent in the ordering. As a consequence of our approach, we provide not only an upper bound of $4^{\frac{k_t}{m}}$ on the number of optimal aggregations, but also the first parameterized bound, $4^{\frac{k_t}{m}}$, on the number of locally-optimal aggregations, and demonstrate that it is tight. Furthermore, since our results rely on a known relation to WEIGHTED DIRECTED FEEDBACK ARC SET, we obtain new results for this problem along the way.

1 Introduction

In the general rank aggregation problem, the goal is to find a single preference list that is as close as possible to a multi-set of preference lists, according to a chosen distance measure. The problem dates back to the 18th century [9,11], when it was raised in the context of fair voting protocols in France; since then it has been applied to such areas as computational social choice, planning problems in artificial intelligence [15], bioinformatics [18], and graph drawing [8]. Here we study KEMENY RANK AGGREGATION [20], where the input preference lists (*votes*) and the output preference list (*optimal aggregation*) are restricted to total orders over the set of elements (*candidates*), the distance between two votes is the number of pairs of candidates ordered differently in the two votes, and the optimal aggregation is at minimum total distance from all votes.

---

* Supported by the Natural Sciences and Engineering Research Council of Canada (NSERC).

F. Dehne, R. Solis-Oba, and J.-R. Sack (Eds.): WADS 2013, LNCS 8037, pp. 512–523, 2013.

KEMENY RANK AGGREGATION is NP-hard for constant even numbers of votes as small as four [3,8,14,17]; therefore, approximations have been studied [1,8,13,14,23]. KEMENY RANK AGGREGATION admits a polynomial-time approximation scheme, based on a reduction to the weighted-directed feedback arc set problem (WDFAS) for special complete digraphs [21].

Since approximate solutions to KEMENY RANK AGGREGATION can violate important properties [12], algorithms to find exact solutions have garnered significant interest. Betzler et al. developed fixed-parameter algorithms with running times of $O(2^n \cdot n^2 m)$ [7], $O(1.53^{k_t} + m^2 n)$ [7] and $O((3k_m+1)! \, k_m \log k_m \cdot mn)$ [6], where n is the number of candidates, m is the number of votes, k_t is the total τ-distance of an optimal aggregation from the votes, and k_m is the maximum pairwise τ-distance of the votes. The idea in the last-mentioned algorithm was later extended to the average pairwise τ-distance of votes, denoted by k_a, and the maximum difference between the positions of a particular candidate in any of the votes, denoted by r_m, yielding bounds of $O(16^{k_a} \cdot (k_a^2 \cdot m + k_a \cdot m^2 \log m \cdot n))$ and $O(32^{r_m} \cdot (r_m^2 \cdot m + r_m \cdot m^2))$ [7]. Simjour [22] considered $\frac{k_t}{m}$ as an average parameter tighter than k_a, and obtained an $O^*(5.823^{\frac{k_t}{m}})$-time algorithm, based on an algorithm for WDFAS in tournaments. Simjour [22] also obtained algorithms of running times $O^*(1.403^{k_t})$ and $O^*(4.829^{k_m})$. Later, a subexponential-time algorithm developed by Alon et al. [2] for WDFAS for tournaments improved the running times with respect to $\frac{k_t}{m}$, k_a, and k_m, to $O(2^{O(\sqrt{\frac{k_t}{m}} \log \frac{k_t}{m})} + n^{O(1)})$ [16]. At about the same time, Karpinski and Schudy [19] reduced KEMENY RANK AGGREGATION to WDFAS for complete digraphs with arc-weights satisfying the probability constraint (the weights of the arcs (a,b) and (b,a) add up to one). Through an elegant analysis, they obtained an improved running time of $O(2^{O(\sqrt{\frac{k_t}{m}})} + n^{O(1)})$. Though most of the parameterized algorithms for KEMENY RANK AGGREGATION have benefited from its connection to WDFAS [16,19,22], details of the reductions differ.

Not much improvement (with respect to $\frac{k_t}{m}$) is expected, since an $O(2^{o(\sqrt{\frac{k_t}{m}})} + n^{O(1)})$-time algorithm for KEMENY RANK AGGREGATION would cause the failure of the Exponential Time Hypothesis [2]. On the other hand, Fernau et al. [16] studied an above-guarantee parameterization of KEMENY RANK AGGREGATION. The reduction to WDFAS results in an $O(2^{O(k_g \log k_g)} + n^{O(1)})$-time algorithm, where k_g is an above-guarantee version of k_t [10]. For an odd number of votes, the algorithm of Karpinski and Schudy [19] runs in time $O(2^{O(\sqrt{k_g})} + n^{O(1)})$. Again, an $O(2^{o(\sqrt{k_g})} + n^{O(1)})$-time algorithm for KEMENY RANK AGGREGATION results in the failure of the Exponential Time Hypothesis [16], thus is very unlikely to exist. In addition, KEMENY RANK AGGREGATION can be reduced to a kernel that includes $2k_t$ votes over at most $2k_t$ candidates [7], and to a partial kernel over at most $\frac{16k_a}{3}$ candidates [4,5].

There are few results on counting and enumeration of optimal aggregations, including those obtainable by adjusting the $O^*(2^n)$-time dynamic programming of Betzler et al. [7] or the subexponential-time algorithm of Karpinski and

Schudy [19] to count the number of optimal aggregations. The only known parameterized bound on the number of optimal aggregations is due to Simjour [22], who gave an $O^*(36^{\frac{k_t}{m}})$-time enumeration algorithm.

Our contributions. Using a refined approach, we improve the running time for enumeration from $O^*(36^{\frac{k_t}{m}})$ [22] to $O^*(4^{\frac{k_t}{m}})$, and show an $4^{\frac{k_t}{m}}$ bound on the number of optimal aggregations. We use the reduction to WDFAS for complete digraphs, exploiting the observation that the arc-weights in all the reduced digraphs satisfy the triangle inequality [23]. Our search tree algorithm, AGGSEARCH, consumes a complete digraph whose arc-weights satisfy the probability and triangle inequality constraints and finds all minimum feedback arc sets of the input graph (sets of arcs whose removal renders the graph acyclic).

The algorithm AGGSEARCH guesses adjacent pairs of minimum feedback arc sets, relying on the fact that all consecutively-ordered vertices in such sets correspond to ($\leq \frac{1}{2}$)-weight arcs. Our algorithm does not use other properties of minimum feedback arc sets; it actually enumerates all locally-minimum feedback arc sets (total orders that are only constrained to have their consecutively-ordered vertices correspond to ($\leq \frac{1}{2}$)-weight arcs). Therefore, our parameterized bound on their number (though restricted to special graph classes) is quite unexpected. Analogously, the bound is carried over to the number of locally-optimal aggregations, defined in Section 2. We are not aware of any parameterized upper bounds on the number of locally-optimal aggregations prior to this bound.

There are instances with 4^k minimum feedback arc sets. Furthermore, all these instances correspond to KEMENY RANK AGGREGATION instances. Consequently, the upper bounds on the numbers of (locally-) minimum feedback arc sets and (locally-) optimal aggregations are asymptotically tight.

2   Definitions

Complete or partial preference lists over a set of candidates U can be represented as binary relations, namely sets of ordered pairs in $U \times U$, where each ordered pair (x, y) in the relation represents the preference of a candidate x over a candidate y. As a benefit, set operations can be used; for instance, the number of preferences common to two lists π_1 and π_2 can be represented as $\pi_1 \cap \pi_2$. Since we reduce our problem to a graph problem, we also treat graph arcs as ordered pairs and sets of arcs as binary relations that consist of the corresponding ordered pairs.

For a binary relation $\rho \subseteq U \times U$, we use $x <_\rho y$ to denote that $(x, y) \in \rho$, that is, that x is preferred over y. The *reverse* of an ordered pair (x, y), denoted $\text{rev}((x, y))$, is the ordered pair (y, x) formed by reversing the first and second elements (the *tail* and *head*, respectively). The preferences opposite to those in a binary relation ρ, its *reverse*, is $\text{rev}(\rho) = \{(y, x) : (x, y) \in \rho\}$. A binary relation ρ is *transitive* if $w <_\rho x$ and $x <_\rho y$ imply $w <_\rho y$; ρ^+ is the transitive binary relation of minimum cardinality that is a superset of ρ. A binary relation ρ is *acyclic* if $\rho \cap \text{rev}(\rho^+) = \emptyset$, and a *total order* over a set U if it is transitive, for any $x <_\rho y$, x is not equal to y and $y \not<_\rho x$, and for any $x, y \in U$, $x \neq y$, either $x <_\rho y$ or $y <_\rho x$. We use $Tot(U)$ to denote the set of total orders over U.

The problem of KEMENY RANK AGGREGATION is defined in terms of a distance measure that describes the degree to which preference lists differ from each other. The τ-*distance* between $\pi_1 \in \text{Tot}(U)$ and $\pi_2 \in \text{Tot}(U)$, denoted by $\tau(\pi_1, \pi_2)$, is the number of pairs in $\pi_1 - \pi_2$, and by extension, the τ-*distance* between π_1 and a multi-set $\mathcal{I}$ over $\text{Tot}(U)$, denoted by $\tau(\pi_1, \mathcal{I})$, is $\sum_{\pi_2 \in \mathcal{I}} \tau(\pi_1, \pi_2)$.

KEMENY RANK AGGREGATION

Input: a multi-set $\mathcal{I}$ of m total orders (*votes*) in $\text{Tot}(U)$ where U is a set of n elements (*candidates*)

Output: an *optimal aggregation* of $\mathcal{I}$ (a total order $\lambda \in \text{Tot}(U)$ that minimizes $\tau(\lambda, \mathcal{I})$)

We use a well-known reduction to WDFAS on complete digraphs [21], where a *feedback arc set* β for a graph G is a subset of the graph arcs whose removal makes the graph acyclic, with *weight* $w_\beta = \sum_{e \in \beta} w_e$.

WDFAS

Input: an arc-weighted directed graph G
Output: a feedback arc set β for G of minimum weight

We use $MF(V, w)$ to denote the set of all minimum feedback arc sets in a complete digraph G on the vertex set V and with the arc-weight function w.

Feedback arc sets in a complete digraph must have many arcs; each must include a total order. The total orders in $\text{Tot}(V)$, for a complete digraph over vertex set V, are exactly the *minimal* feedback arc sets (sets for which the removal of any arc will result in a cycle in the remaining graph); thus since every minimum weight feedback arc set is minimal, $MF(V, w) \subseteq \text{Tot}(V)$.

An instance $\mathcal{I}$ of KEMENY RANK AGGREGATION is reduced to a complete digraph with arc-weights between zero and one. We define $\mathcal{I}_{(a,b)}$ as $\{\pi \in \mathcal{I} : a <_\pi b\}$.

Observation 1. *A total order $\lambda \in \text{Tot}(U)$ is an optimal aggregation of $\mathcal{I}$ if and only if* $\text{rev}(\lambda) \in MF(U, w)$*, where w is the weight function* $w_{(a,b)} = \frac{|\mathcal{I}_{(a,b)}|}{m}$*.*

Proof. This is a consequence of the fact that the τ-distance between any total order $\pi \in \text{Tot}(U)$ and $\mathcal{I}$ is precisely m times the weight of $\text{rev}(\pi)$ in the complete digraph with vertex set U and the arc-weight function $w_{(a,b)} = \frac{|\mathcal{I}_{(a,b)}|}{m}$. $\square$

The weight function satisfies two useful properties, which will be exploited in the analysis of our algorithm (Section 4). a weight function w over $U \times U$ satisfies the probability constraint if $w_{(a,b)} + w_{(b,a)} = 1$ for all pairs $a, b \in U$; we are using $w_{(a,b)}$ to denote the weight assigned to the pair $(a, b) \in U \times U$.

Observation 2. *[23] The weight function* $w_{(a,b)} = \frac{|\mathcal{I}_{(a,b)}|}{m}$ *satisfies the probability constraint and the triangle inequality.*

We can use the arc-weight function to identify pairs of vertices that might be adjacent in minimum feedback arc sets. An ordered pair (x, y) is π-*adjacent* (or *adjacent* when π is implicit) for a total order $\pi \in \text{Tot}(U)$ if $x <_\pi y$ and there is no $w \in U$ such that $x <_\pi w <_\pi y$. We use $\text{adj}(\pi)$ to denote the binary relation consisting of all π-adjacent ordered pairs. For example, let $U = \{1, 2, 3, 4\}$ and $\lambda \in \text{Tot}(U)$ satisfy $1 <_\lambda 2 <_\lambda 3 <_\lambda 4$. Then, the set of λ-adjacent pairs is $\text{adj}(\lambda) = \{(1, 2), (2, 3), (3, 4)\}$.

For a weight function $\{w_{(a,b)} : a, b \in V\}$, we define the binary relations $w_{\leq c}$ and $w_{\geq c}$ as $\{(a, b) : w_{(a,b)} \leq c\}$ and $\{(a, b) : w_{(a,b)} \geq c\}$, respectively. For any $\lambda \in MF(V, w)$, $\text{adj}(\lambda) \subseteq w_{\leq \frac{1}{2}}$, since if $\text{adj}(\lambda)$ includes an arc $e \notin w_{\leq \frac{1}{2}}$, then $(\lambda - e) \cup \text{rev}(e)$ is a feedback arc set whose weight is smaller than λ's weight, contradicting $\lambda \in MF(V, w)$.

Our fixed-parameter algorithm in Section 4 is not merely an enumeration algorithm for KEMENY RANK AGGREGATION; it enumerates all locally-optimal total orders, defined as total orders whose total τ-distances do not decrease after changing the order of an adjacent pair [14]. A closer look at total orders resulting from such a change gives rise to the following equivalent definition [14], analogous to which we define locally-minimum feedback arc sets in digraphs.

Definition 1. *A total order* $\lambda \in \text{Tot}(U)$ *is a* locally-optimal aggregation *for an instance* $\mathcal{I}$ *of* m *total orders of* KEMENY RANK AGGREGATION *if* $\text{adj}(\lambda) \subseteq n_{\geq \frac{m}{2}}$ *for the weight function* $n_{(a,b)} = |\mathcal{I}_{(a,b)}|$.

Definition 2. *A feedback arc set* β *is* locally-minimum *if it is minimal and* $\text{adj}(\beta) \subseteq w_{\leq \frac{1}{2}}$.

A minimal feedback arc set is a locally-minimum feedback arc set if reversing a single arc does not produce a feedback arc set of smaller weight. We use $LF(V, w)$ to denote the set of all locally-minimum feedback arc sets in the complete digraph on the vertex set V and the arc-weight function w.

By the minimality condition, locally-minimum feedback arc sets are forced to be total orders, making them comparable to locally-optimal aggregations.

Observation 3. *A total order* $\lambda \in \text{Tot}(U)$ *is a* locally-optimal aggregation *for an instance* $\mathcal{I}$ *of* m *total orders of* KEMENY RANK AGGREGATION *if and only if* $\text{rev}(\lambda) \in LF(U, w)$, *for the weight function* $w_{(a,b)} = \frac{|\mathcal{I}_{(a,b)}|}{m}$.

3   Ideas Used in the Algorithm

3.1   Branching Based on a Feedback Arc Set

A brute-force search for adjacent pairs of a $\gamma \in LF(V, w)$ can be very inefficient. We use a minimal feedback arc set β (equivalently, a $\beta \in \text{Tot}(V)$) to speed up the search, and show in Theorem 1 that for any β, AGGSEARCH($V, w, \beta, \text{rev}(\beta), \emptyset, \emptyset$) produces every $\gamma \in LF(V, w)$ in the leaves of its search tree. The weight of β affects only the running time: the search tree has at most 4^{w_β} leaves and is computed in times $O(n^\mu \cdot 4^{w_\beta})$, where μ denotes the exponent of matrix multiplication. As a result, $|LF(V, w)| \leq 4^k$, where k is the weight of a minimum feedback arc set in G.

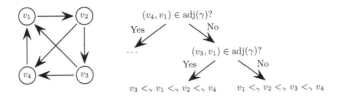

Fig. 1. The first toy example and a decision tree based on adjacent pairs

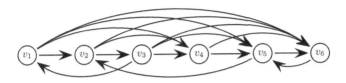

Fig. 2. The second toy example

To give a sense of how branching on adjacent pairs prunes the search space, we consider the graph shown in Fig. 1, along with the decision tree implicit in the algorithm and resulting γ's. For clarity, we have omitted arc weights and have drawn only the arcs in $w_{\leq \frac{1}{2}}$, which must include $\mathrm{adj}(\gamma)$ for any $\gamma \in LF(V, w)$. If $(v_4, v_1) \notin \mathrm{adj}(\gamma)$, v_4 must be ordered last in γ, as no other arc of the form $(v_4, *)$ will remain to be placed in $\mathrm{adj}(\gamma)$. Then, either $(v_3, v_1) \in \mathrm{adj}(\gamma)$, or v_1 must be ordered first in γ, since no arc $(*, v_1)$ will remain. Similar arguments are used to determine the rest of the arcs in $\mathrm{adj}(\gamma)$.

The search for adjacent pairs not in β, $\alpha = \mathrm{adj}(\gamma) - \beta$, is easy if the weight of β is small. The reverse arcs of $w_{\leq \frac{1}{2}} - \beta$, each of which has a weight of at least $\frac{1}{2}$, are all in β. Since the weight of β is small, the number of such arcs must be small, and hence the number of arcs in α, of which $w_{\leq \frac{1}{2}} - \beta$ is a superset.

Still, there are possibly many pairs in $w_{\leq \frac{1}{2}} \cap \beta$ from which to choose the remaining arcs, i.e. $\mathrm{adj}(\gamma) \cap \beta$. In Section 3.2, we will show that all the arcs in γ will be fixed once we figure out those located in a certain region which depends on α. A brute-force search of the region is not very costly, as the triangle inequality on the arc weights ensures that the size of the region is linear in the weight of β. The combination of α and the set of arcs of γ in the region form a concise representation of γ in terms of β, the β-*representation of* γ.

3.2 β-Representations

We use a small example to showcase the basic idea of our representation for a $\gamma \in LF(V, w)$ in Fig. 2: we choose a $\beta \in \mathrm{Tot}(V)$, and draw the vertices from left to right in the order of β (only the arcs in $w_{\leq \frac{1}{2}}$ are shown). When $\alpha = \emptyset$, $\mathrm{adj}(\gamma)$ contains no arcs outside β and hence must adhere to the order in β, that is, $\gamma = \beta$.

For $\alpha = \{(v_5, v_2)\}$, we can be sure that v_1 is ordered first and v_6 is ordered last in γ, but we do not know whether either v_3 or v_4 is ordered before v_2 and v_5. The order will be fixed once we know whether $v_3 <_\gamma v_2$ or $v_2 <_\gamma v_3$, and whether $v_4 <_\gamma v_2$ or $v_2 <_\gamma v_4$. For example, if both $v_3 <_\gamma v_2$ and $v_4 <_\gamma v_2$, then $v_3 <_\gamma v_4$ since otherwise (v_4, v_3) had to be in $\alpha = \mathrm{adj}(\gamma) - \beta$ as well.

Fortunately, not many vertices can be in the same situation as v_3 and v_4. By the triangle inequality, the weight of (v_2, v_3) plus the weight of (v_3, v_5), and in general $w_{(v_2, x)} + w_{(x, v_5)}$ for any vertex x satisfying $v_2 <_\beta x <_\beta v_5$, is at least the weight of (v_2, v_5). On the other hand, $(v_2, v_5) \in \beta$ and $w_{(v_2, v_5)} \geq \frac{1}{2}$ since (v_5, v_2) was initially assumed to be in $\alpha \subseteq w_{\leq \frac{1}{2}} - \beta$. Consequently, the weight of β is at least $\sum_{v_2 <_\beta x <_\beta v_5}(w_{(v_2, x)} + w_{(x, v_5)}) \geq \sum_{v_2 <_\beta x <_\beta v_5} w_{(v_2, v_5)} \geq |\{x : v_2 <_\beta x <_\beta v_5\}| \cdot \frac{1}{2}$. Thus, the number of vertices whose relative orders in γ with respect to v_2 must be determined (like v_3 and v_4) is at most twice the weight of β. We will see how the bounded number of decisions is generalized to arbitrary α's.

The β-representation of $\gamma \in LF(V, w)$ consists of two parts. The first part, α, is the set $\mathrm{adj}(\gamma) - \beta$. For a precise definition of the second part, we define a few terms. An unordered pair $\{x, y\}$ is a β-internal pair of $(a, b) \in \mathrm{rev}(\beta)$ if $x = a$ or $x = b$, and $b <_\beta y <_\beta a$. We use $\mathrm{IP}_\beta(e)$ to denote the set of β-internal pairs of $e \in \mathrm{rev}(\beta)$, and by extension, we use $\mathrm{IP}_\beta(\rho)$ for a binary relation $\rho \subseteq \mathrm{rev}(\beta)$ to denote $\bigcup_{e \in \rho} \mathrm{IP}_\beta(e)$. A binary relation $\rho \in \mathrm{Tot}(U)$ *restricted* to a set of unordered pairs P, denoted as $\rho|P$, is the new binary relation $\{(x, y) \in \rho : \{x, y\} \in P\}$.

Thus, for $\beta \in \mathrm{Tot}(\{v_1, \ldots, v_5\})$ and $v_1 <_\beta v_2 <_\beta v_3 <_\beta v_4 <_\beta v_5$, $\mathrm{IP}_\beta((v_5, v_3)) = \{\{v_3, v_4\}, \{v_5, v_4\}\}$ and $\mathrm{IP}_\beta(\{(v_5, v_3), (v_4, v_1)\}) = \{\{v_3, v_4\}, \{v_5, v_4\}, \{v_1, v_2\}, \{v_4, v_2\}, \{v_1, v_3\}, \{v_4, v_3\}\}$. For $v_1 <_\gamma v_2 <_\gamma v_4 <_\gamma v_5 <_\gamma v_3$, the restriction of $\gamma \in \mathrm{Tot}(\{v_1, \ldots, v_5\})$ to $\mathrm{IP}_\beta((v_5, v_3))$ is $\gamma|\mathrm{IP}_\beta((v_5, v_3)) = \{(v_4, v_3), (v_4, v_5)\}$.

Definition 3. *The β-representation of $\gamma \in LF(V, w)$, for some $\beta \in \mathrm{Tot}(V)$, is (α, δ) where $\alpha = \mathrm{adj}(\gamma) - \beta$ and $\delta = \gamma|\mathrm{IP}_\beta(\alpha)$.*

A locally-minimum feedback arc set can be efficiently reconstructed from its β-representation for an arbitrary $\beta \in \mathrm{Tot}(V)$:

Lemma 1. *If (α, δ) is the β-representation of $\gamma \in LF(V, w)$ for a $\beta \in \mathrm{Tot}(V)$, then $\gamma = \beta - \mathrm{rev}((\alpha \cup \delta)^+) \cup (\alpha \cup \delta)^+$.*

Proof. Since $\beta - \mathrm{rev}((\alpha \cup \delta)^+) \cup (\alpha \cup \delta)^+$ is a total order, it suffices to show that its two subsets $\beta - \mathrm{rev}((\alpha \cup \delta)^+)$ and $(\alpha \cup \delta)^+$ are in γ. The latter is true since α and δ are defined to be subsets of γ and γ is transitive. We prove the former by showing that every $(x, y) \in \gamma - \beta$ is in $(\alpha \cup \delta)^+$. Since γ and β are total orders, $\beta - \gamma$ is a subset of $\mathrm{rev}((\alpha \cup \delta)^+)$, and thus, $\beta - \mathrm{rev}((\alpha \cup \delta)^+)$ is a subset of γ. The proof is by strong induction: assuming the claim is true for every $(x', y') \in \gamma - \beta$ with $y <_\beta y'$, we prove the claim for (x, y).

Drawing the vertices in V on a horizontal line and ordered from left to right consistent with their order of β, suppose that $x = w_1 <_\gamma w_2 <_\gamma \ldots <_\gamma w_\ell = y$, with $\ell \geq 2$, and $(w_i, w_{i+1}) \in \mathrm{adj}(\gamma)$ for all $1 \leq i < \ell$. Fig. 3 demonstrates an example where $w_7 <_\beta w_8 <_\beta \ldots <_\beta w_4 <_\beta w_2$. In traversing the vertices in

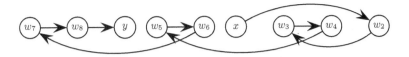

Fig. 3. An example of the case $x <_\gamma y$ and $y <_\beta x$, where the vertices are shown in the order of β from left to right and the ordered pairs in $\mathrm{adj}(\gamma)$ are presented as arcs.

order from w_1 to w_ℓ through the arcs in $\mathrm{adj}(\gamma)$, we use arcs in $\alpha = \mathrm{adj}(\gamma) - \beta$ when we go from right to left; w_ℓ must be to the left of w_1, since $y <_\beta x$. To reach $y = w_\ell$ from $x = w_1$, we must traverse at least one right-to-left arc ending up at y or a vertex to the left of y ((w_6, w_7) in Fig. 3). Since $(x, y) \in \gamma - \beta$, there must exist some $1 \le t < \ell$ such that $(w_t, w_{t+1}) \in \alpha$ with $w_{t+1} \le_\beta y <_\beta w_t$. When $w_{t+1} \ne y$, $\{y, w_{t+1}\} \in \mathrm{IP}_\beta(\alpha)$.

We now prove the induction step. If $(w_t, w_{t+1}) = (x, y)$, then $(x, y) \in \alpha$, and hence $(x, y) \in (\alpha \cup \delta)^+$. Otherwise, we show that $(w_{t+1}, y) \in (\alpha \cup \delta)^+$ if $w_{t+1} \ne y$ and $(x, w_t) \in (\alpha \cup \delta)^+$ if $x \ne w_t$. Together with $(w_t, w_{t+1}) \in \alpha$, these result in $(x, y) \in (\alpha \cup \delta)^+$, as needed to complete the proof.

We first prove that $(w_{t+1}, y) \in (\alpha \cup \delta)^+$ if $w_{t+1} \ne y$. As mentioned above, when $w_{t+1} \ne y$, $\{y, w_{t+1}\}$ is in $\mathrm{IP}_\beta(\alpha)$. Since γ orders w_{t+1} before y, $(w_{t+1}, y) \in \gamma | \mathrm{IP}_\beta(\alpha) = \delta \subseteq (\alpha \cup \delta)^+$.

Next, considering the relative orders of w_t and x, we prove that $(x, w_t) \in (\alpha \cup \delta)^+$ if $x \ne w_t$. For the case in which $w_t <_\beta x$, since γ orders x before w_t, $(x, w_t) \in \gamma - \beta$; therefore, $(x, w_t) \in (\alpha \cup \delta)^+$ by the induction hypothesis. If instead $x <_\beta w_t$, then, $w_{t+1} < x < w_t$, and hence $\{x, w_t\} \in \mathrm{IP}_\beta(\alpha)$. Since γ orders x before w_t, $(x, w_t) \in \gamma | \mathrm{IP}_\beta(\alpha) = \delta \subseteq (\alpha \cup \delta)^+$. $\square$

4 Our Results

Our search tree algorithm AGGSEARCH, shown in Algorithm 1, uses an input total order β to compute every $\gamma \in LF(V, w)$ through recursive construction of its β-representation (α, δ). The β-length of an arc $(a, b) \in \mathrm{rev}(\beta)$, used in the algorithm, is the number of vertices in $\{y : b <_\beta y <_\beta a\}$. A binary relation ρ is an *ordering of* a set of unordered pairs P if both $\rho = \rho | P$ and $|\rho| = |P|$; thus, δ is an ordering of $\mathrm{IP}_\beta(\alpha)$.

Algorithm AGGSEARCH uses an auxiliary parameter σ, initialized to $\mathrm{rev}(\beta)$, which contains the subset of $\mathrm{rev}(\beta)$ for which inclusion in α has not yet been decided. For α to be part of the β-representation of some $\gamma \in LF(V, w)$, the arcs in α must be in $w_{\le\frac{1}{2}}$, since $\alpha = \mathrm{adj}(\gamma) - \beta$ is a subset of $\mathrm{adj}(\gamma)$ and $\mathrm{adj}(\gamma)$ must be in $w_{\le\frac{1}{2}}$. Thus, no further arcs are added to α once $\sigma \cap w_{\le\frac{1}{2}}$ becomes empty (lines 1-5). By that time, δ is an ordering of $\mathrm{IP}_\beta(\alpha)$, since for each arc e inserted in α, all possible orderings of $\mathrm{IP}_\beta(e)$ are added to δ. Hence, the algorithm stops adding arcs to δ as well. Due to Lemma 1, if α and δ now form an β-representation for a $\gamma \in LF(V, w)$, γ must be equal to $(\beta - \mathrm{rev}((\alpha \cup \delta)^+)) \cup (\alpha \cup \delta)^+$. Thus, the algorithm checks if this formula produces a locally-minimum feedback arc

---

Algorithm 1: AGGSEARCH

Require : vertex set V, weight function w, $\beta \in \text{Tot}(V)$, and $\sigma, \alpha, \delta \subseteq V \times V$

1 **if** $\sigma \cap w_{\leq \frac{1}{2}} = \emptyset$ **then**

2    $\gamma \leftarrow (\beta - \text{rev}((\alpha \cup \delta)^+)) \cup (\alpha \cup \delta)^+$;

3   **if** $\gamma \in LF(V, w)$ **then return** $\{\gamma\}$;

4   **else return** $\emptyset$;

5 **end**

6 **else**

7   Select $(u, v) \in \sigma \cap w_{\leq \frac{1}{2}}$ of maximum β-length;

8    $\sigma \leftarrow \sigma - \{(u, v)\}$;

9    $LF \leftarrow \text{AGGSEARCH}(V, w, \beta, \sigma, \alpha, \delta)$;

10    $\alpha \leftarrow \alpha \cup \{(u, v)\}$;

11    $P \leftarrow \{x : u <_\beta x <_\beta v\}$;

12    $\sigma \leftarrow \sigma - \bigcup_{x \in P}\{(u, x), (x, v)\}$;

13    $L \leftarrow \{x \in P : x <_\delta u \text{ or } x <_\delta v\}$;

14    $R \leftarrow \{x \in P : u <_\delta x \text{ or } v <_\delta x\}$;

15   **foreach** $L \subseteq A \subseteq P - R$ **do**

16      $\delta' \leftarrow \delta \cup \bigcup_{x \in A}\{(x, u), (x, v)\} \cup \bigcup_{x \in P - A}\{(u, x), (v, x)\}$;

17      $LF \leftarrow LF \cup \text{AGGSEARCH}(V, w, \beta, \sigma, \alpha, \delta')$;

18   **end**

19   **return** LF;

20 **end**

---

set (line 3). If not, (α, δ) is neither a β-representation for any $\gamma \in LF(V, w)$, nor can it be made into one by adding arcs to α and δ.

For each arc (u, v) in $\sigma \cap w_{\leq \frac{1}{2}}$, the algorithm branches on whether $(u, v) \in \alpha$, removing the arc from σ once the decision is made. In the branch in which $(u, v) \in \alpha$ (lines 10-18), we can also remove all arcs in $\bigcup_{x \in P}\{(u, x), (x, v)\}$, $P = \{x : u <_\beta x <_\beta v\}$ from σ: as (u, v) is in $\text{adj}(\gamma)$ and in γ only one vertex is ordered immediately after u and only one vertex is ordered immediately before v, none of the arcs sharing a head or tail with (u, v) can be in $\text{adj}(\gamma) \supseteq \alpha$. Further branching occurs on the subset $A = \{x \in P : (x, u) \in \gamma\}$ of vertices in P (lines 15-18). The orderings of the vertices in P with respect to u and v, determined by A, are essential in determining $\delta = \gamma | \text{IP}_\beta(\alpha)$ in the β-representation of γ.

We do not want to branch over a pair more than once; one strategy is to consider arcs in order of β-length. Without this selection criterion, if in Fig. 4 (with $\sigma \cap w_{\leq \frac{1}{2}}$ including (u_1, v_1) and (u_2, v_1) such that $v_1 <_\beta u_1 <_\beta u_2$) at line 7 the algorithm selected $(u_1, v_1) \in \sigma \cap w_{\leq \frac{1}{2}}$ to be excluded from α, then further down the search tree, the algorithm could select $(u_2, v_1) \in \sigma \cap w_{\leq \frac{1}{2}}$ to be included in α. This would result in branching twice on (u_1, v_1), once for membership in α and once, at line 15, to decide whether $u_1 <_\delta v_1$ or $v_1 <_\delta u_1$.

Constraining A to include L and exclude R at line 15 avoids another duplicate branching, as otherwise the algorithm could decide on relative orderings of vertices in L and R with respect to u and v after the orderings were already

Fig. 4. Situations in which duplicate decisions could be made over a pair

fixed in δ. In Fig. 4 (where $\sigma \cap w_{\leq \frac{1}{2}}$ includes (u_1, v_1) and (u_2, v_2) such that $v_1 <_\beta v_2 <_\beta u_1 <_\beta u_2$), if $(u_1, v_1) \in \sigma$ is inserted in α at line 10, the algorithm needs to decide whether to include v_2 in A (a decision on the ordering of $\{u_1, v_2\}$) at line 15. Without the constraint on A, $(u_2, v_2) \in \sigma$ could then be inserted in α, necessitating a second decision on the ordering of $\{u_1, v_2\}$ (whether to include u_1 in A).

Removal of the same-head and same-tail arcs from σ (line 12), ordering the arcs in σ in their β-lengths (line 7), and constraining A to include L and exclude R (line 15) all result in less branching.

Theorem 1. *Given a complete digraph on a vertex set V and arc weights $\{w_{(a,b)} : a, b \in V\}$ and $\beta \in Tot(V)$, AGGSEARCH$(V, w, \beta, \mathrm{rev}(\beta), \emptyset, \emptyset)$ returns $LF(V, w)$ in time $O(|V|^\mu \cdot 4^{w_\beta})$, where $\mu < 2.376$ denotes the exponent of matrix multiplication. Furthermore, $|LF(V, w)| \leq 4^{w_\beta}$.*

Proof. Due to space limitations, we provide only a high-level idea of the proof. We prove by strong induction on the cardinality of $\sigma \cap w_{\leq \frac{1}{2}}$ that:

(1) For any ordering δ of $\mathrm{IP}_\beta(\alpha)$, AGGSEARCH$(V, w, \beta, \sigma, \alpha, \delta)$ returns every γ in $LF_{(\beta, \sigma, \alpha, \delta)} = \{\gamma \in LF(V, w) : \alpha \subseteq \mathrm{adj}(\gamma) - \beta \subseteq \alpha \cup \sigma, \text{ and } \delta \subseteq \gamma\}$
(2) If $\sigma \cup \delta$ includes an ordering of $\mathrm{IP}_\beta(\sigma \cap w_{\leq \frac{1}{2}})$, AGGSEARCH$(V, w, \beta, \sigma, \alpha, \delta)$ produces a search tree with at most $4^{w_{\mathrm{rev}(\sigma)}}$ leaves.

Making use of the fact that arcs in $\sigma \cap w_{\leq \frac{1}{2}}$ are selected in order of their β-lengths (line 7), we show that δ is an ordering of $\mathrm{IP}_\beta(\alpha)$ and $\sigma \cup \delta$ includes an ordering of $\mathrm{IP}_\beta(\sigma \cap w_{\leq \frac{1}{2}})$ in all recursive calls originating from AGGSEARCH$(V, w, \beta, \mathrm{rev}(\beta), \emptyset, \emptyset)$; from this we can show $LF(V, w) = LF_{(\beta, \mathrm{rev}(\beta), \emptyset, \emptyset)}$ is returned upon the production of at most 4^{w_β} nodes.

We associate each node v in the search tree with the cost of steps 7-9 or 10-17 performed just before the creation of v plus the cost of steps 1-5 performed at the execution of v. The dominant part is the computation of the transitive closure $(\alpha \cup \delta)^+$ using matrix multiplication at line 2. The time for a node is thus in $O(|V|^\mu)$, yielding $O(|V|^\mu \cdot 4^{w_\beta})$ time overall. □

By Observation 3, KEMENY RANK AGGREGATION instances have at most $4^{\frac{k_t}{m}}$ locally-optimal aggregations.

Corollary 1. *Given a multi-set $\mathcal{I}$ of m total orders in $Tot(U)$ and a total order λ at τ-distance k_λ of $\mathcal{I}$, the set of all locally-optimal aggregations for $\mathcal{I}$ can be found in time $O(m \cdot |U| + 4^{\frac{k_\lambda}{m}} \cdot |U|^\mu)$. Furthermore, $\mathcal{I}$ has at most $4^{\frac{k_t}{m}}$ locally-optimal aggregations, where k_t denotes the minimum τ-distance from $\mathcal{I}$.*

Although $MF(V, w)$ is generally a (small) subset of $LF(V, w)$, the two sets are equal for certain instances, for which Theorem 1's upper bound is tight:

Theorem 2. *For any set $V = \{v_1, v_2, \ldots, v_n\}$ of even cardinality, there exists a weight function w over $V \times V$ that satisfies the triangle inequality and the probability constraint such that $|MF(V, w)| = 4^k$, where k denotes the weight of a minimum feedback arc set in $MF(V, w)$.*

Proof. We consider the following weight function:

$$
w_{(v_i, v_j)} = \begin{cases} 0 & i + 1 < j \text{ or } (i + 1 = j \text{ and } i \text{ is even}) \\ \frac{1}{2} & i + 1 = j \text{ and } i \text{ is odd} \\ 1 - w_{(v_j, v_i)} & \text{otherwise} \end{cases}
$$

It is not hard to see that any minimum feedback arc set must contain all weight-0 arcs. Therefore, elements of $MF(V, w)$ differ only in the ordering of the remaining pairs. All total orders of $\{\{v_1, v_2\}, \{v_3, v_4\}, \ldots \{v_{n-1}, v_n\}\}$ are of equal weight. Since there are $2^{\frac{n}{2}}$ such total orders, each of weight $k = \frac{n}{4}$, the cardinality of $MF(V, w)$ is $2^{2k} = 4^k$ for this instance. $\qquad\square$

As there are KEMENY RANK AGGREGATION instances that reduce to the instances in the proof of Theorem 2, the lower bound also applies to optimal aggregations; the proof is omitted due to space limitations.

Theorem 3. *For any even number m, there exists a multi-set $\mathcal{I}$ of m total orders that has $4^{\frac{k_t}{m}}$ optimal aggregations, where k_t denotes the τ-distance of an optimal aggregation from $\mathcal{I}$.*

5   Concluding Remarks

We gave a tight upper bound on the number of (locally-) optimal aggregations. We emphasize that a $f(\frac{k_t}{m})n^{O(1)}$ upper bound on the number of locally-optimal aggregations is surprising. One future direction for research is the search for a new parameter that is more tuned to the complexity of enumerating all optimal aggregations, rather than locally-optimal aggregations.

References

1. Ailon, N., Charikar, M., Newman, A.: Aggregating inconsistent information: Ranking and clustering. J. ACM 55(5), 1–27 (2008)
2. Alon, N., Lokshtanov, D., Saurabh, S.: Fast fast. In: Albers, S., Marchetti-Spaccamela, A., Matias, Y., Nikoletseas, S., Thomas, W. (eds.) ICALP 2009, Part I. LNCS, vol. 5555, pp. 49–58. Springer, Heidelberg (2009)
3. Bartholdi, J.J., Tovey, C.A., Trick, M.A.: The computational difficulty of manipulating an election. Social Choice and Welfare 6(3), 227–241 (1989)
4. Betzler, N., Bredereck, R., Chen, J., Niedermeier, R.: Studies in computational aspects of voting - a parameterized complexity perspective. In: Bodlaender, H.L., Downey, R., Fomin, F.V., Marx, D. (eds.) Fellows Festschrift 2012. LNCS, vol. 7370, pp. 318–363. Springer, Heidelberg (2012)

5. Betzler, N., Bredereck, R., Niedermeier, R.: Partial kernelization for rank aggregation: Theory and experiments. In: Raman, V., Saurabh, S. (eds.) IPEC 2010. LNCS, vol. 6478, pp. 26–37. Springer, Heidelberg (2010)
6. Betzler, N., Fellows, M.R., Guo, J., Niedermeier, R., Rosamond, F.A.: Fixed-parameter algorithms for kemeny scores. In: Fleischer, R., Xu, J. (eds.) AAIM 2008. LNCS, vol. 5034, pp. 60–71. Springer, Heidelberg (2008)
7. Betzler, N., Fellows, M.R., Guo, J., Niedermeier, R., Rosamond, F.A.: Fixed-parameter algorithms for Kemeny rankings. Theor. Comput. Sci. 410(45), 4554–4570 (2009)
8. Biedl, T.C., Brandenburg, F., Deng, X.: On the complexity of crossings in permutations. Discrete Mathematics 309(7), 1813–1823 (2009)
9. Borda, J.: Mémoire sur les élections au scrutin. Histoire de l'Académie Royale des Sciences (1781)
10. Chen, J., Liu, Y., Lu, S., O'Sullivan, B., Razgon, I.: A fixed-parameter algorithm for the directed feedback vertex set problem. J. ACM 55(5) (2008)
11. Condorcet, M.: Essai sur l'application de l'analyse à la probabilité des décisions rendues à la pluralité des voix. L'imprimerie royale (1785)
12. Conitzer, V., Davenport, A., Kalagnanam, J.: Improved bounds for computing Kemeny rankings. In: AAAI 2006: Proc. of the 21st Nat. Conf. on Artificial Intelligence, vol. 1, pp. 620–626 (2006)
13. Coppersmith, D., Fleischer, L., Rudra, A.: Ordering by weighted number of wins gives a good ranking for weighted tournaments. In: SODA 2006: Proc. of the 17th Annual ACM-SIAM Symp. on Discrete Algorithms, pp. 776–782 (2006)
14. Dwork, C., Kumar, R., Naor, M., Sivakumar, D.: Rank aggregation methods for the web. In: WWW 2001: Proc. of the 10th Int. Conf. on World Wide Web, pp. 613–622 (2001)
15. Ephrati, E., Rosenschein, J.S.: The Clarke tax as a consensus mechanism among automated agents. In: AAAI 1991: Proc. of the 9th Nat. Conf. on Artificial Intelligence, vol. 1, pp. 173–178 (1991)
16. Fernau, H., Fomin, F.V., Lokshtanov, D., Mnich, M., Philip, G., Saurabh, S.: Ranking and drawing in subexponential time. In: Iliopoulos, C.S., Smyth, W.F. (eds.) IWOCA 2010. LNCS, vol. 6460, pp. 337–348. Springer, Heidelberg (2011)
17. Hemaspaandra, E., Spakowski, H., Vogel, J.: The complexity of Kemeny elections. Theor. Comput. Sci. 349(3), 382–391 (2005)
18. Jackson, B.N., Schnable, P.S., Aluru, S.: Consensus genetic maps as median orders from inconsistent sources. IEEE/ACM Trans. Comput. Biol. Bioinformatics 5(2), 161–171 (2008)
19. Karpinski, M., Schudy, W.: Faster algorithms for feedback arc set tournament, Kemeny rank aggregation and betweenness tournament. In: Cheong, O., Chwa, K.-Y., Park, K. (eds.) ISAAC 2010, Part I. LNCS, vol. 6506, pp. 3–14. Springer, Heidelberg (2010)
20. Kemeny, J.G.: Mathematics without numbers. Daedalus 88, 575–591 (1959)
21. Kenyon-Mathieu, C., Schudy, W.: How to rank with few errors. In: STOC 2007: Proc. of the 39th Annual ACM Symp. on Theory of Computing, pp. 95–103 (2007)
22. Simjour, N.: Improved parameterized algorithms for the Kemeny aggregation problem. In: Chen, J., Fomin, F.V. (eds.) IWPEC 2009. LNCS, vol. 5917, pp. 312–323. Springer, Heidelberg (2009)
23. Van Zuylen, A., Williamson, D.P.: Deterministic pivoting algorithms for constrained ranking and clustering problems. Mathematics of Operations Research 34(3), 594–620 (2009)

MapReduce Algorithmics

Sergei Vassilvitskii

Google
Mountain View, CA USA
sergeiv@google.com

Abstract. From automatically translating documents to analyzing electoral voting patterns; from computing personalized movie recommendations to predicting flu epidemics: all of these tasks are possible due to the success and proliferation of the MapReduce parallel programming paradigm. Yet almost ten years after the system was introduced, we still do not have a good understanding of what problems can and cannot be efficiently computed in MapReduce.

In this talk I will give an overview of the MapReduce framework, and explain its connections to both Valiant's Bulk Synchronous Parallel (BSP) model and the classical PRAM model of parallel computing. To demonstrate the power of the MapReduce model I will present the *Sample and Prune* approach that finds an approximate coreset of a manageable size, thereby reducing the problem from the realm of 'Big Data' to that of 'Small Data.'

I will conclude by discussing other considerations that make a large difference when working with MapReduce in practice, but have so far resisted a careful theoretical analysis.

F. Dehne, R. Solis-Oba, and J.-R. Sack (Eds.): WADS 2013, LNCS 8037, p. 524, 2013.
© Springer-Verlag Berlin Heidelberg 2013

The Greedy Gray Code Algorithm

Aaron Williams

Department of Mathematics and Statistics, McGill University
haron@uvic.ca

Abstract. We reinterpret classic Gray codes for binary strings, permutations, combinations, binary trees, and set partitions using a simple greedy algorithm. The algorithm begins with an initial object and an ordered list of operations, and then repeatedly creates a new object by applying the first possible operation to the most recently created object.

1 Introduction

Let $\mathbb{B}(n)$ be the set of n-bit binary strings. The *binary reflected Gray code* **Gray**(n) orders $\mathbb{B}(n)$ so that successive strings have Hamming distance one (i.e., they differ in one bit). For example, the order for $n = 3$ appears below, with overlines denoting the change that creates the next string

$$\mathbf{Gray}(3) = 00\overline{0},\ 00\overline{1},\ 01\overline{1},\ \overline{0}10,\ 11\overline{0},\ 1\overline{1}1,\ 10\overline{1},\ 100. \tag{1}$$

The term *reflected* indicates how the order is created: **Gray**(n) prefixes 0 to each string of **Gray**$(n{-}1)$, and then prefixes 1 to the strings of **Gray**$(n{-}1)$ in reflected order. For example, the top and bottom rows below are $0 \cdot$ **Gray**(3) and $1 \cdot$ reflect(**Gray**(3)), respectively, where $\cdot$ denotes concatenation

$$\mathbf{Gray}(4) = 000\overline{0}, 00\overline{0}1, 001\overline{1}, 0\overline{0}10, 011\overline{0}, 01\overline{1}1, 010\overline{1}, \overline{0}100,$$
$$110\overline{0}, 11\overline{0}1, 111\overline{1}, 1\overline{1}10, 101\overline{0}, 10\overline{1}1, 100\overline{1}, 1000.$$

We can express the construction recursively as **Gray**$(1) = 0, 1$ and for $n > 1$,

$$\mathbf{Gray}(n) = 0 \cdot \mathbf{Gray}(n{-}1),\ 1 \cdot \text{reflect}(\mathbf{Gray}(n{-}1)), \tag{2}$$

where the comma appends the two lists. The above definition uses *global recursion* since it refers to the entire **Gray**$(n{-}1)$ list as one unit. We can instead define the order using *local recursion* by referring to the individual strings in **Gray**$(n{-}1)$. If **Gray**$(n{-}1) = \mathbf{b_1}, \mathbf{b_2}, \ldots, \mathbf{b_{k-1}}, \mathbf{b_k}$ for $k = 2^{n-1}$, then

$$\mathbf{Gray}(n) = \mathbf{b_1} \cdot 0, \mathbf{b_1} \cdot 1, \mathbf{b_2} \cdot 1, \mathbf{b_2} \cdot 0, ..., \mathbf{b_{k-1}} \cdot 0, \mathbf{b_{k-1}} \cdot 1, \mathbf{b_k} \cdot 1, \mathbf{b_k} \cdot 0 \tag{3}$$

where **Gray**$(1) = 0, 1$. In other words, **Gray**(n) can be created by alternately suffixing 0 then 1, and 1 then 0, to successive strings in **Gray**$(n{-}1)$.

Since Frank Gray was granted U.S. Patent 2,632,058 in 1953 [4], his order has been used in numerous applications, with rotary encoders providing a prominent example [12]. The term *Gray code* now refers to minimal change orders of

F. Dehne, R. Solis-Oba, and J.-R. Sack (Eds.): WADS 2013, LNCS 8037, pp. 525–536, 2013.

combinatorial objects. Gray codes are related to efficient algorithms for exhaustively generating combinatorial objects. Knuth recently surveyed combinatorial generation in *The Art of Computer Programming* [6], and included separate subsections on generating tuples, permutations, combinations, partitions, and trees. Although the research area is quite diverse, it is fair to say that it has been dominated by the ideas of recursion and reflection. To demonstrate, we next recount a classic Gray code for combinations and a classic Gray code for permutations.

In the 1980s, Eades and McKay [3] followed Gray's approach to order the k-combinations of an n element set, which can be represented by $\mathbb{B}(n,k)$ the n-bit binary strings with fixed *weight* (i.e., number of 1s) equal to k. The Eades-McKay Gray code is defined using recursion and reflection as follows

$$\mathbf{EM}(n,k) = \mathbf{EM}(n-1,k)\cdot 0, \ \mathsf{reflect}(\mathbf{EM}(n-2,k-1))\cdot 01, \ \mathbf{EM}(n-2,k-2)\cdot 11$$

with $\mathbf{EM}(n,0) = 0^n$, $\mathbf{EM}(n,n) = 1^n$, and $\mathbf{EM}(n,1) = 10^{n-1}, 010^{n-2}, \ldots, 0^{n-1}1$, where exponentiation denotes repetition. For example,

$$\mathbf{EM}(5,3) = \underbrace{11\overline{10}0, 1\overline{10}10, \overline{10}110, 011\overline{10}}_{\mathbf{EM}(4,3)\cdot 0}, \underbrace{\overline{01}101, 1\overline{01}01, 1\overline{10}\overline{01}}_{\mathsf{reflect}(\mathbf{EM}(3,2))\cdot 01}, \underbrace{\overline{10}011, 0\overline{10}11, 00111}_{\mathbf{EM}(3,1)\cdot 11}$$

$$(4)$$

In this order, successive strings differ by a *homogeneous transposition*, meaning that a 1 and 0 can only be interchanged if the intermediate symbols are all 0s. In other words, substrings of the form $00\cdots 01$ and $100\cdots 0$ can be interchanged. Thus, a single s_i changes when the elements of the combination are represented as $1 \le s_1 < s_2 < \cdots < s_k \le n$. For this reason, the order allows pianists to practice all k-note chords while moving only a single finger between chords [3].

Let $\mathbb{P}(n)$ be the permutations of $[n] = \{1, 2, \ldots, n\}$ in one-line notation. For example, $\mathbb{P}(3) = \{123, 213, 213, 231, 312, 321\}$. In the 1960s, researchers considered permutation Gray codes using *adjacent transpositions* (or *swaps*), meaning that two symbols can only be interchanged if they are next to each other. Johnson, Trotter, and Steinhaus all approached the problem using local recursion, and they all rediscovered an order known in the 17th century as *plain changes* [2]. To explain the order, let $\mathsf{zig}(\mathbf{p})$ be the list obtained from $\mathbf{p}$ by applying the following swaps: $(n\ n-1)$, $(n-1\ n-2)$, ..., $(1\ 2)$. For example, $\mathsf{zig}(1234) = 12\overleftarrow{3}4$, $12\overleftarrow{4}3$, $1\overleftarrow{4}23$, 4123 where the arrow shows the movement of 4. Similarly, let $\mathsf{zag}(\mathbf{p})$ apply the following swaps: $(1\ 2)$, $(2\ 3)$, ..., $(n\ n-1)$. Notice that zigs and zags only change the relative order of the last and first symbols, respectively. Thus, we can define a Gray code as follows: If $\mathbf{Plain}(n-1) = \mathbf{p_1}, \mathbf{p_2}, ..., \mathbf{p_{(n-1)!}}$, then

$$\mathbf{Plain}(n) = \mathsf{zig}(\mathbf{p_1}\cdot n), \ \mathsf{zag}(n\cdot \mathbf{p_2}), ..., \mathsf{zig}(\mathbf{p_{(n-1)!-1}}\cdot n), \ \mathsf{zag}(n\cdot \mathbf{p_{(n-1)!}}) \quad (5)$$

where $\mathbf{Plain}(2) = 12, 21$. For example, the following order is $\mathsf{zig}(12)\cdot 3, \mathsf{zag}(21)\cdot 3$

$$\mathbf{Plain}(3) = 123, 132, 312, 321, 231, 231. \quad (6)$$

In this article we propose an alternate method for understanding the aforementioned Gray codes and many others. To illustrate the idea, consider the following method for building a list $\mathcal{L}$ of unique strings in $\mathbb{B}(n)$: Initialize $\mathcal{L}$ to contain 0^n, then repeatedly extend $\mathcal{L}$ by complementing the rightmost possible bit in its last string. For example, if the current list is $\mathcal{L} = 000, 001, 011, 010$ then we examine its last string 010. The rightmost bit cannot be complemented since $01\overline{0} = 011$ is already in $\mathcal{L}$. Similarly, the middle bit cannot be complemented since $0\overline{1}0 = 000 \in \mathcal{L}$. However, the leftmost bit can be complemented since $\overline{0}10 = 110 \notin \mathcal{L}$. Thus, 110 is added to the end of $\mathcal{L}$. The complete list for $n = 3$ is in (1). More generally, we prove that the method always creates **Gray**(n).

As a second example, initialize $\mathcal{L}$ to contain $1^k 0^{n-k}$, then repeatedly extend $\mathcal{L}$ by homogeneously transposing the leftmost possible 1 into the leftmost possible position. For example, if $\mathcal{L} = 11100, 11010, 10110, 01110$ then we examine 01110. The leftmost 1 could be homogeneously transposed into the first position, however $\overline{01}110 = 10110 \in \mathcal{L}$. The middle 1 cannot be homogeneously transposed since it is bordered by 1s. The rightmost 1 can be homogeneously transposed into the last position and $011\overline{10} = 01101 \notin \mathcal{L}$. Thus, 01101 is added to the end of $\mathcal{L}$. The complete list for $n = 5$ and $k = 3$ is in (4). More generally, we prove that the method always creates $\mathbb{B}(n, k)$, and for odd k the order is **EM**(n, k).

As a third example, initialize $\mathcal{L}$ to contain $12\cdots n$, then extend $\mathcal{L}$ by swapping the largest possible symbol once to the left or right. For example, if $\mathcal{L} = 123, 132, 312$ then we examine 312. The 3 cannot swap left since it is in the leftmost position. Similarly, 3 cannot swap right since $3\overset{\leftrightarrow}{1}2 = 132 \in \mathcal{L}$. However, 2 can swap left since $3\overset{\leftrightarrow}{12} = 321 \notin \mathcal{L}$. Thus, 321 is added to $\mathcal{L}$. The complete list for $n = 3$ is in (6). More generally, the method always creates **Plain**(n).

Our "greedy Gray code algorithm" is defined in Section 3 and reinterprets many classic Gray codes. Section 4 discusses these results on binary strings:

1. The binary reflected Gray code complements the rightmost possible bit;
2. Lexicographic order complements the shortest possible suffix;
3. The de Bruijn sequence by Martin [8] shifts in the lowest possible bit.

Section 5 discusses these results for permutations:

4. The plain change order adjacently transposes the largest possible symbol;
5. The pancake flipping order by Zaks [13] reverses the shortest possible prefix;
6. Corbett's rotator graph order [1] rotates the prefix with the first possible length in $n, 2, n-1, 3, \dots$.

Section 6 discusses the following additional results:

7. The Eades-McKay order of combinations homogeneously transposes the leftmost possible 1 into the leftmost possible position when the weight is odd.
8. The Lucas, van Baronaigien and Ruskey order of binary trees [7] rotates the edge with the largest inorder label.
9. Kaye's set partition order [5] moves the largest possible symbol into the leftmost possible subset.

In addition, Section 2 provides an application for our greedy reinterpretations. We conclude the introduction with several clarifications.

- This greedy method is not entirely new. For example, the de Bruijn sequence we discuss here was first defined greedily by Martin in 1934 [8]. However, the term 'greedy' is not common in the literature, nor is it featured in Knuth's 400 page treatise on combinatorial generation [6].

- The greedy method is not suitable for efficiently generating Gray codes since it may have to 'remember' an exponential number of objects relative to their size. However, it can provide a simpler description for previously created Gray codes and a simpler method for discovering new Gray codes (see [9]).

- Recursive constructions are often general results. For example, any swap Gray code of $\mathbb{P}(n-1)$ provides a swap Gray code of $\mathbb{P}(n)$ using (5). In contrast, our greedy method gives only one order. However, this order may illuminate a general recursive principle that leads to an efficient generation algorithm.

In general, the author views the greedy Gray code algorithm as a simple and unified "first step" in understanding and discovering Gray codes.

2 Network Application

Gray codes give Hamilton paths and cycles in well-studied graphs, such as the n-cube (binary reflected Gray code), the permutohedron (plain changes), the rotator graph (Corbett's order), and the pancake graph (Zaks's order). These graphs are used as network topologies, where vertices are computers and two vertices can communicate if they are adjacent (see Siegel [10]). In this section, we illustrate how our greedy algorithms can send messages through these networks.

The *pancake graph* has vertices $\mathbb{P}(n)$ and edges between pairs of vertices that differ by a prefix-reversal. Figure 1 a) illustrates the graph for $n = 4$. In Section 5 we will see that a Hamilton path can be created in this graph from 1234 by repeatedly reversing the shortest possible prefix that gives a new permutation. The order of vertices visited on this Hamilton path is illustrated by Figure 1 b).

Suppose each vertex sets a flag if it has seen a particular message, and each vertex can query the flags of its neighbors. Also assume that the neighbors of a vertex are 'prioritized' by increasing prefix-reversal lengths. Given this scenario, we claim that a message m will propagate from an initial vertex to all other vertices in the pancake graph so long as each vertex runs the following algorithm:

> When a vertex receives m, it sets its flag and passes m to its highest-priority neighbor whose flag is not set.

For example, consider vertex 3214 in Figure 1 b), which is the sixth vertex to receive the message. Once it receives the message, it cannot pass it to its highest-priority neighbor $\overleftrightarrow{23}14$ since this vertex has already seen the message, and similarly it cannot pass it to $\overrightarrow{123}4$. However, it can pass the message to its lowest-priority neighbor $\overrightarrow{4123}$, and at this point its algorithm terminates.

To clarify an important point, we mention that the pancake graph is vertex-transitive, and that our greedy prefix-reversal algorithm generates $\mathbb{P}(n)$ for any initial permutation. Thus, our approach works regardless of where the message originates. Furthermore, the same arguments apply for our greedy algorithms in the n-cube, permutohedron, and rotator graph. In particular, this approach in the rotator graph is much more efficient than the table approach in Corbett [1].

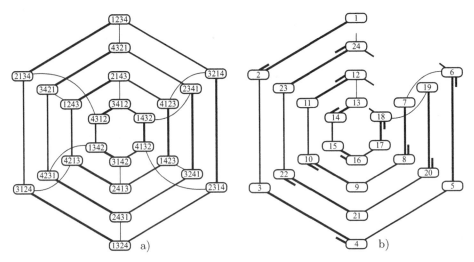

Fig. 1. a) The pancake network for $n = 4$ in which thick, medium, and thin edges are used for prefix-reversals of length two, three, and four, respectively. b) The Hamilton path obtained by greedily reversing the shortest possible prefix starting from 1234, where each partial edge shows a prefix-reversal leading to a previously visited vertex.

3   Greedy Gray Code Algorithm

The *greedy Gray code algorithm* takes as input an object $\mathbf{x} \in \mathbf{X}$ and a prioritized list of operations $\mathcal{O} = o_1, o_2, \ldots, o_k$ where $o_i : \mathbf{X} \to \mathbf{X}$ for all $1 \leq i \leq k$. The algorithm outputs a *greedy object list* $\mathcal{L}$ of distinct objects in $\mathbf{X}$. The list initially contains $\mathbf{x}$, and then is repeatedly extended by one object as follows: If $\mathbf{x}$ is the last object in $\mathcal{L}$, and i is the minimum value such that $o_i(\mathbf{x})$ is not already in $\mathcal{L}$, then $o_i(\mathbf{x})$ is added to the end of $\mathcal{L}$. Greedy$_{\mathcal{O}}(\mathbf{x})$ is *successful* if it generates $\mathbf{X}$. In other words, success occurs if every object of the same type as $\mathbf{x}$ is in $\mathcal{L}$.

Greedy$_{\mathcal{O}}(\mathbf{x})$
1: Initialize list $\mathcal{L}$ to contain the single object $\mathbf{x}$.
2: Let $\mathbf{x}$ be the last object in list $\mathcal{L}$.
3: Let i be minimum such that $o_i(\mathbf{x})$ is not in $\mathcal{L}$. If i does not exist, then return.
4: Add the new object $o_i(\mathbf{x})$ to the end of $\mathcal{L}$.
5: Return to line 2.

Given a prioritized list of operations $\mathcal{O} = o_1, o_2, \ldots$ and an index list $\mathcal{I} = i_1, i_2, \ldots$, we generate a list of objects as follows. Let Apply$_{\mathcal{O}}(\mathbf{x_1}, \mathcal{I})$ be the list $\mathbf{x_1}, \mathbf{x_2}, \ldots$ where $\mathbf{x_{i+1}} = o_{i_k}(\mathbf{x_i})$ for $k = 1, 2, \ldots$. That is, the i_kth operation creates the $(k + 1)$st object from the kth object.

4   Binary Strings

In this section, we give greedy interpretations to three orders of binary strings. Throughout this section we index the bits of a binary string from right-to-left. Thus, if $\mathbf{b} \in \mathbb{B}(n)$, then $\mathbf{b} = b_n b_{n-1} \cdots b_1$ are its individual bits. The *ith bit* of $\mathbf{b}$

uses this right-to-left indexing, so the first bit is the rightmost. A draft of this paper illustrates each order in Table 1 (see the author's website).

4.1 Binary Reflected Gray Code

We first prove the greedy interpretation of the binary reflected Gray code using local recursion. Let bit_i be the operation that complements the ith bit of a binary string. That is, $\text{bit}_i(\mathbf{b}) = b_n \cdots b_{i+1}\overline{b_i}b_{i-1} \cdots b_1$. We prioritize the bit complements from right-to-left in $\text{Bit}_n^\uparrow = \text{bit}_1, \text{bit}_2, \ldots, \text{bit}_n$. (In general, we use lowercase for individual operations and uppercase for prioritized lists of operations, with $\uparrow$ and $\downarrow$ for lists with increasing and decreasing subscripts, respectively.)

Theorem 1. *The greedy Gray code algorithm that complements the rightmost possible bit generates the reflected Gray code. That is,* $\text{Greedy}_{\text{Bit}_n^\uparrow}(0^n) = \mathbf{Gray}(n)$.

Proof. The proof is by induction on n with $\text{Greedy}_{\text{Bit}_1^\uparrow}(0^1) = 0, 1 = \mathbf{Gray}(1)$ for the base case. Inductively assume that

$$\text{Greedy}_{\text{Bit}_{m-1}^\uparrow}(0^{m-1}) = \mathbf{b_1}, \mathbf{b_2}, \ldots, \mathbf{b_{2^{m-1}}} = \mathbf{Gray}(m-1).$$

In particular, $\mathbf{b_1} = 0^{m-1}$ and $\mathbf{b_2} = 0^{m-2}1$. The first four strings generated by $\text{Greedy}_{\text{Bit}_m^\uparrow}(0^m)$ are $0^m, 0^{m-1}1, 0^{m-2}11, 0^{m-2}10 = \mathbf{b_1} \cdot 0, \mathbf{b_1} \cdot 1, \mathbf{b_2} \cdot 1, \mathbf{b_2} \cdot 0$. More generally, suppose $\text{Greedy}_{\text{Bit}_m^\uparrow}(0^m)$ begins

$$\mathbf{b_1} \cdot 0, \mathbf{b_1} \cdot 1, \mathbf{b_2} \cdot 1, \mathbf{b_2} \cdot 0, \ldots, \mathbf{b_{2i-1}} \cdot 0, \mathbf{b_{2i-1}} \cdot 1, \mathbf{b_{2i}} \cdot 1, \mathbf{b_{2i}} \cdot 0 \qquad (7)$$

for some fixed $1 \leq i < 2^{m-1}$. The algorithm cannot apply bit_m to the last string in (7) since $\mathbf{b_{2i}} \cdot \overline{0} = \mathbf{b_{2i}} \cdot 1$ is the second-last string in (7). Therefore, the algorithm can only apply bit_j for some $j < m$. Thus, the next string (if any) generated by the algorithm will end with 0. Observe that the strings ending with 0 in (7) are precisely $\mathbf{b_1} \cdot 0, \mathbf{b_2} \cdot 0, \ldots, \mathbf{b_{2i}} \cdot 0$. Since $\text{Greedy}_{\text{Bit}_{m-1}^\uparrow}(0^{m-1})$ begins by generating $\mathbf{b_1}, \mathbf{b_2}, \ldots, \mathbf{b_{2i}}, \mathbf{b_{2i+1}}$, we know $\text{Greedy}_{\text{Bit}_m^\uparrow}(0^m)$ behaves accordingly. Thus, $\text{Greedy}_{\text{Bit}_m^\uparrow}(0^m)$ follows $\mathbf{b_{2i}} \cdot 0$ by generating $\mathbf{b_{2i+1}} \cdot 0$. Furthermore, the string generated after $\mathbf{b_{2i+1}} \cdot 0$ is $\mathbf{b_{2i+1}} \cdot 1$ since bit_m is the highest priority operation. Therefore, (7) is true when $i+1$ replaces i. Hence, by repeating this argument (7) is true for $i = 2^{m-1}$. Therefore, $\text{Greedy}_{\text{Bit}_m^\uparrow}(0^m)$ and $\mathbf{Gray}(m)$ share the same recursive structure by (3) and (7), which completes the induction. $\square$

4.2 Lexicographic Order

We next give a greedy interpretation to $\mathbf{Lex}(n)$, the *lexicographic order of* $\mathbb{B}(n)$ in which successive strings have decimal value $0, 1, 2, \ldots, 2^n - 1$. For example,

$$\mathbf{Lex}(3) = 00\overline{0}, 00\overline{1}, 01\overline{0}, \overline{01}\overline{1}, 10\overline{0}, 1\overline{0}\overline{1}, 11\overline{0}, 111.$$

Notice that each successive string is obtained by a *suffix complement* suff_i which complements the rightmost i bits. That is, $\text{suff}_i(\mathbf{b}) = b_n b_{n-1} \cdots b_{i+1}\overline{b_i}\overline{b_{i-1}} \cdots \overline{b_1}$

for $\mathbf{b} \in \mathbb{B}(n)$. We prioritize by shortest suffix in $\mathsf{Suff}_n^{\uparrow} = \mathsf{suff}_1, \mathsf{suff}_2, \ldots, \mathsf{suff}_n$. Lexicographic order has the same global recursive definition as the binary reflected Gray code, without the reflection. That is, $\mathbf{Lex}(1) = 0, 1$ and for $n > 1$,

$$\mathbf{Lex}(n) = 0 \cdot \mathbf{Lex}(n{-}1), \; 1 \cdot \mathbf{Lex}(n{-}1). \tag{8}$$

For example, the order below is $0 \cdot \mathbf{Lex}(3)$ followed by $1 \cdot \mathbf{Lex}(3)$

$$\mathbf{Lex}(4) = 000\overline{0}, 00\overline{01}, 001\overline{0}, 0\overline{011}, 010\overline{0}, 01\overline{01}, 011\overline{0}, \overline{0111},$$
$$100\overline{0}, 10\overline{01}, 101\overline{0}, 1\overline{011}, 110\overline{0}, 11\overline{01}, 111\overline{0}, 1111.$$

Theorem 2. *The greedy Gray code algorithm that complements the shortest possible suffix generates lexicographic order. That is,* $\mathsf{Greedy}_{\mathsf{Suff}_n^{\uparrow}}(0^n) = \mathbf{Lex}(n)$.

Proof. Our proof is by induction, with $\mathsf{Greedy}_{\mathsf{Suff}_1^{\uparrow}}(0^1) = \overline{0}, 1 = \mathbf{Lex}(1)$ as the base case. Inductively assume $\mathsf{Greedy}_{\mathsf{Suff}_{m-1}^{\uparrow}}(0^{m-1}) = \mathbf{Lex}(m{-}1)$. Now consider $\mathsf{Greedy}_{\mathsf{Suff}_m^{\uparrow}}(0^m)$. Since suff_m is the lowest-priority operation, the algorithm will begin by creating as many strings as possible using $\mathsf{Suff}_{m-1}^{\uparrow}$. By induction, this produces $0 \cdot \mathbf{Lex}(m{-}1)$, whose last string is $0 \cdot 1^{m-1}$. The greedy algorithm must then apply suff_m to this string to create $\overline{0 \cdot 1^{m-1}} = 1 \cdot 0^{m-1}$ since every string beginning with 0 has already been generated. Now the algorithm again proceeds by creating as many strings as possible using $\mathsf{Suff}_{m-1}^{\uparrow}$ starting from $1 \cdot 0^{m-1}$. By induction, this produces $1 \cdot \mathbf{Lex}(m{-}1)$. Thus, $\mathsf{Greedy}_{\mathsf{Suff}_m^{\uparrow}}(0^m) = 0 \cdot \mathbf{Lex}(m{-}1), \; 1 \cdot \mathbf{Lex}(m{-}1)$, and so our result is true by (8). □

4.3 de Bruijn Sequences

A *de Bruijn sequence* is binary string of length 2^n that contains every string in $\mathbb{B}(n)$ exactly once as a circular substring of length n. Martin [8] showed that a de Bruijn sequence $\mathbf{dB}(n)$ can be built one bit at a time by starting from 0^n and greedily suffixing the largest possible next bit 1 or 0, subject to the condition that the resulting sequence does not contain any substring twice[1]. For example, if the algorithm for $n = 4$ has currently built 00001111011, then Martin's algorithm will not append 1 since the resulting sequence of bits 000011110111 would contain two copies of 0111. Thus, it would append 0. The result of Martin's algorithm for $n = 4$ is $\mathbf{dB}(4) = 0000111101100101$. A de Bruijn sequence is *decoded* by listing its successive substrings of length n. For example,

$$\mathsf{decode}(\mathbf{dB}(4)) = \mathsf{decode}(0000111101100101)$$
$$= 0000, 0001, 0011, 0111, 1111, 1110, 1101, 1011,$$
$$0110, 1100, 1001, 0010, 0101, 1010, 0100, 1000$$

where the final three substrings "wrap around". Successive decoded substrings always differ by a 1-*shift* or a 0-*shift*, meaning that $b_n b_{n-1} \cdots b_1$ is replaced by

---

[1] Martin constructs a sequence of length $2^n + n - 1$ starting from $0^{n-1}1$ whose 2^n non-circular substrings are $\mathbb{B}(n)$. This sequence ends with 0^n, so it is equivalent to $\mathbf{dB}(n)$.

$b_{n-1}b_{n-2} \cdots b_1 1$ or $b_{n-1}b_{n-2} \cdots b_1 0$, respectively. We denote these two operations by shift_1 and shift_0, respectively, and prioritize them as $\mathsf{Shift}_2^{\downarrow} = \mathsf{shift}_1, \mathsf{shift}_0$. This allows us to reinterpret Martin's result using the greedy Gray code algorithm.

Theorem 3 ([8]). *The greedy Gray code algorithm that shifts in the largest possible bit generates decoded strings in Martin's de Bruijn sequence. That is,* $\mathsf{Greedy}_{\mathsf{Shift}_2^{\downarrow}}(0^n) = \mathsf{decode}(\mathbf{dB}(n)).$

We mention that Theorem 2 depends on the initial string. For example, 000 and 001 are the only suitable choices for generating $\mathbb{B}(3)$ in this way.

5 Permutations

In this section, we give greedy interpretations to three permutation orders. Throughout this section we index the symbols of a permutation from left-to-right. Thus, if $\mathbf{p} \in \mathbb{P}(n)$, then $\mathbf{p} = p_1 p_2 \cdots p_n$. A draft of this paper illustrates each order in Table 2 (see the author's website).

5.1 Plain Change Order

The *transposition* $(i\ j)$ interchanges the values in positions i and j of a string. A *swap* is a transposition of the form $(i\ i+1)$. Swaps are also known as *adjacent transpositions*. When considering permutations, we can indicate a specific swap by indicating a value and a direction, instead of a pair of positions. Let swap_{-v} and swap_{+v} be the operations that swap value v one position to the left, or right, respectively. For example, $\mathsf{swap}_{-2}(7654321) = 765\overset{\leftarrow}{4}321 = 7654231$ is a *left swap* of 2, and $\mathsf{swap}_{+2}(7654321) = 765432\overset{\rightarrow}{1} = 7654312$ is a *right swap* of 2. If $\mathbf{p} = p_1 p_2 \cdots p_n \in \mathbb{P}(n)$, then $\mathsf{swap}_{+p_n}(\mathbf{p}) = \mathsf{swap}_{-p_1}(\mathbf{p}) = \mathbf{p}$. In other words, left swapping the first value does not change a permutation, nor does right swapping the last value. We prioritize our swaps by decreasing values, and right before left,

$$\mathsf{Swap}_n^{\downarrow} = \mathsf{swap}_{+n}, \mathsf{swap}_{-n}, \ldots, \mathsf{swap}_{+2}, \mathsf{swap}_{-2}, \mathsf{swap}_{+1}, \mathsf{swap}_{-1}.$$

Note: The relative priorities of swap_{+i} and swap_{-i} do not affect the proof of Theorem 4, so we say that the swaps are prioritized by decreasing value.

Theorem 4. *The greedy Gray code algorithm that swaps the largest possible value generates the plain change order. That is,* $\mathsf{Greedy}_{\mathsf{Swap}_n^{\downarrow}}(12 \cdots n) = \mathbf{Plain}(n).$

Proof. The proof is by induction on n with $\mathsf{Greedy}_{\mathsf{Swap}_2^{\downarrow}}(12) = \overset{\leftarrow}{12}, 21 = \mathbf{Plain}(2)$ as the base case. Inductively assume

$$\mathsf{Greedy}_{\mathsf{Swap}_{m-1}^{\downarrow}}(12 \cdots m{-}1) = \mathbf{p_1}, \ldots, \mathbf{p_{(m-1)!}} = \mathbf{Plain}(m{-}1).$$

In particular, $\mathbf{p_1} = 12 \cdots m{-}1$ and $\mathbf{p_2} = 12 \cdots m{-}3\,m{-}1\,m{-}2$. The first m strings generated by $\mathsf{Greedy}_{\mathsf{Swap}_m^{\downarrow}}(12 \cdots m)$ are

$$1\,2 \cdots m{-}2\,\overset{\leftarrow}{m{-}1\,m}, \ \ 1\,2 \cdots \overset{\leftarrow}{m{-}2\,m}\,m{-}1, \ \ldots, \ m\,1\,2 \cdots m{-}2\,m{-}1 = \mathsf{zig}(\mathbf{p_1} \cdot m)$$

by repeatedly applying swap_{-m}. At this point, the algorithm cannot swap m so it swaps $m-1$ to create $m\,1\,2\cdots\overleftarrow{m-2\,m-1} = m\,1\,2\cdots m-3\,m-1\,m-2$. This is followed by repeatedly applying swap_{+m} as below

$$\overrightarrow{m\,1}\,2\cdots m-3\,m-1\,m-2,\; 1\,\overrightarrow{m\,2}\cdots m-3\,m-1\,m-2,\; ...,\; 1\,2\cdots m-3\,m-1\,m-2\,m = \mathsf{zag}(m\cdot\mathbf{p_2}).$$

More generally, suppose $\mathsf{Greedy}_{\mathsf{swap}_m^\downarrow}(12\cdots m)$ begins

$$\mathsf{zig}(\mathbf{p_1}\cdot m),\mathsf{zag}(m\cdot\mathbf{p_2}),\mathsf{zig}(\mathbf{p_3}\cdot m),\mathsf{zag}(m\cdot\mathbf{p_4}),...,\mathsf{zig}(\mathbf{p_{2i-1}}\cdot m),\mathsf{zag}(m\cdot\mathbf{p_{2i}}) \quad (9)$$

for some fixed $1 \le i < (m-1)!$. Notice that the last string in (9) is $\mathbf{p_{2i}} \cdot m$. The algorithm cannot apply swap_{+m} to $\mathbf{p_{2i}} \cdot m$ since m is in the rightmost position. Similarly, the algorithm cannot apply swap_{+m} since that would result in the second-last string in (9). Thus, the next string (if any) generated by the algorithm will begin with m. Observe that the strings beginning with m in (9) are precisely $m \cdot \mathbf{p_1}, m \cdot \mathbf{p_2}, \ldots, m \cdot \mathbf{p_{2i}}$. Since $\mathsf{Greedy}_{\mathsf{swap}_{m-1}^\downarrow}(12\cdots m-1)$ begins by generating $\mathbf{p_1}, \mathbf{p_2}, \ldots, \mathbf{p_{2i}}, \mathbf{p_{2i+1}}$, we know $\mathsf{Greedy}_{\mathsf{swap}_m^\downarrow}(12\cdots m)$ behaves accordingly. Thus, $\mathsf{Greedy}_{\mathsf{swap}_m^\downarrow}(12\cdots m)$ follows $m \cdot \mathbf{p_{2i}}$ with $m \cdot \mathbf{p_{2i+1}}$. At this point, $m \cdot \mathbf{p_{2i+1}}$ is the first generated string in which the symbols of $[m-1]$ are in the relative order given by $\mathbf{p_{2i+1}}$. Thus, the algorithm continues by generating $\mathsf{zig}(m \cdot \mathbf{p_{2i+1}})$ since swap_{+m} is the highest priority operation. This ends with $\mathbf{p_{2i+1}} \cdot m$, and for similar reasons, the algorithm follows this by $\mathsf{zag}(\mathbf{p_{2i+1}} \cdot \mathbf{m})$. Therefore, (9) is true when $i+1$ replaces i. Hence, (9) is true for $i = (m-1)!$. Therefore, $\mathsf{Greedy}_{\mathsf{swap}_m^\downarrow}(12\cdots m)$ and $\mathbf{Plain}(m)$ share the same recursive structure by (5) and (9), which completes the induction. $\qquad\square$

5.2 Zaks's Pancake Order

Let rev_i be the operation that reverses the first i symbols of a string. Thus, if $\mathbf{p} = p_1 p_2 \cdots p_n$, then $\mathsf{rev}_i(\mathbf{p}) = p_i p_{i-1} \cdots p_1 p_{i+1} p_{i+2} \cdots p_n$. This operation is known as a *prefix-reversal* or a *flip*. The term 'flip' comes from the *pancake problem*: Given a stack of n pancakes of distinct sizes, what is the minimum number of times a waiter must flip over some number of pancakes at the top of the stack in order to sort the pancakes from smallest to largest?

Zaks [13] considered the problem of creating all possible stacks (or permutations) using flips. As he writes, "The poor waiter will be able to generate, in n! such steps, all possible n! stacks". Zaks used global recursion to create his order. For example, $\mathbf{Pan}(3) = \overleftrightarrow{12}3, \overleftrightarrow{21}3, \overleftrightarrow{31}2, \overleftrightarrow{13}2, \overleftrightarrow{23}1, 321$ and $\mathbf{Pan}(4)$ repeats this four times below, with prefix-reversals of length three in between

$$\overleftrightarrow{12}34,\; \overleftrightarrow{21}34,\; \overleftrightarrow{31}24,\; \overleftrightarrow{13}24,\; \overleftrightarrow{23}14,\; \overleftrightarrow{321}4,\; \overleftrightarrow{41}23,\; \overleftrightarrow{14}23,\; \overleftrightarrow{24}13,\; \overleftrightarrow{42}13,\; \overleftrightarrow{12}43,\; \overleftrightarrow{21}43,$$
$$\overleftrightarrow{341}2,\; \overleftrightarrow{43}12,\; \overleftrightarrow{13}42,\; \overleftrightarrow{31}42,\; \overleftrightarrow{41}32,\; \overleftrightarrow{14}32,\; \overleftrightarrow{23}41,\; \overleftrightarrow{32}41,\; \overleftrightarrow{42}31,\; \overleftrightarrow{24}31,\; \overleftrightarrow{34}21,\; 4321.$$

Theorem 5. *The greedy Gray code algorithm that reverses the shortest possible prefix generates Zaks's order. That is,* $\mathsf{Greedy}_{\mathsf{rev}_n^\uparrow}(12\cdots n) = \mathbf{Pan}(n)$.

A new pancake order $\mathbf{Pan'}(n)$ is generated by greedily reversing the longest possible prefix, as prioritized by $\mathsf{Rev}_n^\downarrow$. The reader can refer to the recent article by the author and Sawada [9] for these results.

Theorem 6 ([9]). *The greedy Gray code algorithm that reverses the longest possible prefix generates all permutations. That is,* $\mathsf{Greedy}_{\mathsf{Rev}_n^\uparrow}(12\cdots n) = \mathbf{Pan'}(n)$.

5.3   Corbett's Rotator Order

It is easy to show that $\mathbb{P}(n)$ is not generated by greedily rotating the shortest possible prefix, or the longest possible prefix, for $n \geq 4$. However, we will see that $\mathbb{P}(n)$ can be generated by prioritizing the rotations in a different way. In fact, the Gray code will equal an order given by Corbett in the context of the interconnection network known as the *rotator graph* (see Corbett [1]).

Corbett's order $\mathbf{Rotator}(n)$ is generated with the help of an index sequence $\mathbf{Rotator''}(n)$. The index sequence is defined as follows: $\mathbf{Rotator''}(2) = 2$ and if $\mathbf{Rotator''}(n-1) = r_1, r_2, \ldots, r_{(n-1)!-1}$ then $\mathbf{Rotator''}(n)$ appears below

$$n, ..., n, n+1-r_1, n, ..., n, n+1-r_2, \ldots, n, ..., n, n+1-r_{(n-1)!-1}, n, ..., n.$$

where each $n, ..., n$ denotes n repeated $n-1$ times. For example, $\mathbf{Rotator''}(3) = 3, 3, 2, 3, 3$ and so $\mathbf{Rotator''}(4) = 4,4,4,2,4,4,4,2,4,4,4,3,4,4,4,2,4,4,4,2,4,4,4$. Corbett's order is obtained by applying the sequence as rotations starting from $n\,n{-}1\cdots 1 \in \mathbb{P}(n)$. That is, $\mathbf{Rotator}(n) = \mathsf{Apply}_{\mathbf{Rotator''}(n)}(n\,n{-}1\cdots 1, \mathsf{Rot}_n^\uparrow)$, where $\mathsf{Rot}_n^\uparrow = \mathsf{rot}_1, \mathsf{rot}_2, \ldots, \mathsf{rot}_n$ and rot_1 is included for convenience. For example, the orders for $n = 3$ and $n = 4$ appear below.

$\mathbf{Rotator}(3)$	$\mathbf{Rotator}(4)$
$\overrightarrow{321}, \overrightarrow{213}, \overrightarrow{132},$	$\overrightarrow{4321}, \overrightarrow{3214}, \overrightarrow{2143}, \overrightarrow{1432}, \overrightarrow{4132}, \overrightarrow{1324}, \overrightarrow{3241}, \overrightarrow{2413}, \overrightarrow{4213}, \overrightarrow{2134}, \overrightarrow{1342}, \overrightarrow{3421},$
$\overrightarrow{312}, \overrightarrow{123}, 231$	$\overrightarrow{4231}, \overrightarrow{2314}, \overrightarrow{3142}, \overrightarrow{1423}, \overrightarrow{4123}, \overrightarrow{1234}, \overrightarrow{2341}, \overrightarrow{3412}, \overrightarrow{4312}, \overrightarrow{3124}, \overrightarrow{1243}, 2431.$

Understanding the correctness of Corbett's construction is somewhat tricky, and we refer the reader to [1] and Stevens and Williams [11]. On the other hand, it has a relatively simple greedy interpretation. We prioritize the prefix rotations by *interleaving* the longest and shortest as follows

$$\mathsf{Rot}_n^\updownarrow = \mathsf{rot}_n, \mathsf{rot}_2, \mathsf{rot}_{n-1}, \mathsf{rot}_3, \ldots, \mathsf{rot}_{\lceil \frac{n+1}{2} \rceil}.$$

Due to space restrictions, we omit the proof of Theorem 7.

Theorem 7. *The greedy Gray code algorithm that rotates prefixes with interleaved longest and shortest lengths generates Corbett's order of permutations. That is,* $\mathbf{Rotator}(n) = \mathsf{Greedy}_{\mathsf{Rot}_n^\updownarrow}(12\cdots n)$.

6   Additional Results

In this section, we describe greedy interpretations of additional Gray codes. Formal proofs will appear in the full article. A draft of this paper illustrates each order in Table 3 (see the author's website).

A *k-combination of* $[n]$ is a subset of size k, which we represent by its selected elements $1 \leq s_1 < s_2 < \cdots < s_n \leq n$, or by its incidence vector in $\mathbb{B}(n, k)$ with bitwise indexing from left-to-right. A *homogeneous transposition* $\mathsf{homo}_{i,j}$ transposes the bits in positions i and j only if the bits have opposite values and the intermediate symbols are all 0s. Thus, for a given $\mathbf{b} = b_1 b_2 \cdots b_n \in \mathbb{B}(n, k)$

$$\mathsf{homo}_{i,j}(\mathbf{b}) = \begin{cases} b_1 \cdots b_{i-1}\overline{b_i}b_{i+1} \cdots b_{j-1}\overline{b_j}b_{j+1} \cdots b_n & \text{if } i < j \\ b_1 \cdots b_{j-1}\overline{b_j}b_{j+1} \cdots b_{i-1}\overline{b_i}b_{i+1} \cdots b_n & \text{if } j < i \end{cases}$$

so long as $\{b_i, b_j\} = \{0, 1\}$ and $b_{i+1} \cdots b_{j-1} = 0^{j-i-1}$; otherwise, $\mathsf{homo}_{i,j}(\mathbf{b}) = \mathbf{b}$. In particular, $\mathsf{homo}_{i,i}(\mathbf{b}) = \mathbf{b}$. We prioritize the homogeneous transpositions for a given combination with $1 \leq s_1 < s_2 < \cdots < s_n \leq n$ as follows

$$\begin{aligned} \mathsf{Homo}_n = \;& \mathsf{homo}_{s_1,1}, \; \mathsf{homo}_{s_1,2}, \; ..., \; \mathsf{homo}_{s_1,s_2-1}, && (10) \\ & \mathsf{homo}_{s_2,s_1+1}, \; \mathsf{homo}_{s_1,s_1+2}, \; ..., \; \mathsf{homo}_{s_1,s_3-1}, \; ..., \\ & \mathsf{homo}_{s_n,s_{n-1}+1}, \; \mathsf{homo}_{s_1,s_{n-1}+2}, \; ..., \; \mathsf{homo}_{s_1,n}. \end{aligned}$$

Theorem 8. *The greedy Gray code algorithm that homogeneously transposes the leftmost possible 1 into the leftmost possible position generates all combinations. That is,* $\mathsf{Greedy}_{\mathsf{Homo}_n}(1^k 0^{n-k})$ *generates* $\mathbb{B}(n, k)$. *Furthermore, the order is* $\mathbf{EM}(n, k)$ *when* k *is odd.*

Let $\mathbb{T}(n)$ be the set of binary trees with n vertices, which is enumerated by the nth Catalan number. When modifying a binary search tree, we can use *edge rotations* to keep the tree in balance (see [6]). In the 1990s, Ruskey, van Baronaigien and Lucas [7] showed how to recursively construct a Gray code of $\mathbb{T}(n)$, in which successive trees differ by a single edge rotation. In their Gray code **Tree**(n), they let the *label* of each vertex be its order during an inorder traversal. To describe our greedy interpretation of their Gray code, we *label* an edge between vertices with label u and label v as $\max(uv, vu)$. Given these labels, let $\mathsf{edge}_{i,j}$ denote the operation of rotating the edge with label ij, where $\mathsf{edge}_{i,j}$ has no effect if there is no such edge in the tree. We prioritize the edge rotations by lexicographically largest label as follows

$$\begin{aligned} \mathsf{Edge}_n^{\downarrow} = \;& \mathsf{edge}_{n,n-1}, \; \mathsf{edge}_{n,n-2}, \; ..., \; \mathsf{edge}_{n,1}, \\ & \mathsf{edge}_{n-1,n-2}, \; \mathsf{edge}_{n-1,n-3}, \; ..., \; \mathsf{edge}_{n-1,1}, \; ..., \\ & \mathsf{edge}_{3,2}, \; \mathsf{edge}_{3,1} \\ & \mathsf{edge}_{2,1}. \end{aligned}$$

Theorem 9. *The greedy Gray code algorithm that rotates the edge with the largest possible label generates the Ruskey, van Baronaigien, and Lucas Gray code for binary trees. That is,* $\mathsf{Greedy}_{\mathsf{Edge}_n^{\downarrow}}(1^n 0^n) = \mathbf{Tree}(n)$, *where* $1^n 0^n$ *denotes the binary tree that is a left path from the root.*

A *set partition of* $[n]$ is a collection of disjoint non-empty subsets $S_1, S_2, ...,$ $S_k \subseteq [n]$ with $S_1 \cup S_2 \cup \cdots \cup S_k = [n]$. The disjoint sets are numbered by their minimum elements, so S_1 is the set containing value 1, and S_2 is the set containing the minimum value that is not in S_1, and so on. Let $\mathbb{S}(n)$ denote the set partitions of $[n]$, which is enumerated by the nth Bell number. For example, the following is a set partition of $[6]$ with three subsets $S_1 = \{1, 2, 5\}$, $S_2 = \{3, 6\}$, and $S_3 = \{4\}$: $(\{1, 2, 5\}, \{3, 6\}, \{4\}) \in \mathbb{S}(6)$.

The operation $\mathsf{move}_{i,j}$ *moves* the value i into the jth subset. If j is the only value in its subset, then the operation removes the subset $\{i\}$, and if j is greater than the number of subsets then the operation creates a new subset $\{i\}$. Kaye [5] provided a Gray code for $\mathbb{S}(n)$ in which successive partitions differ by a single move. We denote this Gray code by $\mathbf{Kaye}(n)$, and then show that it has a simple greedy interpretation which prioritizes the operations as follows

$$\begin{aligned}\mathsf{Move}_n = {} & \mathsf{move}_{n,1}, \ \mathsf{move}_{n,2}, \ ..., \mathsf{move}_{n,n}, \\ & \mathsf{move}_{n-1,1}, \ \mathsf{move}_{n-1,2}, \ ..., \mathsf{move}_{n-1,n}, \ ..., \\ & \mathsf{move}_{1,1}, \ \mathsf{move}_{1,2}, \ ..., \mathsf{move}_{1,n}.\end{aligned}$$

Theorem 10. *The greedy Gray code algorithm that moves the largest possible value into the leftmost possible subset generates Kaye's set partition Gray code. That is,* $\mathbf{Kaye}(n) = \mathsf{Greedy}_{\mathsf{Move}_n}(\{1, 2, \ldots, n\})$.

References

1. Corbett, P.: Rotator graphs: An efficient topology for point-to-point multiprocessor networks. IEEE Trans. on Parallel and Distributed Systems 3, 622–626 (1992)
2. Duckworth, R., Stedman, F.: Tintinnalogia (1668)
3. Eades, P., McKay, B.: An algorithm for generating subsets of fixed size with a strong minimal change property. Inf. Proc. Letters 19, 131–133 (1984)
4. Gray, F.: Pulse code communication. U.S. Patent 2,632,058 (1947)
5. Kaye, R.: A Gray code for set partitions. Information Processing Letters 5(6), 171–173 (1976)
6. Knuth, D.E.: The Art of Computer Programming. Combinatorial Algorithms, Part 1, vol. 4. Addison-Wesley (2010)
7. Lucas, J.M., van Baronaigien, D.R., Ruskey, F.: On rotations and the generation of binary trees. Journal of Algorithms 15, 343–366 (1993)
8. Martin, M.H.: A problem in arrangements. Bull. Amer. Math. Soc. 40, 859–864 (1934)
9. Sawada, J., Williams, A.: Greedy pancake flipping. In: Latin-American Algorithms, Graphs and Optimization Symposium, LAGOS 2013 (accepted, 2013)
10. Siegel, J.: Interconnection Networks for Large-Scale Parallel Processing: Theory and Case Studies. McGraw-Hill (1990)
11. Stevens, B., Williams, A.: Hamilton cycles in restricted rotator graphs. In: Iliopoulos, C.S., Smyth, W.F. (eds.) IWOCA 2011. LNCS, vol. 7056, pp. 324–336. Springer, Heidelberg (2011)
12. Wikipedia. Rotary encoder, http://en.wikipedia.org/wiki/Rotary_encoder
13. Zaks, S.: A new algorithm for generation of permutations. BIT Numerical Mathematics 24(2), 196–204 (1984)

Author Index

Ahmed, Mahmuda 1
Ahn, Hee-Kap 13
Alamdari, Soroush 25
Alt, Helmut 13
Arge, Lars 37
Aronov, Boris 49
Asano, Tetsuo 61
Asinowski, Andrei 73
Askalidis, Georgios 85

Bae, Sang Won 13
Bannister, Michael J. 97
Barba, Luis 109
Baumbach, Jan 427
Belazzougui, Djamal 121
Biedl, Therese 25
Bienkowski, Marcin 133
Bille, Philip 146
Biro, Michael 158
Böhmová, Kateřina 170
Bose, Prosenjit 109, 182
Boyar, Joan 195
Bredereck, Robert 207
Buchin, Kevin 219
Buchin, Maike 219
Byrka, Jaroslaw 133

Cabello, Sergio 97
Cardinal, Jean 73
Chan, Timothy M. 25, 231
Chapelle, Mathieu 232
Chen, Danny Z. 244
Chen, Jiehua 207
Chlamtáč, Eden 256
Chowdhury, Iffat 1
Chrobak, Marek 133
Cohen, Nathann 73
Collette, Sébastien 73
Cording, Patrick Hagge 146

Dai, Bang-Sin 268
de Berg, Mark 49
De Carufel, Jean-Lou 109
Demaine, Erik D. 280
Disser, Yann 170

Durocher, Stephane 291
Dvořák, Zdeněk 304

Eppstein, David 97, 316
Epstein, Dror 328

Fischer, Johannes 37
Friggstad, Zachary 256
Fulek, Radoslav 341

Gagie, Travis 121
Georgiou, Konstantinos 256
Geyer, Markus 353
Gibson, Matt 1
Gørtz, Inge Li 146
Golumbic, Martin Charles 328
Goodrich, Michael T. 316
Grant, Elyot 25
Guo, Jiong 427
Gupta, Sushmita 195

Hackl, Thomas 73
Hamann, Michael 365
Hartmann, Tanja 365
Hartung, Sepp 207
Hassidim, Avinatan 378, 390
Hirschberg, Daniel S. 316
Hoffmann, Michael 73, 353
Hon, Wing-Kai 402
Hupp, Philipp 415

Ibragimov, Rashid 427
Immorlica, Nicole 85
Islam, Mohammad Shahedul 1
Iwerks, Justin 158

Jacob, Riko 415
Jampani, Krishnam Raju 25
Jansen, Klaus 439
Jeż, Łukasz 133

Kanj, Iyad A. 451
Kao, Mong-Jen 268
Kaplan, Haim 378
Kaufmann, Michael 353
Keller, Orgad 390
Keshav, Srinivasan 25

Khuller, Samir   475
Kindermann, Philipp   463
Kirkpatrick, David   61
Knauer, Kolja   73
Koehler, Frederic   475
Komusiewicz, Christian   207
Kostitsyna, Irina   158
Kusters, Vincent   353
Kwanashie, Augustine   85

Land, Felix   439
Land, Kati   439
Langerman, Stefan   73
Larsen, Kim S.   195
Lasoń, Michał   73
Lee, D.T.   268
Lee, Lap-Kei   402
Lewenstein, Moshe   390
Liedloff, Mathieu   232
Löffler, Maarten   487, 499
Lubiw, Anna   25

Manlove, David F.   85
Micek, Piotr   73
Mihalák, Matúš   170
Mitchell, Joseph S.B.   158
Mondal, Debajyoti   291
Morgenstern, Gila   328
Mulzer, Wolfgang   487

Navarro, Gonzalo   121
Niedermann, Benjamin   463
Niedermeier, Rolf   207
Nishimura, Naomi   512

Panchekha, Pavel   280
Park, Dongwoo   13
Pathak, Vinayak   25
Pountourakis, Emmanouil   85

Roditty, Liam   390
Roeloffzen, Marcel   49
Rote, Günter   73
Rutter, Ignaz   463

Sach, Benjamin   146
Sadakane, Kunihiko   402
Sanders, Peter   37
Schaefer, Marcus   463
Schulz, André   463
Sgall, Jiří   133
Sherrette, Jessica   1
Simjour, Narges   512
Simons, Joseph A.   499
Sitchinava, Nodari   37
Speckmann, Bettina   49, 219
Staals, Frank   219
Stachowiak, Grzegorz   133
Strash, Darren   499
Suchý, Ondřej   207

Todinca, Ioan   232
Tóth, Csaba D.   341, 353
Tsakalidis, Konstantinos   402
Tůma, Vojtěch   304
Tuval, Omry   378

Ueckerdt, Torsten   73

van Kreveld, Marc   219
van Renssen, André   109, 182
Vassilvitskii, Sergei   524
Verdonschot, Sander   109, 182
Vildhøj, Hjalte Wedel   146
Villanger, Yngve   232
Vind, Søren   146

Wagner, Dorothea   365
Wang, Haitao   244
Widmayer, Peter   170
Williams, Aaron   525
Wilson, David A.   280
Wolff, Alexander   463

Xia, Ge   451

Yang, Edward Z.   280